The Illustrated Book of Wildflowers and Shrubs

The Illustrated Book of Wildflowers and Shrubs

The Comprehensive Field Guide to More Than 1,300 Plants of Eastern North America

Text and Illustrations by
William Carey Grimm

Revised and Updated by
John T. Kartesz

STACKPOLE
BOOKS

Published by
STACKPOLE BOOKS
Cameron and Kelker Streets
P.O. Box 1831
Harrisburg, PA 17105

Printed in the United States of America

Revised Edition
This book was previously published as two volumes: *How to Recognize Shrubs* (1966) and *How to Recognize Flowering Wild Plants* (1968).

10 9 8 7 6 5 4 3 2 1

Cover art by William Carey Grimm

Cover design by Tracy Patterson

Library of Congress Cataloging-in-Publication Data

Grimm, William Carey, 1907–
The Illustrated book of wildflowers and shrubs : the comprehensive field guide to more than 1,300 plants of Eastern North America / William C. Grimm. – 2nd ed.
p. cm.
Includes bibliographical references and index.
ISBN 0-8117-3085-9
1. Wild flowers–East (U.S.)–Identification. 2. Wild flowers–Canada, Eastern–Identification. 3. Shrubs–East (U.S.)–Pictorial works. 4. Shrubs–Canada, Eastern–Pictorial works. I. Title.
QK112.G75 1992
582.1'2'0974–dc20

92-13126
CIP

To
Garvin and Louise Hughes,
Jay and Martha Shuler,
and to my wife,
Ruth Curtis Grimm

Contents

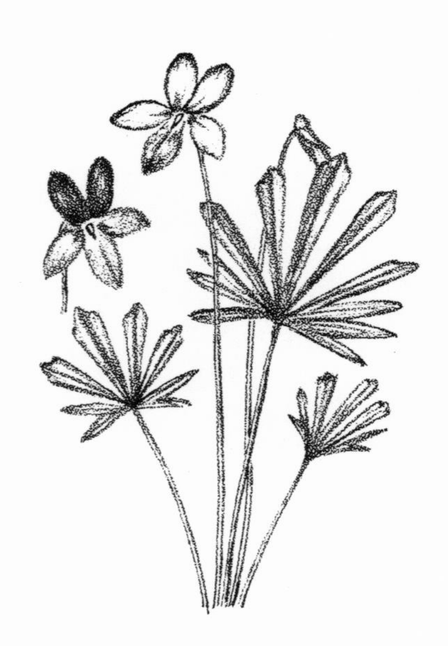

Shrubs

Introduction to the Revised Edition

It is now several decades since the wonderfully illustrated publications of *How to Recognize Shrubs* (1966) and *How to Recognize Flowering Wild Plants* (1968) by William Carey Grimm first appeared, and readers of these works have wondered if indeed the texts would ultimately be revised. As with most works on natural history, time tends to erode the accuracy of scientific nomenclature and taxonomic detail. It is for this reason that I have taken the opportunity to revise the text material and to bring it into current standardization. In revising these works, I have attempted to maintain as much of the original detail as possible, correcting a few minor errors of the original edition, altering the nomenclature, and enhancing some of the plant descriptions to bring them into conformity with current nomenclatural and taxonomic standards.

This edition, like the previous, is intended to provide a convenient method to identify most of the common shrubs and showy wildflowers of eastern United States and Canada. The identification is facilitated by well-illustrated plates and accompanying descriptions, along with dichotomous keys to many of the plant groups. To facilitate this process further, I have added a few generic keys and modified some of the more difficult ones. Although the content of this book remains directed toward the lay public, the descriptive information, especially for the shrubs, should provide valuable diagnostic details for even the most serious students of botany.

John T. Kartesz
January 1993

Introduction

This book is intended to be a popular guide to the wildflowers, native shrubs, and woody vines found east of the Mississippi River, from the Canadian Provinces southward as far as northern Florida and the Gulf Coast region. It therefore covers a much larger area, and includes many more of the species found in the southeastern part of our country, than other popular books published thus far. An attempt has been made to include the flowering plants most apt to attract one's attention, both on trips afoot and while travelling along the highways. However, we have included some which, although local, are of more than casual interest.

No popular book could possibly include all of the shrubs and flowering plants of such a vast area, or even a small part of it. That is the province of the more technical manuals, some of which are listed in the "Books to read for further information" at the end of this book. The choice of the species included here is the author's own, but it has been based upon more than forty years of field experience in various parts of the region covered. In general, only herbaceous plants have been admitted as wildflowers, the only exceptions being a few of the more or less woody vines which have particularly showy flowers. Many of our shrubby plants, such as the rhododendrons and azaleas, have truly spectacular flowers.

An annoying feature of many books has been the separation of illustrations and text on different pages, often without any cross-reference to each other. To remedy this fault, the illustrations of the plants and the text pertaining to them will be found on pages which face each other in this book. To accomplish this it was necessary to make the accounts of the species brief, yet they must include whatever information may be necessary to aid in identification.

The arrangement of the plants in the book follows that found in the technical manuals: by plant families and genera. This seemed to be the most logical arrangement. The illustrations are not all on the same scale for the plants differ greatly in size, and this is often true even among those of the same species. Information about the size of the plant, the size and color of its flowers, its habitat, general range, and other items of interest are given in the text. For wildflowers and flowering shrubs, the time of blooming covers the entire range of the plant. Thus "blooming between April and July" may mean that it begins to bloom in April in the southern part of its range, but it may not bloom until July at the northern extremity or at higher elevations in the mountains.

Colored illustrations are beautiful but frequently they tend to be confusing. This is especially true in the case of photographs of the plant taken in its natural environment, where the leaves of surrounding plants are apt to be taken for those of the plant meant to be portrayed. More often than not, they fail to show points essential for identification. There is often considerable variation in the color of the flowers of many species; and all too often even the best color illustrations are so poorly reproduced that they tend to be very misleading.

Because the book has been prepared for users without any special training in botany, botanical terms have been reduced to a minimum. Nevertheless the usage of some terms could not very well be avoided. These have been explained under *Getting acquainted with flowers* (page 13 through 19) and in the glossary.

Wildflowers

PAGE GUIDE TO PLANT FAMILIES

GETTING ACQUAINTED WITH FLOWERS

The parts of a flower

Many of us who would like to know more about the wild flowers and be able to recognize them have had little or no training in the subject of botany. For this reason it is well to present here a brief introduction to the structure and arrangement of flowers.

Most of the flowers that we ordinarily see are what the botanist would call a *complete flower*: one which has four sets of flower parts or organs—sepals, petals, stamens, and one or more pistils.

The outermost of these sets of organs are the *sepals*. Very often they are green in color and they cover the parts of the flower in the bud stage. Collectively the sepals are known as the *calyx* of the flower. In some flowers the sepals are more or less united and they may form a sort of cup or tube. In this case it is usually spoken of as the *calyx tube*. Usually it has lobes on its rim which represent the free tips of the individual sepals and they are called the *calyx lobes*. The number of calyx lobes therefore tells us how many sepals have been united to form the calyx tube.

The next series of flower parts is made up of the *petals*. They are usually the showy and brightly colored parts of the flower. Collectively the petals are known as the *corolla* of the flower. As in the case of the sepals, the petals may be joined to one another to form a *corolla tube*. The free ends of the individual petals which were united to form the corolla tube are usually seen as *corolla lobes* on its rim. Often at the bases of the petals there are glands which secrete a sweet substance called nectar, from which bees make their honey. The petals actually serve as billboards to attract bees or other insects to the flower.

Sepals and petals taken together are usually spoken of as the *floral envelopes* or the *perianth* of the flower. While the sepals and petals are quite distinct and differently shaped and colored in most flowers, this is not universally true. In most members of the Lily Family, the Amaryllis Family, and in many plants belonging to the Iris Family, the sepals and petals are very much alike in size and color. It is therefore quite convenient just to speak of them as the *perianth parts*. Some flowers actually have but one set or series of floral envelopes, the sepals. In many members of the Buttercup Family these sepals are white or variously colored and look very much like petals. The sepals, in this case, have taken over the usual role of the petals in attracting insects to the flowers.

FLOWER PARTS AND STRUCTURE

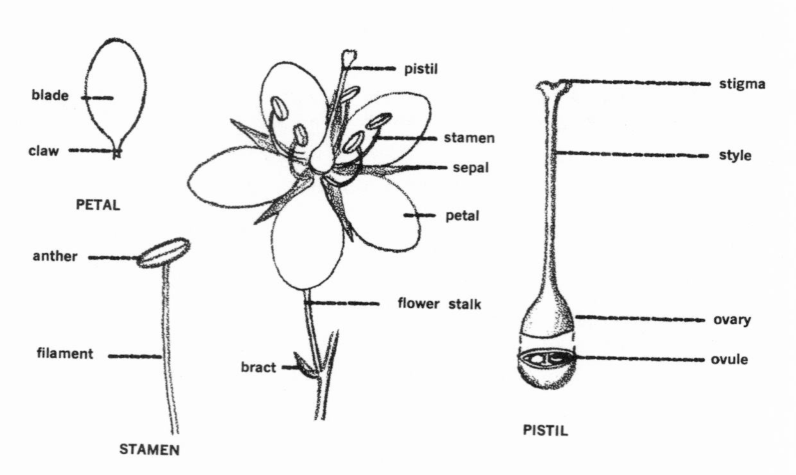

PARTS OF A COMPLETE FLOWER

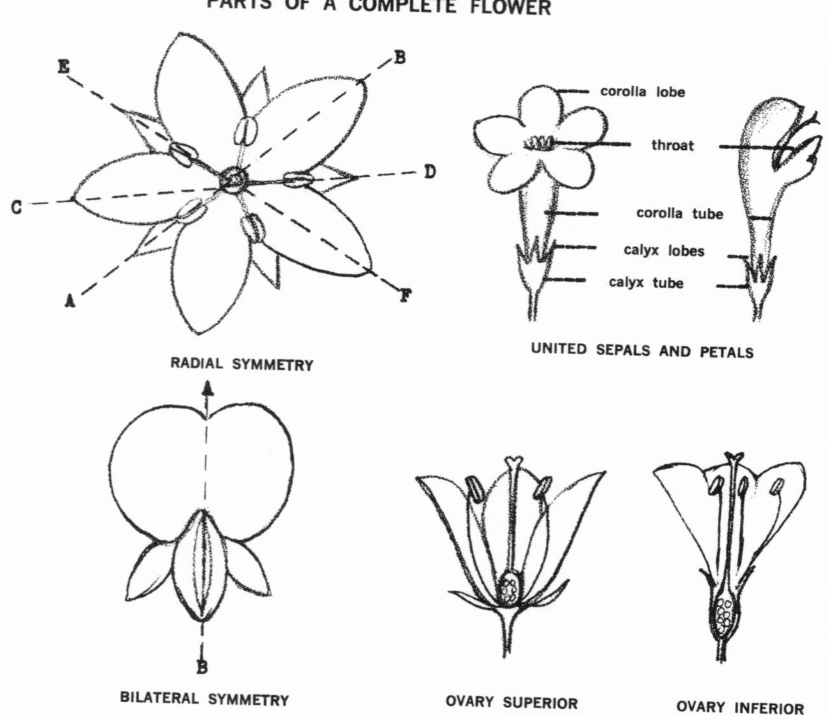

Stamens and pistils are the only flower parts that are actually involved in the production of the fruits and seeds. They are the *sexual organs* of flowers. The *stamens* are the organs just inside of the petals when petals are present. They usually have a slender stalk which is called the *filament* and a box-like compartment at the summit which is known as the *anther*. The anthers contain the grain-like, powdery, and usually yellow or yellowish pollen. Stamens are the male organ of a flower.

In the center of the flower there is a *pistil* or sometimes several pistils. Usually the pistil has a somewhat swollen part at its base which is called the *ovary,* inside of which are one to many small bodies known as *ovules*. The summit of the pistil is commonly enlarged, often knob-like or branched, and usually quite sticky. This part of the pistil is the *stigma*. Between the stigma and the ovary there is usually a stalk-like portion called the *style*. Pistils are the female organs of a flower.

Several things must happen before the flower can produce seeds. First of all, some of the pollen must get from the stamens to the sticky stigma of the pistil. This transfer of the pollen is known as *pollination* and in plants with showy flowers bees or other insects usually perform the task. That is why their flowers have attractive petals, alluring odors, and produce the sweet nectar. Following pollination each pollen grain sends a tube down through the style and into the ovary. There it seeks out one of the ovules. A sperm from the pollen goes down this tube and unites with an egg cell within the ovule, a process which is called *fertilization*. Thereafter the egg cell develops into a tiny plant or *embryo* and the ovule becomes a *seed*. The ovary of the pistil (sometimes with other parts attached) matures into what botanists call a *fruit*.

Various devices are frequently employed to assure that pollen from the stamens will not pollinate the pistil of the same flower. In some flowers the anthers will have shed their pollen before the stigma of the pistil is ready to receive the pollen, or vice versa. In some plants the stamens and pistils are found in separate flowers, or the stamen-bearing and pistil-bearing flowers may even be on separate plants.

The forms of flowers

Flower structure is of primary importance in the identification of plants. In fact, botanists base the classification of plants into species, genera, and families on the basis of their flower structure. While color is sometimes useful in making an identification, it is by no means a reliable characteristic. The same species of plant may have flowers in two or more different colors, and the colors of flowers sometimes change with age. Size, too, is quite variable. A plant may vary considerably in size depending upon environmental factors, especially in the case of annuals.

In the simpler types of flowers all of the flower parts are attached to the *receptacle* or enlarged tip of the stem, and they are all separate and distinct. Good examples are seen in members of the Buttercup Family. A flower in which the ovary of the pistil is free from the surrounding parts is said to be a *superior ovary*. In many flowers the ovary is more or less surrounded by and united with the calyx tube or the receptacle. Such flowers are said to have an *inferior ovary*. The filaments of the stamens in many flowers are attached to the petals or the corolla tube rather than the receptacle.

In many flowers all of the flower parts radiate from the center of the flower and those in each series are more or less alike in size and shape. Like a pie, such a flower could be divided in any number of directions through the center into quite identical halves. Such a flower is said to have a *radial symmetry*. It is often called a *regular flower*.

It should take but one look at a flower of a pea, mint, violet, or orchid to see that it could be divided into two similar halves by cutting it in but one direction—vertically through the center. A flower of this type has a *bilateral symmetry*. It is often called an *irregular flower*. Often when the sepals or petals of a flower are joined together, the calyx or the corolla will be two-lipped. Such flowers are common among the members of the Mint Family but they also occur in many of the other flower families.

Flower clusters or inflorescences

Flowers may be solitary but more often they are arranged in various types of flower clusters or *inflorescences* as shown diagramatically on page 13. A *spike* is a simple inflorescence of stalkless flowers which are arranged along a stem. A *raceme* is similar except that the flowers are stalked. A *head* is a dense cluster of stalkless or very short-stalked flowers at the tip of a stem. In an *umbel* the flowers are on stalks which all arise from the summit of a stem. It is the typical inflorescence of the Parsley Family but it also occurs among the members of other flower families. A *panicle* is a branching flower cluster which may be likened to a compound raceme. The *cyme* is also a branching flower cluster, a more or less forking one in which the central flowers open first, and it tends to be flat-topped. Another flat-topped inflorescence is the *corymb* but in this one it is the outer or marginal flowers which open first.

It is very easy to be fooled unless one is very observant, and what is usually taken to be a flower may really be a flower cluster. In the Arum Family the flowers are quite small and seated on a thick and somewhat fleshy stem called a *spadix*. Usually the spadix is more or less surrounded by a large and often colorful bract called a *spathe*. The "flowers" of the familiar Jack-in-the-pulpit and the cultivated calla-lilies are good examples.

The sunflower or a daisy is not a flower in the ordinary sense, but rather a dense cluster—a head—of many small flowers. What may be mistaken for petals are the marginal *ray flowers* which have a strap-shaped corolla. In the center of the head is a prominent *disk* which contains a large number of small flowers (*disk flowers*) which have a tubular corolla. The entire head is surrounded by an *involucre* made up of greenish bracts which may

TYPES OF INFLORESCENCES

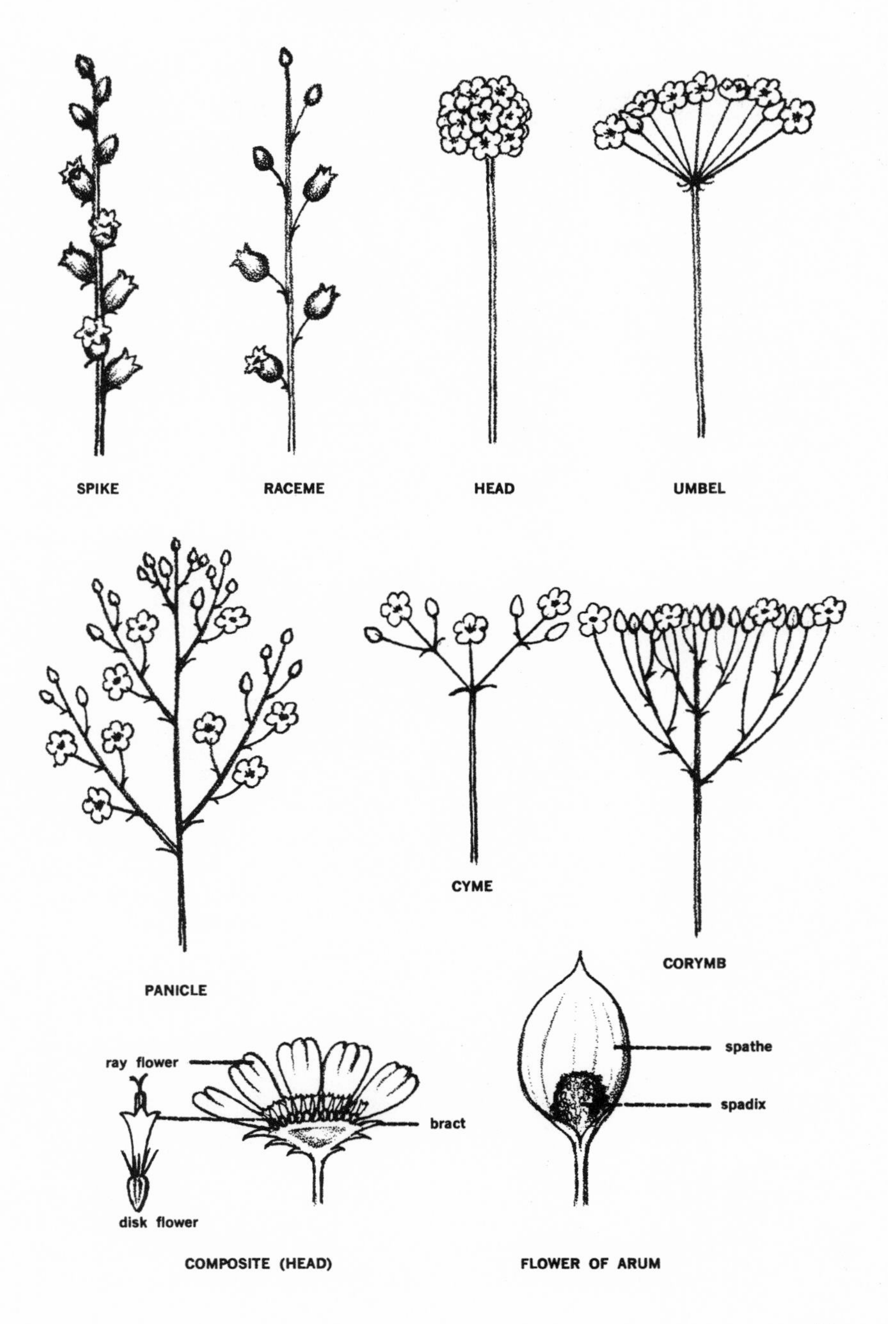

LEAF PARTS AND TYPES OF LEAVES

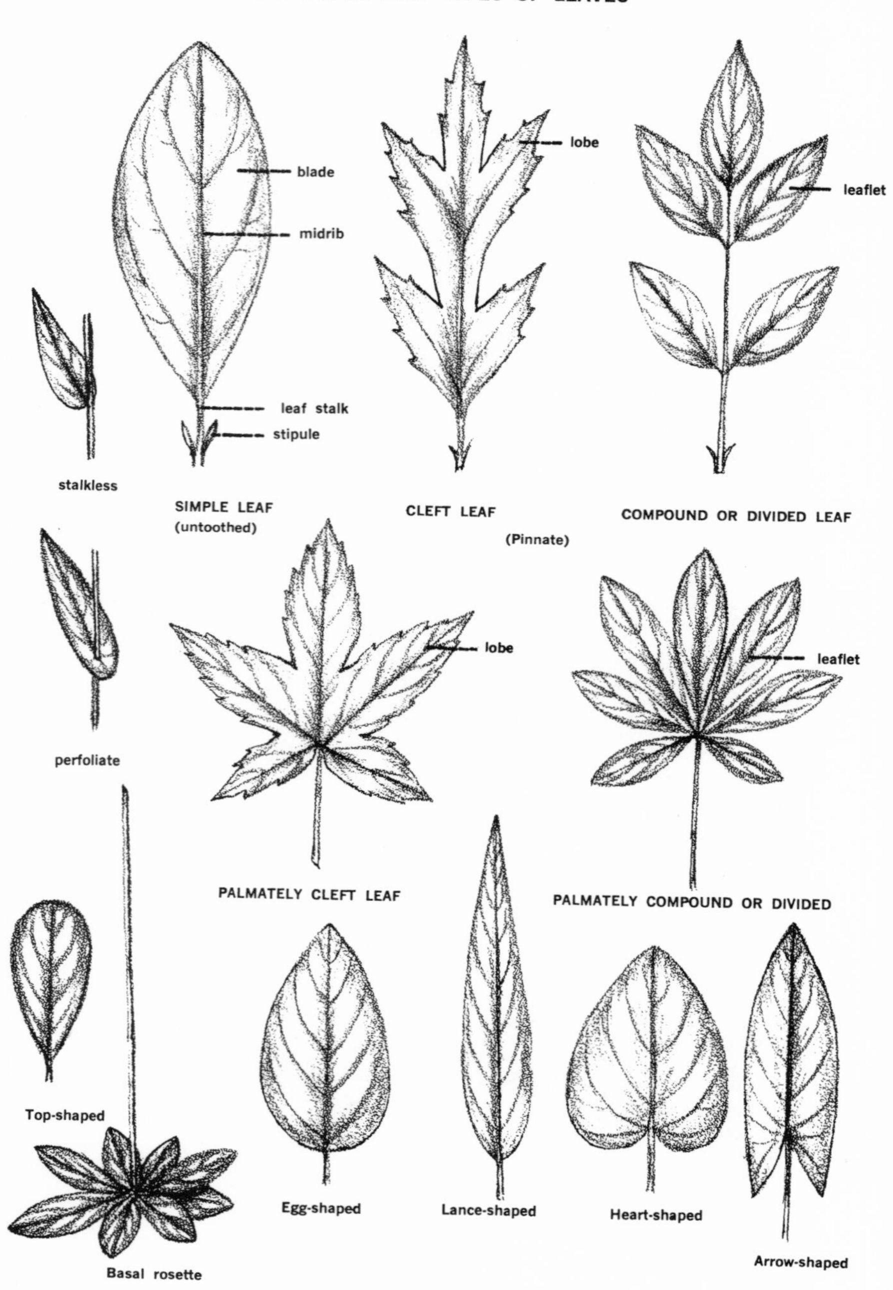
blade
midrib
leaf stalk
stipule
lobe
leaflet
stalkless
SIMPLE LEAF
(untoothed)
CLEFT LEAF
(Pinnate)
COMPOUND OR DIVIDED LEAF
perfoliate
lobe
leaflet
PALMATELY CLEFT LEAF
PALMATELY COMPOUND OR DIVIDED
Top-shaped
Basal rosette
Egg-shaped
Lance-shaped
Heart-shaped
Arrow-shaped

be mistaken for sepals. Sunflowers, daisies, asters, goldenrods, and many other common flowers are called *composites*. They are members of the very large Composite Family.

Types and arrangements of leaves

Leaves show a marked variation in size, shape, and arrangement on the stems. In a few cases plants have leaves which are so distinctive that they may be identified by the leaves alone. Many of the features of leaves are shown on page 14 and it would be well to become acquainted with them.

Most leaves have a flattened and more or less broad portion which is called the *blade*. In a great many there is a prominent central vein or *midrib* from which the veins branch off like the barbs from the shaft of a feather. Such a leaf is said to be *pinnately-veined,* or feather-veined. Other leaves have several main veins which radiate from the base or from the summit of a leaf stalk. This type of leaf is *palmately-veined*. The margin of the leaf blade may be untoothed or *entire*, or it may have teeth of various sizes or shapes. Sometimes a leaf blade is deeply cut or cleft, or it may be divided into smaller leaf-like parts which are called *leaflets*. A leaf in which the blade is in one piece is known as a *simple leaf*. One in which the blade is divided into leaflets is called a *compound leaf*. A leaf may be *pinnately* divided or pinnately compound or *palmately* divided or palmately compound. In many leaves there will be a pair of appendages or *stipules* at the base of the leaf stalk.

WHY PLANTS HAVE SCIENTIFIC NAMES

Have you ever wondered why plants have scientific names, or why common names are not good enough?

To begin with, and this may amaze you, most of our wild plants actually have no common names at all. They never have been given any. On the other hand, every known kind of plant does have a scientific name.

Common names are frequently very confusing. In some instances the same name is used for two or more entirely different plants. Button-snakeroot, for example, is used as a common name for plants in both the Parsley and Composite families. Samson's-snakeroot may be either a gentian or member of the Pea family. A fireweed may be a plant belonging to either the Evening-primrose or the Composite Familes; and a loosestrife could be either a true loosestrife or an entirely different plant of the Primrose Family.

To make things a bit more confusing, a plant may be known by more than one common name. Thus the little member of the Lily Family which bears the scientific name *Erythronium americanum* is variously known as the Trout-lily, Fawn-lily, Yellow Adder's-tongue, and even Dog's-tooth Violet. Often a plant is known by one common name in one part of its range and by other names elsewhere. It is obvious, too, that common names are often misleading in that they do not show the true relationship of a plant. The little lily-like plant just mentioned is most certainly not a violet, nor even remotely related to the true violets.

Scientific names have several advantages. First there can be but one plant known as *Erythronium americanum*. It cannot therefore be confused with any other plant. The first part of its scientific name tells us that it belongs to the genus *Erythronium*. It is the generic name of the plant and this particular genus is a member of the Lily Family. Its relationship is thus firmly established beyond a doubt. The second part of the name—*americanum*—is the Latin adjectival name, the two together comprising the name of the species. In this case it means "American." Scientific names are always in Latin and are the same throughout the world, regardless of language differences in the various countries.

In some cases two or more varieties of the same species of plant may be recognized. *Cypripedium calceolus* is the Yellow Lady's-slipper of the Old World. Botanists now recognize our American plants as a variety of this wide-ranging species. Thus our Yellow Lady's-slipper is known as *Cypripedium calceolus var.* (or variety) *pubescens.*

Naturally the question is going to arise as to why the scientific names do not always agree in even the technical manuals. The rule is that the first published name, accompanied by a valid description of the plant, becomes its scientific name. Thus a great many of the scientific names of our plants have remained the same for a long time, even for more than two centuries. Often, however, researchers find that a plant was given a valid scientific name before one in current usage was given; and in such a case it becomes necessary to make a change in its name. Research is constantly continuing in the taxonomy of plants and for other reasons a name may become invalid. As we learn more about the true relationship of species, a change often becomes necessary in the name. Thus botanists no longer accept *Ascyrum* and *Hypericum* as two distinctly different genera of plants. All species formerly placed in the genus *Ascyrum* are now regarded as species of *Hypericum*. Someday such matters will, of course, be definitely decided, but right now they are in a state of flux.

Scientific names are actually not as dreadful as they may at first seem. We use many of the generic names of plants—names such as *Trillium, Hepatica, Anemone,* and *Chrysanthemum*—without ever thinking, or perhaps even knowing, that they are scientific names. Others such as *Lilium* (lilies), *Viola* (violets), and *Rosa* (roses) come very close to the English equivalents of the generic names. With a little practice one could soon become adept at using scientific names of plants and appreciate their advantages.

USING THIS BOOK TO RECOGNIZE FLOWERS

The following key is provided to the flower families which are covered in this book, and keys to the genera of some of the larger or more difficult families will be found following the discussions of the family characteristics elsewhere in the book.

A key is a guide or aid in helping one to identify a flower without thumbing through the book page by page. Insofar as possible, characteristics which may be seen with the naked eye have been used in the keys, but there were a few cases where this was not feasible. At any rate, a serious student of plants will sooner or later find that a good hand lens or magnifying glass is a very handy piece of equipment.

In using the following key, first decide into which class—monocots or dicots—the plant belongs. Next decide into which group it belongs. Having done this, we proceed to use the sets of numbers beginning with number 1, then 2, 3, etc. until the plant family is found. The numbers are based on two sets of opposing characteristics.

Suppose, for example, we have a plant in which the leaves are distinctly netveined and the flower parts—sepals, petals, and stamens—are in fives. From these characteristics we decide that the plant is a dicot from the description given. As our plant has flowers with both sepals and petals—two sets of floral envelopes—it can't belong in Groups I or II, and as the petals are not united it must belong to Group III. Now we will proceed to number 1 under this group, and take the following steps.

1. This first number 1 reads, *"Ovary of the pistil free from other parts of the flower and apparently situated above them."* As the flowers of our plant obviously have the calyx united with the ovary, we proceed to the second number 1 which reads, *"Ovary of the pistil wholly or partly united with the calyx and apparently situated below the other parts of the flower."* As this fits our flower, we shall proceed to number 2.
2. The first number 2 reads, *"Pistils 2, etc.,"* but as our flower has but a single pistil we will go to the second number 2. Here we find, *"Pistil apparently solitary, completely concealed by and more or less united with the calyx tube."* As this is correct, we shall proceed to number 3.

3. The first number 3 reads, *"Petals and stamens numerous. Plants with fleshy, jointed, flattened, and usually spiny stems."* As none of these characteristics fit our plant, we will proceed to the second number 3. Here we find, *"Petals, sepals, and stamens relatively few."* This seems to fit, so we will go on to number 4.
4. The first number 4 reads, *"Sepals and petals usually in 4's."* As the flower of our plant has 5 petals, we shall go to the second 4. This reads, *"Sepals and petals in 5's. The flowers mostly arranged in umbels."* This is correct and the flowers of our plant are arranged in umbels, so we will go on to number 5.
5. The first number 5 reads, *"Pistil with 2 to 5 styles. Fruit a berry."* Our plant has a pistil with but 2 styles and the fruits do not seem to be berries, so we shall try the second number 5. There we read, *"Pistil with 2 styles, surrounded at their base by a prominent disk. Fruits dry and seed-like."* This description seems to fit our plant perfectly and we find that it is a member of the Carrot Family or *Apiaceae.*

The serious student of flowers should become acquainted with the characteristics of the various flower families. A short discussion of the outstanding characteristics of the various families treated in this book will be found before the accounts of the species. Familiarity with them will make it much easier to identify an unknown flower.

KEY TO THE FLOWER FAMILIES

MONOCOTS

The flower parts of monocots are usually in 3's (rarely in 4's). The leaves have unbranched veins which usually run parallel to each other from the base of the leaf to the tip (Trilliums and some members of the Arum Family are exceptions in that the leaves have branching veins).

GROUP I.
FLOWERS WITHOUT FLORAL ENVELOPES OR SOMETIMES WITH INCONSPICUOUS SEPALS.

1. Flowers (sometimes with small sepals) crowded and seated directly upon a thick and fleshy stalk *(spadix)*, which is often subtended or partly surrounded by a leaf-like structure *(spathe)*.
ARUM FAMILY *(Araceae)* 35
1. Flowers not crowded on a thick and fleshy stalk.

   2. Flowers in ball-like or spike-like heads and arranged along a simple or branching stalk. BUR-REED FAMILY *(Sparganiaceae)* 32
   2. Flowers in small, scaly-bracted, more or less cone-shaped end clusters. Grass-like plants with usually 3-sided stems, each leaf with a basal sheath surrounding the stem.
   SEDGE FAMILY *(Cyperaceae)* 32

3. Flowers in laterally appearing spikes. ACORUS FAMILY *(Acoraceae)* 36
3. Flowers in subrounded, ball-like heads.
BUR-REED FAMILY *(Sparganiaceae)* 32

GROUP II.
FLOWERS WITH A DEFINITE PERIANTH, USUALLY WITH 3 SEPALS AND 3 PETALS.

1. Flowers in a tightly packed head at the end of a long and leafless stem.

   2. Heads of flowers cone-shaped to cylindrical. Petals yellow. Flowers nearly covered by hard, dry, overlapping scales.
   YELLOW-EYED GRASS FAMILY *(Xyridaceae)* 36
   2. Heads of flowers button-shaped or roundish. The flowers small, whitish, with a perianth of dry scales.
   PIPEWORT FAMILY *(Eriocaulaceae)* 36

1. Flowers not in tightly packed heads.

   3. Pistils 6 or more and crowded in heads or ring-like clusters. Stamens often numerous and sometimes in separate flowers. Flowers with 3 green sepals and 3 white or pinkish petals.
   WATER-PLANTAIN FAMILY *(Alismataceae)* 32
   3. Pistils solitary, composed of 3 united carpels and often showing 3 distinct styles or stigmas. Stamens 6 or fewer.

4. Ovary of the pistil free from the perianth parts and seemingly situated above the other parts of the flower.

   5. Sepals and petals different in size or color.

6. Flowers clustered, arising from folded spathes or with leafy bracts at their base. Anthers not much longer than broad and sometimes only 3 of them fertile. Pistil with a single style. Leaves alternate and with bases sheathing the stem. SPIDERWORT FAMILY *(Commelinaceae)* 39
6. Flowers solitary. Anthers much longer than broad. Pistil with 3 styles. Leaves in a whorl of 3 beneath the flower.
TRILLIUMS in LILY FAMILY *(Liliaceae)* 59

5. Sepals and petals very much alike in size and color.

6. Perianth with a bilateral symmetry and somewhat 2-lipped. Stamens 3 or 6, unequal in length and attached to the perianth tube. Flowers usually blue, solitary or in a spike-like cluster subtended by a leafy bract. Aquatic plants. PICKERELWEED FAMILY *(Pontederiaceae)* 39
6. Perianth with a radial symmetry. Stamens 6 (rarely 4) and about the same length; free from or joined to the base of the perianth tube. Flowers solitary or variously clustered. LILY FAMILY *(Liliaceae)* 40

4. Ovary of the pistil more or less united with the perianth parts or tube and thus appearing to be below the other parts of the flower.

7. Flowers with 6 stamens; arranged in showy umbels
LILY FAMILY *(Liliaceae)* 40
7. Flowers with fewer than 6 stamens.

8. Flowers with a radial symmetry or nearly so. Stamens 3.

9. Stamens attached to the base of the 3 larger and inner perianth parts (petals). Flowers woolly on the outside. Plant with a bright red juice. BLOODWORT FAMILY *(Haemodoraceae)* 63
9. Stamens attached to the base of the 3 outer perianth parts (sepals), their filaments sometimes forming a tube about the pistil. Styles united except at the tips, or deeply parted and petal-like. Leaves grass-like or sword-shaped and placed edge to edge.
IRIS FAMILY *(Iridaceae)* 64

8. Flowers very irregular and with a bilateral symmetry. Central petal usually markedly different from the other 2 and forming a "lip." Stamens 1 or 2, united with the style to form a central structure or "column." ORCHID FAMILY *(Orchidaceae)* 68

DICOTS

The flower parts of dicots are usually in 4's or 5's (rarely in 3's), in multiples of these numbers, or indefinite in number. The leaves usually have a network of branching veins (net-veined leaves).

GROUP I. FLOWERS WITHOUT FLORAL ENVELOPES, NEITHER SEPALS NOR PETALS PRESENT.

1. Flowers grouped in a calyx-like cup with prominent glands or petal-like lobes on its rim. Flowers consist of single-stalked stamens or a stalked and 3-lobed pistil with 3 spreading styles. The plants have a milky juice. SPURGE FAMILY *(Euphorbiaceae)* 165
1. Flowers arranged in a long and narrow, recurved cluster. Each flower consisting of 6 to 8 stamens and a 3- or 4-parted pistil.
LIZARD'S-TAIL FAMILY *(Sauraceae)* 82

GROUP II.
FLOWERS WITH BUT ONE SET OF FLORAL ENVELOPES, THE SEPALS. THESE OFTEN COLORED AND PETAL-LIKE.

1. Stamens numerous and indefinite in number. Flower parts all separate and distinct. The pistils usually several or numerous (rarely 1).
BUTTERCUP FAMILY *(Ranunculaceae)* 92
1. Stamens 12 or fewer.

   2. Sepals 3, these united to form a flask-shaped or urn-shaped cup with 3 lobes on the rim. Plants with aromatic rootstocks and more or less heart-shaped leaves.
   BIRTHWORT FAMILY *(Aristolochiaceae)* 85
   2. Sepals 5 or 6 in number.

3. Sepals 5, these united below into a cup with 5 teeth on its rim. Stamens 5. SANDALWOOD FAMILY *(Santalaceae)* 82
3. Sepals 5 or 6 in number, all separate and distinct.

   4. Stamens 10, surrounding the large and somewhat 10-lobed pistil which develops into a purplish-black berry.
   POKEWEED FAMILY *(Phytolaccaceae)* 89
   4. Stamens 6. Sepals often rosy-pink. Stems usually jointed and surrounded by a tube-like sheath above each leaf.
   BUCKWHEAT FAMILY *(Polygonaceae)* 82

GROUP III.
FLOWERS WITH TWO SETS OF FLORAL ENVELOPES, BOTH SEPALS AND PETALS. THE PETALS FREE FROM EACH OTHER AND DISTINCT.

1. Ovary of the pistil free from other parts of the flower and apparently situated above them.

   2. Stamens numerous, their number indefinite.

3. Petals numerous and showing a gradual transition into stamens. Plants aquatic. WATERLILY FAMILY *(Nymphaeaceae)* 90
3. Petals usually 12 or fewer.

   4. Stamens united by their filaments to form a column about the styles of the pistil. MALLOW FAMILY *(Malvaceae)* 166
   4. Stamens all separate and distinct.

5. Sepals 2, falling as the flower opens. Petals usually 4 but sometimes 8 to 12. Plants with a colored or milky juice.
POPPY FAMILY *(Papaveraceae)* 111
5. Sepals 4 or 5 and usually persisting.

   6. Flowers with several to many pistils. Leaves usually toothed, cleft, or divided into leaflets.

7. Stamens attached to the receptacle below the group of pistils. Leaves without stipules. BUTTERCUP FAMILY *(Ranunculaceae)* 92
7. Stamens attached to a disk on the calyx. Leaves usually with prominent stipules. ROSE FAMILY *(Rosaceae)* 131

   6. Flowers with but 1 pistil. Leaves with entire margins.

8. Pistil with 4 or 5 distinct styles. Leaves opposite, often with minute black or transparent dots. ST. JOHN'S-WORT FAMILY *(Hypericaceae)* 169

8. Pistil with a single style or sometimes none. Leaves alternate. ROCKROSE FAMILY *(Cistaceae)* 170

2. Stamens usually no more than twice the number of petals.

9. Flowers with a radial symmetry.

10. Sepals 2. Flowers with 5 petals, 5 stamens, and a solitary pistil with 3 styles. Plants fleshy. PURSLANE FAMILY *(Portulacaceae)* 86

10. Sepals more than 2.

11. Pistil solitary.

12. Sepals and petals not in 5's.

13. Sepals and petals 4, the latter forming a cross. Stamens 6, two of them shorter. MUSTARD FAMILY *(Cruciferae)* 115

13. Sepals 4 or 6. Petals 6 to 9. Stamens as many or twice as many as the petals. BARBERRY FAMILY *(Berberidaceae)* 108

12. Sepals and petals usually in 5's (rarely in 4's).

14. Bog plants with leaves in a basal cluster or rosette.

15. Leaves hollow, pitcher-like or trumpet-shaped. PITCHER-PLANT FAMILY *(Sarraceniaceae)* 120

15. Leaves otherwise. SUNDEW FAMILY *(Droseraceae)* 119

14. Leaves not in a basal rosette; if in a basal cluster not bog plants.

16. Tendril-bearing vines. Flowers with a stalked pistil, 5 pendant anthers, and a fringed crown between the sepals and petals. PASSION-FLOWER FAMILY *(Passifloraceae)* 181

16. Plants not tendril-bearing vines. Flowers otherwise.

17. Leaves deeply cut or divided.

18. Leaves divided into 3 inversely heart-shaped leaflets. WOOD-SORREL FAMILY *(Oxalidaceae)* 158

18. Leaves palmately cut or divided with the divisions variously cut or toothed. GERANIUM FAMILY *(Geraniaceae)* 158

17. Leaves entire.

19. Petals and stamens attached to the calyx tube. LOOSESTRIFE FAMILY *(Lythraceae)* 181

19. Petals and stamens otherwise.

20. Sepals more or less united, at least at the base. Leaves opposite. Stems often swollen at the nodes. PINK FAMILY *(Caryophyllaceae)* 86

20. Sepals not united. Leaves alternate or opposite; small, narrow, and numerous. FLAX FAMILY *(Linaceae)* 161

11. Pistils 2 or more.

21. Pistils 2, more or less united at the base. Sepals and petals 5. Stamens 5 or 10. SAXIFRAGE FAMILY *(Saxifragaceae)* 123

21. Pistils 4 or 5. Sepals and petals 4 or 5. Stamens 8 or 10. Plants fleshy. STONECROP FAMILY *(Crassulaceae)* 123

9. Flowers with a bilateral symmetry.

22. Flowers with 1 or more sacs or spurs.

23. Sepals 5, greenish and distinct from the petals. Petals 5, the lowermost one extended backward as a sac or spur.
VIOLET FAMILY *(Violaceae)* 173

23. Sepals inconspicuous or colored like the petals.

24. Sepals 2, very small. Petals 4, 2 of them large and 2 small; one or more of the large petals extended backward as a sac or spur. Plants with leaf blades divided and often finely cut.
FUMITORY FAMILY *(Fumariaceae)* 112

24. Sepals 3, the lower one extended backward as a spurred sac. Petals 5, all very unequal. Leaves with a 1-piece blade. Plants with very watery stems. JEWELWEED FAMILY *(Balsaminaceae)* 165

22. Flowers without sacs or spurs.

25. Sepals 5, united at the base and distinct from the petals. Stamens usually 10 and often united by their filaments into two groups. Petals 5, sometimes nearly equal but more often the flowers pea-like or papilionaceous. Fruit as legume. PEA FAMILY *(Fabaceae)* 136

25. Sepals 5 but the 2 side ones petal-like and forming wings. Petals 3, grouped together to form a tube, the lower one often crested or fringed at the tip. MILKWORT FAMILY *(Polygalaceae)* 161

1. Ovary of the pistil more or less united with the calyx and apparently situated below the other parts of the flower.

26. Pistils 2, the ovaries partially covered by a cup-like disk.
SAXIFRAGE FAMILY *(Saxifragaceae)* 123

26. Pistil apparently solitary, the ovary completely concealed by and more or less united with the calyx tube.

27. Petals and stamens numerous. Plants with fleshy, jointed, flattened and usually spiny stems. CACTUS FAMILY *(Cactaceae)* 181

27. Petals, sepals, and stamens relatively few.

28. Sepals and petals usually in 4's.

29. Calyx tube urn- or bell-shaped with a narrow neck and 4-lobed rim; united with ovary only near its base.
MELASTOME FAMILY *(Melastomataceae)* 182

29. Calyx tube and ovary completely united.

30. Stamens 4. Style unbranched. Flowers small, crowded in a head which is surrounded by 4 large and petal-like bracts.
DOGWOOD FAMILY *(Cornaceae)* 189

30. Stamens 8. Style with 4 branches at the tip.
EVENING-PRIMROSE FAMILY *(Onagraceae)* 185

28. Sepals and petals in 5's. The flowers usually in umbels.

31. Pistil with 2 to 5 styles. Fruit a berry.
GINSENG FAMILY *(Araliaceae)* 189

31. Pistil with 2 styles, surrounded at the base by a prominent disk. Fruits dry. CARROT FAMILY *(Apiaceae)* 190

GROUP IV.
FLOWERS WITH TWO SETS OF FLORAL ENVELOPES, BOTH SEPALS AND PETALS. THE PETALS MORE OR LESS UNITED.

1. Ovary of the pistil free from other parts of the flower and apparently situated above them.

    2. Flowers with a radial symmetry or very nearly so.

3. Stamens not attached to the corolla.

    4. Petals but slightly united at the base.
    WINTERGREEN FAMILY *(Pyrolaceae)* 199
    4. Petals well united at least toward the base.
    HEATH FAMILY *(Ericaceae)* 200

3. Stamens attached to the corolla.

    5. Anther-bearing stamens 5, these alternating with scale-like sterile ones. PINCUSHION-PLANT FAMILY *(Diapensiaceae)* 200
    5. Anther-bearing stamens not alternating with scale-like sterile ones.

6. Stamens placed opposite the lobes of the corolla.
PRIMROSE FAMILY *(Primulaceae)* 203
6. Stamens alternating with the corolla lobes.

    7. Pistils 2, at least the stigmas united and maturing into pairs of pod-like fruits. Seeds with silky tufts of hairs. Plants usually with a milky juice.

8. Styles of the pistils united. Stamens separate.
DOGBANE FAMILY *(Apocynaceae)* 212
8. Styles of the pistils not united. Anthers of the stamens united in a tube about the disk-like stigma of the pistil.
MILKWEED FAMILY *(Asclepiadaceae)* 212

    7. Pistil solitary or apparently so.

9. Ovary without partitions and containing many ovules.

    10. Leaves opposite and entire. GENTIAN FAMILY *(Gentianaceae)* 204
    10. Leaves alternate or all at the base; either toothed, lobed, or divided. WATERLEAF FAMILY *(Hydrophyllaceae)* 224

9. Ovary of the pistil showing 2 or more partitions in cross-section, or with 4 lobes.

    11. Stamens 2 or 4 in number.

12. Fruit a group of 4 hard seed-like structures.
VERBENA FAMILY *(Verbenaceae)* 228
12. Fruit a 2-celled and many-seeded capsule.
ACANTHUS FAMILY *(Acanthaceae)* 252

    11. Stamens 5 in number.

13. Pistil with a deeply 4-lobed ovary which develops into a group of 4 hard seed-like structures. BORAGE FAMILY *(Boraginaceae)* 227
13. Pistil with an ovary which is not 4-lobed.

    14. Stigma of the pistil with 3 branches.
    PHLOX FAMILY *(Polemoniaceae)* 220
    14. Stigma unbranched or with 2 or 4 branches.

15. Leaves opposite. PINKROOT FAMILY *(Loganiaceae)* 204
15. Leaves alternate.

16. Ovary with 2 chambers, each containing a pair of ovules. MORNING-GLORY FAMILY *(Convolvulaceae)* 219
16. Ovary with 2 chambers, each containing a number of ovules. NIGHTSHADE FAMILY *(Solanaceae)* 239

2. Flowers with a bilateral symmetry.

17. Aquatic plants with minute bladders on the usually thread-like leaves; or plants of wet places without obvious leaves, or with a basal rosette of pale green leaves and solitary spurred flowers on naked stems. BLADDERWORT FAMILY *(Lentibulariaceae)* 251
17. Plants otherwise.

18. Plants (of woodlands) which lack green color and true leaves; they are parasitic on roots of other plants. BROOM-RAPE FAMILY *(Orobanchaceae)* 248
18. Plants with green color and normal leaves.

19. More or less woody vines. BIGNONIA FAMILY *(Bignoniaceae)* 248
19. Plants other than woody vines.

20. Ovary of the pistil deeply 4-lobed, developing into a group of 4 hard, dry, seed-like structures.

21. Anther-bearing stamens 5. Flower clusters unrolling from the tip and straightening as the flowers expand. Leaves alternate. BORAGE FAMILY *(Boraginaceae)* 227
21. Anther-bearing stamens either 2 or 4, if 4 one pair shorter than the other. Leaves opposite. MINT FAMILY *(Labiatae)* 230

20. Ovary of the pistil not deeply 4-lobed.

22. Ovary 1-celled and 1-seeded, in fruit bending sharply downward. VERBENA FAMILY *(Phryma)* 230
22. Ovary otherwise.

23. Ovary developing into 4 hard, dry, seed-like structures. VERBENA FAMILY *(Verbenaceae)* 228
23. Ovary 2-celled, each cell containing a number of small ovules.

24. Seeds supported by short hooks or cup-like expansions of the ovary wall, forcibly thrown by the exploding capsules. ACANTHUS FAMILY *(Acanthaceae)* 252
24. Seeds neither thus supported nor thrown by the exploding capsules. FIGWORT FAMILY *(Scrophulariaceae)* 238

1. Ovary of the pistil united with the calyx tube and apparently situated below the other parts of the flower.

25. Flowers in dense heads which are surrounded by an involucre of bracts.

26. Stamens 4, separate from one another. Stout and prickly plants with opposite leaves. TEASEL FAMILY *(Dipsacaceae)* 259
26. Stamens 5, their anthers united into a ring about the style of the pistil. Heads may have either tubular flowers, flowers with flat and strap-shaped corollas, or both. ASTER FAMILY *(Asteraceae)* 264

25. Flowers not in dense heads.

27. Stamens and pistils in separate flowers. Plants are tendril-bearing vines. GOURD FAMILY *(Cucurbitaceae)* 259

27. Stamens and pistils present in the same flowers.

 28. Stamens 3. Calyx lobes minute or not evident. VALERIAN FAMILY *(Valerianaceae)* 259

 28. Stamens 4 or 5. Calyx lobes usually prominent.

29. Plants with usually opposite leaves.

 30. Flowers with a radial symmetry. Corolla lobes and stamens 4. MADDER FAMILY *(Rubiaceae)* 255

 30. Flowers often with a bilateral symmetry. Corolla lobes and stamens usually 5. HONEYSUCKLE FAMILY *(Caprifoliaceae)* 256

29. Plants with alternate leaves. BLUEBELL FAMILY *(Campanulaceae)* 259

BUR-REED FAMILY (Sparganiaceae)

Bur-reeds are aquatic or marsh plants with long ribbon-like leaves, the bases of which sheath the stems. The flowers are of two kinds—neither of them having sepals or petals—and are arranged in ball-like heads. The one to a few flower heads near the base of the cluster contain the pistil-bearing flowers. Those in the upper part of the cluster are made up of stamen-bearing flowers. The plants get their name from the bur-like heads of pistil-bearing flowers and fruits.

AMERICAN BUR-REED *(Sparganium americanum)*
This is the most widely distributed species of bur-reed. It is often common on mucky or peaty shores and in the shallow water of marshes. The plants grow from 1 to about 3 feet tall and have ball-like fruit heads from ¾ to about an inch in diameter. It blooms between May and September. RANGE: Nfd. to Minn. south to Fla. and Mo.

WATER-PLANTAIN FAMILY (Alismataceae)

Members of this family are marsh or aquatic plants having leaves in a basal cluster. Their flowers are often quite showy and arranged in clusters on scape-like leafless stems. They have 3 greenish sepals, 3 white to pinkish petals and usually more than 6 stamens and pistils. The best known members of the family are the plants called "arrowheads", as many of them have more or less arrow-shaped leaves. Flowers of the sagittarias or arrowheads are usually in whorls of threes. Those in the lowermost whorls have pistils or sometimes both pistils and stamens, but those in the upper whorls have only stamens.

BROAD-LEAF ARROWHEAD *(Sagittaria latifolia)*
This is the commonest, best known, and most widely distributed of all the arrowheads. The leaves are quite variable in width but are usually arrow-shaped. Its 3-petalled white flowers with clusters of golden stamens are quite attractive and may be seen from June to September. They are subtended by thin papery bracts. Also called Wapato or Duck-potato, the tubers of the plant were a food of the Indian.
RANGE: N.S. to B.C. south throughout most of the U.S.

BULL-TONGUE ARROWHEAD *(Sagittaria lancifolia)*
Although recognizable as an arrowhead by its flowers, this species does not have arrow-shaped leaves. Its leaves are pointed at both ends and usually 6 to 24 inches long. It grows in tidal marshes and along streams and ditches near the coast; blooming from June to October. RANGE: Del. south to Fla. and west to Tex.

LANCE-LEAF WAPATO *(Sagittaria lancifolia)*
This species is similar to the preceeding but it is somewhat larger, with leaves up to 2 feet long. It may be readily distinguished, however, by the minute nipple-like projections on both its sepals and bracts. RANGE: S.C. south to Fla. and west to Tex. (Not illustrated)

GRASS-LEAF SAGITTARIA *(Sagittaria graminea)*
The leaves of this species are variable, ranging from lance-shaped to narrow and ribbon-like. It is often common in shallow waters, especially where the botton is sandy.
RANGE: Nfd. to Ont. south to Fla. and Tex.

SEDGE FAMILY (Cyperaceae)

Sedges are grass-like plants which usually grow in moist or wet places. Most of them have 3-sided stems which are sheathed by the bases of the leaves. The flowers are small and quite inconspicuous, being arranged in scaly-bracted and usually cone-shaped clusters. Although there are several genera of sedges, some with numerous species, the following, because of their handsome white bracts, are the only ones likely to be considered as wild flowers.

WHITE-TOPPED SEDGE *(Rhynchospora latifolia)*
This is a conspicuous plant in the wetter pinelands and savannahs of the Southeast from May until after mid-summer. What appear to be petals are really the white leaf-like bracts which surround the cluster of true flowers. They are 1½ to 3 inches long, mostly white but with green tips, and usually more than 7 in number.
RANGE: N.C. south to Fla. and west to Texas.

Another but less showy species is *Rhynchospora colorata*. It usually has less than 7 narrower white bracts, and occurs north to the coast of Virginia.

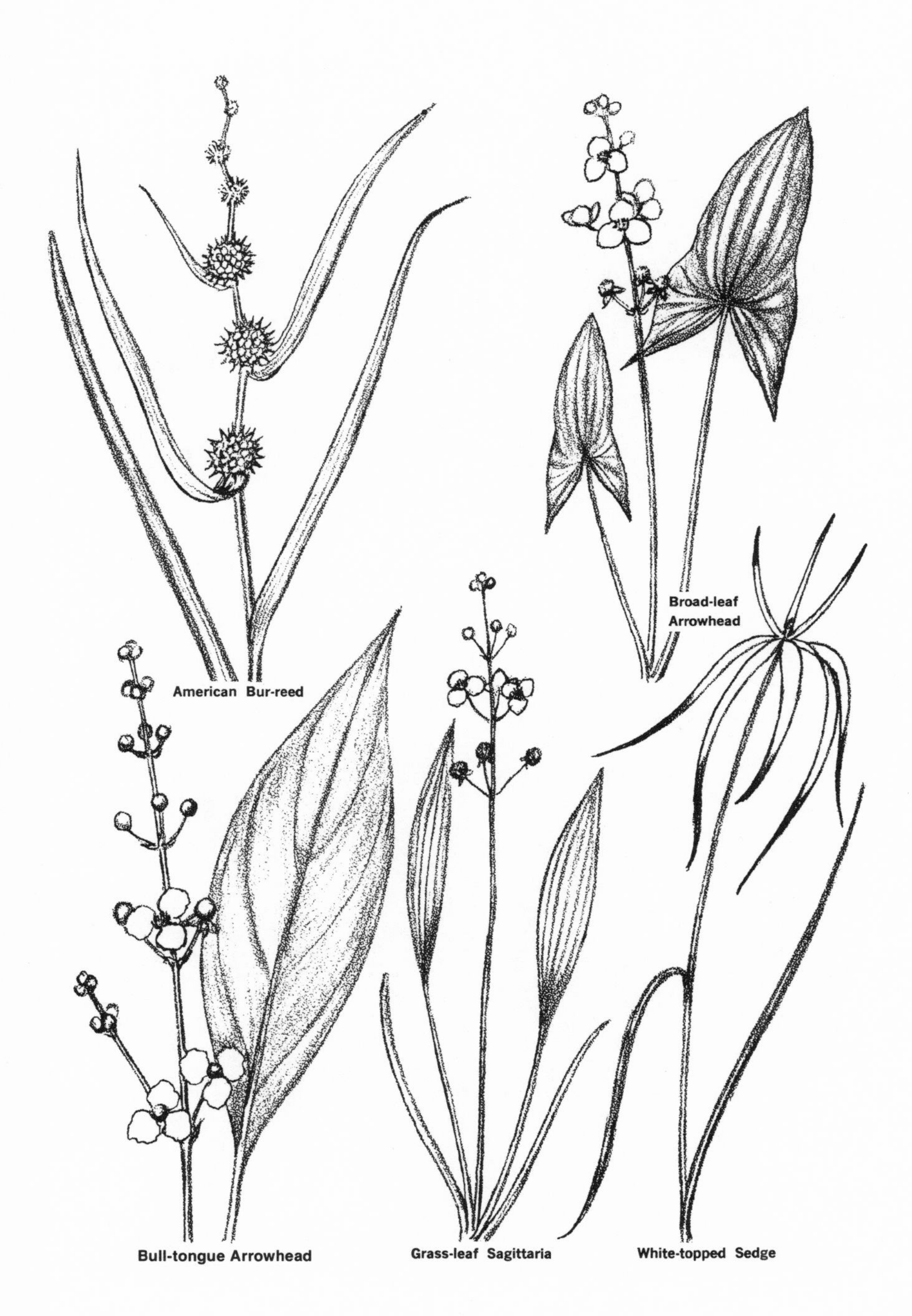
American Bur-reed
Broad-leaf Arrowhead
Bull-tongue Arrowhead
Grass-leaf Sagittaria
White-topped Sedge

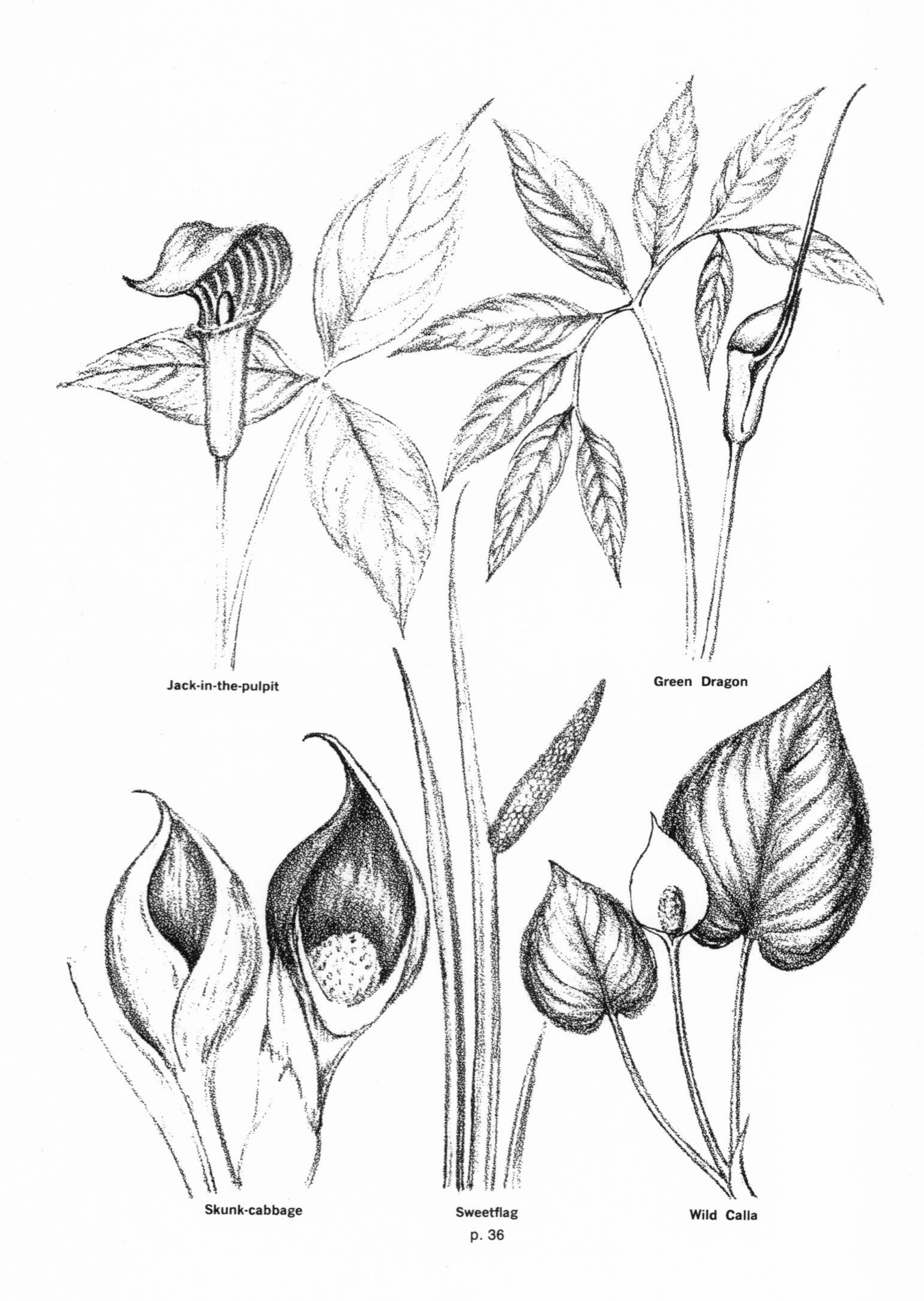
Jack-in-the-pulpit
Green Dragon
Skunk-cabbage
Sweetflag
Wild Calla

p. 36

ARUM FAMILY (Araceae)

The arums have minute flowers usually consisting of either stamens or a pistil—or sometimes both—which are crowded on a fleshy stem called a *spadix*. Usually this spadix is subtended by a bract—often a very large one—which is called a *spathe*. The plants contain minute crystals of calcium oxylate which cause an intense burning sensation when parts of the plants are eaten. The crystals, however, seem to be destroyed by drying or boiling; thus many of these plants were often used as food by the Indians.

JACK-IN-THE-PULPIT *(Arisaema triphyllum)*

This plant is quite familiar to country boys as the "Indian-turnip". Jack is the club-like spadix which bears tiny flowers toward its base. The spathe forms what looks like an old-fashioned canopied pulpit from which Preacher Jack preaches. The hood portion of the spathe may be green or purple, quite often marked with paler or whitish stripes. Jack-in-the-Pulpits are often quite common in rich, moist woods and thickets. The plants grow from 1 to rarely 3 feet tall, and each of the two leaves are divided into 3 leaflets. Towards fall the spathe withers and discloses a cluster of brilliant red, berry-like fruits. The flowering season is between late March and June.

RANGE: N.B. to Man. south to n. Fla. and Miss.

GREEN DRAGON *(Arisaema dracontium)*

A unique plant with usually but one leaf which is divided into from 7 to 15 lance-shaped leaflets or leaf segments. The spadix is prolonged into a long, pointed tail-like portion which projects upward through an opening in the spathe. Less common and much less familiar than the Jack-in-the-pulpit, the Green Dragon grows in low wet woodlands or on flats along streams. The plants are from 1 to 3 feet high, and may be found in flower between May and July.

RANGE: N.H. to Ont. and Minn. south to Fla. and Tex.

SKUNK-CABBAGE *(Symplocarpus foetidus)*

All parts of this plant have a characteristic skunk-like odor when bruised or broken. The large hood-like spathes vary from yellowish-green to reddish-brown and are streaked and spotted with purple. They often appear above the half-thawed ground in late February or March, in wet meadows or swampy woods. These flower spathes are soon followed by big veiny leaves which have heart-shaped bases. They unfurl from tightly wrapped cones as the spathes wither and are very conspicuous throughout the summer.

RANGE: Que. to Man. south to w. N.C., n. Ga. and Tenn.

WILD CALLA *(Calla palustris)*

The Wild Calla or Water-arum grows in cold northern bogs and swamps. It is small plant seldom a foot high; with beautiful dark green, heart-shaped leaves from 2 to 4 inches wide. The nearly flat, snow-white spathe is sometimes 2 inches across and stands behind the spadix on which the real flowers are clustered. It blooms from late April to June or sometimes later. A good plant for bog gardens in regions with cool summers but it must have an acid soil.

RANGE: Nfd. to Alaska south to n. N.J., n. Pa. and Great Lakes region.

GOLDEN-CLUB *(Orontium aquaticum)*

A plant of swamps, pond margins, bogs, and slow-moving streams. The "club" is a spadix from 4 to 8 inches long, tipped with closely packed yellow flowers. Near its base is a small bract-like spathe. The bluish-green leaves are 6 to 12 inches long and often float on the water. When emersed they show a silvery or coppery irridescence and come out perfectly dry. For this reason the plant is locally known as the Never-wet. The bright yellow clubs are produced between late March and May. (Illustrated on page 37.)

RANGE: Mass. to Ky. south to Fla. and Tex.

ARROW-ARUM *(Peltandra virginica)*

This is a common plant of swamps and shallow waters. The rather large arrow-shaped leaves stand a foot or more above the water. Although they look somewhat like the leaves of the broad-leaf arrowheads, they have an entirely different venation. The narrow greenish spathe is wrapped around the spadix which is covered with small flowers. It blooms during May or June. RANGE: Me. to Ont. south to Fla. and Tex.

WHITE-ARUM *(Peltandra sagittifolia)*

This is similar to the preceding species but it has spathes which are whitish and spreading. It grows in boggy places in the coastal plain from N.C. south to Fla. and w. to Miss.

CALAMUS FAMILY (Acoraceae)

SWEETFLAG *(Acorus calamus)*

Rootstocks of the Sweetflag are pleasantly aromatic and in bygone days were often boiled in syrup and eaten as a candy. The spathe in this plant looks very much like the long, narrow, and flattened leaves, being prolonged upward 2 or 3 inches above the outwardly pointing spadix. It grows in wet open places and the flowering season is between May and July. (Illustrated on page 34.)

RANGE: N.S. to Ore. south to S.C., Tenn., Miss., and Tex.

YELLOW-EYED-GRASS FAMILY (Xyridaceae)

Yellow-eyed grasses are rush-like plants with narrow leaves and naked flower stalks which terminate in scaly heads of flowers. The flowers have 3 unequal and brownish sepals, 3 yellow petals, 3 anther-bearing stamens alternating with 3 sterile filaments, and a pistil composed of 3 united carpels. Most of the flower parts are hidden by the large overlapping bracts. There are several species in the Southeast, especially in the coastal plain.

SLENDER YELLOW-EYED-GRASS *(Xyris torta)*

This species has a tuft of deeply-rooted chestnut-brown bulbs from which arise the narrow and often twisted leaves and slender, wiry flower stalks. The latter are much longer than the leaves, from 10 to 30 inches tall, 2-edged above and are often slightly twisted. It blooms in June or July. RANGE: N.J. south to Fla. west to Ark. and e. Texas.

PIPEWORT FAMILY (Eriocaulaceae)

Pipeworts are plants of wet places. They have small flowers crowded in a button-like head at the summit of a long stalk, which is naked except for a sheath at the base. They so remind one of the old-fashioned hatpins that hatpins is one of their common names. Their leaves are narrow and grass-like.

TEN-ANGLED PIPEWORT *(Eriocaulon decangulare)*

This common species has a flower stalk 1 to 3 feet tall which is 10- or 12-angled. The blunt-pointed leaves are as long as, or longer than, the sheath at the base of the flower stalk; and the flowers are in a whitish wooly head about ½ inch in diameter. It blooms from June to October. RANGE: N.J. south to Fla. and west to Tex.

FLATTENED PIPEWORT *(Eriocaulon compressum)*

Has basal leaves sharply pointed and shorter than the sheaths of the flower stalks. It grows in shallow waters of the coastal plain from N.J. south to Fla. and west to Tex. (Not illustrated)

SEVEN-ANGLED PIPEWORT *(Eriocaulon aquaticum)*

Is smaller and often has flower stalks only 1 to 8 inches long, which are 7-angled. It grows from Nfd. to Ont. south to Va. and the Great Lakes; blooming July to September. (Not illustrated)

SHOE-BUTTONS *(Syngonanthus flavidulus)*

Is often called the Yellow Pipewort. Its straw-colored flower heads are on naked stalks 4 to 12 inches tall and arise from a cluster of short, awl-like leaves which are woolly at the base. It grows in wet pinelands and bogs of the coastal plain from N.C. south to Fla. and Ala.; blooming May to October. (Not illustrated)

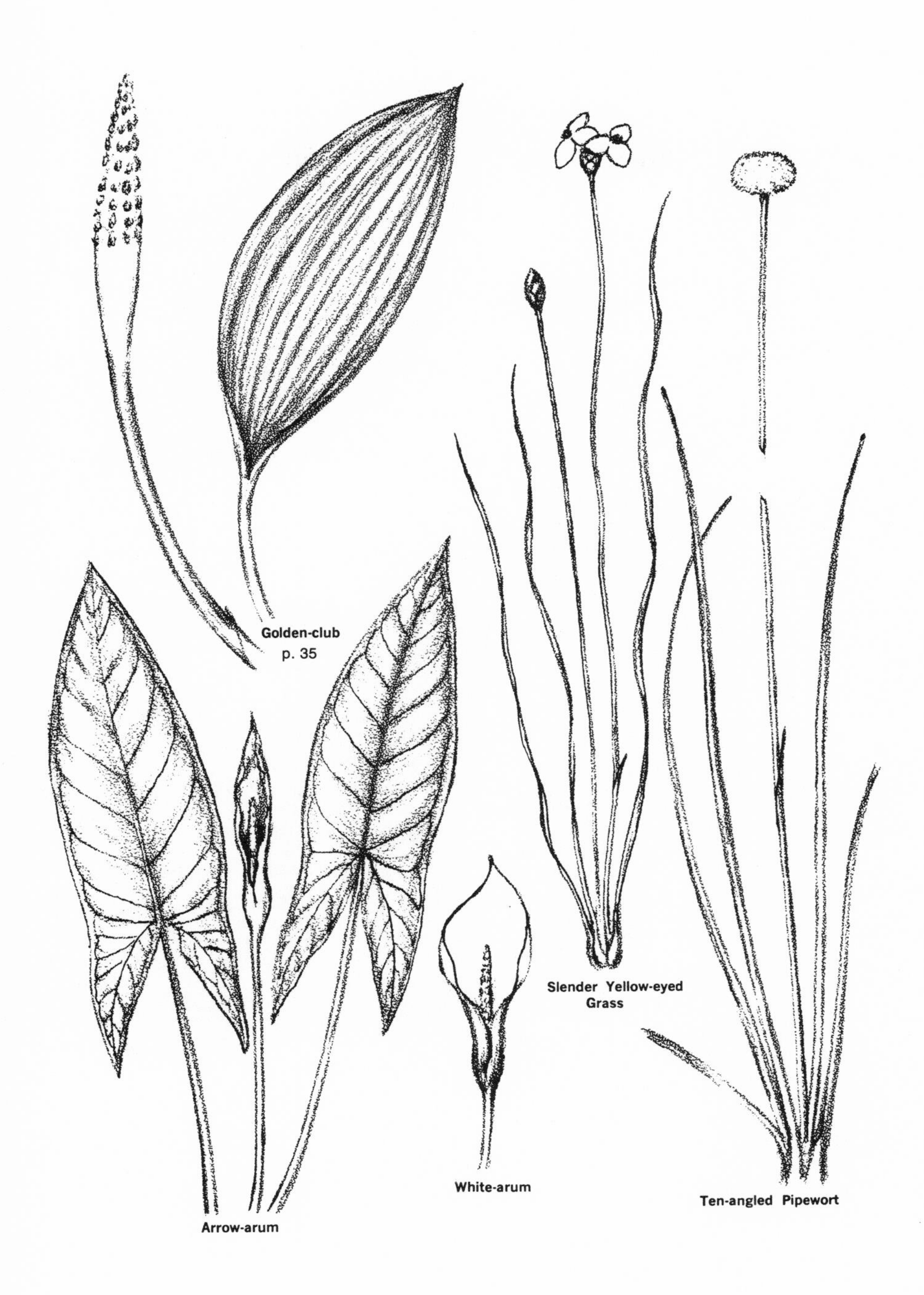

Golden-club
p. 35
Slender Yellow-eyed Grass
Arrow-arum
White-arum
Ten-angled Pipewort

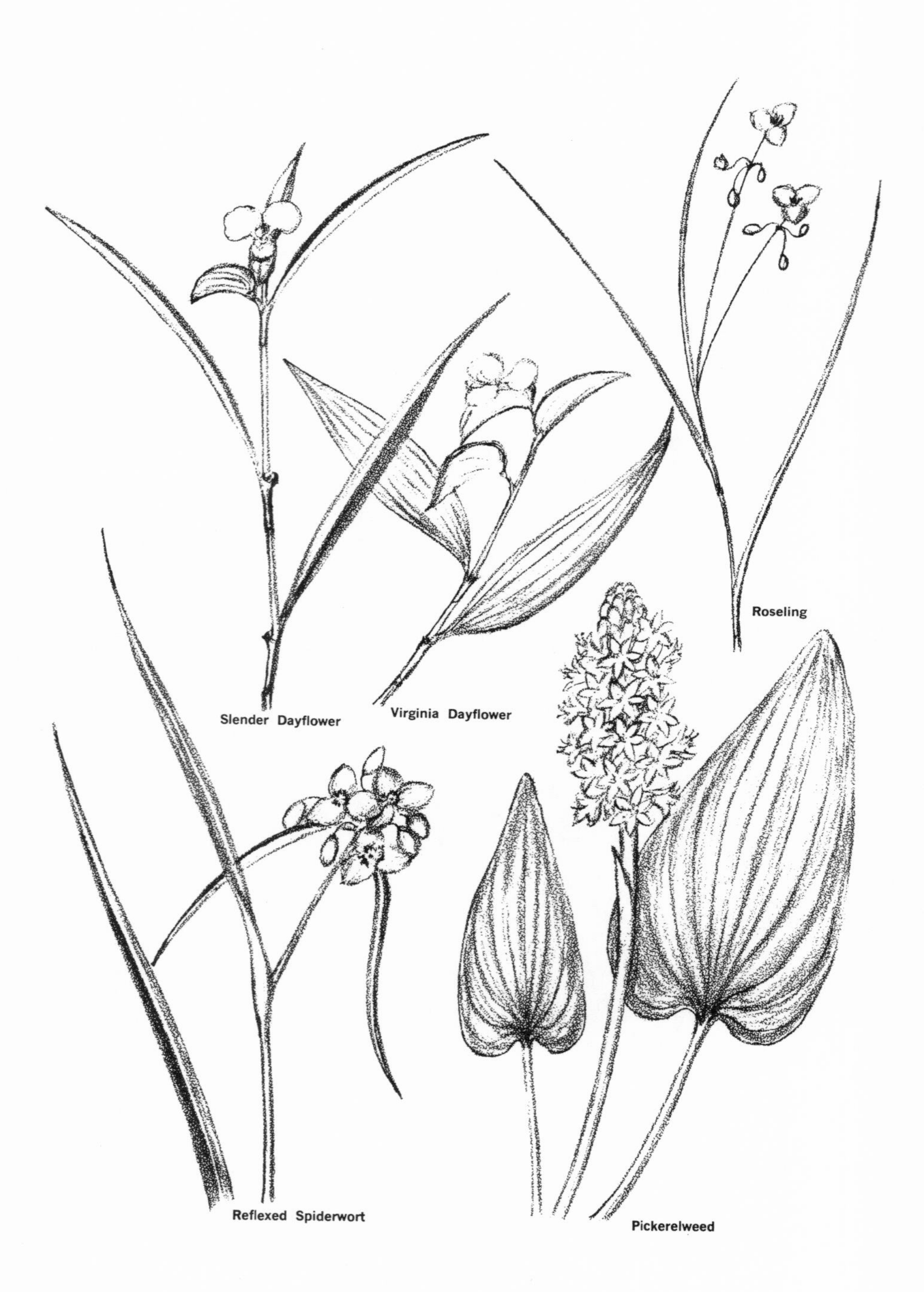
Roseling
Slender Dayflower
Virginia Dayflower
Reflexed Spiderwort
Pickerelweed

SPIDERWORT FAMILY (Commelinaceae)

Members of this family have flowers with 3 greenish or bronzy sepals, 3 petals, 6 stamens, and a 3-parted pistil. Only 1 or 2 of the flowers in the cluster open at a time. They open in the morning and by noon the petals become but blobs of liquid.

SLENDER DAYFLOWER *(Commelina erecta)*

Like the flowers of other dayflowers, those of this species appear from folded, spathe-like bracts; and they have 3 perfect stamens and 3 sterile ones with X-shaped anthers. The flowers of this species show but 2 bright blue petals; the lower one being much smaller, white, and inconspicuous. The narrow leaves are from 2 to 6 inches long, and leaf sheaths which surround the stem at the leaf bases have a spreading green flange. The plant itself grows from a few inches to about 2 feet tall. It is found in dry sandy woods and on rock outcrops; blooming between June and October.

RANGE: Pa. to Kan. south to Fla. and Tex.

VIRGINIA DAYFLOWER *(Commelina virginica)*

Flowers of this species have 3 bright blue petals, the lower one being but slightly smaller. It has broader lance-shaped leaves than the preceding species, and the leaf sheaths are fringed at the top with reddish hairs. The Virginia Dayflower becomes 2 to 3 feet tall, and grows in moist woods and thickets. It blooms between July and October.

RANGE: N.J. to Ill. and Kan. south to Fla. and Tex.

COMMON DAYFLOWER *(Commelina communis)*

This is a widely distributed and weedy plant often found in low moist grounds and waste places. It is a sprawling plant which often roots at the joints of the stem. Its flowers have 2 large blue petals and a smaller white one, and the edges of the spathes are not united near the base. The leaves are broadly lance-shaped to egg-shaped. (Not illlustrated)

REFLEXED SPIDERWORT *(Tradescantia ohiensis)*

Spiderworts have flower clusters which are subtended by long, narrow, leaf-like bracts. The flowers of this one have 3 delicate violet or purple petals and the filaments of the 6 stamens are bearded with hairs. The 3 sepals have tufts of hair only at the tips. The plant is 1 to 2 feet tall and both the stem and narrow leaves are smooth and whitened with a bloom. It grows in dry woods and open places, and blooms between April and July.

RANGE: Mass. to Neb. south to Fla. and Tex. Several other species occur in the eastern United States.

ROSELING *(Callisia rosea)*

This small spiderwort is distinguished by its slender stems, narrow grass-like leaves, and small flowers which have 3 bright rosy-pink petals. It is usually less than a foot tall and often grows in dense clumps in dry sandy pinelands of the coastal plain and piedmont. It blooms between May and July.

RANGE: Va. south to Florida.

ANEILEMA *(Murdannia keisak)*

This is a low, often creeping plant, that resembles the dayflowers. It has small lilac or bluish-purple, 3-petalled flowers with 2 or 3 fertile stamens which are borne in the axils of the upper leaves, which are narrowly lance-shaped and often folded. It grows along stream banks and in marshes; blooming in September or October.

RANGE: Va. south to Ga. (Not illustrated)

PICKERELWEED FAMILY (Pontederiaceae)

PICKERELWEED *(Pontederia cordata)*

The bright lavender-blue flowers of this plant may be seen from May to September, or even later in Florida. There are 2 yellow spots on the middle upper lobe of the corolla but these can be seen only at close range. The leaves vary from egg-shaped to lance-shaped and they may have heart-shaped or wedge-shaped bases. It grows quite commonly on muddy shores and in shallow waters.

RANGE: N.S. to Ont. south to Fla. and Tex.

LILY FAMILY (Liliaceae)

Members of the family typically have flowers with a perianth of 6 similar parts, 6 stamens, and a pistil composed of 3 united carpels. The styles may be separate or united into one style. The ovary of the pistil is separate from any of the other flower parts and it occupies a prominent position in the center of the flower.

The *Bunchlily Family* includes members having a flat perianth of 6 similar parts, usually persisting about the bases of the fruits. The pistil has 3 separate styles and often appears to be 3-parted or 3-horned.

The *Onion Family* includes members whose flowers have a flat or dish-shaped perianth of 6 similar parts, the flowers being arranged in umbels subtended by 2 or 3 thin and dry bracts. There is a single compound pistil with but 1 style and the fruits are few-seeded capsules.

The *Lily Family* includes members whose flowers are flat to bell-shaped, usually quite large and showy, and have 6 similar perianth parts sometimes united near the base. There is a single compound pistil with but 1 style and the fruits are capsules.

The *Lily-of-the-valley Family* includes members with moderate to small-sized, flat to bell-shaped flowers with 6 similar perianth parts sometimes united at the base. There is a single compound pistil with but 1 style and the fruits are berries.

The *Trillium Family* includes members whose 3 petals are more or less distinct from the 3 sepals. The pistil has 3 very prominent stigmas and sometimes a short style and the fruits are berry-like. The plants have their leaves in whorls.

Key to Genera of the Lily Family

1. Ovary wholly or partly inferior.

  2. Perianths hairy externally.

3. Inflorescence 1- to few-flowered. Perianth completely covering ovary. Sepal and petal without woolly tufts. STAR-GRASS *(Hypoxis)*

3. Inflorescence many-flowered. Summit of ovary free from perianth. Woolly tufts at sepal and petal bases. GOLDEN-CREST *(Lophiola)*

  2. Perianth glabrous externally.

4. Scape terminating in 2 to several bracts and an umbel of sessile flowers. Flowers with a conspicuous corona. SPIDER-LILIES *(Hymenocallis)*

4. Scape with a single terminal bract and single flower. Flowers without a corona. ATAMASCO-LILY *(Zephyranthes)*

  1. Ovary superior.

5. Plants with only basal leaves.

  6. Flower stalks forking above; the flowers large and tawny-orange. DAY-LILIES *(Hemerocallis)*

  6. Flower stalks not forking above; the flowers medium or small.

7. Flowers solitary. Leaves mottled. ADDER'S-TONGUES *(Erythronium)*

7. Flowers in clusters.

  8. Styles 3, the pistil sometimes appearing 3-parted or 3-horned.

9. Leaves pointed at the tip. TOFIELDIAS *(Tofieldia)*

9. Leaves blunt at the tip. FLY-POISON *(Amianthium)*

  8. Style 1.

10. Flowers arranged in umbels.

  11. Flower cluster subtended by 2 or 3 papery bracts.

12. Plants with an onion-like odor. ONIONS, LEEKS, ETC. *(Allium)*

12. Plant without an onion-like odor. FALSE GARLIC *(Nothoscordum)*

  11. Flower cluster not subtended by bracts. Leaves broad. CLINTONIAS *(Clintonia)*

10. Flowers in a long and narrow type of cluster.

13. Leaves broad. Flowers bell-shaped, white, and very fragrant. LILY-OF-THE-VALLEY *(Convallaria)*

13. Leaves narrow and grass-like.

14. Flowers white or yellow, perianth cylindrical bell-shaped. COLIC-ROOTS *(Aletris)*

14. Flowers bluish or lavender, the perianth somewhat bilateral. CAMAS *(Camassia)*

5. Plants with leafy stems, leaves of the flower stems sometimes reduced in size.

15. Sepals and petals not alike in size or color. TRILLIUMS *(Trillium)*

15. Sepals and petals much alike in size and color.

16. Flowers large (2 inches or more across); the perianth yellow, orange or red and usually spotted. LILIES *(Lilium)*

16. Flowers much smaller.

17. Leaves in whorls. INDIAN CUCUMBER-ROOT *(Medeola)*

17. Leaves alternate.

18. Styles 3, the pistil sometimes appearing 3-parted or 3-horned.

19. Flowers in a long and narrow type of cluster.

20. Basal leaves broad but those along the flower stalk much smaller and narrow.

21. Flowers white; the stamen-bearing and pistil-bearing ones on separate plants. DEVIL'S-BIT *(Chamaelirium)*

21. Flowers pink, anthers bluish. SWAMP-PINK *(Helonias)*

20. Leaves all long and narrow, those of the flower stem smaller.

22. Leaves very narrow (almost like pine needles) and wiry. Flowers white. TURKEY-BEARD *(Xerophyllum)*

22. Leaves narrow but flat and grass-like. Flowers greenish-white, each perianth segment with a 2-lobed greenish gland about midway from the base. POISON-CAMAS *(Zigadenus)*

19. Flowers in a broad and branching type of cluster.

23. Perianth parts with a pair of greenish glands at the base.

24. Perianth parts with stalk-like bases, in age turning brown. BUNCHFLOWERS *(Melanthium)*

24. Perianth parts without stalked bases, in age turning pink. SOUTHERN-CAMAS *(Zigadenus)*

23. Perianth parts without glands at the base.

25. Flower cluster quite smooth. BUNCHFLOWERS *(Melanthium)*

25. Flower cluster rather downy.

26. Leaves narrow and grass-like. Flowers white turning greenish or bronze. FEATHERBELLS *(Stenanthium)*

26. Leaves broad and prominently plaited lengthwise. Flowers yellowish-green. WHITE-HELLEBORE *(Veratrum)*

18. Style single, the stigmas sometimes 3-parted.

27. Flowers medium-sized, bell-shaped or cylindrical.

28. Flower solitary, or in pairs at the tips of leafy branches.

29. Flowers plain yellow. BELLWORTS *(Uvularia)*

29. Flowers greenish-white, or yellow with purple spots. DISPORUMS *(Disporum)*

28. Flowers arising from the axils of the leaves.

30. Flowers cylindrical bell-shaped, greenish-white, usually 2 or more grouped together. SOLOMON'S-SEALS *(Polygonatum)*

30. Flowers open bell-shaped, greenish-white to rose-purple, usually solitary. TWISTED-STALKS *(Streptopus)*

27. Flowers small, neither bell-shaped nor cylindrical, white to creamy-white and borne in end clusters. FALSE SOLOMON'S-SEALS *(Maianthemum)*

TURKEY-BEARD *(Xerophyllum asphodeloides)*

This plant has a dense basal cluster of slender leaves which might easily be mistaken for a tuft of wiry grass or the needles of a young Longleaf Pine. Sometime in May or June it sends up a flower stalk from 2 to 3 feet tall. Along it are a few scattered leaves and at the top there is a densely cylindrical cluster of white flowers. It grows in dry woodlands, southward chiefly in the mountains.
RANGE: Coastal plain N.J. and Del. south to Va.; mountains Va. to Ga.

WOOD-FEATHERLING *(Tofieldia racemosa)*

In June or July the flowers of this plant often whiten the wet pinelands and savannahs of the southeastern coastal plain. The long, narrow, grass-like leaves are all at the base of the plant. Its flower stalk is from 1 to 2 feet tall and bears a long, narrow, more or less interrupted cluster of small white flowers. Both the upper part of the flower stalk and the pedicels of the individual flowers are roughened by minute glandular hairs.
RANGE: N.J. south to Fla. and west to Tex.

VISCID TOFIELDIA *(Tofieldia glutinosa)*

This is a northern plant which occurs southward only in the high mountains. It closely resembles the preceding species but it has a sticky-hairy stem 6 to about 20 inches tall which is sprinkled with black dots. A plant of boggy places, flowering in July or August.
RANGE: Nfd. to Man. south to N. Eng, the Great Lakes, and in mts. to n. Ga.

WHITE-FEATHERLING *(Tofieldia glabra)*

September and October are the months to look for the slender white flower clusters of the White-featherling. The plant is quite similar to the prceding species but it has a flower stalk which is entirely smooth. It grows in wet pinelands and savannahs of the coastal plain in the Carolinas. (Not illustrated)

DEVIL'S-BIT *(Chamaelirium luteum)*

Another and much more attractive name for this plant is Fairy-wand. Most of the leaves are in a rosette at the base of its slender 2 to 3 foot flower stalk. Along the stalk are just a few smaller and narrower leaves; and toward the summit there is a long, slender cluster of small white flowers. The flowers on a given plant have either stamens or pistils, but not both. It grows in rich moist woods and thickets; blooming between April and June. A good subject for a woodland wild flower garden.
RANGE: Mass. to Ont. south to Fla., Miss. and Ark.

FLY-POISON *(Amianthum muscitoxicum)*

Between May and July this plant sends up a practically leafless flower stalk from 1 to 3 feet tall. On it is a dense cylindrical cluster of flowers which are at first white but later become greenish or purplish. Both leaves and flower stalk arise from a very poisonous bulb-like base. Another name for the plant is Crow-poison. Whether or not flies and crows are affected by it we cannot say, but cattle and sheep are often poisoned by eating the leaves. It grows in a variety of places: dry to moist open woods, mountain balds, bogs, and savannahs.
RANGE: N.Y. and Pa. south to Fla., Miss. and Okla.

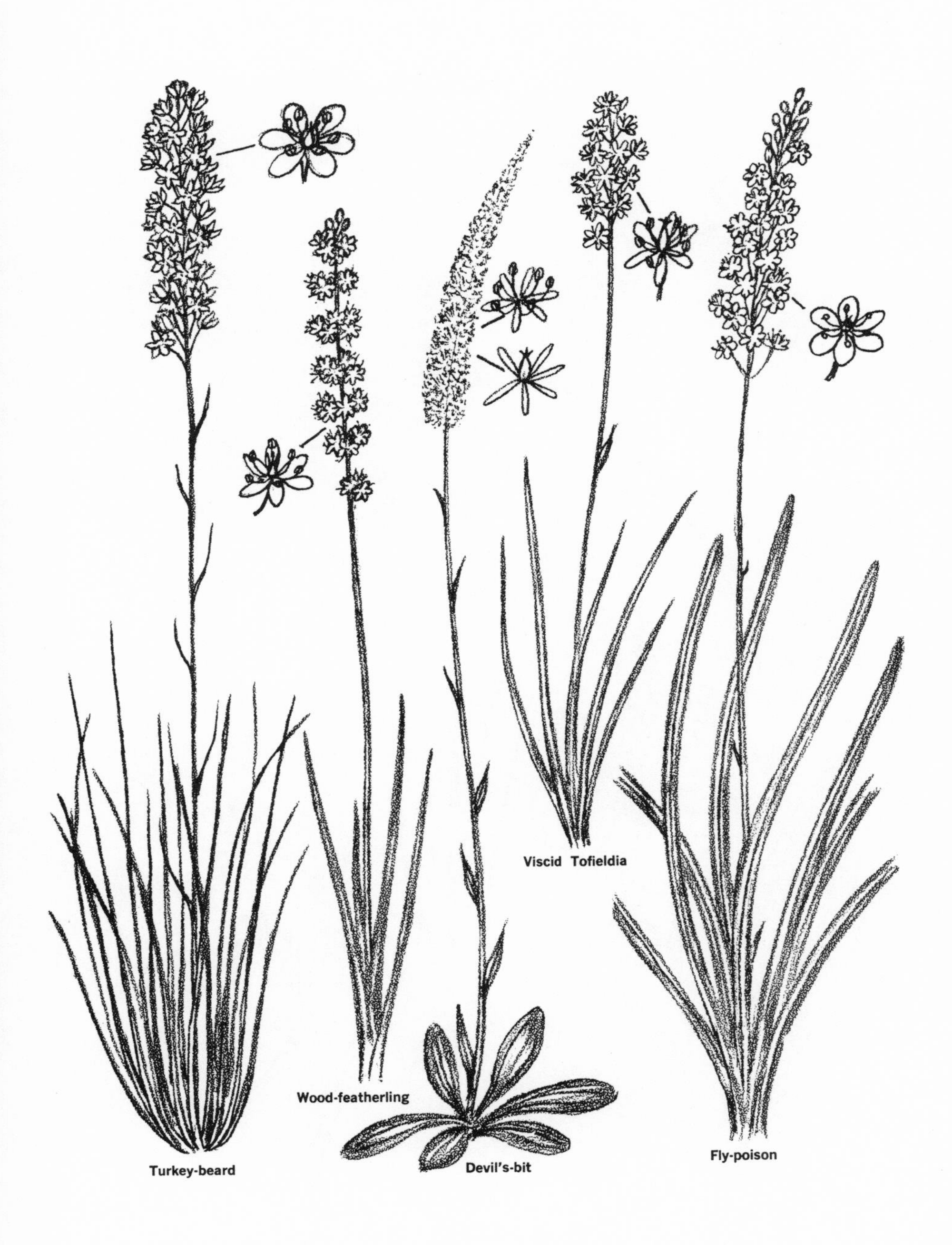
Turkey-beard
Wood-featherling
Devil's-bit
Viscid Tofieldia
Fly-poison

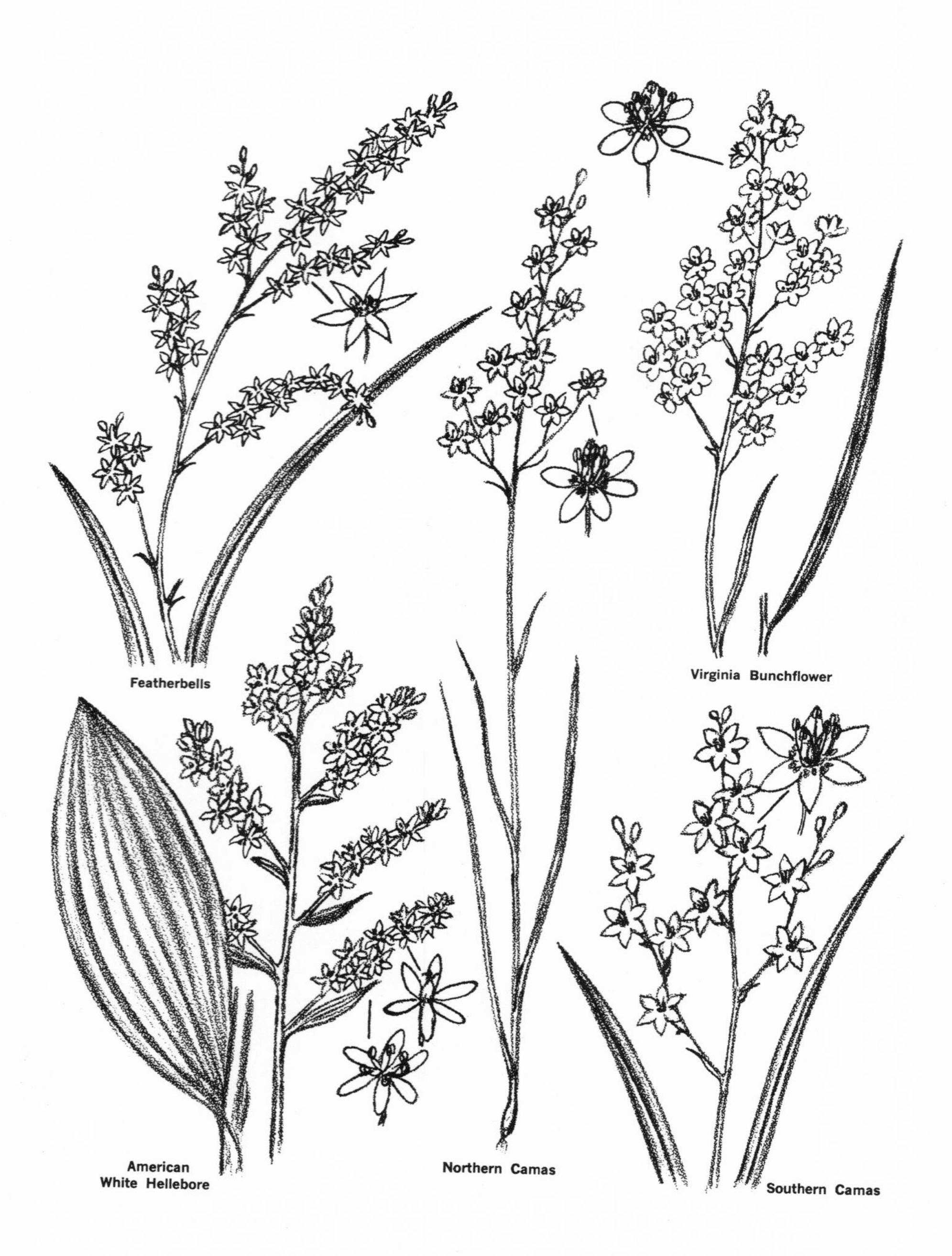
Featherbells
Virginia Bunchflower
American
White Hellebore
Northern Camas
Southern Camas

FEATHERBELLS *(Stenanthium gramineum)*

Often called Feather-fleece, this is an attractive plant with large and more or less drooping clusters of small, starry, white flowers. It grows to a height of 3 to 6 feet and has long, narrow, grass-like leaves which are more numerous and larger toward the base of the plant. This attractive plant grows in moist open woods, meadows, and bogs; and blooms between June and September.

RANGE: Pa. to Ill. and Mo. south to N.C., nw. Fla. and e. Tex.

AMERICAN WHITE HELLEBORE *(Veratrum veride)*

This is often a common and conspicuous plant in swamps and on moist wooded slopes. It has a stout, leafy, more or less downy stem from 2 to 5 feet tall. The large leaves are broadly elliptic, prominently veined and plaited, and have bases which clasp the stem. They are often mistaken for the leaves of some of the lady's-slipper orchids. A large branched cluster of small, downy, yellowish-green flowers is produced between June and August. It is also known as the False Hellebore and Indian-poke.

RANGE: N.B. to Minn. south to Md., n. Ga. and Tenn.

NORTHERN CAMAS *(Zigadenus elegans* var. *glaucus)*

The narrow and grass-like leaves of this plant are chiefly at the base, with only a few smaller ones along the 1 to 3 foot tall flower stalk. The flowers are rather small, creamy-white tinged with green or purplish on the back, and each of the 6 perianth segments has a 2-lobed greenish gland. It grows in moist rocky places and in bogs, blooming between July and September. Another name for this very poisonous plant is White Camas.

RANGE: N.B. to Minn. south to N.Y., w. N.C., and the Great Lakes region.

SOUTHERN CAMAS *(Zigadenus glaberrimus)*

This is a stout leafy-stemmed plant from 2 to 4 feet tall; with a basal cluster of long, narrow, strongly channelled leaves. The flowers are produced in a large, rather open, pyramid-shaped terminal cluster. They are an inch or more across and white when they first open, but they gradually turn pinkish. There is a pair of prominent greenish glands at the base of each of the six perianth segments. It grows in wet coastal plain pinelands, savannahs and bogs; blooming June to August.

RANGE: Va. south to Fla. and west to La.

ST. AGNES'-FEATHER *(Zigadenus densus)*

This plant has white flowers in a dense cylindrical cluster, and each perianth segment has but one gland. Its 2- to 3-foot tall stem is sparingly leafy, the grass-like leaves being chiefly basal. It grows in bogs and savannahs of the coastal plain from N.C. south to Fla. and west to Miss.; blooming between April and June. (Not illustrated)

SMALL-FLOWERED HELLEBORE *(Melanthium parviflorum)*

This plant has a slender leafy stem 2 to 5 feet tall, and lower stem leaves which taper at the base into long stalks. Large clusters of small, smooth, greenish flowers are produced between July and September. It grows in dry deciduous woodlands in the mountains from Va. and W.Va. south to Tenn. and n. Ga. (Not illustrated)

VIRGINIA BUNCHFLOWER *(Melanthium virginicum)*

The rather coarse, narrow, grasslike leaves of this plant are chiefly at the base of 3- to 4-foot tall, rough-hairy stem. Its dime-sized flowers are creamy-white at first but later turn greenish or purplish. Each of the six perianth parts are stalked at the base and have a pair of greenish glands. The flowers are produced in rather large showy clusters between June and August. It grows in moist to wet woods, meadows and bogs.

RANGE: N.Y. to Iowa south to Fla. and Tex.

CRISPED BUNCHFLOWER *(Melanthium latifolium)*

This is a similar but more slender plant with broader and thinner leaves. The six perianth parts of its flowers have wrinkled margins and are pointed rather than blunt at their tips. It grows in moist to dry open woods of the uplands from Conn. and s. N.Y. south to n. Fla.; blooming during July and August. (Not illustrated)

SOUTHERN RED LILY *(Lilium catesbaei)*

This beautiful lily grows in the wet pinelands and savannahs of the southeastern coastal plain. It is 1 to 2 feet tall and along the stem are small, scattered, lance-shaped leaves which point upward. Between July and September each stem bears a solitary flower in which the 6 perianth parts have long and slender stalks. Their blades are yellow toward the base but bright scarlet above, thickly spotted with purple, and the long-pointed tips curve gracefully outward. Also called the Pine Lily, Leopard Lily, and Catesby's Lily. It was named for Mark Catesby (1679-1749), one of the earliest southern naturalists.
RANGE: Va. south to Fla. and west to La.

NORTHERN RED LILY *(Lilium philadelphicum)*

Often called the Wood Lily, this species has a stem 1 to about 3 feet tall on which are from 2 to 6 whorls of lance-shaped leaves. At its summit there may be from 1 to 5 flowers which stand upright. Their 6 perianth parts are distinctly stalked, bright orange-red, and purple-spotted. It grows in dry open woods and clearings and southward in mountain meadows and balds, blooming between June and August.
RANGE: Me. to s. Que. and s. Ont. south to Del., n. Ga. and Ky.

TURK'S-CAP LILY *(Lilium superbum)*

The showy Turk's-cap is the tallest of our native lilies, being from 3 to about 5 feet in height. Along its stem are several whorls of leaves and some scattered ones near the summit. They are broadest about the middle and pointed at both ends. Between June and August the plants produce anywhere from a few to as many as 40 handsome flowers. These are orange-red thickly spotted with purple and the 6 perianth parts are curved backward so their tips often touch. It grows in wet meadows and moist woods, southward chiefly in the mountains.
RANGE: Mass. to Ind. south to N. Fla. and Ala.

CAROLINA LILY *(Lilium michauxii)*

This species closely resembles the Turk's-cap but it is usually less than 3 feet tall and generally has but 1 to 3 flowers. The leaves afford the best distinguishing feature. In this species they are thickish in texture and always broadest toward the bluntly-pointed tip; and they may be scattered or arranged in from 1 to 4 whorls along the stem. Although most common in open woods and on dry rocky slopes in the mountains, it often occurs well down into the coastal plain. It blooms during July and August. This lily was named *michauxii* in honor of Andre Michaux, the noted French botanist who discovered it.
RANGE: Va. and W.Va. south to Fla. and Ala.

CANADA LILY *(Lilium canadense)*

On a stem 1 to 4 feet tall this lily usually has several nodding, bell-shaped, red to yellow flowers which are spotted with purplish-brown. The 6 perianth parts of its flowers have spreading or slightly recurved tips. It usually has several whorls of lance-shaped leaves with roughish margins. This northern species is often common in wet woods and meadows and is found southward along the mountains. It blooms in June or July. Another name for it is Wild Yellow Lily.
RANGE: N.S. to Que. south to Va., n. Fla. and Ala.

GRAY'S LILY *(Lilium grayi)*

This attractive but rather rare lily grows in moist meadows and balds in the southern Appalachians. On a stem 2 to 4 feet tall are usually a few whorls of leaves, or the lower ones may be scattered. The few to sometimes numerous bell-shaped flowers are slightly nodding; and their 6 perianth parts are deep orange-red thickly spotted with purple, with the short-pointed tips but slightly spread. This lily was named for Dr. Asa Gray, the famous American botanist; and it is sometimes called the Roan Lily after Roan Mountain where Dr. Gray discovered it. It blooms during June and July.
RANGE: Mts. of Va. south to w. N.C. and e. Tenn.

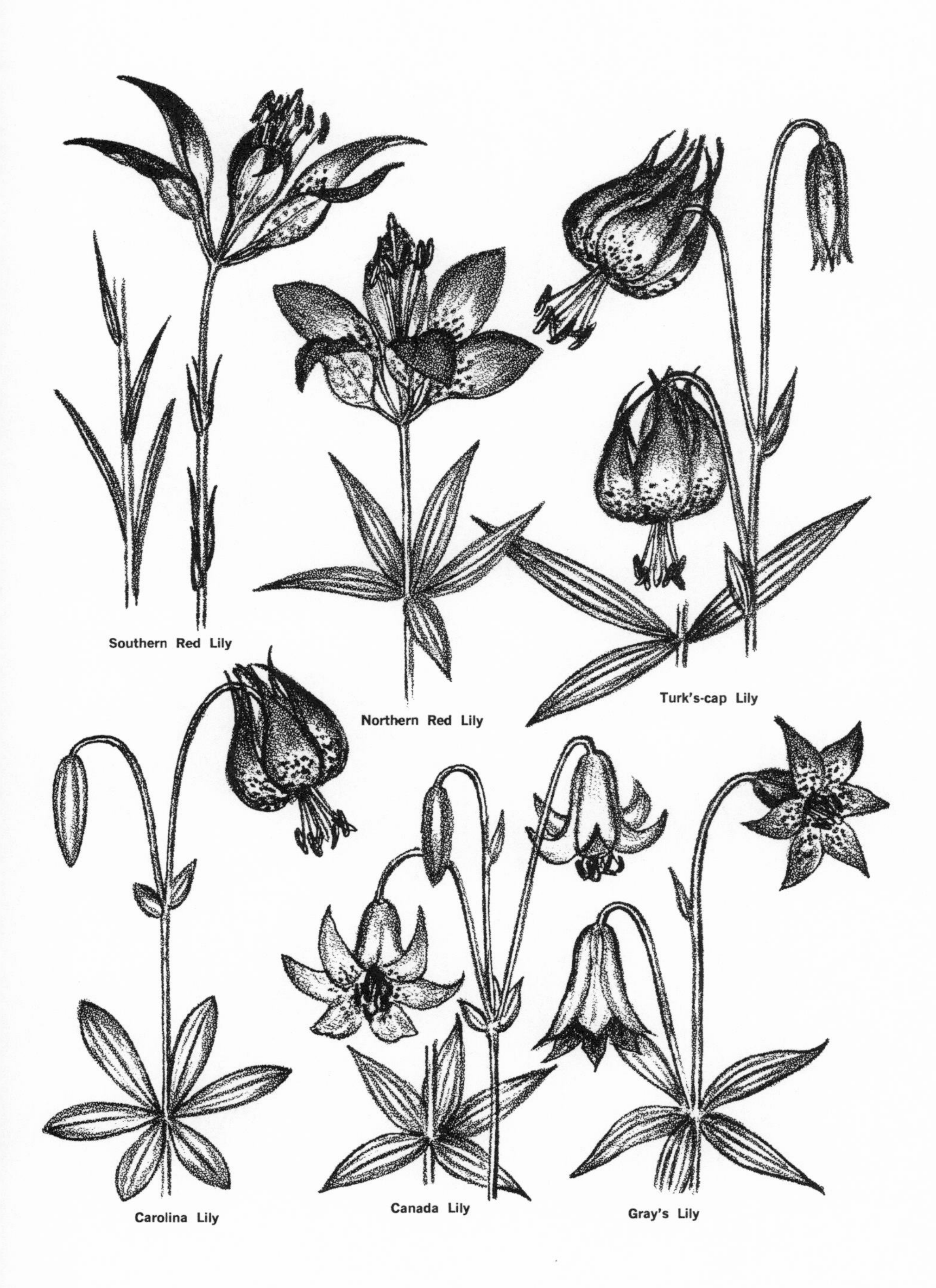
Southern Red Lily
Northern Red Lily
Turk's-cap Lily
Carolina Lily
Canada Lily
Gray's Lily

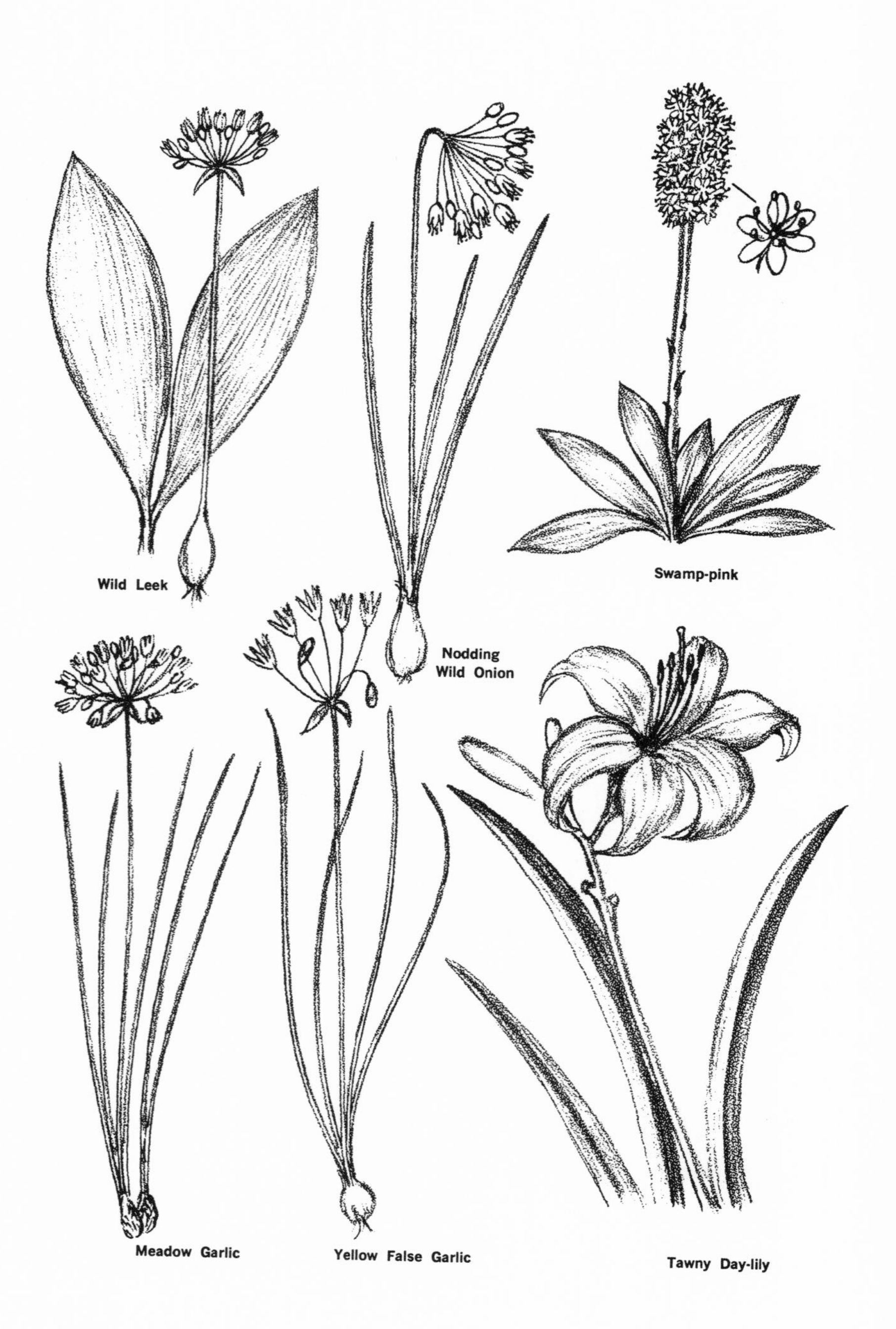
Wild Leek
Nodding
Wild Onion
Swamp-pink
Meadow Garlic
Yellow False Garlic
Tawny Day-lily

WILD LEEK *(Allium tricoccum)*

In the early spring, this plant has broad, flat, elliptical leaves from 6 to 10 inches long but the very definite onion-like odor will aid in identifying it. After the leaves have withered, in June or July, it sends up a naked stalk which bears, at the top, an umbel of greenish-white flowers. The plant grows in rich, moist woodlands. In the southern mountains, where it is often very abundant, the plant is known as the Ramp. The bulbs are eagerly sought by inhabitants of the region, who seem to be very fond of them.
RANGE: N.B. to Que. and Minn.; south to n. Ga., Tenn., and Iowa.

NODDING WILD ONION *(Allium cernuum)*

This species blooms in July or August. It has a flower stalk a foot or more tall which is bent near the top, thus the umbel of lilac-colored or pinkish flowers are in a nodding position. Its flat, soft, long, and grass-like leaves are usually shorter than the stalk which bears the flowers. All of its parts have an onion-like odor. It grows on moist to rather wet, rocky, wooded slopes.
RANGE: N.Y. to B.C. south to Ga., Tex., and N.Mex.

MEADOW GARLIC *(Allium canadense)*

This species has a naked flower stalk from 8 inches to nearly 2 feet tall, topped with an umbel of lilac-pink to whitish flowers. Frequently some or even all of the flowers may be replaced by small bulblets. Its leaves are long and very slender but shorter than the flower stalk. The entire plant has an onion-like odor. It is often abundant in open woods, thickets, and meadows; blooming in May and June.
RANGE: N.B. to Que. and Minn. south to Fla. and Tex.

YELLOW FALSE GARLIC *(Nothoscordum bivalve)*

In its general appearance, the False Garlic resembles the onions and garlics but it has no trace of an onion-like odor. Its flower stalk is 6 to 12 inches tall, leafless, and bears at the summit an umbel of from 6 to 12 starry, greenish-white to yellowish-white flowers. The leaves are narrow and grass-like and do not overtop the flower stalk. It grows in sandy, open woods and open places and blooms in April or May.
RANGE: Va., the Ohio Valley, and Neb. south to Fla. and Tex.

SWAMP-PINK *(Helonias bullata)*

The Swamp-pink, as its name indicates, grows in swamps and bogs. It has a stout, hollow flower stalk from 6 to 15 inches tall with almost scale-like bracts toward the base. Towards the top is a short but dense cluster of pink flowers with lavender-blue stamens. The true leaves are flat, dark green, broadest above the middle, and form a rosette at the base of the flower stalk. It blooms in April and May.
RANGE: Se. N.Y. and N.J. south to Va. and in mts. to n. Ga.

TAWNY DAY-LILY *(Hemerocallis fulva)*

This native of Eurasia long ago escaped from American flower gardens, and it is now so widely naturalized that many think it has always been here. It is often very common along roadsides and in the borders of fields and woods; in fact, it has become a real weed. Unlike the true lilies, the day-lilies have a leafless flower stem which forks repeatedly. Although each flower lasts but a few hours, the plant produces a succession of them from May to about mid-summer. At the base of the plant is a cluster of long, strap-shaped, and channeled leaves. The flowers of this species are quite large and have 6 tawny-orange perianth segments.

PERFOLIATE BELLWORT *(Uvularia perfoliata)*

This plant has a slender forking stem 8 to 20 inches tall; and smooth, pale green leaves which are coated with a thin whitish bloom. The bases of the leaves completely surround the stem, which appears to grow through them. It usually has a solitary pale yellow flower about an inch long, which is shaped like a narrow bell. On the inner surface of the 6 perianth parts there are small grain-like hairs. It grows in moist open woods and blooms in April or May.

RANGE: Mass. to Ont. south to Fla. and La.

LARGE-FLOWERED BELLWORT *(Uvularia grandiflora)*

This species looks very much like a larger edition of the preceding one, but it has a stouter stem 1 to 2 feet tall and its leaves are finely downy beneath and brighter green above. Its flowers are about 1½ inches long, deeper lemon-yellow in color, and the 6 perianth parts are smooth within. It also grows in rich woods, southward only in the mountains, and blooms in April and May.

RANGE: Que. to N.D. south to n.Ga., Miss., Ark. and Okla.

SESSILE-LEAF BELLWORT *(Uvularia sessilifolia)*

The leave of this species are stalkless but their bases do not surround the stem. It has a slender, smooth, forking stem 8 to 16 inches tall. The leaves are pale or somewhat whitened on the lower surface. The 6 perianth parts of the inch-long, pale yellow flowers are smooth within. It likewise grows in rich woods and blooms in April or May. It is also known as Wild-oats.

RANGE: N.B. to N.D. south to n. Fla., Ala. and Mo.

MOUNTAIN BELLWORT *(Uvularia puberula)*

Although similar to the preceding species, this plant has leaves which are bright green on both surfaces; and its stem has minute hairy lines. It commonly grows in clumps in deciduous woodlands and is not confined to the mountains. Blooming in April or May.

RANGE: Va. and W.Va. south to Tenn. and Ga.

FLORIDA BELLWORT *(Uvularia floridana)*

This plant grows in deciduous woods and swamps of the coastal plain from S.C. south to Fla. and west to Ala. (Not illustrated)

YELLOW ADDER'S-TONGUE *(Erythronium americanum)*

Other names given to this little lily family member are Fawn-lily, Trout-lily, and Dog's-tooth Violet. The plant has a pair of elliptic leaves 4 to 6 inches long which are pale green and mottled with purplish-brown. Between them rises a flower stalk which bears a nodding, open bell-shaped, yellow flower about 1½ inches across. Its 6 perianth parts are often tinged with purplish on the back, and usually curve gracefully backward. It grows in rich woods and bottomlands and blooms between March and May. It requires several years for the plants to attain the flowering stage. Younger plants have just one leaf which arises from a deeply buried corm.

RANGE: Ont. to Minn. south to Ga., Ala., Ark. and Tex.

WHITE ADDER'S-TONGUE *(Erythronium albidum)*

This species is similar to the preceding one but it has white to pale bluish or pale pinkish flowers, and its leaves are less mottled and sometimes plain green. It grows in moist woods or on stream banks from Ont. to Minn. south to Ga., Mo. and Okla. (Not illustrated)

EASTERN CAMAS *(Camassia scilloides)*

This is sometimes called Wild-Hyacinth. The plant has a naked flower stalk from 6 inches to almost 2 feet tall. On it is a long cluster of starry, 6-parted, pale blue flowers. The long, narrow, keeled leaves are in a basal cluster, arising from a bulb which the Indians used as food. The Indian name for the plant was "quamash" and from it both the generic name *Camassia* and the common name "camas" have been derived. It grows in wet open woods and meadows and blooms between April and June.

RANGE: Pa. to Wis. south to Ga., Miss., Mo. and Okla.

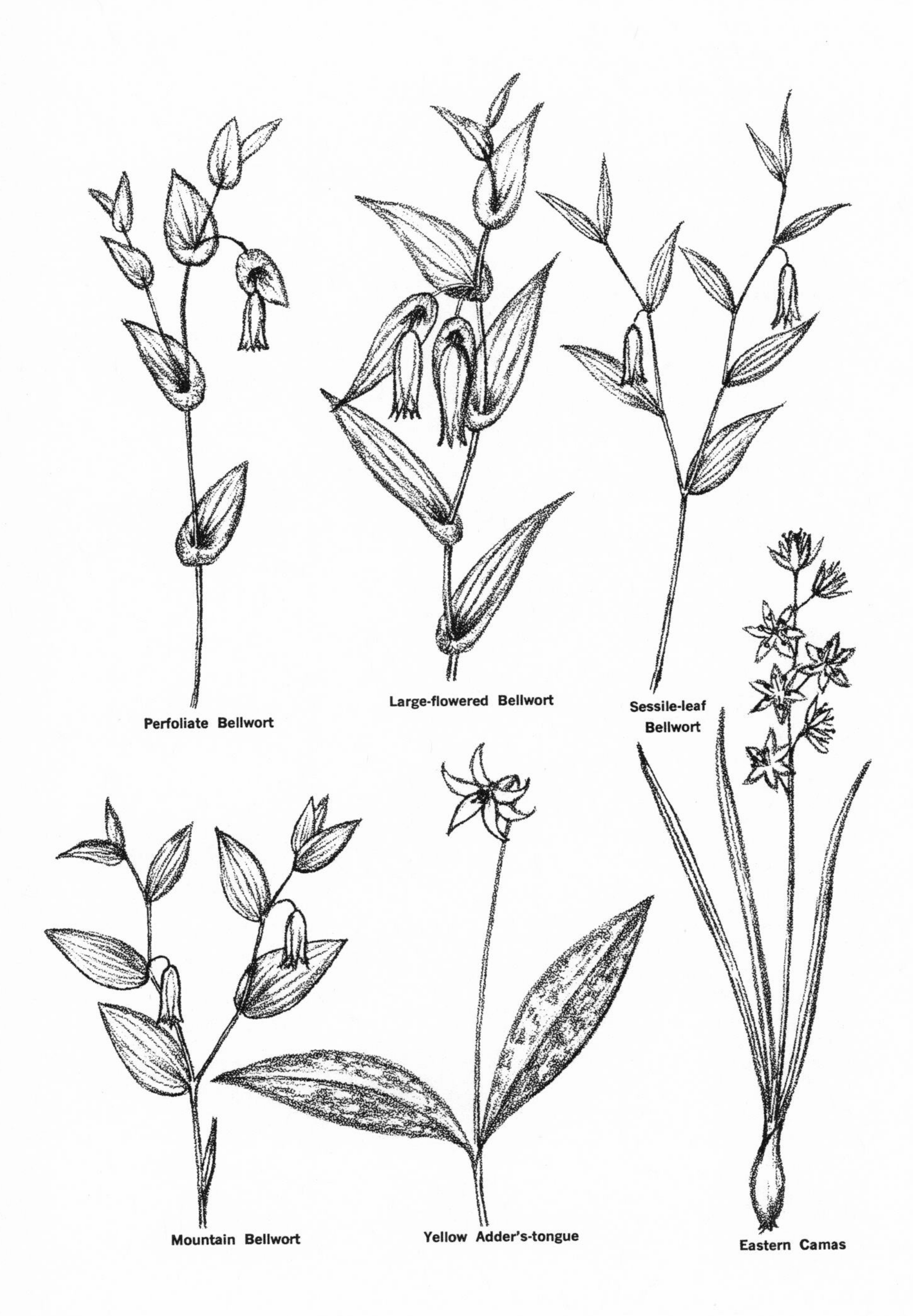
Perfoliate Bellwort
Large-flowered Bellwort
Sessile-leaf Bellwort
Mountain Bellwort
Yellow Adder's-tongue
Eastern Camas

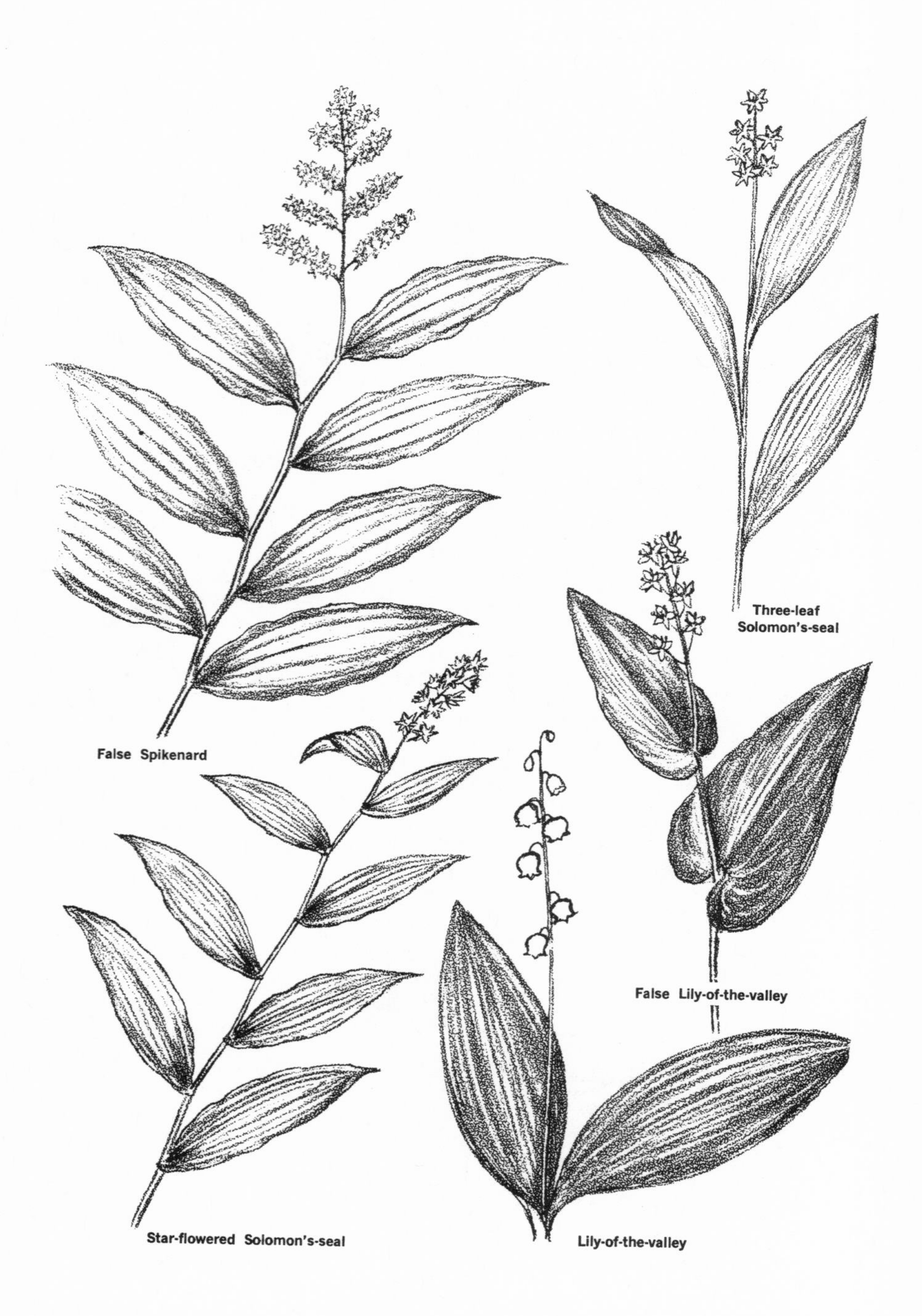
Three-leaf
Solomon's-seal
False Spikenard
False Lily-of-the-valley
Star-flowered Solomon's-seal
Lily-of-the-valley

FALSE SPIKENARD *(Maianthemum racemosa)*

The False Spikenard, or False Solomon's-seal, has an arching, slightly zig-zag stem from 2 to 3 feet tall which arises from a rootstock. Scattered along it are a number of conspicuously veined and almost stalkless leaves. It small white or creamy-white flowers are disposed in a plumy-branched end cluster sometime between April and July. They are followed by round berries which at first are yellowish-white and speckled. Later they become translucent and ruby-red. The plant is often common on rocky wooded slopes. It grows well in a woodland flower garden.

RANGE: Que. to B.C. south to Ga. and Ariz.

STAR-FLOWERED SOLOMON'S-SEAL *(Maianthemum stellata)*

This species has a leafy stem from 8 to about 20 inches tall. The veiny leaves are stalkless and finely downy beneath. Its flowers, while not very numerous, are larger than those of the preceding species and arranged in a simple, rather than branched, end cluster. It blooms during May or June and grows on moist wooded slopes, shores, bluffs, and meadows. The berries at first are greenish with black stripes, finally becoming a bronzy-black.

RANGE: Nfd. to B.C. south to W. Va., the Great Lakes region, Mo., N.M. and Calif.

THREE-LEAF SOLOMON'S-SEAL *(Maianthemum trifolia)*

A small plant usually from 2 to about 8 inches high with from 2 to 4, but usually 3, rather narrow leaves. Its white flowers are in a small, loose, simple end cluster and are produced between late May and August. This is a plant of cold northern bogs and wet woods.

RANGE: Lab. to B.C. south to N. Eng. and the Great Lakes region.

FALSE LILY-OF-THE-VALLEY *(Maianthemum canadense)*

The small white flowers of this plant are unique among members of the Lily Family in that they have a 4-parted perianth and 4 stamens. The plant has an erect stem usually from 3 to 6 inches tall, bearing 2 or 3 smooth leaves which have heart-shaped bases. It has rootstocks which creep extensively and it usually occurs in colonies; in which there are many single, stalked, heart-shaped leaves. The flowers are borne in simple little end clusters above the leaves, sometime between May and July. They are followed by small, round, pale greenish and spotted berries which eventually become a dull red. Other names for the plant are Two-leaf Solomon's-seal and Canada Mayflower. It is often common in damp mossy woods and bogs and not infrequently on the decaying trunks of fallen trees.

RANGE: Lab. to B.C. south to Del., n. Ga., Tenn. and Iowa.

LILY-OF-THE-VALLEY *(Convallaria majuscula)*

This native of our southern Appalachian Mountains is considered by most botanists to be closely related to the European Lily-of-the-valley, which is very common in cultivation. Unlike the European plant *(C. majalis)* our native Lily-of-the-valley is not a colony-forming plant. It has nearly the same broad elliptic leaves, and long clusters of fragrant white flowers which resemble little globe-shaped bells. Our native plants grow on rich, rocky wooded slopes and bloom between April and June.

RANGE: Va. and W. Va. south to w. N.C. and e. Tenn.

GREAT SOLOMON'S-SEAL *(Polygonatum biflorum* var. *commutatum)*

Largest of the Solomon's-seals, this species has a stout, arching stem from 2 to nearly 6 feet tall. Along it are a number of broad, more or less corrugated leaves from 3 to 6 inches long; which have clasping bases. From the leaf axils arise drooping clusters of from 2 to 8 greenish-white flowers a half inch or more long, which are shaped like cylindrical bells. Later these flowers develop into ball-shaped, bluish-black berries. The Great Solomon's-seal grows in rich, moist, woods and swampy thickets; and also along the banks of streams. It blooms in May and June. Solomon's-seals get their name from the seal-like scars which the upright leafy branches leave on the underground, creeping rootstocks. RANGE: N.H. to Man. south to S.C., Mo., and Okla.

HAIRY SOLOMON'S-SEAL *(Polygonatum pubescens)*

This is a much smaller plant with an arching stem from about 1 to 3 feet tall, which grows in woods and thickets. On its slender stem are a number of leaves 2 to about 4 inches long, which are pale and somewhat downy beneath. In their axils hang half-inch, narrowly bell-shaped, greenish-white flowers, which are sometimes solitary but usually in pairs. It blooms in May or June, and the flowers are followed by round, bluish-black berries. RANGE: Que. to Man. south to S.C., Ky., and Iowa.

SMOOTH SOLOMON'S-SEAL *(Polygonatum biflorum)*

Like the preceding species, the Smooth Solomon's-seal usually has its flowers in pairs. The most obvious difference between the two species is in their leaves, those of the Smooth Solomon's-seal being smooth on both surfaces. It is often common in dry to moist, often rocky woods and thickets; blooming in May or June.
RANGE: Ont., Mich., and Neb. south to Fla. and Tex. (Not illustrated)

SESSILE-LEAF TWISTED-STALK *(Streptopus roseus)*

Twisted-stalks get their name from the fact that the stalks of their flowers are abruptly bent or twisted near the middle. Instead of arising from the leaf axils, the slender stalks which bear the open bell-shaped flowers arise to the side or more nearly opposite the leaves. In general appearance, the plants resemble the Solomon's-seals but they have somewhat more zig-zag and often forked stems. This species has a stem from 1 to 2 feet tall. The leaves are deeply corrugated and seated directly on the stem but their bases are not clasping. Between April and July it has pink or rose-purple flowers, which are followed by red berries. It grows in cool, moist woods and the borders of swamps. Another name for it is Rose Mandarin. RANGE: Lab. to Man. south to n. Ga., Tenn. and Minn.

CLASPING-LEAF TWISTED-STALK *(Streptopus amplexifolius)*

From the preceding species, this one can be distinguished by its clasping leaf bases. It has a stem from 1½ to 3 feet tall. The flowers are greenish-white (sometimes dark purple) and the fruits are red berries. It grows in cool, moist woods and thickets; blooming between May and July. It is also known as the White Mandarin.
RANGE: Lab. to Alaska south to N.C., the Great Lakes region, and N. Mex.

HAIRY DISPORUM *(Disporum lanuginosum)*

Disporums are plants similar to the twisted stalks but they have 1 or 2 open bell-shaped flowers at the tips of the stems or their branches. This one has a stem 1 to 2 feet tall which is sparingly branched above. Its leaves are stalkless, pointed at the tip, and rounded to slightly heart-shaped at the base. Their lower surfaces, as well as the younger parts of the stems, are minutely woolly-hairy. The 6 perianth parts of the flowers are about ¾ inch long and greenish-white in color. Smooth red berries follow the flowers which bloom in May or June. This plant grows in rather moist, rich woods. It is also known as the Yellow Mandarin. RANGE: N.Y. to Ont. south to Ga., Ala., and Tenn.

SPOTTED DISPORUM *(Disporum maculatum)*

This species has the younger parts of its stems and lower leaf surfaces roughened by small, stiff, spreading hairs. Its flowers are slightly larger and are yellow spotted with dark purple. The red fruits are rather wrinkled and somewhat hairy. It also grows in rich woodlands, blooming in April or May. Another name for it is Nodding Mandarin
RANGE: Ohio and Mich. south to Ga., Ala., and Tenn. (Not illustrated)

Sessile-leaf
Twisted-stalk
Great Solomon's-seal
Hairy
Solomon's-seal
Clasping-leaf Twisted-stalk
Hairy Disporum

YELLOW CLINTONIA *(Clintonia borealis)*

This plant usually has 3 oval-shaped, somewhat leathery, lustrous leaves with hairy-fringed margins. They are from 5 to about 10 inches long. The greenish-yellow, bell-shaped flowers are borne in a cluster (umbel) of 3 to 8 at the top of a naked stalk 6 to 8 inches tall. Both leaves and flower stalk arise directly from an underground rootstock. The plants bloom in May or June, the flowers being followed by oval-shaped blue berries. The Yellow Clintonia is a plant of cool, moist woods and thickets.

RANGE: Lab. to Man. south to N. Eng., n. Ga., and the Great Lakes region.

WHITE CLINTONIA *(Clintonia umbellata)*

The White Clintonia has a whorl of 3 or 4 leaves very much like those of the preceding species. It differs in having smaller and more numerous (up to 30) white flowers which are minutely speckled with purple and green. The fruits which follow the flowers are ball-shaped black berries. This species is usually more common in rich woods southward. It blooms between May and early July.

RANGE: N. Y. to Ohio south in the mts. to n. Ga. and Tenn.

INDIAN CUCUMBER-ROOT *(Medeola virginica)*

A distinctive plant with a slender stem usually 1 to 2 feet tall, with a whorl of 5 to 9 leaves near the middle and another whorl of 3 smaller ones near the top. From the axils of the latter leaves arise the slender stalks bearing the nodding, greenish-yellow flowers. Both sepals and petals, as well as the 3 slender stigmas of the pistil, curve backward. The plant blooms in May or June. Later the flowers produce dark purple berries. In the ground, at the base of the stem, is a whitish tuber which has a taste very much like that of a cucumber—hence the common name. It grows in rich, moist woodlands.

RANGE: Que. to Ont. and Minn. south to n. Fla. and La.

WHITE COLIC-ROOT *(Aletris farinosa)*

Close to the ground this plant has a star-like cluster of flat, rather thin, yellowish-green, and grass-like leaves. Sometime between May and July or August, it sends up an almost naked stalk usually 1 to about 3 feet tall. Along the upper part of it are scattered white flowers which are shaped like narrow urns, and have a peculiar mealy surface. The plant grows in moist to dry open woods on sandy or peaty soils. It is abundant in the coastal plain pinelands of the Southeast and is also often common in the mountains. Other names for it are White Star-grass and Mealy Starwort.

RANGE: Me. to Ont. and Wis. south to Fla. and Tex.

Aletris obovata is another white-flowered species found from se. S.C. southward to Florida. It can be distinguished by the fact that the flowers are not constricted below the lobes, and the latter are erect rather than spreading. (Not illustrated)

GOLDEN COLIC-ROOT *(Aletris aurea)*

This plant closely resembles the preceding one but it has a more lax cluster of shorter, broader, bell-shaped, orange-yellow flowers. It grows in the wetter pinelands, bogs and savannahs of the coastal plain; blooming during June and July.

RANGE: Md. south to Fla. and west to Tex.

YELLOW COLIC-ROOT *(Aletris lutea)*

This species resembles both of the preceding species and is best distinguished by its narrowly urn-shaped, somewhat mealy, pale yellow to orange flowers. It blooms in late April and May in wet pinelands of the southeastern coastal plain from Va. south to Fla. and west to La. (Not illustrated)

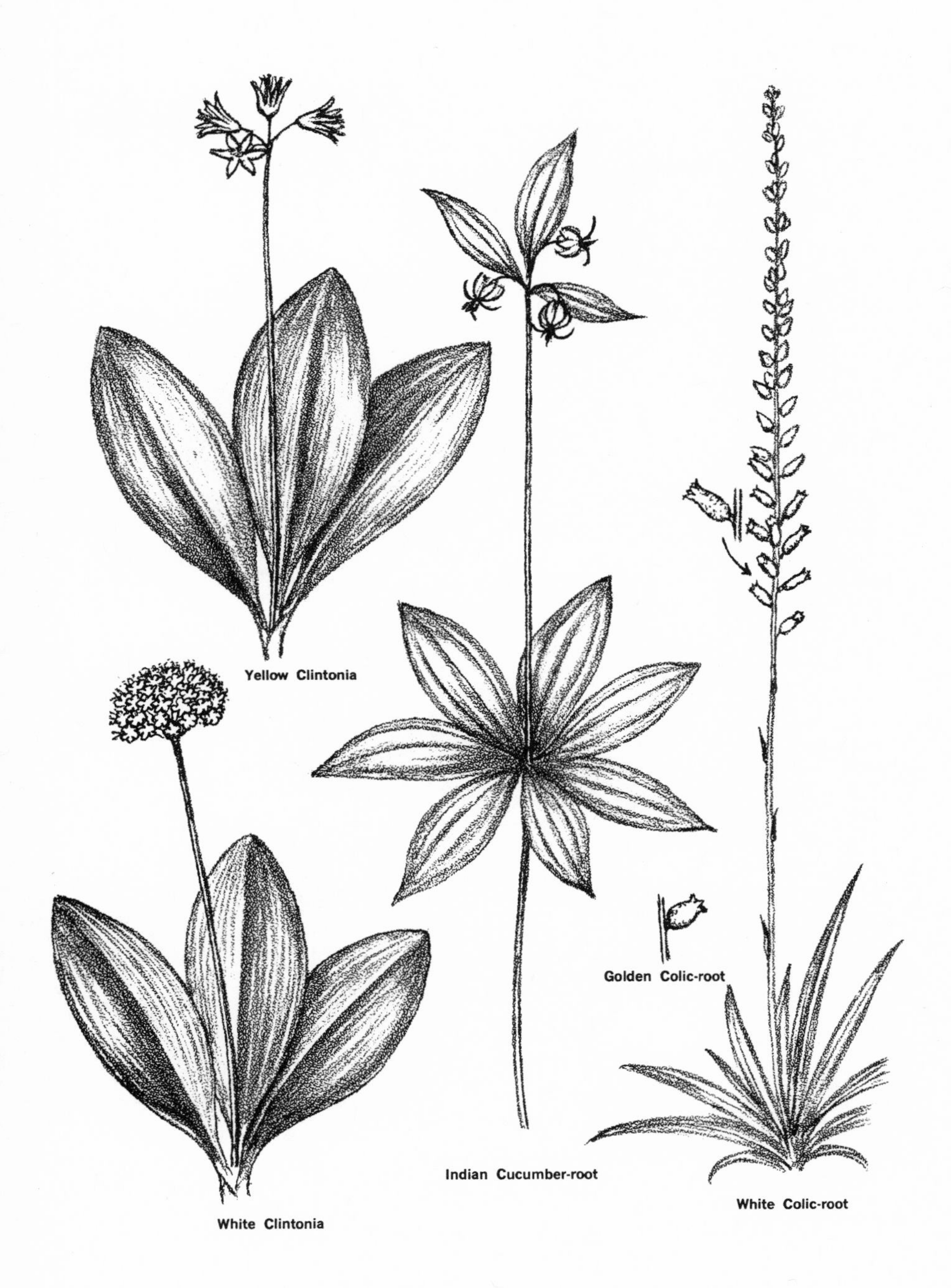
Yellow Clintonia
White Clintonia
Indian Cucumber-root
Golden Colic-root
White Colic-root

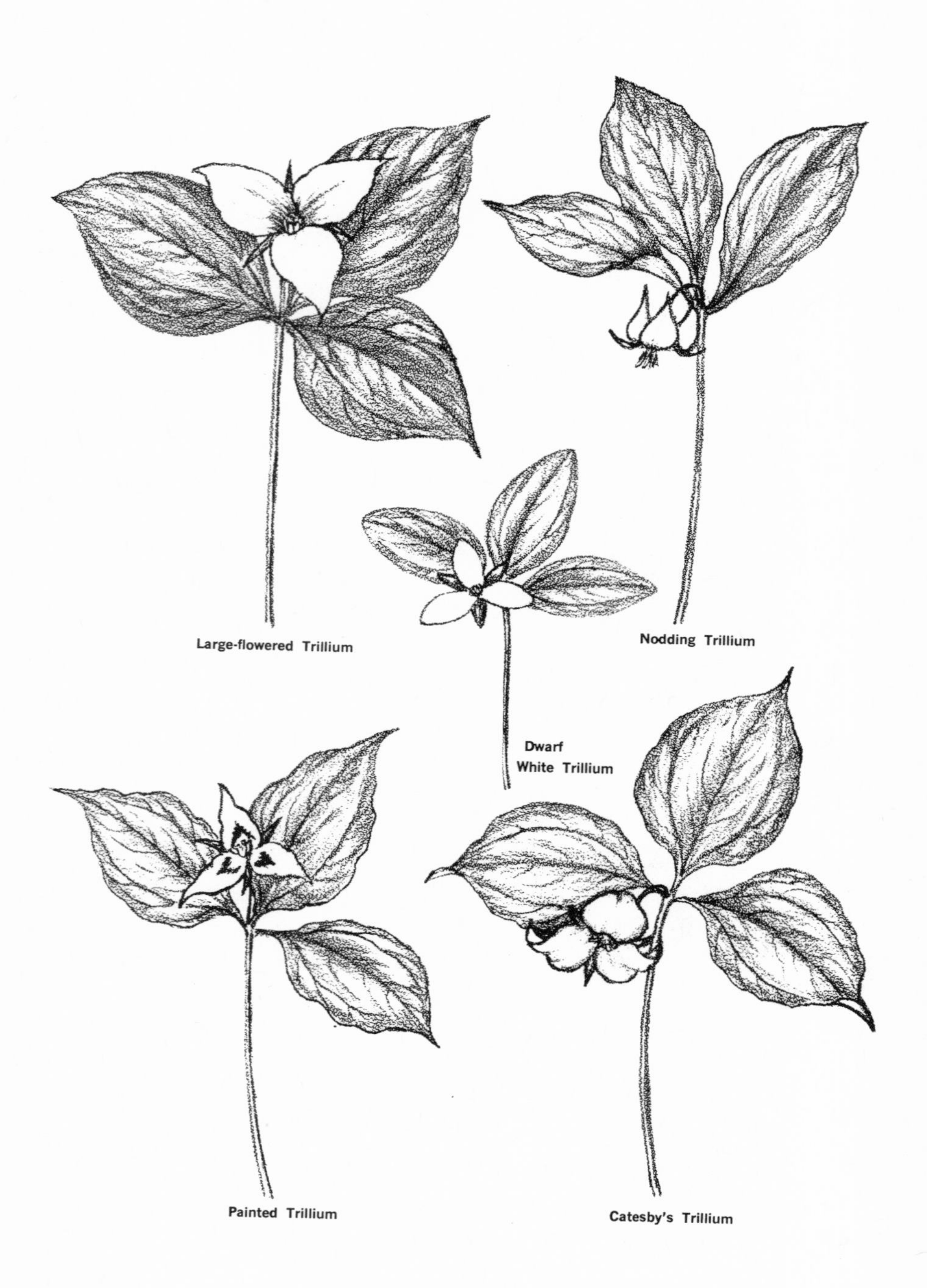
Large-flowered Trillium
Nodding Trillium
Dwarf
White Trillium
Painted Trillium
Catesby's Trillium

Trilliums

Trilliums, or Wake-robins, typically have all of their parts in 3's. At the summit of the stem is a whorl of 3 leaves which are unusual among Lily Family members in that they are net-veined. A solitary flower arises from the center of this whorl of leaves; and they, too, are unusual in that the 3 sepals are narrow and green while the 3 petals are usually much broader and either white or colored. There are 6 stamens and the 3- to 6-angled ovary of the pistil is surmounted by 3 prominent styles.

Some species are highly variable and, especially in the southern Appalachian region, many are difficult to identify. As a group they are among the showiest and best known of our spring wild flowers.

LARGE-FLOWERED TRILLIUM *(Trillium grandiflorum)*

This is one of the most common trilliums, being widespread in rich moist woods and thickets. It has a stout stem 8 to 18 inches tall with 3 large, broadly oval or egg-shaped, nearly stalkless leaves. The flower is 2 to 3½ inches across and stands on an erect or slightly leaning stalk. The 3 broad petals overlap at the base forming a sort of tube. They are white when the flower first opens but turn pink with age. It blooms in April or May.

RANGE: Me. and s. Que. west to Minn. south to n. Ga. and Ark.

NODDING TRILLIUM *(Trillium cernuum)*

Like the flowers of the preceding species, those of this trillium have petals which overlap at the base forming a sort of short tube. A distinctive feature is its purplish anthers and the fact that the flower nods on a slender stalk and hangs below the leaves. The petals are white, creamy-white, or pale pink. It grows in rich moist woods and swamps, blooming between April and June.

RANGE: Nfd. to Que. and Wis. south to n. Ga., Tenn. and Iowa.

DROOPING TRILLIUM *(Trillium flexipes)*

This plant is similar to the Nodding Trillium but is larger and coarser, and the larger flowers have pale anthers. It grows in damp woods and stream bottoms from N.Y. to Minn. south to Md., Tenn. and Mo. (Not illustrated)

DWARF WHITE TRILLIUM *(Trillium nivale)*

Often called the Snow Trillium, this is a small species with a stem 2 to 5 inches tall; with oval, egg-shaped or roundish, stalked leaves from 1 to 2 inches long. The stalked flower, little more than an inch across, has oval-shaped white petals. It grows in rich woods and along streams; blooming between March and early May.

RANGE: w. Pa. to Minn. south to Ky. and Mo.

PAINTED TRILLIUM *(Trillium undulatum)*

This trillium is readily recognized by the V-shaped purplish-pink marks toward the bases of its white, wavy-margined, recurved petals. It is a slender-stemmed plant 5 to about 20 inches tall. The broadly egg-shaped leaves are definitely stalked and are taper-pointed at the tip. It grows in cool wet woods and swamps, blooming between April and June.

RANGE: N.S. to Man. south to N.J., n.Ga., e. Tenn. and Wis.

CATESBY'S TRILLIUM *(Trillium catesbaei)*

This is the pink-flowered trillium so often seen in the piedmont region and lower slopes of the mountains in the Southeast. It has a moderately stout stem 8 to 20 inches tall; and broadly oval leaves with stalked bases and abruptly taper-pointed tips. The flower nods on a slender stalk. A good field mark is the tendency of the sepals to be sickle-shaped; and the stamens have filaments fully as long as the anthers. It grows in rich deciduous woods, blooming in April or May.

RANGE: N.C. and Tenn. south to Ga. and Ala.

LITTLE TRILLIUM *(Trillium pusillum)*

The Little Trillium grows in swamp forests and savannahs of the coastal plain from Va. to S.C. and in s.Mo. and nw. Arkansas. It has a stem 4 to 8 inches tall, with rather narrow oblong-lance-shaped leaves. The white flower turns purplish and is short- to fairly long-stalked. It blooms between March and May. (Not illustrated)

RED TRILLIUM *(Trillium erectum)*

This is one of the commonest and most variable of our trilliums. The flowers are not always red for the petals, varying greatly in width, range from deep purplish red or maroon to pink, greenish-yellow, and even white. The flower stalks may be erect or they may be inclined or even bent downward. The petals, however, do not overlap but spread from the base. Such names as Ill-scented Wake-robin, Stinking-Beth and Stinking-Benjamin attest to the fact that the flowers are ill-scented. Often they are but this is not always true. The plants are commonly 8 to 16 inches tall and have rather broadly oval or somewhat diamond-shaped, stalkless or nearly stalkless leaves. It grows in rich moist woods and thickets, blooming between April and June.

RANGE: N.S. to Ont. south to Del., n. Ga. and n. Ala.

VASEY'S TRILLIUM *(Trillium vaseyi)*

Most botanists consider this a distinct species from *T. erectum.* Its flowers have broad, deep maroon petals and pale anthers; and they are quite sweet-scented. It is sometimes called the Sweet Trillium or Sweet-Beth. A white-flowered form of this is called the Woodland White Trillium *(Trillium simile)* in some manuals. Both grow in rich moist woods from Tenn. and w. N.C. south to nw. S.C. and n. Ga. blooming between April and early June.

SESSILE TRILLIUM *(Trillium sessile)*

Often called the Toadshade Trillium, this species has 3 stalkless, oval or egg-shaped leaves which are strikingly mottled in two or three shades of green. The plant grows from 4 to about 12 inches tall. The stalkless flower has 3 narrow and more or less erect petals from ¾ to 1½ inches long, which vary from maroon or purplish-brown to greenish-yellow. The flowers have a strong odor which is not altogether unpleasant. It grows in rich moist woods and stream bottoms, blooming in April or May.

RANGE: w. N.Y. to Ill. and Mo. south to w. Ga., Miss. and Ark.

HUGER'S TRILLIUM *(Trillium cuneatum)*

Huger's is a striking southeastern trillium which looks like a larger edition of the preceding species. Its flowers have erect maroon-colored petals from 2 to nearly 5 inches long, and they have a fragrance which has been likened to that of crushed strawberries. It grows in rich moist woods from w. N.C. and e. S.C. south to nw. Fla. and Miss. but is most common in the mountains; blooming between late March and May. (Not illustrated.

YELLOW TRILLIUM *(Trillium luteum)*

This striking trillium has a stout stem 8 to 18 inches tall, with broadly oval, stalkless leaves which are beautifully mottled. The stalkless flower has erect lemon-yellow petals from 2 to 3 inches in length, and 3 yellowish-green sepals which spread horizontally. The flowers have a very pleasant odor which suggests that of lemons. It grows in rich moist woods and ravines, blooming in April or May.

RANGE: w. N.C., Ky. and Mo. south to n. Ga., Ala. and Ark.

PRAIRIE TRILLIUM *(Trillium recurvatum)*

A distinguishing feature of this trillium is the slender stalks of its leaves which are ½ to about 1 inch long. The flowers are stalkless but their maroon or reddish-brown to yellowish-green petals are contracted into stalk-like bases. The 3 sepals curve backward between the leaves. This trillium grows in rich woods and blooms during April and May.

RANGE: Ohio and Mich. to Wis. and Neb. south to Ala., Miss. and Ark.

LANCE-LEAF TRILLIUM *(Trillium lanceolatum)*

This species is distinguished by its narrowly oval or lance-shaped, almost stalkless, mottled leaves which top a stem from 4 to 15 inches tall. The stalkless flower has 3 maroon to greenish and erect petals which taper at the base. It grows in deciduous woods and on river bluffs from S.C. and Tenn. south to Fla. and La., blooming in April or May. (Not illustrated)

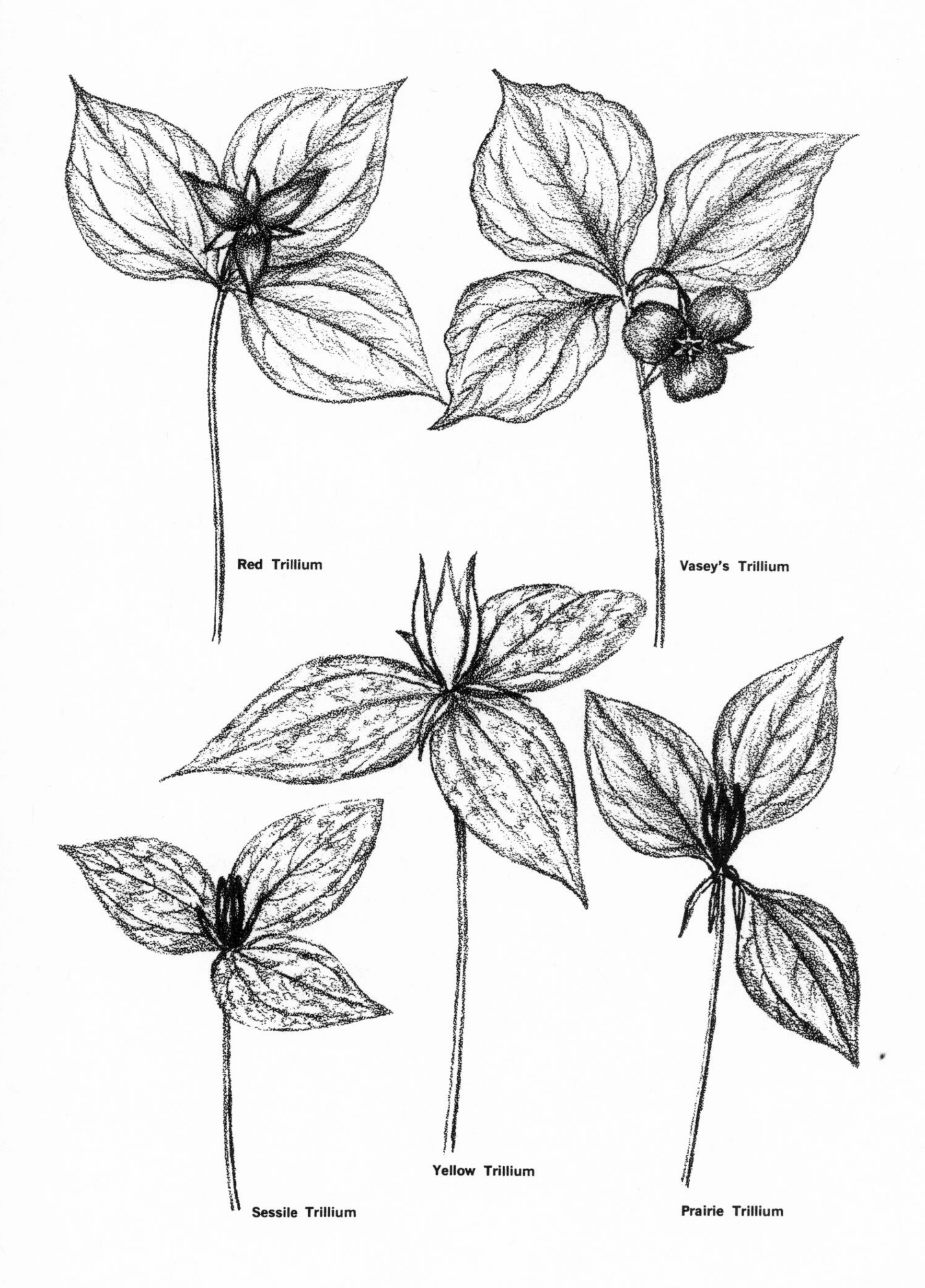
Red Trillium
Vasey's Trillium
Yellow Trillium
Sessile Trillium
Prairie Trillium

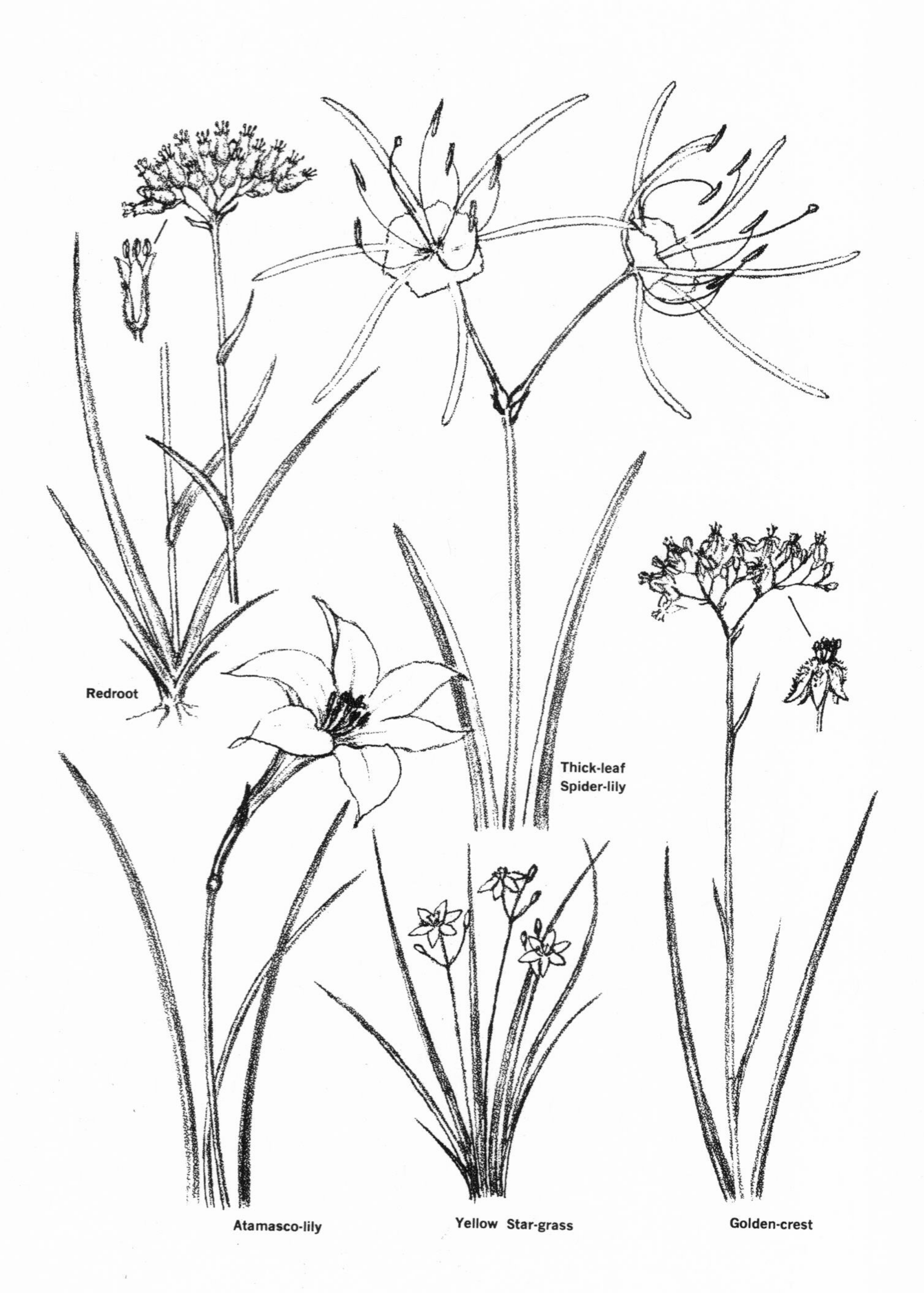
Redroot
Thick-leaf
Spider-lily
Atamasco-lily
Yellow Star-grass
Golden-crest

GOLDEN-CREST *(Lophiola aurea)*

In some ways this plant resembles the Redroot. It has a slightly more leafy stem 1 to 2 feet tall which is densely coated with soft,white, matted wool toward the top. At its summit is a flat-topped cluster of small yellow flowers which are hairy within, and their perianth segments bend backward to expose the 6 stamens. It grows in wet pinelands, bogs and savannahs and blooms in June or July,

RANGE: N.J. south to Fla. and west to Miss.

ATAMASCO-LILY *(Zephyranthes atamasca)*

Also known as the Zephyr-lily and Wild Easter-lily, this plant has rather large lily-like flowers that are usually borne solitary on a naked stalk 8 to 15 inches tall. The flowers are white when they first open but later turn pink. Both the flower stalks and the long, narrow leaves arise from an onion-like bulb. It grows in rich moist woods, swamps, and meadows; blooming between March and May.

RANGE: Va. south to Fla. and west to Miss.

THICK-LEAF SPIDER-LILY *(Hymenocallis floridana)*

At the summit of a naked stalk from 1½ to 2 feet tall, this spider-lily has 2 or 3 large white flowers. The long, green, stalk-like perianth tubes suddenly flare into 6 long and very narrow, petal-like segments. Within them are the 6 stamens, the slender filaments of which are joined toward the base with a cup-shaped white membrane which adds immeasurably to the showiness of the flower. The leaves are all basal, strap-like and up to 20 inches in lenth. This species of spider-lily grows in coastal plain marshes and along the bank of tidewater streams; blooming in May or June.

RANGE: N.C. south to Fla. and west to Ala.

MIDWESTERN SPIDER-LILY *(Hymenocallis caroliniana)*

This species is similar to the preceding species but it usually has from 3 to 6 flowers in a cluster, and blooms between June and September. It grows along the marshy banks of streams from s. Ind., s. Ill and Mo. south to S.C., Fla. and Miss. (Not illustrated)

YELLOW STAR-GRASS *(Hypoxis hirsuta)*

Yellow Star-grasses are small plants with clusters of from 2 to 7 flowers at the summit of a hairy stalk from 2 to 6 inches long. The six-parted perianth is smooth and bright yellow above but usually greenish and hairy beneath, opening to about the size of a dime. Both flower stalks and the narrow grass-like leaves arise from a hard corm. This species grows in dry open woods and meadows, blooming between March and September. Several other species are more commonly found in the southeastern coastal plain.

RANGE: Me. to Man. south to Fla. and Tex.

REDROOT *(Lachnanthes caroliniana)*

This is a plant of sandy swamps and peat bogs, which gets its name from its bright red roots. It has a stem 1 to 2½ feet tall which is woolly toward the top; terminating in a dense and nearly flat-topped cluster of narrowly bell-shaped, dull yellow flowers which are about a half inch long. They are densely woolly outside but smooth within, and have 3 stamens which are longer than the corolla tube. The narrow and grass-like leaves are in a basal cluster, with only a few small ones along the stem. It blooms between June and September in the coastal plain.

RANGE: Mass. and Del. south to Fla. and west to Tex.

Flowers of members of this family have 3 petals and 3 sepals, 3 stamens, and a pistil composed of 3 united carpels but with distinct styles. The bases of the flower parts are joined together and form a tube which is completely united with the ovary of the pistil. They are plants which grow from creeping stems or rootstocks, and have narrow or grass-like leaves which are set end to end.

Wild irises resemble the cultivated kinds and are easy to recognize. They have 3 petal-like sepals which generally curve outward and downward, and 3 smaller petals which are erect. The 3 style branches are petal-like and lie over the sepals, while the 3 stamens are hidden beneath the style branches

BLUE-FLAG IRIS *(Iris versicolor)*

Flowers of the Blue-flag are often seen in wet meadows and marshes between May and August. The violet-blue sepals are whitish toward the base and beautifully veined with a darker purple, while the base of each one shows a yellow or greenish-yellow spot. The 3 petals are about half as large and plain blue-violet.

RANGE: Lab. to Man. south to Va., Ohio, Wis. and Minn.

SOUTHERN BLUE-FLAG IRIS *(Iris virginica)*

This species is quite similar to the preceding but the sepals usually have a brighter yellow and usually downy spot at the base. It is the most common of the tall blue-flowered irises in the South and is often abundant in the borders of swamps, marshes, and roadside ditches. It blooms between April and June, and its range extends from Va. northwestward to Minn. and south to Fla. and Tex. (Not illustrated)

THREE-PETALLED IRIS *(Iris tridentata)*

This iris of the southeastern coastal plain seems to have but 3 instead of the usual 6 perianth parts. Actually the 3 petals are so short they are inconspicuous, but the 3 large sepals are deep violet-blue with a bright yellow blotch at the base. The leaves are quite narrow and are usually gracefully curved. It blooms between April and June in wet pinelands and the borders of swamps.

RANGE: N.C. south to Fla.

Iris hexagona has conspicuously zig-zag stems and long leaf-like bracts at the bases of the spathes from which the flowers appear. As its name indicates it has 6-sided seed pods. It is another iris of the coastal plain swamps and is found from S.C. south to Fla. and west to Ala. (Not illustrated)

SLENDER BLUE-FLAG IRIS *(Iris prismatica)*

This iris has slender flower stalks 1 to 2 feet tall and very narrow leaves. Its pale lavender-blue flowers are about 3 inches across, and the 3 sepals have a yellowish spot at the base. The sharply 3-angled capsules are helpful in distinguishing it. It grows in wet meadows, marshes and on shores; blooming between April and June.

RANGE: Along coast from Me. south to S.C.; in mountains from Va. and Ky. south to n. Ga. and Tenn.

CRESTED DWARF IRIS *(Iris cristata)*

This is a little iris with flower stalks 3 to 6 inches tall. The flowers range from pale to a deep lavender-blue and each of the 3 sepals has a white patch and a 3-ridged, fringed, orange and white crest. The leaves tend to be curved and are ½ to about ¾ of an inch wide. It grows on rich rocky wooded slopes or bluffs, and in ravines; blooming in April or May.

RANGE: Md. to Ind. and Mo. south to n. Ga and Miss.

DWARF IRIS *(Iris verna)*

This is another dwarf species with flower stalks from 1 to 3 inches tall. Its flowers range from a deep to a rather pale violet-blue. The 3 sepals are but little larger than the 3 petals and have a smooth orange-yellow spot at the base. Its leaves are rather stiff, straight-sided, and about ¼ of an inch wide; greatly elongating after the flower fades. It grows in sandy or rocky woods, blooming between March and May. Sometimes it is called the Violet Iris.

RANGE: Md. to s. Pa. south to Fla. and Ala.

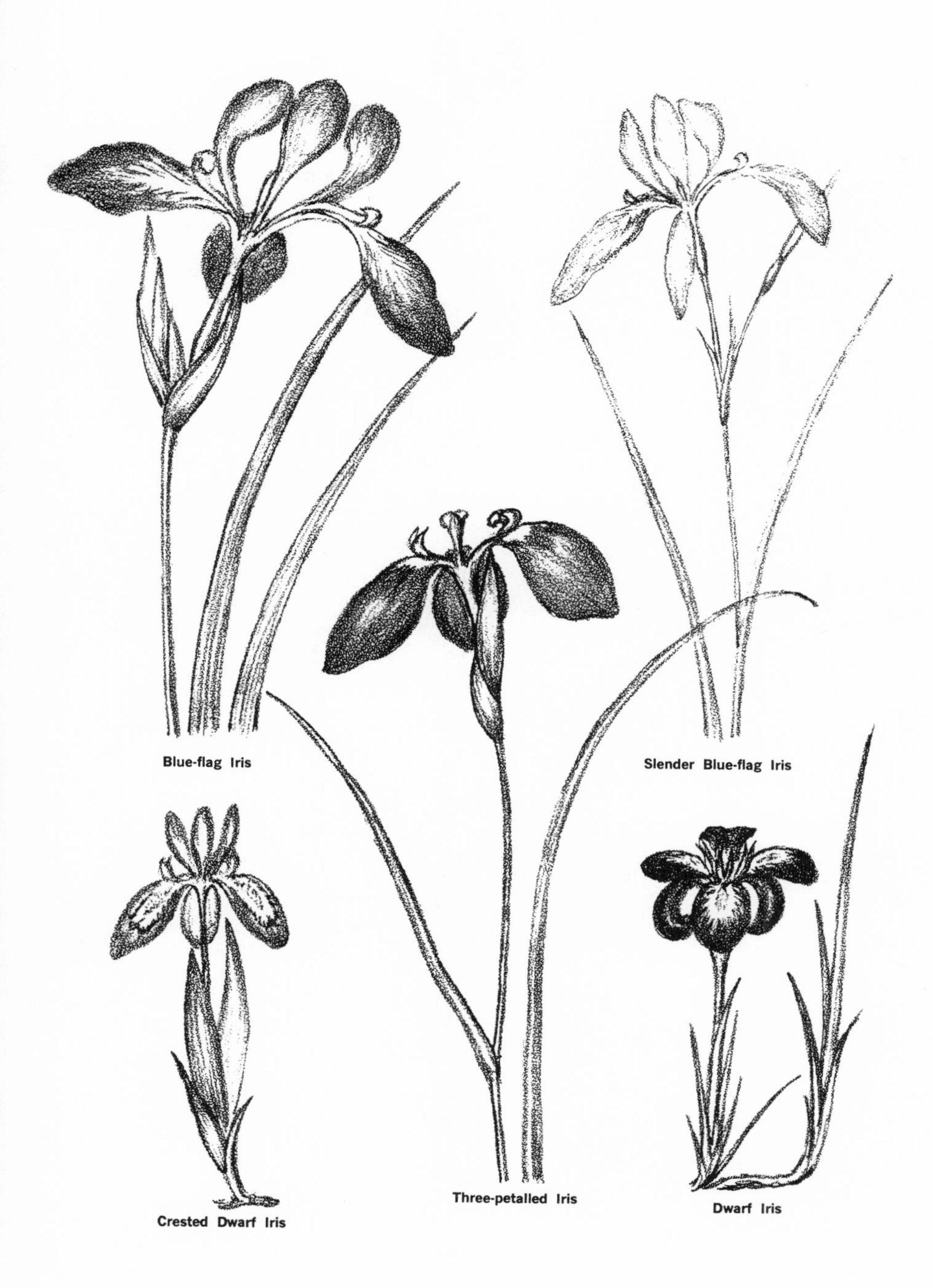
Blue-flag Iris
Slender Blue-flag Iris
Crested Dwarf Iris
Three-petalled Iris
Dwarf Iris

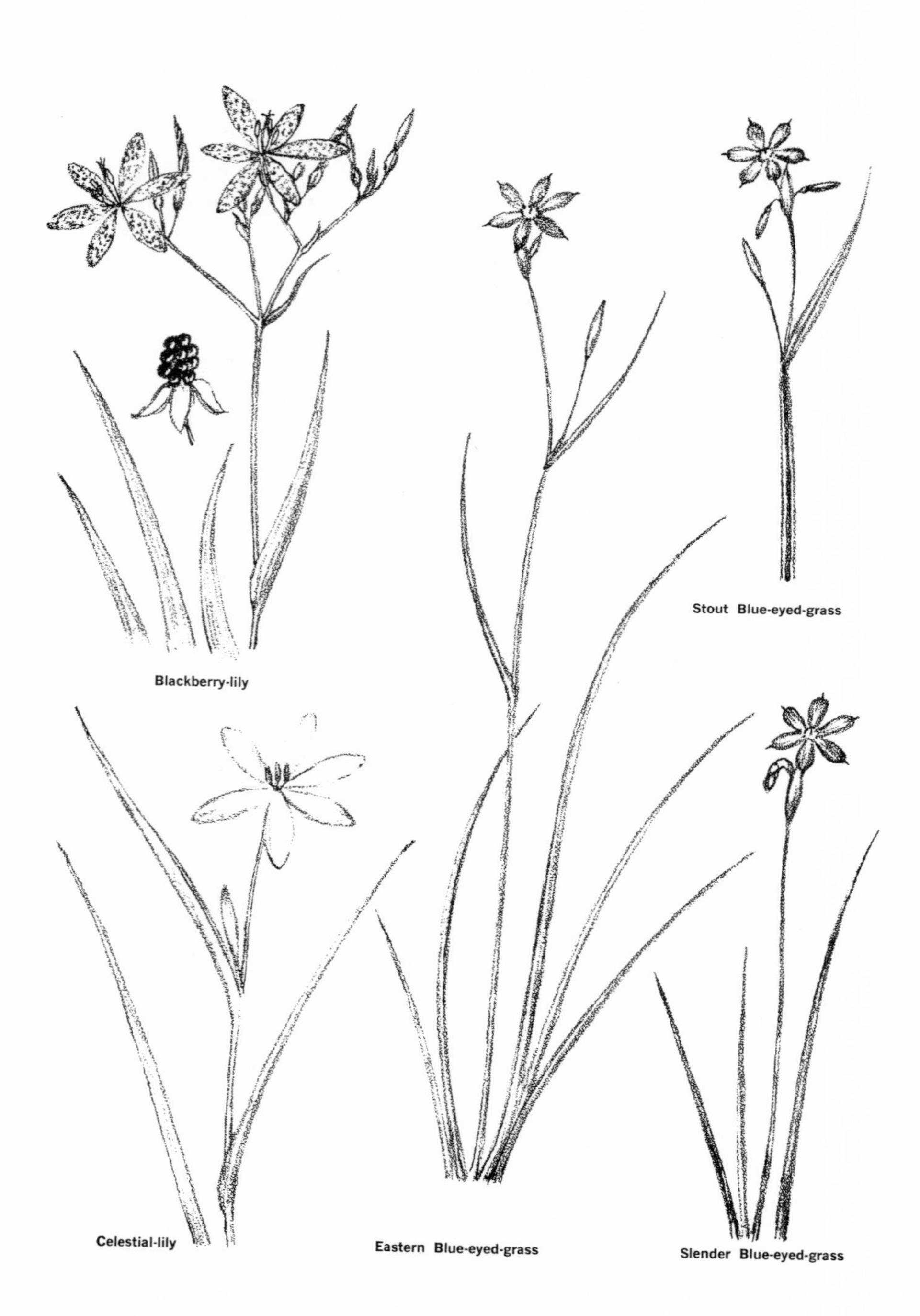
Blackberry-lily
Stout Blue-eyed-grass
Celestial-lily
Eastern Blue-eyed-grass
Slender Blue-eyed-grass

BLACKBERRY LILY *(Belamcanda chinensis)*

The Blackberry-lily has a rather stout, leafless flower stalk from 1½ to about 3 feet tall. It branches above, producing a loose cluster of flowers in which the 6 perianth parts are all alike, orange-yellow, and spotted with crimson or purple. The flower stalks arise from the basal fan-like clusters of leaves which very closely resemble those of the irises. The flowers open between June and August: and are followed by pear-shaped capsules which open to expose a cluster of black, fleshy-coated seeds. The latter are very suggestive of blackberries, hence the common name of Blackberry-lily. This plant, which is a native of Asia, escaped from oldtime flower gardens. It now grows in wayside thickets, open woods, and in waste places throughout much of eastern North America.

CELESTIAL-LILY *(Nemastylis geminiflora)*

Celestial-lilies have stems from 1 to 2 feet tall which bear 3 or 4 long, narrow, folded leaves. This one has very delicate, violet-blue flowers from 1½ to 2½ inches across, which open between late morning and mid-afternoon. Two such flowers appear from each spathe, hence the generic name *geminiflora* which means "twin-flowered". The plant grows on prairies and in open woods or glades; blooming between April and June.

RANGE: W. Tenn. to Kan. south to La. and Tex.

A similar species which grows in the marshes, swamps, and flatwoods of northeastern Florida is *Nemastylis floridana*. It blooms in the fall and its flowers open between mid-afternoon and sundown.

BARTRAM'S IXIA *(Calydorea coelestina)*

This plant is similar to the Celestial-lily but its nodding flowers are violet with a white "eye", or more rarely entirely white. The flowers open early in the morning but they soon wither. It grows in the flatwoods of southern Georgia and northern Florida where it blooms in the spring. This interesting plant was first discovered by the famous Quaker naturalist William Bartram, probably in the spring of 1766, and was described in his "Travels". It was not rediscovered until about 156 years later. (Not illustrated)

BLUE-EYED-GRASSES

About a dozen species of blue-eyed grasses occur in the eastern part of North America. All of them have tufts of narrow or grass-like leaves which stand stiffly erect. The flowering stems are two-edged and the flowers are in small clusters, more or less enclosed by the spathe-like bracts. The 6 perianth parts of the flower are all similar and are usually tipped with a small bristle-like point.

EASTERN BLUE-EYED-GRASS *(Sisyrinchium atlanticum)*

This species is common in moist meadows, marshes, and low woods. It is a pale green or somewhat whitened plant with tufted stems 8 inches to about 2 feet tall. The stems are slender, wiry, and just barely 2-winged. Its flowers are slightly more than a half inch across and are bright blue-violet with a yellow "eye"; blooming March to July.

RANGE: N.S. to Mich. south to Fla. and Miss.

SLENDER BLUE-EYED-GRASS *(Sisyrinchium montanum)*

This species has slender wiry stems 6 to 24 inches tall which have a solitary straw-colored spathe without leafy bracts below it. The violet-blue flowers are about a half inch across and have a yellow "eye". It grows in meadows, fields and open woods; blooming between April and June.

RANGE: Nfd. to B.C. south to Pa., n. Ga., the Great Lakes region, and Colo.

STOUT BLUE-EYED-GRASS *(Sisyrinchium angustifolium)*

The broadly 2-winged stem which usually forks above, and is but slightly longer than the leaves, is helpful in recognizing this species. The violet-blue flowers with a yellow "eye" are a bit more than ½ inch across. It grows in moist meadows and open woods; blooming May to July.

RANGE: Nfd. to Ont. south to Fla. and Tex.

ORCHID FAMILY (Orchidaceae)

The members of this family have highly specialized flowers, many of them apparently being dependent upon a single species of insect to effect their pollination. Seemingly for this reason the flowers are unusually long-lived, but as soon as pollination has taken place the flowers wither.

Orchid flowers have 3, or sometimes only 2, sepals. There are 3 petals. The two side or lateral ones usually form the *wings* while the lower one, which is usually much different from the other two, forms the *lip*. The 1 or 2 stamens are united with the style of the pistil to form the *column*. The pollen grains are very sticky and adhere to one another, forming the 2 pollen masses or *pollinia*. The bases of all of the flower parts are united with the ovary of the pistil, thus this organ appears to be at the very base of the flower (ovary inferior). The fruits of the orchids are capsules which contain a very large number, sometimes even millions, of dust-like seeds.

Many tropical orchids (and those grown in our greenhouses) are epiphytes which grow on the trunks and branches of trees. Only one such species, the Greenfly Orchid, occurs within our range. Aside from this, all of our native orchids are terrestrial plants. Some of them, such as the lady's-slippers, have large and showy flowers, but indiscriminate picking and gathering of the plants is making them increasingly rare. They are very much in need of protection. Others have such small and inconspicuous flowers that only those versed in the subject of botany recognize them as being orchids.

Key to Genera of the Orchid Family

1. Plants with little or no green color; either without leaves or the leaves not present at the time of flowering.
   2. Flower stalks with at least a few scales; the plants entirely leafless. Lip of the flower usually with purple spots and a lobe on either side near the base. CORAL-ROOTS *(Corallorrhiza)*
   2. Flower stalks entirely naked; the leaves not usually present at flowering time.
3. Lip extended backward as a long and slender spur. CRANEFLY ORCHID *(Tipularia)*
3. Lip not extended backward as a spur. PUTTYROOT *(Aplectrum)*
   1. Plants with a green color; the leaves usually present at flowering time.
4. Lip large and pouch-like. LADY'S-SLIPPERS *(Cypripedium)*
4. Lip otherwise.
   5. Lip erect or at the top, at least in the mature flowers.
6. Flowers fairly large, pink or magenta (rarely white), usually 3 to 12 on a stem. Leaves narrow, grass-like. GRASS-PINKS *(Calopogon)*
6. Flowers small, greenish or white, and rather numerous.
   7. Plant with a single oval-shaped leaf. ADDER'S-MOUTH *(Malaxis)*
   7. Plant with 2 or more long and narrow leaves. REIN ORCHID *(Platanthera)*
5. Lip at the bottom of the flower.
   8. Flower solitary, or only 2 or 3 on a stem.
9. Leaves in a whorl of 5. WHORLED-POGONIAS *(Isotria)*
9. Leaves otherwise.
   10. Plant with usually 3 nodding flowers from axils of small, oval-shaped, clasping bracts. NODDING-POGONIA *(Triphora)*
   10. Plant with usually 1 (rarely 2) flower and a single leaf.
11. Leaf broad and oval-shaped. Lip of the flower slipper-shaped. CALYPSO *(Calypso)*
11. Leaf narrow and grass-like. Lip of the flower otherwise.
   12. Flower not subtended by a large leafy bract; the basal leaf appearing after the flower matures. ARETHUSA *(Arethusa)*
   12. Flower subtended by a large leafy bract; the lower leaf well developed before the flower opens.

13. Side petals lying over the trough-shaped lip. Sepals narrow and spreading.
SPREADING-POGONIA *(Cleistes)*

13. Side petals spreading and not covering the spoon-shaped lip. Sepals pink.
ROSE POGONIA *(Pogonia)*

8. Flowers several to numerous.

14. Plant with a pair of leaves near the middle of the stem. TWAYBLADES *(Listera)*

14. Plant with scattered or only basal leaves.

15. Leaves conspicuously mottled with white. RATTLESNAKE-PLANTAINS *(Goodyera)*

15. Leaves otherwise.

16. Lip with a spur at its base.

17. Sepals and petals forming a purple hood above the broad white lip.
SHOWY ORCHID *(Galearis)*

17. Sepals and petals colored alike, the lip sometimes toothed or fringed.
FRINGED ORCHIDS, ETC. *(Platanthera)*

16. Lip without a spur at the base.

18. Plant with 2 oval-shaped basal leaves. TWAYBLADES *(Liparis)*

18. Plant with 2 or more rather narrow or grass-like basal leaves.

19. Plant terrestrial. Flowers stalkless and arranged in 1 or more spiral rows in a slender cluster.
LADY'S-TRESSES *(Spiranthes)*

19. Plant epiphytic. Flowers stalked and arranged in an open cluster.
GREEN-FLY ORCHID *(Epidendrum)*

GREEN-FLY ORCHID *(Epidendrum conopseum)*

This is the only epiphytic orchid to be found north of Florida. It grows on the trunks or branches of live oaks and other trees, along with the Spanish-moss and the resurrection fern. The plant has a bulb-like base and thick roots which serve as holdfasts; and a basal cluster of thickish, leathery, narrow leaves up to 3 inches in length. The flowers are about ¾ inch across, greenish or tinged with purple, and have a delicate fragrance. Several are borne in a cluster on a stalk from 2 to 8 inches long, between July and September.

RANGE: se. N.C. south to Fla. west to La., near the coast.

YELLOW LADY'S-SLIPPER *(Cypripedium pubescens)*

From 1 to 3 flowers are borne on a leafy stem from 8 inches to about 2 feet tall. Often there are several stems in a clump. The prominently veined leaves are hairy and from 3 to 6 inches in length. The flowers have a bright yellow pouch-like lip from ¾ to 2 inches long; and 3 sepals and 2 spirally twisted side petals which are yellowish-green to purplish-brown and often streaked with purple. They are more or less fragrant. Hairs of the plant may cause a dermatitis in sensitive persons. It grows in dry to wet woods and swamps, blooming between April and June.

RANGE: Nfd. and N.S. to B.C. south to N.C., n.Ga., n.La., Tex., Ariz. and Wash.

SMALL WHITE LADY'S-SLIPPER *(Cypripedium candidum)*

This species has a leafy stem 6 to about 15 inches tall which bears 1 or 2 flowers. The lance-shaped, prominently veined leaves are more or less overlapping along the stem. The flowers have a pouch-like lip about ¾ inch long which is white with purple stripes within. The sepals and 2 side petals are greenish-yellow and often streaked with purple. It grows in wet meadows, glades and swamps but is rather local; blooming in May or June.

RANGE: N.Y. west to N.D. south to N.J., Ky. and Mo.

SHOWY LADY'S-SLIPPER *(Cypripedium reginae)*

Often called the Queen Lady's-slipper, this species has a stout, leafy stem from 1 to 2½ feet tall. The large corrugated leaves range up to 7 inches in length and resemble those of the White Hellebore. The flowers are usually solitary and have a pouch-like lip 1½ to 2 inches long which is white but strongly suffused with rose or magenta. The 2 other petals and 2 sepals are white. Hairs of the plant may produce a dermatitis in sensitive persons. It grows in swamps, bogs, and moist wooded slopes; blooming between April and July.

RANGE: Nfd. to Man. south to Pa., w. N.C., e. Tenn., Mo. and N.D.

RAM'S-HEAD LADY'S-SLIPPER *(Cypripedium arietinum)*

The Ram's-head gets its name from its odd flower, the cone-shaped pouch of which is greenish or yellowish at the tip and marked with purple veins. The sepals and 2 side petals are greenish streaked with purple. It has a stem 6 to 12 inches tall on which are from 3 to 5 rather narrow, smooth leaves up to 4 inches in length. The plant grows in wet woods and bogs but it is rather rare, blooming in May or June.

RANGE: Que. to Man. south to N.Y. and Mich.

STEMLESS LADY'S-SLIPPER *(Cypripedium acaule)*

This species is also called the Pink Lady's-slipper and Moccasin-flower and, like other lady's-slippers, it is also known as the Whip-poor-will's-shoe. It is distinctive as the flower stands on a naked stalk 6 to 15 inches tall, at the base of which is a pair of large elliptic leaves. The pouch-like lip is about 2 inches long and pink with deeper colored veins (rarely white). The sepals and 2 side petals are greenish yellow and tinged with purplish. It grows in dry to moist woods and in bogs, blooming between April and July.

RANGE: Nfd. to Alb. south to Ga., Ala., Tenn. and Mo.

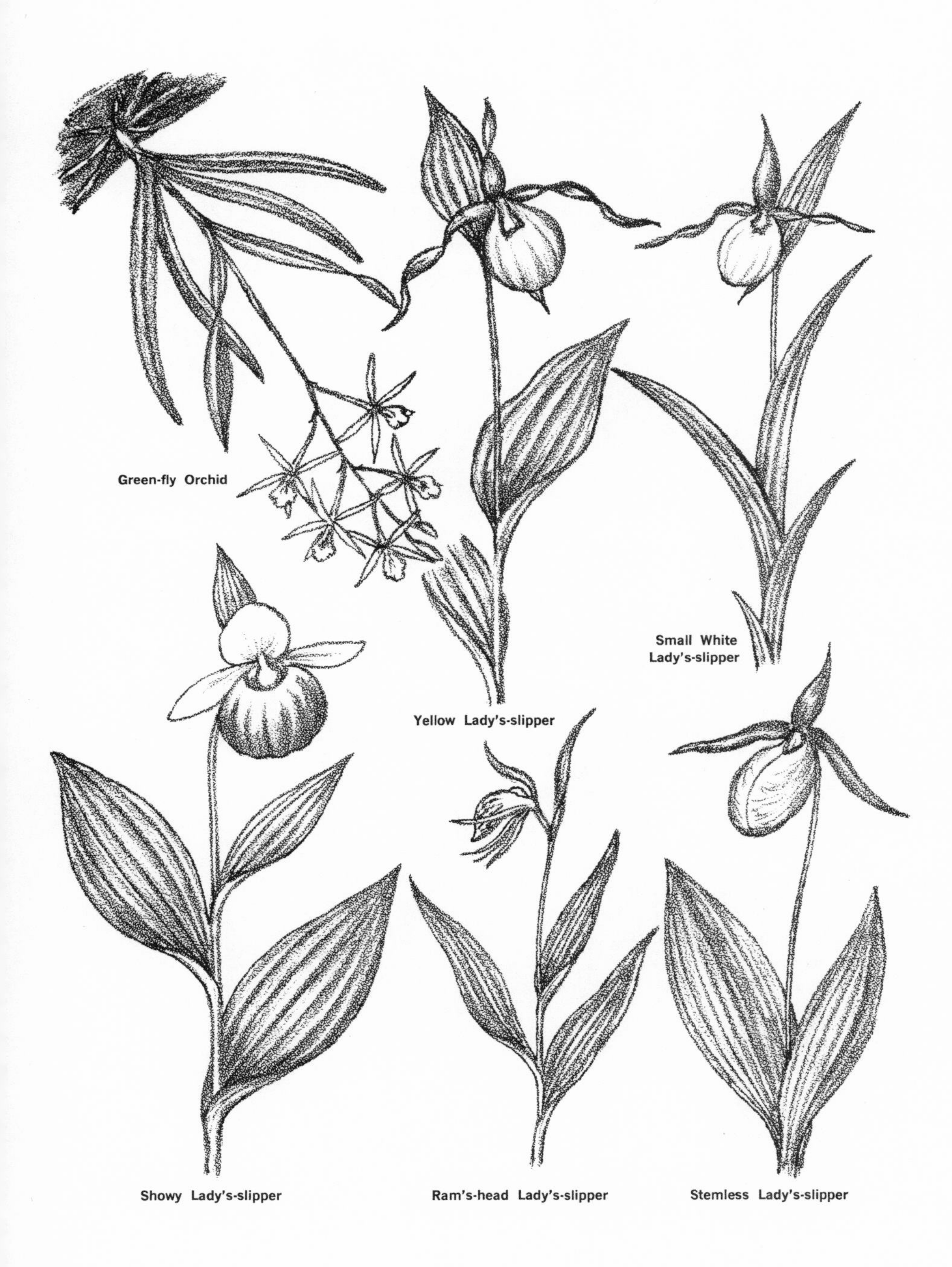
Green-fly Orchid
Yellow Lady's-slipper
Small White
Lady's-slipper
Showy Lady's-slipper
Ram's-head Lady's-slipper
Stemless Lady's-slipper

Showy Orchid
Green Fringed Orchid
Small Purple
Fringed Orchid
Yellow Fringed Orchid
White Fringed Orchid

SHOWY ORCHID *(Galearis spectabilis)*

On a stem from 5 to 9 inches tall, the Showy Orchid displays several, leafy-bracted, attractive flowers. Each one is about an inch long, and from the base of the flat white lip hangs a spur. The other petals and the sepals are joined together to form a lilac or magenta, erect hood. There are 2 smooth, lustrous leaves, from 4 to 8 inches long, at the base of the upright flower stalk. It is found in rich, moist and usually rocky woodlands and it blooms between April and June.

RANGE: N.B. to Ont. south to N.Eng., n. Ga., Tenn., Mo. and Kan.

GREEN FRINGED ORCHID *(Platanthera lacera)*

Another name for this species is Ragged Orchid; and, indeed, its greenish-yellow or greenish-white flowers really look ragged. The flowers are less than an inch long and their fanshaped lip is cut into 3 wedge-shaped lobes which have long-fringed tips. At its base there is a downward pointing spur. The flowers are arranged in a rather loose but long cluster toward the summit of the 1 to 2½ foot stem along which there are some small, pointed leaves. The basal leaves are much larger, some of them often 8 inches in length. This interesting orchid grows in dry to wet meadows, fields, and thickets. It blooms between June and September.

RANGE: Nfd. to Ont. and Minn. south to Fla. and Tex.

SMALL PURPLE FRINGED ORCHID *(Platanthera psycodes)*

On a leafy stem from 1 to 3 feet tall, this orchid has a narrow but quite dense cluster of lilac-pink to rose-purple flowers. Each flower is less than an inch long and its fan-shaped lip is cut into 3 wedge-shaped and fringed lobes. The lip is also projected backward into a downward pointing spur about ¾ of an inch in length. Leaves at the base may be anywhere from 2 to 10 inches long. It grows in moist meadows, thickets, and woods; and blooms between late June and August.

RANGE: Nfd. to Ont. south to N.J., n. Ga., Tenn., and Iowa.

LARGE PURPLE FRINGED ORCHID *(Platanthera grandiflora)*

This orchid is very similar to the preceding species but larger in every respect. Its flowers are an inch or more long, and the spur of the lip is about 1½ inches in length. It grows in cool, moist, rich woods and in meadows; blooming about 2 weeks earlier than the Small Purple Fringed Orchid.

RANGE: Nfd. to Ont. south to N.J., w. N.C., and e. Tenn.

YELLOW FRINGED ORCHID *(Platanthera ciliaris)*

The bright orange-yellow plumes of this common orchid make it one of our most conspicuous wild flowers, and have earned it the name of Orange-plume. It blooms between July and September. The plant has a leafy stem from 1 to 2½ feet tall, the larger leaves toward the base being 4 to 8 inches in length. The flowers have an oblong-shaped lip which is deeply fringed about the margin; and a slender spur as long as, or longer than, the colored and stalk-like ovary. It grows in moist open places, on slopes, and in thickets from the coast to the mountains.

RANGE: Vt. to Ont. and Wis. south to Fla. and Tex.

CRESTED YELLOW ORCHID *(Platanthera cristata)*

This species is similar to, but smaller than, the preceding species. Perhaps the best distinction lies in the length of the spurs. Those of the Crested Yellow Orchid are only about half as long as the stalk-like ovaries. It grows in bogs and in moist to dry meadows and thickets; blooming between July and September.

RANGE: N.J. to Tenn. and Ark. south to Fla. and Tex. (Not illustrated)

WHITE FRINGED ORCHID *(Platanthera blephariglottis)*

The plume-like clusters of snow-white flowers near the summit of a leafy 1 to 3 foot stem will distinguish this orchid. As in the Yellow Fringed Orchid, the slender spur is longer than the stalk-like ovary; but in this species the lip is quite short-fringed, or not fringed at all in *P. integriloba.* It is very partial to wet boggy places and blooms between June and September.

RANGE: Nfd. to Ont. and Mich. south to Fla. and Miss.

WHITE REIN ORCHID *(Platanthera nivea)*

This orchid is often abundant in the wet pinelands and bogs of the Southeast, where it is frequently called the Bog-torch or Snowy Orchid. The plant has a slender stem 10 to 30 inches tall, with a few small leaves and a slender cluster of small white flowers. At the base are a few rather firm, keeled or even longitudinally folded, lance-shaped leaves 4 to 8 inches in length. The lip of the flower is smooth-margined and stands at the top of the flower. It is projected backward as a long, slender spur. The plant blooms in August or September.

RANGE: N.J. south to Fla., west to Tex., and north to Ark.

SMALL SOUTHERN YELLOW ORCHID *(Platanthera integra)*

The small yellow or orange-yellow flowers of this orchid have a lip which is slightly toothed but not fringed. The plant has a stiffly erect stem 1 to 2 feet tall along which are scattered, small leaves. The 1 or 2 lowest ones are larger, lance-shaped, and long-pointed at the tip. Toward the summit of the stem there is a long and narrow flower cluster. It grows in wet pinelands and bogs, blooming in August or September.

RANGE: N.J. to e. Tenn. south to Fla. and Tex.

SMALL GREEN WOOD ORCHID *(Platanthera clavellata)*

One could not call this little orchid very conspicuous. It has a slender stem 8 to 16 inches tall, on which are a few small and bract-like leaves. There is a single large leaf at the base which is 2 to 6 inches long and broadest well above the middle. The small greenish-white flowers have a club-like spur about ½ inch in length, and they are in a narrow cluster. It grows in wet or moist woods and boggy places and blooms during July and August.

RANGE: Mass. to Wis. south to Fla. and Tex.

ROUND-LEAF ORCHID *(Platanthera orbiculata)*

This woodland orchid has a pair of large, round, shiny, pad-like leaves which are 4 to 8 inches wide and usually lie flat on the ground. The flower stalk is 8 to 20 inches tall and usually has several small, bract-like leaves below the narrow flower cluster. Its greenish-white flowers are about an inch long and they have slender dangling spurs. It blooms between June and August.

RANGE: Nfd. to Ont. south to n. Ga., Tenn. and Ill.

HOOKER'S ORCHID *(Platanthera hookeri)*

This plant like the preceding species has big, round, pad-like leaves. The somewhat fleshy, leafless flower stalk is 8 to 15 inches tall; and the flowers are greenish-yellow. The lip tapers to an upturned point in the front and is projected backward as a slender spur. This species grows in rather dry woods, blooming between May and August.

RANGE: Que. and Ont. to Minn. south to W.Va., Ill. and Iowa. (Not illustrated)

FRINGELESS PURPLE ORCHID *(Platanthera peramoena)*

Toward the summit of a stout leafy stem 1 to 3 feet tall, this orchid displays a dense, cylindrical cluster of bright rose-purple or violet-purple flowers. Each one has a lip with 3 wedge-shaped lobes which are but slightly toothed at the tip, and projected backward as a slender spur about an inch in length. Leaves toward the base of the stem are 4 to 8 inches long but they become progressively smaller up the stem. It grows in moist woods, meadows, and along streams; blooming between June and October.

RANGE: N.J. to Ill. and Mo. south to nw. S.C., Ga. and Miss.

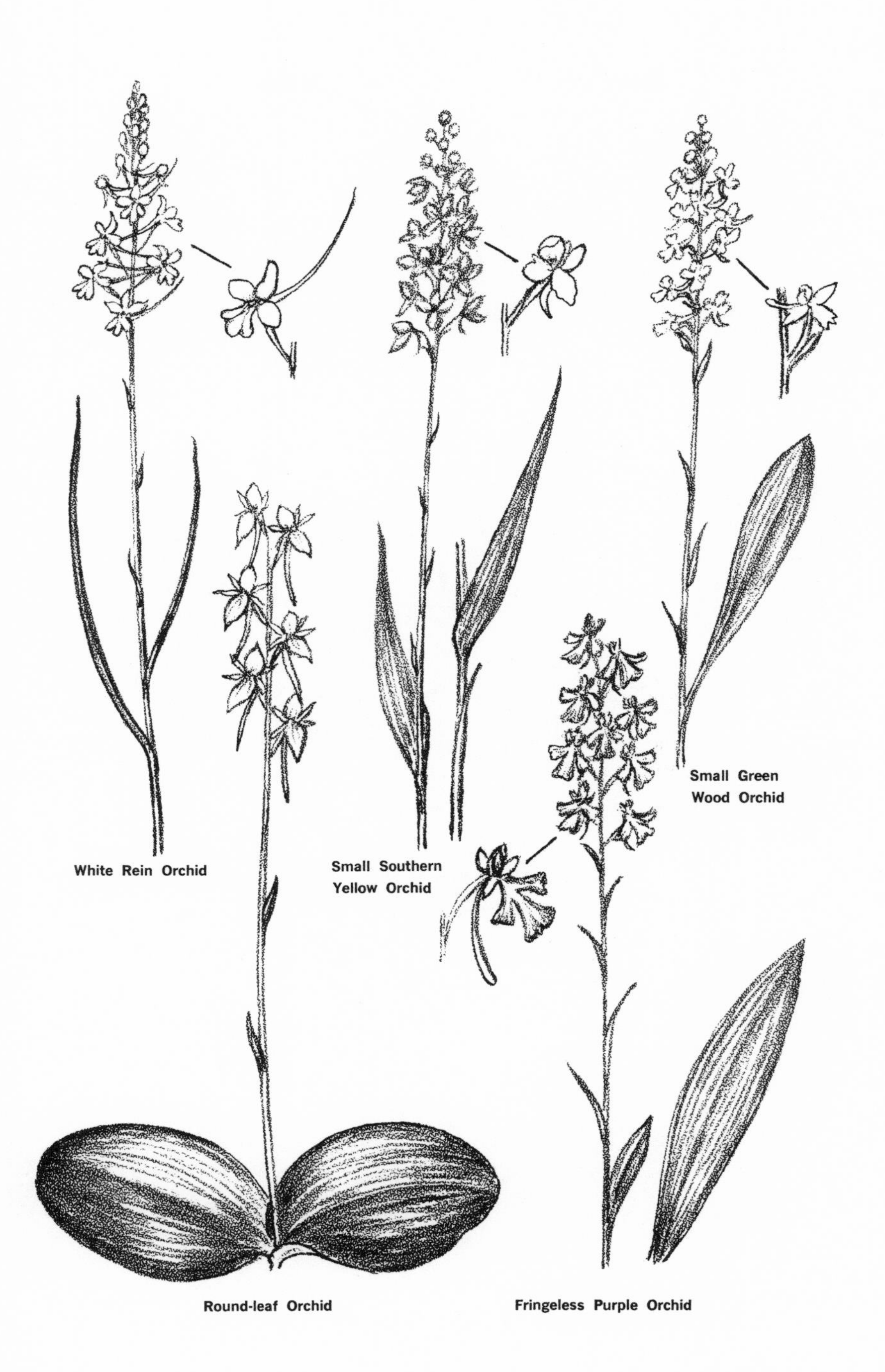
White Rein Orchid
Small Southern Yellow Orchid
Small Green Wood Orchid
Round-leaf Orchid
Fringeless Purple Orchid

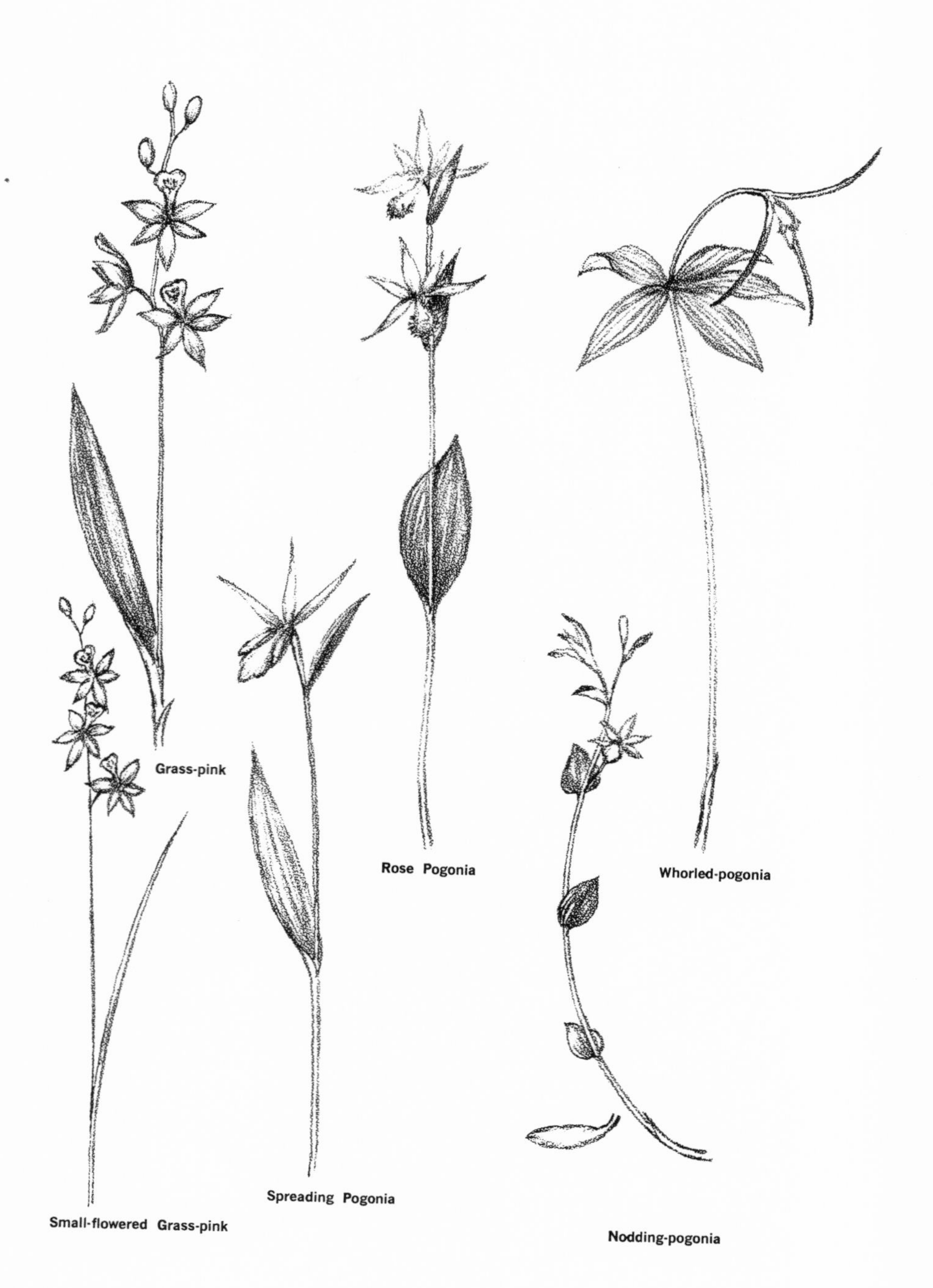
Grass-pink
Rose Pogonia
Whorled-pogonia
Spreading Pogonia
Small-flowered Grass-pink
Nodding-pogonia

GRASS-PINK *(Calopogon tuberosus)*

Several rose-pink to magenta flowers are produced along the 8 to 24 inch tall stem of this plant. As in all grass-pinks the spoon-shaped lip is uppermost and has a tuft of yellow, crimson-tipped, club-shaped hairs. The flowers are an inch or more across and open successively up the stem. There is but one broad grass-like leaf, from 6 to 12 inches long, toward the base. This attractive orchid is quite common in wet meadows, bogs, and southward in the wet pinelands. It blooms between April and August.

RANGE: Nfd. to Que. and Minn. south to Fla. and Tex.

SMALL-FLOWERED GRASS-PINK *(Calopogon barbatus)*

This species of the southern coastal plains has an almost thread-like slender stem 6 to 15 inches tall, with a very narrow grass-like leaf 4 to 10 inches long at the base of the stem. The rose-colored flowers are smaller than those of the preceding species and all open at about the same time. It grows in wet pinelands and savannahs; blooming in April or May.

RANGE: N.C. south to Fla. and west to La.

PALE GRASS-PINK *(Calopogon pallidus)*

This plant closely resembles the preceding species but has pale lilac to whitish flowers. It grows in wet pinelands and bogs of the coastal plain from Va. south to Fla. and west to La., blooming from May to July. (Not illustrated)

SPREADING POGONIA *(Cleistes divaricata)*

At the summit of a 1 to 2 foot stem, this orchid has a solitary, leafy-bracted pink flower an inch to 2 inches long. The trough-shaped lip is veined with purple, lobed at the tip, and crested but not bearded. The other 2 petals lie forward over the lip, while the 3 narrow brownish sepals widely spread. About midway on the stem is a lance-shaped leaf from 2 to 5 inches long. It grows in moist woods, wet meadows and swamps; blooming May to July. Also called Rose Orchid, Rosebud Orchid, and Lady's-ettercap.

RANGE: N.J. south to Fla.; Ky. south to Ga., Tenn. and La.

ROSE POGONIA *(Pogonia ophioglossoides)*

The Rose Pogonia, or Snake-mouth, usually has 1, but sometimes 2, leafy-bracted pink flowers at the top of a slender stem from 10 to 20 inches tall. The flower is an inch to 1½ inches broad, and it has a spoon-shaped lip which is beautifully bearded with yellow-tipped hairs. A solitary and usually broadly lance-shaped leaf 2 to 4 inches long is located about midway on the stem. This orchid grows in wet to moist meadows, thickets, pinelands, and bogs; blooming between May and August.

RANGE: Nfd. to Ont. south to Fla. and Tex.

WHORLED-POGONIA *(Isotria verticillata)*

This orchid of moist woodlands has a stem from 8 to 12 inches tall, on which is a whorl of 5 leaves. The solitary flower at the summit has a greenish-yellow lip which is lobed at the tip, and a ridge-like crest in the center. The 3 wide-spreading sepals are greenish tinged with purple. It blooms in May or June.

RANGE: Me. to Mich. and Mo. south to Fla. and Tex.

SMALL WHORLED-POGONIA *(Isotria medeoloides)*

Like the preceding species, this smaller plant has a whorl of 5 leaves on its 6- to 8-inch stem. At the summit are 1 or 2, inch-long flowers in which the yellow lip is lobed at the tip and bears a broad crest. The other petals and sepals are short and greenish-yellow with a purplish tinge. This orchid grows in dry woodlands but it is quite rare and local. It blooms between May and early July.

RANGE: N.H. and Vt. south to N.C. and Mo. (Not illustrated)

NODDING-POGONIA *(Triphora trianthophora)*

This is a dainty little plant with a fragile stem 3 to 8 inches tall; along which are scattered, small, roundish, clasping leaves. It is often called Three-birds because there are usually 3 flowers nodding from the axils of its upper leaves. They vary from whitish to pink and their lip has 3 greenish lines. It grows in rich hardwood forests, blooming in August or September. Although locally common, it may disappear for a period of several years.

RANGE: Sw. Me. to s. Mich. and Iowa south to Fla. and e. Tex.

ARETHUSA *(Arethusa bulbosa)*

Arethusa has a solitary flower at the summit of a 5 to 10 inch stem; along which are a few, loose, sheathing bracts. Not until the flower matures does the single grass-like leaf protrude from one such sheath, later growing to a length of about 6 inches. The broad tongue-like lip is whitish with conspicuous purple spots, and on it is a crest consisting of 3 rows of fleshy, yellow- and purple-tipped hairs. The other petals and the sepals are magenta-pink. In most places Arethusa is a bog orchid, but nowhere does it grow so abundantly as in the salt marshes along the northern New England coast. It blooms in May or June. It is also known as the Dragon's-mouth and Bog-pink.

RANGE: Nfd. to Ont. and Minn. south to Md., nw. S.C., and the Great Lakes region.

CALYPSO *(Calypso bulbosa)*

The solitary and oddly beautiful flower of Calypso is borne atop a sparsely bracted 3- to 6-inch stem; at the base of which is a broad but pointed, plaited leaf which withers soon after the plant flowers. The lance-shaped sepals and 2 side petals are purplish and spread or ascend over the slipper-shaped, inflated lip. The latter bends downward; its white surface is streaked and spotted with cinnamon-brown and purple within, and in front there are 3 double rows of golden yellow hairs. By many it is considered to have the most beautiful flower of any of our native orchids, and it is often called the Fairy-slipper. It grows in cool, damp, mossy woods and blooms in May or June.

RANGE: Nfd. to Alaska south to N.Y., Mich., Wis., and in mts. to Ariz. and Calif.

DOWNY RATTLESNAKE-PLANTAIN *(Goodyera pubescens)*

This orchid is often called to our attention by its oval-shaped dark green leaves, which have a network of whitish veins. Its small white flowers are arranged in a long, narrow but dense cluster on a downy stalk 6 to 18 inches tall. The leaves are 1 to 2½ inches long. Often common in both moist or dry woods, it blooms during July and August.

RANGE: Me. to Ont. south to Fla., Ala., and Mo.

LESSER RATTLESNAKE-PLANTAIN *(Goodyera repens)*

This species has smaller leaves, ½ to 1 inch long with a few broadly bordered white veins. Its small white flowers are in a more loose, 1-sided flower cluster on a 4 to 10 inch stalk. It grows in cool, moist woods and blooms during July and August.

RANGE: Lab. to Alaska south to N.J., w. N.C., e. Tenn., and S.D. (Not illustrated)

SLENDER LADY'S-TRESSES *(Spiranthes lacera* var. *gracilis)*

About a dozen species of lady's-tresses occur in eastern North America. All of them have rather small white, greenish-white, or yellowish flowers arranged in a more or less spirally twisted spike. This one has small white flowers which have a green spot on the lip. They are arranged in a single and strongly spiraled row on a leafless stalk 8 to 24 inches tall. The leaves are egg-shaped, stalked, and about 1 to 2½ inches long but they may not be present at flowering time. It grows in dry to moist open woods and fields, blooming from late July to October.

RANGE: N.H. to Wis. and Okla. south to Fla. and Tex.

NODDING LADY'S-TRESSES *(Spiranthes cernua)*

This is a widespread and common little orchid; growing in moist to dry fields, thickets, and open woods. It has white flowers about ⅓ of an inch long, which are arranged in 3 or 4 somewhat twisted rows on stalks 6 to 18 inches tall. The grass-like leaves are basal, several inches long, and narrower downward. It blooms from August to October, and the flowers are sweet-scented.

RANGE: N.S. to Que. and Wis. south to Fla. and Tex.

MARSH LADY'S-TRESSES *(Spiranthes odorata)*

Some botanists consider this plant to be a variety of the preceding species. It is a similar but much larger plant growing from 1½ to about 3 feet tall. The flowers are creamy-white and very fragrant. It grows in fresh water marshes and blooms in September or October.

RANGE: Md. and Tenn. south to Fla. and Tex. (Not illustrated)

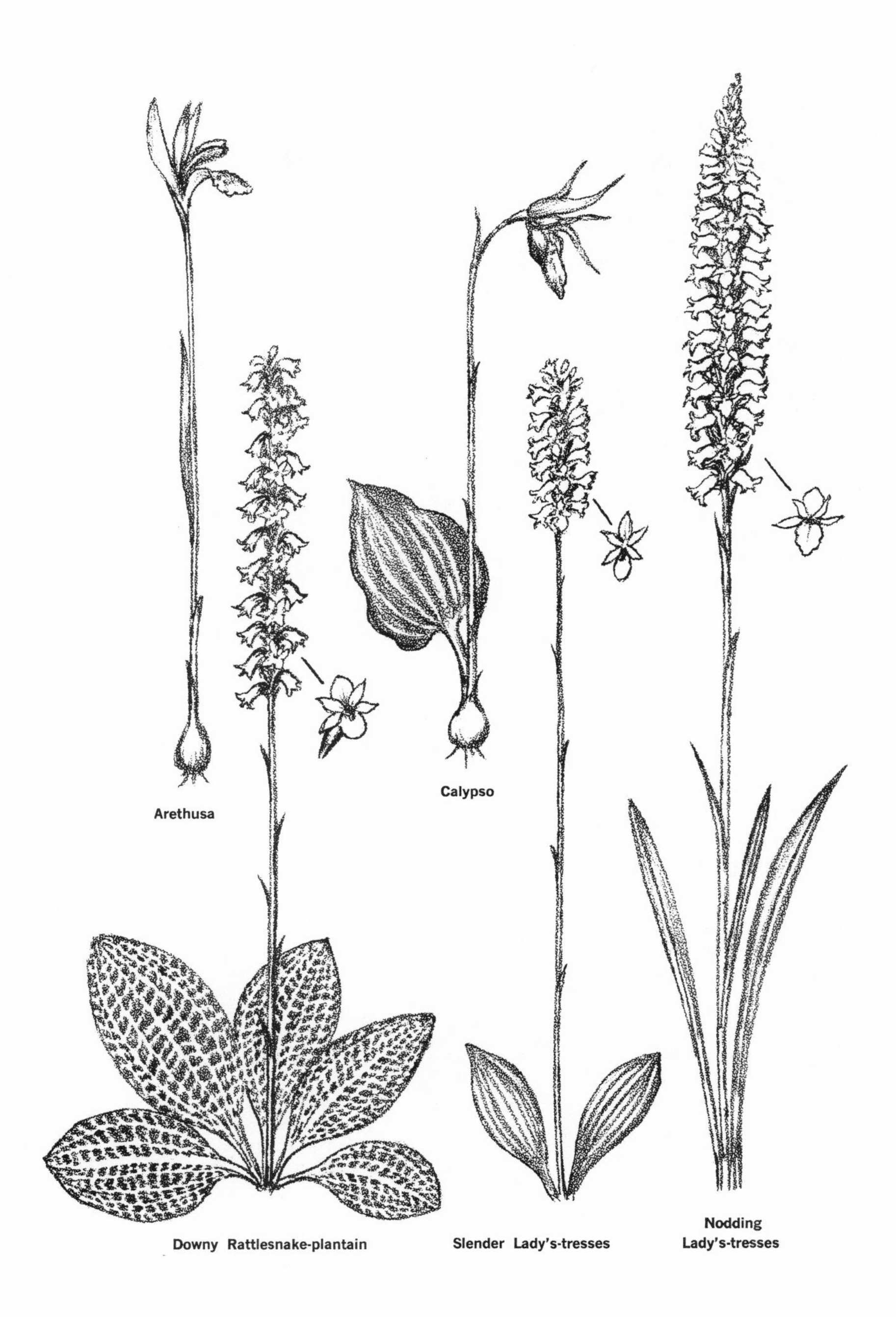
Arethusa
Calypso
Downy Rattlesnake-plantain
Slender Lady's-tresses
Nodding
Lady's-tresses

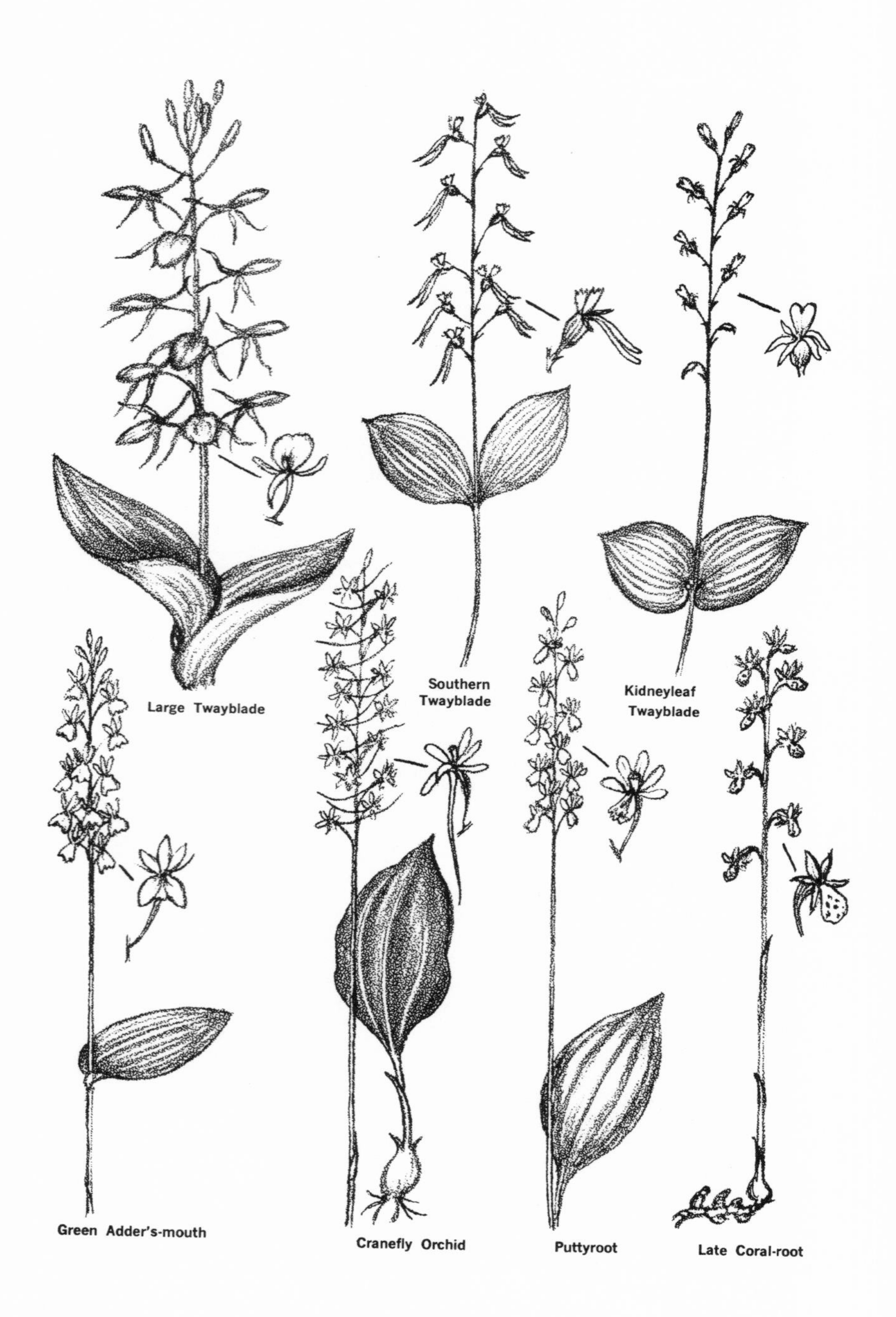
Large Twayblade
Southern
Twayblade
Kidneyleaf
Twayblade
Green Adder's-mouth
Cranefly Orchid
Puttyroot
Late Coral-root

LARGE TWAYBLADE *(Liparis liliifolia)*

Twayblades get their common name from the fact that they have 2 leaves. Those of *Liparis* are oval-shaped, smooth and lustrous, and all at the base of the plant. A number of half-inch greenish-brown flowers are borne on an angled stalk 4 to 10 inches tall. It grows in rich moist woods and blooms between May and early July.
RANGE: N.H. to Minn. south to n. Ga. and Mo.

SOUTHERN TWAYBLADE *(Listera australis)*

On a slender stem 4 to 10 inches tall this plant has a pair of quite firm, narrowly egg-shaped, stalkless leaves ½ to 1½ inches long. The reddish- to greenish-purple flowers are usually less than ½ inch long, with a lip that is split nearly to the base into 2 parallel lobes. It grows in wet woods and bogs, blooming from April to July.
RANGE: Que. and Ont. south to Fla. and La.

HEARTLEAF TWAYBLADE *(Listera cordata)*

This species is similar to the preceding but has thinner heart-shaped leaves ½ to about 1 inch long. The bronzy- or purplish-green flowers have a lip with 2 spreading lobes. It grows in mossy woods or bogs from Nfd. to Alaska south to N.J., w. N.C., Mich., Minn., N.Mex. and Calif.; blooming between late May and August. (Not illustrated)

KIDNEYLEAF TWAYBLADE *(Listera smallii)*

This species has a stem 4 to 12 inches tall which is densely glandular-hairy above. On it is a pair of roundish kidney-shaped leaves ½ to slightly over 1 inch long. The lip of the small greenish flowers is broadly wedge-shaped and deeply notched. It grows in moist mountain woods and thickets, blooming during July and August.
RANGE: Pa. south to n. Ga.

GREEN ADDER'S-MOUTH *(Malaxis unifolia)*

This orchid has a smooth, slender stem 4 to 10 inches tall with a solitary oval-shaped leaf about midway. Its small greenish flowers stand on slender stalks in a cylindrical cluster. At first the lip is lowermost; but, by a twisting of its stalk, the flower becomes inverted so that the lip is at the top. It grows in dry to moist open woods and thickets, blooming between May and August.
RANGE: Nfd. to Sask. south to Fla. and Tex.

CRANEFLY ORCHID *(Tipularia discolor)*

During the fall this orchid produces a single broadly egg-shaped leaf 2 to 3 inches long, which is purple beneath. By flowering time, in July or August, this leaf withers and disappears. The bronzy or tawny flowers have a long slender spur, and they stand on slender stalks which gives the impression of a long-legged insect. A number of the flowers are arranged in a long cluster on a leafless, smooth, tawny stalk 1 to 2 feet tall. It grows in rich woodlands.
RANGE: Mass. to Ind. and Mo. south to Fla. and Tex.

PUTTYROOT *(Aplectrum hyemale)*

This orchid produces a single dark green leaf with paler stripes, from 4 to 6 inches long. It persists until the flower stalk develops in May or June, then it disappears. The purplish-green to yellowish flowers have a white lip spotted with purple, and are arranged in a narrow cluster on a leafless stalk 10 to 20 inches tall. The name of Puttyroot alludes to the putty-like, sticky contents of the whitish corms which are usually in pairs; hence another name for it is Adam and Eve. It grows in rich woods.
RANGE: Que. to Sask. south to Ga., Tenn. and Ark.

LATE CORAL-ROOT *(Corallorrhiza ondontorhiza)*

Coral-roots lack chlorophyll and are yellowish to purplish in color. Their leaves are merely small scales and the underground parts are branching and coral-like. This species has a stalk 4 to 10 inches tall bearing small greenish-purple flowers which have a white, purple-spotted lip. It grows in woodlands, blooming between August and October.
RANGE: Me. to Minn. south to Ga., Miss. and Mo.

LIZARD'S-TAIL FAMILY (Sauraceae)

Flowers of the members of this family have neither sepals nor petals, the 6 to 8 white stamens being the showy part of the flower. They have 3 or 4 pistils which are more or less united at the base. Only the following representative of the family is found in our region.

LIZARD'S-TAIL *(Saururus cernuus)*

The Lizard's-tail is a plant of swamps and shallow waters. It has heart-shaped leaves 3 to 6 inches long which are scattered along the more or less zig-zag 1½ to 3 foot tall stem. The small, fragrant, white flowers are in a long, slender, tapering, and gracefully drooping cluster; blooming between May and September. The extensively creeping root-stocks are aromatic. RANGE: Que. and Ont. south to Fla. and Tex.

SANDALWOOD FAMILY (Santalaceae)

Plants of this family are more or less parasitic on the roots of other plants. The flowers have no petals and the 5 sepals are united to form a cup with a 5-toothed rim. They have 5 stamens and a solitary pistil.

BASTARD-TOADFLAX *(Commandra umbellata)*

This plant has a slender, very leafy, and often a branched stem 6 to 18 inches tall. The thin leaves are stalkless, pale beneath, and from ½ to 1¼ inches long. Its small greenish-white flowers are in clusters at the tip of the stem or its branches. It grows in dry open woods, thickets and fields; blooming between April and June.
RANGE: Me. to Mich. south to Fla. and Ala.

BUCKWHEAT FAMILY (Polygonaceae)

Members of this family may be recognized by the thin tubular sheath about the stem just above each leaf. Their small flowers have 4 to 6 sepals which are often colored and petal-like, no petals, 4 to 6 stamens, and a solitary pistil with 2 or 3 styles. The small, dry, 1-seeded fruits may be flattened or 3-sided.

ARROW-LEAF TEARTHUMB *(Polygonum sagittatum)*

Tearthumbs are well-named as the hooked prickles which line the 4 angles of their stems are capable of scratching the skin. The arrow-shaped leaves of this species are on short stalks and range up to 3 inches in length. Its small greenish to pink flowers are produced between May and late fall. It is often abundant in wet places.
RANGE: Nfd. to Sask. south to Fla. and Tex.

HALBERD-LEAF TEARTHUMB *(Polygonum arifolium)*

This species is similar to the preceding and is best distinguished by its leaves which have the pair of basal lobes pointing outward. It is also common in wet places from N.B. to Ont. and Minn. south to Ga. (Not illustrated)

WATER SMARTWEED *(Polygonum amphibium)*

This is a plant of shallow waters and muddy shores. The sheaths above the leaves have a spreading flange at the top which is rimmed with bristly hairs. Dense clusters of rosy-pink flowers are produced between June and September.
RANGE: Lab. to Alaska south to Pa., Mo., Colo., and Calif.

PENNSYLVANIA SMARTWEED *(Polygonum pensylvanicum)*

A number of smartweeds similar to this one are found in eastern North America, but this is one of the most common and widespread species. It grows to a height of 1 to 3 feet, usually branches freely, and has lance-shaped leaves from 2 to 10 inches long. Showy clusters of pink flowers are produced from July until late fall. The gland-tipped hairs on the stalks of the flower cluster help to distinguish it from similar smartweeds.
RANGE: N.S. to N.D. south to Fla. and Tex.

VIRGINIA KNOTWEED *(Polygonum virginiana)*

Often called Jumpseed as the mature fruits seem to spring from the stalk when touched. This is a woodland plant 2 to 4 feet tall; with egg-shaped pointed leaves 2 to 6 inches long, and small greenish-white flowers widely spaced on a long and slender end stalk. It blooms between July and October. RANGE: Que. to Minn. south to Fla. and Tex.

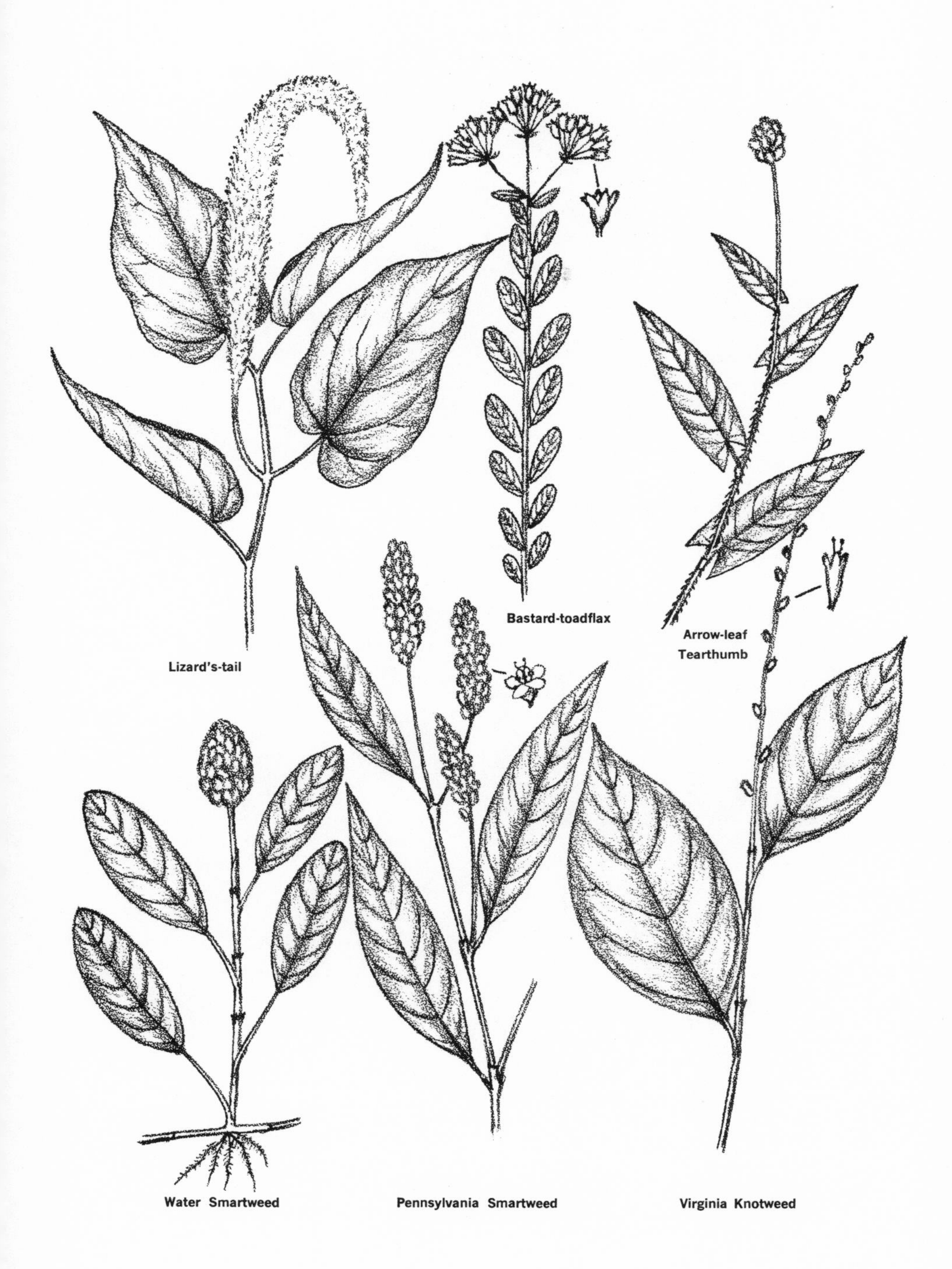
Lizard's-tail
Bastard-toadflax
Arrow-leaf
Tearthumb
Water Smartweed
Pennsylvania Smartweed
Virginia Knotweed

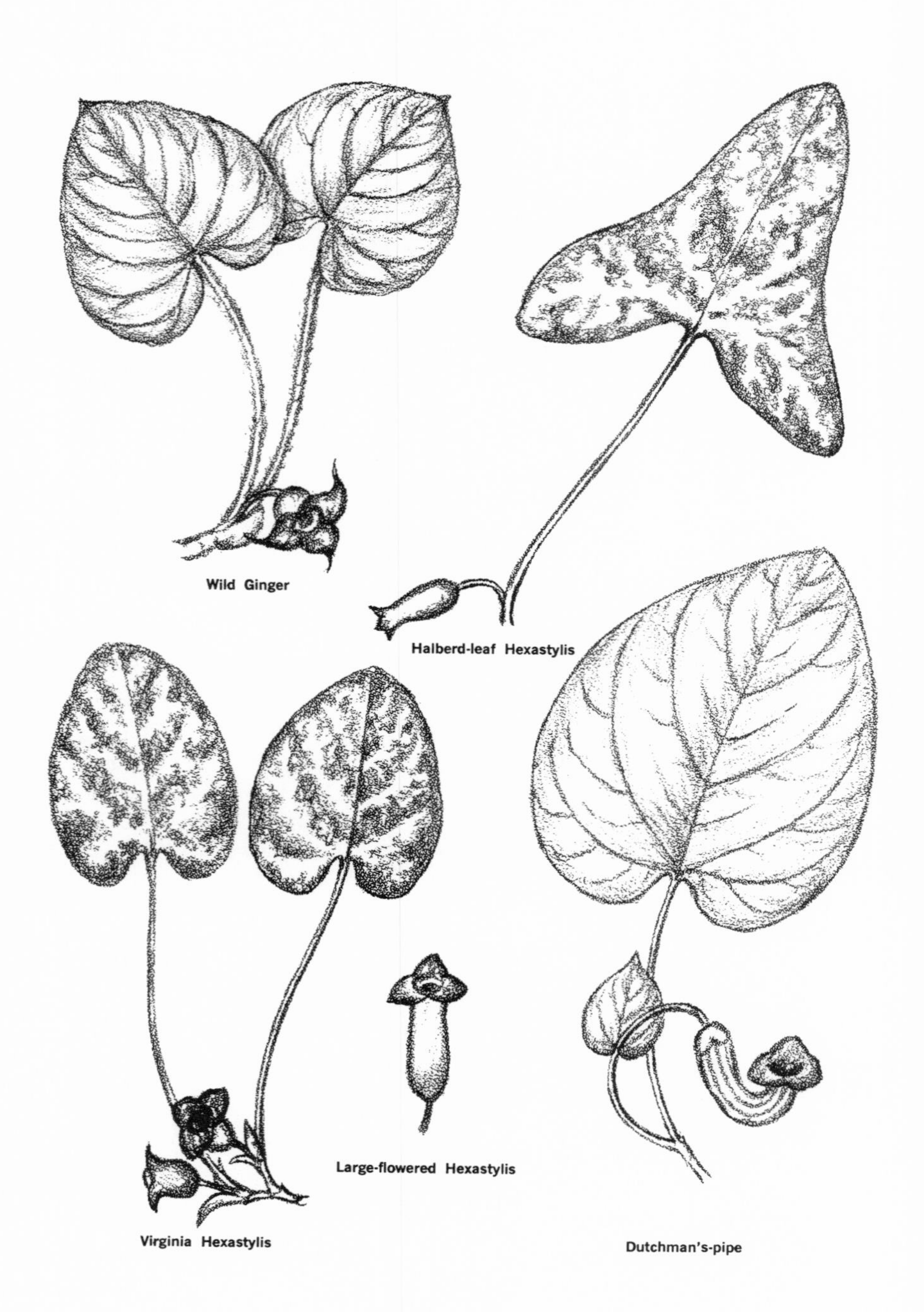
Wild Ginger
Halberd-leaf Hexastylis
Large-flowered Hexastylis
Virginia Hexastylis
Dutchman's-pipe

BIRTHWORT FAMILY (Aristolochiaceae)

Members of this family may be mistaken for monocots as their flower parts are in 3s or multiples of this number. They have 3 sepals which are joined together to form a cup-like structure with 3 lobes on its rim, 6 or 12 stamens, and a pistil which is made up of 6 united carpels, its ovary being more or less united with the calyx tube. The plants have more or less heart-shaped leaves with entire margins, and aromatic rootstocks which often have a ginger-like odor.

WILD GINGER *(Asarum canadense)*

The Wild Ginger is a low hairy plant, with a pair of very veiny leaves at the tip of its creeping underground stem. A solitary flower is produced between the bases of the leaf-stalks and is often hidden among the fallen leaves. It is shaped like a little bell and there are 3 spreading or recurved, pointed, brownish-purple lobes at the summit. Often common on rich and usually rocky wooded slopes, it blooms in April or May. The rootstocks have a ginger-like odor.

RANGE: Que. to Man. south to nw. S.C., Mo. and Kan.

HALBERD-LEAF HEXASTYLIS *(Hexastylis arifolia)*

Like the preceding, these plants have rootstocks with a ginger-like odor; but they have evergreen leaves which are more or less leathery in texture. This species has 1 or 2 such leaves on each branch of its underground stem. They vary from triangular to roundish or arrow-shaped and they are 2 to 5 inches long. Usually they are mottled on the upper surface. The flowers are vase-shaped, contracted at the top, and have 3 slightly spreading lobes at the summit. They are about an inch long. Growing in rich woods, it blooms between April and June.

RANGE: Va., W. Va. and Tenn. south to Fla. and La.

VIRGINIA HEXASTYLIS *(Hexastylis virginica)*

This plant is also known as the Heartleaf and Southern Wild Ginger. It differs from the preceding species in having more roundish or heart-shaped leaves 1½ to 3½ inches long and almost as broad. Its flowers are bell-shaped, about ¾ inch long, and they have 3 broadly-pointed, spreading lobes which are purplish-brown and often mottled. It grows in both sandy and rocky soils and blooms between March and May.

RANGE: E. Va. and W. Va. south to Ga. and Tenn.

LARGE-FLOWERED HEXASTYLIS *(Hexastylis shuttleworthii)*

The flowers of this species are narrow and flask-shaped, and from 1 to 2 inches long. Its leaves are similar in shape to those of the preceding species but they are usually thinner and larger. It grows in rocky woods and ravines in the southern Appalachians, blooming in May or June.

RANGE: Mts. of w. Va. south to n. Ga. and n. Ala.

DUTCHMAN'S-PIPE *(Aristolochia macrophylla)*

The Dutchman's-pipe, or Pipe-vine, is a high-twining vine with heart-shaped leaves 6 to 15 inches across. The flowers are on leafy-bracted stalks in the axils of the leaves. The calyx tube is curved like a Dutch pipe, with a flat and somewhat 3-lobed rim. Its color varies from yellowish-green to brownish-purple. It grows on rich, often rocky wooded slopes and along the banks of streams; blooming in May or June. Often cultivated and sometimes found as an escape outside of its range.

RANGE: Sw. Pa. south in the mts. to n. Ga. and n. Ala.

WOOLLY PIPE-VINE *(Aristolochia tomentosa)*

This species is similar to the preceding but it is readily distinguished by its white-downy young branchlets, lower leaf surfaces, and flowers. The leaves are much smaller, being only 3 to about 6 inches broad. It grows in rich woods and stream bottoms and blooms between May and August.

RANGE: Ill. to Mo. and Kan., south to Fla. and Tex. (Not illustrated)

VIRGINIA SNAKEROOT *(Aristolochia serpentaria)*

The Virginia Snakeroot has an erect, slender, wavy stem from 6 to 18 inches tall which is sparingly branched at the base. Its thin leaves have a heart-shaped base, a pointed tip, and are from 2 to 5 inches long. The dull brown to purplish flowers stand on slender, scaly stalks near the base of the plant and appear between May and July. Beneath the ground, the plant has a knotty rootstock which has an agreeable camphor-like odor and is used medicinally. It grows in rich woodlands.
RANGE: Conn. to Ill. and Kan. south to Fla. and Tex.

PURSLANE FAMILY (Portulacaceae)

The members of the Purslane Family are mostly small, fleshy plants. Their flowers have 2 sepals, usually 5 petals, 5 or more stamens, and a pistil made up of 3 united carpels.

VIRGINIA SPRING-BEAUTY *(Claytonia virginica)*

Spring-beauties have 2 to several flower stems which arise from a deeply buried tuber. In this one the stems are usually 4 to 8 inches tall. Near the middle they have a pair of narrow and rather fleshy leaves from 2 to 4 inches in length. Along the upper part of the stem is a narrow cluster of 5-petalled white or pale pink flowers which have deeper pink veins. They are about a half inch across. The lowermost one opens first, the others open progressively up the stem. This plant is often very common in rich, moist woods and thickets; blooming between March and May. It grows well in a woodland flower garden. RANGE: Que. to Ont. and Minn. south to Ga. and Tex.

CAROLINA SPRING-BEAUTY *(Claytonia caroliniana)*

This species closely resembles the preceding one but it has narrowly egg-shaped or elliptic leaves from 1½ to 3 inches long, and ¼ to 1 inch broad. It grows in rich, moist, open woods and thickets; it blooms between March and May, or later northward. It is also called the Broad-leaved Spring-beauty.
RANGE: Nfd. to Sask. south to w. N.C., e. Tenn., Ill., and Minn.

PINK FAMILY (Caryophyllaceae)

Members of the Pink Family usually have opposite and entire leaves. Their flowers usually have 5 petals and 5 sepals (sometimes only 4), 5 or 10 stamens, and a solitary pistil with from 2 to 5 styles. The fruit is a capsule containing many small seeds.

STAR CHICKWEED *(Stellaria pubera)*

The half-inch, bright white, starry flowers of this plant are often seen on rocky, wooded slopes in the early spring. They have 5 petals which are so deeply notched that there appears to be 10, and there are 10 stamens. The weak stems are 4 to 12 inches tall and have 2 finely hairy lines. On them are pairs of elliptic or oblong leaves from ½ to 2 inches long which are pointed at both ends. Blooming between March and May. It is also known as the Great Chickweed.
RANGE: N.J. to Ill. south to n. Fla. and Ala.

FIRE-PINK *(Silene virginica)*

This is one of our most conspicuous wild flowers. The flowers are an inch or more across and have 5 spreading petals of the most brilliant red, each one being 2-pronged at the tip. The sepals are greenish but tinged with red and united into a sticky tube with a 5-toothed rim. The plant has an ascending, sticky-hairy stem becoming a foot to 2 feet tall and branching above. Along it are widely spaced pairs of narrow leaves, the lower ones being 3 to 5 inches long. The Fire-pink is common in open woods and on rocky slopes. In the South it begins to bloom in late March and it may continue to bloom until June or later. RANGE: N.Y. to Ont. and Minn. south to Ga., Ala., and Ark.

CAROLINA WILD-PINK *(Silene caroliniana)*

This species is a low, tufted plant with sticky-hairy stems from 4 to about 10 inches tall. Its flowers are about an inch across and have 5 spreading petals which are bright pink, and slightly if at all notched at the tip. The paired leaves, and the ones at the base, are 2 to 4 inches long and broadest toward the tip. It grows in dry sandy or rocky woods, blooming between April and June. RANGE: N.H. to Ohio south to S.C. and Tenn.

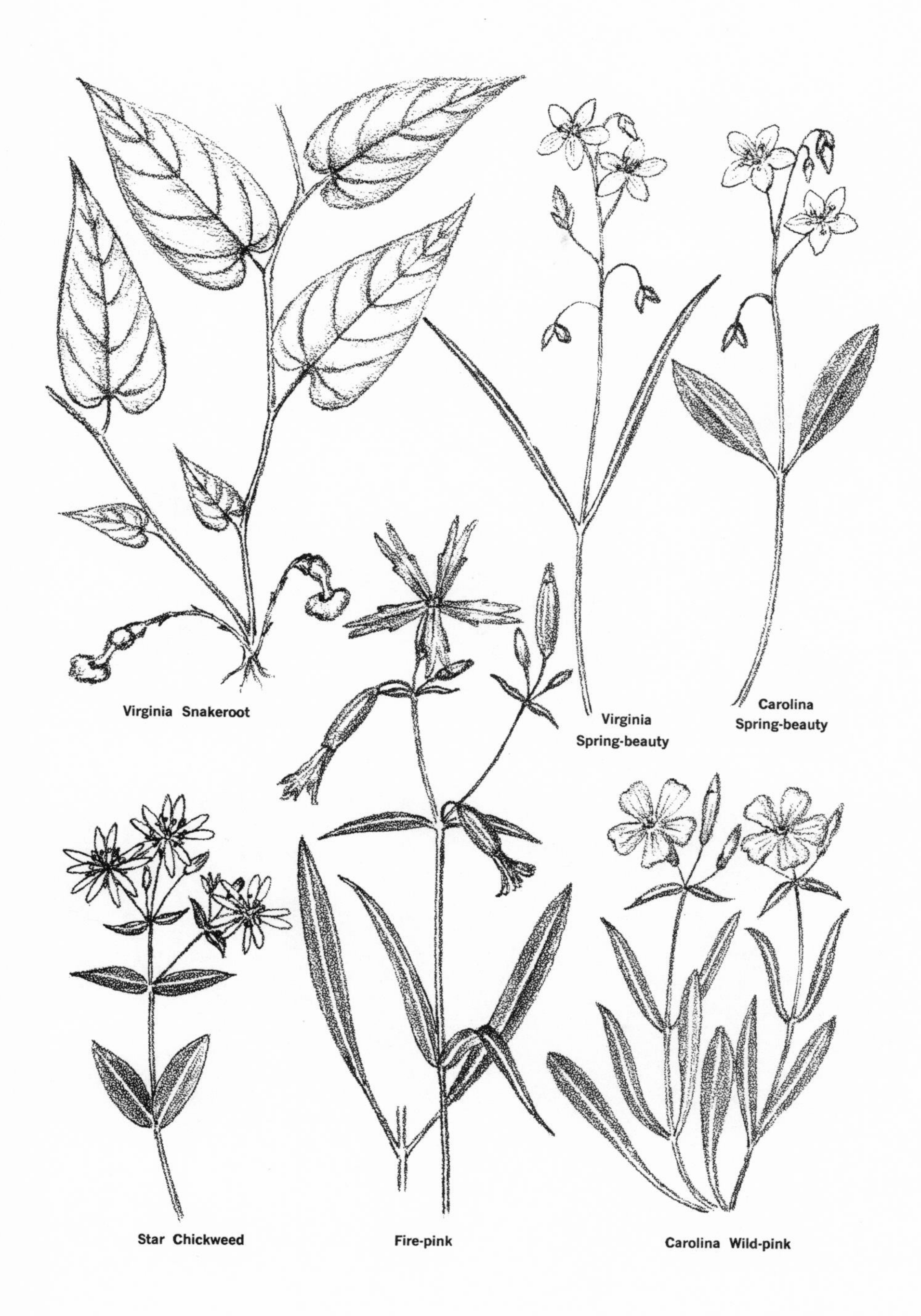
Virginia Snakeroot
Virginia
Spring-beauty
Carolina
Spring-beauty
Star Chickweed
Fire-pink
Carolina Wild-pink

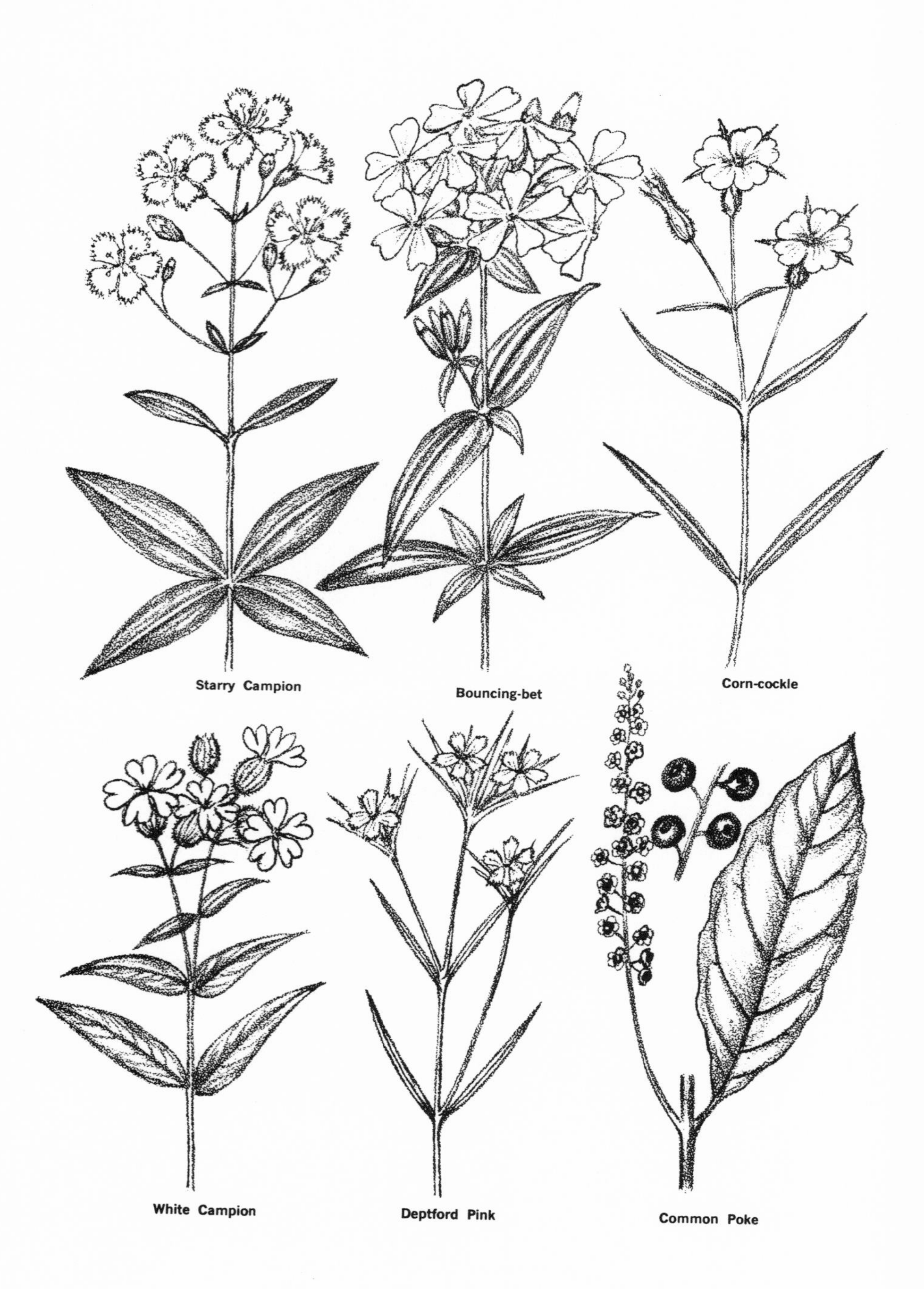
Starry Campion
Bouncing-bet
Corn-cockle
White Campion
Deptford Pink
Common Poke

STARRY CAMPION *(Silene stellata)*

Atop a stem from 2 to 3 feet tall, the Starry Campion has a loose cluster of white flowers. Along the stem below them are whorls of 4 lance-shaped leaves 2 to 4 inches long and minutely downy beneath. The flowers are about ¾ inch across and have an inflated, bell-shaped, usually downy calyx. The 5 petals are delicately fringed at the end. Often common in open woods, clearings, and thickets; it blooms between July and September.

RANGE: Mass. to Minn. south to Ga., Ala., Ark., Okla., and Tex.

WHITE CAMPION *(Silene latifolia* spp. *alba)*

This is another European plant which is widely naturalized in America. It is a loosely branching, sticky-hairy plant from 1 to 2 feet tall. Along the stem are pairs of narrowly egg-shaped or lance-shaped leaves from 1 to 3 inches long. The white or pinkish, fragrant flowers open at dusk and close the following morning, being pollinated by night-flying moths. Their 5 petals are deeply notched at the tip and the pistil has 5 slender styles. Often common by roadsides and in waste places, it blooms between May and September.

BOUNCING-BET *(Saponaria officinalis)*

The Bouncing-bet is a native of Europe but it is now widely naturalized in America. A weed, but a very pretty one, it is often abundant along roadsides, on railroad beds, and in waste places generally. It is a rather coarse but smooth plant from 1 to 2 feet tall; with pairs of broadly lance-shaped, strongly 3- to 5-ribbed leaves which are 2 to 3 inches long. The pale pink or whitish flowers are grouped in quite dense end clusters, and are produced between July and September. Another name for the plant is Soapwort for its juice makes a lather in water.

CORN-COCKLE *(Agrostemma githago)*

This plant is another immigrant from the Old World which has become widely naturalized here. It is a quite slender and silky-hairy plant a foot to nearly 3 feet tall, with pairs of narrowly lance-shaped leaves from 2 to 4 inches long. The showy purplish-pink flowers are 1½ to 2½ inches across. It is frequently found along roadsides, in waste places, and in grain fields; blooming between June and September. The small black seeds often become mixed with grain but must be removed before it is ground into flour as they are poisonous.

DEPTFORD PINK *(Dianthus armeria)*

The Deptford Pink is a stiffly erect, slenderly-branched, finely-hairy plant from 6 to 18 inches tall; with pairs of narrow leaves 1 to 3 inches long. Its small flowers are borne in few-flowered clusters, among long and pointed bracts, at the tips of the stems. Each one is about ½ inch across and the 5 petals are pink or rose with whitish dots. It is an Old World plant which is now widely naturalized; frequently being seen in fields, waste places, and along roadsides. It blooms between May and July.

POKEWEED FAMILY (Phytolaccaceae)

Members of this family have small flowers with 5 sepals, no petals, 10 stamens, and a pistil composed of 10 united carpels. The only representative in our region is the following plant.

COMMON POKE *(Phytolacca americana)*

The Common Poke, or Pokeweed, is a smooth, strong-smelling, branching, reddish-tinged plant from 4 to about 10 feet tall. It has broadly lance-shaped leaves from 5 to 10 inches long which are scattered along the stem and its branches. The small flowers have 5 petal-like white sepals and they are arranged in long and narrow clusters on stalks opposite some of the leaves. As the plant continues to bloom from about mid-summer until late fall, it has flowers and fruits in all stages of development. The mature berries are purplish-black; and as they were once used for making ink, the plant is often called Inkberry. The roots, and perhaps the berries, contain a poisonous substance but the young shoots make wholesome greens. This plant is very common in open woods, thickets, old fields, etc.

RANGE: N.Y. to Que. and Ont. south to Fla. and Tex.

WATERLILY FAMILY (Nymphaeaceae)

Members of this family are aquatic plants with rootstocks creeping in the mud, and long-stalked leaves which commonly float on the surface. The flowers have few sepals, usually numerous petals and stamens showing a gradual transition, and numerous pistils.

AMERICAN LOTUS *(Nelumbo lutea)*

The lotus raises its flowers, and usually its leaves, above the surface of the waters of ponds and slowly moving streams. Its pale yellow flowers are 4 to 8 inches across. The numerous pistils are in pits on the disk-like and elevated receptacle. Later their ovaries become acorn-like fruits which are imbedded on the surface of the disk. The leaves are circular, 1 to 2 feet in diameter, and attached to the leaf stalk in the center. It blooms between July and September. Also called Yellow Lotus and Water-chinquapin.

RANGE: N.Y. and Ont. to Minn. and Iowa south to Fla. and Tex.

SWEET-SCENTED WHITE WATERLILY *(Nymphaea odorata)*

Both the leaves and the flowers of the white waterlily float on the surface of the water. Its flowers are white or pinkish, fragrant, and from 3 to 8 inches across. Their petals are broadest at or near the middle. The leaves are roundish with a V-shaped notch at the base, usually purplish beneath, and from 4 to 10 inches in diameter. It is quite common in ponds and slowly moving streams, blooming between June and September.

RANGE: Nfd. to Man. south to Fla. and Tex.

DWARF WATERLILY *(Nymphaea tetragona)*

Also known as the Small White Waterlily, this species has leaves only 2 to 3 inches across. The white flowers are 1¼ to 3 inches across, with purple lines on the petals. It grows from Me. west to Wash. and northward, blooming between June and September. (Not illustrated)

YELLOW WATERLILY *(Nymphaea mexicana)*

This waterlily has pale yellow flowers 2½ or 3 inches across. The leaves, up to 8 inches broad, are purple or crimson beneath. It is found in the coastal plain from S.C. south to Fla. and west to Tex. (Not illustrated)

YELLOW PONDLILY *(Nuphar lutea* ssp. *advena)*

This plant of pond margins, swamps, and slow-moving streams is often called Spatter-dock. The flowers and leaves float on the water surface or are raised slightly above it. The leaves are egg-shaped or oval with a deep V-shaped notch at the base, and are 5 to 12 inches long. The bright yellow, globe-shaped flowers are 2 to 3 inches across and produced between April and October. RANGE: Mass. to Wis. and Neb. s. to Fla. and Tex.

ARROW-LEAF PONDLILY *(Nuphar sagittifolium)*

This species with arrow-shaped leaves less than half as wide as long is found in the coastal region from Va. south to S.C.

WATER-SHIELD FAMILY (Cabombaceae)

WATER-SHIELD *(Brasenia schreberi)*

The oval-shaped leaves of this plant are attached to stalks near the center and float on the water. They are 1½ to 4 inches long, and both the lower surface and leaf stalk are coated with a sticky jelly. The flowers are dull purple, about ½ inch across, with 3 or 4 sepals and petals and from 12 to 18 stamens. It grows in swamps, ponds, and slow streams and blooms between June and September.

RANGE: N.S. to B.C. south to Fla., Tex. and Ore.

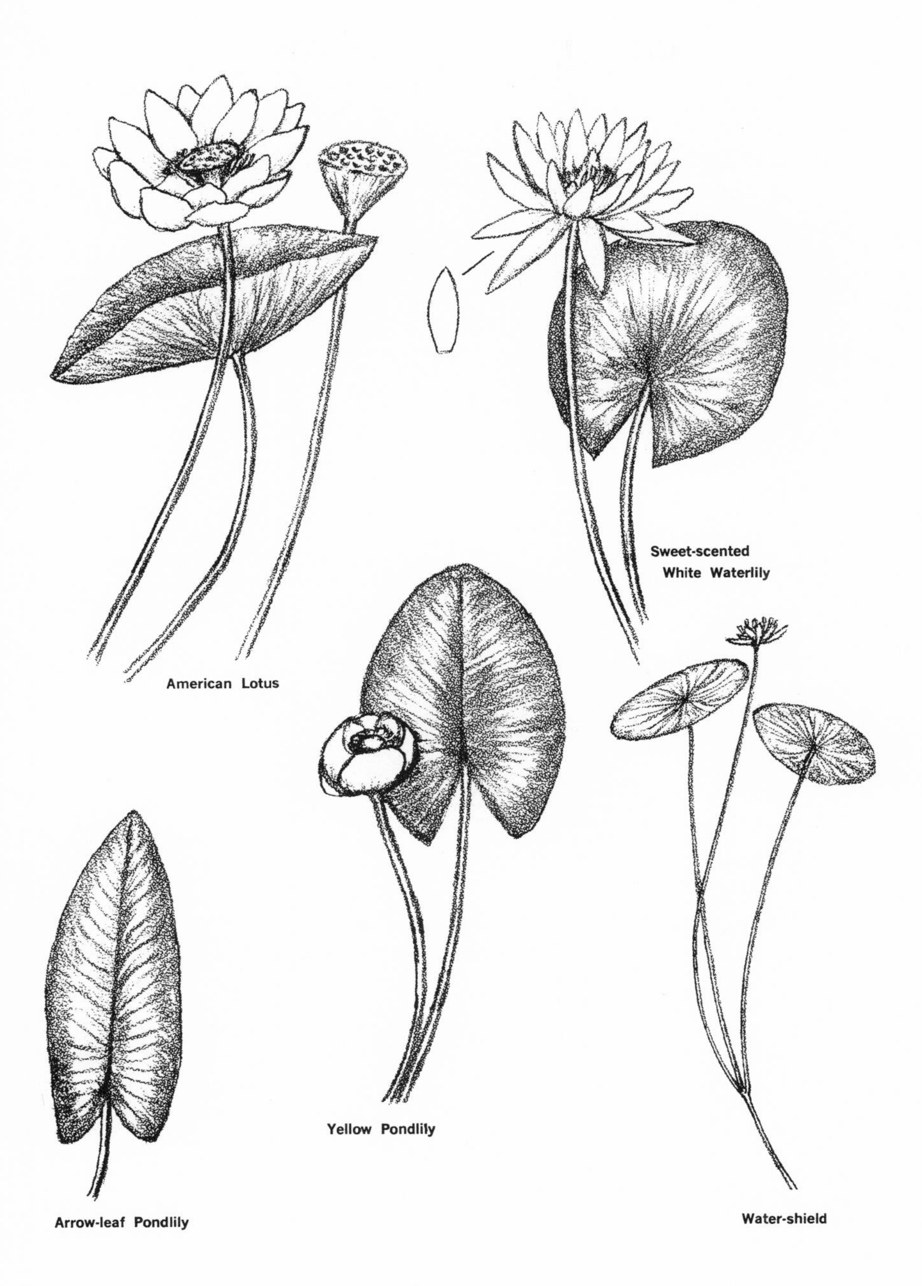
American Lotus
Sweet-scented
White Waterlily
Yellow Pondlily
Arrow-leaf Pondlily
Water-shield

BUTTERCUP FAMILY (Ranunculaceae)

This is quite a large plant family which includes, besides buttercups, many of our most familiar wild flowers. Their flowers have parts which are separate and distinct and they are all attached to a common receptacle. The stamens are usually numerous and indefinite in number. The pistils range from rarely 1 to a great many; in the latter case, being grouped on the knob-like to column-like receptacle in the center of the flower. The petals are usually few (commonly 5) or sometimes none, their place being taken by colorful and petal-like sepals. In some genera which lack petals, the sepals often fall as the flower opens; thus the flowers seem to be without floral envelopes. Most members of the family have flowers with a radial symmetry but those of the larkspurs and monkshoods are bilaterally symmetric.

KEY TO GENERA OF THE BUTTERCUP FAMILY

1. Flowers with a bilateral symmetry.
    2. Upper sepal extended backward as a hollow spur. LARKSPURS *(Delphinium)*
    2. Upper sepal forming a sort of hood or helmet. MONKSHOODS *(Aconitum)*
1. Flowers with a radial symmetry.
    3. Flowers with 2 series of floral envelopes—both sepals and petals.
4. Petals extended backward as hollow spurs. COLUMBINES *(Aquilegia)*
4. Petals ordinary, not spurred.
    5. Flowers with green sepals and yellow or white petals; borne solitary or in loose clusters. BUTTERCUPS *(Ranunculus)*
    5. Flowers with very small white petals, the sepals falling as the flower opens; borne in a long and narrow cluster. BANEBERRIES *(Actaea)*
3. Flowers with but 1 series of floral envelopes—the sepals—which may be showy and petal-like.
    6. Flowers seemingly devoid of floral envelopes, the sepals falling as the flower opens.
7. Flower solitary. GOLDENSEAL *(Hydrastis)*
7. Flowers numerous.
    8. Flowers in long, slender, candle-like clusters. BLACK COHOSH *(Cimicifuga)*
    8. Flowers in open and branching clusters.
9. Plants with slender stems and long-stalked leaves which are divided into many small leaflets. MEADOW-RUES *(Thalictrum)*
9. Plants with stout stems and large palmately lobed and toothed leaves. FALSE BUGBANE *(Trautvetteria)*
    6. Flowers with persistent sepals which are colored and petal-like.
10. Sepals 4. Plants usually vines. CLEMATIS *(Clematis)*
10. Sepals usually 5 or more.
    11. Flowers with yellow sepals.
12. Sepals 5 to 7, greenish-yellow. Leaves deeply cut or divided. GLOBE-FLOWERS *(Trollis)*
12. Sepals usually 5 and bright yellow. Leaves roundish heart-shaped with toothed margins. MARSH-MARIGOLDS *(Caltha)*
    11. Flowers with white, pinkish, blue or lavender sepals.
13. Leaves 3-lobed. Flower with 3 sepal-like bracts beneath the true sepals. HEPATICAS *(Hepatica)*
13. Leaves deeply cut or divided into 3 or more parts. Flowers not with sepal-like bracts beneath the true sepals.
    14. Flowers on short leafless stalks. Basal leaves divided into 3 leaflets. GOLDTHREAD *(Coptis)*
    14. Flower stalks with some leaves or leaf-like bracts.
15. Leaves and bracts deeply cut or divided into 3 or more sharply toothed parts. Flower on a stalk rising well above the leaves or bracts. ANEMONES *(Anemone)*
    16. Styles long and plumose. AMERICAN PASQUE-FLOWER *(Pulsatilla)*
    16. Styles not plumose. THIMBLEWEEDS *(Anemone)*
15. Leaves divided into small roundish leaflets which are somewhat lobed. The bracts of the flower stalk similar.
    17. Plant with several slender-stalked flowers in an end cluster, arising from a whorl of slender-stalked and leaf-like bracts. RUE-ANEMONE *(Thalictrum)*
    17. Plant with scattered leaves each of which is divided into 3 or 9 leaflets; the flowers arising from their axils. FALSE RUE-ANEMONE *(Enemion)*

EARLY BUTTERCUP *(Ranunculus fascicularis)*

Usually the first buttercup to bloom in the spring, the Early Buttercup may flower between late March and May. It grows on wooded hillsides, usually where the soil is quite thin. The plant may be only a few inches tall when the first flowers appear but it later becomes from 6 to 10 inches in height. It is covered with close-pressed, fine, and silky hairs. The bright yellow flowers are almost an inch across, and the leaves have from 3 to 5 cut or lobed divisions. The specific name, *fascicularis,* refers to the cluster (fascicle) of thick, fibrous roots.

RANGE: N.H. to Ont. and Minn. south to Ga. and Tex.

SWAMP BUTTERCUP *(Ranunculus hispidus)*

As its name indicates, the Swamp Buttercup grows in wet or swampy places; and there it may be found in bloom between late April and July. It is a branching plant with smooth to somewhat hairy stems ranging from 1 to about 3 feet tall. The flowers are about an inch across and they have 5 bright yellow petals. Its leaves are long-stalked and divided into 3 lobed and sharply toothed divisions.

RANGE: Que. to Man. south to Md., Ky. and Mo.

SMALL-FLOWERED BUTTERCUP *(Ranunculus abortivus)*

This is not a very showy buttercup for its flowers are barely ¼ inch across. They have tiny yellow petals and somewhat larger sepals which droop. It is a rather smooth plant from 6 to about 24 inches tall. The basal leaves are long-stalked and somewhat heart-shaped or kidney-shaped, but they do not persist very long. Those along the upper part of the stem are quite variable but commonly they have 3 narrow and stalkless segments. It is usually common in low moist woods and thickets where it blooms between late March and June. Another name for it is Kidney-leaf Buttercup.

RANGE: Lab. to Alaska south to Fla. and Tex.

MOUNTAIN BUTTERCUP *(Ranunculus allegheniensis)*

The Mountain Buttercup is quite similar to the preceding species. It can be distinguished from it by the whitish bloom of the plant and also by its achenes which have a strongly recurved or hooked beak. It grows in rich woodlands and blooms between April and June.

RANGE: Mass. and Vt. to Ohio south to nw. S.C. and e. Tenn. (Not illustrated)

ROCK BUTTERCUP *(Ranunculus micranthus)*

This is another small-flowered buttercup similar to both of the preceding species. Usually a smaller plant, it can be distinguished by the fact that it is decidedly hairy. It grows on rocks and in rich, rocky woods, blooming between March and May.

RANGE: Mass. to Ill. and Ark. south to Ga. and Ala. (Not illustrated)

TALL BUTTERCUP *(Ranunculus acris)*

The Tall Buttercup, which is also known as the Common or Meadow Buttercup, is a native of Europe which is now widely naturalized in America. It is the common and weedy buttercup so often seen in fields and meadows or along roadsides, especially in the northeastern United States and Canada. A branching and usually hairy plant from 2 to 3 feet tall, it has deeply 3- to 5-parted leaves in which the divisions are again cut into narrow and toothed lobes. The basal leaves are long-stalked. It has numerous flowers with 5 bright yellow petals, and it blooms between May and August.

CREEPING BUTTERCUP *(Ranunculus repens)*

Like the preceding species, this native of Europe has become a common weed over a large part of North America; and it is commonly found in yards, fields, and in waste places. Its smooth or somewhat hairy trailing stems take root at the nodes, and the plant often grows in rather large patches. Upright stems a few inches high produce bright yellow, 5-petalled flowers about an inch across. The leaves are divided into 3 leaflets which are cleft, lobed, and often mottled. It blooms between May and September.

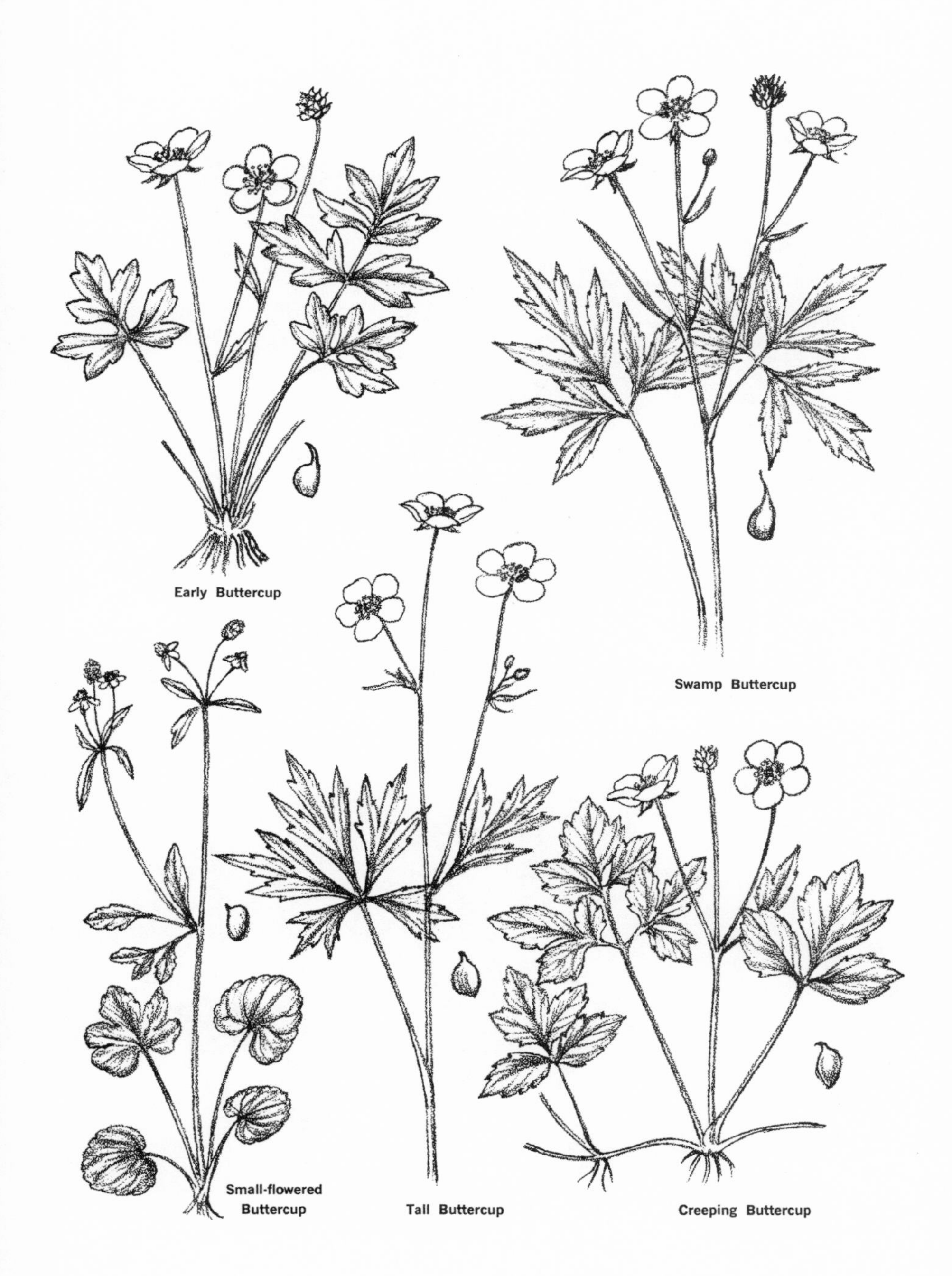
Early Buttercup
Swamp Buttercup
Small-flowered Buttercup
Tall Buttercup
Creeping Buttercup

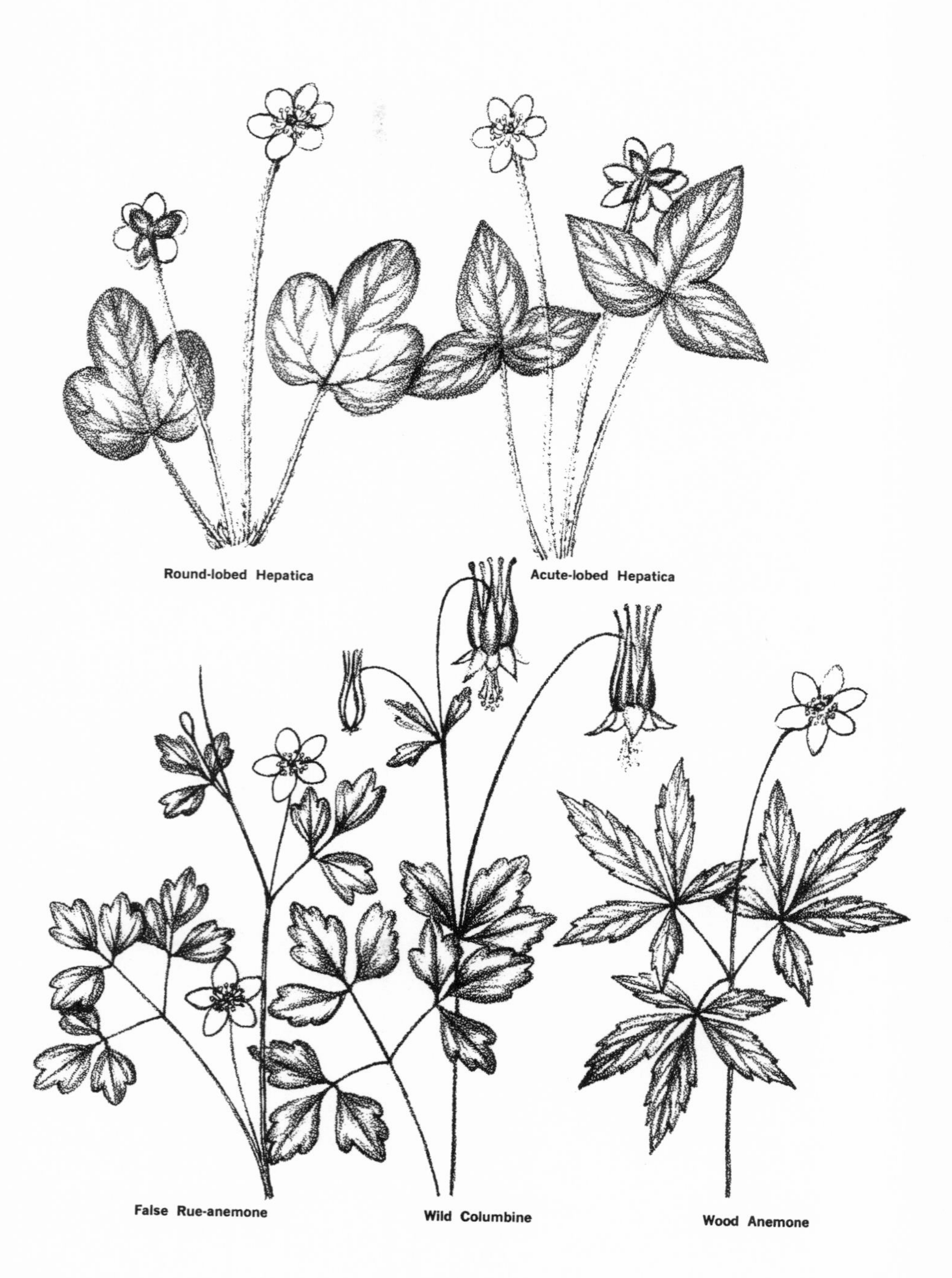
Round-lobed Hepatica
Acute-lobed Hepatica
False Rue-anemone
Wild Columbine
Wood Anemone

ROUND-LOBED HEPATICA *(Hepatica nobilis* var. *obtusa)*

Hepaticas are among the very first wild flowers to bloom in the spring, their blooming season being between March and May. Each flower stands on a silky-hairy stalk a few inches tall. It has no petals but its 5 to 9 sepals are delicate and petal-like; being white, pink, lilac, or even a deep lavender-blue in color. The numerous stamens have whitish or pale anthers; and in the center of the flower there is a little cluster of pistils. Just beneath the petal-like sepals are 3 green bracts which, in this species, have roundish tips. At flowering time the new leaves of the plant have not yet appeared, or are just beginning to put in their appearance. Usually there are some of the leaves of the previous year still present. Although now mostly brown, they are somewhat thick and leathery and they have 3 roundish lobes. The hepatica grows in woodlands, and it makes a good subject for the shady wild flower garden.

RANGE: N.S. to Man. south to Fla., Ala., and Mo.

ACUTE-LOBED HEPATICA *(Hepatica nobilis* var. *acuta)*

This hepatica is very similar to the preceding species; differing from it chiefly in having the 3 bracts below the flowers, and the lobes of its leaves, pointed instead of being roundish. It is also a woodland plant, blooming between March and May.

RANGE: Me. to Minn. south to Ga., Ala., and Mo.

FALSE RUE-ANEMONE *(Enemion biternatum)*

This plant resembles the Rue-Anemone but it has alternate leaves which are divided into either 3 or 9 leaflets. The leaflets are 3-lobed or divided. Its flowers are about ¾ inch across and are borne at the tip of the stem or on stalks arising from the axils of the leaves. They have 5 white, petal-like sepals. Growing in moist woods and thickets, it blooms during April and May.

RANGE: Ont. to Minn. south to Fla. and Tex.

WILD COLUMBINE *(Aquilegia canadensis)*

The Wild Columbine grows in rocky woodlands and on shaded rocks and cliffs, blooming between April and June. It is a smooth plant from 1 to about 2 feet tall. The flowers, like red and yellow bells, nod from the tips of the long and slender stem or its branches. Each one is about 1½ inches long, with 5 petal-like sepals and 5 true petals which are prolonged backward as hollow spurs. Nectar is secreted in the ball-like tips of these spurs. The numerous stamens protrude beyond the mouth of the flower. The ovaries of its 5 pistils mature into slender, dry, pod-like fruits which contain many shiny, black seeds. Its leaves are compound, divided into 3s, and the leaflets are more or less 3-lobed and bluntly toothed. It blooms between April and July, and it is quite easy to grow in the wild flower garden.

RANGE: Que. and Ont. to Wis. south to Fla. and Tex.

WOOD ANEMONE *(Anemone quinquefolia)*

The Wood Anemone, or Windflower, has a stem from 4 to 9 inches tall. At its summit is a whorl of 3 leaves, each of which is divided into from 3 to 5 narrow and sharply toothed segments, and a solitary flower. The basal leaves are similar to the ones on the flower stem but they are either not present or just starting to grow at flowering time. The flower is about an inch across and it has from 5 to 7 white, petal-like sepals. It grows in open woods, thickets, and clearings; blooming between April and June.

RANGE: Que. to Man. south to S.C., Tenn., and Iowa.

MOUNTAIN ANEMONE *(Anemone lancifolia)*

This is one of several species of *Anemone* popularly called Thimbleweeds because their achenes are crowded on a long and thimble-shaped receptacle. The Mountain Anemone bears quite a resemblance to the Wood Anemone but it is a somewhat stouter and taller plant. It has a stem from 6 to 14 inches tall which has a whorl of 3 leaves and a solitary flower at the summit. Each of the 3 leaves is divided into 3 narrowly egg-shaped leaflets which have rather shallowly toothed margins. The flower is an inch or a little more across and it has from 4 to 7 white petal-like sepals. It grows in damp woods and thickets in the mountains, blooming between April and June.

RANGE: S. Pa. south to Ky. and n. Ga.

CAROLINA ANEMONE *(Anemone caroliniana)*

The solitary flower of the Carolina Anemone is from an inch to 1½ inches broad and it has from 10 to 20 white, pink, or violet-colored, petal-like sepals; so many, in fact, that it has the appearance of a daisy. The plant has a stem from 4 to about 10 inches tall, on which are 3 stalkless and deeply cut leaves. The stalked basal leaves are divided into 3 parts which are, in turn, deeply cut and variously toothed. It grows in open woods, fields, and on prairies; blooming in April or May.

RANGE: N.C. to Ind. and Wis. south to Fla. and Tex.

TALL ANEMONE *(Anemone virginiana)*

The Tall Anemone has a stout and hairy stem from 2 to 3 feet tall. On it is a whorl of usually 3 stalked and 3-parted leaves, which are variously cleft and sharply toothed. From this whorl of leaves arise the long stalks which end in a solitary flower. The flower is a little more than an inch across and has 5 greenish-white, petal-like sepals. The basal leaves are long-stalked but similar to those of the flower stem. In this species the achenes are crowded in a woolly, cylindrical head an inch or more long. The Tall Anemone grows in usually dry and rocky open woods and thickets; blooming between June and August.

RANGE: Me. to Minn. south to Ga., Tenn., and Kan.

LONG-FRUITED ANEMONE *(Anemone cylindrica)*

In general this species resembles the preceding one. It has a stem from 1 to 2 feet tall which is quite densely hairy, on which is a whorl of from usually 5 to 9 stalked leaves. The achenes are crowded in a narrowly cylindrical head from an inch to 1½ inches long. It grows in dry open woods and on prairies, blooming between May and July.

RANGE: Me. to Alb. south to N.J., the Great Lakes region, Mo., N.Mex., and Ariz.

ROUNDLEAF ANEMONE *(Anemone canadensis)*

This species has a somewhat hairy stem from 1 to 2 feet tall on which are stalkless, broad, and sharply toothed or cut leaves. The basal ones are similar but long-stalked. The flowers are an inch to 1½ inches across and have 5 white, petal-like sepals. The achenes are crowded in ball-like heads. It grows in damp open woods and thickets, blooming between May and July.

RANGE: Lab. to B.C. south to N.J., W.Va., Mo., and N.Mex.

PASQUE-FLOWER *(Pulsatilla patens)*

The beautiful Pasque-flower has a silky-hairy stem from 4 to about 16 inches tall, but at flowering time it is considerably shorter. Its leaves are basal, cut into numerous narrow divisions, and also covered with silvery-silky hairs. The flower is about 2 inches across and has from 5 to 7 large petal-like sepals varying from white to lavender-blue or purple in color. It grows on prairies and on dry, exposed slopes; blooming between April and June. May be grown in a sunny or lightly shaded spot in the flower garden.

RANGE: Mich. to B.C. south to Ill., Mo., Tex., N.Mex., Utah, and Wash.

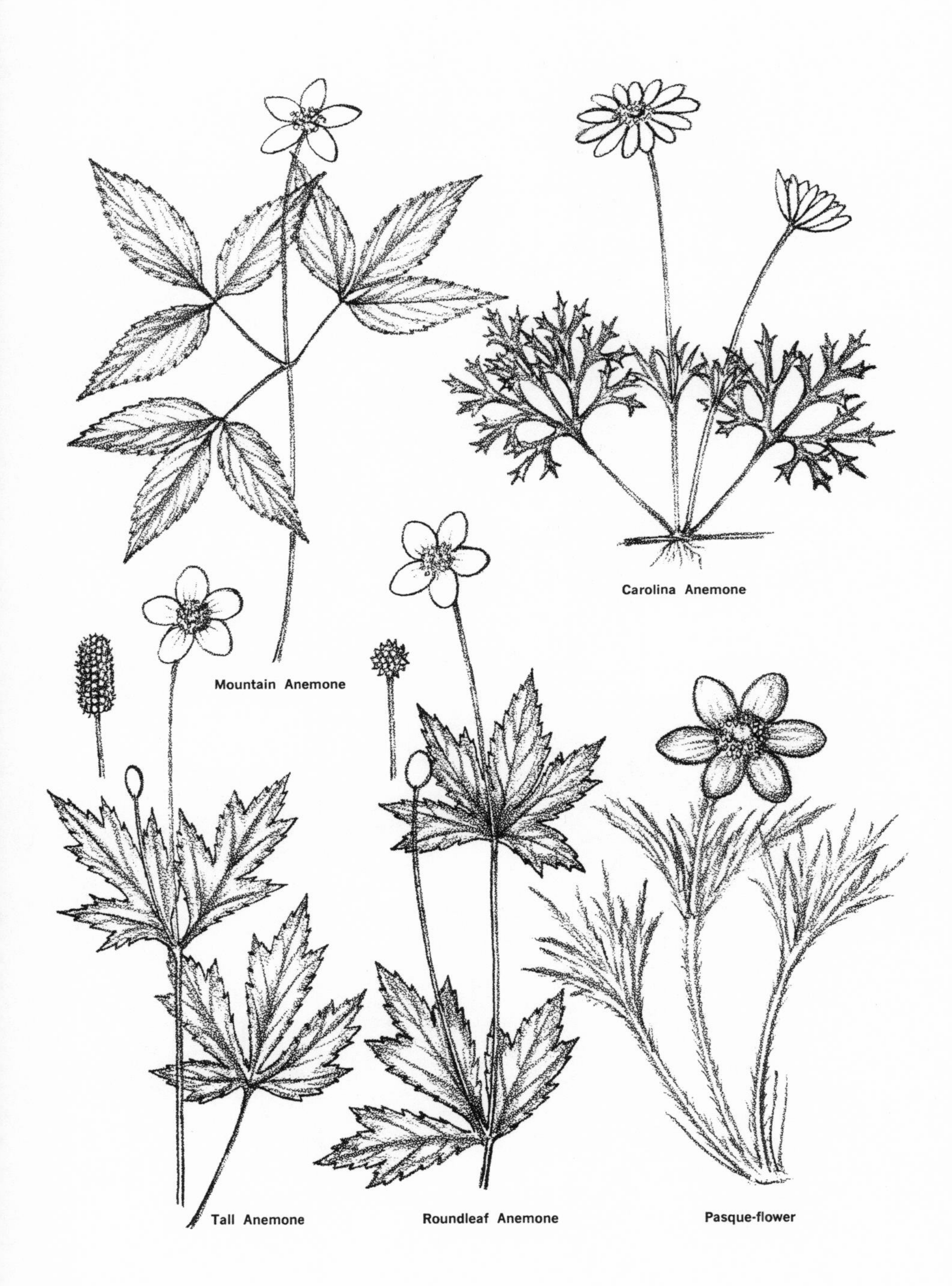
Carolina Anemone
Mountain Anemone
Tall Anemone
Roundleaf Anemone
Pasque-flower

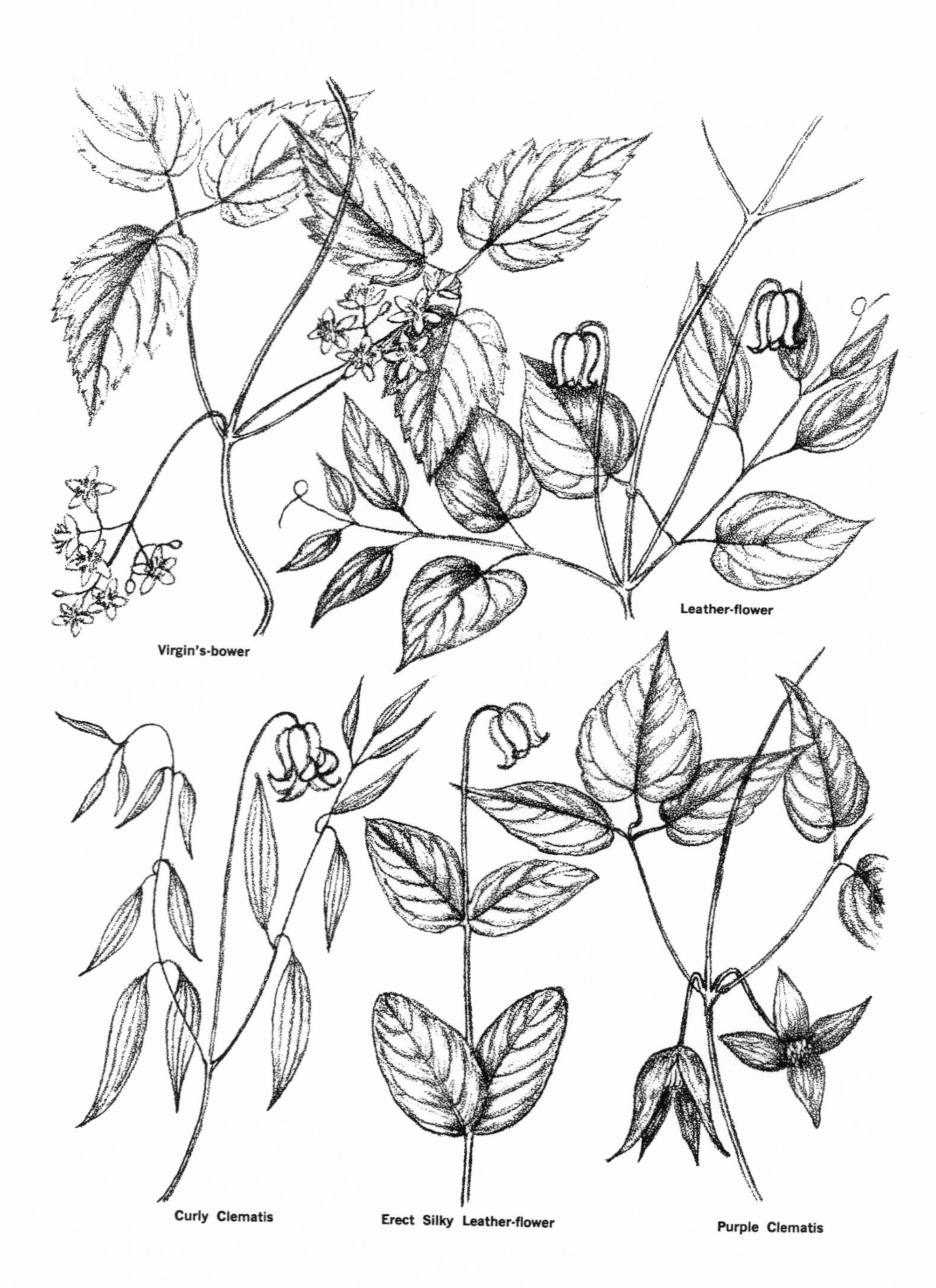
Virgin's-bower
Leather-flower
Curly Clematis
Erect Silky Leather-flower
Purple Clematis

VIRGIN'S-BOWER *(Clematis virginiana)*

The Virgin's-bower is a vine that commonly climbs over other vegetation on the borders of swamps, along streams, and in moist wayside thickets. Between July and September, it produces numerous, showy clusters of white flowers. Each one is a half to ¾ of an inch broad, with 4 or 5 petal-like sepals and tassel-like stamens. In the autumn the clusters of achenes with their long and plume-like tails are just as attractive as the flowers which preceded them. The opposite leaves are divided into 3 egg-shaped leaflets which have a few sharp teeth on their margins. Like other species of *Clematis,* it climbs by means of its leafstalks which act like tendrils. It can be grown in the garden on a fence or trellis.
RANGE: N.S. to Man. south to Ga., La., and Kan.

LEATHER-FLOWER *(Clematis viorna)*

Leather-flowers have bell-shaped flowers which nod from the ends of long stalks arising from the axils of the leaves. They have 4 thick, somewhat leathery, dull purplish sepals in which the tips are recurved. The opposite leaves of this climber are divided into from 3 to 7 egg-shaped to lance-shaped, bright green leaflets. It grows in rich woods and thickets, blooming between May and August. As in other species of *Clematis,* the flowers are followed by clusters of achenes with long, plume-like tails.
RANGE: Pa. to Ill. and Iowa south to Ga. and Tex.

CURLY CLEMATIS *(Clematis crispa)*

The Curly Clematis is also known as the Marsh Leather-flower and the Blue-jasmine. It has attractive bluish-purple flowers an inch or more long, shaped like bells but with the wavy-edged sepals spreading from about the middle. Each one nods from a slender stalk at the tip of a branch. The leaves are divided into from 5 to 9 rather thin, lance-shaped or narrowly egg-shaped leaflets. This species grows in wet woods and about the borders of swamps, blooming between April and August.
RANGE: Se. Va., s. Ill., and Mo. south ot Fla. and Tex.

ERECT SILKY LEATHER-FLOWER *(Clematis ochroleuca)*

Instead of being a vine, this leather-flower is an erect branching plant from 1 to 2 feet tall. It has pairs of simple, egg-shaped, stalkless leaves ranging from 1½ to 2 inches in length. New growth and the lower surfaces of the leaves are silky-hairy but they eventually become smooth. The flowers are bell-shaped, dull yellowish or purplish, silky on the outside, and about an inch long. Each one nods at the end of a slender stalk. It grows in woods, thickets, and on rocky slopes, blooming during April or May. Another name for it is Curly-heads.
RANGE: Se. N.Y. and se. Pa. south to Ga.

PURPLE CLEMATIS *(Clematis occidentalis)*

The showy flowers of the Purple Clematis, or Purple Virgin's-bower, are from 2 to 3 inches broad. They have 4 large, veiny, bluish-purple, petal-like sepals. Some of the outer stamens have the appearance of being small petals. Each flower is on a stalk arising from a leaf axil. The leaves of this trailing or climbing plant are opposite and divided into 3 egg-shaped or heart-shaped leaflets, sometimes with a few teeth on their margins. It grows in open woods and on rocky slopes, blooming during May or June.
RANGE: Que. to Man. south to Md., w. N.C., Ohio, and Iowa.

MARSH-MARIGOLD *(Caltha palustris)*

The Marsh-marigold, or Cowslip, is a smooth plant with a stout, hollow, forking stem from 1 to 2 feet tall. It has flowers an inch to 1½ inches broad, which have 5 or 6 petal-like sepals of the brightest yellow. The roundish, heart-shaped leaves are 2 to 6 inches across and have low teeth on their margins. The young leaves are very popular as spring greens, and in the Northeast they are often sold in the markets. As its name implies, the plant grows in swamps and in wet open woods and meadows. The flowering season is between April and June, or even later in the Far North.

RANGE: Lab. to Alaska south to N.C., Tenn., Iowa, and Neb.

AMERICAN GLOBE-FLOWER *(Trollius laxus)*

Like the preceding species, the Globe-flower grows in swamps and wet meadows or thickets, but it is rather rare and local in occurrence. It has a slender, smooth stem 1 to 2 feet tall. The leaves are divided into from 5 to 7 radiating and cut-toothed segments. They are from 2 to 4 inches wide and all but the uppermost ones are stalked. The flowers are usually solitary, 1 to 1½ inches across, and have from 5 to 7 pale greenish-yellow petal-like sepals. It blooms in April or May.

RANGE: Conn. to Pa. and Mich.

DWARF LARKSPUR *(Delphinium tricorne)*

At the time of flowering, in April or May, the stems of this plant are usually 6 to 12 inches high, but later they become much taller. The larkspurs are unusual among the members of this family in that their flowers have a bilateral symmetry. Those of this species are 1 to 1½ inches long, usually a deep blue-violet or variegated with white, and arranged in a rather loose, cylindrical cluster. As in other larkspurs there are 5 small, irregular, petal-like sepals and the uppermost one is prolonged into a hollow spur. There are also 4 petals but they are quite small, the upper pair having spurs which are enclosed within the spur of the sepal. The leaves are cut into 5 radiating divisions which are themselves cleft and sharply toothed. The Dwarf Larkspur grows in rich woods and on rocky slopes, blooming in April or May.

RANGE: Pa. to Minn. and Neb. south to Ga., Ala., Ark., and Okla.

TALL LARKSPUR *(Delphinium exaltatum)*

This species of larkspur has a slender, leafy stem from 2 to 5 feet tall; bearing toward the summit a long, narrow, but rather dense cluster of lavender-blue to whitish flowers. Each one is about ¾ of an inch long and the upper sepal has a spur which is almost straight. The leaves are deeply cleft into from 3 to 5 narrow divisions which are 3-cleft at the summit. The Tall Larkspur grows in rich woods or thickets and on rocky slopes. Its flowering season extends from July to September.

RANGE: Pa. and Ohio south to N.C. and Miss.

CAROLINA LARKSPUR *(Delphinium carolinianum)*

The Carolina Larkspur usually has a simple, slender stem from 1 to 2 feet tall which is glandular-hairy on the upper portion. Its leaves are deeply 3- to 5-parted and the divisions are cut into very narrow lobes. The flowers are about an inch long, deep blue or violet in color, and have rather long and upcurved spurs. It grows in dry open woods, fields, sandhills, and prairies and blooms during May or June.

RANGE: Va. to Ky., Mo., and Okla. south to Ga. and Tex.

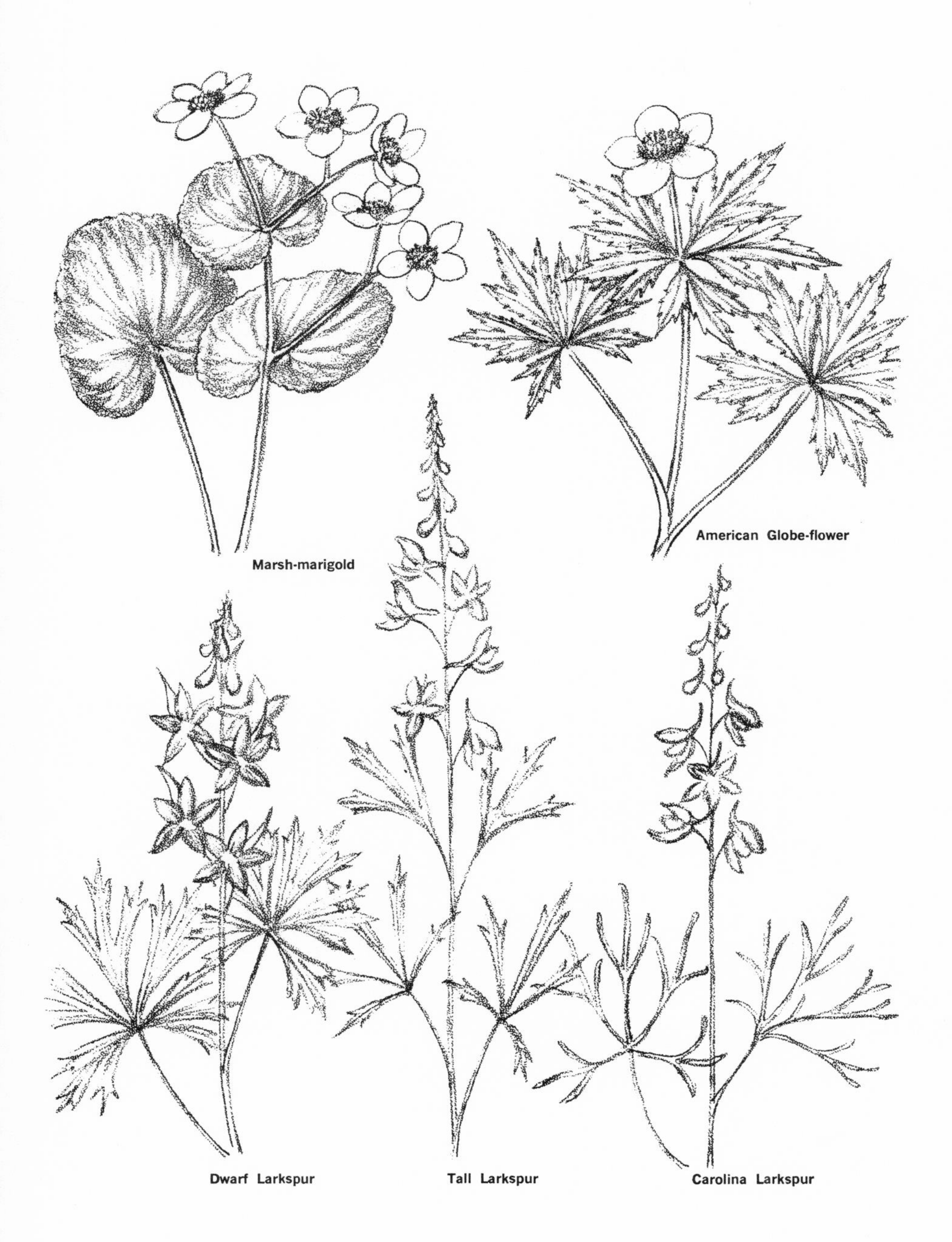
Marsh-marigold
American Globe-flower
Dwarf Larkspur
Tall Larkspur
Carolina Larkspur

BLACK COHOSH *(Cimicifuga racemosa)*

The Black Cohosh, or Black Snakeroot, is one of the most conspicuous plants in our wooded areas during the summer months; for it is then that it displays its long, slender, candle-like clusters of small white flowers. On close inspection, one will find that each flower consists largely of a tassel-like group of white stamens with a pistil in the center. Some of the outermost stamens resemble small petals but there are no true petals, and the sepals drop off as the flowers open. The pistil develops into a small dry pod. The Black Cohosh has a somewhat slender stem from 3 to 6 feet tall, on which are 2 or 3 large compound leaves. The leaflets are mostly egg-shaped, rounded or somewhat pointed at the base, and are cleft or sharply toothed on the margin. It blooms between June and September. The rootstocks are used medicinally.

RANGE: Mass. to Ont. south to Ga., Tenn., and Mo.

HEARTLEAF COHOSH *(Cimicifuga rubifolia)*

This plant of the damp southern Appalachian woodlands is almost identical with the Black Cohosh, except for its leaves. They are divided into usually fewer and larger leaflets, and at least the end ones are deeply heart-shaped at the base. It is generally regarded to be merely a variety of the Black Cohosh, although some botanists have considered it to be a distinct species.

RANGE: Mts. of Va., N.C., and Tenn.

AMERICAN BUGBANE *(Cimicifuga americana)*

This plant also closely resembles the Black Cohosh and to identify it one must examine either the flowers or the pod-like fruits which follow them. The flowers of this species usually contain 3 or more pistils (the Black Cohosh usually has only 1); and the resulting pods are stalked and in groups of 3 or more. Also known as the Summer Cohosh and Mountain Bugbane, it grows only in rich woods in the mountain region; blooming during August or September.

RANGE: Pa. south to Ga. and Tenn.

WHITE BANEBERRY *(Actaea pachypoda)*

The White Baneberry is far more likely to attract one's attention when in fruit than in flower. The plant, in general, closely resembles the Black Cohosh but it grows only a foot or two high. Its large leaves are ternately compound with a number of stalked, sharply toothed leaflets. The flowers very much resemble those of the Black Cohosh but they are in shorter, cylindrical clusters. Those of the baneberries actually have petals but they are so small one has to look very carefully to see them. The sepals fall away as the flowers open. In the center of the flower is one pistil with a stalkless stigma seated at the top. It develops into a somewhat egg-shaped and snow-white berry, about ¼ of an inch in diameter, with a purple spot at the end. Each berry stands on a thick and bright red stalk, a number of them being displayed in a cylindrical cluster. The White Baneberry grows in rich woods and thickets. The flowering season is in May and June; the berries are showy between July and October. It is also called the White Cohosh and Doll's-eyes.

RANGE: N.S. to Man. south to Ga., La., and Okla

RED BANEBERRY *(Actaea rubra)*

This species very closely resembles the preceding but it has bright red berries, or sometimes ivory-white ones, on slender rather than stout stalks. It grows in woods and thickets, blooming between May and July. The berries are present between August and October. Another name for it is Snakeberry.

RANGE: Lab. to B.C. south to N.J., W.Va., S.D., Colo., Utah, and Oregon.

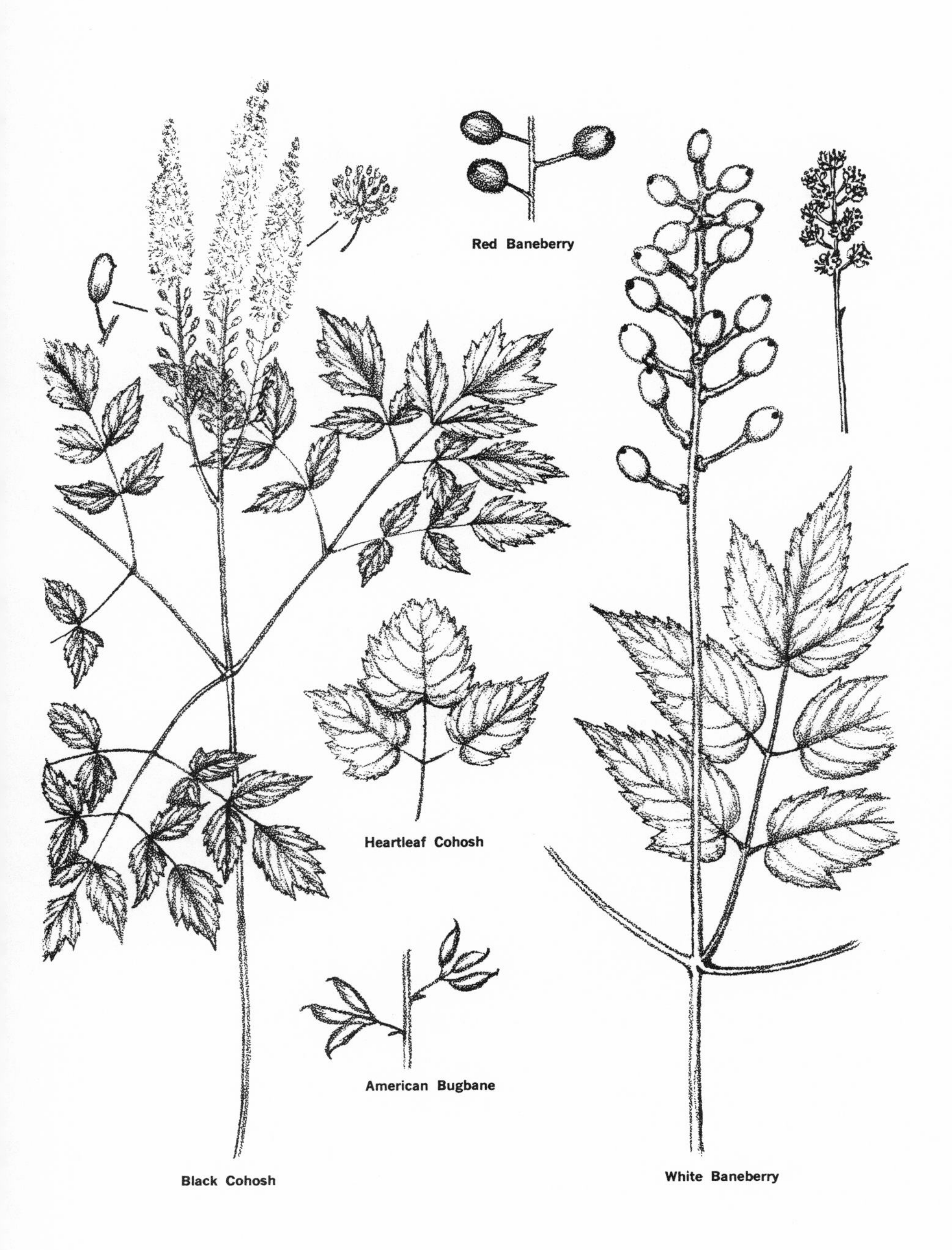
Red Baneberry
Heartleaf Cohosh
American Bugbane
Black Cohosh
White Baneberry

Wild Monkshood
Goldthread
False Bugbane
Goldenseal

WILD MONKSHOOD *(Aconitum uncinatum)*

The Wild Monkshood has a slender, leafy, rather weak, branching stem from 2 to about 3 feet tall; with violet-blue flowers clustered at the ends of the branches. The flowers are about 1 inch across and have a bilateral symmetry. They have 5 very irregular, petal-like sepals, the upper one being much larger and shaped like a hood or helmet. The petals are small and inconspicuous, 2 of them being covered by the hooded sepal. The leaves are deeply divided into from 3 to 5 cut-toothed lobes and they, like the rest of the plant, are smooth or nearly so. The Wild Monkshood is also known as the Wild Wolfsbane. It grows in moist woods, thickets, and on rocky slopes; blooming between August and October.

RANGE: Pa. to Ind. south to Ga. and Ala.

NEW YORK MONKSHOOD *(Aconitum noveboracense)*

This species is similar to the preceding but the stem, at least in the flower clusters, is quite hairy. It grows in rich woods, shady ravines, and on damp slopes; blooming in June and July.

RANGE: Se. N.Y. to Wis. and Iowa. (Not illustrated)

TRAILING WOLFSBANE *(Aconitum reclinatum)*

The Trailing Wolfsbane has a trailing stem from 2 to 8 feet long. It has white to yellowish flowers which are almost an inch long, the hooded sepal being cone-shaped and lying nearly or quite horizontal. The leaves are 3- to 7-cleft and the wedge-shaped divisions are toothed and often deeply cut toward the tips. The lower leaves are roundish and 6 to 8 inches broad; and all but the upper ones are stalked. It grows in mountain woods, blooming between June and September. Also called Trailing Monkshood. (Not illustrated)

RANGE: Va. and W.Va. south to Ga.

GOLDTHREAD *(Coptis trifolia)*

The Goldthread gets its name from its threadlike and bright yellow rootstocks. It is a small plant 3 to 6 inches high, with evergreen leaves which are divided into 3 more or less rounded and toothed lobes. The white flowers are on slender stalks and they have 5 to 6 petal-like sepals. Petals are also present but they are very small and club-shaped. It grows in cold, damp woods and bogs; blooming between May and July.

RANGE: Lab. to Man. south to N.J., e. Tenn., the Great Lakes region, and Iowa.

FALSE BUGBANE *(Trautvetteria carolinensis)*

This plant has a branching stem from 2 to 3 feet tall along which are alternate, deeply-lobed, and sharply toothed leaves. The basal ones are 6 to 8 inches across, 5- to 11-lobed, and long-stalked. The white flowers are borne in clusters at the ends of the branches. They have 3 to 5 sepals which drop off as the flowers open, thus each flower consists of a tassel of white stamens and a head of several pistils. The False Bugbane is also known as Tassel-rue. It grows along the banks of streams and in ravines, flowering between June and August.

RANGE: Sw. Pa. to Mo. south to nw. Fla.

GOLDENSEAL *(Hydrastis canadensis)*

Orangeroot and Yellow Puccoon are other names often given to the Goldenseal; but all three names refer to its thick, knotted, and bright yellow rootstock. In the spring it sends up a simple hairy stem about a foot high; and on it is a pair of large, 5- to 7-lobed, sharply toothed leaves, and there is a similar roundish leaf at the base which is long-stalked. In summer it becomes from 5 to 8 inches broad. From the center of the pair of stem leaves arises a stalk on which is the solitary greenish-white flower. It has a few sepals which promptly drop off, leaving a cluster of stamens and pistils. The latter develop into a little cluster of red berries. The Goldenseal grows in rich woods and blooms in April or May. The rootstock is used medicinally and in many places it has been exterminated by collectors.

RANGE: Vt. to Minn. and Neb. south to Ga., Ala., and Ark.

RUE-ANEMONE *(Thalictrum thalictroides)*

The Rue-anemone is a delicate little plant with a very slender stem 4 to 8 inches high. At the summit it bears a whorl of thin, pale green, 3-lobed, leaf-like bracts together with several slender-stalked flowers. The flowers are about ¾ inch across and have 5 to 10 white or pale-pink, petal-like sepals. The basal leaves are ternately compound and the leaflets very much resemble the bracts of the flower stalk, but they are not present when the plant blooms in April or May. It grows in rich open woodlands.

RANGE: Me. to Minn. south to Fla., Miss., Ark., and Okla.

EARLY MEADOW-RUE *(Thalictrum dioicum)*

The Early Meadow-rue is a smooth, slender-stemmed plant from 1 to 2 feet tall. Its drooping flowers are yellowish-green, or often tinged with purplish, and are borne in a loose terminal cluster. The stamens and the pistils are not only in different flowers but on different plants. The leaves are divided into a number of thin, delicate, roundish leaflets which have from 5 to 9 rounded teeth on their margins. Flowering in April or May, it is quite common in rich, rocky woods and shady ravines.

RANGE: Me. and Que. to Ont., Minn., and S.D. south to Ga., Ala., and Mo.

MOUNTAIN MEADOW-RUE *(Thalictrum clavatum)*

This little meadow-rue differs from all of our other species in that its flowers always contain both stamens and pistils. The flowers are white, long-stalked, and in a relatively few-flowered end cluster. Their stamens have club-shaped, petal-like filaments and the pistils stand on little stalks. The Mountain Meadow-rue is 6 inches to about 2 feet tall. Its leaves are biternately divided into 9 thin, oval- or egg-shaped leaflets which are usually 3-lobed. It is found in wooded ravines and along streams in the mountains. Another name for it is Lady-rue. It blooms in May or June.

RANGE: Va., W. Va. and Ky. south to n. Ga. and nw. Ala.

TALL MEADOW-RUE *(Thalictrum pubescens)*

The Tall Meadow-rue has a stout, smooth or somewhat hairy stem 3 to 8 feet tall. Some of the white flowers have only stamens, while others have both pistils and a few stamens. The stamens have club-shaped filaments. Its leaves are large and ternately divided into a number of leaflets which usually have 3 lobes. It is the common tall meadow-rue of moist meadows, thickets, and sunny swamps; flowering between June and August.

RANGE: Nfd. to Ont. south to Ga. and Tenn.

WAXY-LEAF MEADOW-RUE *(Thalictrum revolutum)*

This one can be distinguished by the tiny waxy particles on the lower surfaces of its leaflets. Its stem and flowers are often purplish. The plant's strong odor suggested another common name, Skunk Meadow-rue. It grows in open woods, thickets, meadows, and prairies; blooming between May and July.

RANGE: Mass. to Ont. south to Fla., Ala., and Mo. (Not illustrated)

PURPLE MEADOW-RUE *(Thalictrum dasycarpum)*

This is another tall meadow-rue which grows in wet meadows or thickets and swamps. The 4- to 7-foot stem is often purplish, the leaflets commonly downy beneath, and the stamens of its flowers have slender filaments. It blooms between late May and July.

RANGE: Ont. to B.C. south to Ohio, La., N.Mex., and Ariz. (Not illustrated)

BARBERRY FAMILY (Berberidaceae)

Flowers of the members of this family have 4 to 6 sepals, 6 to 9 petals, the stamens as many, or twice as many, as the petals, and a solitary pistil.

BLUE COHOSH *(Caulophyllum thalictroides)*

The leaves are divided into a number of leaflets with from 3 to 5 lobes. Blue Cohosh is a smooth, whitened plant from 1 to 3 feet tall. About midway on the stem there is a leaf divided into 3 stalked divisions, each of which bears about 9 leaflets. Above this is the stalked cluster of purplish-green flowers. Each flower is about ½ inch across and it has 6 sepals, 6 small gland-like petals, 6 stamens and a single pistil. The 2 ovules within the ovary of the pistil burst the ovary wall as they grow; and they mature as a pair of large, stalked, blue, and berry-like seeds which look like small blue grapes. Another name is Papoose-root. It grows in rich, moist woods; blooming in April or May.

RANGE: N.B. to Man. south to S.C., Tenn., and Mo.

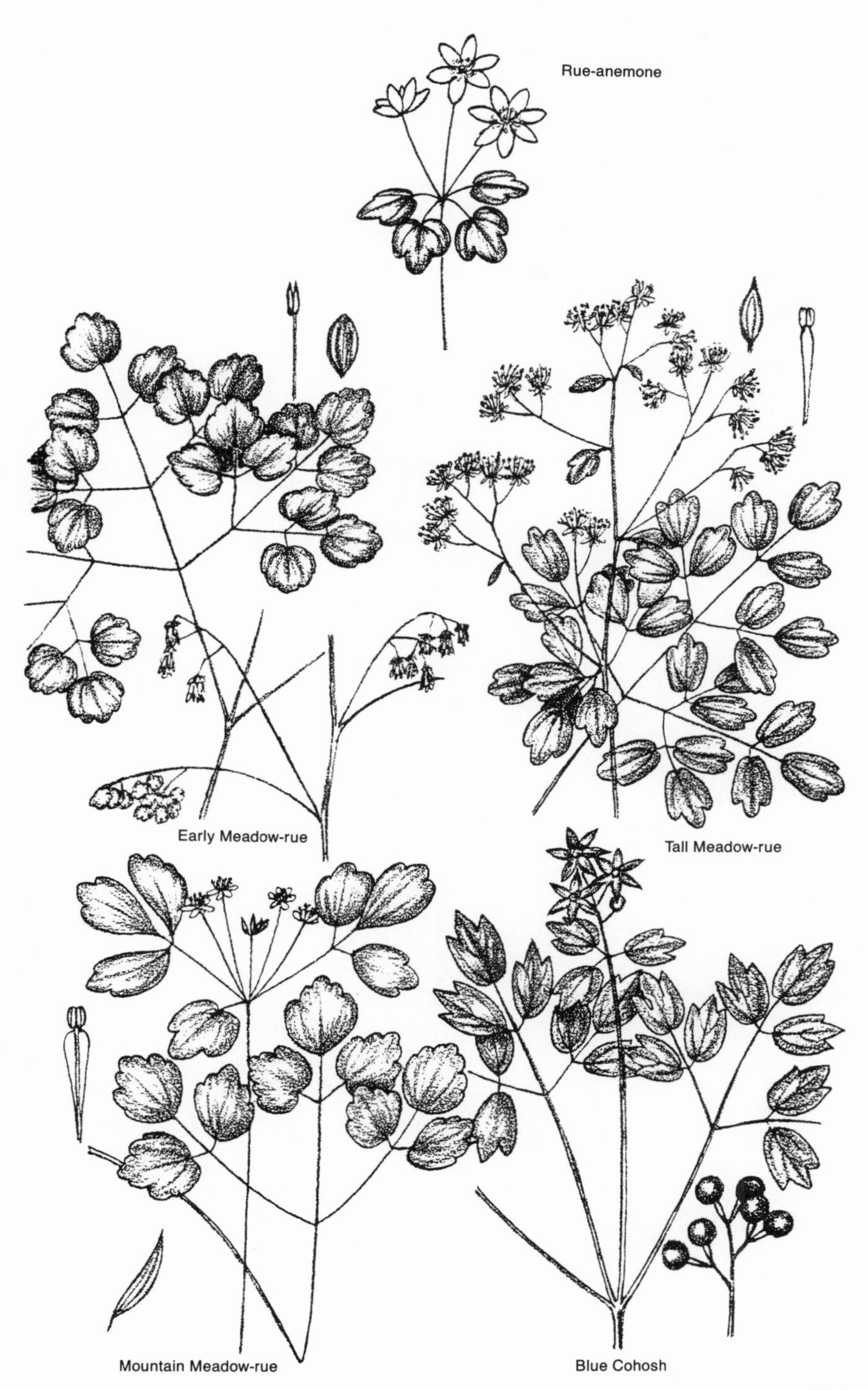
Rue-anemone
Early Meadow-rue
Tall Meadow-rue
Mountain Meadow-rue
Blue Cohosh

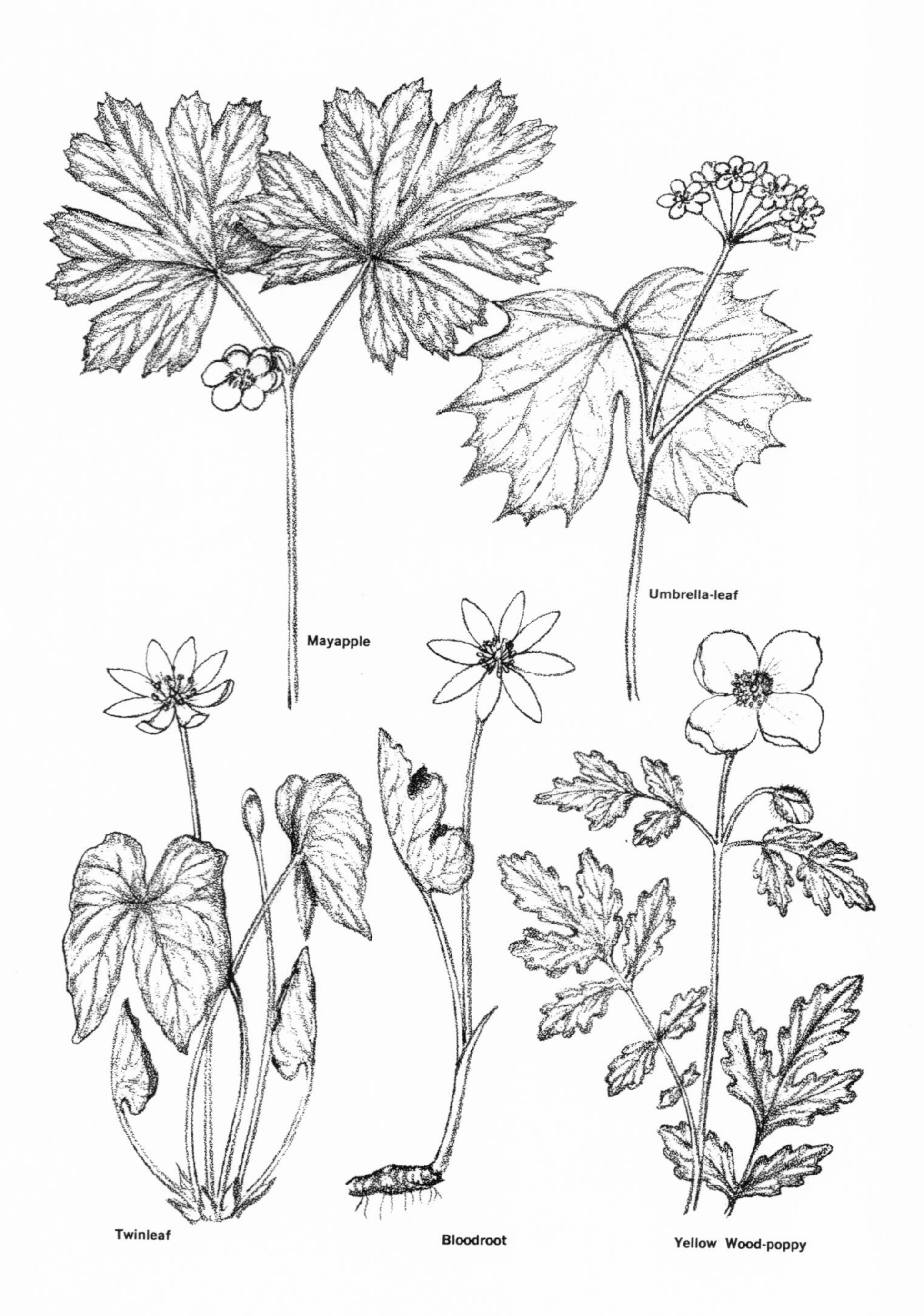
Mayapple
Umbrella-leaf
Twinleaf
Bloodroot
Yellow Wood-poppy

MAYAPPLE *(Podophyllum peltatum)*

The Mayapple is readily recognizable by the pair of large, stalked, umbrella-like leaves on its foot-high stem. They are attached to their stalks near the middle; and deeply divided into from 5 to 7 lobes, each of which is 2-cleft at the end and coarsely toothed. In the fork of the two leafstalks hangs a solitary flower an inch or more across. The 6 sepals are shed as the flower opens but the 6 to 9 large, waxy-white petals remain; and there are from 12 to 18 stamens with bright yellow anthers, and a solitary pistil. The latter develops into a large berry, about the size and color of a small lemon, which is edible and enjoyed by some. Non-flowering plants have but a single umbrella-like leaf at the top of their stem. Mayapples commonly grow in colonies in rich open woods, thickets, and clearings; blooming between April and June. Their rootstocks contain a poisonous substance but they are used medicinally. Another name often given it is Mandrake, but this name properly belongs to an Old World plant of no relationship to the Mayapple. RANGE: Que., Ont., and Minn. south to Fla. and Tex.

UMBRELLA-LEAF *(Diphylleia cymosa)*

This plant gets its name from the single, large, umbrella-like leaf of the non-flowering plants. It is attached to the stem near the center of the leaf. Flowering plants have a pair of similar but smaller leaves which are deeply cleft into two divisions, coarsely toothed, and attached to their stalks near the leaf margin. The white flowers are about ¾ inch across and borne in clusters. They have 6 sepals which are soon shed, 6 oval-shaped petals, 6 stamens, and a pistil. The latter develops into a roundish or oval-shaped, blue berry about ½ inch long. It grows in rich woods, often along streams, in the southern mountains; blooming in April or May. RANGE: Western Va. south to n. Ga.

TWINLEAF *(Jeffersonia diphylla)*

So-called because its leaves are so deeply divided that they seem to have 2 blades, the Twinleaf is very easy to recognize. At flowering time, in April or May, the leaves are only partly developed and 2 halves may be folded together. When fully mature, the leaves are 3 to 6 inches long and from 2 to 4 inches wide. Each flower is on a smooth stalk 6 to 8 inches in length. About an inch across, they have 4 sepals which soon fall, 8 white petals, and 8 stamens. The pistil becomes a many-seeded pod. The Twinleaf grows in rich, moist woods but it is not very common. Its generic name *Jeffersonia* honors our illustrious scientist-president, Thomas Jefferson.
RANGE: N.Y. and Ont. to Wis. south to Va. and Ala.

POPPY FAMILY (Papaveraceae)

Members of the Poppy Family usually have a milky juice which may be whitish, yellowish, or reddish. Their flowers are usually quite large and showy and have a pair of sepals which are quickly shed, 4 petals or petals in multiples of 4, numerous stamens, and a solitary pistil maturing as a many-seeded capsule.

BLOODROOT *(Sanguinaria canadensis)*

In early spring—between March and early May—the bright white flowers of the Bloodroot appear above the fallen leaves in our woodlands. As they open, their 2 sepals fall away; but the 8 to 12 petals spread into a blossom 1½ to 2 inches across. Each flower is accompanied by an enveloping leaf, which has not yet unfolded. Later this 5- to 9-lobed and bluntly toothed leaf may become 6 or more inches in width, persisting until about mid-summer. The Bloodroot gets its name, and another name of Red Puccoon, from the bright orange-red juice of its thick rootstock. It is easily grown in the woodland flower garden. RANGE: Que. to Man. south to Fla. and Tex.

YELLOW WOOD-POPPY *(Stylophorum diphyllum)*

Sometimes called the Celandine-poppy, this plant has a stem 12 to 18 inches tall, with a pair of deeply cut leaves and 2 to 4 flowers at the top. Leaves with similar side-lobes but with long stalks occur at the base. The flowers have 2 hairy sepals which soon drop, and 4 bright yellow petals. The plant grows in rich, moist woodlands and blooms between March and May. It has a yellow juice.
RANGE: Pa. to Wis. south to Va., Tenn., and Mo.

Members of this family have flowers with a bilateral symmetry. They have 2 small sepals; 4 petals, one or both of the outer ones being spurred at the base; 6 stamens in two groups of 3 each; and a solitary pistil composed of 2 united carpels. The plants have a watery juice and compound dissected leaves.

MOUNTAIN-FRINGE *(Adlumia fungosa)*

Also known as the Allegheny-vine, this plant with long and slender stems climbs by means of its coiling leaf stalks. The leaves are divided into many thin, delicate, usually lobed leaflets which are pale beneath. It has pinkish or purplish flowers shaped like narrow hearts, and borne in drooping clusters on stalks arising in the axils of the leaves. The Mountain-fringe grows on damp, rocky, wooded slopes and is sometimes common on recently burned areas. It blooms between June and October.
RANGE: Que. to Ont. and Minn. south to N.Eng. and in mts. to N.C. and Tenn.

WILD BLEEDING-HEART *(Dicentra exima)*

The native Wild Bleeding-heart is sometimes called Turkey-corn and Staggerweed, and in cultivation it is known as the Plumy Bleeding-heart. It is a somewhat whitened, smooth plant from 10 inches to about 2 feet high. The pink to flesh-colored flowers are about ¾ inch long and they are produced between April and September. The leaves are all basal, ternately divided, and cut into numerous oblong segments. It grows in rocky woods and on rock cliffs in the mountains. RANGE: N.Y. and W.Va. south to n. Ga. and e. Tenn.

DUTCHMAN'S-BREECHES *(Dicentra cucullaria)*

This well-known wild flower gets its name from the resemblance of its flowers to a Dutchman's baggy breeches, hanging upside down along the nodding stalks which are 5 to 10 inches long. They are about ¾ inch long, white or faintly pinkish, and their 2 broad spurs are spread like the letter V. The leaves are all at the base and are ternately divided and cut into numerous narrow segments which are pale beneath. Both the flower stalks and the leaves arise from a knobby-scaled bulb. The Dutchman's-breeches grows in rich woods and on rocky slopes; blooming during April or May.
RANGE: N.S. to S.D. south to Ga., Ala., and Kan.

SQUIRREL-CORN *(Dicentra canadensis)*

Leaves of the Squirrel-corn are just like those of the Dutchman's-breeches, but the flowers afford a fine point of distinction. Those of the Squirrel-corn have 2 short and rounded spurs which point upward. They are a little more than ½ inch long, white or pale pinkish, rather heart-shaped, and arranged along an arching stalk 6 to 8 inches high. The plant gets its name from the yellow tubers which occur on its rootstocks. It grows in rich woods and on rocky slopes, blooming in April or May.
RANGE: Que. to Minn. south to Va., N.C., n. Ga., and Mo.

PINK CORYDALIS *(Corydalis sempervirens)*

This is a more or less erect, smooth, and whitened plant 5 to 20 inches tall, with slender and branching stems. Its flowers, like those of other species of Corydalis, are slender and have only 1 petal spurred. They are about ½ inch long, pink or purplish with a yellow tip, and arranged in clusters at the tips of the branches. The leaves are much divided and bluntly lobed, the upper ones being stalkless. It grows on rocky slopes and in recent clearings, blooming between May and September. Another name for it is Pale Corydalis.
RANGE: Nfd. to Alaska south to Ga., Tenn., Minn., and Mont.

YELLOW CORYDALIS *(Corydalis flavula)*

This species is similar to the preceding but it has yellow flowers which have a short, rounded spur. It grows in rocky woods, blooming in April or May.
RANGE: Conn. to Ont. and Minn. south to Va., Tenn., La., and Kan. (Not illustrated)

GOLDEN CORYDALIS *(Corydalis aurea)*

The bright yellow flowers of this species are almost ½ inch long and have a slightly curved spur nearly half the length of the flower. It grows in open woods, on rocky slopes, and on shores; blooming between May and July. Also called Scrambled Eggs.
RANGE: Que. to Alaska south to N.Y., W. Va., Mo., Ariz., and Calif. (Not illustrated)

Wild Bleeding-heart
Mountain-fringe
Dutchman's-breeches
Squirrel-corn
Pink Corydalis

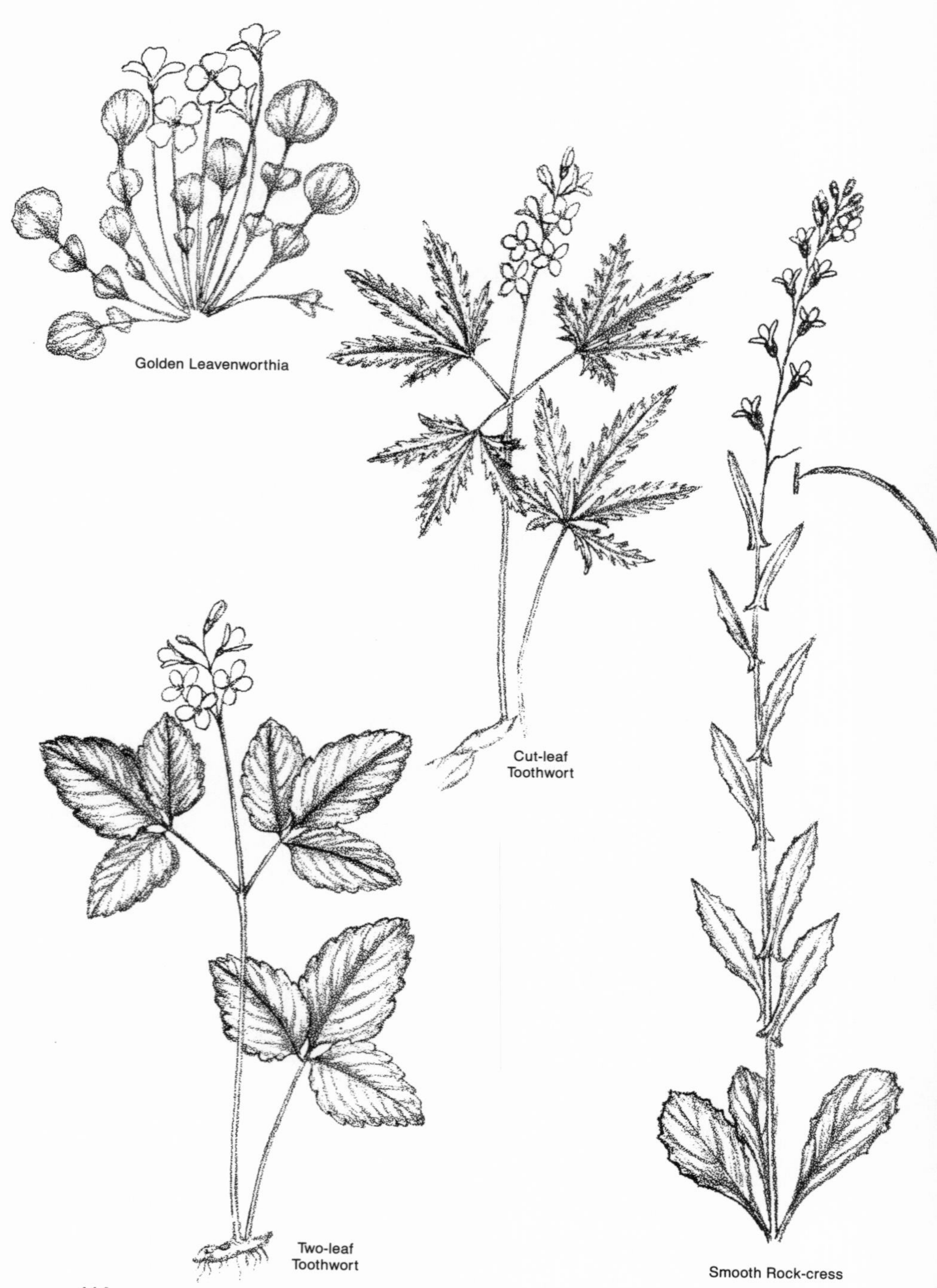
Golden Leavenworthia
Cut-leaf
Toothwort
Two-leaf
Toothwort
Smooth Rock-cress

Flowers of the members of this family have 4 sepals; 4 petals which spead out flat to form a symmetrical cross; 6 stamens, 2 of them shorter than the other 4; and a single pistil composed of 2 united carpels. The latter develops into a flattened or narrow pod containing several seeds.

GOLDEN LEAVENWORTHIA *(Leavenworthia aurea)*

Leavenworthias are small plants with basal tufts of leaves which have side lobes. In this species they are divided into a relatively few, large, and blunt lobes. The flowers are not quite ½ inch across and usually solitary on leafless stalks 1½ to about 4 inches in length. They have 4 white or purplish petals which are yellow at the base. The Golden Leavenworthia grows in rocky places; blooming in March or April.

RANGE: Tenn. to Ark. south to Ala. and Tex.

MICHAUX'S LEAVENWORTHIA *(Leavenworthia uniflora)*

This species differs from the preceding in that the leaves are divided into a number of angled and somewhat toothed lobes or segments. It generally has several but slightly smaller flowers on leafless stalks 2 to 6 inches tall. The plant grows in dry rocky woods or sometimes on wet rocks; blooming in March or April.

RANGE: Ohio and Ind. to Mo. south to Ala. and Ark. (Not illustrated)

CUT-LEAF TOOTHWORT *(Cardamine concatenata)*

On a smooth or somewhat hairy stem 8 to 15 inches tall, this toothwort has a whorl of 3 leaves; each of them being divided into 3 narrow and sharply toothed or lobed segments. The basal leaves are similar but they are seldom present at flowering time, between March and May. The flowers are about ½ inch across and have 4 white or lavender tinged petals. Several of them occur along a stalk which stands above the leaves. Beneath the ground the plant has a jointed rootstock which readily separates into inch-long tubers shaped like sweet-potatoes. Being slightly peppery, they afford a quite pleasant nibble. It is often common in rich, moist woods and on rocky slopes. Another name for it is Pepperroot.

RANGE: Vt. and Que. to Minn. and Neb. south to Fla. La., and Kan.

TWO-LEAF TOOTHWORT *(Cardamine diphylla)*

Another name for this plant is Crinkleroot; a name suggested by its long, continuous, crinkled rootstock. It has a smooth stem from 6 to 12 inches tall on which is a pair of leaves, both of them divided into 3 broadly egg-shaped and bluntly toothed leaflets. The basal leaves are similar but they are long-stalked. Along the stem, and above the 2 leaves, is a loose cluster of white flowers with 4 white petals. This toothwort is usually quite common in rich, moist, rocky woods and thickets; blooming between April and June.

RANGE: N.S. and Ont. south to S.C., Ky., and Mich.

SMOOTH ROCK-CRESS *(Arabis laevigata)*

A number of rock-cresses occur in eastern North America, but this species is one of the more conspicuous ones. It has a smooth, whitened, leafy stem from 1 to 3 feet tall. The pale, narrow leaves clasp the stem by their eared bases and their margins may have a few teeth. The basal leaves taper to a stalked base and are toothed and sometimes deeply cut. The small whitish flowers are followed by slender pods 3 to 4 inches long, which arch outward and downward. The plant is quite common in rich, rocky woods and slopes; blooming between March and June.

RANGE: Que. to Minn. south to Ga., Ala., Ark., and Okla.

SICKLE-POD *(Arabis canadensis)*

This rock-cress is similar to the preceding but it is somewhat hairy, the stem leaves do not have clasping bases, and their margins are toothed. The slender sickle-shaped pods bend abruptly downward. It grows in rich, rocky woods and on banks; blooming between April and June.

RANGE: Me. to Ont., Minn., and Neb. south to Ga. and Tex. (Not illustrated)

SLENDER TOOTHWORT *(Cardamine angustata)*

This toothwort has a slender, smooth or somewhat downy stem 6 to 14 inches tall. On it is a pair of leaves which are divided into 3 very narrow and sparingly toothed or untoothed segments. The basal leaves are divided into 3 more or less egg-shaped leaflets which are bluntly toothed or slightly lobed. Along the stem above its 2 leaves is a loose cluster of flowers, each of them about ¾ inch across and with 4 pinkish or pale purplish petals. The rootstock separates into tubers like those of the Cut-leaf Toothwort. It grows in rich, moist and usually rocky woods; blooming in April or May.

RANGE: N.J. to Ohio south to n. Ga., and Tenn.

LARGE TOOTHWORT *(Cardamine maxima)*

One might mistake this toothwort for the Two-leaf Toothwort for the leaves of the two are quite similar. This one, however, usually has 3 leaves which are widely spaced along the stem. Its flowers are about ¾ inch across and they have 4 white or pale purplish petals. The rootstock is jointed but does not separate easily into tubers. It grows on wooded slopes and streambanks but it is rather local, blooming in March or April.

RANGE: Me. to Wis. south to W. Va. and Tenn.

BULBOUS CRESS *(Cardamine bulbosa)*

This smooth, leafy-stemmed plant has a knobby tuber at the base of its 6- to 18-inch stem, hence the name Bulbous Cress. It is also known as the Spring Cress. Between March and June it produces quite showy clusters of white flowers which are about a half-inch across. The leaves along the stem vary from lance-shaped to roundish and they may or may not have wavy-toothed margins. The basal ones are roundish to heart-shaped, often angled, and long-stalked. It is quite common in wet woods and meadows.

RANGE: N.H. to Ont. and Wis. south to Fla. and Tex.

PURPLE CRESS *(Cardamine douglasii)*

The Purple Cress is quite similar to the preceding species. It differs principally in having a somewhat hairy stem and flowers tinged with lavender or rose. It grows quite generally in wet woods, swamps, and about springs; usually blooming about 2 weeks earlier than the Bulbous Cress.

RANGE: Conn. to Ont. and Wis. south to Va., Tenn., and Mo. (Not illustrated)

PURPLE ROCKET *(Iodanthus pinnatifidus)*

This plant has a slender, smooth, leafy stem from 1 to 3 feet tall which branches above. The lower leaves are divided into a large egg-shaped end segment below which there are 2 to 6 pairs of side lobes, all of them with more or less sharply toothed margins. The leaves along the upper part of the stem are much smaller and have bases which clasp the stem. Numerous flowers about ⅓ inch across, with 4 light violet petals, are arrayed in frequently branched end clusters. The Purple Rocket commonly grows on the flood plains of streams and on adjoining slopes; blooming in May or June.

RANGE: Pa. to Minn. south to Ala. and Tex.

COMMON WINTER-CRESS *(Barbarea vulgaris)*

The Winter-cress, or Yellow Rocket, is a common weed which is often abundant in fields, meadows, and wet woods. It is a smooth, branching plant with leafy stems from 1 to 2 feet tall. The lower leaves are stalked and have a terminal segment much larger than the 1 to 4 pairs of side ones. Those along the upper portion of the stem are stalkless and often have clasping bases. The bright yellow, 4-petalled flowers are about ⅓ inch across; and are borne in rather dense, cylindrical clusters at the end of the stem and its branches; blooming from April to June or later. It is a native of Europe.

EARLY WINTER-CRESS *(Barbarea verna)*

This species quite closely resembles the preceding one, and like it is a native of Europe. It is now widespread in eastern North America as far south as Florida. All of its leaves are deeply lobed, the larger basal ones having from 4 to 8 pairs of side lobes below the larger terminal one. Perhaps the best field mark, however, is its sharply 4-sided pods. It has bright yellow flowers between late March and May. Another name for it is Belle Isle Cress. (Not illustrated)

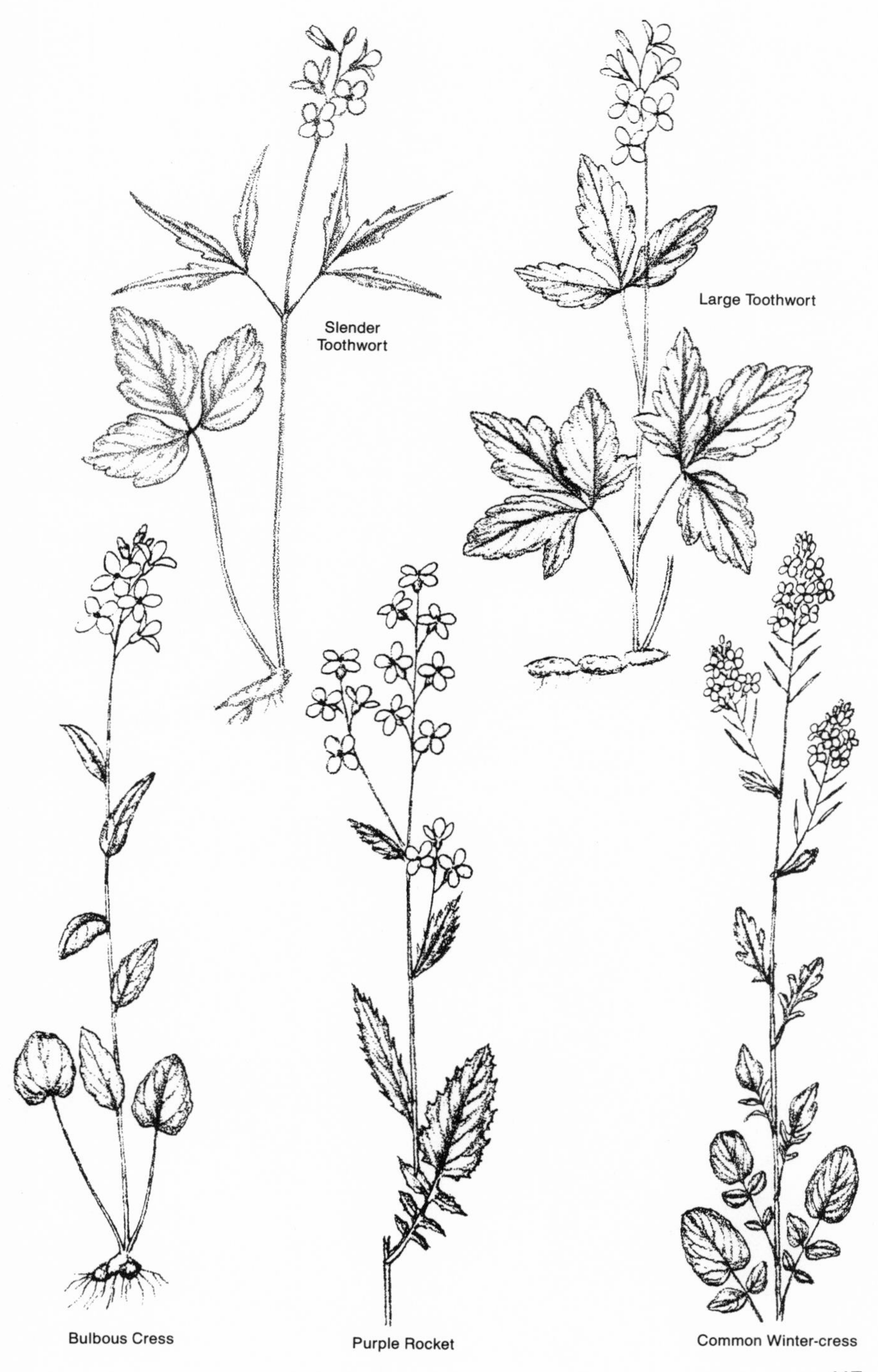
Slender
Toothwort
Large Toothwort
Bulbous Cress
Purple Rocket
Common Winter-cress

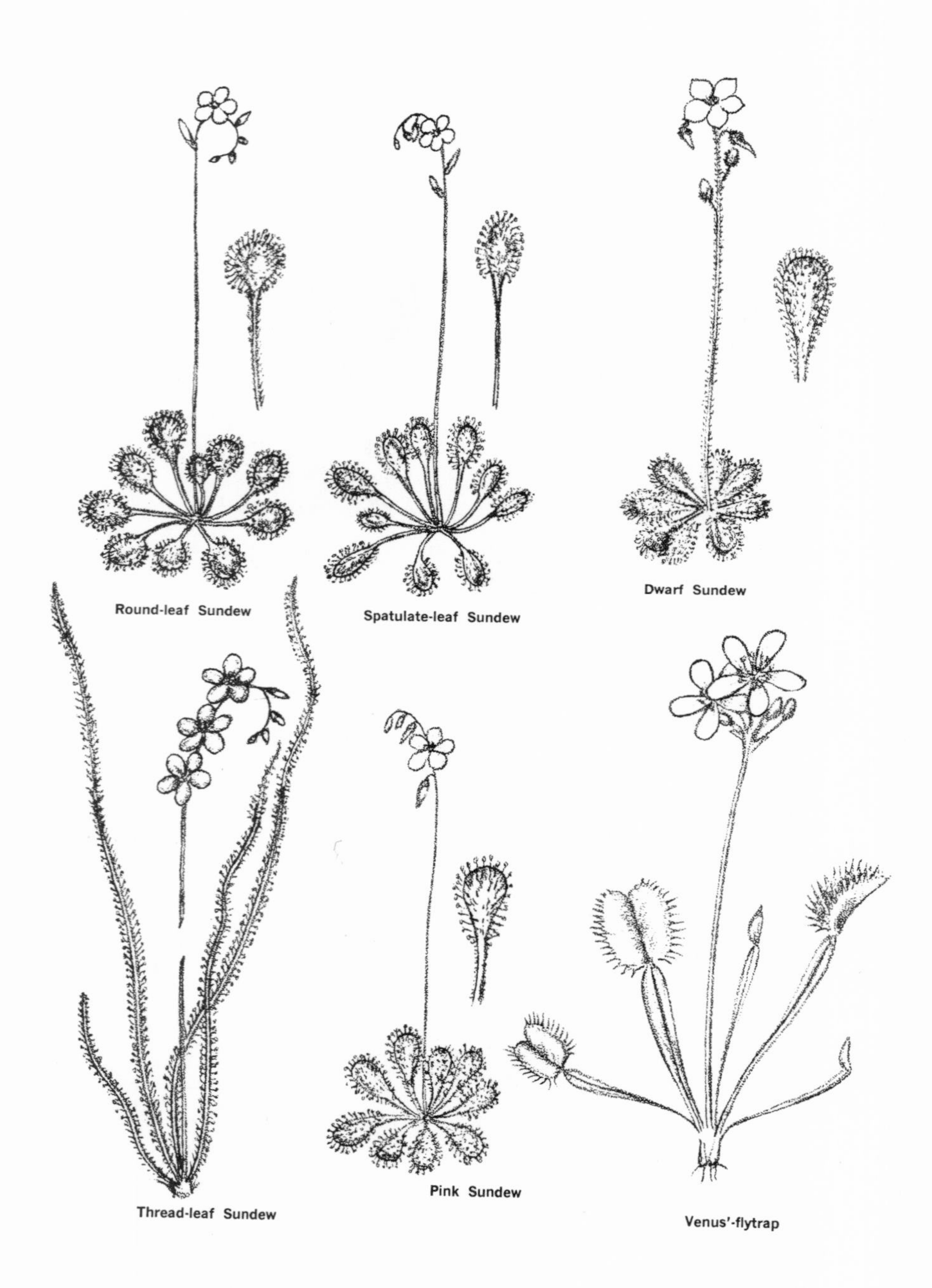
Round-leaf Sundew
Spatulate-leaf Sundew
Dwarf Sundew
Thread-leaf Sundew
Pink Sundew
Venus'-flytrap

Sundews are small insectivorous plants found in bogs and other wet places. Their leaves have gland-tipped red hairs which exude little droplets of a sticky fluid. In the sunlight, these little droplets sparkle like dewdrops, hence the name given to the plants. Small insects trapped on this natural "flypaper" are gradually digested by the plant. The flowers have 5 sepals, 5 petals, 5 stamens, and a pistil with usually 3 styles. The leaves are in a basal rosette.

ROUND-LEAF SUNDEW *(Drosera rotundifolia)*

This is the most common and widespread species of sundew. The leaves have disk-shaped blades on slender, minutely hairy stalks from ½ to 2 inches long. Several white flowers about ¼ inch across are borne on a slender, smooth stalk from 3 to 10 inches high. In bogs and on wet sands; blooming between June and August.

RANGE: Lab. to Alaska south to Fla., Ala., Minn., Mont., and Calif.

SPATULATE-LEAF SUNDEW *(Drosera intermedia)*

The leaves of this species have oval-shaped blades up to ¾ inch long on slender, smooth stalks from ½ to 1½ inches long. The several white flowers are about ¼ inch across and are borne on a slender, smooth stalk from 2 to 8 inches high. It grows in bogs and on wet sands, blooming between June and August.

RANGE: Nfd. to Ont. south to Fla. and Tex.

DWARF SUNDEW *(Drosera brevifolia)*

The Dwarf Sundew has top-shaped leaves ¼ to about ⅝ inch long which have very short and glandular-hairy stalks. Its 1 to 3 white flowers are almost ½ inch across and are borne on a slender, glandular-hairy stalk from 2 to 3½ inches high. It grows in wet, sandy, coastal plain pinelands in the Southeast; blooming in May and June.

RANGE: Va., Tenn., and Ark. south to Fla. and Tex.

THREAD-LEAF SUNDEW *(Drosera filiformis)*

This sundew is unmistakable as it has very narrow leaves from 6 to 15 inches in length, which are covered with gland-tipped hairs. Its lavender flowers are ⅜ to ¾ of an inch across and are borne on a slender, smooth stalk from 8 to 20 inches tall. It grows on wet sands in the coastal plain, blooming between June and September. Another name for it is Dew-thread.

RANGE: Mass. to N.J. and S.C. south to Fla. and west to La.

PINK SUNDEW *(Drosera capillaris)*

The Pink Sundew has top-shaped leaves from ½ to 1¼ inches long which taper into short, broad, glandular-hairy stalks. Its pink flowers are about ¼ inch across; several of them being borne on a very slender, smooth stalk from 2½ to 6 inches tall. It grows in peaty bogs and on wet sands, blooming between May and July.

RANGE: Coastal plain se. Va. south to Fla. and west to Tex.; also on the Cumberland Plateau in Tenn.

VENUS'-FLYTRAP *(Dionaea muscipula)*

This unique insectivorous plant is the only representative of its family. Its leaves, which are in a basal rosette, are 2 to 5 inches long. The leaf stalk has a pair of wings which taper gradually to the base; and at its summit there is a roundish, 2-lobed blade with a marginal fringe of bristly hairs. The upper surface of the blade is red or reddish and on each of the 2 lobes there are 3 sensitive hairs which serve to trigger the trap. An insect touching these hairs causes the two halves of the blade to snap shut. and the marginal bristles then fit together like the fingers of one's folded hands. Glands on the leaf surface then begin to secrete a fluid which digests the softer parts of the trapped insect. The white flowers are almost ¾ of an inch across and are clustered at the summit of a smooth stalk from 4 to 12 inches tall. They have 5 sepals, 5 petals, 10 to 20 stamens, and a pistil composed of 5 united carpels. The Venus'-flytrap is restricted to bogs in southeastern North Carolina and eastern South Carolina. It blooms during May or June.

Pitcher-plants have hollow leaves with a wing down one side and a lid-like hood at the summit. These leaves contain a watery fluid in which insects drown and are digested. Their flowers are solitary and nod from the summit of a long, naked stalk. They have 5 persistent sepals with 3 bracts at their base; 5 fiddle-shaped petals; a number of stamens; and a solitary pistil which has a conspicuous umbrella-like end, bearing the 5 hooked stigmas on its underside. All grow in wet or boggy places.

PURPLE PITCHER-PLANT *(Sarracenia purpurea)*

This pitcher-plant really has pitcher-shaped leaves which are 4 to 10 inches long, with a rather broad wing and an erect hood which is covered with backward-pointing bristles. The reclining or ascending pitchers are often beautifully veined with purple. Its flower is more or less globe-shaped, about 2 inches across, usually a deep purplish-red, and stands on a stalk 10 to 20 inches tall. It is the only species of pitcher-plant found north of the southern part of Virginia. It blooms between April and August. Sometimes it is called the Huntsmans-cap or Sidesaddle-flower.

RANGE: Nfd. to Sask. south to Fla. and La.

PARROT PITCHER-PLANT *(Sarracenia psittacina)*

The pitcher-shaped leaves of this pitcher-plant are only 2 to 6 inches long. They have a broad wing and a domed hood which resembles the head and bill of a parrot; the opening being on the under side. Its pitchers are beautifully marked with white spots and purple veins. The dark purplish flowers are 1¼ to 2 inches across and each one stands on a stalk 8 to 16 inches tall. It grows in wet coastal plain pinelands and bogs, blooming in April or May.

RANGE: S. Ga. and Fla. west to La.

HOODED PITCHER-PLANT *(Sarracenia minor)*

This species has erect, trumpet-shaped leaves 8 to 16 inches tall, with an arching dome-like hood which covers the open top of the trumpet-like leaf. Toward the summit they have numerous translucent, window-like spots; and they are often veined with purple. The yellow flowers are about 2½ inches across and stand on stalks 6 to 12 inches tall. It grows in wet coastal plain pinelands and bogs, blooming between April and June.

RANGE: N.C. south to Fla.

SWEET PITCHER-PLANT *(Sarracenia rubra)*

The leaves of this species are very slender, trumpet-shaped, erect, and 6 to 16 inches tall. They have a narrow wing down the side and an erect or leaning pointed hood ½ to 1¼ inches wide which is usually veined with purple. Its reddish-purple or maroon flowers are 1½ to 2 inches across and stand on stalks 10 to 18 inches tall; and they have a delightful fragrance. It grows in wet pinelands and bogs of the coastal plain, and also along streams in the mountains; blooming in April or May.

RANGE: N.C. south to Fla. and west to Miss.

TRUMPETS *(Sarracenia flava)*

This distinctive species has trumpet-shaped leaves 1½ to 4 feet tall, which have a narrow wing and an erect hood 3 to 4½ inches broad. The latter is often yellow or strikingly veined with red or purple. Its yellow flowers are 3 to 4½ inches across and stand on stalks from 1½ to 3 feet tall. It grows in bogs and savannahs, chiefly in the coastal plain; blooming in March or April. Also known as the Trumpet-leaf or Huntsman's-horn.

RANGE: Se. Va. south to Fla. and west to Ala.

PURPLE TRUMPET *(Sarracenia leucophylla)*

This species resembles the preceding in having tall trumpet-like leaves. Those of this species, however, are white above and beautifully veined with purple. Its large flowers are purplish-red. It grows is sandy bogs in the coastal plain from sw. Ga. and nw. Fla. west to Miss. (Not illustrated)

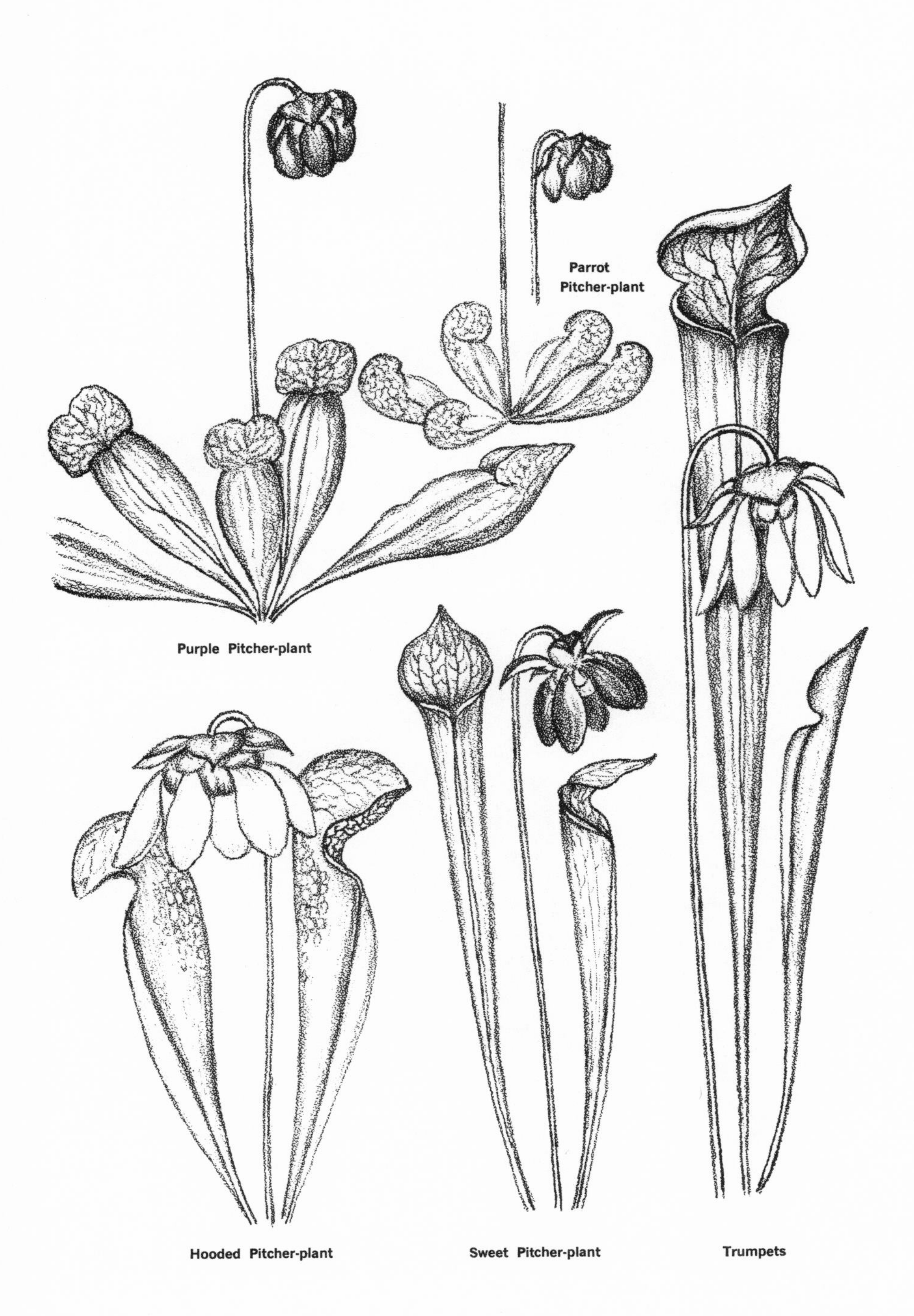
Parrot
Pitcher-plant
Purple Pitcher-plant
Hooded Pitcher-plant
Sweet Pitcher-plant
Trumpets

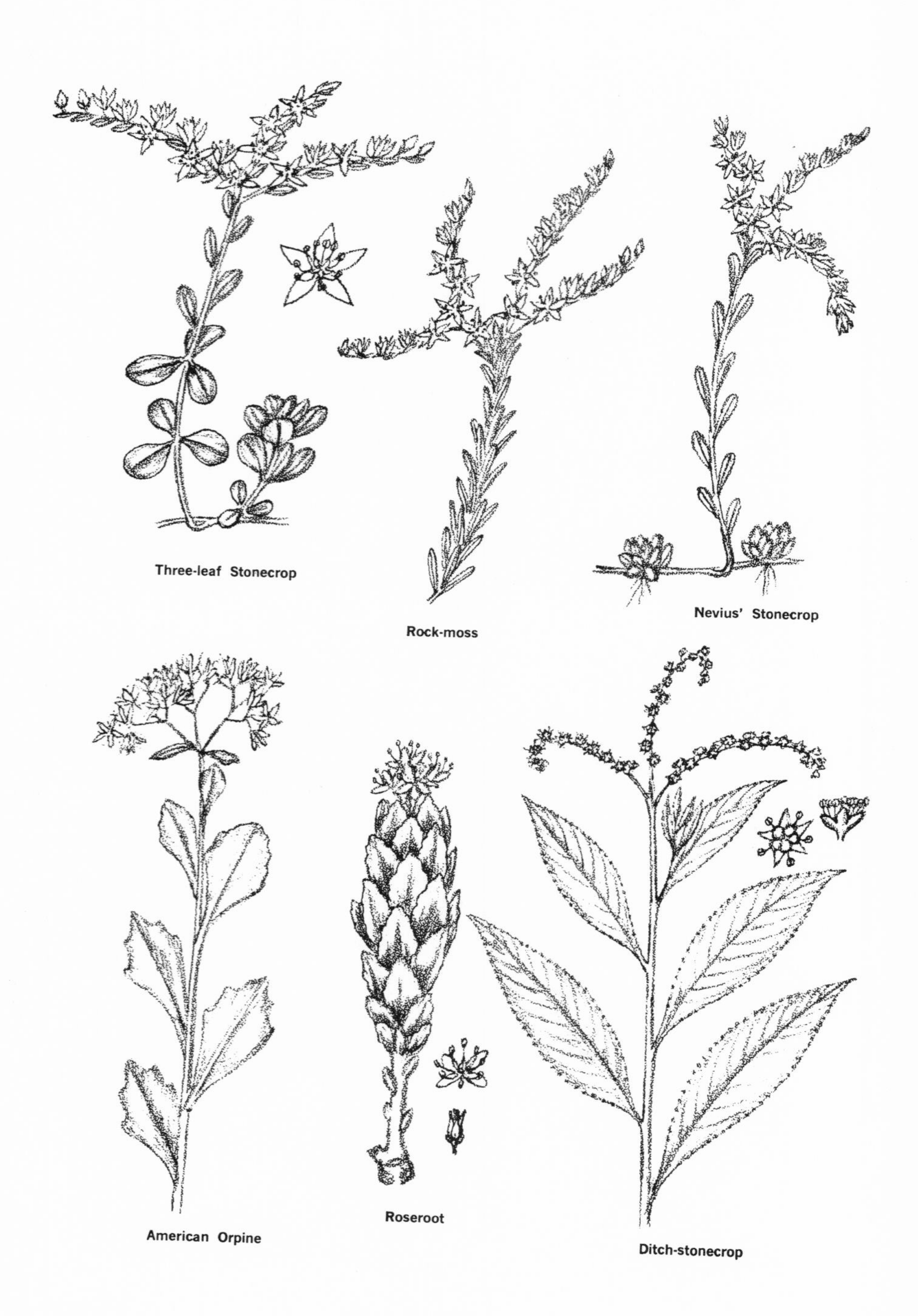
Three-leaf Stonecrop
Rock-moss
Nevius' Stonecrop
American Orpine
Roseroot
Ditch-stonecrop

STONECROP FAMILY (Crassulaceae)

The members of the Stonecrop Family are mostly small and fleshy plants. Their flowers have 4 or 5 sepals which are united at the base, 4 or 5 petals, 8 or 10 stamens, and a group of 4 or 5 separate pistils.

THREE-LEAF STONECROP *(Sedum ternatum)*

The Three-leaf Stonecrop is a small, succulent plant with a creeping stem and ascending flowering branches from 3 to 6 inches high. Its lower leaves are top-shaped and arranged in whorls of 3's, but those on the upper portion of the stem are scattered singly. Its flowers are almost ½ inch across, starry in appearance, with 5 narrow white petals and twice as many stamens which have dark-colored anthers. They suggest another common name of Pepper-and-salt. This plant is usually common on moist cliffs and rocky, wooded slopes; blooming between April and June. RANGE: N.Y. to Mich. south to Ga. and Tenn.

ROCK-MOSS *(Sedum pulchellum)*

This is a small, succulent plant with a creeping or ascending stem from 4 to 12 inches long. It has numerous, densely crowded leaves ¼ to 1 inch long with slightly clasping bases. The flower cluster has from 4 to 7 spreading or recurved branches which are crowded with starry, 5-petalled, white to rose-pink flowers about ½ inch across. Also known as the Mountain-moss and Widow's-cross, it grows on dry to moist rocks and on rocky soils; blooming in May or June.

RANGE: W.Va. to Ill. and Kan. south to Ga., Ala., Ark., and Tex.

NEVIUS' STONECROP *(Sedum nevii)*

This is another small succulent with a creeping stem and ascending flowering branches from 3 to 5 inches high. Its leaves are narrowly top-shaped, spirally arranged, and somewhat spreading. Those of the sterile shoots are very densely crowded. The flower cluster has 3 branches crowded with starry, 5-petalled, white flowers about ½ inch across. It grows on moist rocks in the mountains, blooming in May and June.

RANGE: Western Va. and W.Va. south to N.C. and Ala.

AMERICAN ORPINE *(Sedum telephioides)*

The American Orpine is an erect plant seldom over 10 inches tall; with thick and fleshy, oval or top-shaded, somewhat toothed leaves from 1 to 2 inches long. It has a flat-topped or slightly dome-shaped end cluster of pale pink flowers from 2 to 4 inches broad. The 5-petalled flowers are about ⅜ inch across, blooming during August or September. The entire plant is smooth, purplish, and is whitened with a bloom. It grows on dry rock outcrops and cliffs. Another name for it is Wild Live-forever.

RANGE: W. N.Y. to Ill. south in the mts. to Ga.

ROSEROOT *(Sedum rosea)*

The Roseroot is a peculiar little fleshy plant with a thick root which is fragrant when bruised. It has ascending or erect, leafy stems from 4 to about 12 inches high. The leaves are pale, oval-shaped, sometimes toothed on the margin, from ½ to 1½ inches long, and are closely crowded on the stems. Some plants have yellowish flowers with usually 8 stamens; others have purplish ones with usually 4 pistils. It grows in rocky places, blooming between May and August.

RANGE: Arctic regions south to Me., N.Y., ne. Pa., and Roan Mt., N.C.

DITCH-STONECROP *(Penthorum sedoides)*

This is a smooth plant with an erect, usually branched, somewhat angled stem from 1 to 2 feet tall. It has scattered, elliptic to lance-shaped leaves 2 to 4 inches long, which are finely saw-toothed on the margin. Its flowers are small and yellowish-green; with 5 sepals, no petals, and 10 stamens. It grows in low wet places, swamps, and along stream banks and ditches; blooming between July and October. It is also known as the Virginia-stonecrop. RANGE: N.B. to Ont., Wis., and Neb. south to Fla. and Tex.

SAXIFRAGE FAMILY (Saxifragaceae)

Members of this family usually have flowers with 5 sepals which are more or less united at the base; 5 petals; 5 or 10 stamens; and either a solitary pistil or a pair of pistils with the ovaries at least partly joined together.

EARLY SAXIFRAGE *(Saxifraga virginiensis)*

When the Early Saxifrage begins to bloom in late March or early April, it has a sticky-hairy flower stalk only 4 to 6 inches high. It continues to bloom for several weeks, often into June, becoming branched and up to a foot high. At the base of the flower stalk it has a rosette of rather fleshy, oval or egg-shaped leaves from 1 to 3 inches long which are often purplish-beneath. Their margins may have low and blunt teeth or be coarsely toothed or scalloped. Similar but smaller leaves subtend the branches of the flower stalk. The flowers are about ¼ inch across and have 5 oblong or top-shaped white petals and 10 stamens with bright yellow anthers. This is our most common and widespread species of saxifrage, growing on wet to dry rock ledges and rocky slopes.

RANGE: N.B. to Ont. and Minn. south to Ga., Tenn., and Mo.

CAROLINA SAXIFRAGE *(Saxifraga caroliniana)*

This is a sticky-hairy plant with a cluster of basal leaves, and a flower stalk from 6 to 18 inches tall which branches above. Its leaves are 1 to 2½ inches long, with an oval or roundish blade contracted to a long, hairy leaf stalk; and from 6 to 10 coarse teeth on each side of the margin. Its flowers are almost ½ inch across and have 5 white petals with 2 yellow spots near the base. Also known as Gray's Saxifrage, it grows on moist, rocky slopes or cliffs in the southern mountains; blooming between May and July.

RANGE: Western Va., w. N.C., and e. Tenn.

LETTUCE SAXIFRAGE *(Saxifraga micranthidifolia)*

The Lettuce Saxifrage has a more or less sticky-hairy flower stalk from 1 to about 3 feet tall, with a cluster of big leaves at the base. The latter are narrowly top-shaped, rather thin, coarsely and sharply toothed on the margin, and from 6 to about 12 inches in length. The flowers are small and have 5 white petals. It grows on wet rocks and along the banks of streams in the mountains, blooming during May and June. Also known as Mountain-lettuce, as the leaves are used as a green by many mountain people.

RANGE: Pa. and W.Va. south to Ga. and Tenn.

SWAMP SAXIFRAGE *(Saxifraga pensylvanica)*

As its name indicates, the Swamp Saxifrage grows in swamps, bogs, and wet meadows. It has a stout and somewhat sticky-hairy flower stalk from 1 to 3½ feet tall, with a rosette-like cluster of leaves at its base. The leaves are oval to narrowly top-shaped, thick and somewhat leathery in texture, and from 4 to about 12 inches long. Its small flowers vary from greenish-white to yellowish or purplish and have 5 narrow petals. The flowering season is between April and June.

RANGE: Me. to Minn. south to Va., w. N.C., Ill., and Mo.

MICHAUX'S SAXIFRAGE *(Saxifraga michauxii)*

Michaux's Saxifrage grows in crevices on the face of rocky cliffs and on sunny, wet rocks in the southern Appalachians. In typical saxifrage fashion, it has a sticky-hairy flower stalk from 6 to about 20 inches tall with a rosette-like cluster of leaves at the base. Its leaves are narrowly top-shaped, from 3 to 7 inches long, and have coarse, sharp teeth on their margins. Among saxifrages, the flowers of this one are unusual in that the white petals are not all alike. There are 3 large ones which are heart-shaped at the base and have a pair of yellow spots, and 2 smaller ones which have tapering bases and no spots. It blooms between June and August.

RANGE: Western Va. and W.Va. south to n. Ga. and e. Tenn.

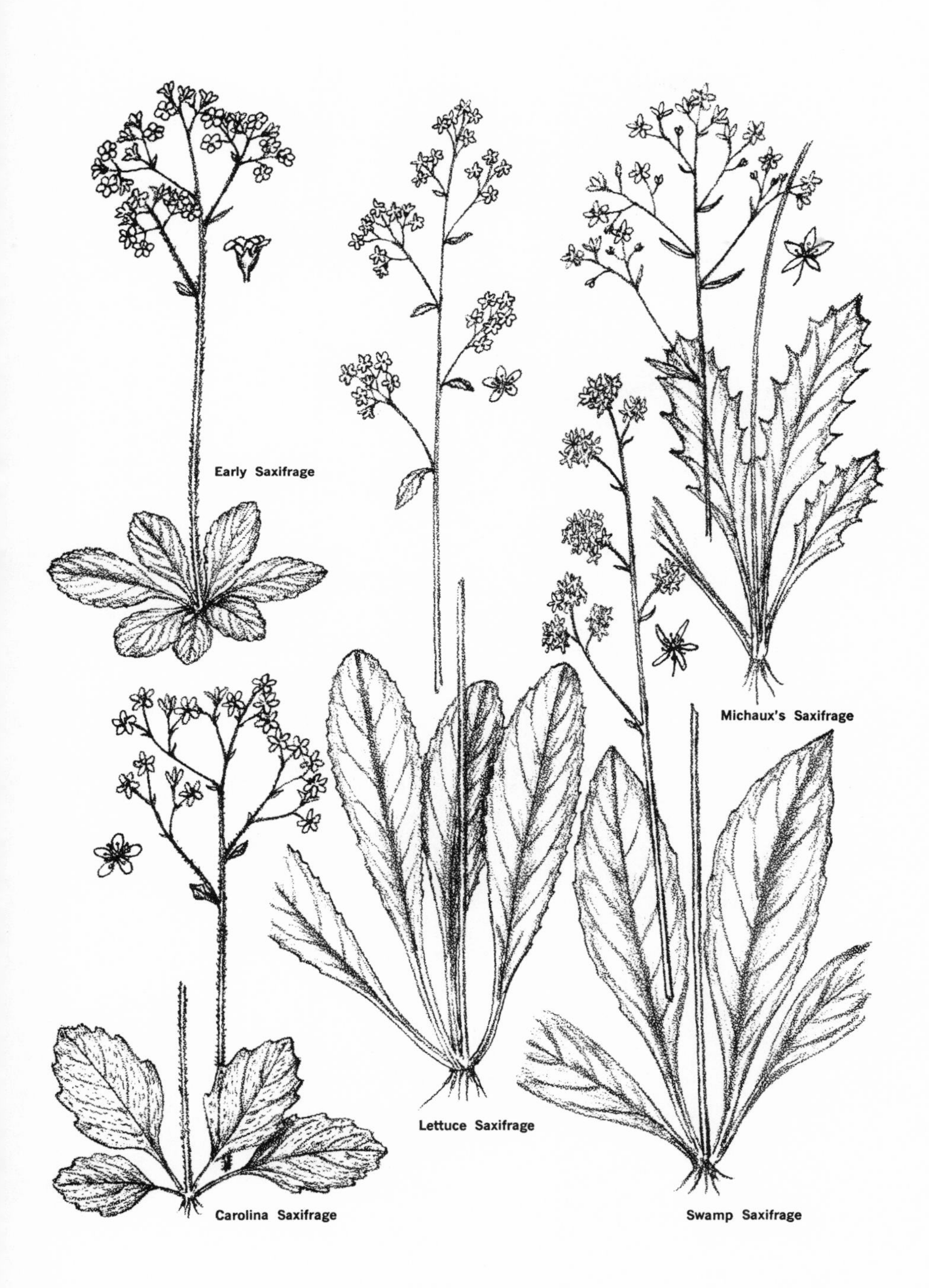
Early Saxifrage
Michaux's Saxifrage
Lettuce Saxifrage
Carolina Saxifrage
Swamp Saxifrage

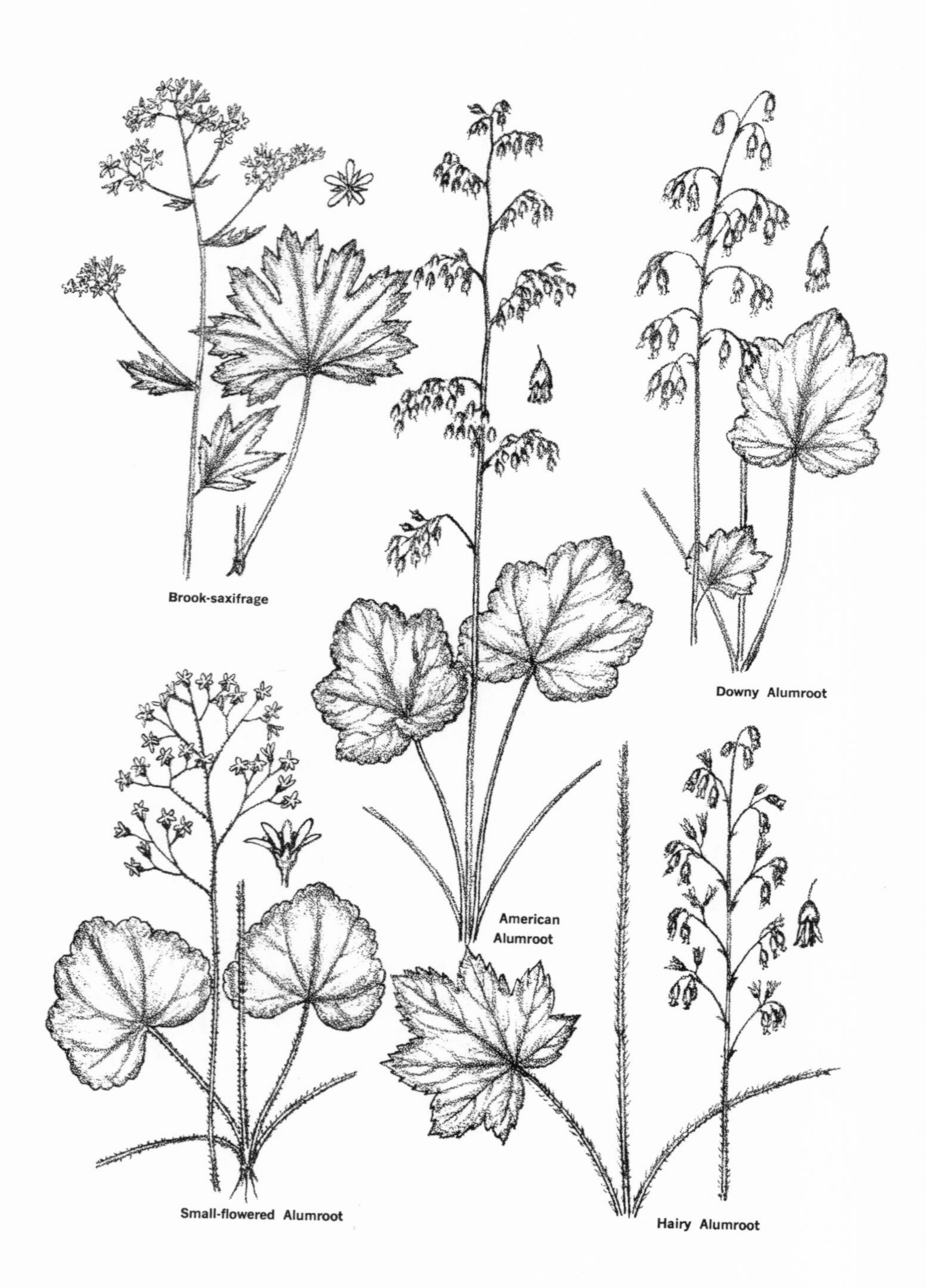
Brook-saxifrage
Downy Alumroot
American
Alumroot
Small-flowered Alumroot
Hairy Alumroot

BROOK-SAXIFRAGE *(Boykinia aconitifolia)*

The Brook-saxifrage is a rather stout-stemmed, erect plant from 1 to 2 feet tall. Most of its leaves are basal and they, as well as the ones on the lower part of the stem, are 5- to 7-lobed, cut-toothed, and long-stalked. Both the stem and the leaf stalks are glandular-hairy. The small flowers clustered at the upper part of the stem have 5 white petals and 5 stamens. It grows in rich woods, on wet rocks, and along the banks of streams, chiefly in the mountains. The flowering season is during June and July. Another name often given it is Aconite-saxifrage, due to the resemblance of its leaves to those of the monks-hoods.

RANGE: Va., W.Va., and Ky. south to Ga. and Ala.

AMERICAN ALUMROOT *(Heuchera americana)*

This is our most common and widespread species of alumroot, being found quite generally in dry woods and on rocky slopes. It has a smooth or sparingly hairy, rather stout stem from 1 to 3 feet tall. The leaves are chiefly basal. They are roundish heart-shaped with from 7 to 9 rounded and bluntly toothed lobes, smooth or nearly so, 3 to 4 inches wide; and have long, slender stalks. The small purplish or reddish-tinged flowers have a cup-like calyx. 5 minute petals, and 5 stamens which extend well beyond the calyx cup. They are very numerous and arranged in a long, loose, several-branched cluster which overtops the leaves. The flowering season extends from April to June. It is often called the Rock-geranium.

RANGE: Conn. to Ont. and Mich. south to Ga., Ala., and Okla.

DOWNY ALUMROOT *(Heuchera pubescens)*

The Downy Alumroot has a rather stout stem 1 to 3 feet tall which usually has a few leaves and is glandular-hairy at least near the summit. Most of the leaves are basal. They are 2 to 4 inches wide, roundish heart-shaped, and the 5 to 7 lobes have abruptly pointed teeth on their margins. The flowers are greenish or purplish-tinged and have a bell-shaped calyx about ¼ inch long which is glandular-downy. The purplish petals are a trifle longer than the calyx cup. This species grows in rich woods and in crevices among rocks in our mountains, blooming between May and July.

RANGE: Pa. and Md. south to Ky. and w. N.C.

SMALL-FLOWERED ALUMROOT *(Heuchera parviflora)*

Both this and the following species have small flowers in which the 5 petals are 2 or 3 times as long as the bell-shaped calyx cup. In this one the petals are usually white. It usually produces 2 or more slender, hairy flowering stems from 6 to 18 inches tall. The leaves are quite thin, kidney-shaped, broader than long, 2 to 5 inches wide, and have very low rounded lobes. They are commonly whitish-downy beneath and also have stalks which are quite hairy. It grows on moist rocks and shaded cliffs, blooming between July and September.

RANGE: W.Va. to Ill. south to N.C., Ala., and Mo.

HAIRY ALUMROOT *(Heuchera villosa)*

The Hairy Alumroot has rusty-hairy flowering stems 1 to 2½ feet tall, and often rusty-hairy leaf stalks. The small flowers have white or pink petals which are 2 to 3 times as long as the bell-shaped calyx. Its leaves are chiefly basal, long-stalked, 2 to 4½ inches wide more or less hairy, and have 7 to 9 triangular and sharply toothed lobes. It grows on damp rocks and rich, rocky wooded slopes; blooming between June and August.

RANGE: W.Va. and s. Ohio south to Ga., Ala., and Mo.

FOAMFLOWER *(Tiarella cordifolia)*

The Foamflower, or False Miterwort, is a low plant which spreads by means of slender runners and forms little colonies. It usually has a leafless flower stalk 6 to 12 inches tall, with a narrow cluster of small white flowers along the upper portion. Occasionally the flower stalks do have a pair of small leaves, in which case one will have to look closer at the flowers to distinguish it from the true miterwort. The flowers have 5 small petals which taper into stalked bases, and 10 very conspicuous long stamens. The leaves are roundish to egg-shaped, heart-shaped at the base, and have from 3 to 7 rather shallow but sharply toothed lobes. They are 2 to 4 inches broad and have long stalks. The entire plant is more or less hairy. It grows in cool, moist, rich and often rocky woodlands; blooming between April and June.

RANGE: N.B. to Ont. and Mich. south to N.Eng., w. N.C. and e. Tenn.

WHERRY'S FOAMFLOWER *(Tiarella cordifolia* var. *collina)*

This variety is quite similar to the preceding one but the plants are not stoloniferous. It also has more narrow flower clusters and leaf blades which are mostly longer than broad. Of more southern distribution than the preceding, it grows in similar situations from Va. and Tenn. south to Ga. and Miss.; blooming in April or May. (Not illustrated)

TWO-LEAF MITERWORT *(Mitella diphylla)*

The Two-leaf Miterwort has a flower stalk 10 to 18 inches tall with a pair of quite stalkless, heart-shaped, 3-lobed, and toothed leaves about the middle. Above them is a slender cluster of small, short-stalked, starry flowers in which the 5 white petals are delicately fringed. The 10 stamens are short and in the mouth of the cup-like calyx. The basal leaves are broadly egg-shaped, heart-shaped at the base, 3- to 5-lobed, and sharply toothed. The entire plant is minutely hairy. It grows in rich and often rocky woodlands, blooming in April or May. Another name is Bishop's-cap.

RANGE: N.H., Que., and Minn. south to w. S.C., Tenn., Miss., and Mo.

NAKED MITERWORT *(Mitella nuda)*

This is a little plant of cold northern woods and bogs, spreading by means of runners. The slender flower stalks are 3 to 7 inches tall, either leafless or with a single small leaf. The basal leaves are roundish kidney-shaped, deeply and doubly blunt-toothed but not lobed, and 1 to 1½ inches wide. It has greenish-yellow flowers; blooming between May and August.

RANGE: Lab. to Mackenz., south to New Eng., Pa., N.D., and Mont. (Not illustrated)

EASTERN GRASS-OF-PARNASSUS *(Parnassia glauca)*

This is a smooth plant with a basal cluster of roundish leaves which are often slightly heart-shaped at the base, 1 to 2½ inches long, and long-stalked. Like other species of *Parnassia* the solitary flower is on a long stalk on which there is a stalkless leaf between the middle and the base. The flower is 1 to 1½ inches across and has 5 sepals, and 5 broadly oval white petals with greenish veins. There are 5 anther-bearing stamens which alternate with the petals; and 5 sterile stamens (staminodia) with 3 prongs usually shorter than the stamens, and placed at the bases of the petals. It grows in wet meadows and thickets, blooming between July and October.

RANGE: Nfd. to Man. south to Pa., the Great Lakes region, Iowa, and S.D.

LARGE-LEAF GRASS-OF-PARNASSUS *(Parnassia grandifolia)*

This is similar to the preceding species but coarser, with leaves 2 to 4 inches long. The flowers are 1 to 2 inches across; the staminodia having 3 to 5 prongs which are very slender and much longer than the stamens. It grows on moist soil or on wet rocks; blooming between August and October.

RANGE: Va., W. Va., Tenn. and Mo. south to Fla. and Tex.

KIDNEY-LEAF GRASS-OF-PARNASSUS *(Parnassia asarifolia)*

The petals of the flowers in this species are abruptly contracted at the base into a distinct stalk or claw; and the staminodia are 3-pronged and shorter than the stamens. The inch-broad flower is at the summit of a stalk 10 to 20 inches tall; and the stem leaf is nearly round, almost surrounding the stalk. The basal leaves are kidney-shaped, 2 to 4 inches broad and long-stalked. It grows in wet woods, bogs, and on rocky banks in the mountains; blooming between August and October.

RANGE: Western Va. and W.Va. south to n. Ga. and n. Ala.

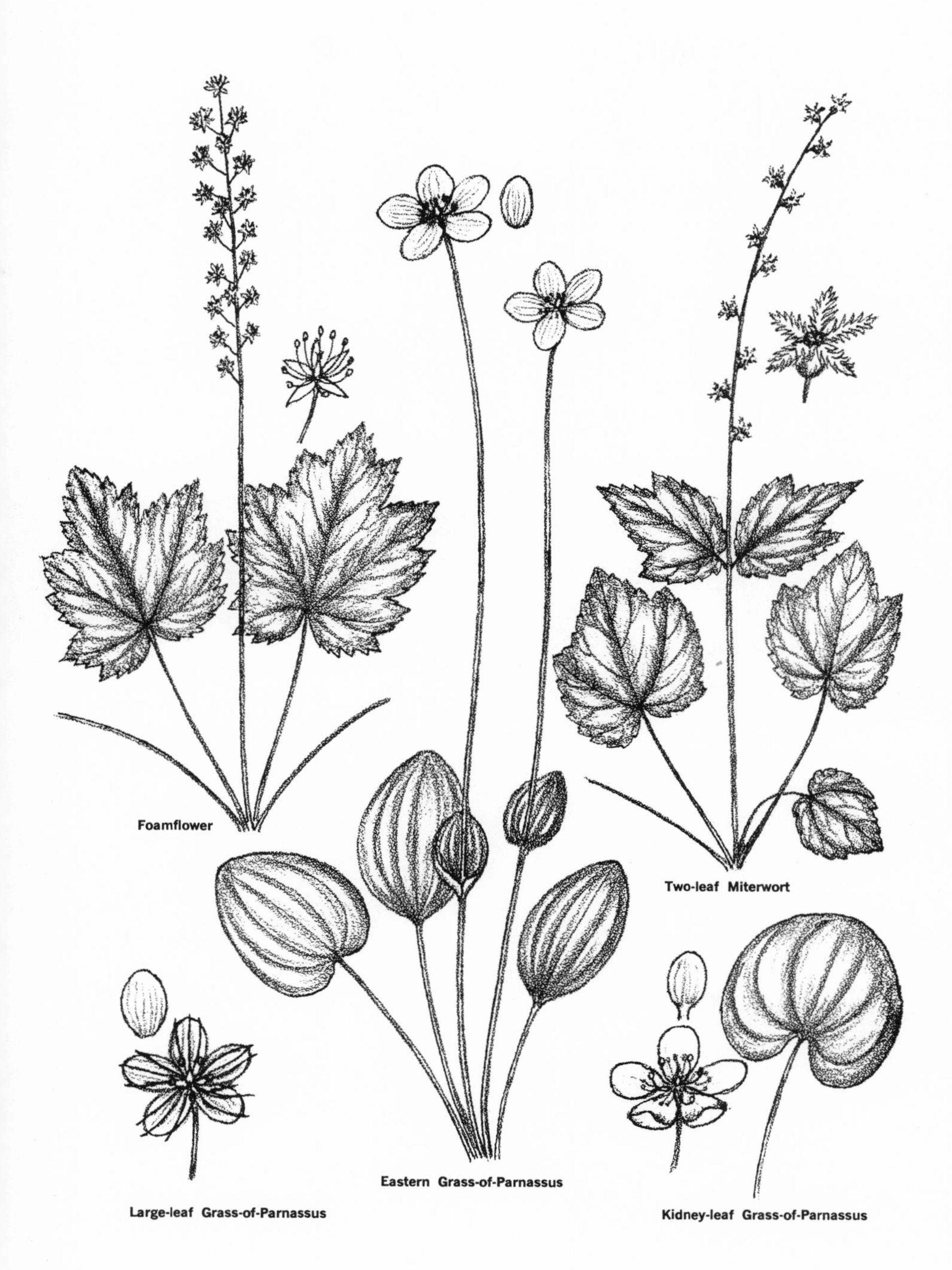
Foamflower
Two-leaf Miterwort
Eastern Grass-of-Parnassus
Large-leaf Grass-of-Parnassus
Kidney-leaf Grass-of-Parnassus

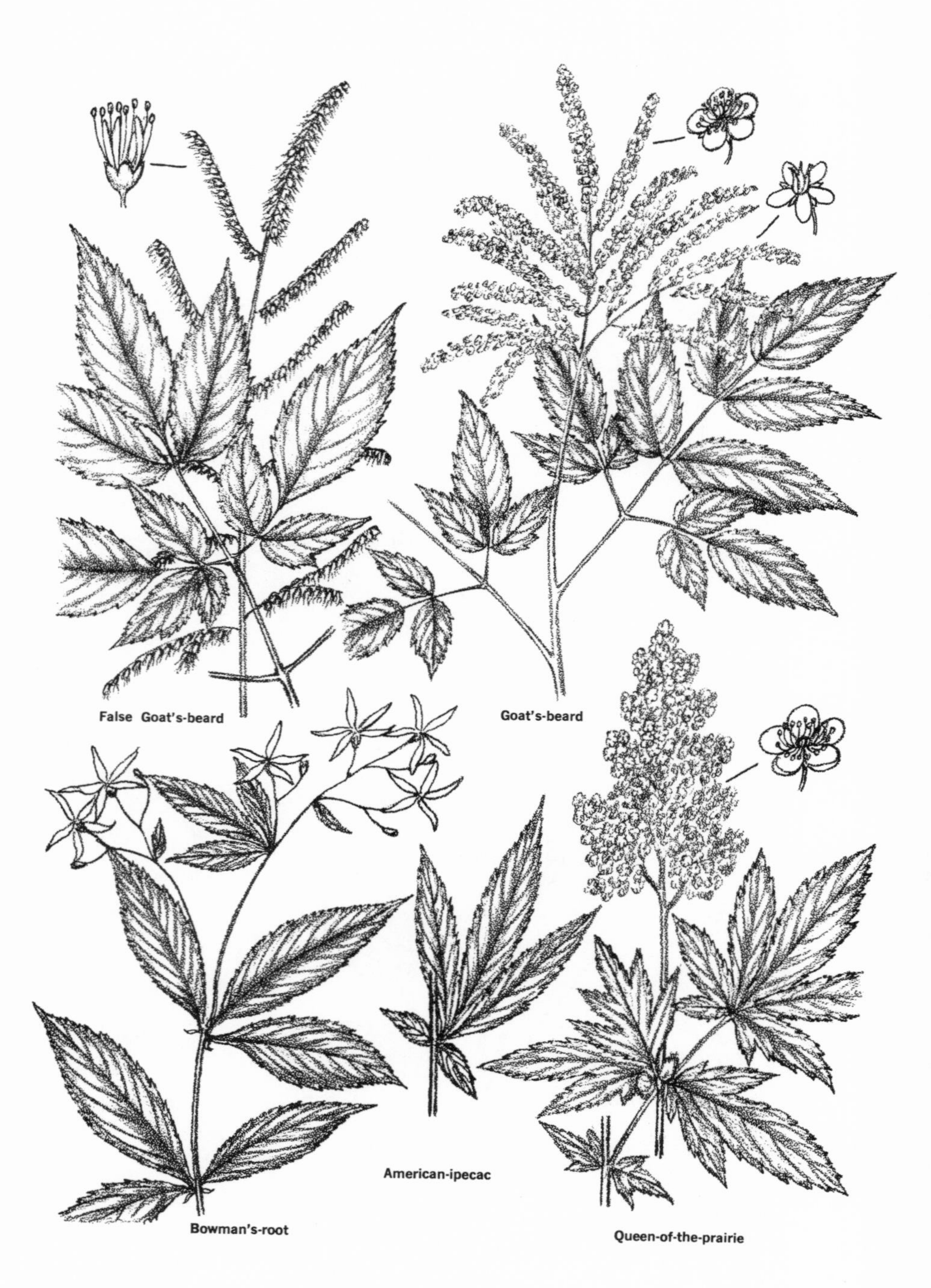
False Goat's-beard
Goat's-beard
Bowman's-root
American-ipecac
Queen-of-the-prairie

FALSE GOAT'S-BEARD *(Astilbe biternata)*

This is a large, somewhat hairy plant from 3 to 6 feet tall; with large leaves, sometimes 2 feet across, which are divided ternately 2 or 3 times into a large number of leaflets. These leaflets are 2 to 5 inches long, more or less egg-shaped or often heart-shaped, with sharply toothed and sometimes lobed margins. The small flowers are yellowish-white, with 4 or 5 narrow petals, from 8 to 10 stamens which have long filaments, and 2 pistils. Although the individual flowers are small, they are crowded together on the numerous branches of the inflorescence; forming a showy, pyramid-like cluster often a foot in length. It grows in mountain woodlands, blooming between May and July.

RANGE: Western Va. and W.Va. south to Tenn. and n. Ga.

ROSE FAMILY (ROSACEAE)

This family to which the roses, the blackberries and raspberries, and most of our common fruit trees belong is a large and varied one. In general, its members have flowers with a radial symmetry with 5 sepals united at the base and borne on a disk-like expansion of the receptacle, usually 5 free petals, numerous stamens, and 1 or more pistils. The leaves are usually alternate, often compound, and nearly always have a pair of prominent stipules.

GOAT'S-BEARD *(Aruncus dioicus)*

This plant quite closely resembles the preceding species in size and in leaf characteristics. Its flowers are likewise arranged in similar large, showy, plume-like clusters. Each flower is less than ¼ inch across and has 5 oval-shaped petals. Those of some plants have about 15 stamens with long, slender filaments. On other plants they have a group of 3 pistils but only tiny rudiments of stamens. When in bloom between May and July, this plant is very conspicuous in rich woods, and ravines.

RANGE: Pa. to Iowa south to Ga., Ala., and Okla.

BOWMAN'S-ROOT *(Porteranthus trifoliatus)*

The Bowman's-root, or Indian-physic, is often conspicuous in rich, open woods and thickets when it is in flower between May and July. It grows from 2 to 3 feet high and its slender and often somewhat twisted branches terminate in loose clusters of showy flowers. They are about an inch across, with a reddish calyx cup and 5 strap-shaped white or pale pinkish petals. The leaves are stalkless and divided into 3 sharply toothed leaflets 2 to 3 inches in length. There is a pair of small stipules at the base of each leaf.

RANGE: N.Y. to Ont. and Mich. south to Ga. and Ala.

AMERICAN-IPECAC *(Porteranthus stipulatus)*

This plant closely resembles the Bowman's-root but its leaves are short-stalked, its leaflets somewhat more deeply cut or toothed, and there is a pair of large leaf-like stipules at the base of each leaf. The latter are so large that the leaves appear to have 5 rather than just 3 leaflets. It grows in woods, thickets, and on rocky slopes; blooming between May and July.

RANGE: N.Y. to Ill., Mo. and Kan. south to Ga. and Tex.

QUEEN-OF-THE-PRAIRIE *(Filipendula rubra)*

Moist thickets, meadows, and prairies to the west of the Appalachians seem to have been the original home of this plant; but it has escaped from gardens eastward. It is a smooth plant 2 to 6 feet tall with an angled or grooved, leafy stem. Its small but numerous flowers are grouped in a dense, showy, end cluster and bloom between June and August. Each flower is about ¼ inch across and has 5 pink petals and a large cluster of stamens. The leaves are deeply cut or divided and sharply toothed. Those along the upper portion of the stem have but 3 divisions. The larger ones toward the base have from 5 to 7 divisions or leaflets; the end one being much the larger, 4 to 8 inches broad, and cleft into 7 to 9 radiating lobes.

RANGE: Pa. to Mich. and Iowa south to Ga. and Ky. Escaped from cultivation eastward into N.Y., N. Eng., and N.S.

WILD STRAWBERRY *(Fragaria virginiana)*

Like our cultivated strawberries, the Wild Strawberry is a low plant with a very short stem. From it grow the leaves, the flower clusters, and the slender runners on which new plants are produced. The long-stalked leaves are divided into 3, elliptic or oval shaped, sharply toothed leaflets which are usually 2 to 4 inches long. The flowers are about ¾ inch across and have 5 oval-shaped white petals, a large number of stamens with yellow anthers, and numerous pistils on a cone-shaped receptacle in the center of the flower. This receptacle develops into an egg-shaped red "berry" with the small fruits (achenes) imbedded in pits on its surface. This native species grows in open woods, fields, and on grassy slopes; blooming between April and June.

RANGE: Nfd. to Alb. south to Ga., La., and Okla.

AMERICAN WOOD STRAWBERRY *(Fragaria vesca* var. *americana)*

This is a more slender, paler green, and less hairy plant than the preceding species. It has thinner leaflets and more narrowly egg-shaped or cone-shaped "berries", with the achenes or "seeds" on its surface rather than in pits. This wild strawberry is usually found on rocky, wooded slopes; blooming in May or June. It is sometimes called the Sow-teat Strawberry.

RANGE: Nfd. to Alb. south to N.Eng., w. Va., w. N.C., Ill., Mo., Neb., and N.Mex.

BARREN-STRAWBERRY *(Waldsteinia fragarioides)*

The Barren- or Dry-strawberry is a low, smooth or slightly hairy plant which also has creeping rootstocks. It resembles a strawberry plant and is often mistaken for one. The long-stalked leaves are divided into 3 broadly wedge-shaped and cut-toothed leaflets from 1 to 2 inches long. The flowers are about ½ inch across and have 5 bright yellow petals, a number of stamens, and 2 to 6 pistils which are within a cup-shaped calyx with 5 lobes on its rim. In the variety *parviflora*, the petals are shorter or barely as long as the calyx lobes. It grows in woods, thickets, and clearings; blooming between April and June.

RANGE: N.B. to Ont. and Minn. south to n.Ga., Tenn., Ind., and Mo.

BARREN-STRAWBERRY *(Waldsteinia lobata)*

This plant resembles the short-petalled variety of the preceding species, but its leaves are merely 3-lobed instead of being divided into 3-leaflets. It grows on river banks from the coast inland from Ga. north to N.C. (Not illustrated)

THREE-TOOTHED CINQUEFOIL *(Sibbaldiopsis tridentata)*

The Three-toothed Cinquefoil is a low, tufted, woody-based, evergreen plant which has extensively creeping rootstocks. Its leaves are divided into 3 wedge-shaped leaflets, ½ to about 2 inches long, which are 3-toothed at the broad summit; hence its name. The flowers are about ⅓ inch across and have 5 white petals and numerous stamens. They are in few-flowered clusters atop leafy or leafy-bracted stems a few inches high. It grows in dry, open, rocky places; in the south at high elevations in the mountains. The flowering season is between June and September.

RANGE: Lab. to Machenz. south to N.Eng., Ont., N.D. and in mts. to n. Ga.

COMMON CINQUEFOIL *(Potentilla canadensis)*

A large number of cinquefoils occur in eastern North America. Many are native plants and a few have been introduced from Europe. Some are erect plants with leafy stems and they have leaves which are variously divided. None are more common or better known than this creeping plant of dry open places. Its leaves are divided into 5 radiating and wedge-shaped leaflets which are sharply toothed on the margin, at least above the middle. The flowers are about ⅓ inch across and have 5 broadly top-shaped, bright yellow petals and a number of stamens. They are borne singly on slender stalks which arise from the axils of the leaves. Another name for the plant is Five-finger. It blooms between March and June.

RANGE: N.S. to Ont. south to Ga., Tenn., and Mo.

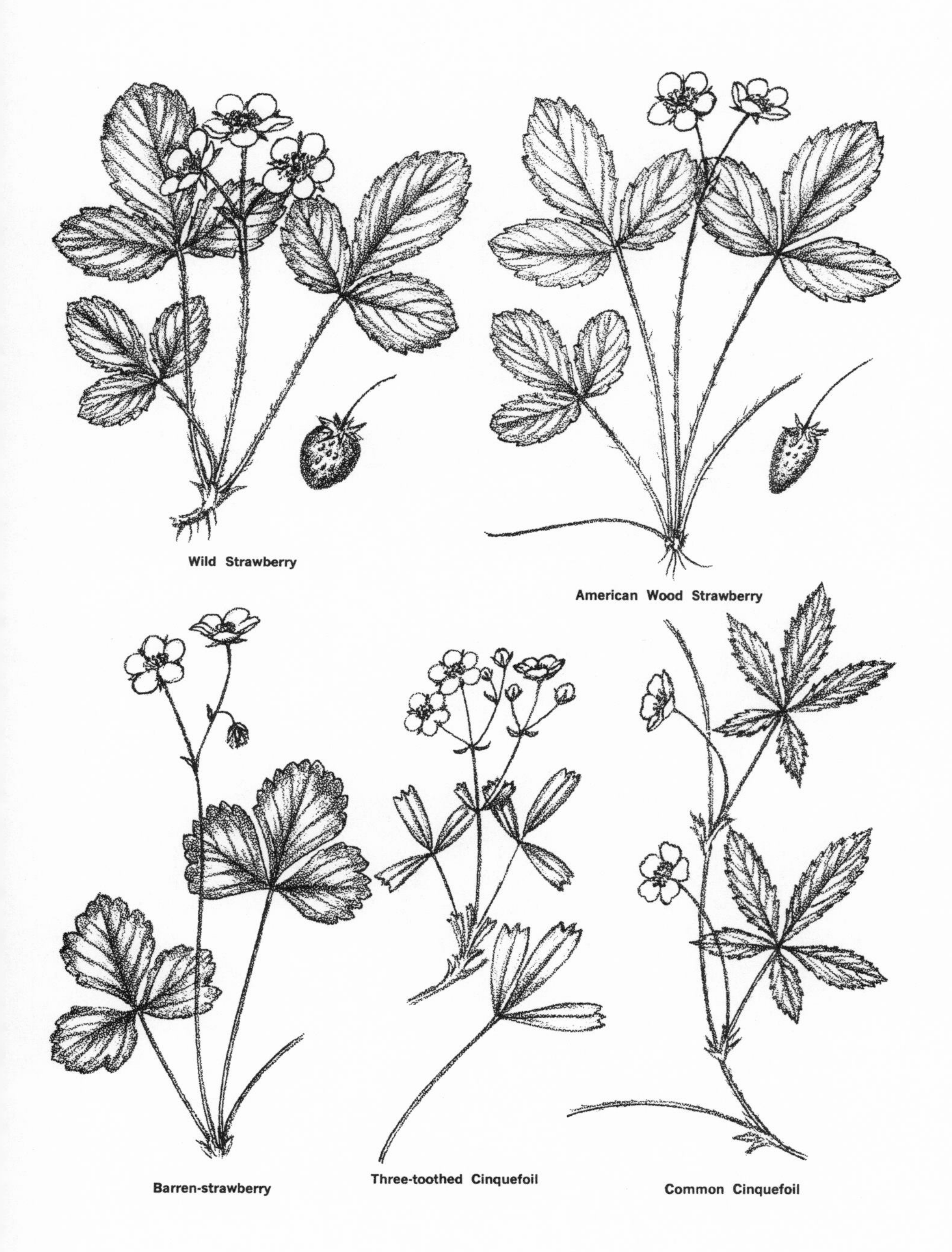
Wild Strawberry
American Wood Strawberry
Barren-strawberry
Three-toothed Cinquefoil
Common Cinquefoil

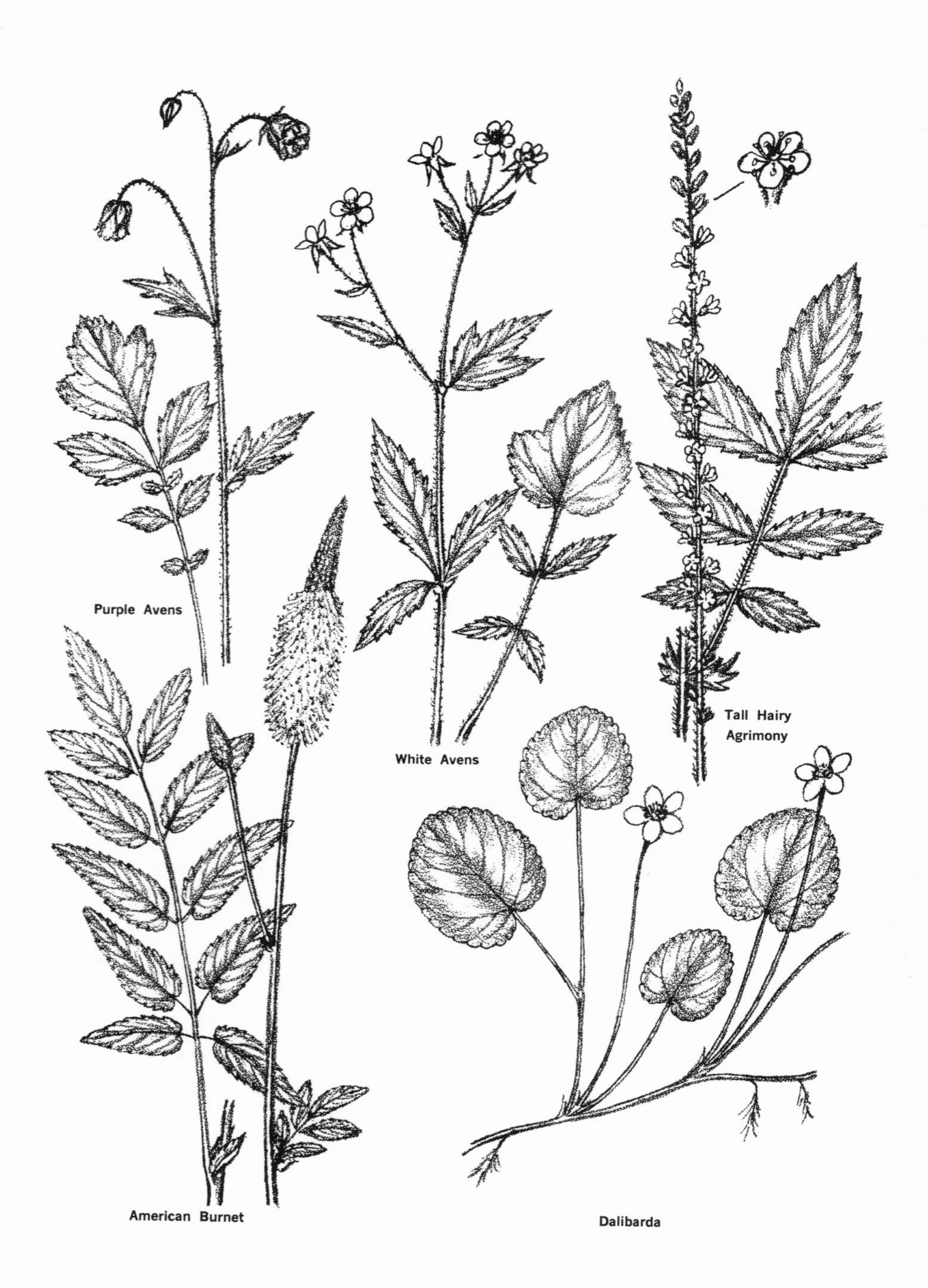
Purple Avens
White Avens
Tall Hairy
Agrimony
American Burnet
Dalibarda

PURPLE AVENS *(Geum rivale)*

The Purple or Water Avens is at home in wet meadows, swamps, and bogs; and there it blooms between May and August. The plant has an erect, downy, and nearly simple stem from 1 to 3 feet tall; along which there are a few 3-parted and toothed leaves. The leaves at the base are much larger, with a large segment at the end and several smaller ones along the sides. Its flowers are about an inch wide and nod at the tips of slender stems. They have a bell-shaped calyx, 5 erect petals, and are usually purple or purplish in color.

RANGE: Lab. to B.C. south to N.J., W.Va., Minn. and N.Mex.

WHITE AVENS *(Geum canadense)*

The White Avens is usually common in open woods and thickets; blooming between June and August. It has an erect, downy or smoothish, branching stem from 1½ to 2½ feet tall; with mostly 3-parted, sharply toothed leaves. The basal leaves have a large terminal leaflet with usually 2 pairs of smaller ones below it, all of which are sharply toothed and sometimes lobed. The flowers are about ½ inch broad and have 5 rather small white petals.

RANGE: N.B. to Minn. south to Ga. and Tex.

ROUGH AVENS *(Geum virginianum)*

This plant closely resembles the preceding species but its stem and leaf stalks are bristly-haired. Its flowers are also somewhat smaller and the petals are creamy-white to greenish-yellow. It grows in dry woods and thickets and on rocky banks from Mass. to Ind. south to S.C. and Tenn; blooming between June and August. (Not illustrated)

TALL HAIRY AGRIMONY *(Agrimonia gryposepala)*

The agrimonies, of which there are a number of species in eastern North America, have a top-shaped hypanthium or calyx-tube with many hooked bristles at the summit. They also have leaves which are pinnately divided, often with small segments between the larger one. This species has a leafy stem usually 2 to 4 feet tall which is minutely glandular and has scattered spreading hairs. The small flowers have 5 yellow petals and between 5 and 15 stamens. They are arranged in a long and very slender cluster and are produced between June and August. Its leaves usually have 7 of the larger leaflets which are coarsely and sharply toothed. It grows in the borders of woods and thickets.

RANGE: N.S. to Que. and N.D. south to S.C., Tenn., and Kan.

AMERICAN BURNET *(Sanguisorba canadensis)*

This is a decidedly attractive plant with a smooth, erect, simple or branching stem from 1 to 5 feet tall. The small white flowers are arranged in dense, showy, terminal clusters from 2 to about 6 inches long, blooming from the base upward. Each flower has 4 whitish sepals and 4 stamens which have very long white filaments. The leaves are divided into from 7 to 15 oval or egg-shaped leaflets with sharply toothed margins. It grows in marshes and in wet meadows or thickets, blooming between May and September. Another name for it is Great Burnet.

RANGE: Lab. to Mich. south to N.J., w. N.C., and Ill.

DALIBARDA *(Dalibarda repens)*

The Dalibarda is a dainty little plant which grows in moist woods and bogs. It has creeping and densely tufted stems or rootstocks from which arise slender stalks bearing a solitary flower. This flower is about ½ inch across and has 5 white petals and numerous stamens, but it usually produces no fruit or seeds. There are other flowers which look like mere buds on short, curved stalks and they produce the little, dry, seed-like fruits. The leaves are roundish heart-shaped, bluntly toothed on the margin, and are somewhat downy on both surfaces. From 1 to 2 inches wide and long-stalked, they might easily be mistaken for the leaves of a violet. Some other names for it are False-violet, Dewdrop, and Robin-run-away. It blooms between June and August.

RANGE: N.S. and Que. to Ont. south to Conn. w. N.C., and the Great Lakes region.

PEA FAMILY (Fabaceae)

This is a very large plant family which includes a great many important food, forage, and ornamental plants. They are popularly referred to as "legumes", a legume technically being the peculiar bilaterally symmetric pod which is characteristic of the members of the family. This pod is developed from a simple pistil, and at maturity it usually splits into two parts sometimes with considerable force. The flowers vary from a regular or nearly regular form to the decidedly bilateral or papilionaceous (meaning butterfly-like) flowers seen in the typical peas. Botanists sometimes divide the family into 3 separate families, or into the following subfamilies.

Mimosoideae. Members of this subfamily have small flowers with a radial symmetry which are arranged in dense and often ball-like heads. The petals are often united into a 4- or 5-lobed cup with the stamens protruding far beyond it. This subfamily is sometimes called the Mimosa Family.

Caesalpinioideae. Members of this subfamily have flowers with an almost radial symmetry or ones which are somewhat bilateral. They have 5 sepals united at the base, 5 petals, and usually 10 separate stamens. This subfamily is sometimes called the Cassia Family.

Papilionoideae. Members of this subfamily have papilionaceous flowers in which the upper petal forms a broad banner or *standard;* the 2 side ones are more or less parallel and form the *wings;* and the 2 lower ones are united into a *keel,* which encloses the stamens and the pistil. The 5 sepals are more or less united and the 10 stamens are united by their filaments into 1 or 2 groups. This subfamily includes most of the family members.

KEY TO GENERA OF THE PEA FAMILY

1. Plants with undivided (simple) leaves.
    2. Flowers blue or lilac-purple. Leaves mostly over 2 inches long. — LUPINES *(Lupinus)*
    2. Flowers yellow. Leaves usually smaller.
3. Leaves stalkless or nearly so. — RATTLEBOXES *(Crotalaria)*
3. Leaves long-stalked. — DOLLARWEED *(Rhynchosia)*
    1. Plants with divided (compound) leaves.
4. Leaflets 3.
    5. Plants with twining and climbing stems.
6. Keel of the flowers spirally twisted or strongly incurved.
    7. Keel spirally twisted. — WILD BEAN *(Phaseolus)*
    7. Keel merely strongly incurved. — SAND-BEANS *(Strophostyles)*
6. Keel of the flowers otherwise.
    8. Flowers large, the standard often an inch or more broad.
9. Standard with a spur on back near the base. — SPURRED BUTTERFLY-PEA *(Centrosema)*
9. Standard very large, notched at the tip but not spurred. — BUTTERFLY-PEA *(Clitoria)*
    8. Flowers smaller, the standard ½ inch or less broad.
10. Plant brownish-hairy. — HOG-PEANUT *(Amphicarpa)*
10. Plant at most minutely hairy. — MILK-PEAS *(Galactia)*
    5. Plants with erect or reclining stems.
11. Margins of the leaflets with minute teeth.
    12. End leaflet stalked. Flowers in long and narrow clusters. Plant very fragrant when crushed. — SWEET-CLOVERS *(Melilotus)*

12. End leaflet not stalked. Flowers in ball-shaped, egg-shaped, or cylindrical heads. CLOVERS *(Trifolium)*

11. Margins of the leaflets untoothed.

13. Filaments of the stamens all separate and distinct.

14. Pods inflated or plump. FALSE INDIGOS *(Baptisia)*

14. Pods flattened. THERMOPSIS *(Thermopsis)*

13. Stamens united by their filaments into 1 or 2 groups.

15. Flowers 1½ to 2 inches long, the corolla slender and bright red. CORAL-BEAN *(Erythrina)*

15. Flowers otherwise; mostly quite pea-like.

16. Leaflets with small stipules on their stalks.

17. Pods jointed and separating into flat 1-seeded segments which adhere to clothing, etc. TICK-TREFOILS *(Desmodium)*

17. Pods merely flattened and pea-like. MILK-PEAS *(Galactia)*

16. Leaflets without small stipules.

18. Flowers yellow, solitary or in few-flowered clusters; their calyx tube stalk-like. PENCIL-FLOWERS *(Stylosanthes)*

18. Flowers whitish to purplish, in many-flowered clusters.

19. Leaves and pods with small dark or translucent dots. SCURF-PEAS *(Orbexilum)*

19. Leaves and pods not dotted. BUSH-CLOVERS *(Lespedeza)*

4. Leaflets more numerous than 3.

20. Leaves palmately divided, with radiating leaflets.

21. Leaflets 4. Flowers yellow. BRACTED ZORNIA *(Zornia)*

21. Leaflets 7 to 11. Flowers blue. LUPINES *(Lupinus)*

20. Leaves pinnately divided into numerous leaflets.

22. Leaves ending in a tendril.

23. Wing petals adhering to the keel. Style bearded near the summit. Leaves with rather small stipules. VETCHES *(Vicia)*

23. Wing petals not adhering to the keel. Style bearded along its inner side. Leaves with large stipules. VETCHLINGS *(Lathyrus)*

22. Leaves not ending in a tendril.

24. Flowers not at all pea-like or papilionaceous.

25. Flowers pinkish and in dense ball-shaped heads. Plants prostrate and prickly. SENSITIVE-BRIERS *(Schrankia)*

25. Flowers yellow, the corolla often with a nearly radial symmetry. WILD SENNAS *(Senna)*

24. Flowers pea-like or papilionaceous.

26. Plants twining and climbing.

27. Leaflets 5. Flowers brownish-purple, the keel coiled. GROUND-NUT *(Apios)*

27. Leaflets 9 to 15. Flowers lilac-colored. Plant woody. WISTERIAS *(Wisteria)*

26. Plants erect or prostrate.

28. Plant prostrate. Flowers white, turning red. GOAT'S-RUE *(Tephrosia)*

28. Plants erect.

29. Flowers creamy-white and purple. GOAT'S-RUE *(Tephrosia)*

29. Flowers yellow. Plants quite tall.

30. Standard petal red on the back. Pods very slender and 5 or more inches long. SESBAN *(Sesbania)*

30. Standard petal yellow on both sides. Pods plump and bladder-like, 2 to 3 inches long. BAGPOD *(Glottidium)*

CAT-CLAW SENSITIVE-BRIER *(Schrankia nuttallii)*

Sensitive-briers have long, slender, trailing or sprawling stems often 2 to 4 feet in length. They, as well as the leaf stalks and those of the flower clusters—and even the fruit pods—are well-armed with small and hooked prickles. The small pink to rose-purple flowers are in puffy ball-shaped heads which are borne on long stalks arising from the leaf axils. The sensitive leaves are doubly compound, each of the primary leaf divisions being divided again into from 8 to 16 small leaflets. In this species the leaflets show conspicuous veins on the lower surface. It has a densely prickly and narrow pod 2 to 3½ inches long which is short-pointed at the tip. Quite common in dry fields, open woods, and pinelands; it blooms between June and September.

RANGE: Ill., Neb. south to Ala. and Tex.

SMALL-LEAF SENSITIVE-BRIER *(Schrankia microphylla)*

This eastern species is very similar to the preceding one but it has somewhat smaller and narrower leaflets without obvious veins on the lower surface. The slender and prickly pods are 3 to 6 inches long and have long-pointed tips. It grows in dry sandy open woods, thickets, or fields; flowering between June and September.

RANGE: Va. and Ky. south to Fla. and Tex.

WILD SENNA *(Senna marilandica)*

The wild, or American, Senna has an erect, simple or sparingly branched stem 3 to 5 feet tall; with clusters of numerous flowers arising from the axils of the upper leaves. Its flowers are about ¾ inch across and have 5 yellow petals, 7 anther-bearing stamens, and 3 sterile ones. The leaves are pinnately divided into from 6 to 10 pairs of lance-shaped to narrowly egg-shaped leaflets from 1 to 2 inches long. There is a conspicuous cone-shaped gland near the base of the leaf stalk. It grows in dry thickets and along roadsides, blooming during July and August.

RANGE: Pa. to Iowa and Kan. south to Fla. and Tex.

MEDSGER'S WILD SENNA *(Senna hebecarpa)*

This wild senna is similar to the preceding species but the gland on its leaf stalks is stalked and club-shaped. The segments of its pods are about as long as broad, while those of the preceding species are much shorter than they are broad. It grows in moist open woods or in stream bottoms from N.Eng. to Wis. and south to w. N.C. and Tenn.; blooming during July and August. (Not illustrated)

LOW SENNA *(Senna obtusifolia)*

This species is also called Sicklepod on account of its slender, 4-sided, and curved pods which are 4 to 6 inches long. It is a smooth plant 1½ to 2 feet high; its leaves having from 2 to 3 pairs of top-shaped leaflets. There is a long gland on the stalk between or just above the lowest pair of leaflets. It grows in rich soil, often along streams, from Pa. to Mich. and Kan. south to Fla. and Tex.; blooming between July and September. (Not illustrated)

PARTRIDGE-PEA *(Chamaecrista fasciculata)*

Also known as the Golden Cassia or Large-flowered Sensitive-plant, this is a branching plant 1 to 2 feet high. Its leaves are pinnately divided into 10 to 15 pairs of small leaflets which fold together when the leaf is touched. The flowers are 1 to 1½ inches across and have 5 bright yellow petals often with reddish-purple spots at the base. Four of the 10 stamens have yellow anthers and 6 have purple ones. It grows in open thickets, meadows, and along roadsides; blooming between July and September.

RANGE: Mass. to Ont. and Wis. south to Fla. and Tex.

WILD SENSITIVE-PLANT *(Chamaecrista nicitans)*

The Wild Sensitive-plant resembles the preceding species but it grows only 6 to 15 inches high; its flowers are short-stalked and less than ½ inch across, and they have but 5 stamens. Its sensitive leaves have from 9 to about 20 pairs of narrow leaflets. This species grows on dry and sandy soils and also blooms from July to September.

RANGE: Mass. to Ill. and Kan. south to Ga. and Tex.

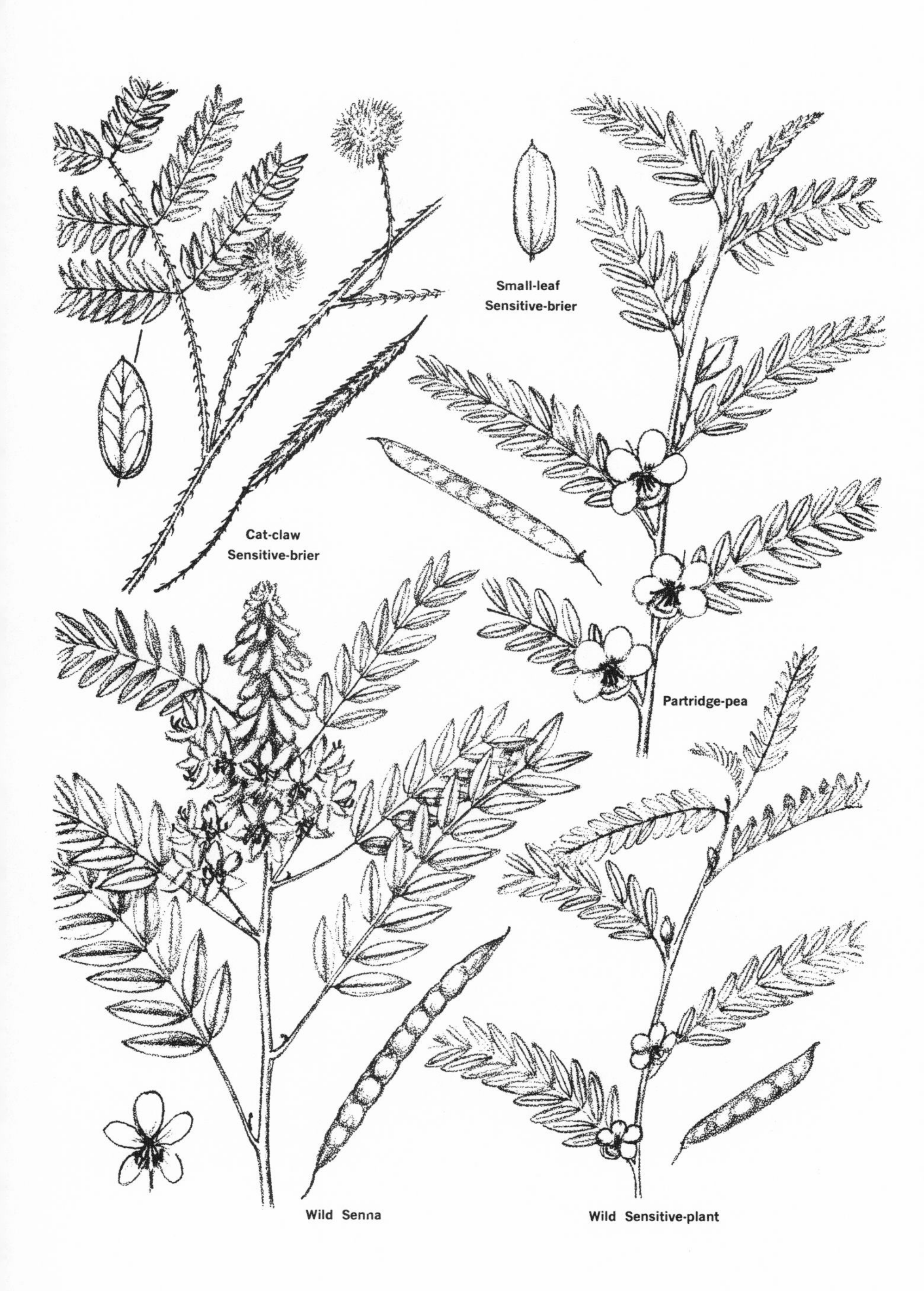
Small-leaf
Sensitive-brier
Cat-claw
Sensitive-brier
Partridge-pea
Wild Senna
Wild Sensitive-plant

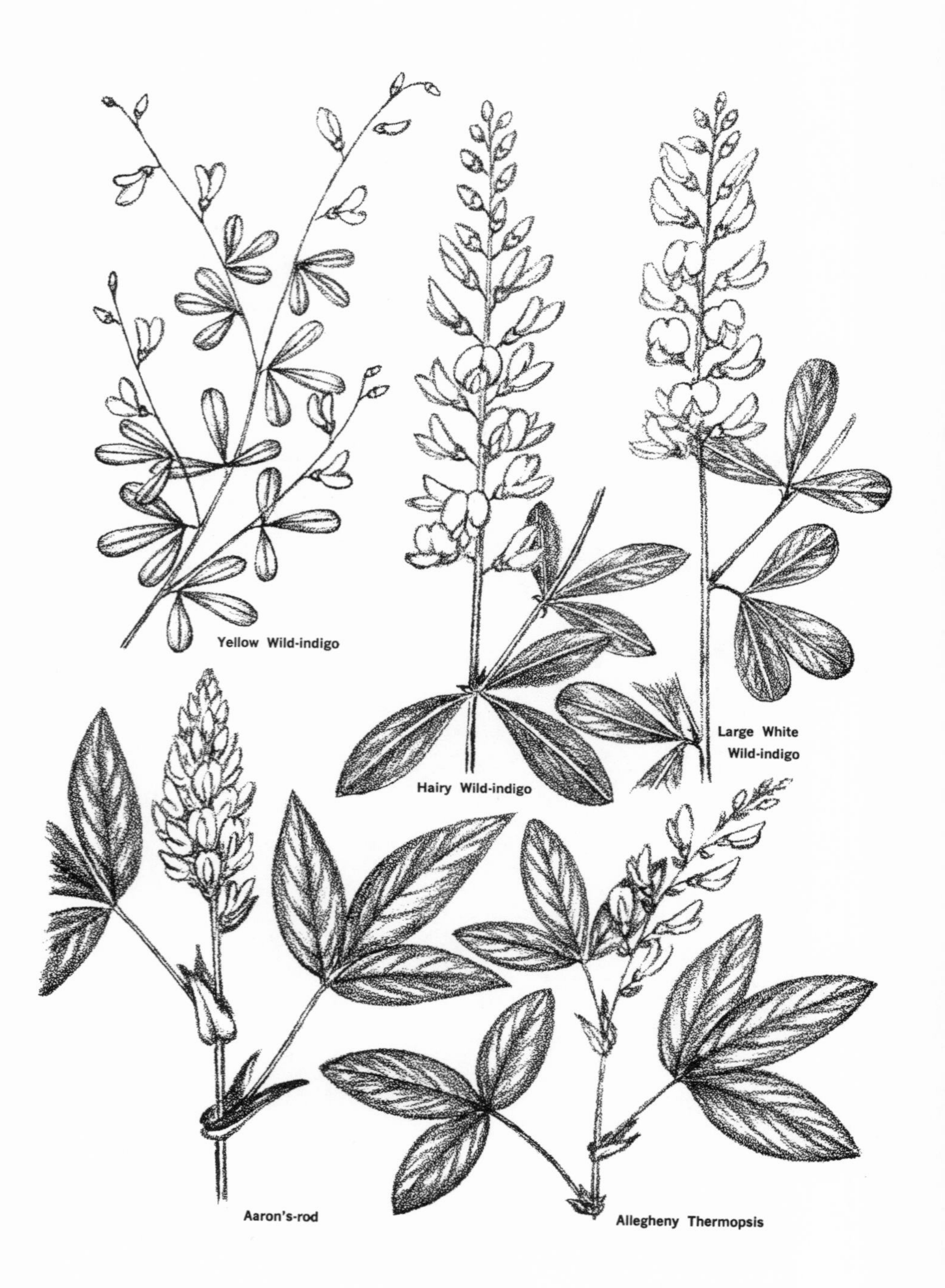
Yellow Wild-indigo
Hairy Wild-indigo
Large White
Wild-indigo
Aaron's-rod
Allegheny Thermopsis

YELLOW WILD-INDIGO *(Baptisia tinctoria)*

Some of the wild indigos, including this one, were once used as a substitute for the true indigo in dyeing cloth. This species is a smooth, slender, bushy-branched plant 2 to 3 feet high. Its yellow flowers are ½ inch or less long and in numerous but few-flowered clusters at the ends of the branches; blooming between May and September. The short-stalked leaves are divided into 3 wedge-shaped leaflets from ½ to 1½ inches long. Usually a common plant in dry open woods, clearing, etc. It is used to a limited extent in medicine.

RANGE: Me. to Ont. and Minn. south to Fla. and La.

HAIRY WILD-INDIGO *(Baptisia cinerea)*

This is an erect, branching plant 2 to 3 feet tall which may be softly hairy or merely downy on the younger growth. It has many inch-long yellow flowers arranged in narrow clusters up to 10 inches in length. Its leaves are stalkless or nearly so and divided into 3 lance-shaped to narrowly top-shaped leaflets 2 to 4 inches in length; with a pair of prominent stipules at the base of each leaf. It grows in dry sandy pinelands and on sand-hills in the coastal plain blooming in April or May.

RANGE: se Va. south to S.C.

LARGE WHITE WILD-INDIGO *(Baptisia alba* var. *macrophylla)*

The large white flowers which are about an inch long, and the stalked, drooping pods serve to identify this species. It is a smooth plant 2 to 4 feet tall with ascending branches; and stalked leaves which are divided into 3 narrowly top-shaped leaflets 1 to 2¼ inches long. Blooming between May and July, the flowers are arranged in rather loose clusters up to a foot long. It grows in rich woods, and on prairies and river-banks.

RANGE: Ont. to Wis. and Neb. south to La. and Tex.

WHITE WILD-INDIGO *(Baptisia alba)*

In this species the white flowers are a little more than ½ inch long, and the pods which follow them stand erect. It is a smooth plant 1 to 3 feet tall with rather spreading slender branches. There are numerous flowers in a narrow and long-stalked cluster; blooming during May or June. The leaves are stalked and have 3 leaflets from 1 to 2 inches in length. It grows in dry woods and pinelands, from the coast to the mountains.

RANGE: N.C. and Tenn. south to Fla. (Not illustrated)

BLUE WILD-INDIGO *(Baptisia australis)*

This is our only species of wild indigo with blue flowers. They are about ¾ inch long and violet-blue in color. It is a smooth and branching plant 3 to 4 feet tall, with stalked leaves divided into 3 narrowly top-shaped leaflets 1 to 2½ inches in length. It grows in rich woods and along streams, blooming in May or June; and is frequently cultivated.

RANGE: Pa. to s. Ind. south to Ga. and Tenn. (Not illustrated)

AARON'S-ROD *(Thermopsis villosa)*

Members of the genus *Thermopsis* resemble the wild indigos but they have flattened rather than plump or inflated pods. Aaron's-rod is a somewhat woolly-hairy plant 1 to 3 feet tall with spreading branches. It has yellow flowers nearly an inch long which are in crowded, long, and narrow erect clusters. The leaves are divided into 3 ellipitic or top-shaped leaflets 2 to 4 inches long; and they have large, leafy stipules at the bases of the leafstalks. It grows in dry open woods and along roadsides, blooming between May and July. Often cultivated as Carolina Thermopsis.

RANGE: N.C. and Tenn. south to Ga.

ALLEGHENY THERMOPSIS *(Thermopsis mollis)*

This is an erect, finely downy, branched plant 2 to 3 feet tall. Its yellow flowers are about ⅔ inch long and are in long but rather loose clusters. Its leaves are divided into 3 elliptic to top-shaped leaflets from 1 to 3 inches in length, and there is a pair of prominent stipules at the base of the leafstalk. It grows in dry woods and on rocky ridges in the mountains, blooming in May or June. Another name given it is Bush-pea.

RANGE: Western Va. and Tenn. south to n. Ga. and n. Ala.

WILD LUPINE *(Lupinus perennis)*

This is the only lupine found north of the Carolina coastal plain. Usually it has several erect or ascending stems from 1 to 2 feet tall. Between April and June these terminate in elongate clusters of showy flowers which are about ⅔ inch long and usually lavender-blue to purplish-blue in color. Occasional plants have flowers which are pinkish or white. The leaves are divided into from 7 to 11 leaflets which radiate from the summits of the long leaf stalks like the spokes of a wheel. The leaflets are narrowly top-shaped and from 1 to 2 inches in length. This plant grows in dry open woods, thickets, and on sandhills. It is often called the Sundial Lupine.

RANGE: Me. to Ont. and Minn. south to Fla. and La.

LADY LUPINE *(Lupinus villosus)*

This is a striking plant of the dry pinelands and sandhills of the southeastern coastal plain. The stalked leaves are 1-foliate with lance-shaped or elliptic blades from 2 to 6 inches long, and are so densely covered with silky hair that they appear quite a grayish-green. The plant has several radiating stems or branches which are more or less prostrate, but turn up at the ends and terminate in dense clusters of flowers from 4 to 7 inches in length. The flowers are about ¾ inch long and are deep lilac to purple, with a very dark purple or red spot in the center of the standard petal. They are followed by pods so densely covered with silvery-gray hairs that they suggest pussy willows. It blooms in April or May.

RANGE: e. N.C. south to Fla. and west to Miss.

BLUE SANDHILL LUPINE *(Lupinus diffusus)*

While this lupine closely resembles the preceding species, its stems and leaves have shorter and less spreading hairs. Its flowers are a bright deep blue with a white or cream-colored spot on the standard petal, and they often appear a few weeks earlier than those of the Lady Lupine. It also grows in dry woods and on sandhills of the coastal plain.

RANGE: e. N.C. south to Fla. and west to Miss.

RABBIT-BELLS *(Crotalaria rotundifolia)*

The slender trailing stems of this little plant often form mats up to 2½ feet across. Scattered along the stems are short-stalked, oval-shaped leaves from ½ to slightly more than 1 inch long. The 2 to 6 yellow flowers are borne on slender erect stalks from 3 to 6 inches in length, appearing between May and August. They are followed by inflated pods, sometimes an inch long, which look like miniature punching-bags. When mature and dry the many seeds within them become loose and rattle. Sometimes it is called the Prostrate Rattlebox. It grows in dry sandy pinelands, chiefly in the coastal plain.

RANGE: se. Va. south to Fla. and west to La.

RATTLEBOX *(Crotalaria sagittalis)*

This plant has a simple to bushy-branched hairy stem seldom more than a foot tall. Its leaves are oval or lance-shaped, nearly stalkless, and 1 to 2½ inches long. Each leaf usually has a pair of arrow-shaped stipules at its base and these extend down the stems as wings. The 2 to 4 yellow flowers are on slender stalks 1 to 3 inches in length, and are followed by bag-like pods similar to those of the preceding species. It grows in dry open woods and clearings, blooming between June and September.

RANGE: Mass. to Wis. south to Fla. and Tex.

SHOWY CROTALARIA *(Crotalaria spectabilis)*

This species is a tall branching plant with showy yellow flowers and large top-shaped leaves, which has escaped from cultivation and is often seen in fields and along roadsides from Va. and Ky. south to Fla. and La. It blooms between July and September. (Not illustrated)

Wild Lupine
Lady Lupine
Blue
Sandhill Lupine
Rabbit-bells
Rattlebox

Buffalo Clover
Red Clover
Crimson Clover
White Clover
Alsike Clover

BUFFALO CLOVER *(Trifolium reflexum)*

This native clover is a smooth to somewhat hairy plant 10 to 20 inches tall. Its flowers have a deep pink or rose-colored standard petal but the rest of the flower is white or a very pale pink. Each flower is about ½ inch long and they are crowded in globe-shaped heads 1 to 2 inches across. The 3 leaflets are oval to broadly top-shaped, with a paler green V-shaped spot and minutely toothed margins. Rather large, egg-shaped and pointed stipules are present at the bases of the leaf stalks. This clover grows in open woods, fields, and along roadsides; blooming between April and August.

RANGE: N.Y. and Ont. to Iowa south to Fla. and Tex.

RED CLOVER *(Trifolium pratense)*

Although introduced from Europe as a forage plant, the Red Clover is now widely naturalized in North America. It is a more or less hairy, usually branched plant, from 6 to about 24 inches tall. The purplish-pink flowers are in dense, roundish to somewhat egg-shaped heads about an inch in diameter, with a pair of leaves at the base. Each of the 3 oval or top-shaped leaflets have toothed margins and have a prominent pale V-shaped spot. It is common in fields, waste places and along roadsides; blooming between April and October.

WHITE CLOVER *(Trifolium repens)*

This is another immigrant from Europe which is now common in fields, lawns, and along roadsides throughout most of North America. It is a creeping plant with long, smooth, mat-forming stems. The inversely heart-shaped leaflets are minutely toothed on their margins; and the white or pinkish flowers are in globe-shaped, long-stalked heads about ¾ inch in diameter. It blooms between April and October.

RUNNING BUFFALO CLOVER *(Trifolium stoloniferum)*

This plant is a native creeping species found in open woods and on prairies from W.Va. to S.D. south to Ky., Mo. and e. Kan. Its white flowers are tinged with purple and are in heads an inch or more across, on stems often 4 or more inches tall. The stalks bearing the flowers have a pair of leaves toward the summit. It blooms between May and August. (Not illustrated)

CRIMSON CLOVER *(Trifolium incarnatum)*

This is still another immigrant from Europe which is now naturalized in waste places, fields, and along roadsides. It is an erect, branching, soft-hairy plant 6 to 24 inches tall; with crimson or scarlet flowers in narrowly cone-shaped heads 1 to 2½ inches in length. It is sometimes called the Italian Clover. In bloom from April to July.

ALSIKE CLOVER *(Trifolium hybridum)*

Often called the Alsatian Clover, this European species is now widely naturalized in fields and along roadsides throughout much of North America. A smooth plant, it has erect or ascending, branched stems often 1 to 2 feet tall. Its flowers are white or pink-tinged but turn brown when old, very fragrant; and in long-stalked, globe-shaped heads about ¾ inch across. The broadly oval to top-shaped leaflets have minutely toothed margins. It blooms between April and October.

CAROLINA CLOVER *(Trifolium carolinianum)*

This is a native species with somewhat downy, tufted, ascending or reclining stems 3 to 8 inches in length. Its purplish flowers are in roundish heads about ¼ inch across; and the 3 leaflets are broadly oval or top-shaped, slightly notched at the top, and minutely toothed on the margin. It grows in fields and on rocks from N.C. to Mo. and se. Kan. south to Fla. and Tex. (Not illustrated)

VIRGINIA CLOVER *(Trifolium virginicum)*

This species is a tufted hairy plant with white flowers in heads about an inch across; and long-stalked leaves which have 3 very narrow leaflets. It grows on shady slopes in the mountains from s. Pa. to W.Va., Md. and Va.; blooming in May or June. (Not illustrated)

YELLOW SWEET-CLOVER *(Melilotus officinalis)*

The two sweet-clovers are smooth and branching plants from 3 to 6 feet tall; with numerous, small, flowers arranged in long and narrow clusters arising from the axils of the leaves. They get their name of "sweet-clover" from the sweet-scented, vanilla-like odor the plants give off when they are crushed or dried. Both species are natives of Europe which have become widely naturalized here in America; being found in fields, waste places, and along roadsides. They bloom almost continuously between May and October. This species is distinguished by its yellow flowers. The leaves are divided into 3 narrowly top-shaped leaflets which are toothed on the margin and rounded at the tip.

WHITE SWEET-CLOVER *(Melilotus alba)*

This species is practically identical to the preceding one but it is readily distinguished by its white flowers. The 3 leaflets of its leaves are also notched at the tip. (Not illustrated)

GOAT'S-RUE *(Tephrosia virginiana)*

Among the other names given to this plant are Devil's-shoestring, Cat-gut, Wild Sweet-pea, and Dolly Varden. It is quite an attractive, more or less whitish silky-hairy plant with an erect stem 1 to 2 feet tall. The showy flowers are about ¾ inch long. They have a yellow standard petal which is commonly flushed with pink, and a rose-pink keel. The leaves are pinnately divided into from 14 to 28 narrowly oblong leaflets. It grows in dry open woods, thickets, or fields in usually sandy soils; blooming between May and August.

RANGE: Mass. to Ont., Wis. and Okla. south to Fla. and Tex.

FEW-FLOWERED GOAT'S-RUE *(Tephrosia hispidula)*

Instead of being erect, the stems of this species are reclined upon the ground and range between 1 and 2 feet in length. It is a somewhat hairy or smoothish plant and its leaves are pinnately divided into from 7 to 19 lance-shaped leaflets. The half-inch long flowers are white when they first open but change to red, and are in few-flowered and long-stalked clusters arising from the axils of the leaves. It grows in dry, sandy, coastal plain pinelands; blooming between May and August.

RANGE: Va. south to Fla. and west to La.

AMERICAN VETCH *(Vicia americana)*

One of our few native vetches, this species is found in moist thickets and meadows where it blooms between May and July. It is a rather smooth plant which, like most members of the genus, climbs by means of a tendril at the tip of its leaves. The leaves are pinnately divided into from 8 to 14 leaflets which are egg-shaped or elliptic, blunt but usually tipped with a minute point, and ¾ to 1¼ inches long. Its flowers are about ¾ inch long, bluish-purple, and in rather loose and rather 1-sided clusters. Another name for it is Purple Vetch.

RANGE: Que. to Alaska south to Va., Kan., N.Mex. and Ariz.

CAROLINA VETCH *(Vicia caroliniana)*

The flowers of this native vetch are about ½ inch long, white, with a blue tip on the keel. The long and narrow clusters have from 8 to 20 flowers, and it blooms between April and June. It is a smooth or slightly hairy plant with slender stems 2 to 3 feet long. Its leaves are divided into from 10 to 18 narrowly oblong leaflets which are blunt but with a minute point at the tip, and ½ to ¾ inch long. It grows in rich woods, thickets, and along the banks of streams. Also known as the Pale Vetch.

RANGE: N.Y. and s. Ont. to Minn. south to Ga., La. and Okla.

The following are among the more common and widespread introduced species of vetch naturalized in our region.

COMMON VETCH *(Vicia sativa)* has usually 2 bluish-purple or rose-colored flowers about an inch long in the axils of the leaves. (Not illustrated)

COW VETCH *(Vicia cracca)* has blue to purple flowers in a dense 1-sided cluster. It is also called Canada-pea.

HAIRY VETCH *(Vicia villosa)* is similar to the preceding but more spreading and hairy, with violet to whitish flowers.

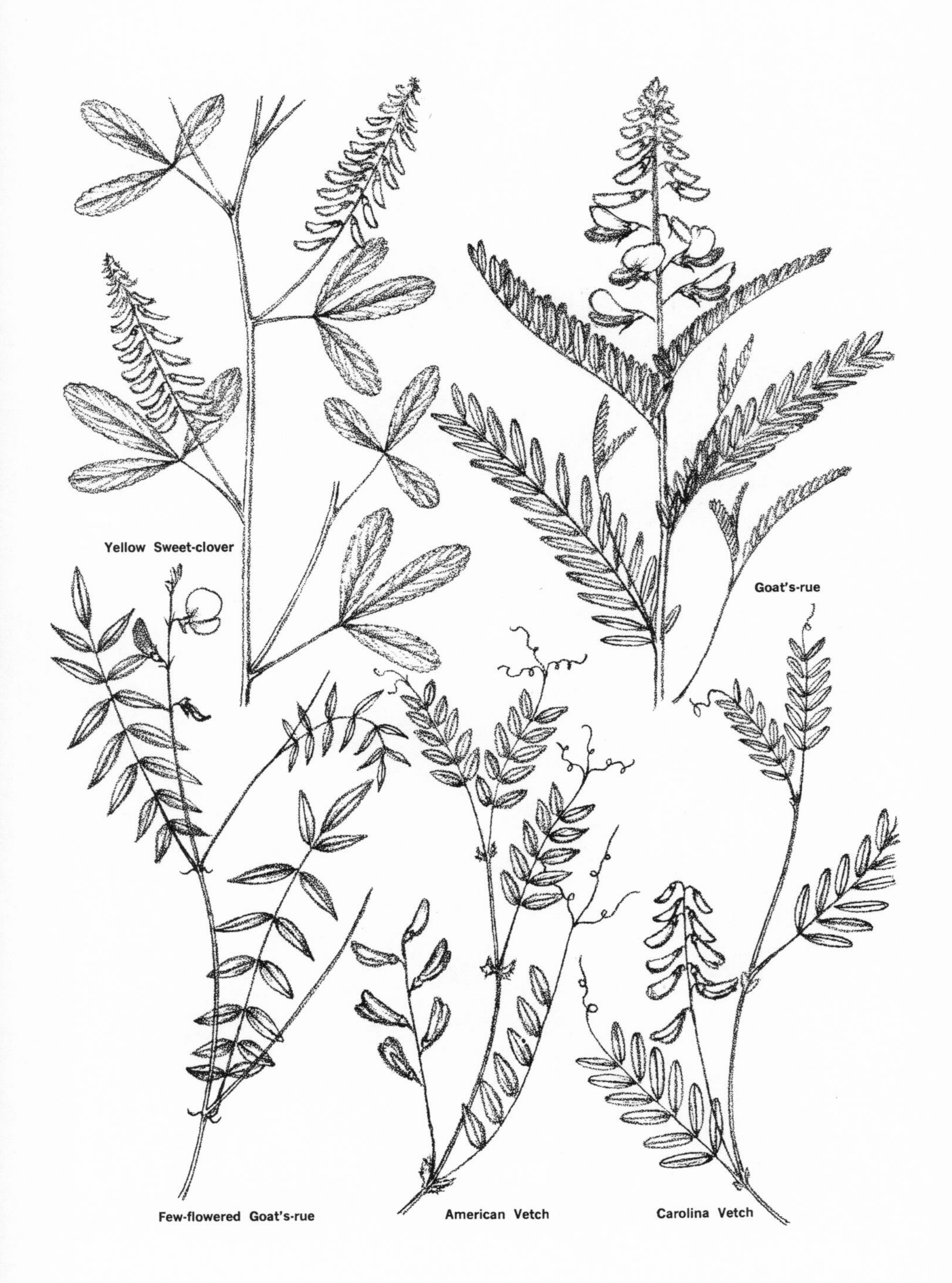
Yellow Sweet-clover
Goat's-rue
Few-flowered Goat's-rue
American Vetch
Carolina Vetch

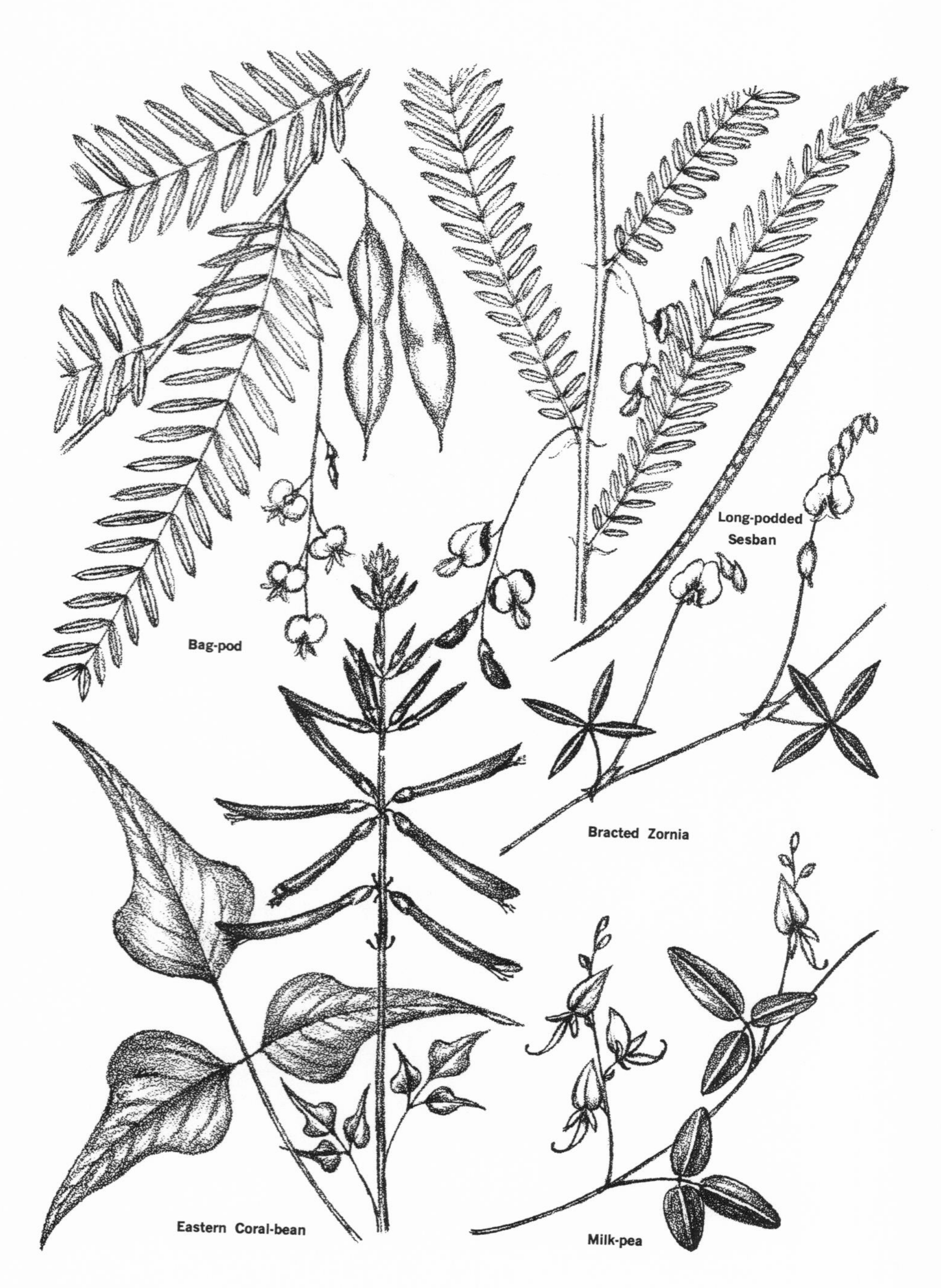
Long-podded
Sesban
Bag-pod
Bracted Zornia
Eastern Coral-bean
Milk-pea

BAG-POD *(Glottidium vesicarium)*

The Bag-pod is a branching plant from 4 to about 12 feet tall, with leaves pinnately divided into 24 to 50 narrowly oblong leaflets. Between July and September it has yellow, or sometimes dark purple flowers about ½ inch long which are borne in open, drooping clusters from the axils of the leaves. It is often conspicuous in thickets, or along roadsides in the coastal plain. The 2-seeded, bladdery pods remain on the dry brown stalks during the winter. Another name for it is Bladder-pod.

RANGE: E. N.C. south to Fla. and west to Tex.

LONG-PODDED SESBAN *(Sesbania exaltata)*

Most of the 15 or so members of this genus grow in tropical lands. This one is a smooth, branching plant from 3 to 10 feet in height. Its leaves are pinnately divided into 24 to 70 narrowly oblong leaflets an inch or less in length. The flowers are yellow, often spotted with red, and the back side of the standard petal is largely red. They are about ¾ inch long and borne in long-stalked, few-flowered clusters which arise from the axils of the leaves. The flowers are followed by very slender bean-like pods from 6 to 12 inches long. It grows in low wet fields, moist thickets, and about the borders of swamps; blooming between June and September.

RANGE: N.C., Mo., and Okla. south to Fla. and Tex; occasionally adventive north to se. Pa. and se. N.Y.

BRACTED ZORNIA *(Zornia bracteata)*

The Bracted Zornia has slender, wiry, much-branched stems up to 3 feet in length which lie flat on the ground. The plant often forms broad carpets in the dry, sandy pinelands or along sandy roadsides in the coastal plain. Its leaves are divided into 4 dark green, lance-shaped leaflets from ½ to 1 inch long which are spread like the letter X. At the base of the leaf stalks there are large stipules which taper to a tail-like base, and the flowers appear from pairs of similarly shaped but broader bracts on long stalks arising from the axils of the leaves. The flowers are bright yellow, the broad standard with usually some red at the base. It blooms between May and September.

RANGE: Se. Va. south to Fla. and west to Tex.

EASTERN CORAL-BEAN *(Erythrina herbacea)*

Between May and July, the brilliant red flowers of this plant are conspicuous in coastal plain pinelands, thickets, and the borders of woods; the long and open clusters being held aloft on leafy stems from 2 to 4 feet tall. Its flowers are slender-looking and about 2 inches long, consisting of a long and narrow standard which turns up only toward the tip. Stamens protrude slightly beyond the end of the standard but the wing and keel petals are quite small. These flowers are followed by bean-like pods which have conspicuous constrictions between the bright red seeds. The leaves are long-stalked and divided into 3 rather triangular leaflets. At the base there is a broad and roundish lobe on each side, above which they taper to a long-pointed tip. Some of the other names given this plant are Cardinal-spear, Cherokee-bean, and Firecracker-plant; the latter because the color and shape of its flowers suggest firecrackers.

RANGE: e. N.C. south to Fla. and west to Tex.

MILK-PEA *(Galactia regularis)*

The Milk-peas, in spite of their name, have nothing milky about them. They are plants with slender, prostrate, or climbing stems. Most species, like this one, have their leaves divided into 3 leaflets; and have violet-purple flowers. The leaves of this species have 3 narrowly egg-shaped or elliptic leaflets from ½ to 1½ inches long, with stalks shorter than the ones bearing the flowers. It grows in open woods, thickets, and roadsides in dry sandy soils; blooming between June and August.

RANGE: N.Y. south to Fla. and west to La.

ELLIOTT'S MILK-PEA *(Galactia elliottii)*

This species is unusual in that its somewhat evergreen leaves have from 7 to 9 leaflets, and its flowers are white. It grows in pinelands from N.C. and Tenn. south to Fla. (Not illustrated)

SPURRED BUTTERFLY-PEA *(Centrosema virginianum)*

This is a trailing or climbing plant with minutely rough-hairy stems from 2 to 4 feet in length. It has clusters of from 1 to 4 flowers on stalks which arise from the axils of the leaves. They are violet colored and have a big, broad standard an inch or more across. On its back, and near the base, is a spur-like projection from which the plant gets its name. The leaves are divided into 3 narrowly egg-shaped to lance-shaped leaflets from 1 to 2 inches long. It grows in dry sandy open woods and fields; blooming between June and August.

RANGE: N.J. to Ky. and Ark. south to Fla. and Tex.

BUTTERFLY-PEA *(Clitoria mariana)*

The showy, 2-inch long, pale lavender-blue or violet flowers of the Butterfly-pea are really impressive. They owe their beauty, however, to the large, rounded standard which is notched at the tip, for the keel and wing petals are quite short and far from conspicuous From 1 to 3 such flowers are borne on stalks arising from the axils of the leaves between June and August. Late in the season, small and bud-like flowers are produced; and, although hardly noticeable they form pods with viable seeds. The Butterfly-pea has trailing but seldom climbing stems from 1 to 3 feet in length; and stalked leaves which are divided into 1-to 2-inch egg-shaped or lance-shaped leaflets. It grows in dry open woods, thickets, and pinelands.

RANGE: se. N.Y. to W. Va. and Iowa south to Fla. and Tex.

SAND-BEAN *(Strophostyles umbellata)*

The Sand-bean has slender, branching, and trailing stems 1 to 5 feet in length which are minutely hairy near the tips. Its flowers are about ½ inch long and are borne in a close cluster at the summit of a stalk from 4 to 8 inches tall. They are pink at first but become yellowish or tawny in fading. The keel is noticably curved. Its leaves are divided into 3 egg-shaped leaflets. The Sand-bean grows in dry sandy woods, thickets, and fields; blooming between June and September. It is also called the Pink Wild-bean.

RANGE: N.Y. to s. Ill. and Mo. south to Fla. and Tex.

TRAILING WILD-BEAN *(Strophostyles helvula)*

This species has trailing or twining and climbing stems from 2 to 8 feet long which are usually somewhat hairy. Its flowers are about ½ inch long and are borne in compact clusters at the summit of a stalk from 2 to 7 inches long. They are pink or purplish and turn greenish with age. The keel is strongly curved. Its leaves are divided into 3 leaflets which are often somewhat 3-lobed. It grows in moist sandy thickets and on shores; blooming between June and September.

RANGE: Mass. to Ont., Wis. and S.D. south to Fla. and Tex.

AMERICAN WISTERIA *(Wisteria frutescens)*

One can readily recognize this woody-stemmed climber as a wisteria, for it closely resembles the cultivated ones. It has rather dense clusters of lilac-purple flowers from 2 to about 5 inches in length, and it blooms in April or May. Its leaves are from 4 to 8 inches long and are pinnately divided into from 9 to 15 narrowly egg-shaped or lance-shaped leaflets, from 1 to 2 inches in length. It grows about the borders of swamps and along the banks of streams.

RANGE: Va. to Ill. and Mo., south to Fla. and Tex.

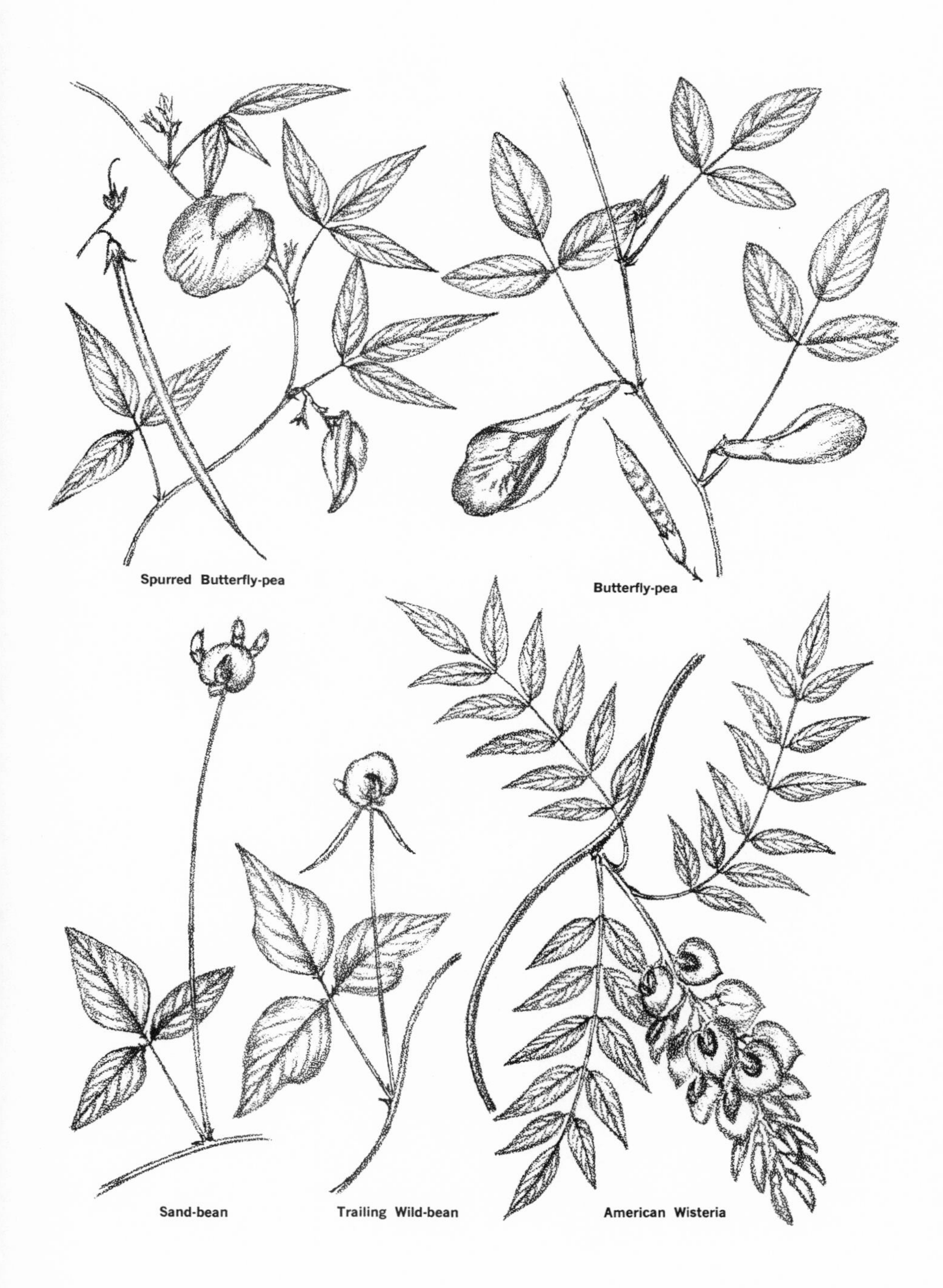
Spurred Butterfly-pea
Butterfly-pea
Sand-bean
Trailing Wild-bean
American Wisteria

Sampson's Snakeroot
Ground-nut
Beach-pea
Hog-peanut
Wild Bean

SAMPSON'S SNAKEROOT *(Orbexilum pedunculatum* var. *psoralioides)*

Another name for this erect, slender-stemmed, sparingly branched plant is the Congo-root. It grows a foot to 2½ feet tall, producing a dense cluster of bracted, small, lilac-purple flowers toward the end of a long stalk. As the cluster gradually grows longer, the flowers become more spread apart. Its leaves are divided into 3 lance-shaped but rather blunt-tipped leaflets, each from 1½ to 3 inches long. Growing in dry open woods, fields, and clearings; it blooms between May and July.
RANGE: Va. to Ohio, Ill., Mo., and Kan. south to Ga. and Tex.

GROUND-NUT *(Apios americana)*

The Ground-nut is a slender-stemmed, twining plant which often climbs over other vegetation to a height of several feet. It has purplish-brown, fragrant flowers about ½ inch long which are in compact clusters on stalks arising from the axils of the leaves. The keel is strongly upturned. Its leaves are divided into 5 or 7 egg-shaped or broadly lance-shaped leaflets. The "nuts" are tubers which occur in necklace-like strings underground, and were a source of food for the Indians. Also known as the Wild-bean or Potato-bean, the plant is quite common in moist thickets; blooming between June and August.
RANGE: N.B. to Minn. and Colo. south to Fla. and Tex.

BEACH-PEA *(Lathyrus japonicus* var. *maritimus)*

One might detect a kinship between this and other species of *Lathyrus* and the sweet-peas of our gardens. The Beach-pea is a somewhat fleshy plant with angled, branching, and trailing stems 1 to 2 feet in length. Its leaves are pinnately divided into 6 to 12 broadly egg-shaped or oval leaflets. They have branching tendrils at the leaf tip; and a pair of very large, leaf-like, often more or less arrow-shaped stipules at the base of the leaf stalk. The flowers are purple, violet and purple, or bluish, about ¾ inch long, and are in rather compact clusters on stalks arising from the axils of the leaves. It grows on sandy and gravelly beaches and shores; blooming between June and August.
RANGE: Arctic region south to coast of N.J. and the Great Lakes.

VEINY-PEA *(Lathyrus venosus)*

The Veiny-pea has strongly 4-angled, reclining, ascending, or climbing stems to 3 feet in length. Its leaves are pinnately divided into from 8 to 14 egg-shaped or oval leaflets which are usually very veiny beneath. The stipules are decidedly smaller than the lowest leaflets and shaped like an arrowhead split lengthwise. The flowers are between ½ and ¾ inch long, purple, with 5 to 19 in a stalked cluster. It grows in rich woods, thickets, and along streams; blooming between April and June.
RANGE: Que. to Sask. south to Ga., La., and e. Tex. (Not illustrated)

HOG-PEANUT *(Amphicarpa bracteata)*

Often called the Wild-peanut, this plant has twinging stems more or less clothed with brownish hairs. Purplish or sometimes whitish flowers, about ½ inch long, are produced in nodding clusters on stalks arising from the axils of the upper leaves. They form curved pods containing a few seeds. But like the true peanut, the plant also has small bud-like flowers near the ground; and these mature their 1-seeded pods beneath its surface. The leaves are divided into 3 egg-shaped leaflets and have small stipules at the bases of the leaf stalks. It grows in rich, moist woods and thickets; blooming from July to September.
RANGE: Me. to Que., Man. and Mont. south to Fla. and Tex.

WILD BEAN *(Phaseolus polystachios)*

A close relative of our garden beans, the Wild Bean has lilac or purple flowers about ⅓ inch long in which the keel is spirally coiled. They are produced between July and September in narrow and long-stalked clusters; and are followed by somewhat curved and flattened, bean-like pods 1½ to 2½ inches long. The leaves are quite bean-like in appearance and have 3 broadly egg-shaped leaflets with pointed tips. It grows in dry open woods and thickets.
RANGE: N.J. to W. Va., Ill., Iowa and Neb. south to Fla. and Tex.

The Tick-trefoils, or Tick-clovers, are common plants in woods, thickets, and clearings. They get their name from the fact that the leaves are divided into 3 leaflets; and the fruits, which are called "loments", are made up of several one-seeded joints which separate very easily at maturity, adhering to one's clothing. There are two dozen or more species of these plants in eastern North America.

SMOOTH TICK-TREFOIL *(Desmodium laevigatum)*

This plant has an erect, simple or branched stem from 2 to 4 feet tall, which is smooth or sometimes minutely downy. It has rose-pink or purple flowers about ½ inch long, arranged in a large, branched cluster at the summit. The leaves are divided into 3 egg-shaped leaflets from 1½ to 3½ inches long. It grows in dry sandy woods and clearings, blooming from June to August or September. The fruits usually have 3 or 4 joints.
RANGE: N.Y. to Ind. and Mo. south to Fla. and Tex.

PROSTRATE TICK-TREFOIL *(Desmodium rotundifolium)*

This species has a slender and trailing stem from 2 to 6 feet in length which may be downy or softly white-hairy. It has purple flowers about ⅓ inch long, arranged in rather loosely-branched clusters. The leaves are divided into 3 roundish leaflets from 1 to 2 inches in length, and have prominent stipules at the bases of the leaf stalks. The fruits are 3- to 5-jointed. It grows in dry woods, blooming between June and September. Often called Dollarleaf.
RANGE: Mass. to Ont. and Mich. south to Fla. and Tex.

DILLEN'S TICK-TREFOIL *(Desmodium perplexum)*

This species is an erect plant from 2 to 3 feet tall and somewhat hairy or downy. It has purple flowers about ⅓ inch long which are in a loosely-branched end cluster. The leaves are divided into 3 egg-shaped or oval leaflets 1½ to 4 inches long. The fruits are 2- to 4-jointed. It grows in woods and thickets, blooming between June and September.
RANGE: Me. to Wis. south to Ga. and Tex.

NAKED-FLOWERED TICK-TREFOIL *(Desmodium nudiflorum)*

This tick-trefoil gets its name from the fact that there are no leaves on the flowering stem which usually overtops the one bearing the leaves. The leaves are clustered at the summit of a stem generally less than a foot high. They have 3 leaflets from 1 to 3 inches long; the end one usually somewhat diamond-shaped, the others rather unevenly egg-shaped. Its rose-purple flowers are about ⅓ of an inch long. The fruits are 2- to 4-joined and are not constricted along the top edge. It grows in rather dry but rich woodlands, blooming between June and August.
RANGE: Me. to Que. and Minn. south to Fla. and Tex.

PANICLED TICK-TREFOIL *(Desmodium paniculatum)*

This species has an erect, slender, smooth or minutely downy stem from 2 to 3 feet tall. It has rose-purple flowers about ⅓ inch long which are in a much-branched end cluster. The leaves are divided into 3 lance-shaped leaflets from 1 to 2 inches long, being noticeably narrower than those of most tick-trefolds. The fruits are 4- to 6-jointed. It grows in the borders of dry woods and in clearings, blooming between June and September.
RANGE: N.H. to Ont., Iowa and Neb. south to Fla. and Tex.

SESSILE-LEAF TICK-TREFOIL *(Desmodium sessilifolium)*

The leaves of this species are almost stalkless and have 3 very narrow, blunt-tipped leaflets 1 to 3 inches long. The plant is quite hairy, the stem having small hooked hairs. Its flowers are small and purplish. The fruits being 2- or 3-jointed. It grows in dry sandy soils.
RANGE: Mass. to Mich., Ill., s. Mo. and e. Kan. south to e. S.C., W.Va., La. and e. Tex. (Not illustrated)

HAIRY SMALL-LEAF TICK-TREFOIL *(Desmodium ciliare)*

The name of this species is quite descriptive. It is a hairy plant up to about 5 feet tall. The leaves have 3 egg-shaped or elliptic leaflets usually less than an inch long. Its flowers are very small. The fruits are 2- or 3-jointed. It grows in dry sandy woods and clearings.
RANGE: Mass. to N.Y., Ohio, Mich. and Mo. south to Fla. and Tex. (Not illustrated)

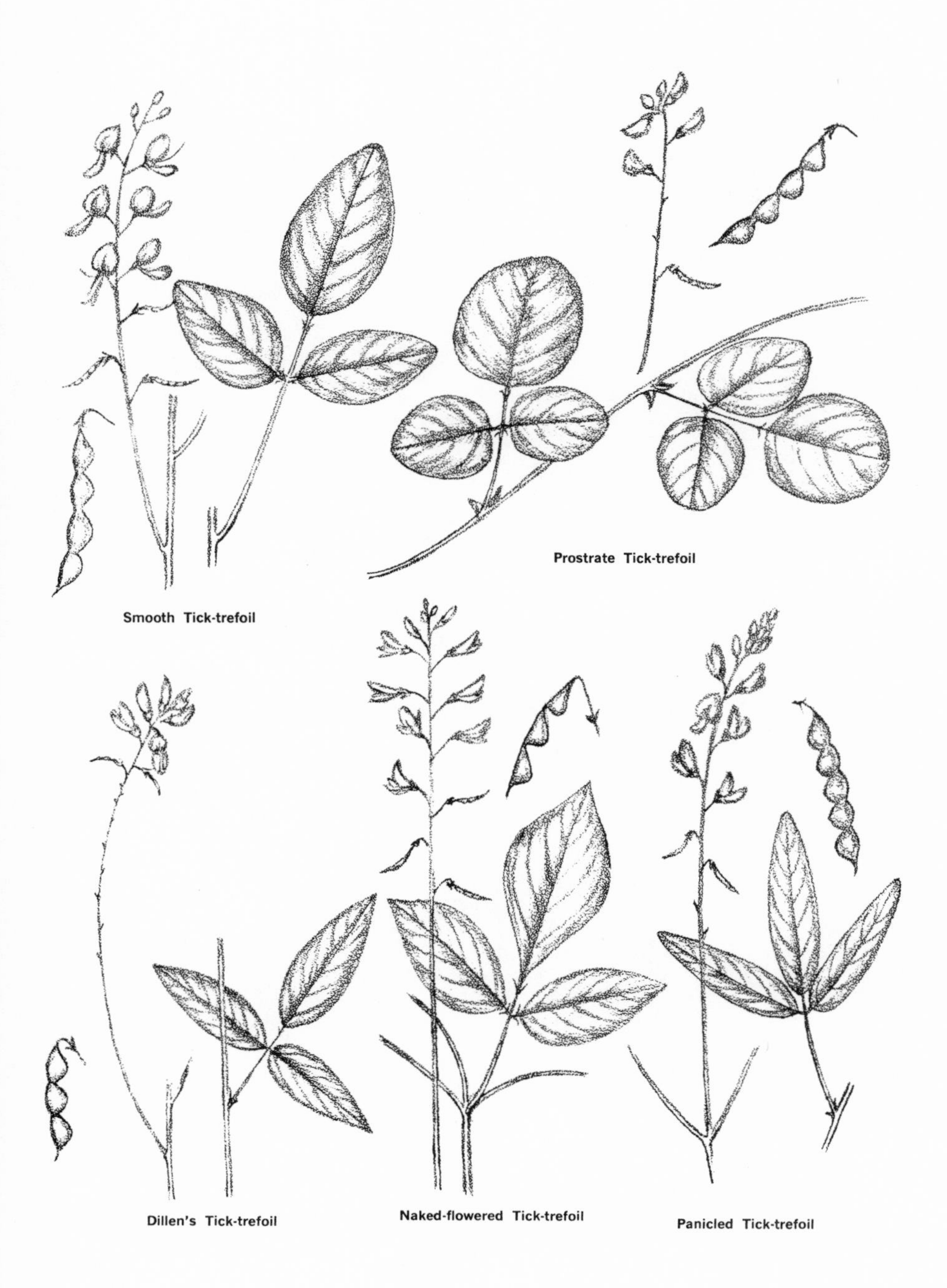
Smooth Tick-trefoil
Prostrate Tick-trefoil
Dillen's Tick-trefoil
Naked-flowered Tick-trefoil
Panicled Tick-trefoil

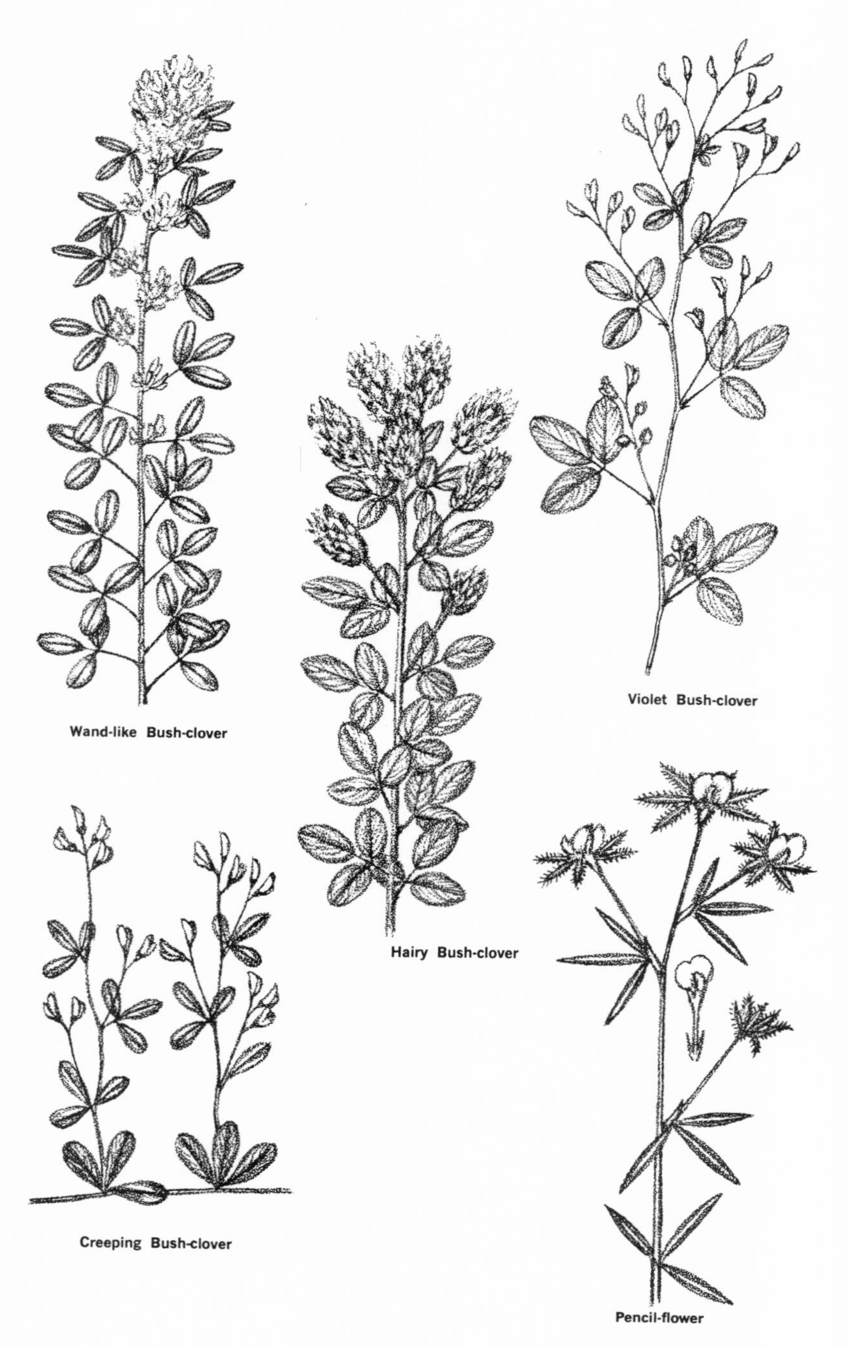

Wand-like Bush-clover

Hairy Bush-clover

Violet Bush-clover

Creeping Bush-clover

Pencil-flower

WAND-LIKE BUSH-CLOVER *(Lespedeza intermedia)*
This is a stiffly erect, smooth or somewhat hairy plant with a simple or branched stem from 1 to 3 feet tall. It has small violet-purple flowers in short-stalked clusters in the axils of the leaves, and generally crowded toward the top of the stem. Its leaves are divided into 3 elliptic leaflets which are about ¾ inch long and very dark green above. It grows in dry open woods and thickets, blooming between July and September.
RANGE: Me. to Ont. and Wis. south to Fla. and Tex.

HAIRY BUSH-CLOVER *(Lespedeza hirta)*

The Hairy Bush-clover is a densely hairy or silky-hairy plant with an erect, usually branched stem from 2 to 4 feet tall. Its small flowers are whitish or yellowish-white with a purplish base, and are crowded in cylindrical heads on stalks which are longer than the leaves. The latter are divided into 3 oval, top-shaped, or roundish leaflets from ½ to 2 inches in length. It grows in dry open woods, thickets, and on slopes; blooming between July and October. RANGE: Me. to Ont. south to Fla. and Tex.

VIOLET BUSH-CLOVER *(Lespedeza violacea)*

The Violet Bush-clover has rather slender, upright or spreading, bushy-branched stems from 1 to 2½ feet tall. The violet-purple flowers are about ⅓ inch long and are borne in loose and slender-stalked clusters. Its leaves are divided into 3 thin, oval or elliptic leaflets which are ½ to 2 inches in length. It grows in dry woods, thickets, and clearings; blooming between July and September.
RANGE: N.H. to Wis. and Kan. south to Fla. and Tex.

CREEPING BUSH-CLOVER *(Lespedeza repens)*

This species of bush-clover has trailing stems from 6 to 24 inches long which are smooth or minutely hairy. Between May and September it has small violet-purple flowers, which are in loose clusters on slender stalks longer than the leaves. The latter are divided into 3 oval or top-shaped leaflets from ¼ to ¾ of an inch long. It is found in dry open woods, thickets, and fields. RANGE: Conn. to Wis., Iowa, and Kan. south to Fla. and Tex.

TRAILING BUSH-CLOVER *(Lespedeza procumbens)*

The stems of this little bush-clover are also low and trailing, but it is a somewhat stouter and softly hairy plant. Its leaves and flowers ar quite similar to those of the preceding species. It grows in dry sandy or rocky woods and clearings, blooming between August and October. RANGE: Mass. to Wis., Iowa, and Kan. s. to Fla. and Tex. (Not illustrated)

PENCIL-FLOWER *(Stylosanthes biflora)*

Pencil-flowers get their name from the peculiar stalk-like calyx tubes of their flowers. This species has a wiry stem from 6 to 20 inches tall, which is either simple or with a few stiffly ascending branches. It has a few orange-yellow flowers about ⅓ of an inch long; which are grouped at the tips of the stiff stem or its branches, and surrounded by bristly-margined leaflets. The leaves lower on the stem are divided into 3 lance-shaped leaflets ¾ to 1½ inches long and with smooth margins. It grows in dry woods, thickets, and fields; blooming between June and September.
RANGE: se. N.Y. to s. Ill. and Kan. south to Fla. and Tex.

DOLLAR-WEED *(Rhynchosia reniformis)*

This is an erect, hairy little plant from 3 to about 9 inches high. Its slender-stalked leaves have 1-pieced blades; which are roundish or somewhat broader than long, often slightly heart-shaped at the base, and from 1 to 2 inches long. The yellow flowers are usually less than ½ inch long and in fairly dense end clusters. It grows in sandy open woods, pinelands, and fields; blooming all summer.
RANGE: se. Va. south to Fla. and west to La. (Not illustrated)

GERANIUM FAMILY (Geraniaceae)

Members of this family have flowers with 5 sepals, 5 petals, 10 stamens, and a pistil composed of 5 united carpels which matures into a capsule splitting at the base into 5 parts.

WILD GERANIUM *(Geranium maculatum)*
On a hairy stem usually 1 to 2 feet tall, the Wild Geranium has a pair of deeply 5-parted and sharply toothed leaves 3 to 6 inches wide. Other ones at the base are similar but long-stalked. Between April and June it has lavender-purple flowers an inch or more across. This is one of our most common woodland wild flowers and it sometimes is found in open areas. It is also known as the Spotted Cranesbill. The slender capsules have a fancied resemblance to a crane's bill.
RANGE: Me. to Man. south to Ga., Tenn., Mo., and Kan.

HERB-ROBERT *(Geranium robertianum)*
This smaller relative of the Wild Geranium is a strongly scented plant with a weak, branching, and sparsely hairy stem about a foot long. It has purple flowers about ½ inch across, borne in pairs on long axillary stalks. The leaves are divided into 3 or 5 deeply cut divisions. It grows in rocky woods and ravines, blooming between May and October. RANGE: Nfd. to Man. south to Md., W.Va., and Ill.

CAROLINA CRANESBILL *(Geranium carolinianum)*
This species is a bushy-branched, hairy plant 5 to 15 inches high; readily recognizable as a *Geranium* by its deeply cleft and toothed leaves with spreading divisions; and the long, bill-like capsules. Small, pale-purple flowers in 4- to 12-flowered clusters are produced between May and July. It is found in dry woods, thickets, waste places, etc. and in many places it is a common weed.
RANGE: Mass. to Wyo., Idaho, and s. B.C. south to Fla. and Calif. (Not illustrated)

WOOD-SORREL FAMILY (Oxalidaceae)

Members of this family all have long-stalked leaves with blades divided into 3 inversely heart-shaped leaflets. The flowers have 5 sepals, 5 petals, 10 stamens, and a pistil composed of 5 united carpels which matures into a slightly 5-lobed or 5-sided capsule. The 5 styles are distinct.

COMMON WOOD-SORREL *(Oxalis montana)*
This low and creeping plant is usually common in cool, moist northern and mountain woodlands. The solitary flowers are about ¾ inch across, borne on 2- to 6-inch stalks which usually overtop the leaves. Their 5 white petals are veined with pink and are usually deeply notched at the tip. Also called the White or True Wood-sorrel and Wood-shamrock, it blooms between May and August.
RANGE: Nfd. to Man. south to N. Eng., w. N.C., e. Tenn. and the Great Lakes region.

VIOLET WOOD-SORREL *(Oxalis violacea)*
Both leaves and flower stalks of this species arise from scaly-coated bulbs. Usually 3 to 12 flowers are clustered on leafless stalks from 4 to 8 inches high. They are about ⅜ inch across and have rose-purple or violet petals. The leaves have shorter stalks than the flowers. It grows in rich or alluvial woods, thickets, meadows, and on slopes; blooming in April and May and August to October.
RANGE: Mass. to Minn. south to Fla. and N. Mex.

UPRIGHT YELLOW WOOD-SORREL *(Oxalis stricta)*
This is a grayish-green and somewhat whitish-hairy plant with leafy, erect or ascending stems 5 or 6 inches tall. The flowers are about ⅔ inch across and the yellow petals are often red at the base. It grows in dry open woods and in fields, blooming between May and October.
RANGE: N.S. to B.C. south to Fla. and Tex.

GREAT YELLOW WOOD-SORREL *(Oxalis grandis)*
The leafy stalk of this species is 1 to 3 feet tall, its yellow flowers about ¾ in across, and its leaflets 1 to 2 inches broad. The latter usually have purplish margins. It grows in woods and on shady slopes; blooming in May and June.
RANGE: Pa. to Ill. south to Ga. and Ala. (Not illustrated)

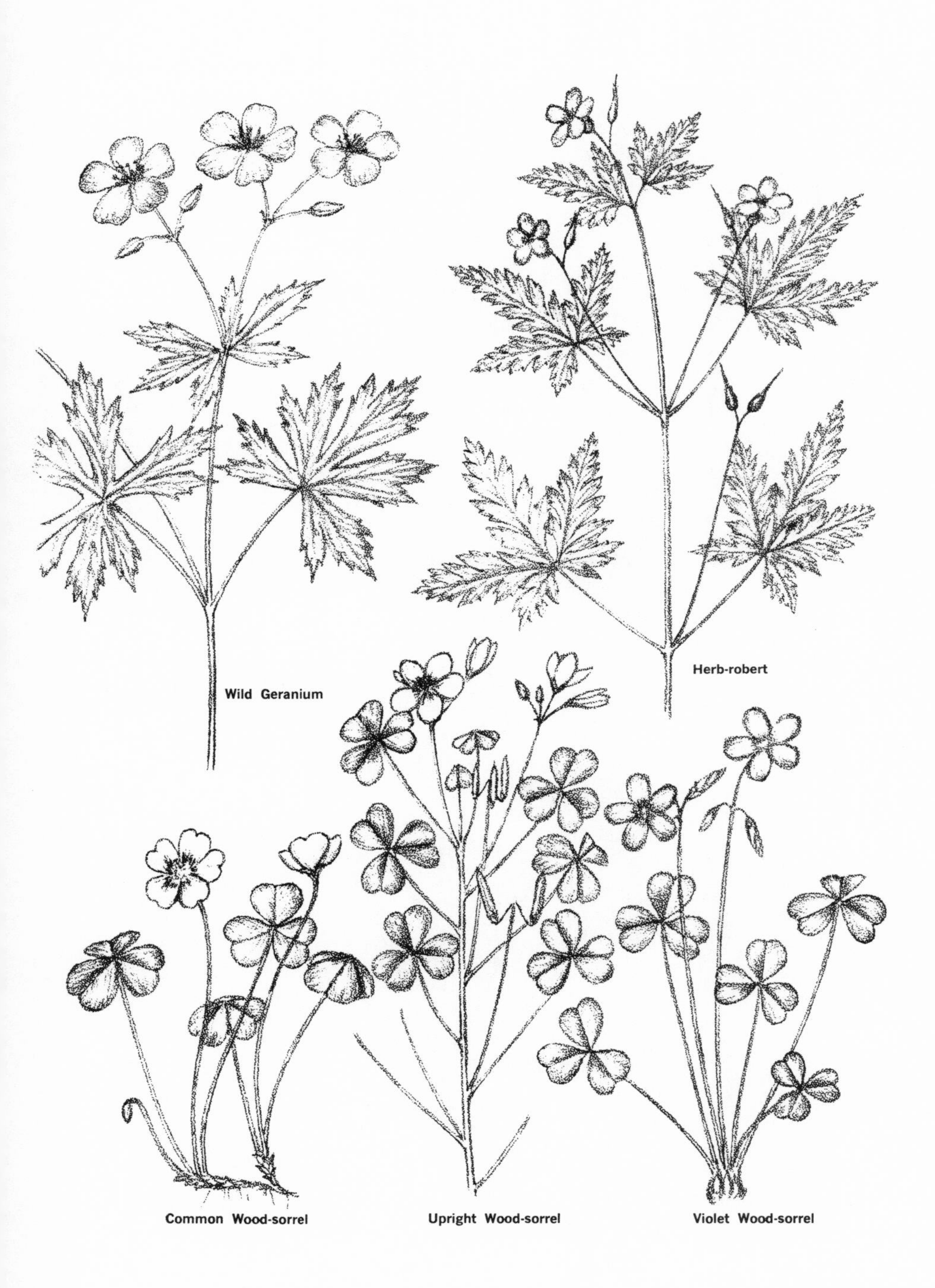
Wild Geranium
Herb-robert
Common Wood-sorrel
Upright Wood-sorrel
Violet Wood-sorrel

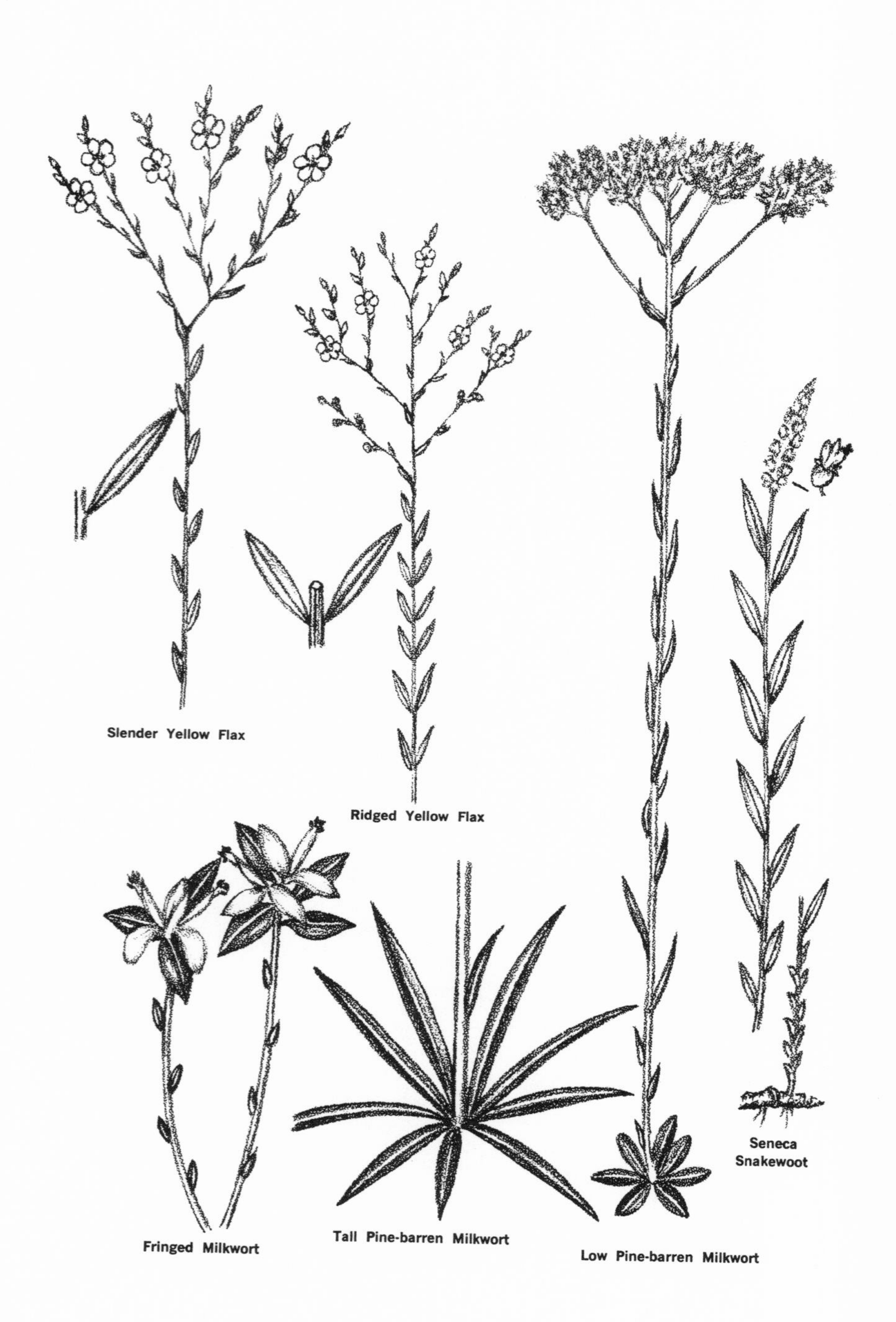
Slender Yellow Flax
Ridged Yellow Flax
Seneca
Snakewoot
Fringed Milkwort
Tall Pine-barren Milkwort
Low Pine-barren Milkwort

FLAX FAMILY (Linaceae)

Members of the Flax Family usually have flowers with 5 sepals, 5 petals, 5 stamens with the filaments united at the base, and a pistil composed of 5 carpels with their ovaries united but the styles separate.

SLENDER YELLOW FLAX *(Linum virginianum)*

Flaxes are very slender-stemmed plants with numerous, small, simple, and narrow leaves. This species is 1 to 2 feet tall, with ascending thread-like branches toward the top. Its flowers are about ⅜ inch across and have 5 yellow petals. The scattered lance-shaped or elliptic leaves are ¼ to ½ inch long. It grows in rich upland or alluvial woods; blooming between June and October. RANGE: Mass. to Ont. south to Ga. and Ala.

RIDGED YELLOW FLAX *(Linum striatum)*

From the preceding species, this one can be distinguished by the sharp angles and ridges which run down its stem from the leaf bases. The leaves on the lower part of the stem are also in pairs. Its flowers are somewhat less than ¼ inch across, blooming between June and October. It grows in moist woods, bogs, and wet sandy places.
RANGE: Mass. to Mich., Mo. and Okla. south to Fla. and Tex.

GROOVED YELLOW FLAX *(Linum sulcatum)*

This species has grooved and wing-angled stems but its leaves are very narrow and all scattered; and the flowers are about ½ inch across. It grows in dry open places, blooming between July and September.
RANGE: Mass. to Man. south to Ga., Ala., Ark., and Tex. (Not illustrated)

MILKWORT FAMILY (Polygalaceae)

Members of this family have flowers with a bilateral symmetry. They have 5 sepals, the 2 lateral ones being petal-like; 3 petals which form a tube, the lower one having a fringed or crested tip; 6 or 8 stamens; and a pistil consisting of 2 united carpels.

FRINGED MILKWORT *(Polygala paucifolia)*

Also known as the Flowering Wintergreen, Gay-wings, and Bird-on-the-wing, this milkwort has very showy, rose-purple flowers about ¾ of an inch long, with a beautifully fringed crest. The plant has slender creeping rootstocks which, in the spring, send up stems 3 to 6 inches high; bearing from 1 to 4 of the pretty flowers and a few small, egg-shaped leaves, with smaller scale-like leaves down along the stem. It grows in rich woods and on rocky slopes, blooming between April and July.
RANGE: N.B. to Man. south to Va., n. Ga., Ill., and Minn.

LOW PINE-BARREN MILKWORT *(Polygala ramosa)*

This plant is very conspicuous in the low, wet, coastal plain pinelands when it blooms between June and September. It has a smooth, erect stem, or stems, from 6 to about 16 inches high, along which are numerous small and narrow leaves. At the top it has a flat-topped cluster of small sulfur-yellow flowers which may be from 3 to 6 inches across. At the base of the plant there is a rosette of broader, top-shaped leaves up to about an inch in length. RANGE: Del. south to Fla. and west to La.

TALL PINE-BARREN MILKWORT *(Polygala cymosa)*

In a general way, this plant is quite similar to the preceding species, but it has a much stouter stem from 2 to about 3 feet tall. The cluster of leaves at the base of the plant are narrowly lance-shaped or grass-like and from 2 to 3 inches in length. Its flowers are no larger than those of the low species and they are in only slightly bigger flat-topped clusters. It also grows in the wet coastal plain pinelands and cypress ponds, blooming between May and August. RANGE: Del. south to Fla. and west to La.

SENECA SNAKEROOT *(Polygala senega)*

The Seneca Snakeroot is a smooth plant with several stems 6 to 15 inches tall, which arise from a thick crown and hard, woody rootstock. Each stem bears a number of narrowly lance-shaped leaves from ½ to 2½ inches long; and small white or whitish flowers in a dense, cone-shaped, 1- to 2-inch end cluster. It grows in dry rocky woodlands and blooms between April and July. Its rootstock is used medicinally.
RANGE: Que. to Alb. south to Me., n. Ga., Tenn., Ark. and S.D.

YELLOW MILKWORT *(Polygala lutea)*

The bright orange-yellow flower heads of this milkwort are conspicuous in coastal plain bogs and wet pinelands between April and October. It is the only milkwort having flowers of this color, which makes its identification easy. It is a smooth plant with several erect or ascending stems from 6 to about 12 inches high. Along them are numerous narrow leaves from ¾ to 1½ inches long. Other names for it are Orange Milkwort and Yellow Bachelor's-button.

RANGE: se. N.Y. south to Fla. and west to La.

CROSS-LEAF MILKWORT *(Polygala cruciata)*

This species has a simple or branched, 4-sided, smooth stem from 4 to 16 inches tall. The leaves are in whorls of 4's, thus forming crosses. They are narrow and ½ to 1½ inches long. The small flowers may be purplish or greenish-white and grouped in dense, cylindrical or barrel-shaped end clusters; blooming between June and October. It grows in bogs and in wet meadows and pinelands.

RANGE: Me. to Minn. south to Fla. and Tex.

FIELD MILKWORT *(Polygala sanguinea)*

The Field or Purple Milkwort is common in fairly moist fields, meadows, and open woods. It is a smooth little plant with a simple or branched stem 6 to about 15 inches high; along which are numerous, scattered, narrow leaves from ¾ to 1½ inches long. Its small flowers vary from a bright rose-purple to greenish and white; and are borne in dense, round to oval-shaped heads from ¼ to ½ inch in diameter. The flowering season is between June and August.

RANGE: N.S. to Ont. and Minn. south to S.C., Tenn., La., and Okla.

WHORLED MILKWORT *(Polygala verticillata)*

This little milkwort has a very slender stem from 6 to 16 inches high which is often branched. It has very narrow leaves which are arranged in whorls of from 3 to 7. The small purplish, greenish, or whitish flowers are in dense cone-shaped clusters at the ends of slender stalks. It grows in moist to dry fields and open woods on more or less sterile soils, blooming between June and September.

RANGE: Me. to Ont. and Minn. south to Fla. and Tex.

RACEMED MILKWORT *(Polygala polygama)*

The numerous simple, erect stems of this milkwort are 4 to 18 inches high; and on them are numerous, rather crowded, and usually narrowly top-shaped leaves from ⅓ to 1½ inches in length. The rose-pink to purple flowers are almost ¼ inch long and arranged in loose but narrow end-clusters from 1 to 4 inches long. It grows in open woods, fields, and meadows but practically always in dry and sandy soils. The flowering season is during June and July.

RANGE: N.S. to Man. south to Fla. and Tex.

PINK MILKWORT *(Polygala incarnata)*

This milkwort has a very slender, stiff, somewhat whitened stem between 5 and 16 inches tall. Its leaves are relatively few, widely scattered, very narrow, and from ⅛ to ½ inch long. The pinkish or flesh-colored flowers have a slender tube formed by the bases of the true petals, about ¼ inch long and far surpassing the wing-like sepals. They are arranged in a dense terminal cluster and the fruits often persist below the tuft of flowers. Another name often given it is Procession-flower. The plant grows in dry fields and meadows, blooming between June and October.

RANGE: N.Y. and Ont. to Wis., Iowa and Neb. south to Fla. and Tex.

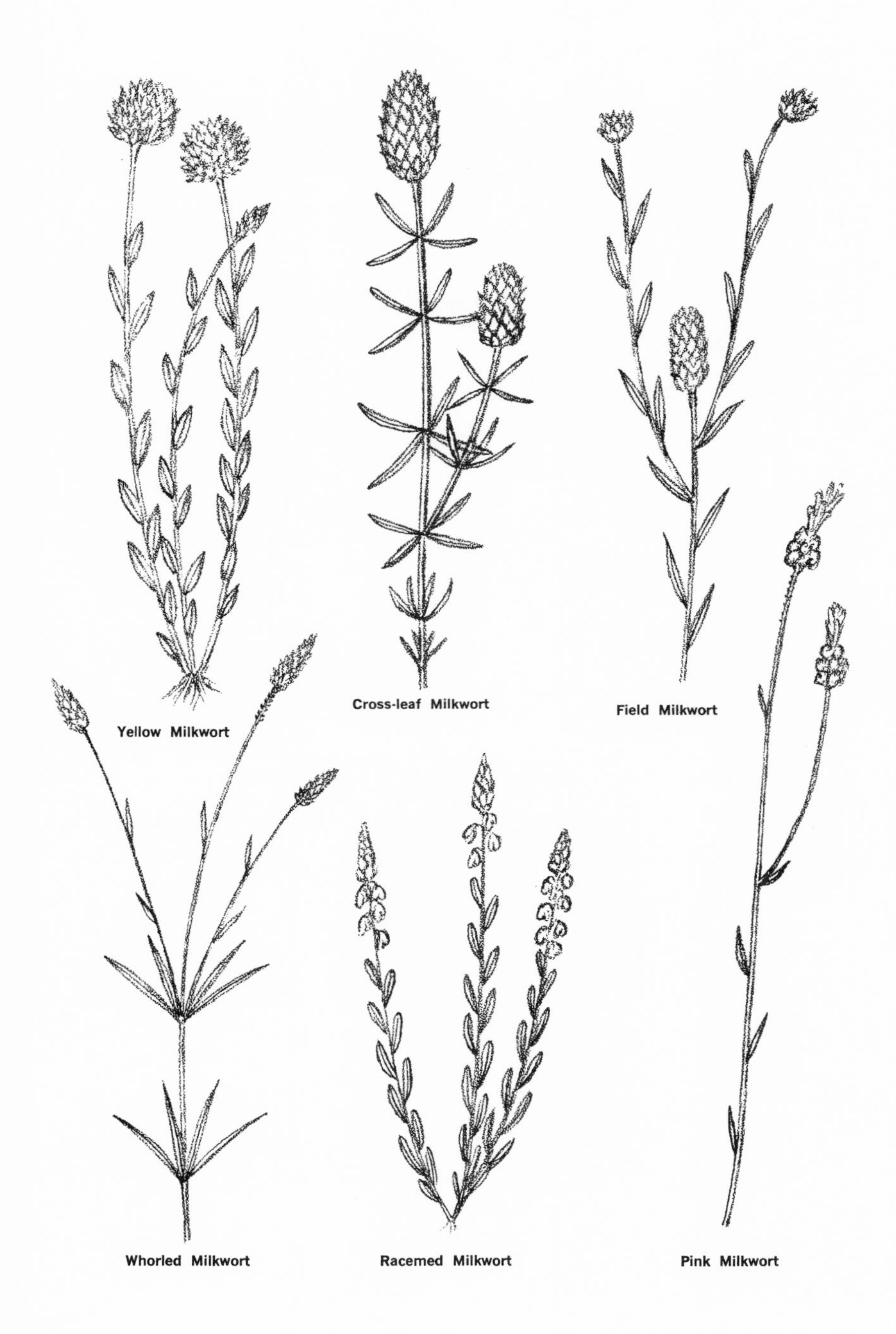
Yellow Milkwort
Cross-leaf Milkwort
Field Milkwort
Whorled Milkwort
Racemed Milkwort
Pink Milkwort

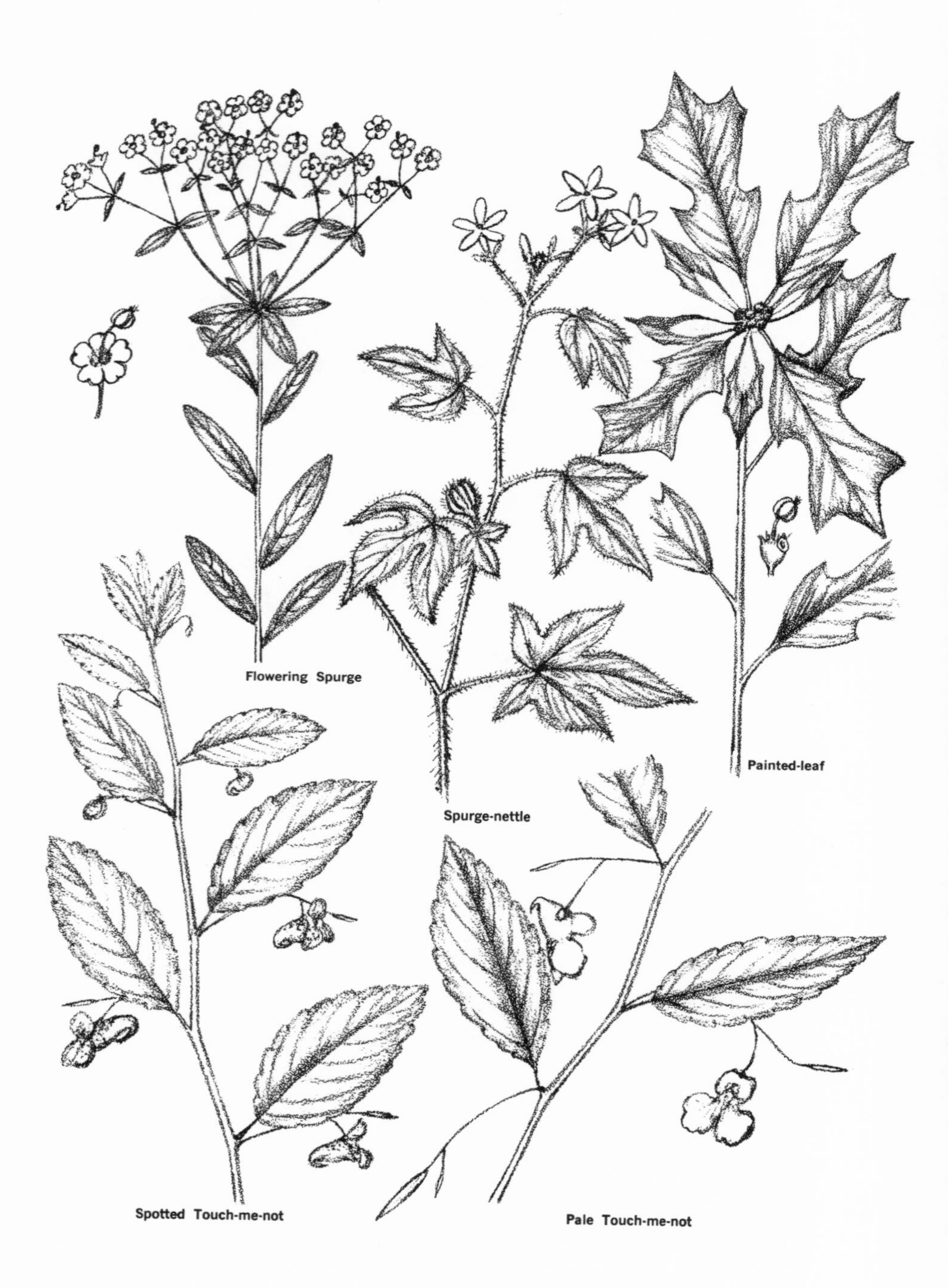
Flowering Spurge
Spurge-nettle
Painted-leaf
Spotted Touch-me-not
Pale Touch-me-not

SPURGE FAMILY (Euphorbiaceae)

This is a large and varied family of plants, most of them occurring in the warmer regions of the world. Many of them have a milky juice. Some have a calyx which may be petal-like. In the true spurges *(Euphorbia)* the flowers consist of only a stalked stamen, or a stalked pistil consisting of 3 united carpels. These flowers are borne within a cup-like involucre which may have prominent glands or corolla-like lobes on its rim. The stamen-bearing and pistil-bearing flowers are always in separate involucres.

FLOWERING SPURGE *(Euphorbia corollata)*

This is a smooth plant with leafy stems from 1 to 3 feet tall, with a branching flower cluster at the summit. What appear to be 5-petalled white flowers are not flowers at all, but really cup-shaped structures which contain the real flowers. The white "petals" are lobes on the rim of the cup, and each one has a yellowish-green gland at the base which might be mistaken for a stamen. The numerous leaves are stalkless and from 1 to 2 inches long. Like other true spurges, the plant has a milky juice. It grows in dry open woods, thickets, fields, and by roadsides. The flowering season is between May and September. RANGE: N.Y. and Ont. to Minn. and Neb. south to Fla. and Tex.

SPURGE-NETTLE *(Cnidoscolus stimulosus)*

This bristly plant of dry sandy pinelands, fields, and sandhills has such other descriptive names as Bull-nettle, Tread-softly, and Finger-rot. It is a branching plant from 6 inches to about 2 feet high, covered with bristle-like and stinging hairs. The stamen-bearing flowers have a corolla-like calyx with 5 spreading, petal-like lobes. The ones which bear the pistils are greenish and not so showy. The leaves are roundish heart-shaped, deeply 3- to 5-lobed, toothed on the margin, from 2½ to 10 inches broad, and are long-stalked. The flowering season is between March and August.

RANGE: Va. south to Fla. and west to Tex.

PAINTED-LEAF *(Euphorbia heterophylla)*

It is easy to see that this plant is a close relative of the familiar Christmas Poinsettia. Its leaves are variable, commonly oval- to fiddle-shaped and coarsely toothed, and from 2 to 5 inches long. Those surrounding the cup-shaped structures containing the flowers have bright red blotches at the base. As in the Poinsettia, there are prominent glands on the flower cups. It grows in open woods, thickets, and waste places usually with a moist sandy soil; flowering between June and September. Other names are Fiddler's Spurge and Annual Poinsettia. It is frequently cultivated under the latter name.

RANGE: Va., Ind., Minn. and S.D. south to Fla. and Tex.

JEWELWEED FAMILY (Balsaminaceae)

Members of this family have flowers with a bilateral symmetry. They have 3 sepals, the lower one extended backward as a spur; 5 petals, all very unequal; 5 stamens; and a pistil composed of 5 united carpels. The fruit is a capsule which splits explosively into 5 spirally twisted parts at maturity, hurling the seeds in all derections.

SPOTTED TOUCH-ME-NOT *(Impatiens capensis)*

This is the smooth, branching, watery-stemmed plant so often met with in moist woods, swamps, and in springy places. It grows from 2 to 5 feet tall and has alternate, thin, elliptic to egg-shaped leaves with coarsely toothed margins. They are from 1½ to about 3½ inches long and are noticeably stalked. The orange flowers are more or less spotted with red, about ¾ inch long, and have a tail-like spur. Usually they are in pairs on slender stalks arising from the leaf axils and are produced continuously between June and September. Other names for it are Jewelweed and Snapweed.

RANGE: Nfd. to Alaska south to Fla., Ala., Ark., and Okla.

PALE TOUCH-ME-NOT *(Impatiens pallida)*

This species is quite similar to the preceding one but somewhat larger and stouter; the canary-yellow to creamy-white flowers being about an inch long, and either unspotted or sparingly spotted with reddish-brown. It is often common in wet or springy woods and in moist, shaded ravines; blooming between July and September. Pale Snapweed or Pale Jewelweed are other names sometimes given to it.

RANGE: Nfd. to Sask. south to Ga., Tenn., Mo., and Kan.

Members of the Mallow Family have flowers with 5 sepals which are united at the base, 5 petals and numerous stamens united by their filaments to form a hollow column about the style, or styles, of the compound pistil.

SWAMP ROSE-MALLOW *(Hibiscus moscheutos)*

Between June and September, this plant of marshes and wet spots produces a succession of showy flowers along its 4- to 7-foot-tall stem. They are 4 to 6 inches across and their white or creamy white or pink petals have purplish or dark crimson bases. The flower stalks are united for part of their length with the leaf stalks, thus appearing to grow out of the latter. Its leaves are 3 to 8 inches long, narrowly egg-shaped or lance-shaped, and toothed on the margin. It is also known as the Mallow-rose and Wild-cotton.

RANGE: Ont. and Mass., W. Va. to Ohio and Ind. south to Fla. and Ala.

HALBERD-LEAF ROSE-MALLOW *(Hibiscus laevis)*

Leaves of this rose-mallow are arrow-shaped with the basal lobes pointing outward, and from 4 to 5 inches long. It is a smooth plant 3 to 5 feet tall; its 3- to 5-inch flowers being pink or flesh-colored with a reddish-purple center. Growing in wooded swamps and on wet stream banks, it blooms between June and August.

RANGE: Pa. to Minn. and Neb. south to Fla. and Tex.

ROUGH ROSE-MALLOW *(Hibiscus aculeatus)*

This is a very rough-hairy plant 3 to 6 feet tall with leaves deeply 3- to 5-lobed or parted. Its flowers are yellow or cream-colored with a reddish-purple base. It grows in woods and pinelands in the coastal plain from S.C. south to Fla. and west to La. (Not illustrated)

RED HIBISCUS *(Hibiscus coccineus)*

This is a smooth plant up to 10 feet tall with leaves divided into from 5 to 7 narrow and toothed lobes. The crimson or deep red flowers are often 6 inches across. It grows in swamps near the coast from Ga. and Fla. west to Tex. (Not illustrated)

COASTAL-MALLOW *(Kosteletzyka virginica)*

The Coastal-mallow is an erect, branching, more or less downy or sometimes roughish-hairy plant 1½ to 4 feet tall; growing in salt or brackish marshes and on shores. The leaves are egg-shaped to somewhat arrow-shaped, often 3-lobed, always coarsely toothed, and 2 to 5 inches long. It has pink flowers 1½ to 2½ inches across which bloom between July and October. Other names for it are Saltmarsh-mallow and Seashore-mallow.

RANGE: se. N.Y. and Del. south to Fla. and west to Tex.

SOUTHEASTERN POPPY-MALLOW *(Callirhoe papaver)*

There are several species of poppy-mallows and most of them are plants of the mid-west. This one has smoothish or somewhat hairy, reclining stems 8 to 24 inches long. Its leaves are 3- to 5-parted or lobed, with narrow but untoothed segments. It has solitary reddish-purple flowers from 1 to 2½ inches broad which bloom between June and August. The plant grows in sandy or rocky woods, glades, and on prairies.

RANGE: Ga. to Mo. south to Fla. and Tex.

Pink Swamp Rose-mallow

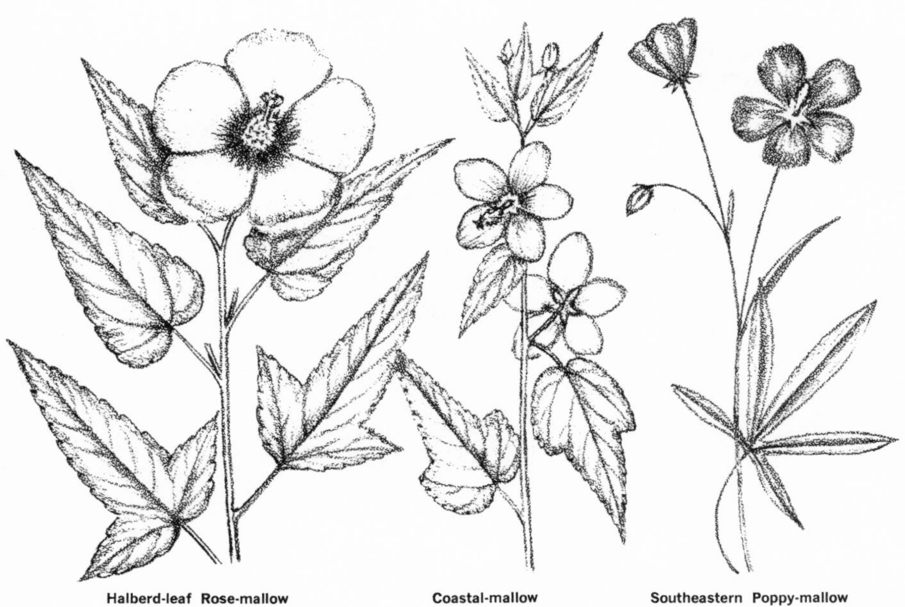

Halberd-leaf Rose-mallow Coastal-mallow Southeastern Poppy-mallow

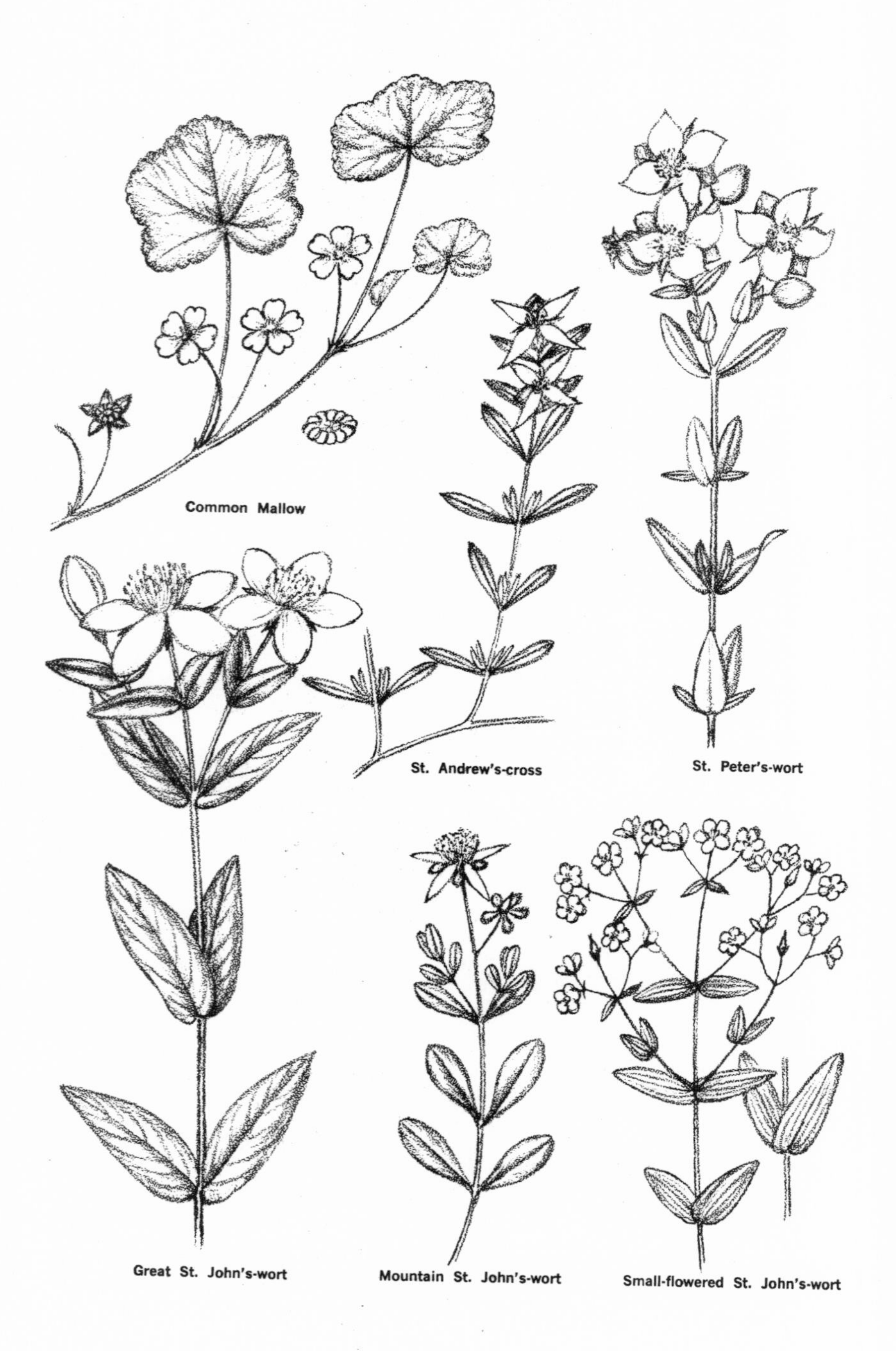
Common Mallow
St. Andrew's-cross
St. Peter's-wort
Great St. John's-wort
Mountain St. John's-wort
Small-flowered St. John's-wort

COMMON MALLOW *(Malva neglecta)*

Another common name for this little trailing plant is Cheeses, as its rings of small fruits suggest a tiny cheese and have a mildly pungent taste. Its flowers are pale lilac or whitish, about ½ inch across, and are clustered in the axils of the leaves; blooming over a long period between April and October. The leaves are roundish heart-shaped, shallowly lobed, bluntly toothed, from 1 to 3 inches wide, and have long stalks. A native of Europe, it has become widely naturalized here in America; growing in yards, barnyards, and waste places.

ST. JOHN'S-WORT FAMILY (Hypericaceae)

Members of this family have opposite, entire leaves with tiny translucent or black dots. Their flowers have 4 or 5 sepals and a like number of petals, usually numerous stamens, and a pistil composed of from 2 to 5 united carpels which are indicated by the number of styles.

ST. ANDREW'S-CROSS *(Hypericum hypericoides)*

This is a woody-based little plant with ascending, flattened and 2-edged, often branching stems from 5 to 10 inches high. Its flowers are about ¾ across and have 2 broadly egg-shaped sepals and 2 very much smaller and narrow ones. The 4 yellow petals form an oblique cross. The pistil is composed of 2 carpels and there are 2 distinct styles. It develops into a pod enclosed by the pair of large sepals. The leaves are oblong or top-shaped and ½ to 1½ inches long. This plant grows in dry sandy or rocky woods and thickets; blooming between May and August.

RANGE: Mass. to Ill. and Kan. south to Fla. and Tex.

ST. PETER'S-WORT *(Hypericum crux-andreae)*

The St. Peter's-wort is a woody-based plant with a simple or sparingly branched, 2-edged stem from 1 to 2 feet high. Its flowers are ¾ to an inch across and have 4 broad, bright yellow petals; a pair of large sepals and a pair of smaller ones; and a pistil with 3 or 4 styles. The leaves are oval or oblong, whitened beneath, from ¾ to 1½ inches long, and have heart-shaped bases which clasp the stem. It grows in moist to dry, sandy, open woods and in fields; blooming between June and October.

RANGE: se. N.Y. and N.J. to Ky. south to Fla. and Tex.

GREAT ST. JOHN'S-WORT *(Hypericum ascyron)*

This plant has a stout, erect, branching stem from 2 to 5 feet tall; its branches being 2- to 4-angled. The flowers are 1 to 2 inches across and have 5 bright yellow petals, and a pistol with 5 styles. Its leaves are narrowly egg-shaped, from 2 to 5 inches long, and have somewhat clasping bases. It grows on rich, rocky, wooded slopes and along the banks of streams; blooming between July and September.

RANGE: Me. to Man. south to N.J., Md., Ill., Mo. and Kan.

MOUNTAIN ST. JOHN'S-WORT *(Hypericum buckleyi)*

On rocky mountain summits and cliffs in the Southern Appalachians, the Mountain St. John's-wort blooms between June and August. It is a trailing or straggling little plant from 4 to 12 inches high, with pairs of top-shaped or elliptic leaves from ¼ to ¾ of an inch long. Its flowers, between ¾ and an inch across, have 5 yellow petals about twice as long as the spoon-shaped sepals, and a pistil with 3 styles.

RANGE: w. N.C. south to n. Ga.

SMALL-FLOWERED ST. JOHN'S-WORT *(Hypericum mutilum)*

This is one of several species of St. John's-worts having small flowers less than ¼ inch across. This one has branching, slender, weak stems 6 inches to 2½ feet tall; and oblong or egg-shaped leaves ½ to an inch long, usually with a clasping base and 5 main veins. The numerous flowers are about 3/16 inch across, with 5 yellow petals and about 10 stamens. They are arranged in rather large, loose, leafy-bracted end clusters; blooming between June and October. It grows in low, moist, open woods and meadows.

RANGE: N.S. to Ont. and Wis. south to Fla. and Tex.

DOTTED ST. JOHN'S-WORT *(Hypericum punctatum)*

This is a smooth and sparingly-branched plant from 1½ to 3 feet tall, with a crowded flower cluster at the summit. The flowers are about ½ inch across and the 5 yellow petals have several rows of small black dots. The pistil has 5 styles. Its leaves are 1 to 3 inches long, oblong in shape, rounded at the tip, and more or less clasping at the base. They are liberally sprinkled with both translucent and black dots. This is a common plant in open woods, thickets, and fields; blooming between June and September.

RANGE: Que. to Ont. and Minn. south to Fla. and Tex.

COMMON ST. JOHN'S-WORT *(Hypericum perforatum)*

A native of Europe, this St. John's-wort is now widely naturalized here in America and generally abundant in fields, waste places, and along roadsides. It somewhat resembles the preceding species but it is usually less tall, much more branched, and the yellow petals of the flowers are black-dotted only on their margins. The pistil has but 3 styles. It blooms between June and September. (Not illustrated)

CANADA ST. JOHN'S-WORT *(Hypericum canadense)*

This is a slender-stemmed, branching plant from 6 to 20 inches high. It has very narrow leaves with 1 to 3 main veins, and they are ½ to 2 inches long. The small flowers are ¼ inch or less across and have 5 yellow petals and 5 to 10 stamens. It grows in swampy places, on wet rocks, or on moist banks; blooming between July and September.

RANGE: Nfd. to Man. south to Ga., Ala., Ill. and Iowa.

MARSH ST. JOHN'S-WORT *(Triadenum virginicum)*

The Marsh St. John's-wort has a smooth, simple or branched stem 1 to 2½ feet tall. Its leaves are 1 to 3 inches long, egg-shaped or oblong, rounded at the tip, clasping at the base, black-dotted, and often whitish beneath. The flowers are about ⅔ inch across, pink to greenish-purple, and have 9 stamens. The latter are united by the bases of their filaments into 3 groups which alternate with 3 large orange-colored glands. It grows in swamps, bogs, and wet sandy places; blooming between July and September.

RANGE: Nfd. to Man. south to Fla. and Tex.

LARGER MARSH ST. JOHN'S-WORT *(Triadenum tubulosum)*

This is usually a larger and more branched plant than the preceding. It can be distinguished by its leaves which are short-stalked or almost stalkless but not clasping.

RANGE: Md. and W.Va. to Ind. and Mo. south to Fla. and Tex. (Not illustrated)

ROCKROSE FAMILY (Cistaceae)

Members of this family have flowers with 5 sepals, 2 of which are small or sometimes lacking; 3 or 5 petals; numerous stamens; and a solitary pistil with 1 style, or the style sometimes absent. The plants usually have alternate leaves.

FROSTWEED *(Helianthemum canadense)*

The Frostweed branches near the base into several grayish-downy stems from 8 to 16 inches high. Usually each one has a solitary flower between May and July, with 5 bright yellow petals and an inch to 2 inches across. Later there are smaller bud-like flowers which are rather inconspicuous but produce most of the seeds. The leaves are lance-shaped or inversely lance-shaped and broadest near the tip, stalkless or nearly so, and from ½ to 1½ inches long. In late autumn ice crystals shoot from the bases of the stems, hence the common name. It grows in rocky open woods and on sandy barrens.

RANGE: N.S. to Que. and Wis. south to Ga., Miss. and Mo.

CAROLINA ROCKROSE *(Helianthemum carolinianum)*

The Carolina Rockrose is a plant of the drier coastal plain pinelands. In May or June it usually has 1 or 2 showy flowers 1 to 1½ inches across, with 5 bright yellow petals. As in the preceding species, it also produces small bud-like flowers which are quite inconspicuous. The plant has a hairy stem from 2 to about 10 inches tall on which there are a few scattered leaves, most of its leaves being in a basal cluster. They are elliptical or inversely lance-shaped and from ¾ to 1½ inches long.

RANGE: e. N.C. south to Fla. and west to Tex.

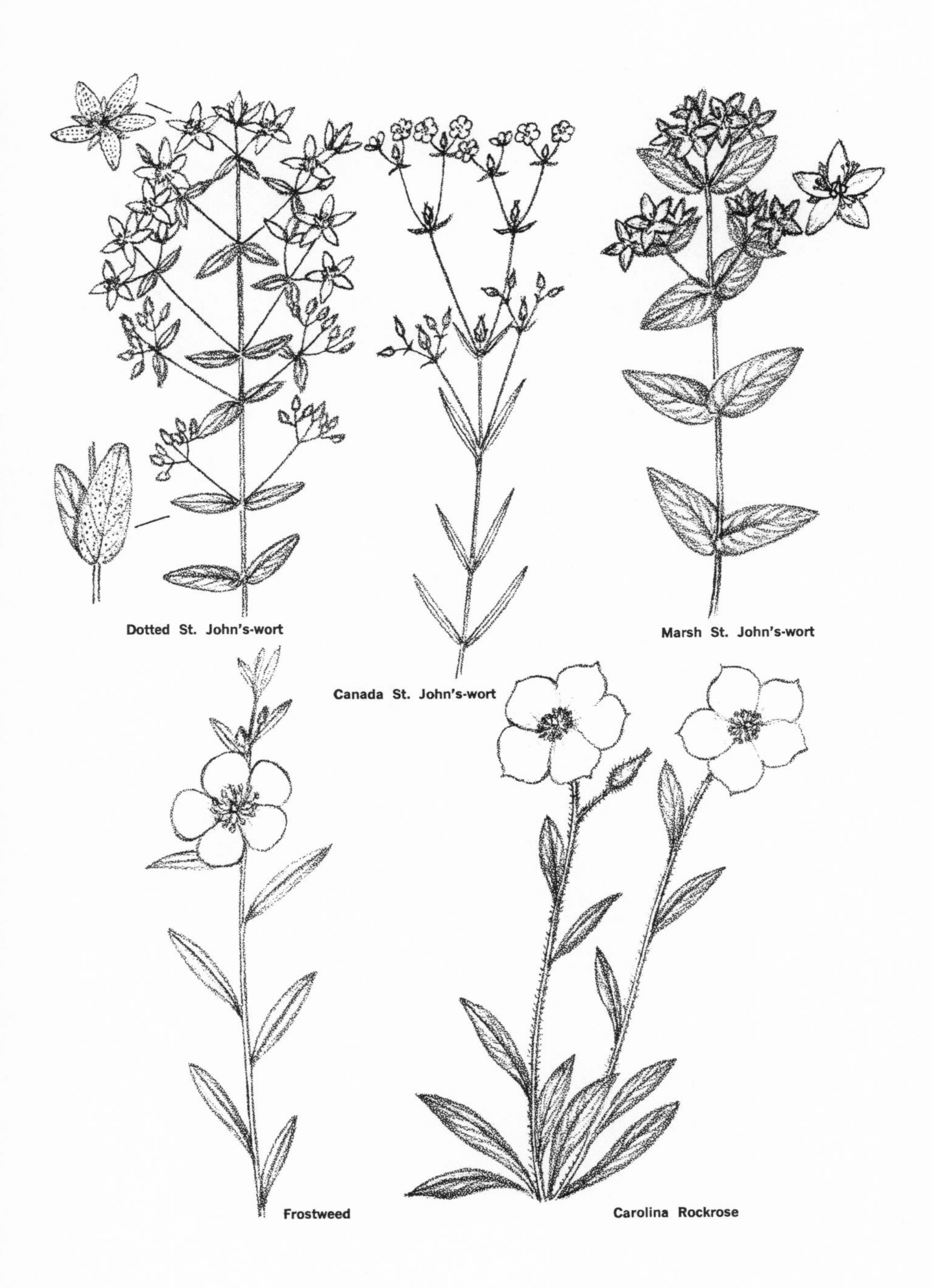
Dotted St. John's-wort
Canada St. John's-wort
Marsh St. John's-wort
Frostweed
Carolina Rockrose

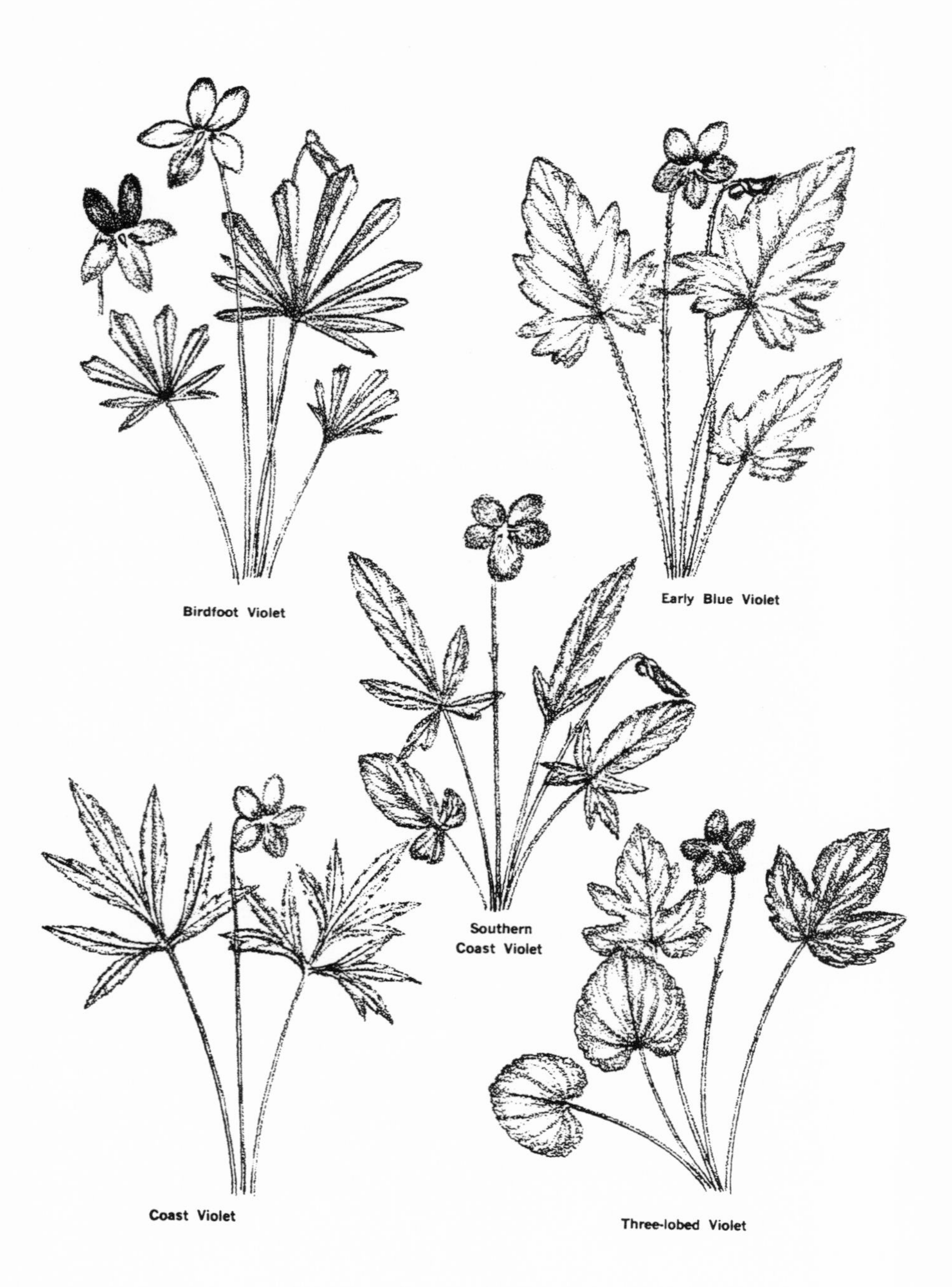
Birdfoot Violet
Early Blue Violet
Southern
Coast Violet
Coast Violet
Three-lobed Violet

Members of this family are among our most common and best known wild flowers. Although they are easy to recognize as being violets, it is often very difficult to identify the numerous species. This is especially true of the so-called "stemless" blue violets, among which there are many confusing hybrids.

The flowers have a bilateral symmetry. There are 5 sepals; 5 petals, the lower one being extended backward as a hollow sac or spur; 5 stamens in which the anthers form a cone-shaped cluster about the base of the pistil; and a solitary pistil with a club-shaped style and stigma. Besides the showy flowers, many violets also have bud-like flowers which produce large quantities of seeds. The fruit is a capsule containing numerous seeds.

BIRDFOOT VIOLET *(Viola pedata)*

This is certainly one of our most striking species of violet, and one that is very easy to recognize. Its showy flowers are often an inch or more across. Usually all 5 petals are lilac-colored but in one variety the 2 upper petals are dark violet. The stamens are a bright orange-yellow. Its name comes from the leaves which are deeply cut into narrow and radiating segments. Both leaves and flowers arise from a short, erect, fleshy rootstock. It grows on dry, sunny, sandy, rocky, or clayey banks and open woods; blooming between March and June.

RANGE: Mass. to Ont., Minn. and Kan. south to Ga. and Tex.

EARLY BLUE VIOLET *(Viola palmata)*

From March to May this violet blooms on dry but rich wooded hillsides. It is best distinguished by its leaves which have from 5 to 11 lobes or segments, the middle one being largest and all variously toothed and cleft. The violet-purple flowers are about an inch across. Both they and the leaves arise from a thick, erect, fleshy rootstock and both have stalks which are somewhat hairy.

RANGE: N.H. to Ont. and Minn. south to Fla. and Miss.

COAST VIOLET *(Viola brittoniana)*

Often the earliest leaves of this violet are merely toothed, but the ones which follow are 3-parted and the segments are cut into 2 to 4 narrow and pointed lobes. The flowers are a rich violet with a conspicuous white throat and are an inch or more across. Both the flowers and leaves arise from a thick, erect, fleshy rootstock and have stalks which are quite smooth. This one grows in moist sandy or peaty soils, blooming between April and June.

RANGE: Coast of s. Me. south to coast and mts. of N.C.

SOUTHERN COAST VIOLET *(Viola septemloba)*

In spite of its scientific name, the leaves of this violet are not always 7-lobed. They are very variable, ranging from some that are just heart-shaped to others which are variously lobed or divided. The flowers are an inch across and usually well above the leaves. They are a deep blue-violet with a broad white center, and all 3 of the lower petals have tufts of hair at the base. Both flowers and leaves arise from a thick, erect, fleshy rootstock and have stalks which are sparingly hairy. It is the common blue violet of the sandy coastal plain pinelands; blooming from late March to May.

RANGE: Va. south to Fla. and west to La.

THREE-LOBED VIOLET *(Viola palmata* var. *triloba)*

The earliest leaves, as well as some of those developed later in the summer, are heart-shaped with toothed margins. Those produced in between may be 3- to 5-lobed with the middle segment always the largest. The flowers are about an inch across, deep violet, and all 3 of the lower petals have a hairy tuft at the base. Both flowers and leaves arise from a thick, erect, fleshy rootstock and have stalks which are somewhat hairy. It grows in rich but dry woodlands, blooming from late March to May.

RANGE: Pa. to Ill., Mo. and Okla. south to Fla. and Tex.

MARSH BLUE VIOLET *(Viola cucullata)*

As its name implies, this violet grows in wet meadows, bogs, and springy places. Its violet-blue flowers are darker toward the center, about an inch broad, and are on long stalks which overtop the leaves. The side petals have hairy tufts containing many club-shaped hairs. The leaves are heart-shaped with rather bluntish teeth on the margin, and they are smooth or very nearly so. The plant is usually tufted with several crowns; and both flowers and leaves arise from a thick, erect, fleshy rootstock. It blooms between April and July.

RANGE: Nfd. to Ont. and Minn. south to Va., n.Ga., Ark. and Neb.

COMMON BLUE VIOLET *(Viola sororia)*

This is undoubtedly the commonest of our blue violets; growing in open woods, meadows, dooryards, and along roadsides everywhere. Its rich violet-colored flowers are about an inch across, whitish toward the center, and are on smooth stalks no longer than the leaves. Both of the side petals are "bearded" but none of the hairs are club-shaped. Its smooth leaves are broadly heart-shaped and by midsummer may become 5 inches wide. Both flowers and leaves arise from a thick and fleshy rootstock. It blooms between late February and June. Also called Hooded Blue Violet.

RANGE: Minn. to Me. south to Fl. and west to Ks.

ARROW-LEAF VIOLET *(Viola sagittata)*

The leaves of this violet are shaped like narrow arrow-heads, tapering to a pointed tip; usually with 1 or more pairs of lobes or coarse teeth at the base. At maturity they become from 1½ to 4 inches long. They are usually smooth but sometimes finely hairy, or with a hairy fringe on the margin. The flowers are violet-purple with a white center, about ¾ inch across, and are on stalks about as long as the leaves; both arising from a stout, fleshy rootstock. It grows on moist banks and in fields and upland woods; blooming between late March and June.

RANGE: Mass. to Minn. south to Ga. and Tex.

ROUND-LEAF YELLOW VIOLET *(Viola rotundifolia)*

This is our only "stemless" violet with yellow flowers. It has roundish or oval leaves with a wavy-toothed margin. At flowering time, between March and May, they are only about an inch wide and are finely hairy. Later they become 2 to 4 inches broad, thickish, quite smooth, and lie flat on the ground. The flowers are about ½ inch across, bright yellow, the 3 lower petals being veined with brown. Both leaves and flowers arise from a long, stout, jagged rootstock. Look for it in cool, moist, rich woods or on banks; southward in the mountains.

RANGE: Me. to Ont. south to Del., Pa., n. Ga. and e. Tenn.

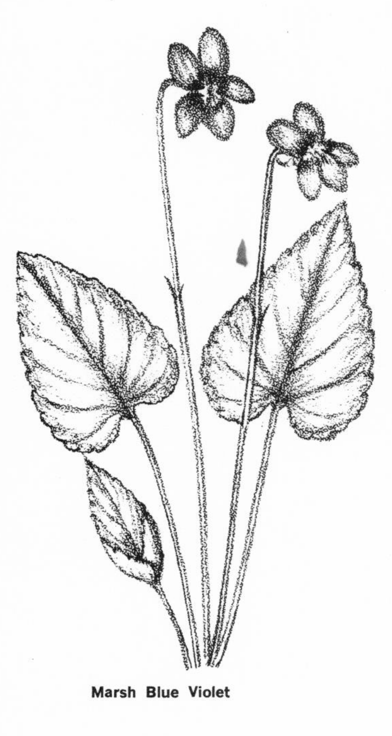

Marsh Blue Violet

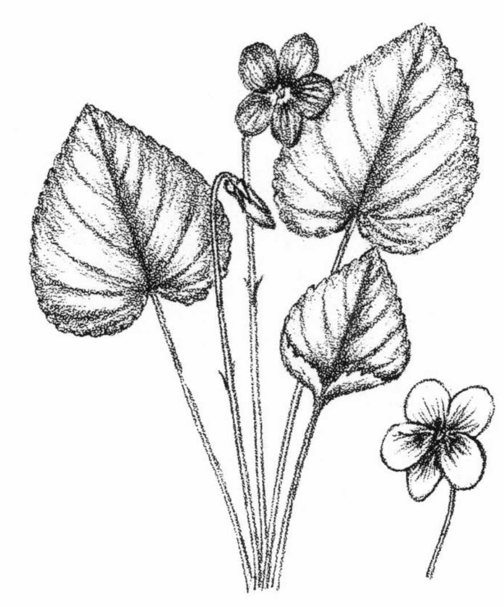

Common Blue Violet

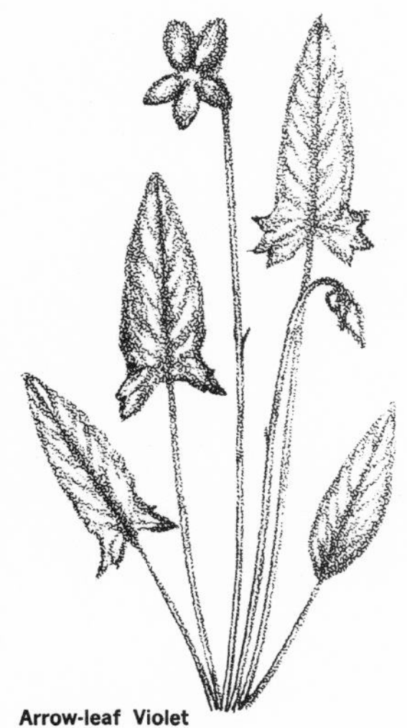

Arrow-leaf Violet

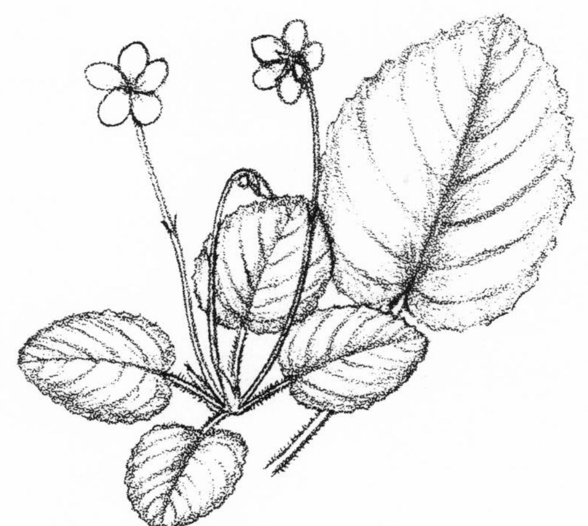

Round-leaf Yellow Violet

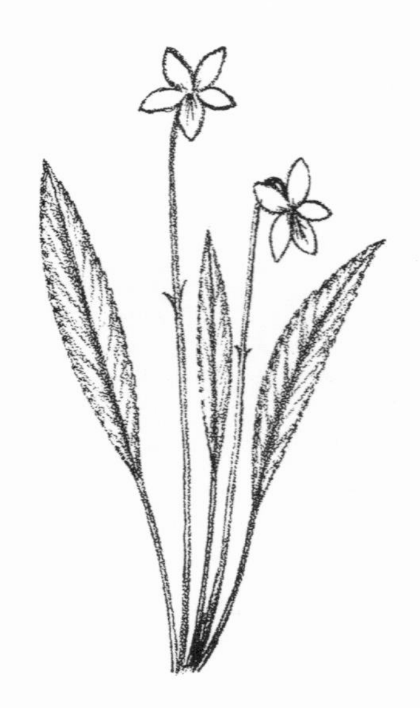

Lance-leaf Violet

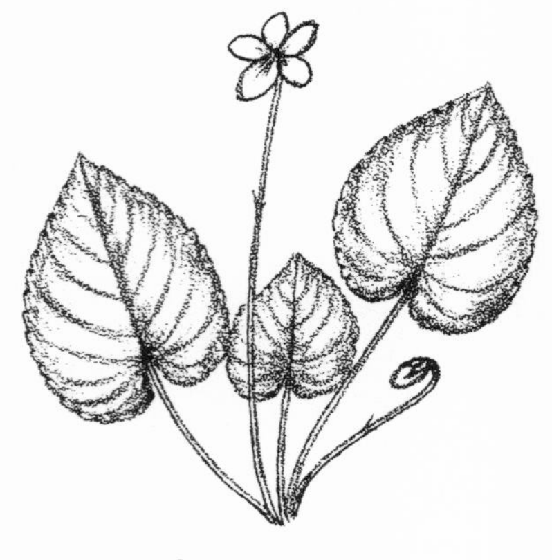

Sweet White Violet

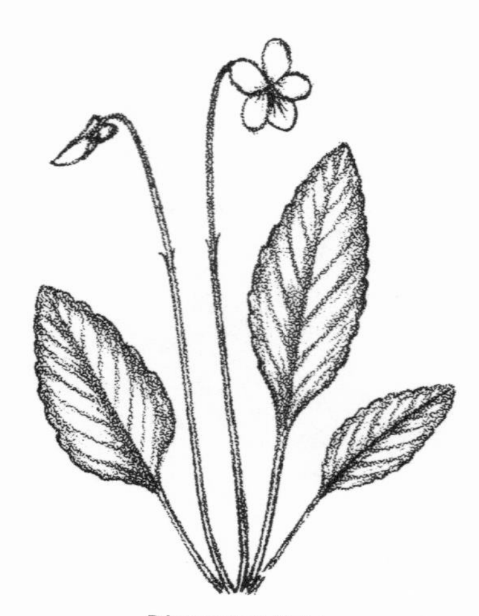

Primrose-leaf Violet

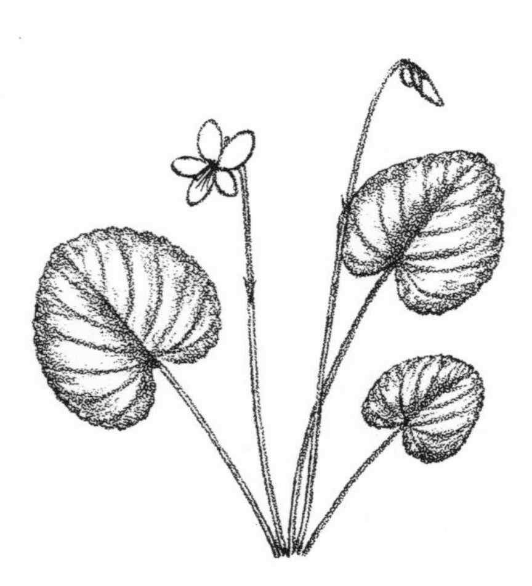

Northern White Violet

SWEET WHITE VIOLET *(Viola blanda)*

Expect to find this little white-flowered violet in cool, moist woods and ravines, where it blooms in April or May. Its fragrant flowers are about ½ inch across. The 3 lower petals are veined with purple; the 2 upper ones plain but often twisted or bent backward. The leaves are heart-shaped, 1 to 2 inches wide, the sinus at the base being very narrowly V-shaped. They are smooth except for a few small and widely scattered hairs on the upper surface. Both leaves and flowers usually have reddish-tinged stalks and arise from a slender rootstock.

RANGE: N.H. and Que. to Minn. south to Md., n.Ga., Tenn., Ohio, Ill. and Wis.

LARGE-LEAF WHITE VIOLET *(Viola blanda* var. *palustriformis)*

While this variety resembles the preceding species, it is a coarser plant; its leaves having a wrinkled-veiny appearance and being downy beneath. The 2 side petals of its flowers have hairy tufts at the base, and the flowers are slightly, if at all, fragrant. It grows in wet or moist woods, thickets, and clearings; blooming between April and June.

RANGE: Nfd. to Ont. south to N.Y., e. Tenn., the Great Lakes region, and N.D. (Not illustrated)

NORTHERN WHITE VIOLET *(Viola macloskeyi* var. *pallens)*

The flowers of this species resemble those of the preceding one and are very fragrant. Its leaves are ½ to 2½ inches wide, entirely smooth, heart-shaped, and have a very broadly V-shaped basal sinus. It grows in wet or springy woods, bogs, and along stream banks; blooming between April and July.

RANGE: Lab. to Alaska south to Pa., mts. of Ga. & Tenn. Ill., Iowa, N.D. and Mont.

LANCE-LEAF VIOLET *(Viola lanceolata)*

This violet can be recognized very easily by its long and narrow, lance-shaped leaves which have low and rounded teeth on the margin. The leaves are commonly from 2 to 6 inches long, but sometimes as much as a foot in length. The white flowers are a half inch or a little more across, the 3 lower petals having conspicuous purple veins. Both leaves and flowers arise from a slender rootstock, and the plants spread by means of runners. Look for this violet in wet meadows, marshy places, and bogs. It blooms between March and July.

RANGE: N.S. to Que., Minn. and Neb. south to Fla. and Tex.

PRIMROSE-LEAF VIOLET *(Viola × primulifolia)*

This little white-flowered violet may be recognized by its egg-shaped or elliptic leaves which taper at the base into their stalks. They have rather inconspicuous teeth on the margin and are more or less hairy. The flowers are about ½ inch across and the 3 lower petals are veined with purple. Both flowers and leaves arise from a slender rootstock, and the plants spread freely by means of runners. It grows in moist open woods, clearings, and meadows; blooming between March and June. This is a hybrid of *V. lanceolata* and *V. macloskeyi.*

RANGE: N.S. to Que., Minn. and Okla. south to Fla. and Tex.

CANADA VIOLET *(Viola canadensis)*

This is a smooth or slightly hairy plant with 1 or more leafy stems usually between 8 and 14 inches tall. The flowers have white petals which are tinged with lilac on the back. They are about ¾ inch across and on slender stalks arising from the axils of the heart-shaped leaves. This violet grows in cool, moist, rich woodlands; blooming between April and July.

RANGE: N.H. and Que. to Mont. south to Md., nw. S.C., n. Ga., Tenn., Iowa, S.D., Colo. and Utah.

HALBERD-LEAF VIOLET *(Viola hastata)*

On a slender stem 4 to 10 inches tall, this violet has from 2 to 4 narrowly triangular, taper-pointed leaves. Its yellow flowers are about ½ inch across and on slender stalks arising from the axils of the leaves. The petals are tinged with violet on the back. It grows in rich, often rocky woods; blooming late March to May.

RANGE: Pa. and Ohio south to Fla. and Ala.

STRIPED VIOLET *(Viola striata)*

The large and conspicuously fringed stipules afford a good field mark in identifying this violet. It usually has several smooth, angled, leafy stems from 6 to 12 inches long; and heart-shaped leaves 1 to 1½ inches broad. The flowers are white or creamy-white, about ¾ inch across, and the 3 lower petals are conspicuously veined with purple. They are on slender stalks arising from the axils of the leaves; and bloom between March and June. It is often common in low, moist woods and meadows, and along streams. Other names given it are Pale Violet or Cream Violet.

RANGE: N.Y. to Ont. and Wis. south to Ga., Tenn. and Ark.

SMOOTH YELLOW VIOLET *(Viola pubescens)*

This violet has smooth, leafy stems 6 to 12 inches tall with from 1 to 3 basal leaves. The latter, and the few stem leaves are smooth, heart-shaped, and from 2 to 4 inches wide. The yellow flowers are about ¾ inch across, their 3 lower petals being veined with purple. It grows in moist woods and on cool rocky slopes; blooming in April or May.

RANGE: Me. to Minn. south to Ga., Ala., Ark. and Okla.

LONG-SPURRED VIOLET *(Viola rostrata)*

The flowers of this violet are lilac-colored and have a slender spur about ½ inch long, or fully as long as the petals. It is a smooth little plant with several leafy stems 4 to 8 inches high. The flowers are on slender stalks which arise from the axils of the heart-shaped leaves. It grows in moist rich woodlands and on rocky slopes, blooming between April and June.

RANGE: Vt. to Que. and Wis. south to n. Ga. and n. Ala.

AMERICAN DOG VIOLET *(Viola conspersa)*

This is a small, smooth plant with branching and leafy stems 3 to 6 inches high at flowering time between late March and June. The flowers are about ½ inch across, pale violet, and have a blunt-tipped spur about half as long as the petals. It grows in moist woods and meadows.

RANGE: Que. to Minn. south to Md., n. Ga., n. Ala. and Tenn. (Not illustrated)

FIELD PANSY *(Viola bicolor)*

This little annual has smooth, slender, branching stems 3 to 8 inches high; and mostly top-shaped leaves an inch or less long, with large and deeply cut stipules at their base. Its flowers are about ½ inch across, creamy-white to pale blue or lilac, the 3 lower petals with purple veins. Naturalized from Europe.

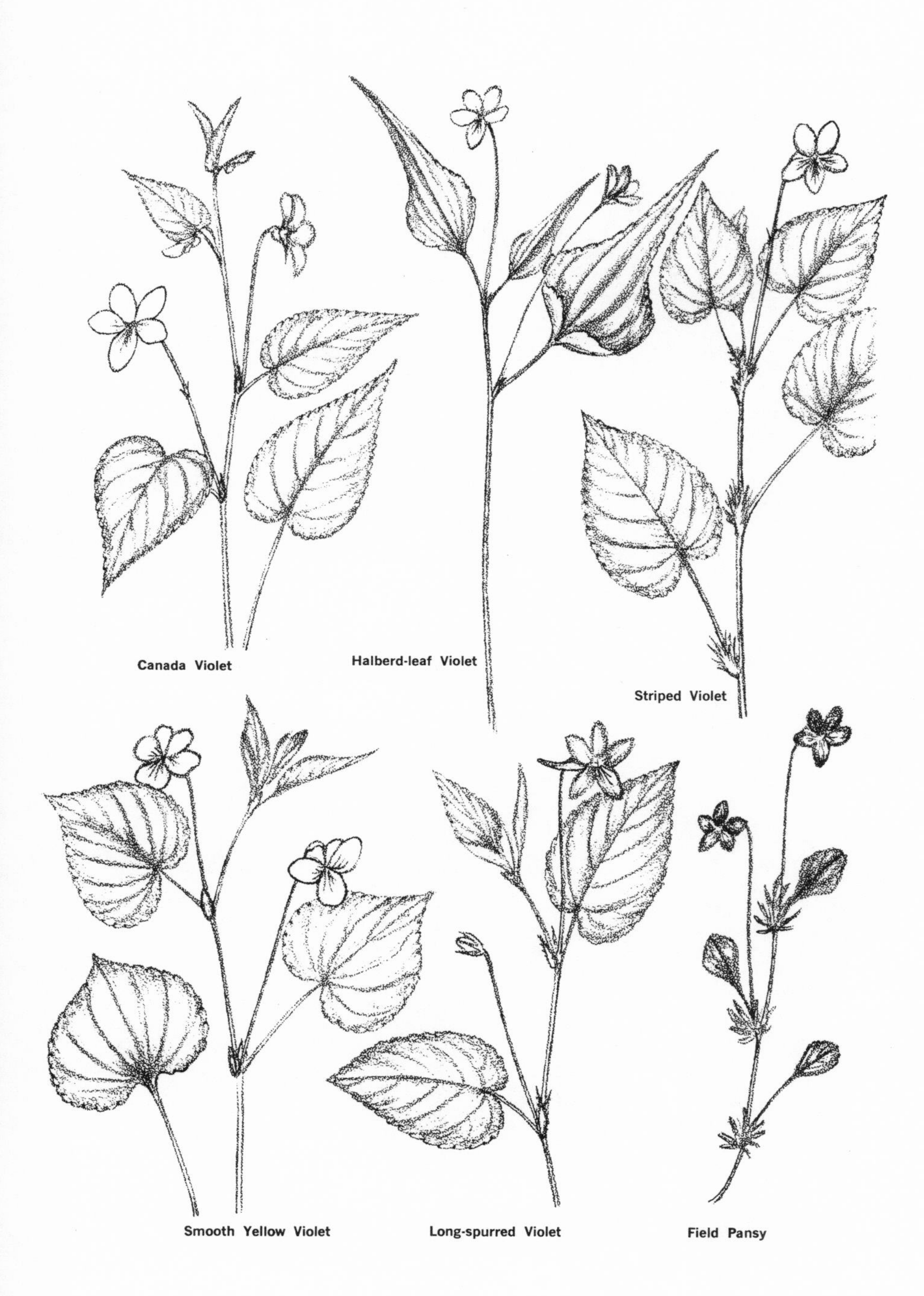
Canada Violet
Halberd-leaf Violet
Striped Violet
Smooth Yellow Violet
Long-spurred Violet
Field Pansy

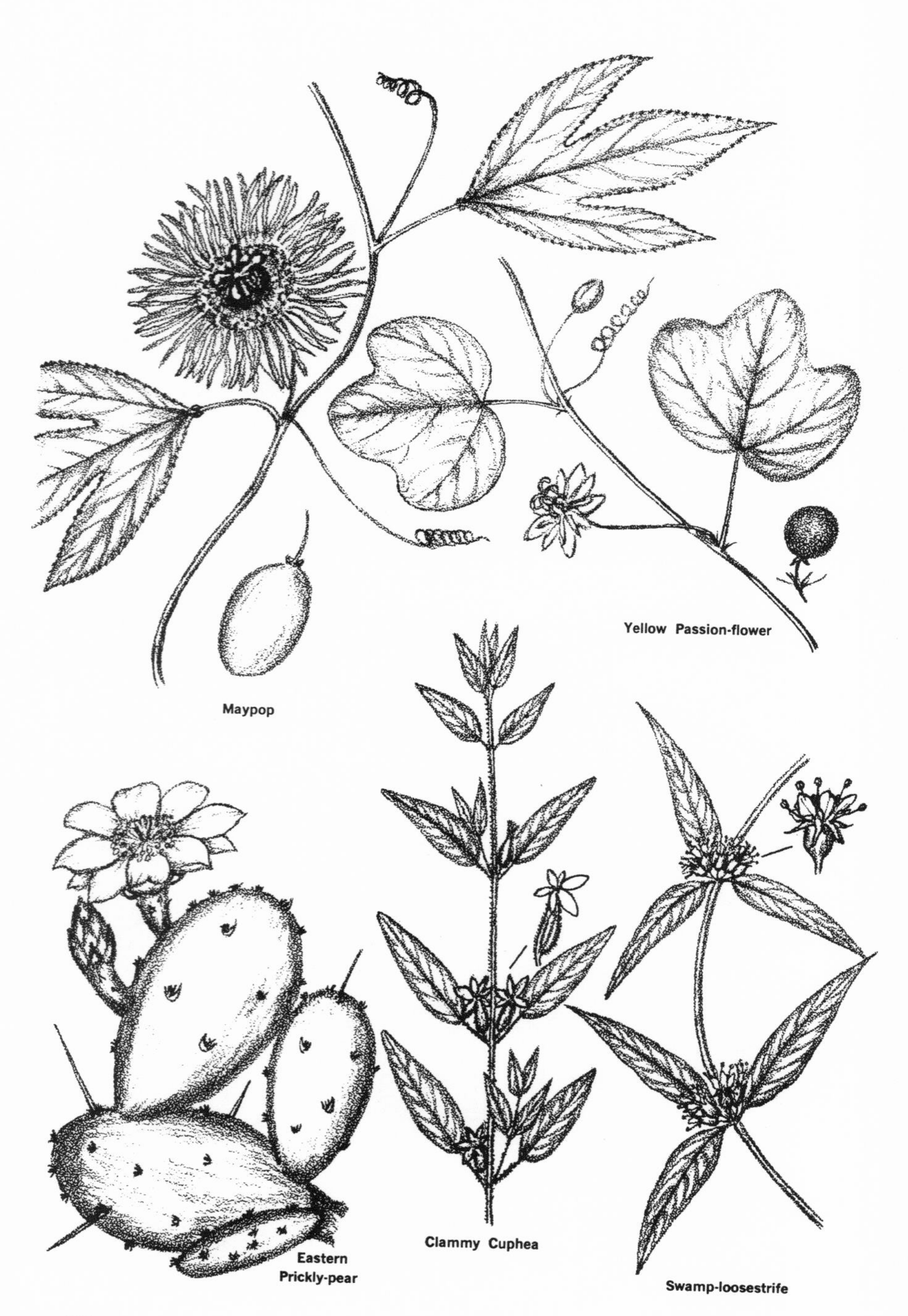
Maypop
Yellow Passion-flower
Eastern
Prickly-pear
Clammy Cuphea
Swamp-loosestrife

PASSION-FLOWER FAMILY (Passifloraceae)

Our members of this family have flowers with 5 sepals united at the base, the throat of the calyx bearing a "crown" of numerous fringe-like segments; 5 petals which are insterted on the calyx; 5 stamens, their filaments united to form a tube about the long stalk of the ovary but the large anthers separate. The fruits are many-seeded berries.

MAYPOP *(Passiflora incarnata)*

Also known as the Purple Passion-flower, this trailing or climbing vine grows abundantly in open woods, thickets, and dry fields in the South. Its leaves are 3-lobed, toothed on the margin, from 3 to 5 inches wide, and have stalks 1 to 2 inches long with a pair of prominent glands at their summits. The odd-looking lavender flowers are 2 to 3 inches broad and have a very striking fringed "crown". They bloom between May and August and are followed by egg-shaped fruits about 2 inches long. When ripe they are yellowish and edible. Passion-flowers get their name from a fancied resemblance of the flower parts to various implements of the crucifixion.

RANGE: Md. to s. Ohio, Ill., Mo. and Okla. south to Fla. and Tex.

YELLOW PASSION-FLOWER *(Passiflora lutea)*

This is a slender-stemmed trailing or climbing vine which grows in thickets and the borders of woods. It has leaves which are much broader than long, shallowly 3-lobed, untoothed on the margin, and 2 to 3 inches wide. Their stalks are ½ to 1½ inches long and have no glands at the summit. The greenish-yellow flowers are usually less than an inch across and bloom between June and September. They are followed by roundish, dark purple fruits about ½ inch in diameter.

RANGE: Pa. to Ill., Mo. and Kan. south to Fla. and Tex.

CACTUS FAMILY (Cactaceae)

Members of the Cactus Family have flowers in which the several to many sepals and petals have their bases joined to the ovary; numerous stamens which are inserted on the inside of the cup or tube formed by the union of the sepals and petals; and a pistil with 1 style.

EASTERN PRICKLY-PEAR *(Opuntia humifusa)*

Prickly-pears have fleshy, flat, jointed stems which are green. The leaves are small and scale-like and are soon shed; but in their axils there are clusters of barbed hairs and occasional slender spines. The flowers are 2 to 3 inches across, yellow sometimes with a red star-shaped "eye", and have from 8 to 12 petals. They are followed by pear-shaped, dull purplish-red, fleshy fruits 1 to 1½ inches long. It grows in dry sandy and rocky places, blooming between May and July. Another name for it is Indian-fig, the fruits being edible. RANGE: Mass. to Minn. south to Fla., Ala., Mo. and Okla.

LOOSESTRIFE FAMILY (Lythraceae)

Members of this family have flowers with 4 to 6 sepals united into a tube; 4 to 6 petals and 4 to 12 stamens which are both inserted on the calyx tube; and a solitary pistil which is within but not united with the calyx tube.

CLAMMY CUPHEA *(Cuphea viscosissima)*

This is an erect, branched, very sticky-hairy little plant 6 to 20 inches tall; often very common in dry fields. It has broadly lance-shaped and slender stalked leaves from 1 to 1½ inches long. The purple flowers borne in the axils of the leaves are not very large and have 6 petals of varying size. It blooms between July and October. Another name for it is Blue Waxweed. RANGE: N.H. to Ill. and Iowa south to Ga. and La.

SWAMP-LOOSESTRIFE *(Decodon verticillatus)*

This plant with angled and arching stems 2 to 8 feet long is often common in swamps and other shallow waters. The submerged portions of the stems are thickened and spongy, and the tips often bend over and take root. It has lance-shaped leaves 2 to 5 inches long which are usually in whorls of 3. Pinkish-purple flowers about an inch long are clustered in the axils of the upper leaves, blooming from July to September. They have 5 petals, 5 long-stalked stamens and 5 shorter ones. It is sometimes called the Water-willow. RANGE: Me. to Ont. and Minn. south to Fla. and La.

This is a large family of chiefly tropical plants. Our members of the family—the Meadow-beauties—are small plants with opposite 3- to 7-ribbed leaves. Their flowers have 4 sepals united to form an urn-shaped or vase-like tube with 4 lobes at the summit; 4 somewhat oblique, showy, and delicate petals; 8 stamens; and a pistil composed of 4 united carpels but with 1 long style and 1 stigma. The petals and sepals are attached near the rim of the calyx tube. The flowers open early in the morning and the petals are usually dropped by mid-day.

VIRGINIA MEADOW-BEAUTY *(Rhexia virginica)*

Deergrass is another name often given this attractive plant of bogs and moist, open, sandy places. It is a more or less hairy plant 1 to 2 feet tall, with a simple or branched stem which is 4-sided and has narrow wings. The leaves are oval or egg-shaped, 1 to 2 inches long, and have small bristly teeth on their margins. Its bright purple flowers are an inch or so across and have a glandular-hairy calyx tube. They bloom between May and September. RANGE: N.S. to Ont. south to Ga., Ala., Tenn. and Mo.

CILIATE MEADOW-BEAUTY *(Rhexia petiolata)*

This meadow-beauty has a smooth square stem from 1 to 2 feet tall which is simple or but slightly branched above. Its leaves are oval or egg-shaped, ½ to 1 inch long, and have a bristly fringed margin. The flowers are violet-purple, 1 to 1½ inches across, and are stalkless or very nearly so. They have a calyx tube which is quite smooth. This species grows in wet pinelands in the Southeast, blooming between June and September.
RANGE: se. Va. south to Fla. and La.

MARYLAND MEADOW-BEAUTY *(Rhexia mariana)*

The Maryland Meadow-beauty has a hairy, roundish or slightly 4-sided stem from 1 to 2 feet tall which is simple or somewhat branched above. Its leaves are lance-shaped or elliptic, 1 to 1½ inches long, bristly toothed on the margin, and are narrowed at the base to very short stalks. The flowers vary from pale purplish to whitish, are about an inch across, and have a glandular-hairy calyx tube shaped like a long-necked vase. It grows in coastal plain pinelands, swamps, and other moist sandy places; blooming between May and September.
RANGE: se. Mass. and e. Pa. south to Fla.; inland from Va. and Ky. south to Ga. and La.

SAVANNAH MEADOW-BEAUTY *(Rhexia alifanus)*

This, the largest and certainly the most beautiful of our meadow-beauties, is often abundant in the wetter coastal plain pinelands and savannahs of the Southeast. There it blooms between May and September. It is a smooth plant, more or less whitened by a waxy bloom, with roundish, slender, simple or sparingly branched stems from 1 to about 3 feet tall. The pairs of 1- to 3-inch lance-shaped leaves point upward. Its flowers are a bright rose-purple, about 2 inches across, and have a calyx tube bristling with reddish glandular hairs. RANGE: e. N.C. south to Fla. and west to La.

YELLOW MEADOW-BEAUTY *(Rhexia lutea)*

Of all our meadow-beauties, this is the only one having yellow flowers. It is a sparsely hairy plant with a 4-sided and branching stem from 6 to 10 inches tall; with pairs of narrow, pale green, and bristly margined leaves an inch or less long. The flowers are an inch or less across, with a calyx tube shaped like a narrow-necked vase and stamens with relatively short anthers. It grows in wet coastal plain pinelands, blooming between April and July. RANGE: e. N.C. south to Fla. and west to La.

AWN-PETALLED MEADOW-BEAUTY *(Rhexia aristosa)*

This meadow-beauty is distinguished by the little tail-like tips on its magenta or rose-purple petals. It has a slender, smooth, 4-sided stem 1 to 2 feet tall which is sometimes branched above. The leaves are ¾ to 1½ inches long, narrow, stalkless, and may have a few inconspicuous teeth. It grows in sandy swamps and wet coastal plain pinelands, blooming between June and September.
RANGE: N.J. and Del. south to Ga. and Ala. (Not illustrated)

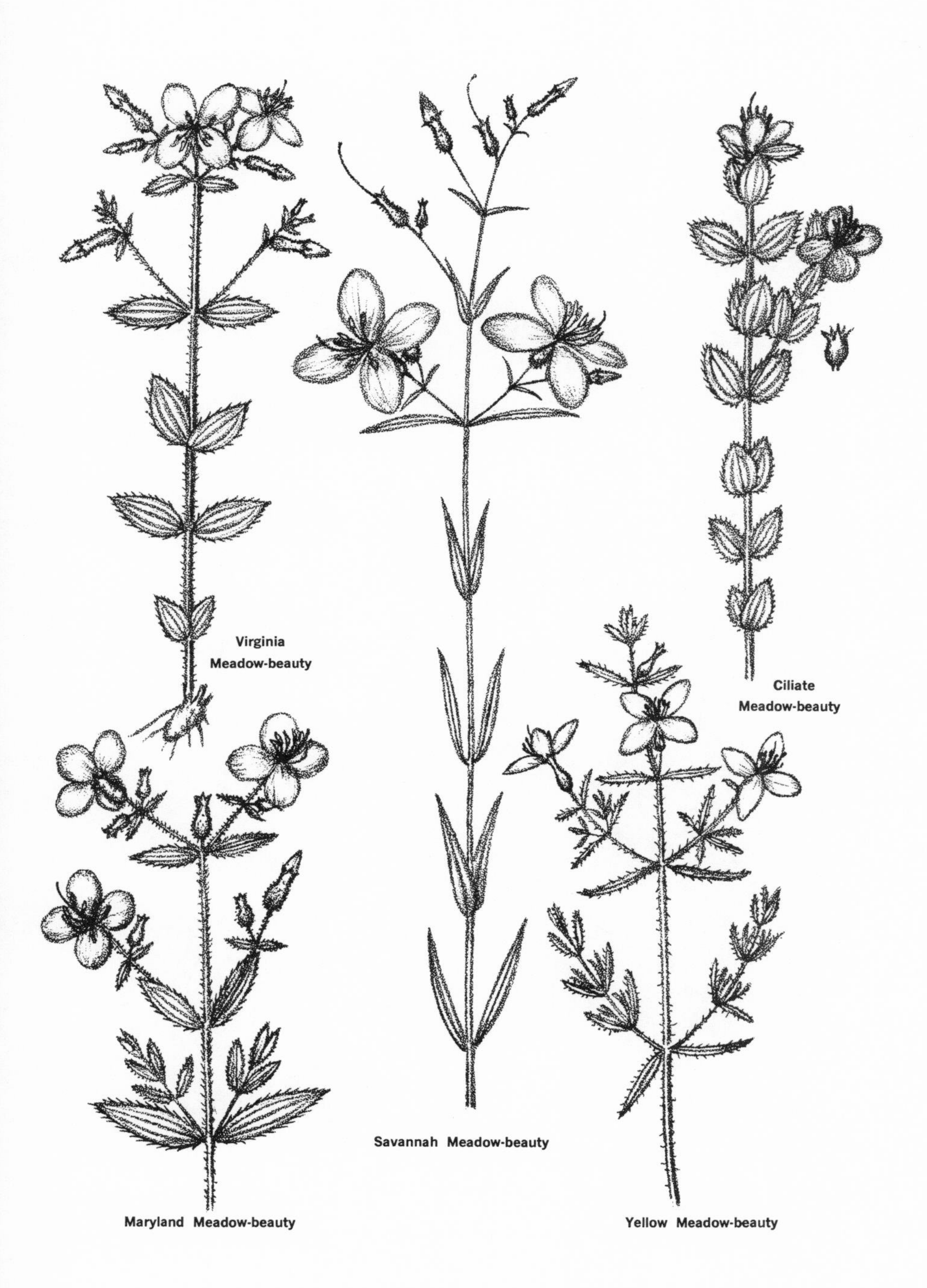
Virginia
Meadow-beauty
Ciliate
Meadow-beauty
Savannah Meadow-beauty
Maryland Meadow-beauty
Yellow Meadow-beauty

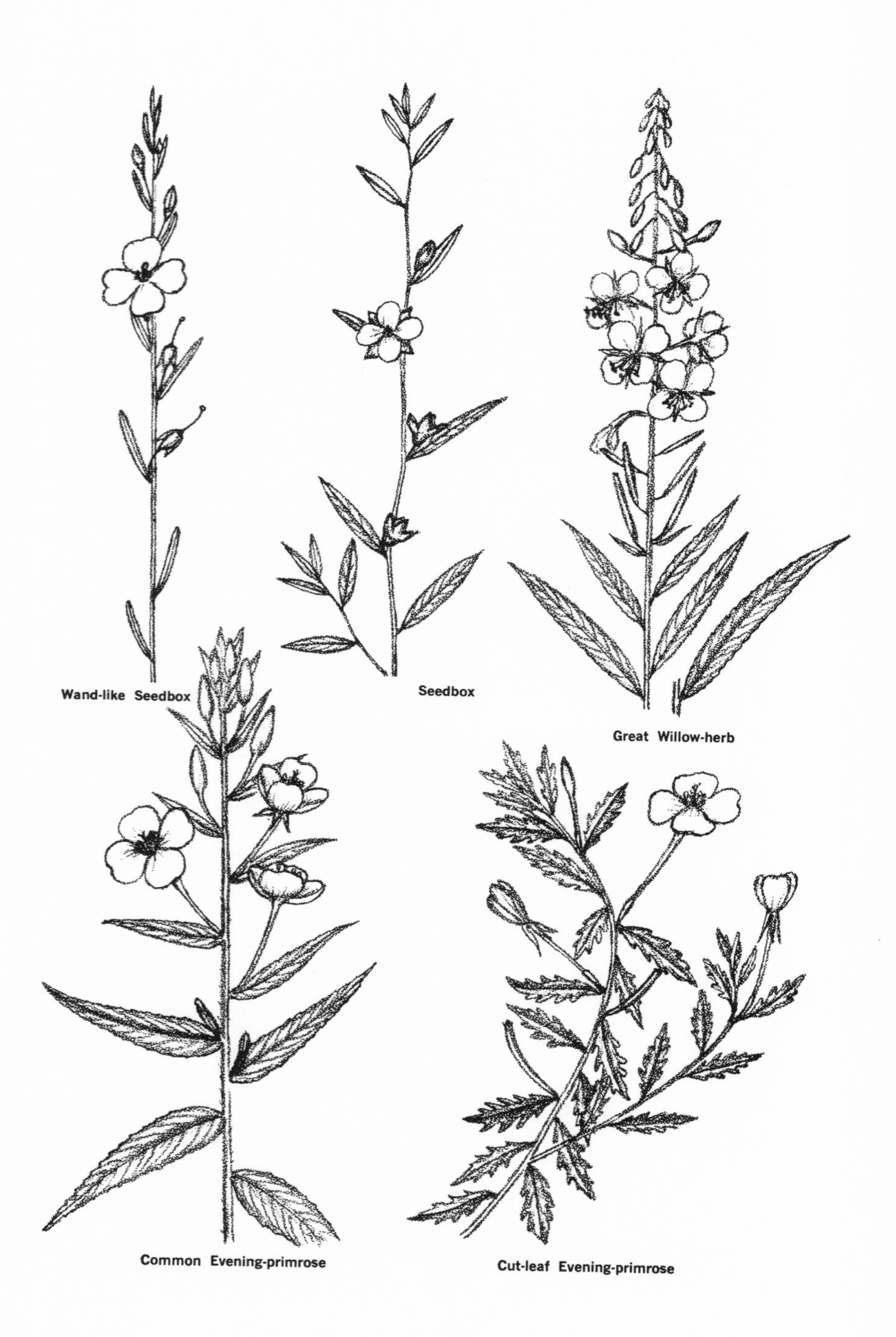
Wand-like Seedbox
Seedbox
Great Willow-herb
Common Evening-primrose
Cut-leaf Evening-primrose

Members of the Evening-primrose Family have flowers with usually 4 sepals which are united with the ovary of the pistil, and with each other to sometimes form a long tube; 4 petals, attached to the top of the calyx tube; 4 or 8 stamens, also attached to the calyx tube; and a solitary pistil consisting of 2 or 4 united carpels.

WAND-LIKE SEEDBOX *(Ludwigia virgata)*

Seedboxes have the 4 calyx lobes, and their 4 petals and stamens seated on the summit of the ovary, which later becomes a little box-like capsule full of tiny seeds. This one has an erect, simple or branching, smooth or minutely downy stem from 1 to 2½ feet tall. Its narrow, stalkless leaves are ¾ to about 1½ inches long and point upward. The flowers are about an inch across and the 4 yellow petals are much longer than the calyx lobes, which turn abruptly backward soon after the petals fall. It grows in wet coastal plain pinelands and bogs, blooming between June and September.

RANGE: e. Va. south to Fla.

SEEDBOX *(Ludwigia alternifolia)*

While similar in general appearance to the preceding species, this plant has lance-shaped leaves 1½ to 4 inches long which taper at the base to a short stalk. Its flowers are a little smaller and their 4 yellow petals are about the same length as the calyx lobes. The latter remain erect or spreading after the petals fall. It grows in marshes, wet meadows, and swamps; blooming between May and October.

RANGE: Mass. to Ont., Iowa and Kan. south to Fla. and Tex.

HAIRY SEEDBOX *(Ludwigia hirtella)*

This species is very similar to the preceding except that it is hairy throughout. It grows in wet pinelands and swamps from N.J. and Ky. south to Fla. and Tex. (Not illustrated)

GREAT WILLOW-HERB *(Epilobium angustifolium)*

The Great Willow-herb, or Fireweed, is one of some 20 species found in northeastern North America. It is an erect, smooth, leafy-stemmed plant 2 to 6 feet tall; with inch-wide, 4-petalled, bright purple or magenta flowers in a long terminal cluster. The leaves are almost stalkless, willow-like and 2 to 6 inches long. It grows in open woods, clearings, and in burned-over areas; blooming between July and September. The slender pods which follow the flowers contain numerous seeds with a tuft of silky hairs at one end.

RANGE: Lab. to Alaska south to Md., w. N.C., the Great Lakes region, S.D., Ariz. and Calif.

COMMON EVENING-PRIMROSE *(Oenothera biennis)*

This common plant has an erect, stout, more or less hairy, often red-tinged, leafy stem 2 to 6 feet tall. The flowers, which open in the evening and close the following morning, are 1 to 2 inches across; with 4 pale yellow petals atop the long and slender calyx tube. The lance-shaped leaves are stalkless, elliptic or lance-shaped, wavy-toothed on the margin and 1 to 6 inches in length. It grows in dry open places and along roadsides, blooming between June and September.

RANGE: Lab. to Alaska south to Fla. and Tex.

CUT-LEAF EVENING-PRIMROSE *(Oenothera laciniata)*

The smooth to somewhat hairy, branching stems of this often weedy plant sprawl close to the ground; and the numerous leaves are deeply toothed or lobed and mostly 1 to 2½ inches long. Its flowers are ¾ to 1½ inches across and have 4 pale yellow or yellowish-white petals. They are borne in the axils of the leaves and turn reddish in fading. It is a common plant in fields, waste places and along roadsides; blooming between March and October.

RANGE: N.J. to N.D. south to Fla. and Tex.

SEASIDE EVENING-PRIMROSE *(Oenothera humifusa)*

This species is a similar but hoary plant which grows on coastal sand dunes from N.J. south to Fla. and west to La.; blooming between May and October. (Not illustrated)

SHOWY EVENING-PRIMROSE *(Oenothera speciosa)*

The original home of this attractive little plant was on the prairies of the midwest, but it has escaped from cultivation and now grows wild in fields and along roadsides in the Southeast. It forms colonies by means of creeping underground stems which send up 6- to 12-inch stems bearing leaves and flowers. The nodding flower buds open into pink or white, 4-petalled flowers from 1½ to about 3 inches across; the flowering season being between May and July. Its leaves are lance-shaped, 1 to 3 inches long, and have wavy-toothed margins.

RANGE: Mo. and Kan. south to Tex.; naturalized from Va. and Ill. s. to Fla. and La.

COMMON SUNDROPS *(Oenothera fruticosa)*

This is quite a variable plant and botanists recognize several varieties. As a rule it has erect or ascending and usually branched stems from 1 to 3 feet high which may be hairy or nearly smooth. Its leaves range from narrowly lance-shaped or narrowly egg-shaped to oblong, and the margins may be untoothed or wavy-toothed. They are usually 1 to 4 inches long. The flowers are an inch to 2 inches across and have 4 inversely heart-shaped, bright yellow petals and a slender calyx tube; the ovary being distinctly club-shaped. As the flowers open during the daytime, the name "sundrops" is quite appropriate. It is common in dry to moist open woods, fields, and meadows; blooming between April and August. RANGE: N.H. to Mich. south to Fla., La. and Okla.

SMALL SUNDROPS *(Oenothera perennis)*

The Small Sundrops is a little plant with slender stems a few inches to nearly 2 feet tall, along which are small and narrow leaves. Usually it has a basal rosette of top-shaped leaves from 1 to 2 inches long. Its yellow flowers are in leafy-bracted end clusters. They are about ¾ inch across and open during the daytime. It grows in dry to moist fields and meadows, blooming between May and August; from Nfd. to Man. south to e. Va., n. Ga., nw. S.C., Tenn. and Mo. (Not illustrated)

BIENNIAL GAURA *(Gaura biennis)*

This is an erect, much-branched, leafy plant 2 to 5 feet tall; with a hairy or downy stem and flowers in long, narrow end clusters. The flowers are about ½ inch across, white at first but turning pinkish, with 4 petals turned upward and 8 stamens which are directed downward. The lance-shaped leaves have a few widely spaced teeth on the margin, and are 2 to 4 inches in length. It grows in dry thickets, fields and along roadsides; blooming between June and October. RANGE: Que. to Minn. south to n. Ga., Tenn. and Mo.

NARROW-LEAF GAURA *(Gaura angustifolia)*

This species, found in the coastal plain from N.C. south to Fla. and west to Tex., has very narrow leaves and capsules which are sharply angled. (Not illustrated)

ENCHANTER'S-NIGHTSHADE *(Circaea lutetiana* var. *canadensis)*

The Enchanter's-nightshades are woodland plants with opposite, rather long-stalked, wavy-toothed, egg-shaped leaves. The small white flowers are in long and narrow end clusters. They have but 2 petals which are so deeply notched that they appear to be four, and a pair of stamens. The fruits are oval-shaped pods covered with hooked bristles, and they often hitch a ride on one's clothing. This species has leaves from 2 to 4 inches long, rather firm in texture, dark green, and usually rounded at the base; and the stem is 8 inches to 3 feet tall. It blooms between June and August.

RANGE: N.S. to s. Ont. and N.D. south to Ga., Tenn., Mo. and Okla.

SMALLER ENCHANTER'S-NIGHTSHADE *(Circaea alpina)*

This is a smaller plant with leaves 1 to 2½ inches long which are heart-shaped at the base; and its rootstocks are thickened and tuber-like. It grows in cool moist woods and bogs and blooms between June and September.

RANGE: Lab. to Alaska south to N.Y., n.Ga., e.Tenn., the Great Lakes region, S.D., Colo., Utah and Wash.

CANADA ENCHANTER'S-NIGHTSHADE *(Circaea × intermedia)*

This plant is a hybrid of *C. alpina* and *C. lutetiana*. It is similar to the preceding species, with heart-shaped leaves 1½ to 4 inches long which are more coarsely toothed. It grows from Lab. to Que. and Minn. south to Me., Conn., and the mts. of Va. and W. Va.

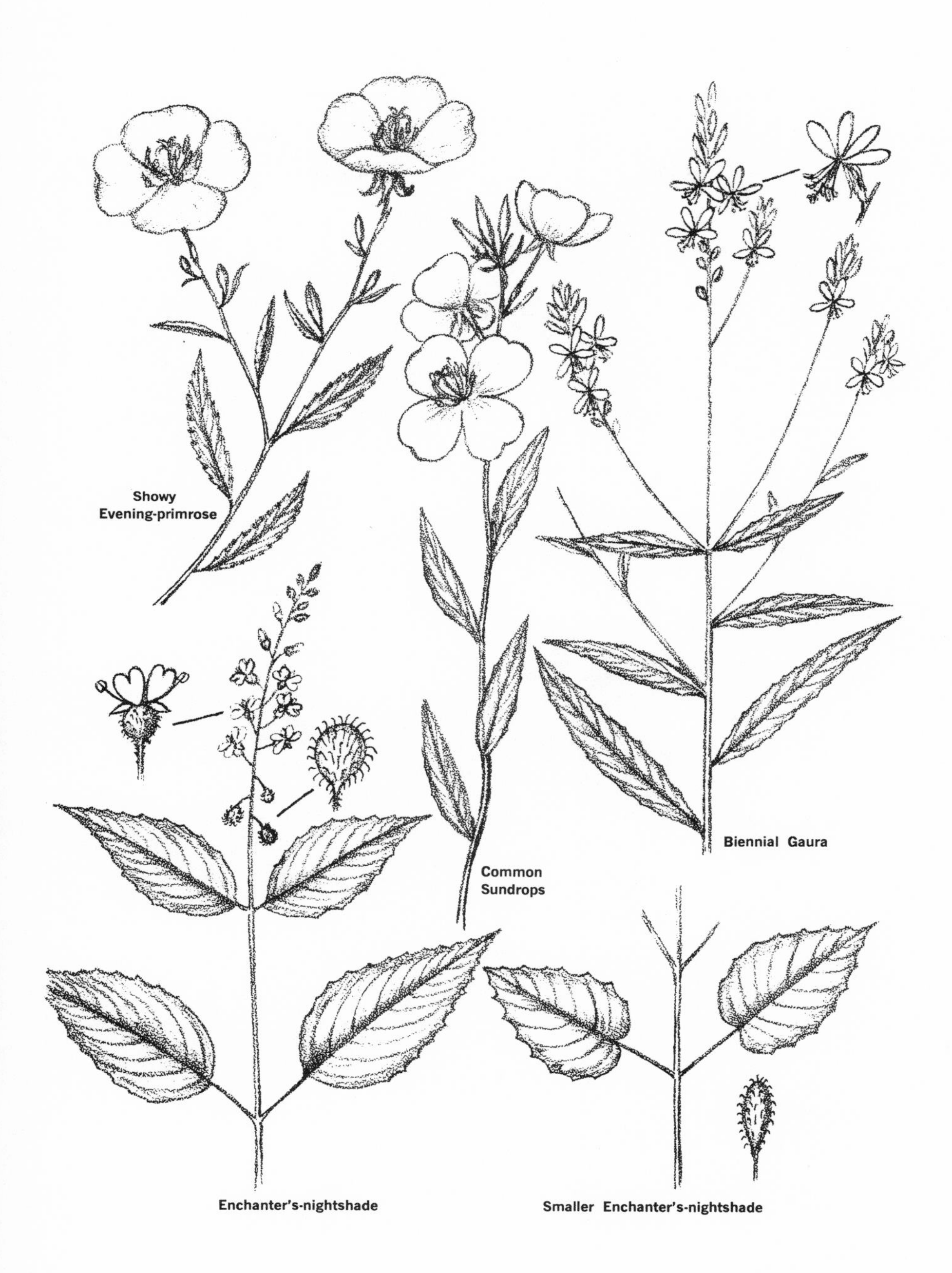
Showy
Evening-primrose
Common
Sundrops
Biennial Gaura
Enchanter's-nightshade
Smaller Enchanter's-nightshade

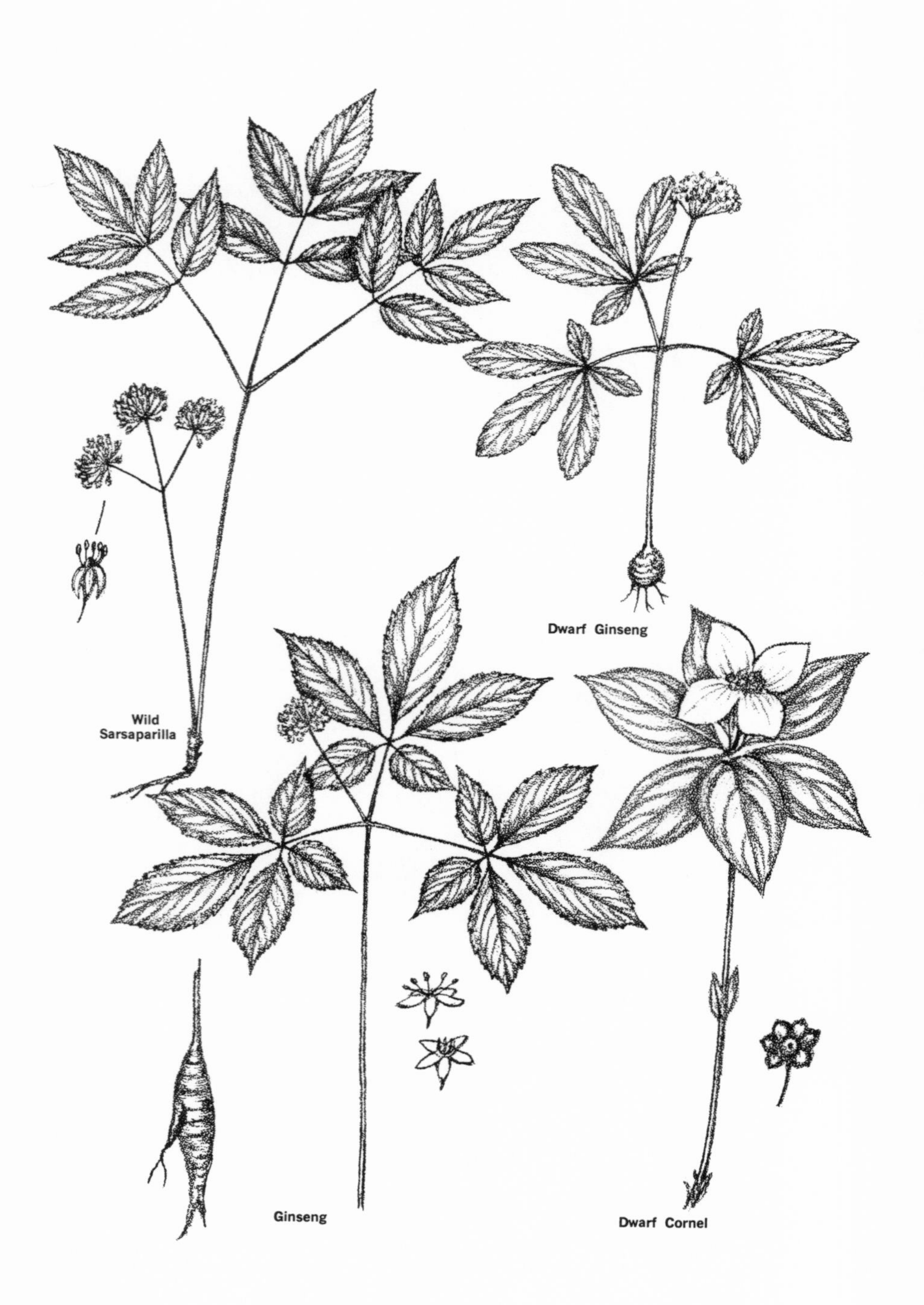
Dwarf Ginseng
Wild
Sarsaparilla
Ginseng
Dwarf Cornel

GINSENG FAMILY (Araliaceae)

Members of this family have small flowers in umbels. Our species have 5 sepals united with the ovary which has a disk at the top, 5 petals, 5 stamens, and a pistil composed of 2 to 5 united carpels with the same number of distinct styles. The fruit is a berry.

WILD SARSAPARILLA *(Aralia nudicaulis)*

This plant has but one long-stalked leaf which is divided into 3 parts; each of which are again divided into usually 5 egg-shaped or oval, sharply toothed leaflets 2 to 5 inches long. The naked flower stalk usually has 3 clusters of small greenish-white flowers; which are followed by round, purplish-black berries. Beneath the ground the plant has a creeping aromatic rootstock. It grows in rich moist woods, blooming from May to July.

RANGE: Nfd. to Man. south to Va., n. Ga., e. Tenn. and Mo.

AMERICAN SPIKENARD *(Aralia racemosa)*

This species is a coarse, herbaceous plant 4 to 10 feet tall. Its leaves are 2 to 3 feet in length and breadth. They are divided into 3 parts which are again pinnately divided into a number of heart-shaped leaflets 2 to 6 inches in length. The numerous umbels of small white flowers are in a large terminal cluster. It is often common in rich moist woods from N.B. to Man. south to n.Ga., Miss., Mo. and Kan.; blooming from June to August. (Not illustrated)

BRISTLY SARSAPARILLA *(Aralia hispida)*

This plant resembles the preceding but it is somewhat woody and bristly toward the base, and the bases of the leaflets are not heart-shaped. It grows in rocky or sandy open woods from Nfd. to Man. south to N.J., W.Va., Ill. and Minn.; blooming June to August. (Not illustrated)

GINSENG *(Panax quinquefolius)*

This is an erect plant 8 to 15 inches tall, bearing 3 leaves and a solitary umbel of 6 to 20 small yellowish-green flowers at the summit. The leaves are divided into 5 radiating leaflets which are 2 to 5 inches long and sharply toothed on the margin. The flowers are followed by bright crimson berries. Ginseng or "Sang" grows in rich, moist deciduous forests, and blooms between May and August. Excessive gathering of its rootstocks has eliminated it in many places.

RANGE: Que. to Man. south to Ga., Tenn. and Okla.

DWARF GINSENG *(Panax trifolius)*

Dwarf Ginseng has a stem 3 to 8 inches tall which bears 3 leaves and a solitary umbel of small white flowers. Each leaf is divided into 3 to 5 stalkless leaflets which are 1 to 1½ inches long and blunt at the tip. The flowers are followed by yellowish berries. The plant grows from a roundish pungent tuber about ½ inch in diameter, which gives it the common name of Groundnut. It grows in rich moist woods and blooms between May and August.

RANGE: N.S. to Minn. south to n. Ga., Tenn., Iowa and Neb.

DOGWOOD FAMILY (Cornaceae)

Most members of this family are trees or shrubs. The small flowers have 4 or 5 sepals united with the ovary, 4 or 5 petals, usually 4 stamens, and a solitary pistil which consists of 2 united carpels. The fruits are fleshy but have a large 2-seeded stone.

DWARF CORNEL *(Cornus canadensis)*

Often called the Bunchberry, this plant has a slender stem which runs underground and sends up erect branches from 2 to 10 inches tall. On them is what appears to be a whorl of usually 6 oval-shaped leaves, from 1 to 3 inches long, and what appears to be a solitary flower. Actually it is a dense little cluster of small greenish-yellow or sometimes purplish flowers, and the flower cluster is surrounded by usually 4 petal-like white bracts. Roundish bright red fruits follow the flowers which bloom between May and July. It grows in cool moist woods and bogs.

RANGE: Lab. to Alaska south to Md., W.Va., the Great Lakes region, S.D., N.Mex. and Calif.

CARROT FAMILY (Apiaceae)

All but a very few members of this large family have their flowers arranged in umbels. Usually the umbels are compound; the primary branches or rays of the flower cluster bearing the umbels at their tips. The flowers are small and have the 5 sepals united with the ovary. The 5 petals and 5 stamens are attached to the disk-like summit of the ovary which surrounds the 2 styles of the pistil. The pistil consists of 2 united carpels which mature as a pair of dry 1-seeded fruits, at maturity separating at the bottom but remaining joined at the tip to the stalk which rises between them. They have 5 primary ribs and sometimes 4 additional secondary ones, and through the walls run tubes containing a usually aromatic oil. Many members of the family are plants used for food or flavorings but some members are extremely poisonous.

KEY TO GENERA OF THE CARROT FAMILY

1. Flowers in dense heads — ERYNGOS *(Eryngium)*
1. Flowers in umbels.
    2. Leaves simple, on long stalks arising from a creeping underground stem.
3. Leaves roundish or kidney-shaped. Umbels with several flowers. — WATER-PENNYWORTS *(Hydrocotyle)*
3. Leaves egg-shaped. Umbels only 2- to 4-flowered. — MARSH-PENNYWORT *(Centella)*
    2. Leaves compound, the blades divided into leaflets.
4. Plants with yellow (sometimes purplish) flowers.
    5. Margins of the leaflets untoothed. — YELLOW PIMPERNEL *(Taenidia)*
    5. Margins of the leaflets toothed.
6. Central flower in each umbel stalkless. Fruits with narrow and rounded ribs. — ALEXANDERS *(Zizia)*
6. Central flower in each umbel stalked like the others. Fruits with winged ribs. — MEADOW-PARSNIPS *(Thaspium)*
    4. Plants with white or whitish flowers.
7. Umbels 2- to 4-flowered, each umbel with a ternately divided leaf at its base. — HARBINGER-OF-SPRING *(Erigenia)*
7. Umbels many-flowered.
    8. Bracts of the primary umbel large and pinnately divided into narrow segments. Fruits bristly. — CARROTS *(Daucus)*
    8. Bracts of the primary umbel rather small or absent.
9. Fruits very slender, several times as long as broad. Plants with a licorice-like odor. — SWEET-CICELIES *(Osmorhiza)*
9. Fruits not more than twice as long as broad.
    10. Leaflets often 6 inches or more broad. Plant stout and very woolly-hairy. — COW-PARSNIP *(Heracleum)*
    10. Leaflets much smaller. Entire plant not woolly-hairy and often quite smooth.
11. Leaves merely pinnately compound, the leaflets narrow and evenly toothed. — WATER-PARSNIPS *(Sium)*
11. Leaves doubly or even triply compound.
    12. Uppermost leaves reduced to large sheath-like stalks, or with blades much shorter than the sheaths. — ANGELICAS *(Angelica)*
    12. Uppermost leaves otherwise.
13. Leaflets thin and deeply cut-toothed, cleft, or dissected. Primary umbel with small bracts at its base. Fruits with wavy ribs. — POISON-HEMLOCK *(Conium)*
13. Leaflets firm and merely toothed. Primary umbel without bracts at its base. Fruits with broadly rounded ribs. — WATER-HEMLOCK *(Cicuta)*

BUTTON-SNAKEROOT *(Eryngium yuccifolium)*

The Button-snakeroot, Eryngo, or Rattlesnake-master is a rather unique plant with a stout, stiffly erect, and smooth stem from 2 to 6 feet tall. At the base it has a cluster of long, narrow, stiff, and bristly-margined leaves which range from 6 inches to sometimes nearly 3 feet in length. The few and much smaller ones on the stem are similar but have clasping bases. Its flowers are whitish and crowded into globe-shaped heads from ½ to an inch across. This plant grows in dry to moist open woods or thickets and on prairies, blooming between June and August.

RANGE: N.J. to Mich., Minn. and Kan. south to Fla. and Tex.

WILD CARROT *(Daucus carota)*

This weedy ancestor of our cultivated carrots is a native of Europe which is now widely naturalized here in North America. It is frequently abundant in old fields, waste places, or along roadsides; blooming between May and September. The plant has a very bristly-hairy stem from 1 to 3 feet tall, and leaves which are pinnately divided and deeply cut into innumerable narrow segments. Its many umbels of small white flowers are disposed into a lacy-looking and flat-topped cluster usually 3 or 4 inches broad, most often with a solitary and deep purple floret in its center. Before the flower cluster is in full bloom, and again as the fruits form, it is hollow and its shape suggests that of a bird's nest. In fact it is often called the Queen Anne's-lace or Bird's-nest.

AMERICAN CARROT *(Daucus pusillus)*

This native American carrot is apt to be passed by as a poor specimen of the preceding species, for it looks very much like it. It is a smaller plant usually 6 inches to 2 feet high, and it has a minutely bristly stem and even more finely dissected leaves. It grows in dry fields and on barren hillsides from S.C. to Mo., se. Kan. and B.C. south to Fla. and Calif.; blooming in April and May. (Not illustrated)

MANY-FLOWERED WATER-PENNYWORT *(Hydrocotyle umbellata)*

Water-pennyworts get their name from their simple, roundish leaves; and the fact that they grow in wet places. This one has leaves ½ to 2 inches across which are attached to their stalks in the center of the leaf. Both the leaf and the flower stalks arise from creeping stems and are 2 to 6 inches long. The umbels of tiny white flowers appear between June and September. It grows in wet areas, swamps, and on shores.

RANGE: Mass. to Mich. south to Fla. and Tex.

AMERICAN WATER-PENNYWORT *(Hydrocotyle americana)*

This species has its leaves attached to the stalks at a heart-shaped base; and the flowers are in small, short-stalked clusters at the bases of the long-stalked leaves. It grows in wet meadows and woods from Nfd. to Minn. south to Md., w. N.C. and e. Tenn.; blooming between June and September. (Not illustrated)

Hydrocotyle bonariensis is a southern species with shield-shaped leaves and umbels in a branching cluster. It grows among beach dunes and in wet sandy areas in the coastal plain from N.C. south to Fla. and west to Tex.; blooming April to September. (Not illustrated)

OVATE-LEAF WATER-PENNYWORT *(Centella asiatica)*

The stems of this plant creep through the wet sand about the margins of pools, and from them the long-stalked leaves and flower clusters arise. The leaves are egg-shaped, 1 to 2 inches long, with a wavy-toothed margin and heart-shaped base. Flower clusters of 2 to 4 small flowers and a pair of bracts are produced between June and August.

RANGE: Del. south to Fla. and west to Tex.

HARBINGER-OF-SPRING *(Erigenia bulbosa)*

This is a dainty and smooth little plant 3 to 8 inches tall, with 1 or 2 finely dissected leaves and small umbels of white flowers with brown anthers. Both the leaves and the flower stalk arise from a small round tuber. It grows in rich deciduous woods and thickets, blooming between late February and April. Another name for it is Pepper-and-salt.

RANGE: w. N.Y. to Ont. and Wis. south to w. N.C., Ala., Miss. and Mo.

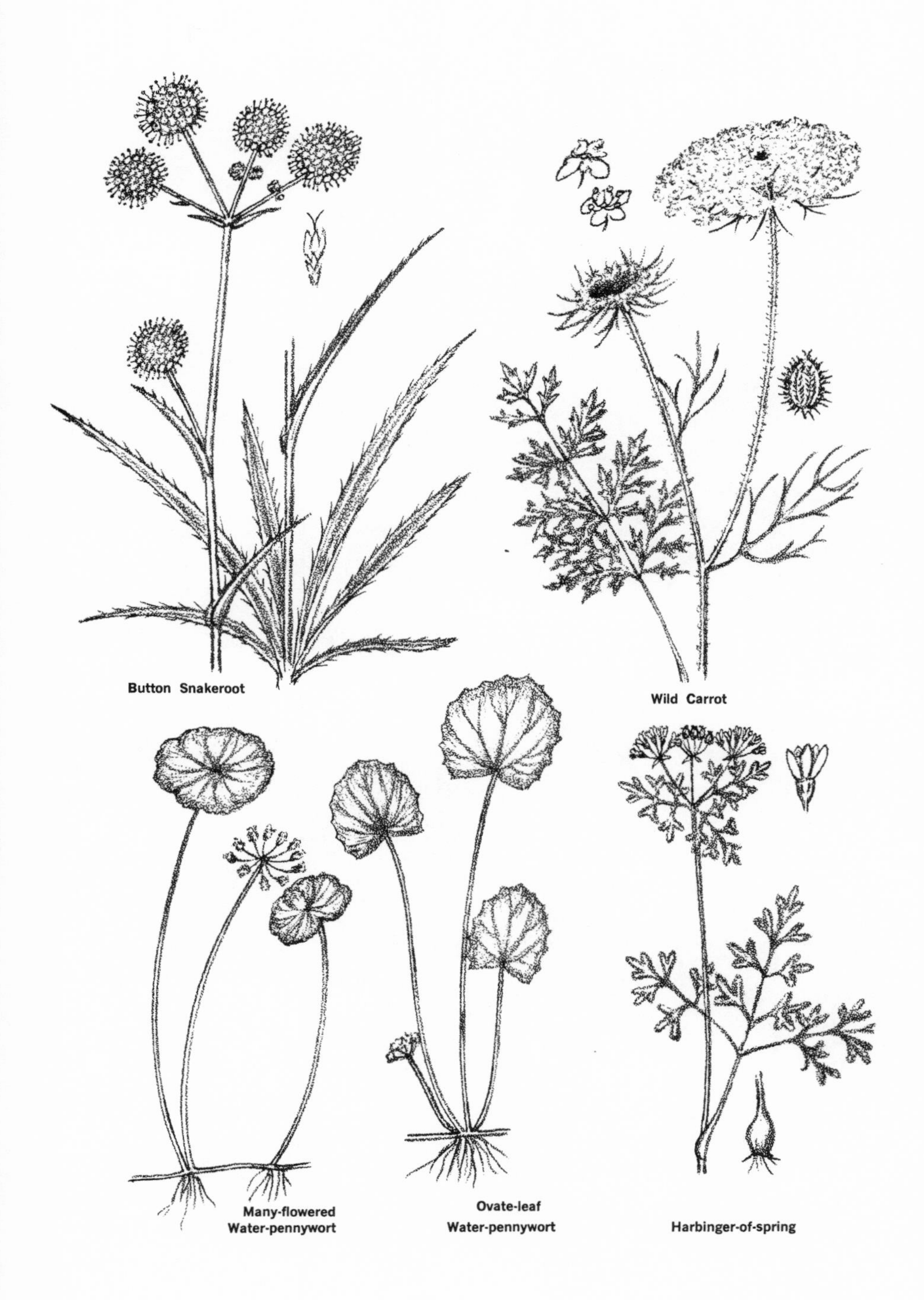
Button Snakeroot
Wild Carrot
Many-flowered
Water-pennywort
Ovate-leaf
Water-pennywort
Harbinger-of-spring

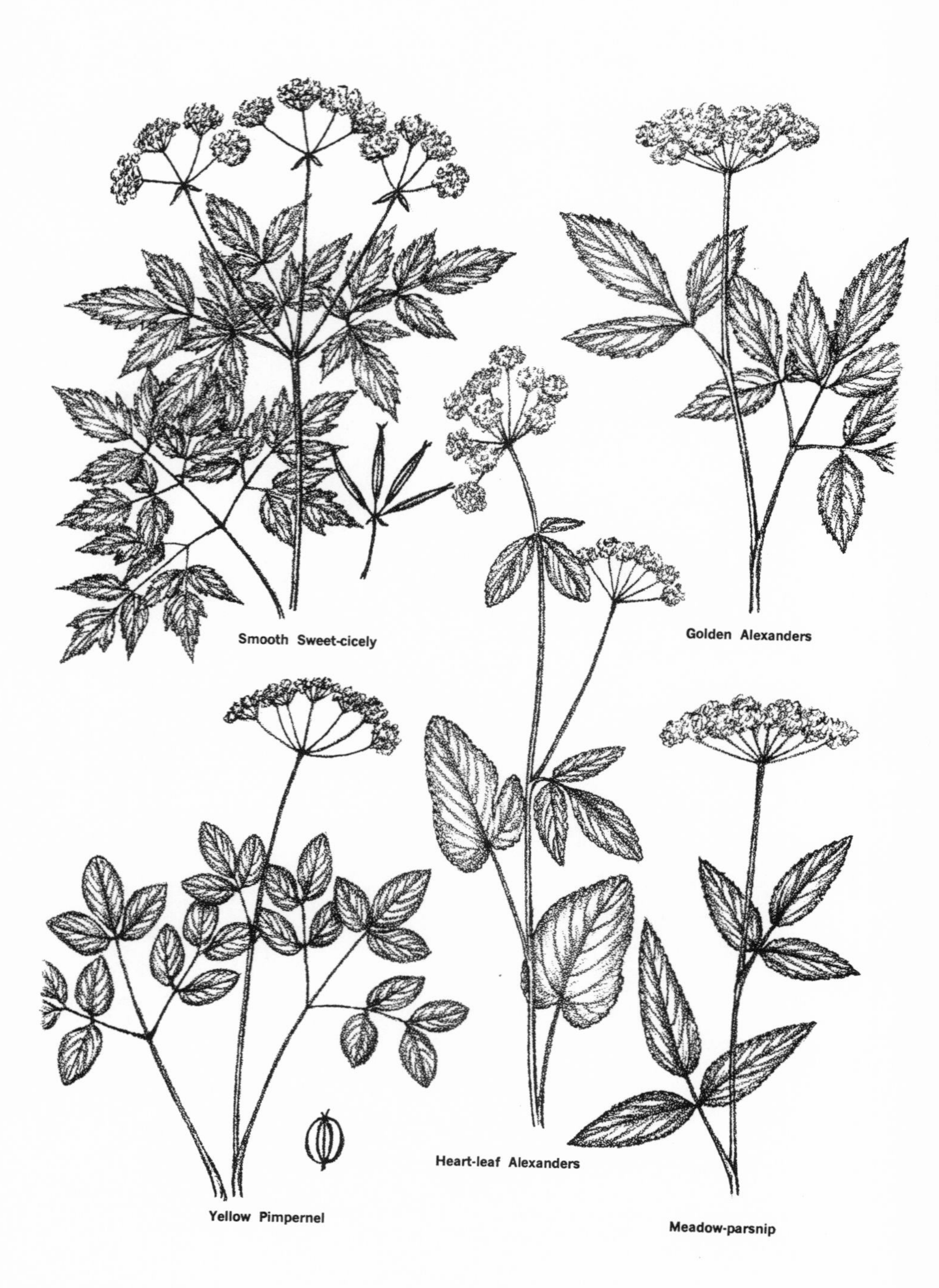
Smooth Sweet-cicely
Golden Alexanders
Heart-leaf Alexanders
Yellow Pimpernel
Meadow-parsnip

SMOOTH SWEET-CICELY *(Osmorhiza longistylis)*

The sweet-cicelies get their name from the pleasant licorice- or anise-like odor of their carrot-like roots. This species has a stem from 1 to about 3 feet tall which is usually smooth except at the points where the leaves are attached. The leaves are divided into 3 segments which are again divided, being quite fern-like in their appearance. Its small white flowers are in few-flowered umbels, blooming April to June. They are followed by slender fruits which taper at both ends and are bristly along the ribs. Those of this species are tipped with a pair of styles about ⅛ inch long. It grows in rich woodlands and on wooded slopes. Sometimes called Sweet-myrrh and Aniseroot.

RANGE: N.S. to Alb. south to e. Va., n. Ala., and e. Tex.

HAIRY SWEET-CICELY *(Osmorhiza claytonii)*

This plant very closely resembles the preceding species but it is quite hairy or downy throughout. The slender fruits have very short styles, not more than 1⁄16 inch long. It grows in woods and on wooded slopes, blooming between April and June.

RANGE: Que. to Sask. south to w. S.C., n. Ga., Ala., Ark. and Kan. (Not illustrated)

YELLOW PIMPERNEL *(Taenidia integerrima)*

The Yellow Pimpernel is an erect, smooth, slender-stemmed plant 1 to 3 feet tall, which is more or less whitened with a bloom. Its leaves are ternately divided into oval, egg-shaped, or lance-shaped leaflets ½ to an inch long which have untoothed margins. The small yellow flowers are in several long-stalked umbels, blooming April to June. It is a plant of dry rocky or sandy woods and thickets; southward only in the piedmont and mountains. RANGE: Que. to Minn. south to Ga. and Tex.

HEART-LEAF ALEXANDERS *(Zizia aptera)*

This is an erect and usually smooth plant from 1 to 2 feet tall. At the base it has long-stalked leaves which are egg-shaped to roundish, heart-shaped at the base, bluntly toothed on the margin and sometimes lobed, and 2 to 3 inches long. Those upward along the stem are short-stalked and divided into 3 leaflets. The small yellow flowers are in several umbels, the central flower in each umbel being stalkless. It grows in wooded bottomlands and meadows, blooming between April and June. The plant closely resembles the Meadow-parsnip.

RANGE: R.I. to Minn. and B.C. south to n. Fla., Miss., Mo., Col., Utah, and Ore.

GOLDEN ALEXANDERS *(Zizia aurea)*

Golden Alexanders is a smooth, erect plant 1 to 2 feet tall. Its leaves are all ternately divided and sometimes redivided into lance-shaped or egg-shaped, finely-toothed leaflets 1 to 2 inches long. The several umbels of small yellow flowers form a more or less flat-topped cluster. The central flower in each umbel is stalkless. It grows in wet woods and meadows, blooming between April and June. RANGE: Que. to Sask. south to n. Fla. and Tex.

BEBB'S ZIZIA *(Zizia trifoliata)*

This species has somewhat leathery textured and more coarsely toothed leaflets than the preceding, and the umbel has very slender primary rays. It is found from Va. and W. Va. south to n. Fla. and Ala., blooming in April or May. (Not illustrated)

MEADOW-PARSNIP *(Thaspium trifoliatum)*

One variety of this plant has greenish or purplish flowers and is called the Purple Meadow-parsnip or Purple Alexanders. More common and widespread is the variety which has yellow flowers. Both are smooth erect plants 1 to 2½ feet tall. The long-stalked basal leaves may be undivided, or divided into 3 egg-shaped to lance-shaped, toothed leaflets 1 to 2 inches long. All of the flowers in the umbels are stalked. It grows in open woods and thickets, blooming between April and June. It closely resembles the Heart-leaf Alexanders. RANGE: N.Y. to Minn. south to Fla. and La.

HAIRY-JOINTED MEADOW-PARSNIP *(Thaspium barbinode)*

This plant has its basal leaves more dissected and has whitish hairs surrounding the stems at the nodes. It grows in rich woods and along streams from N.Y. to Minn. south to Fla., Miss. and Okla. (Not illustrated)

COW-PARSNIP *(Heracleum maximum)*

This is a conspicuous woolly-hairy plant 4 to 8 feet tall, with a stout grooved stem and very large leaves. Between June and August it displays umbels of white or purple-tinged flowers 6 inches to nearly a foot across. Although it has a rank odor the plant is not poisonous. It grows in moist open places and thickets.

RANGE: Lab. to Alaska south to n. Ga., Tenn., Mo., N. Mex. and Calif.

WATER-HEMLOCK *(Cicuta maculata)*

The roots of water hemlocks resemble small sweet-potatoes, have an odor similar to parsnips, and are deadly poisonous. This species, often called the Spotted Cowbane, Musquash-root and Beaver-poison, has a stout erect stem 3 to 6 feet tall which is usually streaked or spotted with purple. The lower leaves are often a foot long and pinnately divided, or redivided, into a number of lance-shaped, sharply toothed leaflets 1 to 5 inches long. The small white flowers are in umbels 2 to 5 inches across. It grows in swamps, wet meadows and thickets; blooming between May and August.

RANGE: N.B. to Man. south to Fla. and Tex.

BULB-BEARING WATER-HEMLOCK *(Cicuta bulbifera)*

This is a slender plant 1 to 3 feet tall, with very narrow leaflets, and bearing small bulbs in the axils of the upper leaves. It grows in wet places from Nfd. to B.C. south to Va. Ind., Neb., Mont. and Ore. (Not illustrated)

WATER-PARSNIP *(Sium suave)*

The Water Parsnip has an erect, branching stem 2 to 6 feet tall which is longitudinally furrowed. Its leaves are pinnately divided into 5 or more very narrow, lance-shaped, sharply and evenly toothed leaflets 1½ to 5 inches long. The lower ones are long-stalked and submersed leaves are often finely dissected. Its small white flowers are in umbels 2 to 4 inches across. It grows in swamps, wet meadows, and on muddy shores; blooming between June and September.

RANGE: Nfd. to B.C. south to Fla., La. and Calif.

POISON-HEMLOCK *(Conium maculatum)*

This is a European plant, the one used by the ancient Greeks for putting prisoners to death. It is a dangerous plant for all parts of it are poisonous, and it is now widely naturalized here in bottomlands, waste places, etc. It is an erect, much-branched, smooth plant with purple-spotted and hollow stems 2½ to 5 feet tall. Its large leaves are divided and redivided into numerous, thin, egg-shaped leaflets which are sharply toothed and deeply cut. The small white flowers are in umbels from 1 to 3 inches across, blooming between June and August. The fruits are rather roundish, somewhat flattened, and have prominent wavy ribs.

HAIRY ANGELICA *(Angelica venenosa)*

This is a rather slender-stemmed plant from 2 to about 6 feet tall, which is somewhat hairy toward the summit. The lower and basal leaves are often a foot long and divided into rather thickish, oval to lance-shaped leaflets 1 to 2 inches long, with toothed margins. Those on the upper part of the stem are reduced to sheathing leaf stalks, with or without small blades. The small white flowers are in umbels 2 to 4 inches across, blooming between July and September. It grows in dry woods, thickets, and clearings.

RANGE: Conn. to Mich. and Ill. south to Fla., Ala., Miss. and Mo.

PURPLE-STEMMED ANGELICA *(Angelica atropurpurea)*

This is a stouter plant with a purple or purple-blotched, smooth stem 3 to 10 feet tall. The leaf stalks have large, swollen, and veiny sheathing bases; and the umbels are 4 to 10 inches broad. It grows in swamps and wet bottomlands from Lab. to Wis. south to Md., W. Va., Ohio and Ill. Blooming May to September. (Not illustrated)

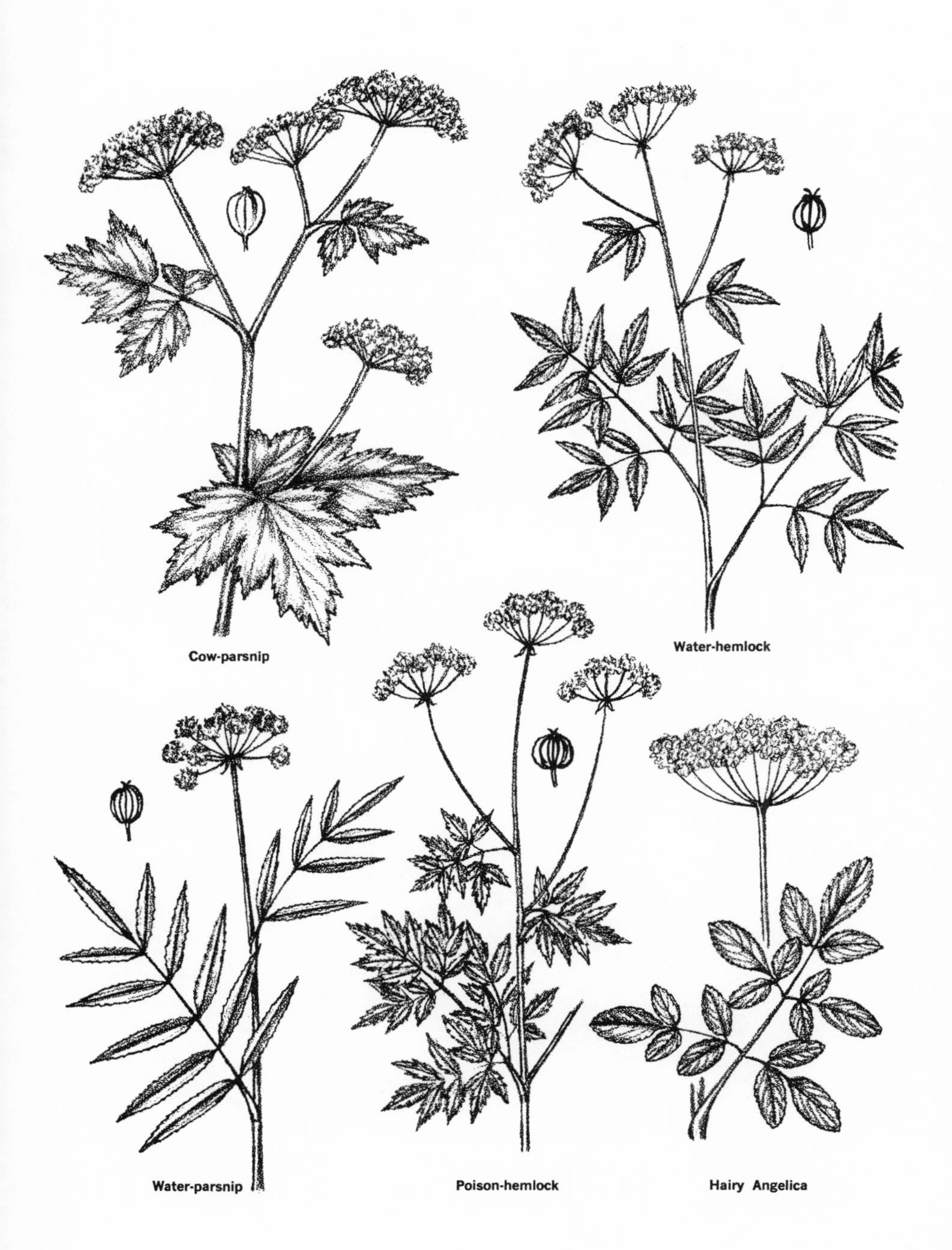
Cow-parsnip
Water-hemlock
Water-parsnip
Poison-hemlock
Hairy Angelica

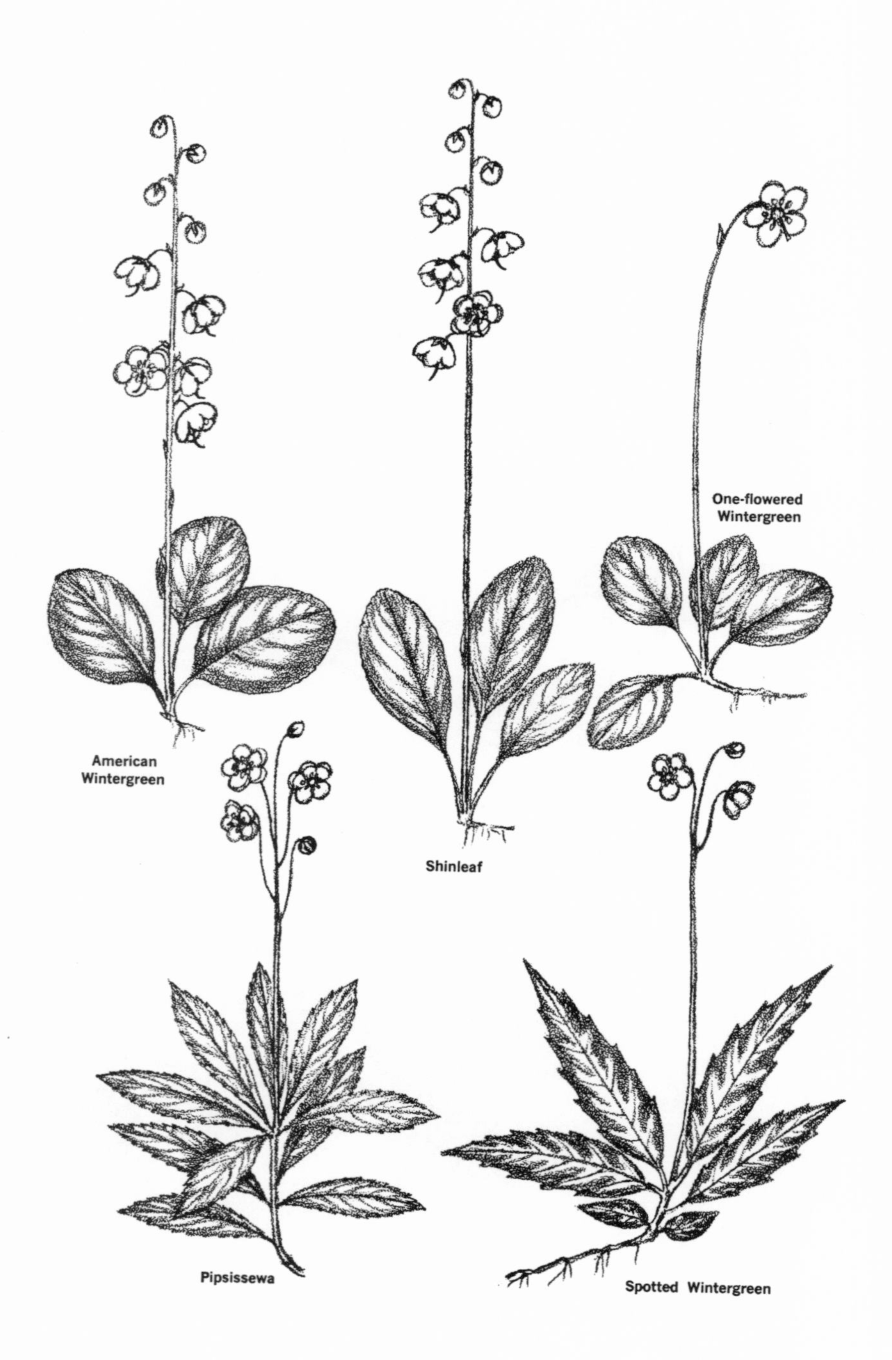
One-flowered
Wintergreen
American
Wintergreen
Shinleaf
Pipsissewa
Spotted Wintergreen

WINTERGREEN FAMILY (Pyrolaceae)

Members of this family have flowers with a radial or slightly bilateral symmetry. They have 5 sepals, somewhat united at the base; 5 petals, barely united at the base; 10 stamens; and a solitary pistil consisting of 5 united carpels, with 1 style and stigma. The fruits are capsules containing a large number of very small seeds. The name "wintergreen" refers to the evergreen leaves.

AMERICAN WINTERGREEN *(Pyrola americana)*

On an erect flower stalk 6 to 20 inches tall, this plant has a few scaly bracts and a narrow cluster of nodding white flowers. Each one is about ⅔ inch across and they are quite fragrant. At the base of the flower stalk are several oval or roundish leaves which are somewhat leathery in texture, lustrous above, obscurely toothed on the margin, and from 1 to 3 inches long. It grows in rather dry woods and clearings, blooming between May and August. RANGE: N.S. to Ont. south to N.C., Ky., Ind., and Wis.

SHINLEAF *(Pyrola elliptica)*

The Shinleaf has an erect, naked flower stalk 5 to 10 inches tall; on which is a narrow cluster of nodding, white, fragrant flowers about ⅔ inch across. At the base of this stalk are several leaves which are oval or elliptic, rather thin in texture, dull green above, obscurely toothed, and from 1 to 3 inches in length. It grows in dry to fairly moist, rich woods; blooming between June and August.

RANGE: Nfd. to B.C. south to W. Va., Ohio, Ill., Iowa, S.D. and N. Mex.

PINK WINTERGREEN *(Pyrola asarifolia)*

The Pink Wintergreen has distinctive pale pink to crimson flowers; and heart-shaped, leathery, lustrous leaves. It grows in rich, wet woods and swamps from Nfd. to Yuk. Ter. south to N.Y., n. Ind., Wis., S.D. and N. Mex. (Not illustrated)

ONE-SIDED WINTERGREEN *(Orthilia secunda)*

This species has whitish or greenish yellow flowers in a dense 1-sided cluster; and elliptic or egg-shaped, somewhat leathery, lustrous leaves which are usually pointed at the tip. It grows in dry or moist woods and thickets and bogs from Nfd. to Alaska south to Va., the Great Lakes region, Iowa, S.D., N. Mex. and Calif. (Not illustrated)

ONE-FLOWERED WINTERGREEN *(Moneses uniflora)*

This plant can be distinguished from the *pyrolas* by its solitary flower at the tip of its erect 2- to 6-inch stem. It is about ¾ inch across, white or pink-tinged, and quite fragrant. Below the flower the stalk bears only 1 or 2 small bracts. The leaves at the base of it are roundish or egg-shaped, rather thin, obscurely toothed, and from ½ to 1¼ inches long. It grows in cool, mossy woods and in bogs, blooming between June and August. RANGE: N.S. to Ont. south to n. Ga., Ohio, Ill. and Minn.

PIPSISSEWA *(Chimaphila umbellata)*

Our Pipsissewa, or Prince's-pine, is a variety of a plant found throughout the northern portion of the Northern Hemisphere. It is a little plant with erect leafy flowering stems 6 to 12 inches high, arising from extensively creeping underground stems. Its leaves are nearly whorled, 1 to 2½ inches long, bright green and shining, sharply toothed, and distinctly broadest toward their tips. The white or pale pinkish flowers are ½ to ⅔ inch across and in an umbel-like end cluster. The Pipsissewa grows in dry woodlands, blooming between May and August. RANGE: N.S. to Ont. south to Ga., Ohio, Ill. and Minn.

SPOTTED WINTERGREEN *(Chimaphila maculata)*

The Spotted Wintergreen, or Spotted Pipsissewa, is similar in habit and stature to the preceding species. Its leaves, however, are broadest toward the base, egg-shaped or lance-shaped, from 1 to 3 inches long and have sharp but rather widely spaced teeth on the margin. Even in winter, the plant is brought to our attention by the white mottling along the veins of its otherwise dark green leaves. Between May and August a few white or pinkish flowers, about ¾ inch across, are borne in an end cluster. It usually grows in rather dry rich woodlands.

RANGE: N.H. to Ont. and Mich. south to Ga., Ala. and Tenn.

INDIAN-PIPE *(Monotropa uniflora)*

This is a strange and ghostly-looking plant which grows in dimly lighted, rich, moist woodlands. Containing no chlorophyll, the entire plant is waxy-white, or sometimes pinkish, and in drying becomes blackish. A solitary flower about ¾ inch long, with 4 or 5 petals and 10 stamens, nods at the end of a scaly stalk from 4 to 10 inches tall. The flower stalks are usually in clusters, arising from a ball-like mass of roots. The plant is a saprophyte, obtaining its nourishment from the decaying vegetable matter in the soil. Also known as the Corpse-plant, it blooms between June and October.
RANGE: Nfd. to Alaska south to Fla., Tex. and Calif.

PINE-SAP *(Monotropa hypopithys)*

The Pine-sap is a somewhat hairy or downy plant with stalks and flowers ranging from lemon-yellow to tawny or red, and darkening in drying. It has several flowers, each about ¾ inch long, toward the top of a scaly stalk 4 to 14 inches in height. Growing in rich woodlands, it blooms between May and October. Another name for it is False Beech-drops. RANGE: Nfd. and Que. to Ont. south to Fla. and La.

HEATH FAMILY (Ericaceae)

Members of this family are chiefly woody plants. Their flowers have 4 or 5 sepals, united toward the base; 4 or 5 petals, which are more or less united; 8 to 10 stamens; and a solitary pistil consisting of 5 to 10 united carpels.

TEABERRY *(Gaultheria procumbens)*

The Teaberry has slender stems which creep on or beneath the surface of the ground, sending up erect branches 2 to 6 inches high which bear the leaves and flowers. Its urn-shaped white flowers are about ⅜ inch long and usually solitary in the leaf axils. They are followed by globular, bright red, berry-like fruits about ⅜ inch in diameter. The leaves are elliptic, oval, or top-shaped, ¾ to 2 inches long, somewhat leathery, and have low bristle-tipped teeth on their margins. The plant has an oil of wintergreen odor and taste. It grows in woods and thickets, blooming between June and August. Other names for it are Mountain-tea, Checkerberry, and Spicy Wintergreen.
RANGE: Nfd. to Man. south to Ga., Ala., Wis. and Minn.

TRAILING ARBUTUS *(Epigaea repens)*

Trailing Arbutus is a prostrate or trailing plant with branches 6 to 15 inches long; and evergreen, veiny, oval to roundish leaves 1 to 3 inches long, which are heart-shaped at the base. Its fragrant whitish or pink flowers are about ⅝ inch long, and are in small axillary or end clusters; blooming between late February and May. Also known as the Mayflower and Ground-laurel, it grows in sandy or rocky woods and banks.
RANGE: Lab to Sask. south to Fla. and Miss.

PINCUSHION-PLANT FAMILY (Diapensiaceae)

Members of this family have flowers with 5 sepals, united at the base; 5 petals, united toward the base; 5 stamens alternating with 5 sterile ones or staminodia, either attached to the corolla or joined with one another to form a tube; and a pistil composed of 3 united carpels, with a single style.

GALAX *(Galax urceolata)*

Galax has long-stalked, roundish heart-shaped, thickish, shiny leaves 1 to 3 inches across, with small bristly teeth on their margins. Bright green in summer, they become bronzed or reddish during the winter. The small white flowers are arranged in a narrow cluster on a naked flower stalk 10 to 18 inches tall, blooming May to July. It grows chiefly in mountain and upland woods. Another name for it is Beetleweed.
RANGE: Va. and W. Va. south to n. Ga. and n. Ala.

SHORTIA *(Shortia galacifolia)*

The Shortia, or Oconee-bells, is an interesting little plant which spreads by means of short runners and often forms large colonies. It has a very short stem with long-stalked, shiny, oval or roundish, wavy-toothed leaves ¾ to 2½ inches long. The white or pinkish flowers are about an inch across, with 5 irregularly toothed petals, and are on naked stalks 3 to 6 inches long; blooming during March and April. It grows in wooded ravines and along streams in the foothills and lower slopes of the mountains from sw. Va. south to n. Ga.

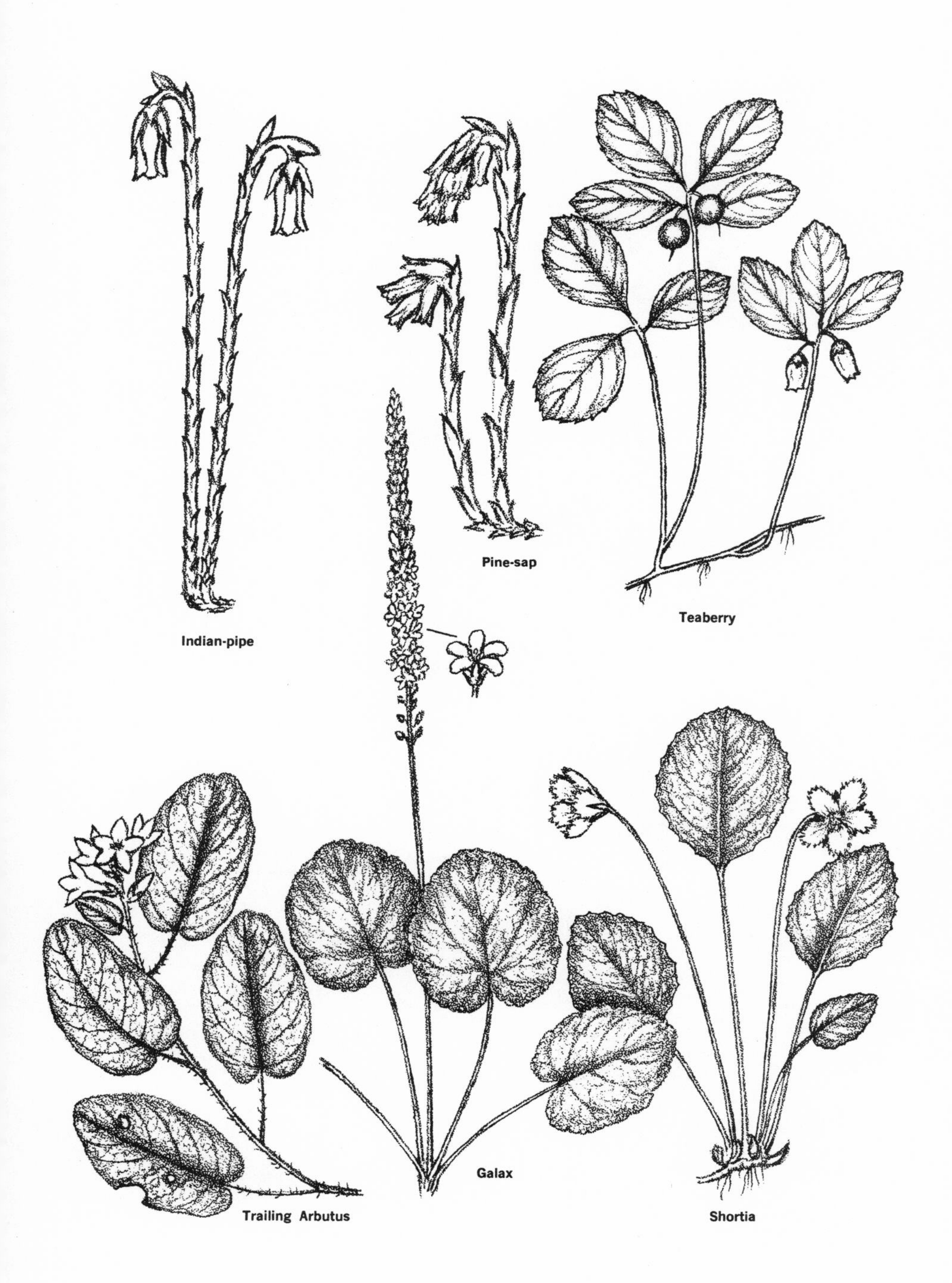
Indian-pipe
Pine-sap
Teaberry
Galax
Trailing Arbutus
Shortia

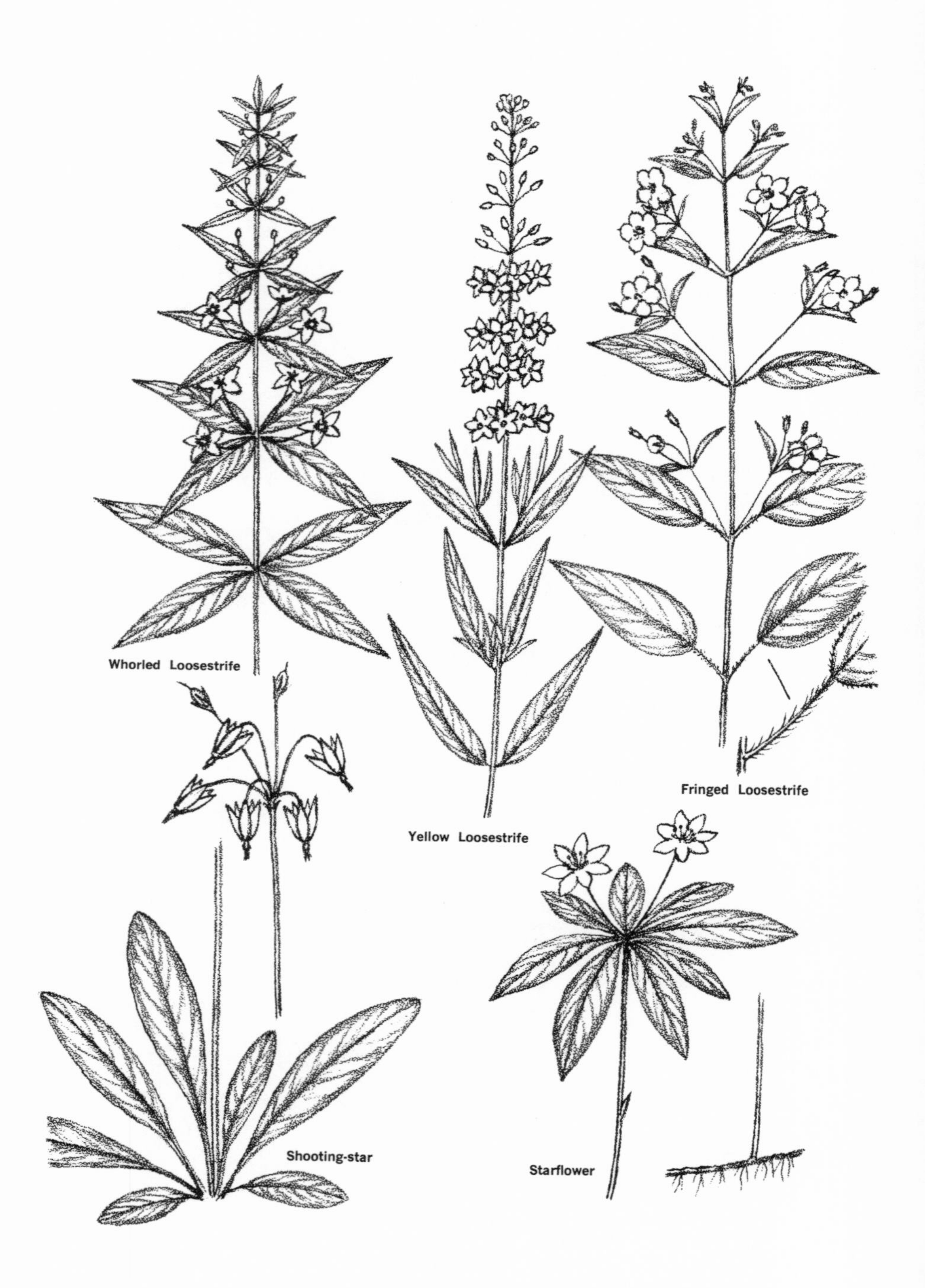

Whorled Loosestrife
Yellow Loosestrife
Fringed Loosestrife
Shooting-star
Starflower

PRIMROSE FAMILY (Primulaceae)

Members of the Primrose Family have flowers with 5 to 8 sepals which are united at the base to form a sort of cup; 5 to 8 petals, somewhat united at the base; 5 to 8 stamens which are attached to the bases of the petals; and a single pistil with a 1-chambered ovary, the seeds being attached to a stalk in its center.

WHORLED LOOSESTRIFE *(Lysimachia quadrifolia)*

This common plant of open woodlands and thickets has a simple, smooth or sparingly hairy stem from 1 to 2 feet tall. Along it are several whorls of usually 4 lance-shaped, stalkless leaves from 1½ to 3½ inches long. The star-like, 5-petalled, yellow flowers are about ½ inch across and have a reddish center. They are on thread-like stalks arising from the axils of the leaves; blooming between May and August. It is also known as the Four-leaf Loosestrife.

RANGE: Me. to Ont. and Wis. south to Ga., Ala., Tenn. and Ill.

YELLOW LOOSESTRIFE *(Lysimachia terrestris)*

Often called Swamp-candles, this loosestrife has a simple or branched stem 8 inches to about 2 feet tall, which ends in a long and narrow cluster of flowers. Each flower is about ⅓ inch across, star-like and 5-petalled, and yellow with purplish streaks or dots. It grows in swamps and low wet woods, blooming between May and August.

RANGE: Nfd. to Ont. and Wis. south to Ga., Ky. and Iowa.

FRINGED LOOSESTRIFE *(Lysimachia ciliata)*

The Fringed Loosestrife has an erect, simple or branched stem from 1 to 4 feet tall. On it are pairs of narrowly egg-shaped leaves from 2 to 6 inches long, which have slender leaf stalks which are fringed with hairs—hence the name of the plant. Its flowers are about ¾ inch across, the 5 yellow petals being ragged-toothed and with a minute point at the summit. They bloom between June and August, nodding on slender stalks at the tip of the stem or its branches.

RANGE: N.S. to B.C. south to Fla. and Tex.

SOUTHERN LOOSESTRIFE *(Lysimachia tonsa)*

This species resembles the preceding species but the leaf stalks are not hairy-fringed. It grows in dry upland woods and rocky slopes from Va., Ky. and Ark. south to Ga. and Ala.; blooming between May and July. (Not illustrated)

LANCE-LEAF LOOSESTRIFE *(Lysimachia lanceolata)*

This plant has lance-shaped leaves which taper at both ends and often have no evident leaf stalks. The flowers are similar to those of the preceding species. It is a slender-stemmed plant which grows in dry to moist open woods from Pa. to s. Mich. and Wis. south to Fla. and La. (Not illustrated)

SHOOTING-STAR *(Dodecatheon meadia)*

The Shooting-star or American Cowslip is a smooth plant with a basal cluster of narrowly elliptic or top-shaped leaves 3 to 10 inches long, their bases often marked with red. The distinctive flowers are about an inch long and have 5 lilac, pink or white petals which point backward; the stamens forming a beak-like cone in the center of the flower. From a few to sometimes a hundred or more of the flowers are arranged in an umbel, and nod from the summit of a naked flower stalk 8 inches to 2 feet tall. It grows in rich moist woods, meadows, prairies and cliffs; blooming between late March and June.

RANGE: D.C. to w. Pa., Wis. and Alb. south to n. Ga., La. and Tex.

STARFLOWER *(Trientalis borealis)*

Also known as the Chickweed-wintergreen, this little plant has a creeping rootstock which sends up erect branches 3 to 9 inches tall. These have a whorl of leaves at the summit; and 1 or 2 white, usually 7-petalled flowers about an inch across, which stand on long and thread-like stalks. The leaves vary considerably in size, from 1½ to about 4 inches in length, and are lance-shaped, pointed at both ends, and are stalkless or nearly so. It grows in moist woods, thickets, and on hummocks in swamps; blooming between May and July

RANGE: Lab. to Sask. south to Va., Ohio, Ill. and Minn.

PINKROOT FAMILY (Loganiaceae)

Members of this family have flowers with 4 or 5 partly united sepals; 4 or 5 petals partly united to form a tube; 4 or 5 stamens, united with the corolla tube; and pistil composed of 2 united carpels, with 1 to 4 style branches or stigmas. They typically have opposite leaves with untoothed margins.

YELLOW-JESSAMINE *(Gelsemium sempervirens)*

The Yellow-jessamine grows most profusely in the woods and thickets in the coastal regions of the South, where it often begins to bloom in late February. Elsewhere it may be as late as April or early May before the showy, trumpet-shaped, and very fragrant yellow flowers appear. They are 1 to 1½ inches long, with 5 rounded lobes, and have a slightly bilateral symmetry. The plant is a climber with slender, smooth, shining stems often 20 feet in length. It has pairs of lance-shaped, short-stalked leaves 1½ to 3 inches long which remain on the vine over winter. Also called the Carolina-jessamine and Evening-trumpet flower, it has found favor in cultivation. Its rootstocks are used medicinally, and children have been poisoned by sucking nectar from its blossoms.

RANGE: Va. and Ark. south to Fla. and Tex.

INDIAN-PINK *(Spigelia marilandica)*

This is a smooth plant with a simple erect stem from 6 to 18 inches tall. It has pairs of stalkless, egg-shaped to broadly lance-shaped, pointed leaves from 2 to 4 inches long; and a terminal cluster of showy flowers which bloom during May or June. They are shaped like slender trumpets 1½ to 2 inches long, with 5 rather small lobes at the summit. On the outside they are bright red or scarlet, but inside they are yellow. Several are borne in a 1-sided cluster. Also known as the Carolina Pinkroot or Wormgrass, its rootstocks have medicinal properties. It grows in rather rich and moist woodlands.

RANGE: Md. to Ohio, Ind., Mo. and Okla. south to Fla. and Tex.

GENTIAN FAMILY (Gentianaceae)

Members of the Gentian Family have flowers with from 4 to 12 partially united sepals; 4 to 12 partially or often well united petals; stamens as many as the petals or corolla lobes and attached to them; and a solitary pistil composed of 2 united carpels, the ovary having a single chamber which contains many ovules. They typically have opposite leaves with untoothed margins.

LARGE MARSH-PINK *(Sabatia dodecandra)*

Also known as the Sea-pink, this plant has attractive flowers 1½ to 2½ inches across, usually pink but sometimes white with a bright yellow "eye", and from 8 to 12 petal-like corolla lobes. They bloom between July and September. The plant has a simple or sparingly branched stem 1 to 2 feet high, on which are pairs of lance-shaped or narrower leaves and the terminal flowers. The basal leaves are larger, 1½ to 3 inches long, and are rather spoon-shaped. It grows on the sandy borders of pools and in brackish or salt marshes, blooming between June and August. RANGE: Conn. south to Fla. and west to La.

MANY-PETALLED MARSH-PINK *(Sabatia gentianoides)*

Very attractive when in flower, this species has a simple or branched stem 1 to 2 feet high; on which are pairs of rather stiff, very narrow, stalkless and pointed leaves from 1½ to 4 inches long. Its flowers are 1½ to 2 inches across, lilac to purplish-pink, with 8 to 12 petal-like corolla lobes but no yellow "eye"—just the yellow stamens in the center. They differ from the flowers of most marsh-pinks in having curved rather than coiled anthers. It grows in bogs and savannahs in the southern coastal plain, blooming during July and August. RANGE: se. N.C. south to Fla. and west to Tex.

SEA-PINK *(Sabatia stellaris)*

The Sea-pink has starry 5-petalled deep pink to white flowers 1 to 1½ inches across, with a central yellow "star" which is bordered with red. They are usually numerous but solitary at the tips of the branches. The plant has a roundish or slightly 4-angled stem from 6 inches to about 20 inches tall; and narrowly lance-shaped or top-shaped leaves from ½ to 2 inches long. It grows in salt and brackish marshes along the coast; blooming between July and October.

RANGE: Mass. south to Fla, and west to La.

Yellow-jessamine
Indian-pink
Large Marsh-pink
Many-petalled Marsh-pink
Sea-pink

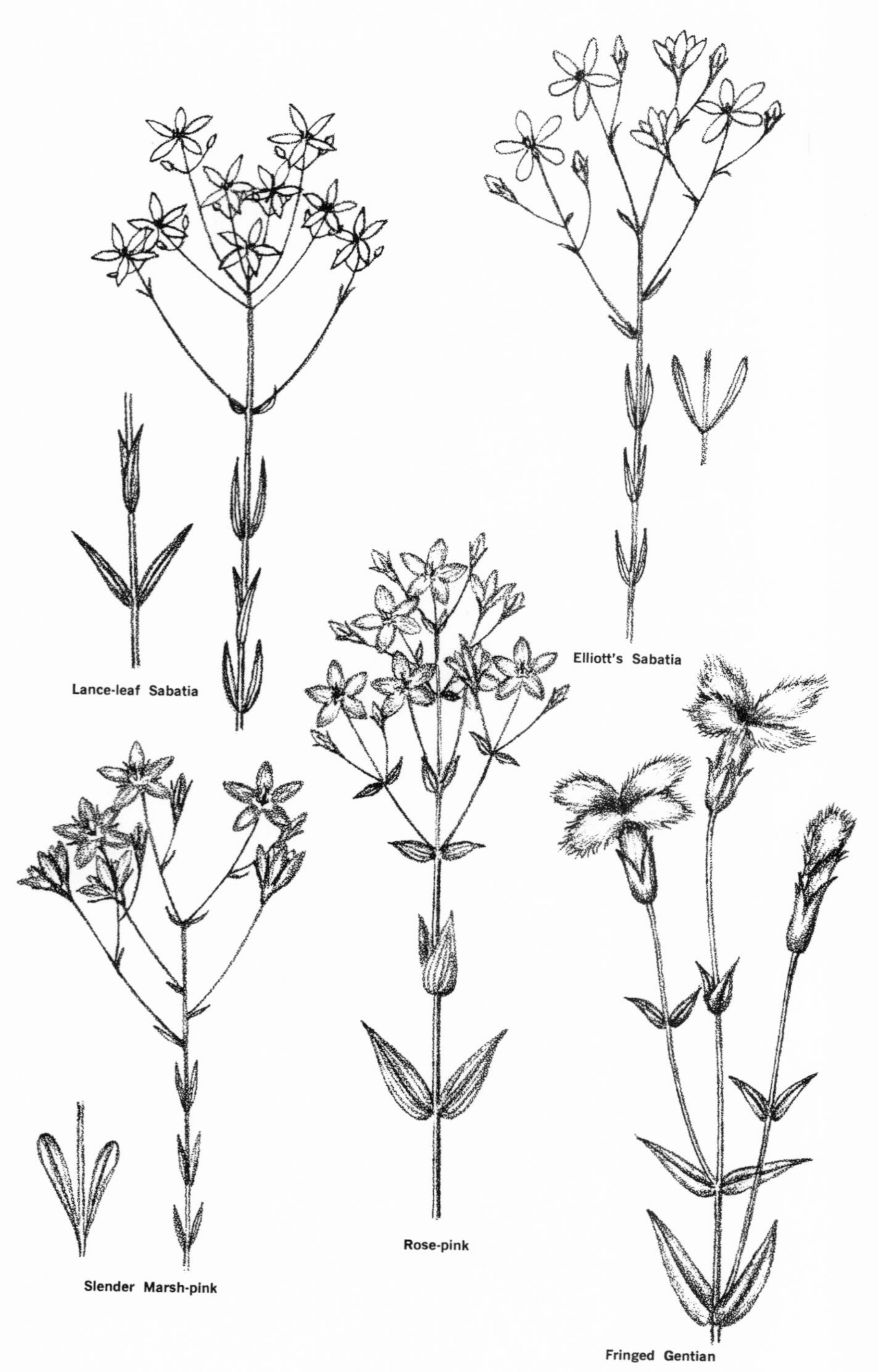
Lance-leaf Sabatia
Elliott's Sabatia
Rose-pink
Slender Marsh-pink
Fringed Gentian

LANCE-LEAF SABATIA *(Sabatia difformis)*

This is a smooth plant with a somewhat 4-angled and slender stem from 1 to about 3 feet tall. On it are pairs of narrowly lance-shaped leaves 1 to 2 inches long. Its flowers are white, ¾ to an inch across, and have 5 petal-like corolla lobes. They are quite numerous and arranged in a terminal flower cluster which has opposite, or forking branches. It is often exceedingly abundant in the wet coastal plain pinelands and savannahs, blooming between July and September.
RANGE: N.J. south to Fla.

ELLIOTT'S SABATIA *(Sabatia brevifolia)*

Elliott's Sabatia is a smooth plant with a roundish or slightly ridged stem from 1 to 2 feet high; with pairs of very narrow leaves ½ to nearly an inch long. The flowers are white or cream-colored, about ¾ inch across, and have 5 petal-like corolla lobes. They are quite numerous but arranged in a terminal flower cluster which has alternate branches. It grows in coastal plain pinelands, blooming September and October.
RANGE: se. S.C. south to Fla. and Ala.

SLENDER MARSH-PINK *(Sabatia campanulata)*

The Slender Marsh-pink has pale crimson, pink, or occasionally white flowers about an inch across, with 5 petal-like corolla lobes and a yellow "eye". It is a smooth plant with a very slender stem, or stems, from 1 to 2 feet high; with pairs of often very narrow leaves 1 to 1½ inches long, tapering to the tip from below the middle. The flowers are generally numerous and the flower cluster has alternate branching. It grows in savannahs and wet sandy or boggy places, blooming between June and August.
RANGE: Coast from Mass. south to Fla. and La.; inland from Va. and Ind. south to Ga. and Ala.

ROSE-PINK *(Sabatia angularis)*

This is our most common and best known species of *Sabatia* for it grows in open woodlands, thickets, meadows and marshes over a wide range. Known also as the Bitter-bloom, it has a much branched and sharply 4-angled stem from 1 to about 3 feet tall. The paired leaves are egg-shaped, ¾ to 1½ inches long, and have heart-shaped clasping bases. The numerous flowers are rose-pink, rarely white, about an inch across, with 5 petal-like corolla lobes and a greenish-yellow star-shaped "eye". The branching of the flower cluster is opposite. This is one of our more attractive mid-summer wild flowers. It blooms in July and August.
RANGE: N.Y. and Ont. to Wis. and Mo. south to Fla., La. and Okla.

FRINGED GENTIAN *(Gentianopsis crinita)*

The Fringed Gentian has violet-blue flowers about 2 inches long, with the 4 petals united below into a bell-shaped tube and with the free ends conspicuously fringed. It has a stem 1 to 3 feet tall, with pairs of lance-shaped leaves 1 to 2 inches long on the stem and spoon-shaped ones at the base. It grows in cool, moist, open woods and meadows; blooming in September and October. RANGE: Me. to Man. south to Ga., Ind. and Iowa.

SMALLER FRINGED GENTIAN *(Gentianopsis procera)*

This species is similar to the preceding but is only 6 to 18 inches tall, and the very narrow leaves show only a prominent midrib. The flowers are 1 to 2 inches long and the corolla lobes are much shorter fringed. It grows in wet or boggy places from w. N.Y. and s. Ont. to Alaska south to Ohio, Iowa and N.D. (Not illustrated)

PINE-BARREN GENTIAN *(Gentiana autumnalis)*

This is a distinctive plant which has a solitary flower at the tip of a slender, simple or few-forked stem 6 to 18 inches tall. Along the stem are a number of pairs of rather thick and very narrow leaves from 1 to 2 inches in length. The flower has an almost lily-like, funnel-shaped corolla about 2 inches long, which is bright indigo-blue or rarely lilac or whitish, and is often spotted with greenish or brown within. The corolla lobes are much longer than the small fringed lobes between them. It grows in savannahs and moist, sandy coastal plain pinelands; blooming from September to early December. It is also known as the One-flowered Gentian.
RANGE: N.J. south to S.C.

CLOSED GENTIAN *(Gentiana andrewsii)*

The club-shaped flowers of this and the next species do not open, nor do they show prominent corolla lobes at the tip. If spread open, the pale appendages between the narrower corolla lobes show a fringed margin in this species. The deep violet-blue (rarely white) flowers are 1 to 2 inches long, and clustered in the axils of the upper leaves. It grows in moist open places and thickets, blooming between August and October. Often called the Blind or Bottle Gentian. RANGE: Mass. and Que. to Man. south to Ga. and Ark.

CLOSED GENTIAN *(Gentiana clausa)*

This species is very similar to the preceding. It has porcelain blue flowers which turn violet-blue. The membranes between the corolla lobes are 2- to 3-cleft but not fringed, and are no wider than the corolla lobes. It grows in rich moist woods and meadows from Me. to Minn. south to w. N.C., Tenn. and Mo. (Not illustrated)

STRIPED GENTIAN *(Gentiana villosa)*

This species has slender, smooth stems 6 inches to 2 feet tall; with pairs of egg-shaped leaves ½ to 2 inches long, which are stalkless and have clasping bases. Its flowers are 1 to 3 inches long, club-shaped, greenish-white to purplish-green with purple stripes, and have 5 corolla lobes which become erect. It grows on wooded slopes and stream banks, blooming from August to November. Also called Samson's Snakeroot.
RANGE: N.J. and Pa. south to Fla. and La.

SOAPWORT GENTIAN *(Gentiana saponaria)*

This plant has a corolla open only slightly at the summit and the pleats or membranes are nearly or quite equal to the corolla lobes. It grows in wet places and moist woods from N.Y. to Ind. and Minn. south to Fla. and Tex., blooming September to November. (Not illustrated)

CATESBY'S GENTIAN *(Gentiana catesbaei)*

This gentian of the wet coastal plain pinelands and savannahs is readily identified by its bell-shaped, deep blue or violet-purple flowers which are about 2 inches long. It has a simple or sparingly branched stem 8 inches to 2 feet tall, with pairs of egg-shaped or lance-shaped leaves 1 to 2½ inches in length. It blooms from September to November.
RANGE: Del. south to Fla.

DOWNY GENTIAN *(Gentiana puberulenta)*

This species has a minutely downy stem 8 to 18 inches tall on which are many pairs of narrowly lance-shaped leaves. Its flowers are 1½ to 2 inches long, violet-blue, and have a funnel-shaped corolla. It grows on prairies and sandy ridges.
RANGE: N.Y. to Ont. and Minn. south to Ga. and Kan.

STIFF GENTIAN *(Gentianella quinquefolia)*

Also called Agueweed and Gall-of-the-earth, this species has stiffly erect, wing-angled stems 6 inches to 2 feet tall. The paired leaves are egg-shaped, ½ to 2 inches long, stalkless, and with clasping bases. Its flowers are about ¾ inch long, funnel-shaped, violet-blue to lilac, and have no pleats between the 5 triangular lobes. They are borne in both end and axillary clusters and bloom between August and October. It grows in dry to moist open woods, and along stream banks and roadsides.
RANGE: Me. to Ont. south to Fla., La. and Mo.

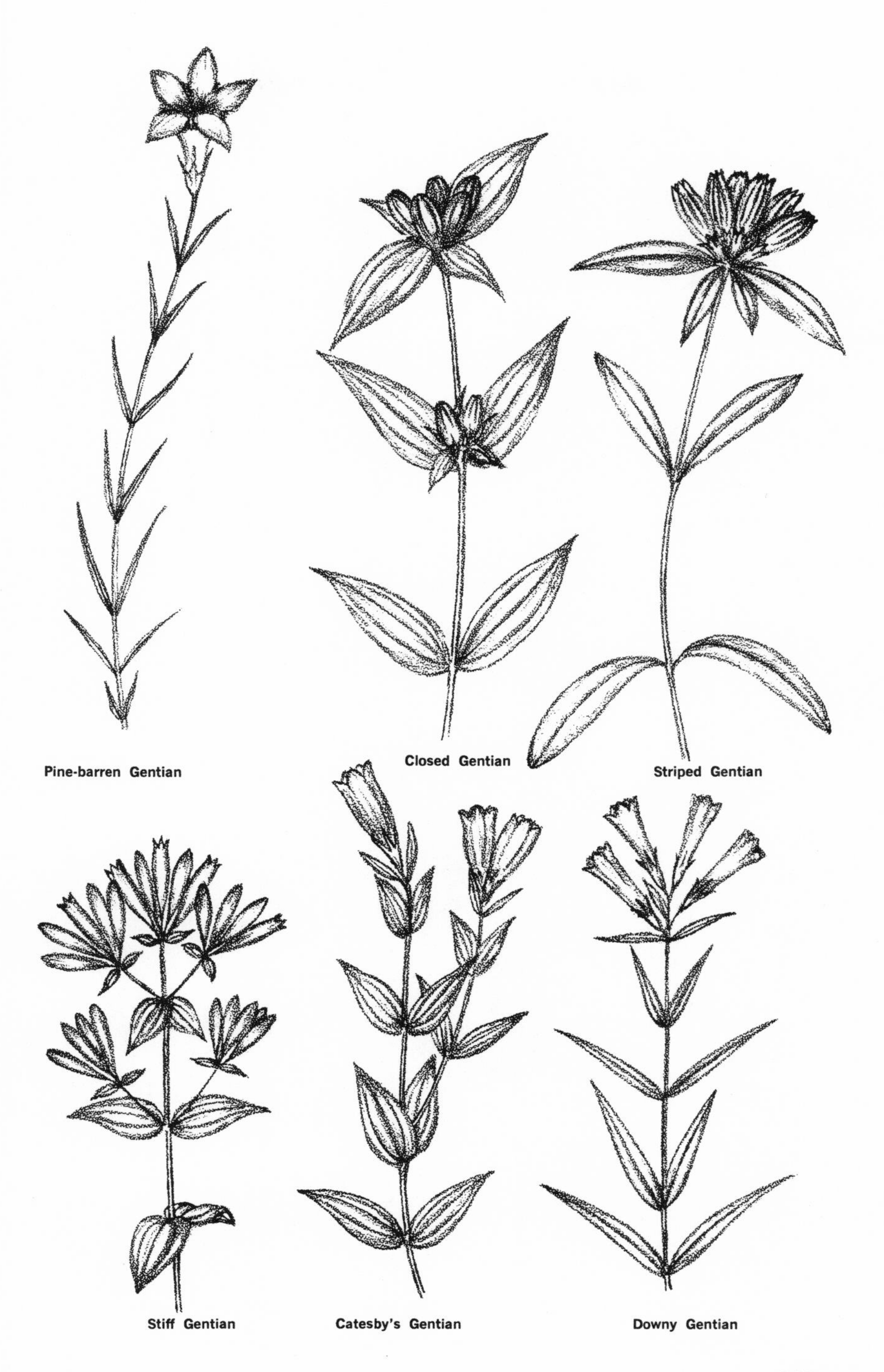
Pine-barren Gentian
Closed Gentian
Striped Gentian
Stiff Gentian
Catesby's Gentian
Downy Gentian

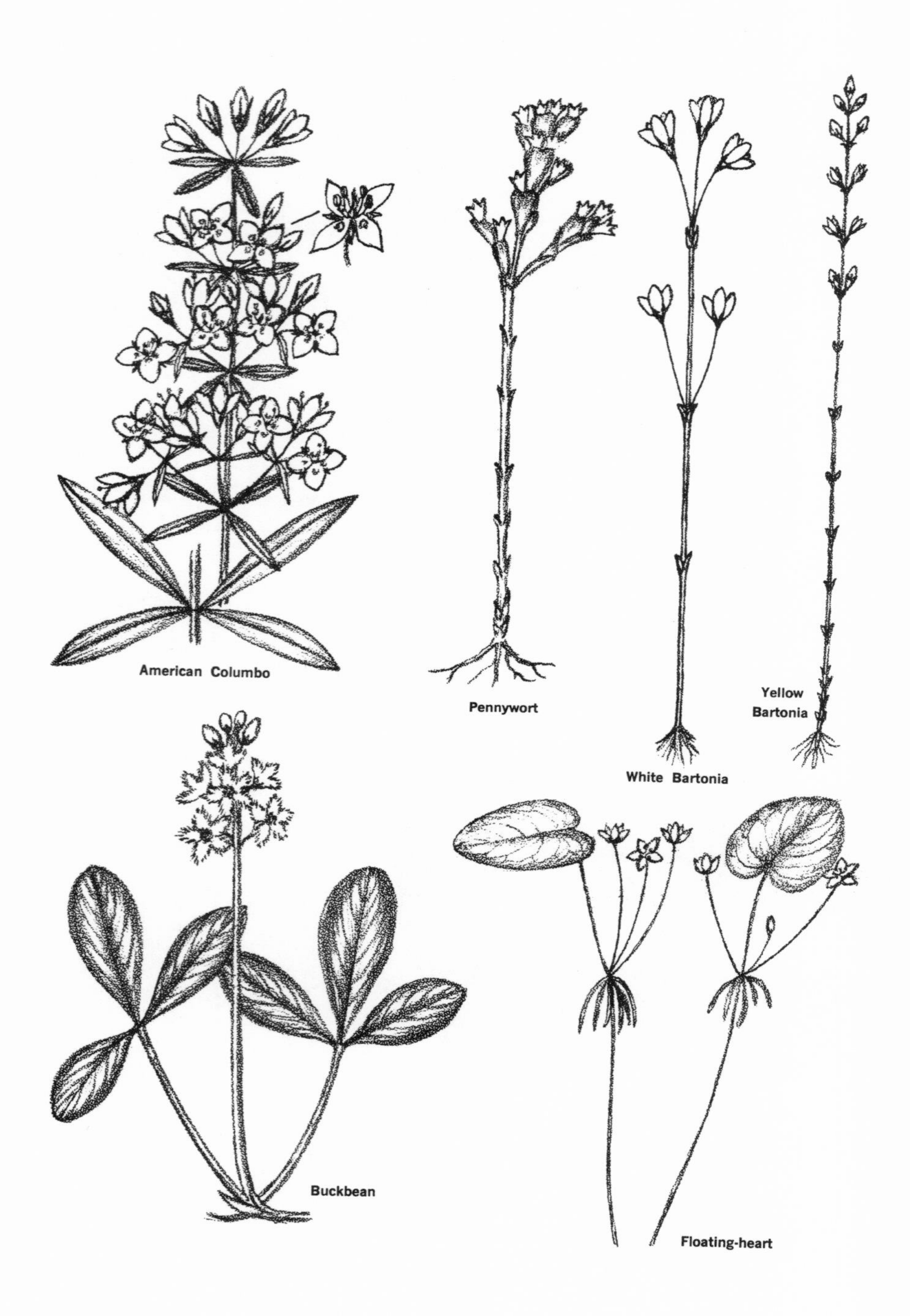
American Columbo
Pennywort
Yellow
Bartonia
White Bartonia
Buckbean
Floating-heart

AMERICAN COLUMBO *(Frasera carolinensis)*

Pyramid-plant is another name for this tall member of the Gentian Family. It is a smooth plant with a stout stem 3 to 6 feet high, which is terminated by a branched flower cluster which may be 1 to 2 feet long. Each flower is about an inch across, yellowish-white dotted with brownish-purple; with a large, round, fringed gland on each of the 4 petals. The lance-shaped leaves are 3 to 6 inches long, mostly in whorls of 4, and the ones toward the base of the plant are broader toward the tip. It grows in dry open woods, thickets, and meadows; blooming in May or June.
RANGE: N.Y. and Ont. to Wis. south to Ga. and La.

PENNYWORT *(Obolaria virginica)*

This is a smooth, fleshy, purplish-green plant from 3 to 6 inches high. Near the summit of the stem there are pairs of top-shaped leaves which have 1 to 3 flowers in their axils. Lower on the stem there are 2 to 6 pairs of small bracts in place of leaves. The flowers are about ½ inch long, white to dull purplish, and have 4 lobes at the summit. It grows in rich, moist woods and thickets, getting much of its nourishment from the decaying humus. Flowering season is between March and May.
RANGE: N.J. to Ill. south to Fla. and Tex.

WHITE BARTONIA *(Bartonia verna)*

This plant has a very slender, yellowish-green stem 2 to 15 inches high; with a few widely-spaced, scale-like leaves. Its flowers are white, about ⅓ inch long, 4-petalled, and loosely clustered on the upper part of the stem. It grows in moist pinelands, prairies, and lake shores in the coastal plain; blooming during March and early April.
RANGE: e. N.C. south to Fla. and west to La.

YELLOW BARTONIA *(Bartonia virginica)*

Yellow Bartonia has a fine, wiry, yellowish-green stem 4 to 15 inches high, usually simple but sometimes with a few ascending branches above. Along it are many pairs of small, awl-like scales in place of leaves. Its flowers are small, yellowish, and in the axils of bracts on the upper part of the stem. It grows in moist to fairly dry open woods and on grassy banks; blooming between July and October.
RANGE: N.S. to Que. and Minn. south to Fla. and La.

PANICLED BARTONIA *(Bartonia paniculata)*

This one differs from the preceding Bartonias in having a branched cluster of slender stalked flowers, and scattered rather than paired scale-like leaves. It grows in wet peaty or sandy places from Mass. to Ky. and Ark. south to Fla. and La. (Not illustrated)

BUCKBEAN FAMILY (Menyanthaceae)

BUCKBEAN *(Menyanthes trifoliata)*

The Buckbean has a thick, scaly, creeping stem from which its leaves and flower stalks arise. The former are divided into 3 elliptic or narrowly top-shaped, untoothed leaflets 1½ to 3 inches long; on leaf stalks 2 to 10 inches in length. Its white flowers are about ½ inch across, and the 5 petals have glistening hairs on their upper surface. They are clustered on stalks 3 to 12 inches long and bloom between May and July. Bogbean and Marsh-trefoil are other names often given this plant of shallow waters and boggy places.
RANGE: Lab. to Alaska south to Del., W.Va., the Great Lakes region, Mo., Neb. and Wyo.

FLOATING-HEART *(Nymphoides cordata)*

At first glance, this plant may be mistaken for a miniature water-lily. It grows in ponds and the waters of slowly moving streams with its roots in the bottom mud. Between June and September, it sends up to the surface a long stalk which produces a single 1- to 2-inch long heart-shaped leaf, several small white flowers, and a cluster of tuber-like roots.
RANGE: NFD. to Ont. south to Fla. and La.

LARGE FLOATING-HEART *(Nymphoides aquatica)*

This is a much coarser plant with leaves 2 to 6 inches across, and larger flowers about ¾ inch wide. It grows in ponds and sluggish streams of the coastal plain from N.J. south to Fla.; blooming between May and August. (Not illustrated)

DOGBANE FAMILY (Apocynaceae)

Members of this family have flowers with 5 partly united sepals; 5 petals united to form a corolla tube; 5 stamens with anthers converging around the stigma; and 2 pistils which have their styles and stigmas united and appearing as one, maturing as a pair of long and slender pods. The seeds very often have long, silky hairs. Most species belonging to the family have a milky juice.

BLUE-DOGBANE *(Amsonia tabernaemontana)*

The Blue-dogbane, or Willow Amsonia, is a smooth or slightly downy plant with clustered stems from 1 to 3 feet high. It has scattered lance-shaped or narrowly egg-shaped leaves 2 to 4 inches long. In April or May, it produces rather large but open end clusters of numerous pale blue flowers about ¾ of an inch across. The slender corolla tube abruptly expands into 5 narrow and pointed lobes, giving the flowers a starry appearance. The plant grows in rich, moist woods and along the banks of streams. It is often cultivated and has become naturalized northward and eastward of its normal range.

RANGE: Va. to s. Ill., Mo. and Kan. south to Ga. and Tex.

SPREADING DOGBANE *(Apocynum androsaemifolium)*

The Spreading Dogbane has a stem which forks repeatedly into wide-spreading, mostly smooth, and usually reddish branches. It attains a height of between 1 and 4 feet. Its leaves are paired, egg-shaped or oval, short-stalked, 2 to 4 inches long, and are pale and either smooth or downy beneath. The flowers are about ⅓ inch across, bell-shaped, and rosy-pink. It grows in dry, open woods, thickets, and fields; blooming between June and August. RANGE: Nfd. to Alska south to n. Ga., Ohio, Ill., Ark., Neb. and N.Mex.

INDIAN-HEMP *(Apocynum cannabinum)*

This is a common but less attractive looking plant than the preceding species, usually with a rather erect main stem and ascending, greenish side branches. It has smaller, more urn-shaped, greenish-white flowers which bloom between May and August. The plant apparently got its name from the use of the fibers of its stems by the Indians. It grows in open places and thickets from Que. to Alb. and Wash. south to Fla., Tex. and Calif. (Not illustrated)

MILKWEED FAMILY (Asclepidaceae)

Members of this family have flowers with 5 sepals, barely united at the base; 5 petals, united at the base; 5 stamens which are united at the base and have their anthers adhering to the stigma; and a double pistil, the 2 ovaries having a common stigma and maturing as a pair of pod-like fruits containing seeds usually with silky hairs. Most members of the family have opposite leaves and a milky juice.

In the milkweeds proper, the flowers have 5 reflexed corolla lobes and a central crown of 5 trowel-shaped hoods which are attached to the filaments of the stamens. Within each hood there is usually a slender horn that curves up and over the stigma. The flowers have a remarkable adaptation to insect pollination.

BUTTERFLYWEED *(Asclepias tuberosa)*

This is the beautiful orange-flowered milkweed so often seen in dry fields, on rocky open slopes, and along the roadside. It is a roughish-hairy plant with reclining, ascending, or even somewhat erect stems 1 to 2 feet high, which usually branch toward the summit. The numerous leaves are lance-shaped or narrowly egg-shaped, stalkless or very short-stalked, 1 to 4 inches long, and usually scattered singly along the stems. The flowers are orange-yellow to orange-red but vary considerably in their brilliance. Although it is a milkweed, the plant does not have a milky juice. It blooms between May and September. Other names for it are Yellow Milkweed and Pleurisy-root.

RANGE: Vt. and Ont. to Minn., Neb. and Colo. south to Fla., Tex. and Ariz.

LANCE-LEAF MILKWEED *(Asclepias lanceolata)*

Also known as the Few-flowered Milkweed, this is a smooth plant with usually a simple stem 2 to 4 feet tall. Along it are very widely spaced pairs of narrow or lance-shaped leaves from 4 to 10 inches in length. There are from 1 to 3 umbels at the summit of the stem, each having from 5 to 12 flowers with bright orange-yellow hoods and a deep red corolla. It grows on the borders of swamps and in wet pinelands from Del. south to Fla. and west to Tex.

Blue-dogbane
Spreading Dogbane
Butterflyweed
Lance-leaf Milkweed

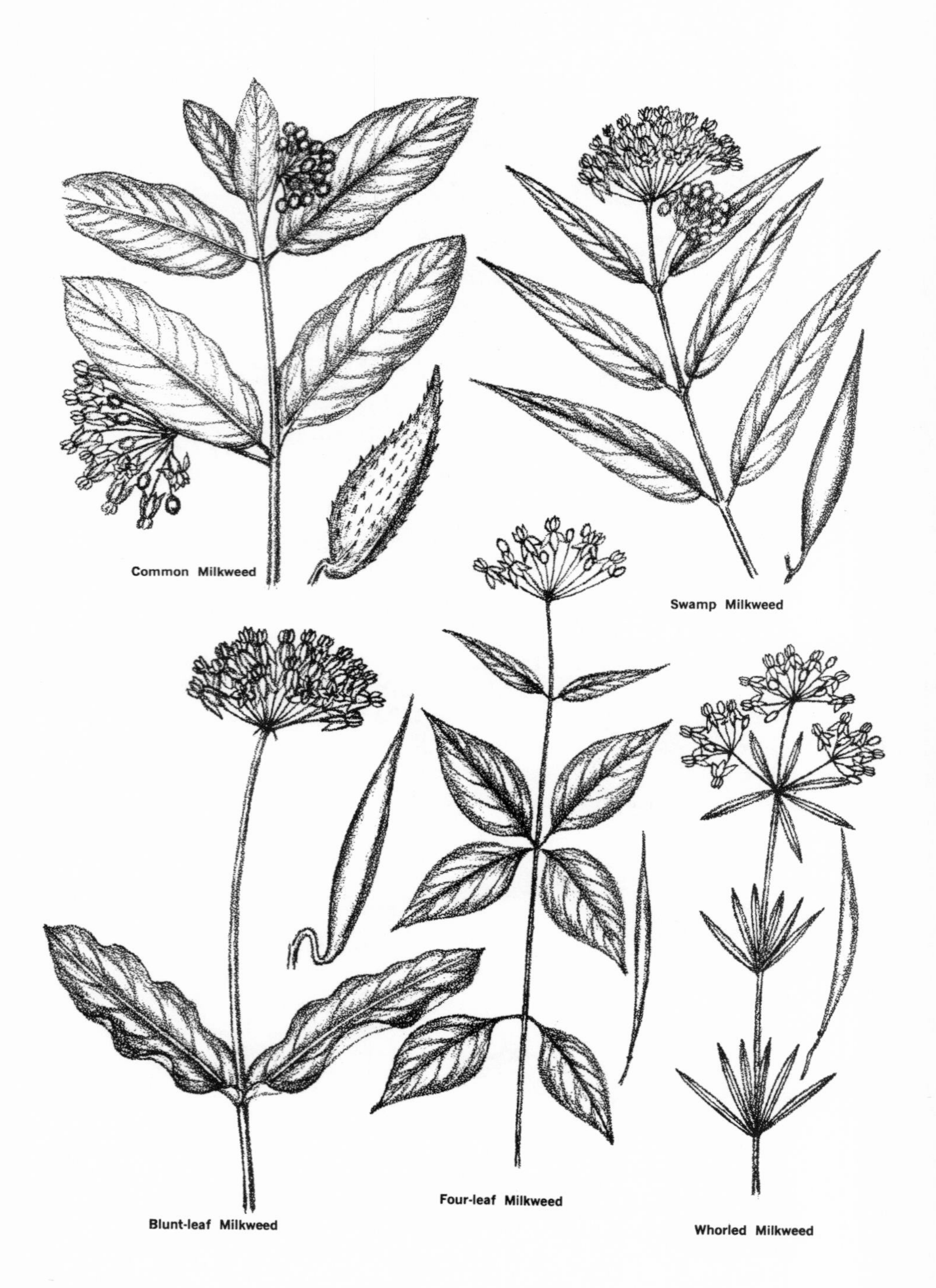
Common Milkweed
Swamp Milkweed
Blunt-leaf Milkweed
Four-leaf Milkweed
Whorled Milkweed

COMMON MILKWEED *(Asclepias syriaca)*

The Common Milkweed has a stout, usually simple stem from 3 to 5 feet in height. Its paired leaves are oblong or oval, more or less rounded at both ends or abruptly pointed at the tip, softly downy beneath, and from 3 to 8 inches long. The dull lavender to greenish-white flowers are about ⅓ inch long and in rather densely crowded umbels a few inches across. This is our commonest species of milkweed, growing in dry fields, thickets, and along roadsides everywhere. It blooms between June and August.
RANGE: N.B. to Sask. south to Ga., Tenn., Mo. and Kan.

SWAMP MILKWEED *(Asclepias incarnata)*

This is another common milkweed and, as its name indicates, it grows in swamps and other wet places. It has a slender and often branched stem from 2 to 4 feet in height; and numerous pairs of leaves which are lance-shaped, taper-pointed, from 3 to 6 inches long and either smooth or downy beneath. They are very short-stalked or practically stalkless. Its flowers are about ¼ inch long, pink or rose-colored, and in many-flowered umbels. The flowering season extends from July to September.
RANGE: N.S. to Man. and Wyo. south to Ga., Tex. and N.Mex.

RED MILKWEED *(Asclepias rubra)*

This is a smooth plant usually with a simple stem 1 to 4 feet tall; and pairs of broadly lance-shaped to egg-shaped leaves 2 to 7 inches long which taper from the rounded or heart-shaped base to a long and slender tip. It has from 1 to 4 many-flowered umbels of purplish-red flowers, blooming in June and July. The plant grows in wet pinelands and bogs in the coastal plain from se. N.Y., N.J. and e. Pa. south to Fla. and west to Tex. (Not illustrated)

BLUNT-LEAF MILKWEED *(Asclepias amplexicaulis)*

The Blunt-leaf Milkweed is a rather smooth, pale green and somewhat whitened plant 2 to 3 feet tall. It has stalkless elliptic or oblong leaves 3 to 5 inches long which have broad and clasping bases, rounded and abruptly short-pointed tips, and conspicuously wavy margins. The dull purple or greenish-purple flowers are about ½ inch long, in a usually solitary and densely-flowered umbel which terminates a long stalk. It grows in dry and usually sandy fields, blooming between May and July.
RANGE: Mass. to Minn. and Neb. south to Fla. and Tex.

FOUR-LEAF MILKWEED *(Asclepias quadrifolia)*

This dainty little milkweed grows in dry open woods, blooming between May and July. On a usually unbranched, slender stem 1 to 2 feet tall, it has 1 or 2 whorls of leaves near the middle and a pair of leaves both above and below the whorled ones. All are quite thin, egg-shaped to lance-shaped, definitely stalked and from 2 to 6 inches long. The flowers have pale pink corolla lobes and white hoods. They are in from 1 to 4 rather small terminal umbels.
RANGE: N.H. to Ont. and Minn, south to n. Ga., Ala., Ark., and Kan.

WHORLED MILKWEED *(Asclepias verticillata)*

The Whorled Milkweed gets its name from its very narrow leaves which are arranged along the stem in several whorls of from 3 to 7. They are stalkless, ¾ to 2¼ inches long, and have the margins rolled. The plant has a slender, simple or sparingly branched stem from 1 to 2½ feet tall. Its flowers are about ¼ inch long and have greenish-white corolla lobes and white hoods. They are in several small umbels, blooming between June and September. It grows in dry woods and fields.
RANGE: Mass. to Ont. and Sask. south to Fla. and Tex.

THIN-LEAF MILKWEED *(Asclepias perennis)*

This is another of our smaller milkweeds, one which grows in the wet woods and along stream banks. It has a slender, ascending or somewhat erect stem 1 to 3 feet high; often rooting near the base and somewhat downy above. Its leaves are paired, rather lance-shaped to narrowly egg-shaped, thin in texture, 2 to 6 inches long, and have slender leaf stalks. The flowers are small, white, and have hoods much shorter than the needle-shaped horns. They are in rather small umbels, blooming between June and August. RANGE: S.C. to s. Ind., s. Ill. and se. Mo. south to Fla. and Tex. (Not illustrated)

PURPLE MILKWEED *(Asclepias purpurascens)*

The Purple Milkweed has a moderately stout, smooth or minutely downy, and usually simple stem 2 to 3 feet tall. It has pairs of elliptic or egg-shaped, short-stalked leaves 3 to 8 inches long which are minutely downy beneath. The flowers are about ⅓ inch long, deep purple, and in several many-flowered umbels; blooming between May and July. It grows in dry to moist open woods, thickets, and fields.

RANGE: N.H. and Ont. to Minn. and S.D. south to N.C., Tenn., Miss. and Okla.

POKE MILKWEED *(Asclepias exaltata)*

The Poke or Tall Milkweed is a smooth plant, usually with a simple stem 3 to 6 feet tall; and pairs of rather thin, egg-shaped to broadly lance-shaped leaves 4 to 9 inches long, which taper at both ends and are long-stalked. Its flowers are about ⅔ inch long, with greenish or greenish-purple corolla lobes and white hoods. They are in rather large but loosely-flowered and usually nodding umbels; blooming between June and August. It grows in rich open woods and thickets.

RANGE: Me. to Minn. south to Ga., Ky., Ill. and Iowa.

WHITE MILKWEED *(Asclepias variegata)*

The White Milkweed has a moderately stout, usually simple stem 1 to 3 feet tall, which is somewhat downy on the upper portion. It has pairs of egg-shaped or broadly lance-shaped leaves 3 to 6 inches long, which are definitely stalked and rounded to broadly pointed at both ends. The flowers are about ⅓ inch long, white with a purplish center, and their roundish hoods have a half-moon-shaped horn with a horizontal point. They are arranged in 1 to 4 many-flowered umbels, blooming between May and July. It grows in open woods and thickets.

RANGE: Conn. and se. N.Y. to Ill., Mo. and Okla. south to Fla. and Tex.

SANDHILL MILKWEED *(Asclepias humistrata)*

This unusual milkweed grows in the driest of sandy places: chiefly in coastal plain pine and oak woods, and on sandhills. It is a pale green and smooth plant which is whitened by a bloom. The often purplish stem is prostrate, 1 to 3 feet long, and the paired leaves stand almost vertically. They are egg-shaped, pointed at the tip, stalkless, and from 2 to 5 inches long. Both surfaces of the leaves are alike—pale green and strikingly veined and bordered with white or coral-pink. The flowers are about ⅓ inch long and have grayish or greenish-purple corolla lobes and white hoods. They are arranged in rather loosely-flowered umbels and bloom in May or June. It cannot possibly be mistaken for any other species of milkweed. RANGE: e. N.C. south to Fla. and west to Miss.

GREEN MILKWEED (*Asclepias viridiflora*)

The flowers of this milkweed are greenish ½ inch or less long, and the hoods have no horns. They are arranged in quite dense and practically stalkless umbels in the axils of the leaves; blooming between June and August. The plant has reclining or erect stems 1 to 2½ feet long, along which are many pairs of oval to lance-shaped or very narrow leaves which may be smooth or slightly downy. It grows in dry woods, openings, and on prairies from Mass. to Ont. and Man. south to Fla., Tex. and N.Mex. (Not illustrated)

LARGE-FLOWERED ANGLEPOD *(Matelea obliqua)*

The anglepods are vining plants with opposite heart-shaped leaves. Six species besides this one occur in the eastern United States. The heart-shaped leaves of this species are 2 to 8 inches long, downy beneath, and have rather long downy leaf stalks. Its flowers are ½ to ⅔ inch across and have 5 narrow corolla lobes which are greenish and downy on the back and reddish-purple above. They are arranged in somewhat flat-topped and branching clusters, blooming between May and July. It grows in rather dry but rich thickets, woods borders, and on rocky slopes. The pods have small fleshy protuberances.

RANGE: Pa. to Ill. south to Ga., Tenn. and Mo.

SANDVINE *(Cynanchum laeve)*

This is also a vine but it has pairs of egg-shaped leaves which are heart-shaped at the base and pointed at the tip. It has small whitish flowers which are bell-shaped with 5 nearly erect corolla lobes, and are in stalked axillary clusters; blooming from July to September. It grows along stream banks from e. Pa. to Ill. and Neb. south to w. Ala. and Tex. (Not illustrated)

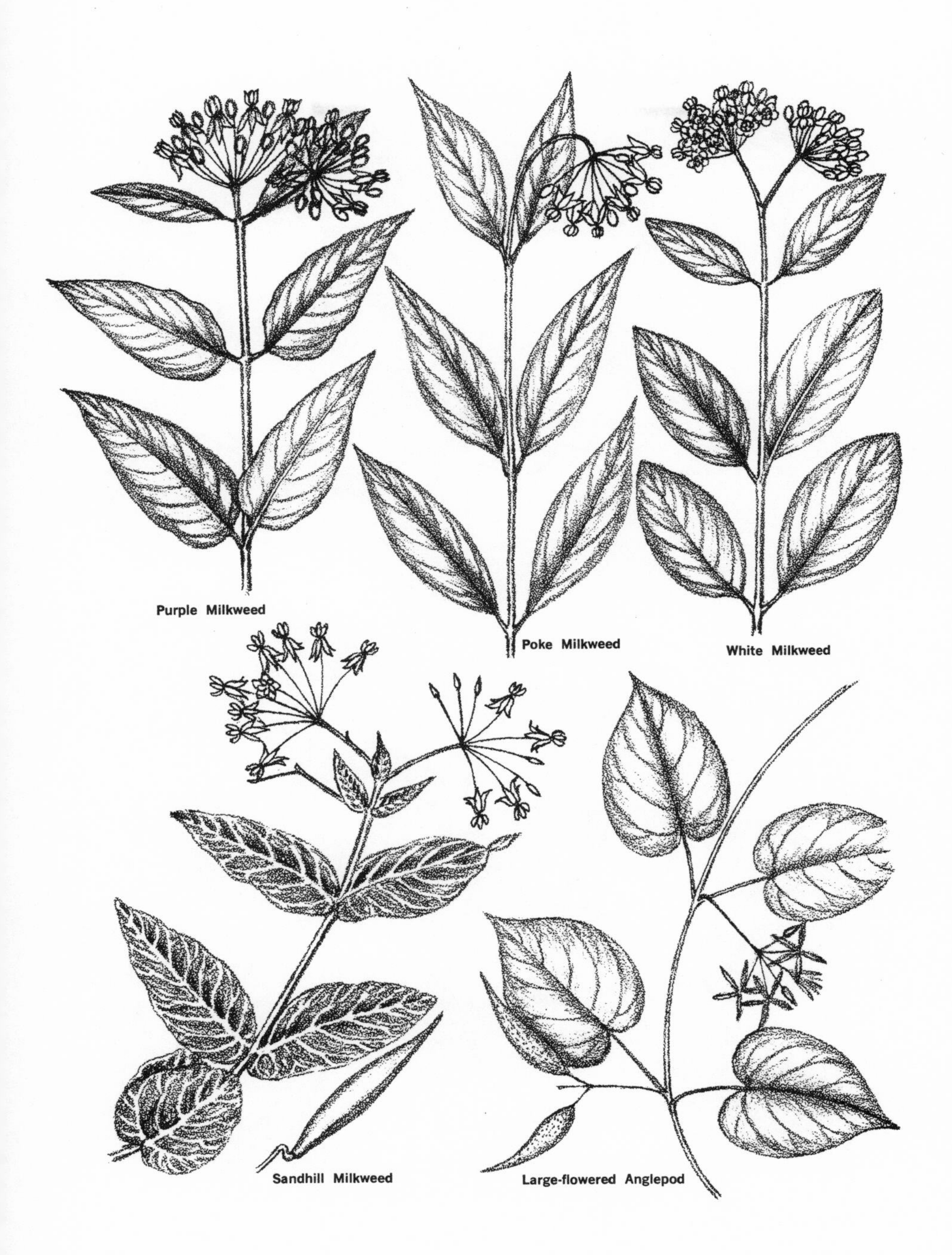
Purple Milkweed
Poke Milkweed
White Milkweed
Sandhill Milkweed
Large-flowered Anglepod

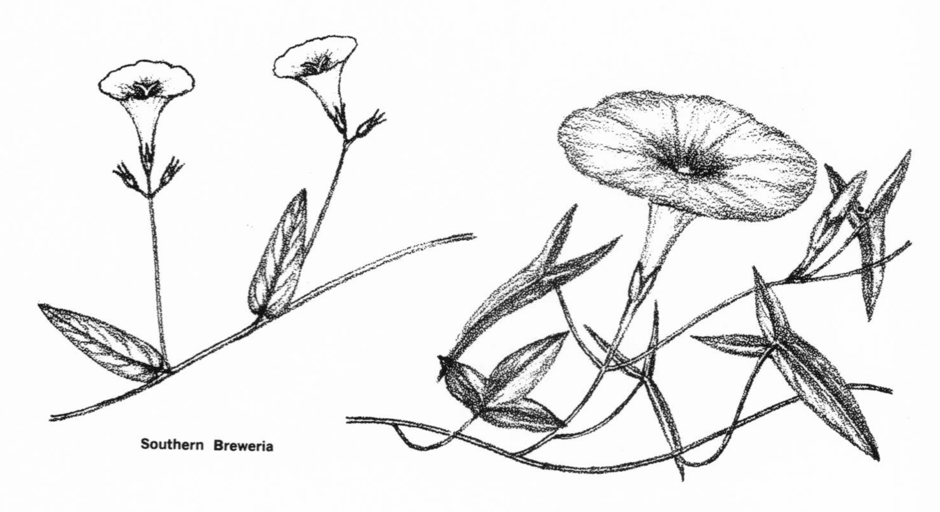

Southern Breweria

Arrow-leaf Morning-glory

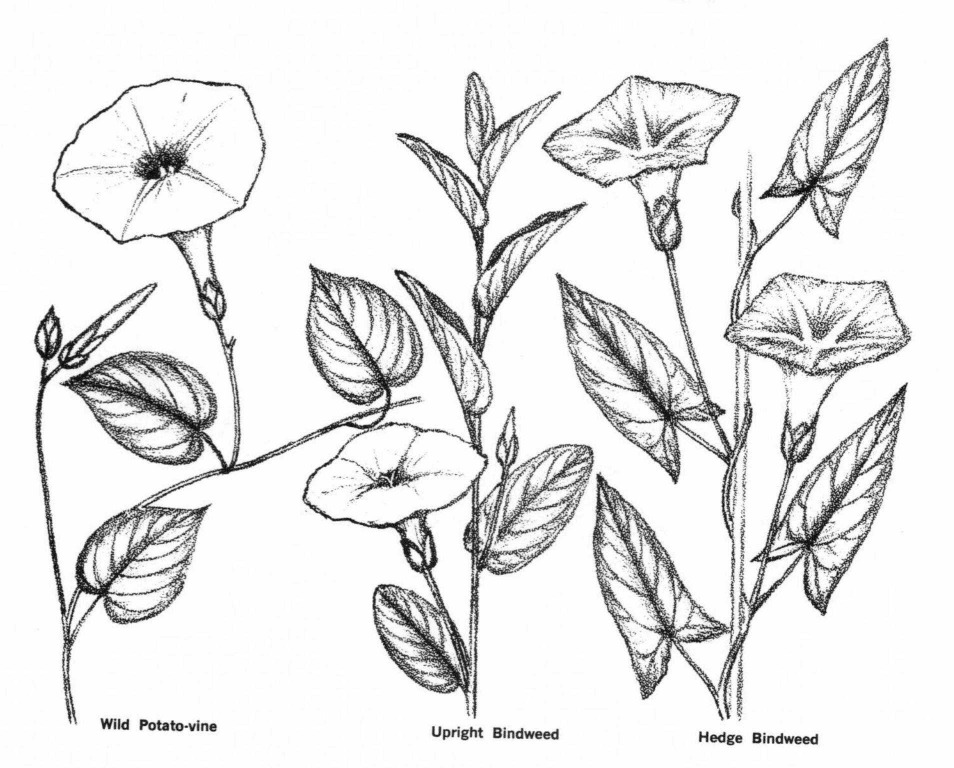

Wild Potato-vine

Upright Bindweed

Hedge Bindweed

This is chiefly a family of twining and climbing or trailing plants. The flowers usually have 5 (sometimes 4) sepals, united at the base; a like number of petals united to form a funnel-shaped or sometimes bell-shaped corolla; 5 (sometimes 4) stamens which are attached to the corolla tube; and a pistil consisting of 2 united carpels, with from 1 to 4 styles or style-branches. The fruits are capsules with 1 or 2 seeds in each of the chambers.

SOUTHERN BREWERIA *(Stylisma humistrata)*

This is a little plant which grows in dry, sandy coastal plain pinelands. It has a slender, trailing, often branched stem 1 to 2 feet long; and scattered erect or ascending, short-stalked, oblong or elliptic leaves 1 to 2 inches long. Stalks bearing from 1 to 7 white, trumpet-shaped, ¾-inch flowers arise from the leaf axils and overtop the leaves. They have a style with 2 branches. It blooms in June to August.

RANGE: e. Va. south to Fla. and west to Tex.

WATER BREWERIA *(Stylisma aquatica)*

Similar to the preceding, this species is a hairy plant with pink or purple flowers. It grows in wet pinelands and about pond margins from se. Va. south to Fla. and west to Tex. (Not illustrated)

ARROW-LEAF MORNING-GLORY *(Ipomea sagittata)*

This native morning-glory has attractive rosy-pink flowers about 3 inches across. Its leaves are quite variable but most often arrow-shaped with the 2 basal lobes pointing outward, and usually 1½ to 3 inches long. It creeps over the ground or climbs in bushes on the borders of coastal sand dunes, swamps, and marshes; blooming between July and September. RANGE: e. N.C. south to Fla. and west to Tex.

WILD POTATO-VINE *(Ipomea pandurata)*

The slender, smooth or minutely downy trailing or slightly twining stems of this plant arise from an enormous root often weighing as much as 15 to 30 pounds. Its leaves are heart-shaped, 2 to 6 inches long; and often have a purplish midrib, stalks, and margin. The flowers are white with a purple blotch in the center and are about 3 inches across. It grows quite commonly in dry fields, thickets, and along roadsides; blooming between June and September. Another name frequently given it is Man-of-the-earth.

RANGE: Conn. to Ont. and Mich. south to Fla. and Tex.

UPRIGHT BINDWEED *(Calystegia spithamaea)*

The flowers of bindweeds resemble those of the morning-glories but they have 2 stigmas, and usually have large bracts surrounding the calyx. This one is a smooth or downy plant with an erect or ascending stem 6 to 12 inches high, sometimes feebly twining toward the tip. Its oval-shaped leaves are short-stalked or nearly stalkless and 1 to 2 inches long. From 1 to 4 flowers are produced in the axils of the lowermost leaves. They are white and between 1½ and 2 inches long. It grows in dry sandy or rocky open woods and fields, blooming between May and August. It is also known as the Low Bindweed. RANGE: Me. and Que. to Ont. and Minn. south to Va., n. Ga., n. Ala. and Iowa.

HEDGE BINDWEED *(Calystegia sepia)*

The Hedge Bindweed has smooth or somewhat hairy, trailing or twining stems up to 10 feet long; and triangular or arrow-shaped leaves from 2 to 5 inches long, the 2 basal lobes being rather short and blunt. The white or pink trumpet-shaped flowers are 2 to 3 inches across, and on long stalks arising from the leaf axils. At the base they have a pair of large, heart-shaped bracts. It is often abundant in fields, thickets, waste places, and along roadsides; blooming between May and September. It is often called the Wild Morning-glory. RANGE: Nfd. to B.C. south to Fla., Tex. and N.Mex.

FIELD BINDWEED *(Convolvulus arvensis)*

This immigrant from Europe is now widely naturalized here in America, and is often a troublesome weed in cultivated fields. It has a slender, usually trailing stem 1 to 2½ feet long; arrow-shaped leaves from 1 to 2 inches long; and pink or white flowers an inch or less across, which lack bracts at the base. It is also known as the Small Bindweed. (Not illustrated)

DODDER FAMILY (Cuscutaceae)

COMMON DODDER *(Cuscuta gronovii)*

Dodders are parasitic plants with thread-like, twining and climbing, yellow to bright orange stems on which the leaves are reduced to scales. The small, waxy-white, 4- or 5-parted flowers have globe- or bell-shaped corollas, and are usually in dense clusters. The Common Dodder has its flower parts in 5's. It usually occurs in low grounds and is parasitic on a wide variety of plants. It blooms between August and October or later.

RANGE: Que. to Man. south to Fla., Tex. and Ariz.

PHLOX FAMILY (Polemoniaceae)

Members of this family have flowers with 5 united sepals, 5 petals united to form a tube with 5 spreading lobes above; 5 stamens attached to the corolla tube; and a pistil composed of 3 united carpels, with a 3-chambered ovary and a 3-branched style. The fruits are capsules.

GREEK-VALERIAN *(Polemonium reptans)*

This is a native plant with branching, smooth or sparsely hairy stems 6 to 15 inches long. Its leaves are divided into from 3 to 15 lance-shaped or elliptic leaflets. The flowers are about ½ inch across, light blue-violet; with their stamens included within the somewhat bell-shaped, deeply 5-lobed corolla. It grows in rich woods and bottomlands, blooming between April and June.

RANGE: N.Y. to Minn. south to Ga., Miss., Mo. and Okla.

AMERICAN JACOB'S-LADDER *(Polemonium vanbruntiae)*

This is a taller plant with bluish-purple flowers almost an inch across, with the stamens extending well beyond the bell-shaped corolla. The basal leaves are 6 to 12 inches long and have 11 to 17 narrow leaflets. Those on the upper part of the stem are much smaller and have 3 to 7 leaflets. It grows in bogs, marshy meadows, and along streams; blooming between May and July.

RANGE: Vt. and N.Y. south to Md. and W.Va.

WILD BLUE PHLOX *(Phlox divaricata)*

This phlox has slender, upright, more or less hairy stems 6 inches to 1 foot tall; with widely-spaced pairs of lance-shaped or narrowly egg-shaped leaves 1 to 2 inches long. The pale bluish to lilac flowers have a slender corolla tube with 5 spreading lobes which are often notched at the tip; and the stamens are completely within the tube. Often called the Wild Sweet-William, it is often common in open woods and on rocky slopes; blooming from April to June. Often cultivated as a border plant.

RANGE: Vt. and Que. to Minn. and Neb. south to Fla. and Tex.

HAIRY PHLOX *(Phlox amoena)*

The Hairy Phlox has tufted, simple, slender and more or less hairy stems 6 to 18 inches tall; with pairs of lance-shaped or narrowly egg-shaped leaves from ½ to 2 inches long, which tend to stand erect and are rather bluntly pointed. The purplish or pink flowers have 5 narrowly top-shaped lobes which are rarely notched; and the stamens are included within the corolla tube. It grows in dry woods and on rocky slopes and banks; blooming between April and June.

RANGE: N.C. and Ky. south to Fla. and Miss.

SWORD-LEAF PHLOX *(Phlox buckleyi)*

This rather rare phlox has a sticky-hairy stem 4 to 10 inches tall, with pairs of mostly lance-shaped leaves. At the base it has a tuft of evergreen, sword-shaped leaves up to 6 inches in length. The bright purple flowers have corolla tubes and flower stalks densely glandular, and the anthers of the stamens reach the summit of the corolla tube. It grows in open woods and on shaly slopes in western Va. and W.Va., blooming between May and July. (Not illustrated)

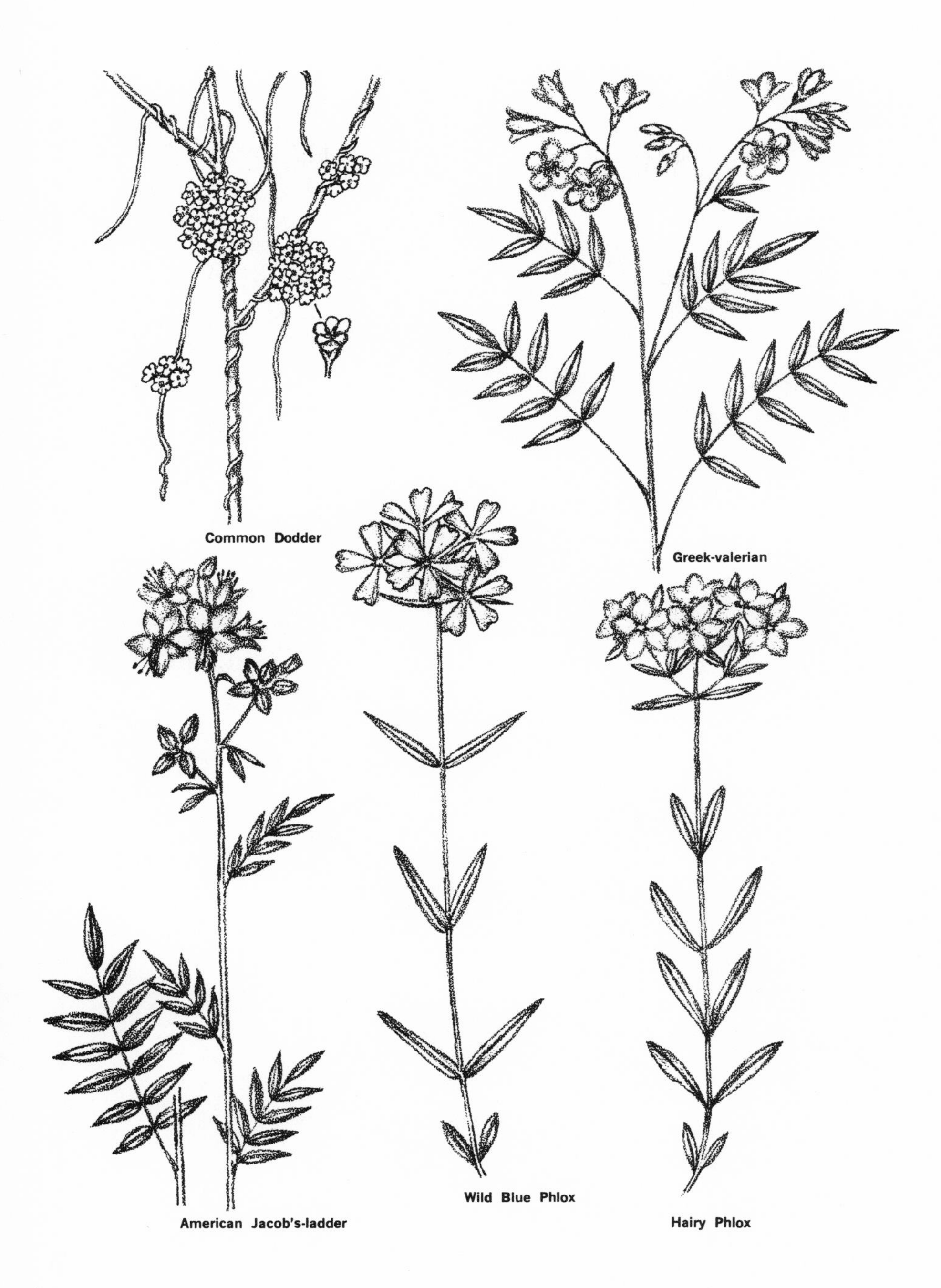

Common Dodder
Greek-valerian
American Jacob's-ladder
Wild Blue Phlox
Hairy Phlox

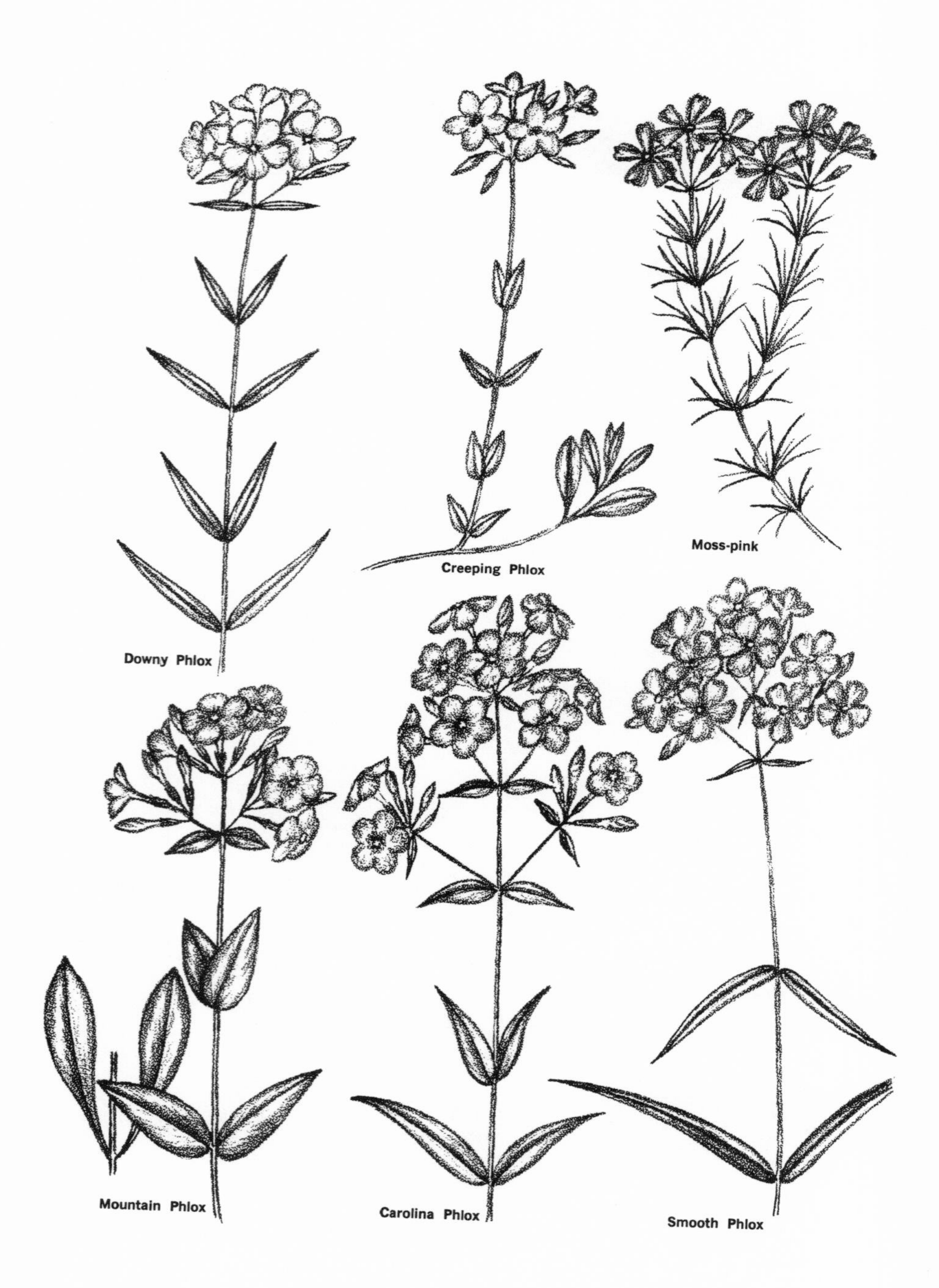
Downy Phlox
Creeping Phlox
Moss-pink
Mountain Phlox
Carolina Phlox
Smooth Phlox

DOWNY PHLOX *(Phlox pilosa)*

This phlox is quite similar to the Hairy Phlox, and most apt to be confused with it. It is a slightly higher plant which is softly hairy or downy. The leaves are more spreading, 1 to 4 inches long, and taper to sharp-pointed tips. Another difference is found in the usually glandular and sticky-hairy calyx of this species. It grows in dry open woods, thickets, sandhills and prairies; blooming between April and June. It is also called Prairie Phlox. RANGE: Ont. to Sask. south to Fla. and Tex.

CREEPING PHLOX *(Phlox stolonifera)*

The Creeping Phlox produces creeping leafy stems which often form sizeable colonies of the plant. It has upright flowering stems 4 to 10 inches high with a few pairs of lance-shaped or narrowly oblong leaves up to ¾ inch long. Those of the sterile shoots are often 1 to 3 inches long, top-shaped, and taper into stalks. The flowers are bright pink or violet-purple, about ¾ inch across, and have rounded corolla lobes. A few of the orange-yellow anthers usually protrude from the summit of the corolla tube. It grows in moist woods and on flats along streams, chiefly in the mountain region; blooming between April and June. RANGE: Pa. and Ohio south to n. Ga. and Tenn.

MOSS-PINK *(Phlox subulata)*

This is a low and mat-forming plant with trailing or creeping stems; and numerous pairs of very narrow, almost needle-like leaves ½ to 1 inch long. Its flowers are about ½ inch across and usually rose-pink or purplish (rarely white), the spreading corolla lobes commonly being deeply notched at the tip. It grows in dry, rocky or sandy, open woods and rocky slopes; blooming between April and June. The plant is widely cultivated, in the South generally under the name of Thrift.
RANGE: N.Y. and Ont. to Mich. south to N.J., w. N.C. and e. Tenn.

TRAILING PHLOX *(Phlox nivalis)*

This plant resembles the preceding one in habit and general appearance. It differs in that the corolla lobes of its flowers are slightly if at all notched at the tip, but they often have some small ragged looking teeth. The stamens in this species are well included within the corolla tube, while those of the preceding one protrude slightly from its opening. This is chiefly a plant of dry woods and pinelands of the coastal plain and piedmont, from s. Va. south to Fla. It blooms between March and May. (Not illustrated).

MOUNTAIN PHLOX *(Phlox latifolia)*

The flowering stalks of this phlox arise from reclining leafy stems. They are usually slender, quite smooth, 1 to 2 feet tall; and have from 3 to 7 pairs of leaves which are 1 to 2 inches long, rounded at the base, and pointed at the tip. The lower ones are somewhat larger and narrowed to a stalked base. The flowers are about ¾ inch across, pink to reddish-purple, and have rounded corolla lobes. It grows in moist to rather dry open woods, thickets, and meadows; blooming during May and June.
RANGE: Pa. to Ind. south to S.C. and Tenn.

CAROLINA PHLOX *(Phlox carolina)*

This phlox has tufted, minutely downy to hairy stems from 1 to 3 feet tall; with from 5 to 12 pairs of lance-shaped to narrowly egg-shaped leaves 1½ to 4 inches long. The flowers are almost an inch across, deep to pale reddish-purple, and have 5 rounded corolla lobes. It grows in moist or dry open woods, chiefly in the mountains; blooming in May to July. RANGE: Md. to Ky. and s. Ind. south to Fla. and Miss.

SMOOTH PHLOX *(Phlox glaberrima)*

The Smooth Phlox resembles the preceding species but it is smooth or very nearly so. It has slender stems 1 to 4 feet tall and usually from 10 to 20 pairs of lance-shaped or even narrower leaves from 1½ to 6 inches long. Its flowers are similar but the corolla lobes tend more to be notched at the tip. It grows in moist open woods and meadows, blooming between April and June.
RANGE: Va. to Ky., south to N.C. and Ala.

WILD SWEET-WILLIAM *(Phlox maculata)*

This phlox has fairly smooth and usually purple-spotted stems 1 to 3 feet tall; with 7 or more pairs of lance-shaped or narrowly egg-shaped leaves, rounded at the base, and 2 to 5 inches long. Its flowers are about ⅔ inch across, pinkish-purple or rarely white, and are in narrowly cylindrical end clusters. It grows in moist meadows, bottomlands, and along the banks of streams; blooming between June and September.

RANGE: Conn. and N.Y. to Que. and Minn. south to Md., w. S.C., Tenn. and Mo.

FALL PHLOX *(Phlox paniculata)*

The Fall Phlox has smooth or somewhat downy stems 2 to 4 feet tall; with 14 or more pairs of short-stalked or stalkless, oblong to broadly lance-shaped, and conspicuously veined leaves from 2 to 6 inches long. Its flowers are about ¾ inch across, pinkish-purple to white, and have lobes shorter than the somewhat downy corolla tube. They are arranged in a large, dense, pyramidal end cluster; blooming between July and October. It grows in rich open woods, thickets, and on stream banks. Often cultivated as the Garden or Perennial Phlox, it has escaped beyond its original range.

RANGE: N.Y. to Iowa south to S.C., n. Ga., n. Miss. and Ark.

BROADLEAF PHLOX *(Phlox amplifolia)*

This phlox resembles the preceding species in having pyramidal flower clusters and veiny leaves. It has 7 to 15 pairs of broader leaves which are roughish above; stems which are glandular-hairy above; and flowers with smooth rather than downy corolla tubes. It grows on rocky wooded slopes and the banks of streams, often high in the mountains from sw. Va. to Ind. and Mo. south to w. N.C., Tenn. and Ala. Flowering during July and August. (Not illustrated)

WATERLEAF FAMILY (Hydrophyllaceae)

Members of the Waterleaf Family have flowers with 5 barely united sepals; 5 petals, united to form a tubular or bell-shaped corolla; 5 stamens with filaments more or less united with the corolla tube; and a solitary pistil consisting of 2 united carpels, the style having 2 branches.

VIRGINIA WATERLEAF *(Hydrophyllum virginianum)*

Often a very common plant in rich, moist woodlands, the Virginia Waterleaf is made conspicuous by its usually mottled leaves which are pinnately divided into 5 or 7 cut-toothed segments. It is a smoothish or sparingly hairy plant 1 to 2 feet high with a few leaves on the slender flowering stems, the lower ones being 6 to 10 inches long and long-stalked. Its bell-shaped flowers are white to lavender or purple, about ⅓ inch long; with narrow, bristly, spreading sepals; and stamens with hairy filaments longer than the corolla tube. They are in fairly dense end clusters and bloom between May and August.

RANGE: Que. to Man. south to Va., Tenn. and Kan.

LARGE-LEAF WATERLEAF *(Hydrophyllum macrophyllum)*

This plant is similar to the Virginia Waterleaf but it is quite bristly-hairy. Its leaves are pinnately divided into from 9 to 13 cut-toothed divisions. The flowers are white, blooming between May and July. It grows in woodlands from Va. to Ill. south to N.C. and Ala. (Not illustrated)

APPENDAGED WATERLEAF *(Hydrophyllum appendiculatum)*

This is a bristly-hairy plant 1 to 2 feet high. Its stem leaves are palmately 5-lobed and coarsely toothed and the lower ones are divided. The name comes from the tiny reflexed appendages between the narrow and bristly-hairy sepals of its flowers; which have white to purplish, bell-shaped corollas. It grows in rich, moist woodlands; blooming in May or June. RANGE: Ont. to Minn. south Tenn. and Kan.

CANADA WATERLEAF *(Hydrophyllum canadense)*

The Canada Waterleaf is a nearly smooth plant 1 to 2 feet high which has roundish leaves 2 to 10 inches across. The leaves are palmately 5- to 9-lobed and have coarse, sharp teeth. Its flowers resemble those of the other waterleafs but the sepals are quite smooth, and the corollas are usually white. It grows in rich, moist woodlands; blooming from May to July. RANGE: Mass. to Ont. south to Ga., n. Ala. and Mo.

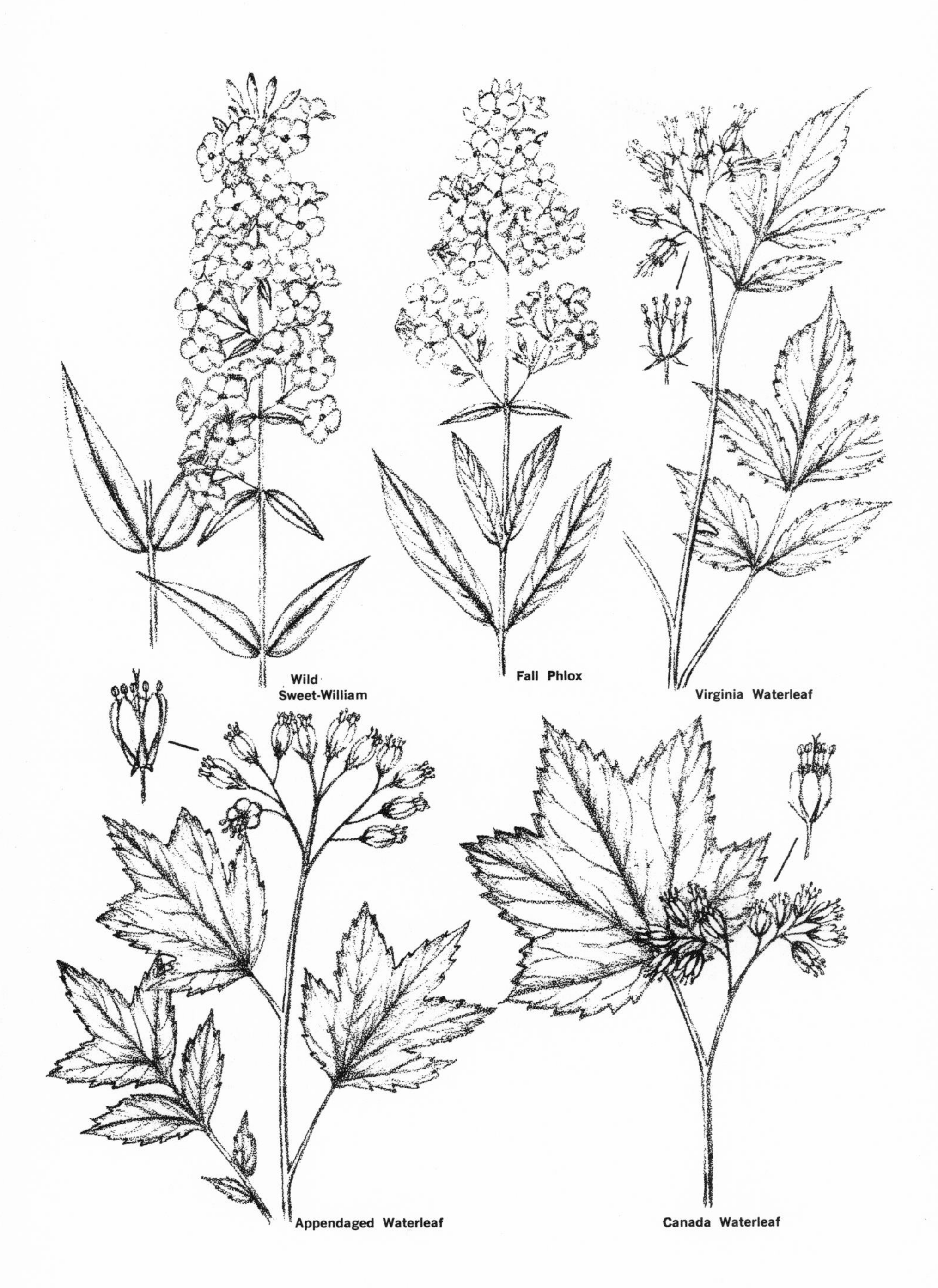
Wild
Sweet-William
Fall Phlox
Virginia Waterleaf
Appendaged Waterleaf
Canada Waterleaf

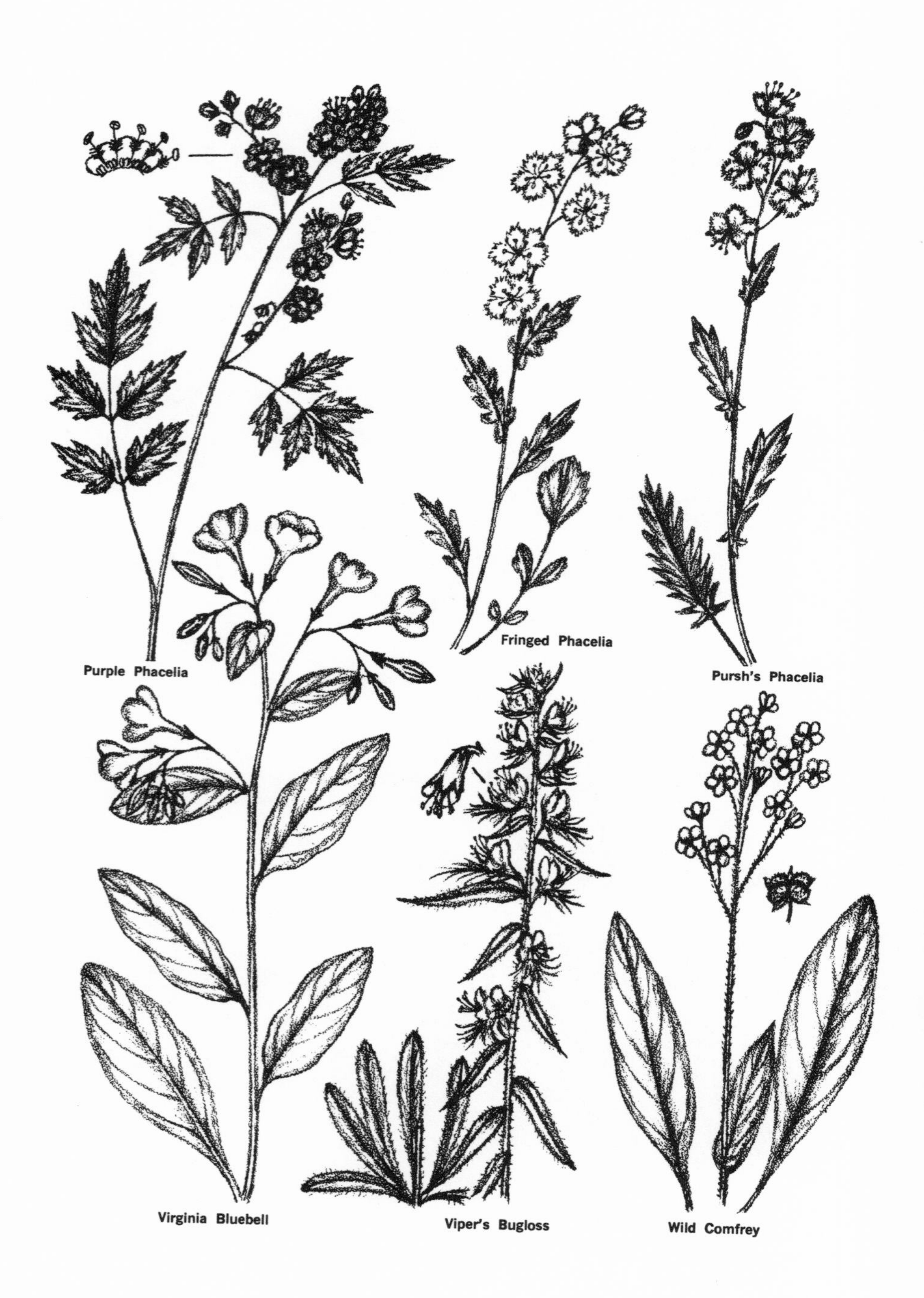
Purple Phacelia
Fringed Phacelia
Pursh's Phacelia
Virginia Bluebell
Viper's Bugloss
Wild Comfrey

PURPLE PHACELIA *(Phacelia bipinnatifida)*

This phacelia has a hairy, branching stem 1 to 2 feet tall; with long-stalked leaves pinnately divided into 3 to 7 sharply toothed or deeply cut segments. The numerous bell-shaped, violet-blue flowers are about ½ inch across, and have projecting stamens with hairy filaments. It grows in moist woods or on rocky slopes and stream banks, blooming between April and May. Also called Loose-flowered Phacelia and Fern-leaf Scorpion-weed.

RANGE: Va. to Ill. and Iowa south to Ga., Ala. and Ark.

FRINGED PHACELIA *(Phacelia fimbriata)*

The stems of this phacelia are weak, seldom over 6 inches tall, and have some spreading hairs. Its leaves are cut into 5 to 9 unequal, rather blunt-tipped segments. The lower ones are 2 to 4 inches long and slender-stalked but those along the stem are much smaller and stalkless. Its flowers are white to pale lilac, about ½ inch across, with lilac-colored anthers and the corolla lobes strongly fringed. It grows in rich woods and along streams in the mountains, blooming in April and May.

RANGE: w. Va. south to w. N.C. and e. Tenn.

PURSH'S PHACELIA *(Phacelia purshii)*

This is a finely hairy plant with usually branching stems 6 to 18 inches tall. The leaves are 1½ to 3 inches long and cut into 9 to 15 usually pointed segments, the lower ones being stalked. Its flowers are lavender-blue with a broad white "eye", about ½ inch across, and have the corolla lobes short-fringed. It grows in rich moist woods and meadows, blooming May or June.

RANGE: Pa. to Ill. and Wis. south to n. Ga., Ala., Tenn. and Okla.

SMALL-FLOWERED PHACELIA *(Phacelia dubia)*

The flowers of this species are lilac-blue to white, the corolla lobes are unfringed, and the stamen stalks are hairy. Its stem is 6 to 16 inches tall and has scattered entire or few-cleft small leaves, but the basal ones are 1 to 2 inches long and divided into 3 to 7 segments. It grows in rich woods and rocky slopes or cliffs, blooming from April to June. RANGE: Del. to N.Y. and Ohio south to Ga. and Miss. (Not illustrated)

BORAGE FAMILY (Boraginaceae)

Members of this family have flowers with 5 sepals, united at the base; 5 petals united to form a tube; 5 stamens, with filaments attached to the corolla tube; and a pistil with a deeply 4-lobed ovary, which matures as a group of 4 seed-like nutlets. The flower clusters are usually spirally coiled, straightening out as the flower buds open.

VIRGINIA BLUEBELL *(Mertensia virginica)*

Also known as the Virginia Cowslip and Roanoke-bells, this is one of our most beautiful wild flowers of spring. It is a smooth and pale green plant 1 to 2 feet high, with elliptic or egg-shaped leaves 2 to 5 inches long. The nodding, trumpet-shaped flowers are about an inch long, pink when in the bud but becoming a bright lavender-blue or bright blue when fully open. It grows in rich moist woods, on rocky slopes, or along streams; blooming between March and May.

RANGE: Ont. to Minn. south to Va., Ala., Ark. and Kan.

VIPER'S BUGLOSS *(Echium vulgare)*

Blueweed is another name for this bristly-hairy European immigrant which is now thoroughly at home here in America. Between June and September, its bright blue, ¾-inch long flowers with their long red stamens are very conspicuous in old fields, waste places, and by the wayside. The plant grows from 1 to 2½ feet tall and has many narrowly oblong or lance-shaped leaves 1 to 6 inches long, the ones on the upper part of the stem being small and stalkless.

WILD COMFREY *(Cynoglossum virginianum)*

The Wild Comfrey is a hairy plant with a simple stem 1½ to 2½ feet tall, the few leaves on the upper part being small and with clasping bases. The basal leaves are 4 to 12 inches long, oval-shaped, and taper into stalks. Its flowers are about ⅜ inch across and pale lilac to white, blooming between April and June. It grows in rather rich open woods and thickets. RANGE: Conn. to Ill. and Mo. south to Fla. and Tex.

Members of this family have flowers with 4 or 5 unequal sepals united into a tube; 4 or 5 petals united to form a slender tube with a flaring 4- or 5-lobed rim; usually 4 stamens, in pairs of different lengths; and a pistil with a more or less 4-lobed ovary, maturing as 4 seed-like nutlets. The leaves are nearly always opposite.

BLUE VERVAIN *(Verbena hastata)*

The Blue Vervain has an erect, 4-sided, rough-hairy stem 3 to 5 feet tall, which usually branches above. On it are pairs of narrowly egg-shaped to lance-shaped leaves 1½ to 5 inches long, which are roughish-hairy and have coarsely toothed margins. The small violet-blue flowers, hardly more than ⅛ inch across, are arranged in several long and very slender clusters at the tip of the stem and its branches. It is usually common in moist fields, meadows, and along streams; blooming between June and September.
RANGE: N.B. to B.C. south to Fla., Tex. and Calif.

WHITE VERVAIN *(Verbena urticifolia)*

This plant is similar to the preceding one but it has more lax end clusters of small white flowers. It grows in fields, thickets, and the borders of woods from Que. to Ont. and S.D. south to Fla. and Tex. (Not illustrated)

HOARY VERBENA *(Verbena stricta)*

This is densely pale-hairy plant with a quite roundish, simple or sparingly branched stem 1 to 4 feet tall. The leaves are stalkless or nearly so, sharply-toothed, and 2 to 4 inches long. Its flowers are deep blue or purple, a bit over ¼ inch across; and in dense, narrow, blunt-tipped, and practically stalkless clusters. It grows in dry open places from N.Y. and Ont. to Mont. south to Tenn., Ark., Okla., Tex. and N. Mex. Introduced in the Northeast. It flowers June to September. (Not illustrated)

MOSS VERBENA *(Glandularia pulchella)*

The Moss Verbena is a native of South America now naturalized in the Southeast. It is a prostrate or sprawling, hairy little plant with pairs of very finely dissected leaves; and flat-topped clusters of showy rose-purple, pink, or white flowers with 5-lobed corollas about ¼ inch across. It is commonly seen blooming in the sands along highways in the coastal plain between June and September, from e. N.C. south to Fla. and west to La.

LARGE-FLOWERED VERBENA *(Glandularia canadensis)*

This is a smooth or hairy-stemmed plant 8 to 20 inches high, with ascending branches. The leaves are 1 to 3 inches long and are coarsely toothed, deeply cut, or 3-lobed. Its flowers are about ½ inch across, flesh-colored to reddish-purple, and the 5 corolla lobes are notched at the tip. The flower cluster tends to be somewhat flat-topped. This verbena grows in dry sandy or rocky thickets, fields, and on prairies; blooming between April and October. RANGE: Va. to Ill., Iowa, Kan. and Colo. south to Fla. and Tex.

CAROLINA VERBENA *(Stylodon caroliniana)*

The Carolina Verbena is a slender stemmed, roughish-hairy plant 1 to 2½ feet tall; with widely spaced pairs of narrowly elliptic or lance-shaped leaves above and spoon-shaped ones toward the base. They are ¾ to about 3½ inches long. Its flowers are almost ¼ inch across, pink purplish, or white; and arranged in a long and narrow cluster, blooming between May and September. It grows in open sandy woods and thickets, chiefly in the coastal plain. RANGE: N.C. south to Fla. and west to La.

FOG-FRUIT *(Phyla lanceolata)*

The Fog-fruit is a smooth or sparingly hairy plant with slender, trailing stems; and pairs of elliptic or lance-shaped, coarsely toothed leaves 1 to 3 inches long. The flowers are very small, pink, bluish, or white, and in scaly-bracted heads on long stalks from the leaf axils. It grows in wet places and on streambanks, flowering between May and September. RANGE: N.J. to Ont., Minn. and Neb. south to Fla. and La.

CAPEWEED *(Phyla nodiflora)*

This is a similar but downy plant growing in low grounds. Its leaves are ½ to 2½ inches long and broadest with a few sharp teeth toward the rounded tip. The small flowers are bluish to nearly white. RANGE: Va. to Mo. and Okla. s. to Fla. and Tex. (Not illustrated)

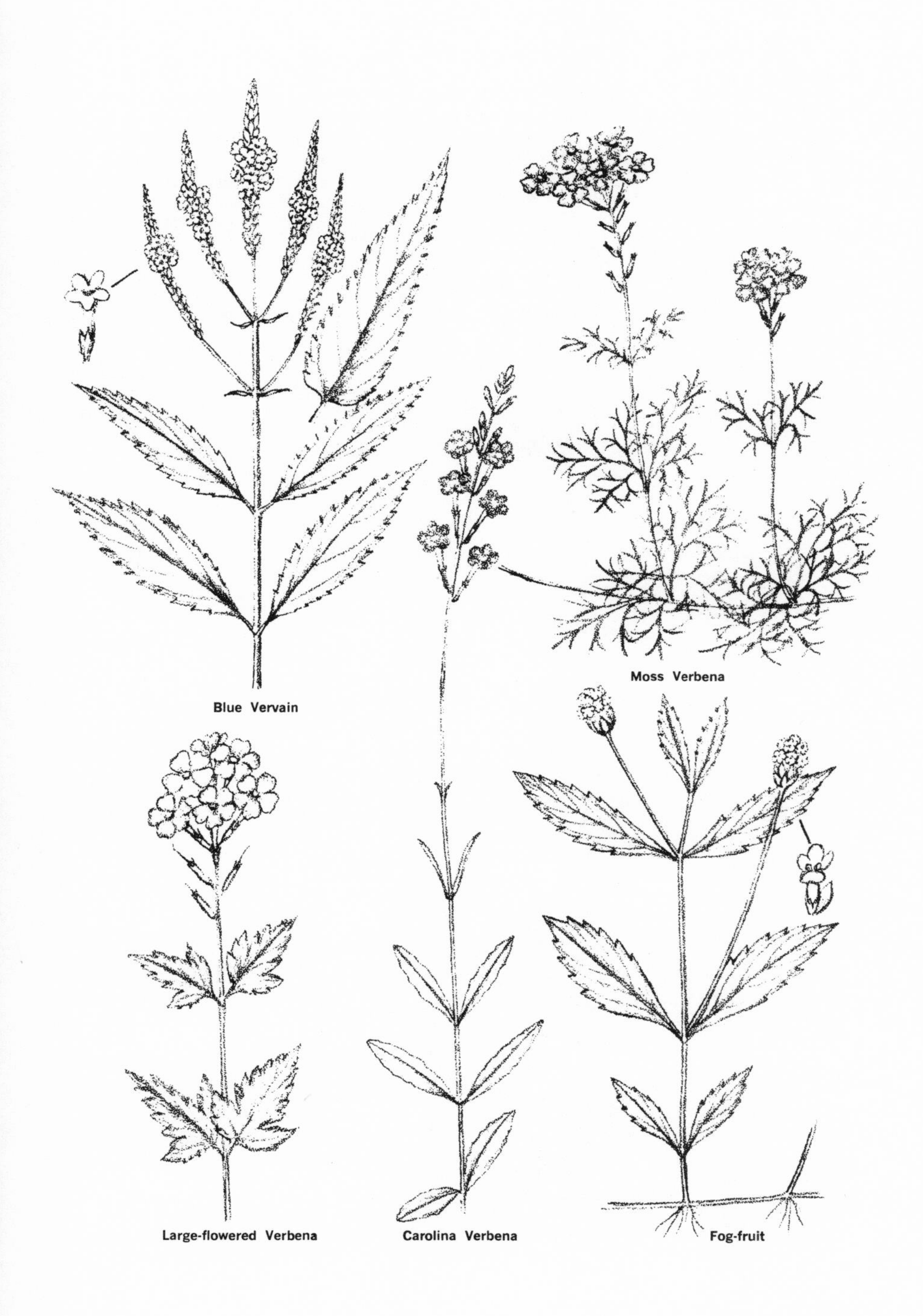
Blue Vervain
Moss Verbena
Large-flowered Verbena
Carolina Verbena
Fog-fruit

LOPSEED *(Phryma leptostachya)*

This is a smooth or somewhat minutely downy plant with a branching stem 1½ to 3 feet tall. Its leaves are in widely-spaced pairs; egg-shaped, coarsely toothed, 2 to 6 inches long, and all but the upper ones are long-stalked. The white to lavender flowers are about ⅓ inch long and are arranged in pairs in a very long and slender cluster, blooming between June and August. After the flower fades, the calyx bends abruptly downward against the stalk and the fruits develop within it, hence the common name given to the plant. It grows in rich woodlands and in thickets.

RANGE: N.B. to Man. south to Fla. and Tex.

MINT FAMILY (Labiatae)

Plants belonging to the Mint Family have opposite leaves and usually, but not always, 4-sided stems. Many of them also have a distinctive and aromatic odor when crushed or bruised. Their flowers have a bilateral symmetry with 4 or 5 united sepals, 4 or 5 petals united to form a tube, and usually with a 2-lipped corolla; 4 stamens, in 2 pairs of different length, or sometimes only a single pair; and a pistil which has a decidedly 4-lobed ovary, which matures as 4 seed-like nutlets.

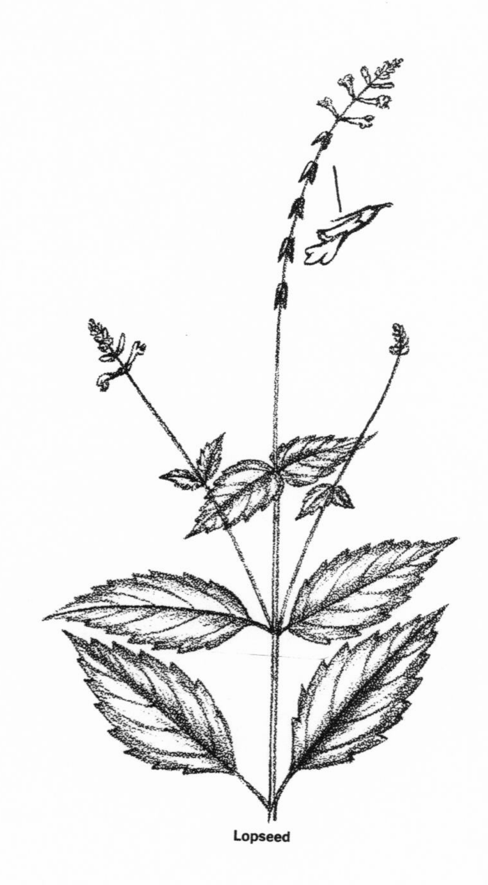

Lopseed

KEY TO GENERA OF THE MINT FAMILY

1. Flowers with but 2 anther-bearing stamens.
    2. Lower lip of the corolla fringed. Flowers in a large, open, branched end cluster which overtops the leaves. HORSE-BALM *(Collinsonia)*
    2. Lower lip of the corolla not fringed. Flowers in dense heads or narrow end clusters.
3. Corolla large, usually an inch or more long.
    4. Flowers in terminal heads or with several head-like clusters in the axils of the upper leaves; calyx regular and 5-toothed. MONARDAS *(Monarda)*
    4. Flowers in elongate and often interrupted terminal clusters; calyx distinctly 2-lipped. SAGES *(Salvia)*
3. Corolla less than ½ inch long; calyx 2-lipped, with bristle-pointed teeth. WOOD-MINTS *(Blephilia)*
    1. Flowers with 4 anther-bearing stamens.
5. Lower lip of the corolla fringed. Flowers in a large, open, branched end cluster which overtops the leaves. STONEROOT *(Collinsonia)*
5. Lower lip of the corolla not fringed.
    6. Calyx with a cap-like hump on the upper side. SKULLCAPS *(Scutellaria)*
    6. Calyx otherwise.
7. Stamens very long, describing an arch or curve extending well beyond the corolla. BLUE-CURLS *(Trichostema)*
7. Stamens otherwise.
    8. Flowers in the axils of large rounded bracts and in a dense cylindrical end cluster. Leaves untoothed. HEAL-ALL *(Prunella)*
    8. Flowers not in a dense cylindrical cluster. Leaves usually toothed.
9. Plants prostrate or creeping.
    10. Flowers few, in a 1-sided end cluster. MEEHANIA *(Meehania)*
    10. Flowers few and in the axils of the regular leaves. GROUND-IVY *(Glechoma)*
9. Plants more or less erect.
    11. Corolla apparently without an upper lip, split on the upper side and the lower lip seemingly 5-lobed. GERMANDER *(Teucrium)*
    11. Corolla nearly regular or distinctly 2-lipped.
12. Calyx deeply 4-toothed. Leaves heart-shaped. SYNANDRA *(Synandra)*
12. Calyx 5-toothed.
    13. Corolla large, ¾ inch or more long; flowers in a long and narrow end cluster. FALSE DRAGONHEAD *(Physostegia)*
    13. Corolla usually less than ½ inch long; flowers in dense axillary or end clusters.
14. Corolla nearly regular with more or less erect lobes; flowers in axillary clusters or narrow and interrupted end clusters. MINTS *(Mentha)*
14. Corolla decidedly 2-lipped; flowers in axillary or sometimes flattened end clusters. MOUNTAIN-MINTS *(Pycnanthemum)*

BLUE-CURLS *(Trichostema dichotomum)*

This is a minutely downy or sometimes sticky-hairy plant with a slender and branching stem from 6 inches to 2 feet high; and lance-shaped or oblong, untoothed leaves from 1 to 3 inches long. Its violet-blue flowers are very distinctive as the long blue or violet filaments of the stamens describe an arch which extends well beyond the corolla. It grows in dry sandy open woods, fields, and on slopes; blooming between August and October. Another name often given it is Bastard-pennyroyal.

RANGE: S. Me. to Mich. and Mo. south to Fla. and Tex.

LARGER SKULLCAP *(Scutellaria integrifolia)*

Skullcaps get their name from a peculiar little cap-like projection on the upper side of the calyx. A dozen or more species of them occur in the eastern United States north of Florida, and this species is one of the more showy ones. It is a somewhat hoary-downy or hairy plant with an erect, simple or branched stem from 6 inches to 2 feet high. The leaves of the upper part of the stem are narrow, untoothed, 1 to 2 inches long, and stalkless or very short-stalked. Those toward the base are slender-stalked, egg-shaped to roundish, and have toothed margins. Its purplish-blue flowers are often whitish underneath and almost an inch long. Also known as the Hyssop Skullcap, it grows in open woods, thickets, and clearings; blooming between May and July.

RANGE: Mass. to Ky. and Mo. south to Fla. and Tex.

SHOWY SKULLCAP *(Scutellaria serrata)*

The Showy Skullcap has a simple or branched stem 1 to 2 feet tall, which is quite smooth or sometimes minutely downy above. Its egg-shaped leaves are sharply toothed, stalked, and 2 to 4 inches in length. The flowers are violet-blue, about an inch long, and bloom in May or June. It grows in rather dry but rich open woods, thickets, and clearings. RANGE: N.J. and s. N.Y. to Tenn. and Mo. south to S.C. and Ala.

LYRE-LEAF SAGE *(Salvia lyrata)*

The Lyre-leaf Sage is a somewhat hairy plant with a simple or sparingly branched and erect stem 1 to 2 feet tall, on which are pairs of small bract-like leaves and a terminal flower cluster. The basal leaves are stalked, 3 to 8 inches long, and vary from ones with merely wavy-toothed margins to others which are pinnately lobed, usually with a large terminal segment. Its bright blue flowers are about an inch long and occur in several whorls along the upper part of the stem, blooming between April and June. It grows in rather dry but rich open woods, pinelands, and along roadsides.

RANGE: Conn. to Ill., Mo. and Okla. south to Fla. and Tex.

BLUE SAGE *(Salvia azurea)*

The Blue Sage is a pale and minutely downy plant 1 to 4 feet tall with leaves narrowly oblong to lance-shaped or narrower, 1½ to 4 inches long, and with wavy or shallowly toothed margins. It has azure blue to white flowers about ¾ inch long. It grows in coastal plain pinelands and sandhills from N.C. south to Fla. and west to Tex. (Not illustrated)

RED SAGE *(Salvia coccinea)*

This, our only native sage with red or scarlet flowers is a softly-hairy, leafy-stemmed plant from 1 to about 2 feet tall. It has egg-shaped and bluntly toothed leaves 1 to 2½ inches long which are on slender stalks. The flowers are ¾ to about an inch long. It grows in sandy hammocks, waste places, and along roadsides in the coastal plain from S.C. south to Fla. and west to Tex. and Mex. (Not illustrated)

SELF-HEAL *(Prunella vulgaris)*

This plant which is also known as Heal-all or Carpenter's-weed, is common everywhere in open places. It is a low plant commonly with several branches and may be but a few inches to nearly 2 feet high. The egg-shaped to lance-shaped leaves are 1 to 4 inches long, slender-stalked, and often have some low teeth on the margin. Its small, bluish to lavender or whitish flowers are in cylindrical-shaped heads; in the axils of large, greenish or purplish-tinged, bristly-fringed bracts and bloom between May and October. The narrow-leaved variety found from Nfd. to Alaska south to se. N.Y., w. N.C., Tenn., Kan., N. Mex., Ariz. and Calif. is believed to be a native plant. The broad-leaved form has evidently been introduced from Europe.

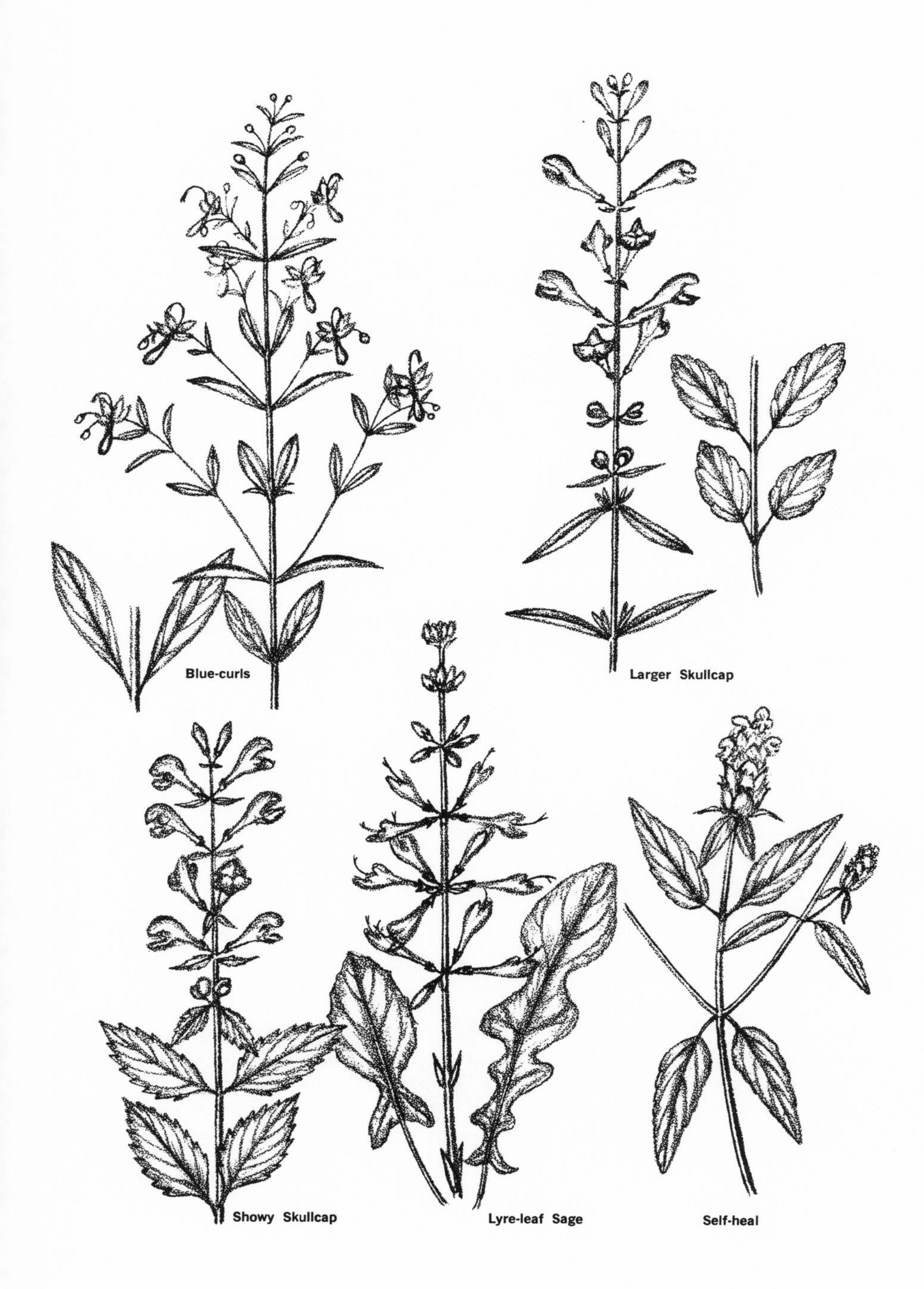
Blue-curls
Larger Skullcap
Showy Skullcap
Lyre-leaf Sage
Self-heal

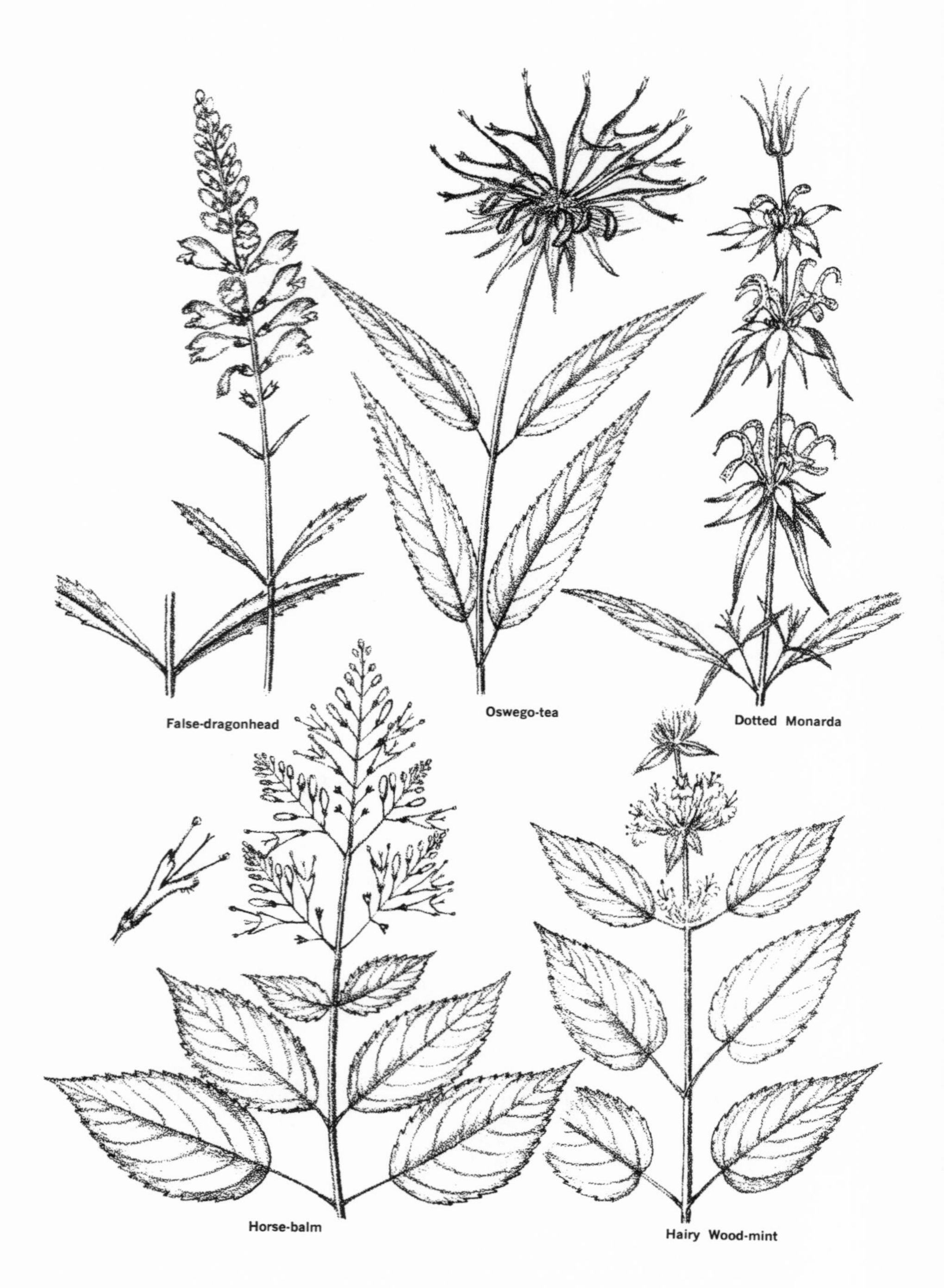
False-dragonhead
Oswego-tea
Dotted Monarda
Horse-balm
Hairy Wood-mint

FALSE DRAGONHEAD *(Physostegia virginiana)*

This plant has a smooth, erect, often branching stem 1 to 4 feet tall; with sharply-toothed lance-shaped leaves 1½ to 5 inches long. The pale purple or rose flowers are about an inch long, and borne in showy terminal spikes. The corolla is swollen at the throat, the upper lip being domed and the lower one spreading and 3-lobed. It is often called the Obedient-plant as the flowers tend to stay put when bent from their normal position. It grows in wet meadows, bogs, and along streams; blooming July to October. RANGE: Que. to Minn. south to Fla. and La.

FEW-FLOWERED DRAGONHEAD *(Physostegia purpurea)*

This species has fewer flowers in a looser spike and usually fewer than 10 pairs of bluntly-toothed leaves below the stem branches. It grows in river swamps and marshes from se. Va. south to Fla. and west to Miss., blooming May to July. (Not illustrated)

OSWEGO-TEA *(Monarda didyma)*

Often called the Bee-balm, this plant is well known for its bright red or scarlet flowers which are borne in showy heads at the summit of the stem or its branches. They are 1½ or 2 inches long and have a narrow, ascending upper lip and a somewhat drooping and broader lower one with 3 short lobes. It grows from 2 to 3 feet tall and has lance- to egg-shaped, sharply-toothed, stalked leaves 3 to 6 inches in length. It is often common in rich moist woods and bottomlands, blooming July to September.
RANGE: N.Y. to Mich. south to n. Ga. and Miss.

BASIL-BALM *(Monarda clinopodia)*

This plant is similar to the preceding but it has smaller whitish to pink, dark-spotted flowers, below which are white or partly white leafy bracts. It grows in woods from N.Y. to Ill. south to Md., n. Ga. and Tenn.; blooming between late May and Sept. (Not illus.)

WILD BERGAMOT *(Monarda fistulosa)*

This species has lilac to pinkish flowers with a hairy tuft near the tip of the upper lip, and leafy bracts which are often pinkish at the base. It grows in open woods and meadows from Me. to Sask. south to Ga. and Tex., blooming from June to September (Not illustrated)

DOTTED MONARDA *(Monarda punctata)*

Also known as Horsemint, this species has a stem 2 to 3 feet tall; with narrowly lance-shaped, shallowly toothed leaves which are pointed at both ends. The upper ones and the bracts of the flower cluster are lilac-pink to whitish. Th inch-long yellow flowers are spotted with purple and have a very slender, arching upper lip. It grows in open sandy woods and fields, blooming between late July and September.
RANGE: N.J. to Minn. south to Fla., Tex. and Ariz.

HORSE-BALM *(Collinsonia canadensis)*

Richweed and Stoneroot are other names for this nearly smooth plant with a branching stem 2 to 5 feet tall. Its long-stalked, coarsely-toothed, egg-shaped leaves are 6 to 10 inches long, and have a lemon-like odor when crushed. The flowers are light yellow, about ½ inch long, and have 2 stamens which protrude well beyond the corolla. It grows in rich moist woods, blooming between July and September.
RANGE: Mass. to Ont. and Wis. south to Fla., Miss. and Ark.

STONEROOT *(Collinsonia verticillata)*

Stoneroot has a slender, finely downy, often purplish stem 4 to 20 inches tall; with an apparent whorl of 4 thin, coarsely-toothed leaves at the summit below the flower cluster. The tawny or purplish flowers have 4 protruding stamens and a fringed lower lip. It grows in rich woods from s. Va. to Tenn. south to Ga., blooming late April to June. (Not illustrated)

HAIRY WOOD-MINT *(Blephilia hirsuta)*

This wood-mint has a downy or hairy stem 1½ to 3 feet tall, and sharply toothed, long-stalked leaves 2 to 4 inches long. The small white or pale lavender flowers are purple-spotted. It grows in moist and rocky woods, blooming between June and August.
RANGE: Que to Minn. south to w. N.C. Tenn., Mo. and Tex.

MOUNTAIN-MINT *(Pycnanthemum incanum)*

A number of species of mountain-mints occur in the eastern United States, some of them in the coastal region. This wide-spread species is apt to attract attention by the conspicuously whitened leaves or bracts associated with the flower clusters. It is a more or less downy plant with a branching stem 1½ to 3 feet tall. The egg-shaped to broadly lance-shaped leaves are 1½ to 3 inches long, with rather widely spaced marginal teeth. Its flowers, about ¼ inch long, are whitish to rose-pink or purple and have small purple spots. It grows in dry open woods, thickets and field; blooming between June and August.

RANGE: N.H. to Ill. south to Ga. and Miss.

FIELD BASIL *(Clinopodium vulgare)*

This mint has a slender, simple or branched, more or less hairy stem 6 inches to 2 feet tall. The leaves are egg-shaped to elliptic, short-stalked, 1 to 2½ inches long, and sometimes toothed on the margin. Its small flowers are lilac-pink to whitish and in dense clusters in the axils of the upper leaves. It grows in woods, thickets and fields; blooming between June and September.

RANGE: Nfd. to Man. south to N.C., Tenn., Ind., Wis. and Minn.

AMERICAN WILD MINT *(Mentha arvensis)*

The true mints, including the familiar Peppermint and Spearmint, are members of the genus *Mentha*. Although several species of them are found in eastern North America, only this species of world-wide distribution is a native plant. The others are immigrants from Europe. It is a smooth to somewhat downy plant with a slender, simple or branched stem 6 inches to 2 feet tall. The leaves are narrowly elliptic or lance-shaped, stalkless, sharply toothed on the margin, and 1½ to 3 inches long. The flowers are small, lilac-pink to purplish, and clustered in the axils of the leaves. It grows in woods, thickets and fields; blooming between July and October.

RANGE: Nfd. to Alaska south to Md., W. Va., Mo., N. Mex., and Calif.

AMERICAN GERMANDER *(Teucrium canadense)*

Also called the Wood-sage, this plant has an erect, simple or branched, finely hairy stem 1 to 3 feet tall. The leaves are lance-shaped or narrowly egg-shaped, coarsely toothed, short-stalked, grayish-downy beneath, and 1½ to 5 inches long. Its purplish-pink to cream-colored flowers are about ⅔ inch long, the lower lip with a large central lobe. They are arranged in long, narrow terminal clusters. It grows in moist woods and thickets and in marshes, blooming between June and September.

RANGE: N.S. to Minn. and Neb. south to Fla. and Tex.

GROUND-IVY *(Glechoma hederacea)*

This little native of Europe is widely naturalized in North America and is often abundant in moist woods, thickets, yards, waste places and along roadsides. Sometimes called Gill-over-the ground, it is a smooth or minutely downy plant with extensively creeping stems and short ascending branches. The leaves are roundish, heart-shaped at the base, ½ to 1½ inches wide, and coarsely and bluntly toothed. Its lavender to purplish-blue Flowers are about ½ inch long, in small axillary clusters, and bloom between late March and July.

MEEHANIA *(Meehania cordata)*

Meehania is a native plant most likely to be confused with the Ground-ivy for it also has trailing stems and similar but usually larger heart-shaped leaves. Its somewhat larger flowers are in end clusters, each with a small egg-shaped bract at the base. It grows in rich wooded coves and slopes from Pa. to Ill. South to w. N.C. and Tenn., blooming in May or June (Not illustrated)

HENBIT *(Lamium amplexicaule)*

Henbit is a European plant widely naturalized in fields, lawns and waste places; blooming between March and June. The whorls of purplish flowers are subtended by pairs of stalkless leaves, but the lower leaves are roundish, coarsely and bluntly toothed, and long-stalked. (Not illustrated)

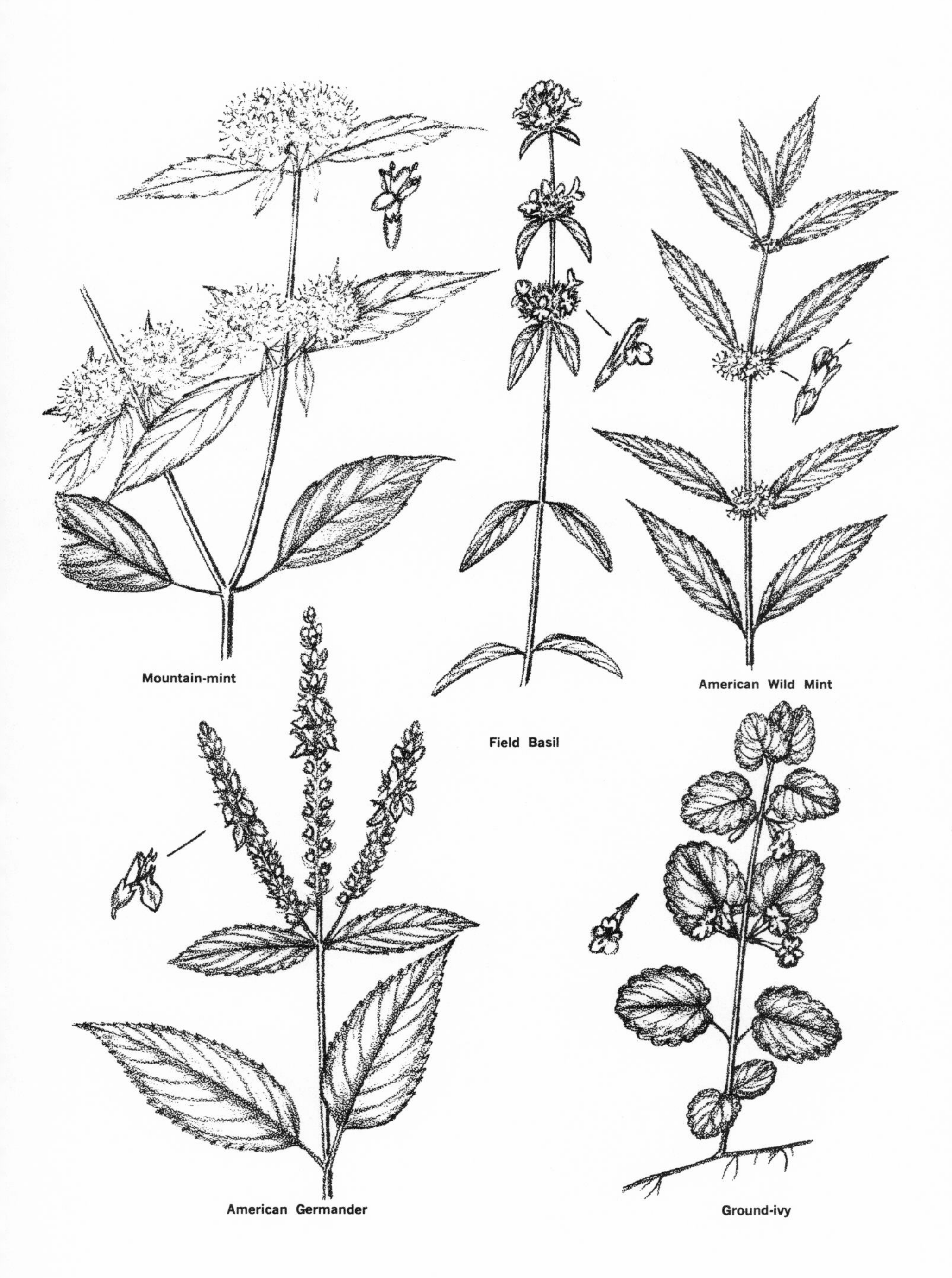
Mountain-mint
Field Basil
American Wild Mint
American Germander
Ground-ivy

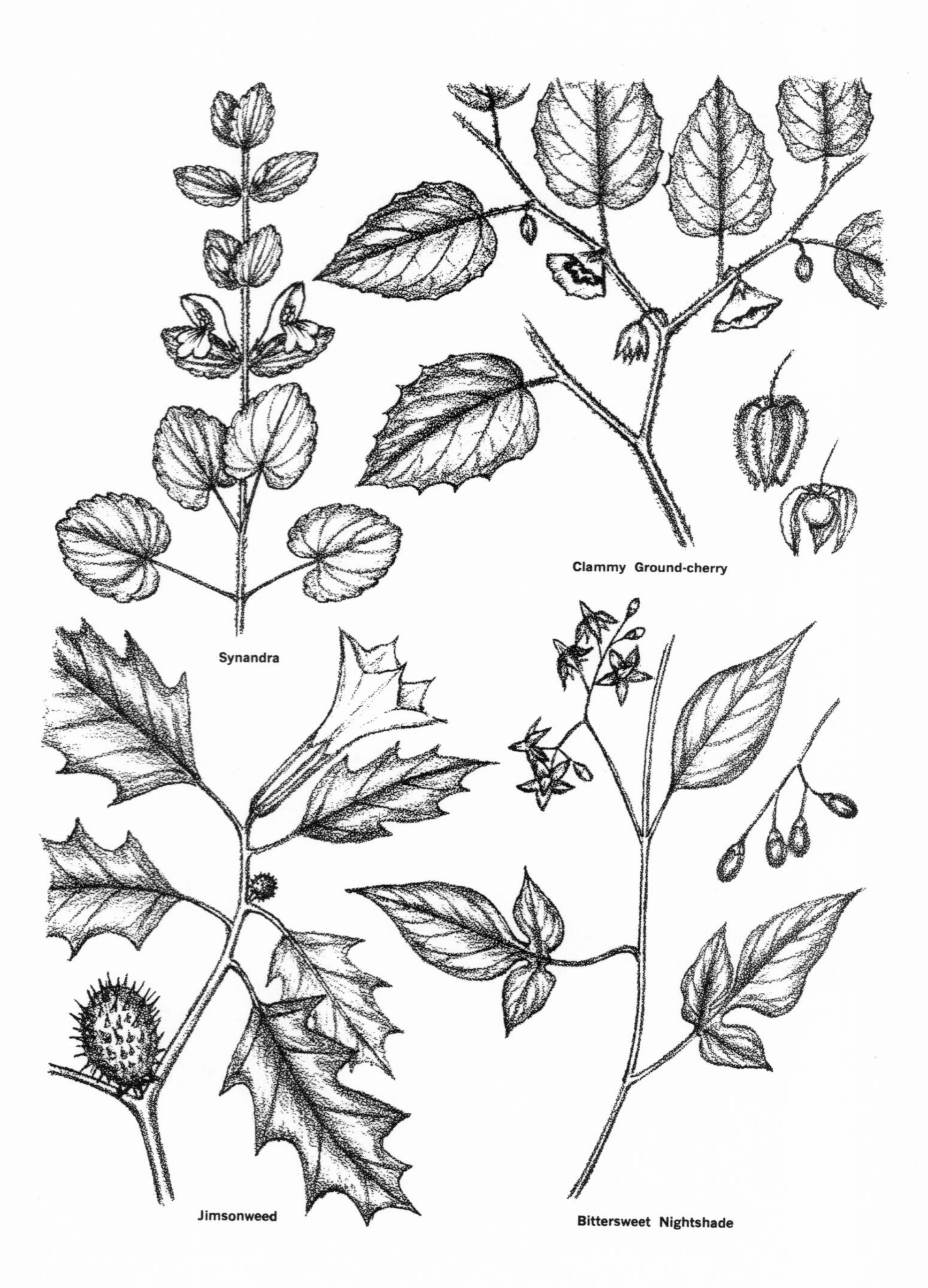
Clammy Ground-cherry
Synandra
Jimsonweed
Bittersweet Nightshade

SYNANDRA *(Synandra hispidula)*

This plant has a rather slender, hairy, ascending or erect stem 1 to 2½ feet high. The leaves on the upper part of it are egg-shaped to lance-shaped and stalkless. Those toward the base are egg-shaped to roundish, long-stalked, 2 to 4 inches long, and are bluntly toothed. Its flowers are 1 to 1½ inches long, yellowish-white with purple lines on the lower lip, and solitary in the axils of the upper leaves. It grows on damp wooded slopes and stream banks, blooming in May or June.

RANGE: Va. and W.Va. to Ill. south to w. N.C. and Tenn.

NIGHTSHADE FAMILY (Solanaceae)

Members of this family usually have alternate leaves, and flowers with 5 more or less united sepals; 5 petals which are united into a star-like, bell-shaped, or trumpet-shaped corolla; 5 stamens which are inserted on the corolla tube; and pistil which usually has a 2-chambered ovary. The fruits are either berries or capsules containing a number of small seeds.

CLAMMY GROUND-CHERRY *(Physalis heterophylla)*

There are a number of species of ground-cherries in our region, this being one of the more common ones. It is a sticky-hairy plant with often branching stems 1½ to 3 feet high, sometimes with long, spreading, and jointed hairs. Its leaves are egg-shaped, usually broadly rounded or heart-shaped at the base, more or less wavy-toothed on the margin, and 2 to 4 inches long. The flowers are shaped like a broad open bell about ¾ inch across, and are yellowish with a purplish-brown center. They are followed by small, round, yellow berries enclosed in the bag-like calyx. It grows in open woods and clearings, blooming between May and September.

RANGE: N.B. to Sask. south to Ga., Okla. and Tex.

VIRGINIA GROUND-CHERRY *(Physalis virginiana)*

This is another common species, with egg-shaped leaves which are pointed at both ends. It has red berries and the calyx which surrounds them has a deep depression at the stem end. Conn. to s. Ont. and s. Man. south to Fla. and Tex. (Not illustrated)

JIMSONWEED *(Datura stramonium)*

Also known as the Thorn-apple and Stramonium, this is a smoothish plant with a stout, forking, widely branched, often purple-tinged stem 1 to 5 feet tall. The rather thin, egg-shaped, irregularly toothed leaves are 3 to 8 inches long. It has trumpet-shaped white, lavender or violet flowers 3 to 4 inches long, with 5 pointed lobes on the corolla rim. The flowers are followed by erect spiny capsules. It is a native of Asia now widely naturalized in fields and waste places, blooming from July to September.

BITTERSWEET NIGHTSHADE *(Solanum dulcamara)*

This widely naturalized native of Europe is a somewhat woody vine which often has flowers, green berries, and bright red ripe ones present at the same time. The leaves, 2 to 4 inches long, are egg-shaped and often have a pair of ear-like lobes at the base. Its flowers, about ½ inch across, have 5 pointed violet or purple corolla lobes and a cone-shaped group of yellow stamens. It blooms between May and September.

HORSE-NETTLE *(Solanum carolinense)*

This species is a weedy native plant with slender yellowish prickles on its stem and midribs of its leaves. It is an erect plant 1 to 2 feet high; with lavender or whitish, 5-lobed flowers nearly an inch wide; followed by round yellow berries about ¾ inch in diameter. It grows in dry fields, waste places and along roadsides; blooming between May and July. Vt. to Ont., Iowa and Neb. south to Fla. and Tex. (Not illustrated)

BLACK NIGHTSHADE *(Solanum ptychanthum)*

Black Nightshade is a smooth bushy plant up to 2 feet tall; with long-stalked, egg-shaped leaves of a thin texture and from 1 to 4 inches long. Its white or purple tinged flowers are borne in 2- to 4-flowered umbels and are followed by lustrous black berries ¼ to ⅜ inch across. It grows in dry open woods, fields and along roadsides from Me. to N.D. south to n. Fla. and e. Tex. (Not illustrated)

FIGWORT FAMILY (Scrophulariaceae)

Members of this family have flowers with 4 or 5 more or less united sepals; 4 or 5 petals, more or less united and most often forming a 2-lipped corolla; from 2 to 5 (usually 4) stamens attached to the corolla tube, 1 or 2 of them often without an anther; and a pistil consisting of 2 united carpels. The fruit is a 2-chambered pod containing numerous small seeds.

MOTH MULLEIN *(Verbascum blattaria)*

The Moth Mullein usually has a simple, smooth, slender stem 2 to 4 feet tall. Its leaves are egg-shaped to lance-shaped, toothed or sometimes cleft; those along the upper part of the stem being broad-based, stalkless or somewhat clasping, and ½ to 2½ inches long. The flowers are about an inch across, white or yellow, and have the filaments of the stamens bearded with violet hairs. They are rather widely spaced along the upper portion of the stem, blooming between June and September. The plant is a native of Europe but it is now common in fields, waste places, and along roadsides here in America.

COMMON MULLEIN *(Verbascum thapsus)*

During the first year this plant produces a large rosette of big, grayish-green, flannel-like leaves which are often conspicuous in fields, waste places, and by the wayside. The second year it sends up a wand-like leafy stem from 2 to 7 feet tall, which ends in a dense and cylindrical flower cluster. Stem and leaves alike are densely woolly-hairy; the leaves being elliptical, pointed at both ends, 4 to 12 inches long, and tapering into winged stalks. The flowers are about ¾ inch across, the 5-lobed yellow corolla being almost regular. It blooms between June and September, only a few flowers being open at any time. A native of Europe, it is widely naturalized in America.

YELLOW TOADFLAX (*Linaria vulgaris*)

Butter-and-eggs is another name often given this plant, which was originally introduced into this country from Europe as a garden flower. It now grows in fields, waste places, and along roadsides almost everywhere. Usually it has several, smooth, very leafy stems from 1 to 2 feet high, which end in a long cluster of flowers. The scattered leaves are very narrow, stalkless, and ½ to 1½ inches long. The attractive flowers are about an inch long, bright yellow, and have a prominent orange protuberance or palate on the lower lip; as well as a long, slender, and curved spur. The flowering season is between May and October.

BLUE TOADFLAX *(Nuttalanthus canadensis)*

The Blue Toadflax is a smooth plant with very slender ascending stems 8 inches to about 2 feet tall; with numerous, scattered, very narrow leaves ½ to 1½ inches long. Usually there are trailing offshoots at the base of the plant. The flowers are ¼ to ½ inch long, blue-violet with a white palate, and have a rather short spur. It grows in dry sandy or clayey fields and along roadsides, blooming between April and September. In the Southeast it is very abundant and often colors old fields blue in the early spring. Another name for it is Old-field Toadflax.

RANGE: N.S. to B.C. south to Fla., Tex. and Calif.

BLUE-EYED MARY *(Collinsia verna)*

The Blue-eyed Mary has slender, weak, sometimes minutely downy stems from 6 inches to 2 feet tall. The upper leaves are lance-shaped to egg-shaped, stalkless, partly clasping at the base, toothed on the margin, and ½ to 2 inches long. The lower ones are egg-shaped to roundish and have slender stalks. Its attractive flowers are ½ inch or more long, the 2-lobed upper lip being white and the 3-lobed lower one violet-blue. They are on slender stalks arising from the leaf axils, apparently often in whorls of from 4 to 6. It grows in rich, moist, open woodlands and thickets; blooming between April and June.

RANGE: N.Y. to Wis. and Iowa south to W.Va., Ky., Ark. and Kan.

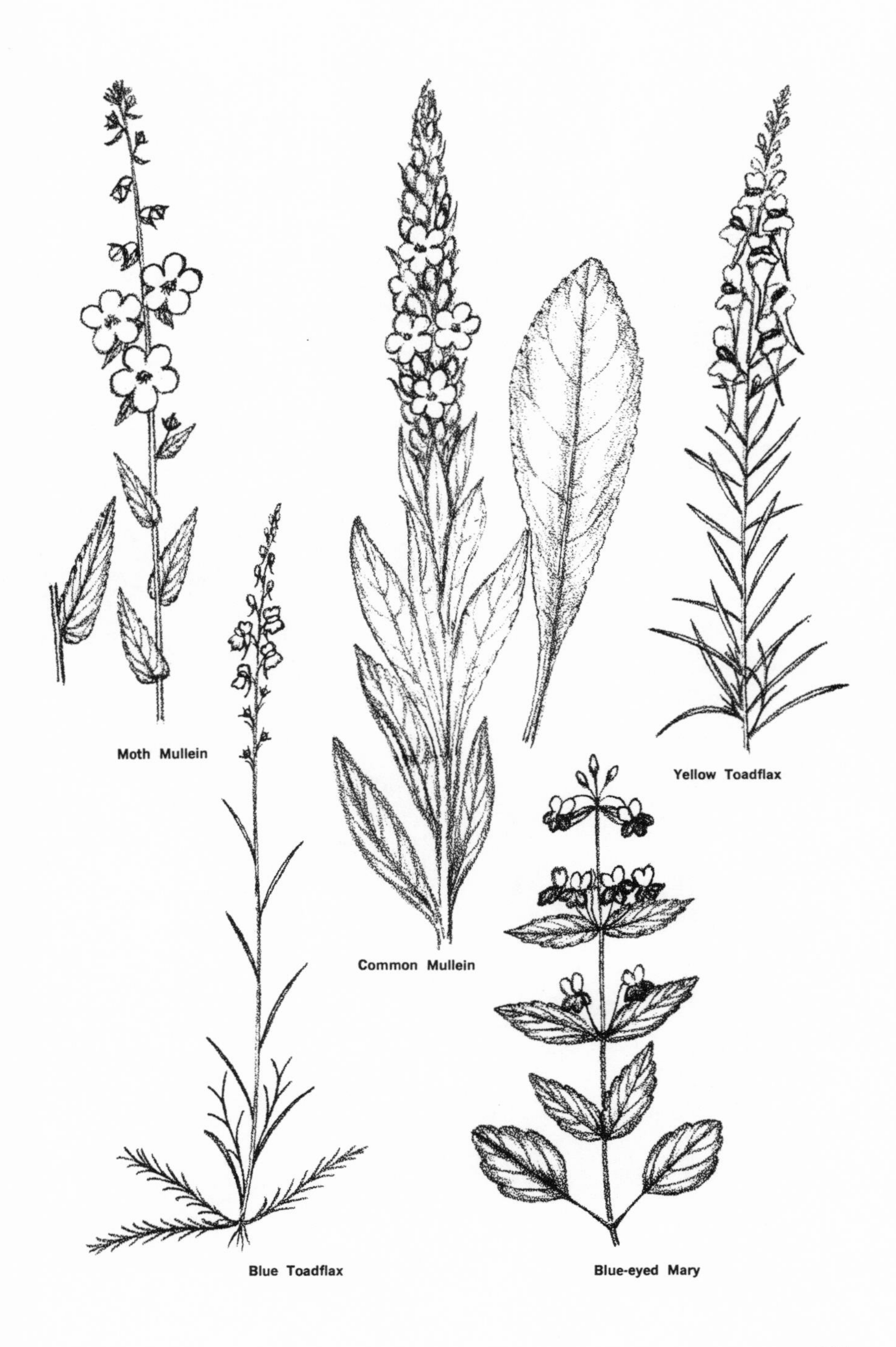
Moth Mullein
Common Mullein
Yellow Toadflax
Blue Toadflax
Blue-eyed Mary

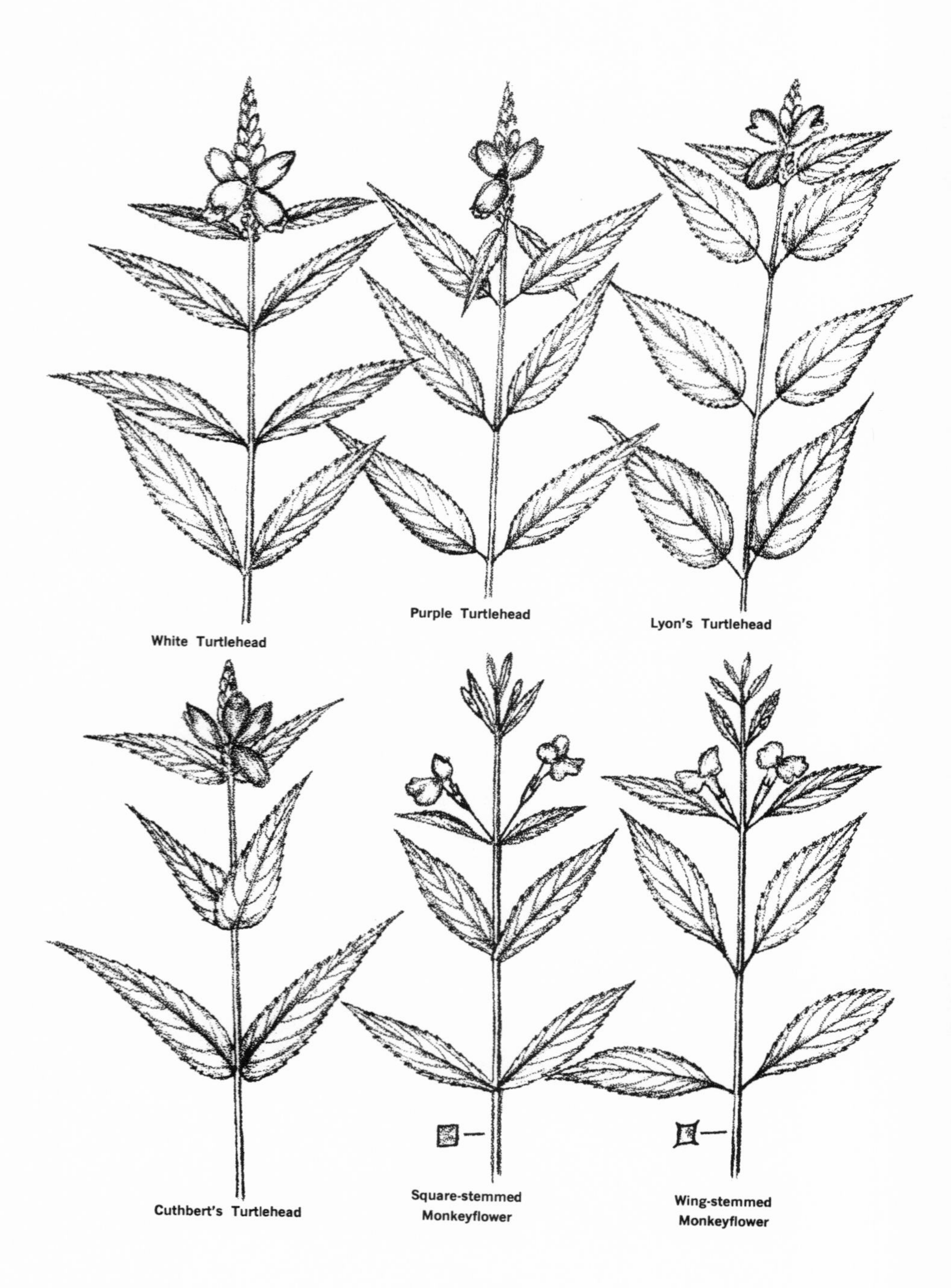
White Turtlehead
Purple Turtlehead
Lyon's Turtlehead
Cuthbert's Turtlehead
Square-stemmed
Monkeyflower
Wing-stemmed
Monkeyflower

WHITE TURTLEHEAD *(Chelone glabra)*

The turtleheads get their name from a fancied resemblance of their flowers to the head of a turtle. Those of this species are about an inch long and usually white, though they are quite often tinged with pink or purple near the tip. It is a smooth plant with a slender erect stem 1 to 3 feet tall; and the opposite leaves are lance-shaped, sharply toothed, stalkless or nearly so, and from 3 to 6 inches long. The flowers of this, and other species, are arranged in narrow but dense end clusters; blooming between August and October. It grows in swamps, along streams, and in other wet places.

RANGE: Nfd. to Ont. and Minn. south to Ga., Ala. and Mo.

PURPLE TURTLEHEAD *(Chelone obliqua)*

This is also a smooth plant 1 to 2 feet tall; with sharply toothed and rather broadly lance-shaped leaves 2 to 6 inches long. They taper at the base into slender stalks ¼ to ½ inch long. The flowers are about an inch long, deep pink to rose-purple and the lower lip is bearded with yellow hairs. It grows in wet woods and swamps; blooming between August and October.

RANGE: Md. to Tenn. south to Ga. and Miss.; Ind. to s. Minn. south to ark.

LYON'S TURTLEHEAD *(Chelone lyonii)*

This turtlehead can be distinguished by its egg-shaped leaves which are 3 to 7 inches long, rounded at the base, taper-pointed at the tip, sharply toothed, and have slender stalks up to 1½ inches long. Its flowers are about an inch long, deep pink to rose-purple; the corolla having a sharp ridge on its back and a lower lip bearded with deep yellow hairs. It grows in wet woods and along streams in the southern mountains; blooming between July and September.

RANGE: Va. south to w. N.C., nw. S.C. and e. Tenn.

CUTHBERT'S TURTLEHEAD *(Chelone cuthbertii)*

Cuthbert's Turtlehead may be distinguished by its sharply toothed, lance-shaped leaves which are rounded to somewhat heart-shaped at the base and completely stalkless. Its flowers are a deep violet-purple, the corolla having broad darker lines on the inside of the lower lip. It grows in wet woods and thickets; blooming in July to September.

RANGE: Uplands of N.C. into the coastal plain of se. Va.

SQUARE-STEMMED MONKEYFLOWER *(Mimulus ringens)*

The monkeyflowers received their name from the fancied resemblance of their 2-lipped corollas to a grinning face. This species is a smooth plant with a branching 4-sided or square stem 1 to 3 feet tall. The lance-shaped to narrowly oblong leaves are 2 to 4 inches long, sharply toothed, and stalkless. Its flowers are about an inch long and have slender stalks usually 1 to 2 inches in length; the corolla being a light violet-blue. It grows in wet meadows, swampy places, and along streams; blooming between June and September.

RANGE: N.S. to Man. south to Ga., La., Tex. and Colo.

WING-STEMMED MONKEYFLOWER *(Mimulus alatus)*

This species is similar to the preceding one but the 4-sided stems are more or less winged on the angles. Its leaves have very evident stalks and its pale blue-violet or pinkish flowers have stalks which are shorter, or no longer than the leaf stalks. It grows in similar situations, blooming between July and October.

RANGE: Conn. to s. Ont., Iowa and Neb. south to Fla. and Tex.

MUSKFLOWER *(Mimulus moschatus)*

The Muskflower is really a monkeyflower with yellow flowers. It has a weak, hairy or somewhat sticky-hairy stem which lies partly flat on the ground, rooting at the lower nodes, but sometimes ascending to a height of about a foot. The leaves are paired, short-stalked, egg-shaped, sometimes toothed, and 1 to 2 inches long. They may be rounded or somewhat heart-shaped at the base. The corolla is rather open, yellow, often striped with red in the throat, about ¾ inch long, and there are 2 densely hairy lines below the lower lip. It grows in the wet margins of ponds and streams, blooming between June and September.

RANGE: Nfd. to Ont. south locally to Mass., N.C., W.Va. and Mich. (Not illustrated)

HAIRY BEARD-TONGUE *(Penstemon hirsutus)*

Beard-tongues are so-called because their flowers have 5 stamens, the fourth one lacking an anther but having a hairy, or bearded, filament. This species has a slender, grayish-downy stem 1 to 3 feet tall, often with gland-tipped hairs in the flower cluster. It has pairs of lance-shaped, elliptic, or narrowly top-shaped leaves 2 to 4 inches long, with sharply toothed margins; the upper ones being stalkless, the lower ones with stalks. The flowers are about an inch long, the slender, dull purple or violet corolla having whitish lobes, and its throat closed by an upward arching lower lip. It grows in dry, rocky, open woods and in fields; blooming between May and July.

RANGE: Que. to Ont. and Wis. south to Va. and Tenn.

SMOOTH BEARD-TONGUE (*Penstemon laevigatus*)

The Smooth Beard-tongue has slender stems 1½ to 3 feet tall which are smooth except in the terminal flower cluster, where they are glandular-hairy. Its leaves are quite firm, lance-shaped to narrowly egg-shaped, 3 to 6 inches long, their margins usually with inconspicuous teeth above the middle. Those on the upper part of the stem are stalkless and slightly clasping; the lower ones tapering into somewhat winged stalks. Its flowers are about an inch long, light violet-purple to white; the corolla expanded above the middle and with an open throat. It grows in dry open woods, fields, and along roadsides; blooming during May and June. RANGE: N.J. and Pa. south to Fla. and Miss.

GRAY BEARD-TONGUE *(Penstemon canescens)*

This species has a slender, gray-downy stem from 1 to 3 feet tall. The egg-shaped to broadly lance-shaped leaves are 2 to 4 inches long; the upper ones are stalkless with rounded to heart-shaped bases and have sharply toothed margins; the lower ones are broader and stalked. The flowers are about an inch long, pale to rather deep violet-purple; the corolla being abruptly swollen about the middle, and with grooves and darker lines in the open throat. It grows in dry woods and on rocky slopes, chiefly in the mountains; blooming from May and July.

RANGE: Pa. to Ind. south to se. Va., n. Ga. and n. Ala.

SOUTHERN BEARD-TONGUE *(Penstemon australis)*

This is chiefly a coastal plain plant, growing in sandy pinelands and dry oak woods; and blooming in May or June. It is distinguished by its flowers which have a creamy corolla with a reddish-purple lower lip. The stems are downy and the upper leaves are nearly or quite toothless. Se. Va. south to Fla. and west to Ala. (Not illustrated)

INDIAN-PAINTBRUSH *(Castilleja coccinea)*

Also known as the Scarlet Painted-cup, this is a somewhat hairy, simple-stemmed plant 8 to 15 inches tall; with the upper stem leaves, or bracts, deeply 3- to 5-cleft and tipped with brilliant red (sometimes pink or white). The flowers are not very conspicuous, about an inch long, greenish-yellow, and in the axils of the colored bracts. It grows in moist meadows, thickets, along roadsides and on prairies; flowering between April and July. A parasite on the roots of other plants.

RANGE: N.H. to Man. south to Fla., La. and Okla.

CULVER'S-ROOT *(Veronicastrum virginicum)*

Culver's-root has a smooth, slender stem 2 to 6 feet tall; with lance-shaped, toothed, short-stalked leaves 3 to 6 inches long, arranged in whorls of from 3 to 9. Its small flowers have a tubular, white or purplish corolla; and are crowded in dense, long, narrow end clusters. It grows in rich moist woods, thickets, and meadows; blooming between June and September. RANGE: Mass. to Man. south to Fla. and Tex.

COMMON WOOD-BETONY *(Pedicularis canadensis)*

This is a hairy plant with a cluster of simple stems 6 to 18 inches high. The lance-shaped leaves are so deeply cut into toothed lobes that they appear almost fern-like, the larger ones being 3 to 5 inches long. Its flowers are about ¾ inch long, the corolla yellow and reddish and with a long and arching upper lip, and are borne in short, densely-bracted end clusters. It grows in open woods, thickets, and clearings; blooming between April and June. Another name is Lousewort.

RANGE: Me. to Que. and Man. south to Fla. and Tex.

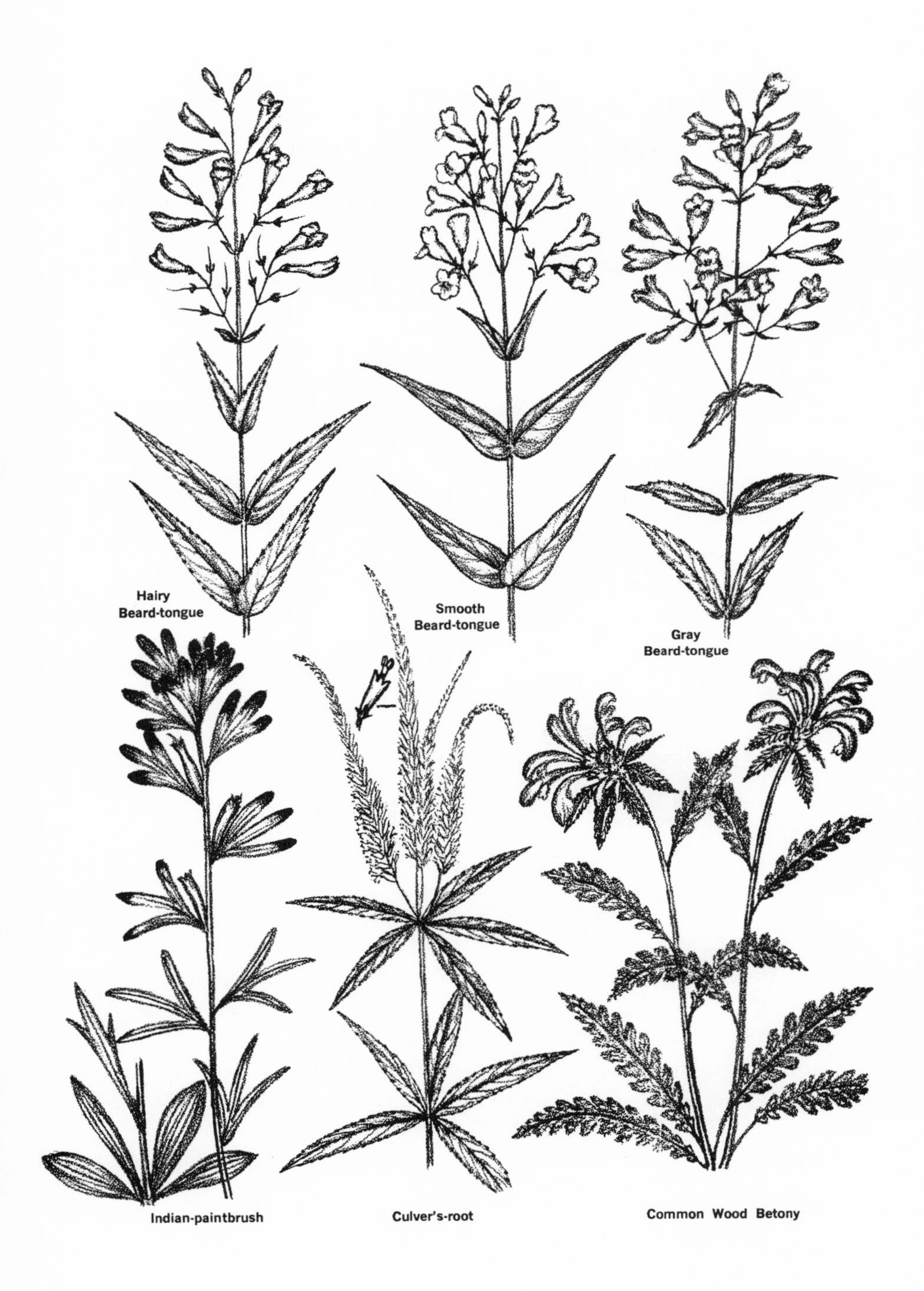
Hairy
Beard-tongue
Smooth
Beard-tongue
Gray
Beard-tongue
Indian-paintbrush
Culver's-root
Common Wood Betony

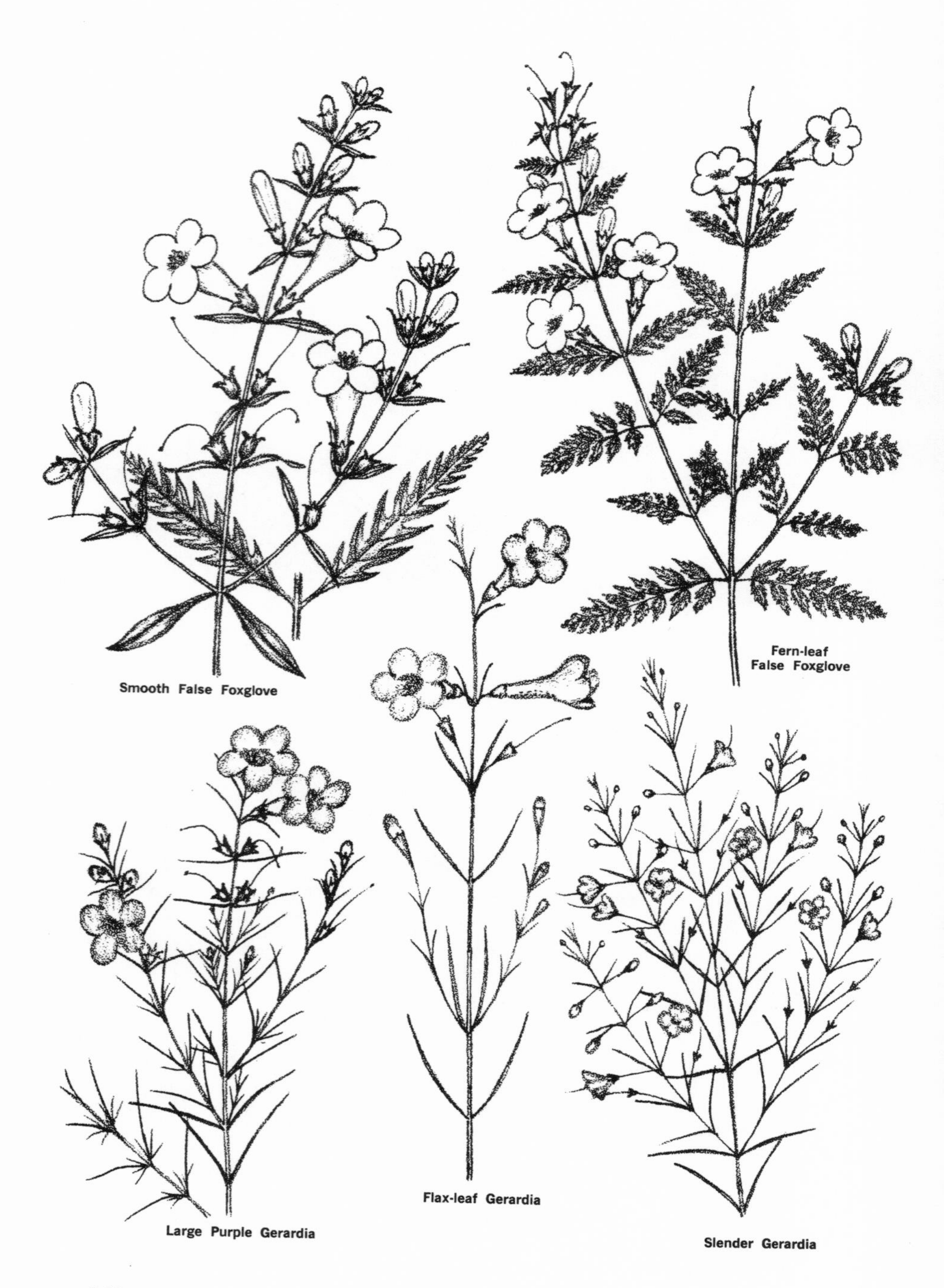
Smooth False Foxglove
Fern-leaf
False Foxglove
Flax-leaf Gerardia
Large Purple Gerardia
Slender Gerardia

SMOOTH FALSE FOXGLOVE *(Aureolaria flava)*

The yellow-flowered false foxgloves are parasitic on the roots of oak trees. This species is characterized by usually branched and often purplish stems 3 to 6 feet tall, which are more or less whitened with a bloom. Its leaves are elliptic to lance-shaped; the lower ones 4 to 6 inches long and pinnately cleft, the upper ones smaller and either toothed or untoothed. The flowers have a slightly bilateral, trumpet-shaped corolla 1½ to 2 inches long, with roundish lobes. It grows in dry to moist woods, blooming between July and September. RANGE: Me. to Minn. south to Fla. and La.

DOWNY FALSE FOXGLOVE *(Aureolaria virginica)*

This species resembles the preceding one but it has a simple or sparingly branced stem 2 to 4 feet tall, which is grayish-downy. It grows in dry open woods and thickets, blooming between May and August. RANGE: N.H. to Mich. south to Fla. and La. (Not illustrated)

ENTIRE-LEAF FALSE FOXGLOVE *(Aureolaria laevigata)*

This species has a simple or sparingly branched smooth, green stem 1 to 3 feet tall; and even the lower leaves are commonly untoothed. The flowers are somewhat smaller, with corollas just a little over an inch long. It grows in dry to moist woods and thickets, chiefly in the mountains, blooming in August and September.
RANGE: Pa. and Ohio south to Tenn. and Ga. (Not illustrated)

FERN-LEAF FALSE FOXGLOVE *(Aureolaria pedicularia)*

This is a more or less sticky-hairy plant with a much-branched and very leafy stem 1 to 4 feet high. The leaves are all deeply cut, sharply-toothed, and almost fern-like; from 1 to about 3 inches long. The flowers have a trumpet-shaped corolla about 1¼ inches long, which is yellow but commonly purple-tinged. It grows in dry woods and thickets, blooming in September or October. RANGE: Me. to Minn. south to Ga., Ky. and Ill.

COMB-LEAF FALSE FOXGLOVE *(Aureolaria pectinata)*

While similar to the preceding species, this one is usually 1 to 2 feet tall; with leaves even more sharply cut and very often purple-tinged. It grows in sandy woods, blooming from May to September. RANGE: Va. to Ky. and Mo. s. to Fla. and La. (Not illustrated)

LARGE PURPLE GERARDIA (*Agalinis purpurea*)

A number of species of *Agalinis* with pink or purplish flowers occur in the eastern United States, but few of them have such large flowers as the present species. Those of the Large Purple Gerardia average about an inch in length; and are rose-purple, with darker spots within the throat. It is a smooth or slightly roughish plant with a slender branching stem 1 to 2½ feet high; the very narrow leaves being 1 to 1½ inches long, often with smaller ones clustered in their axils. It grows in moist fields, thickets, and boggy places; blooming between August and October.
RANGE: Me. to Minn. and Neb. south to Fla. and Tex.

FLAX-LEAF GERARDIA *(Agalinis linifolia)*

This species also has large flowers, an inch or more long; rose-purple with deeper purple spots on the lower side. It has a smooth, slender, simple or sparingly branched stem 1 to 3 feet tall; with pairs of nearly erect, very narrow but thickish leaves 1 to 2 inches long. It grows in wet pinelands and about the margins of ponds in the coastal plain, blooming during August and September.
RANGE: Del. south to Fla. and west to La.

SLENDER GERARDIA *(Agalinis tenuifolia)*

The Slender Gerardia is a smooth plant with a slender and much-branched stem 6 inches to 2 feet tall; the numerous pairs of very narrow but flattened leaves being ½ to 1¼ inches long. Its flowers are light rose-purple or rarely white, smooth and spotted within the throat, about ⅔ inch long, and are on slender stalks arising in the axils of the upper leaves. It grows in dry to moist woods, thickets, and fields; blooming between August and October. RANGE: Me. to Minn. south to Ga. and La.

THREAD-LEAF GERARDIA *(Agalinis setacea)*

This is similar to the preceding species but its leaves are thread-like (not flattened), and its small flowers are downy. It grows in dry sandy woods, pinelands, and openings from se. N.Y. south to Ga. and Ala. (Not illustrated)

BIGNONIA FAMILY (Bignoniaceae)

Members of this family are mostly woody plants with opposite leaves, and flowers with a bilateral symmetry. Their flowers usually have 5 sepals, more or less united; 5 petals, united to form a tubular and often somewhat 2-lipped corolla; 4 or 5 stamens; and a solitary pistil consisting of 2 united carpels. The fruit is a capsule, usually with numerous winged seeds.

CROSS-VINE *(Bignonia capreolata)*

The pith in the stems of this high-climbing woody vine is shaped like a Maltese-cross, hence its common name. Its paired leaves are divided into 2 egg-shaped or oblong leaflets which are heart-shaped at the base, 3 to 7 inches long, and have the leaf stalk extended beyond them as a branched tendril. Often there is a pair of small leaves at the base of the leaf stalk which resemble stipules. The flowers are long-stalked, trumpet-shaped, red on the outside and orange or yellow within, and about 2 inches long; with a cup-like calyx. They are borne in axillary clusters, blooming during April and May. It grows in rich moist woods and in swamps.
RANGE: Md. and W.Va. to Ohio, Ill. and Mo. south to Fla. and Tex.

TRUMPET-CREEPER *(Campsis radicans)*

The Trumpet-creeper climbs by means of rows of aerial rootlets on its stems. Its paired leaves are divided into from 7 to 11 egg-shaped leaflets which are coarsely and sharply toothed and from 1½ to 3 inches long. The bright orange or reddish-orange, trumpet-shaped flowers are about 3 inches long; and borne in terminal clusters between June and September. It grows in moist woods, thickets, and along fencerows. Although often cultivated as an ornamental vine, it produces a dermatitis similar to that of Poison-ivy in some persons; giving rise to another common name, Cow-itch.
RANGE: N.J. to W.Va., Ill. and Iowa south to Fla. and Tex.

BROOM-RAPE FAMILY (Orobanchaceae)

All members of this family are parasitic plants, without green color and with only scale-like leaves. Their flowers have 4 or 5 partly united sepals; 4 or 5 united petals, the corolla usually irregular or 2-lipped; 4 stamens which are attached to the corolla tube; and a solitary pistil consisting of usually 2 united carpels, with the ovary 1-chambered. The fruits are capsules.

SQUAWROOT *(Conopholis americana)*

This peculiar plant is usually found in the woods, at the bases of oak trees. It forms large rounded knobs on the roots of the trees from which it receives its nourishment; and produces groups of stout, brownish or yellowish, scaly stems 3 to 10 inches high. They look very much like slender pine cones. The tubular flowers on these stems are yellowish, about ½ inch long, and are in the axils of the scales; blooming between April and July. Another name for it is Cancer-root. RANGE: N.S. to Wis. south to Fla. and Miss.

BEECH-DROPS *(Epifagus virginiana)*

In woods, under the beech trees, one often sees this parasitic plant with slender, purplish or yellowish-brown stems and ascending branches. They are 6 to 15 inches high and have scales instead of leaves; and they arise from a thick, scaly base. It produces its flowers between September and November; the upper ones being largest, about ½ inch long, whitish with a band of purplish-brown on the upper side, but they are sterile. The smaller flowers produce the seeds. RANGE: N.S. to Ont. south to Fla. and La.

ONE-FLOWERED BROOM-RAPE *(Orobanche uniflora)*

This plant has a scaly stem which creeps beneath the surface, around the roots of the plants on which it is parasitic. It sends up whitish, naked flower stalks 3 to 8 inches high; each one bearing a solitary creamy-white to lilac-colored flower about ¾ inch long. The tubular corolla has 2 bearded, yellow folds in the throat and it is minutely downy on the outside. It grows in rich, moist woodlands and thickets where it is parasitic on the roots of various plants; blooming between April and June. Other names for it are Pale Broom-rape and Naked Broom-rape.
RANGE: N.B. to Que. and Mont. south to Fla. and Tex.

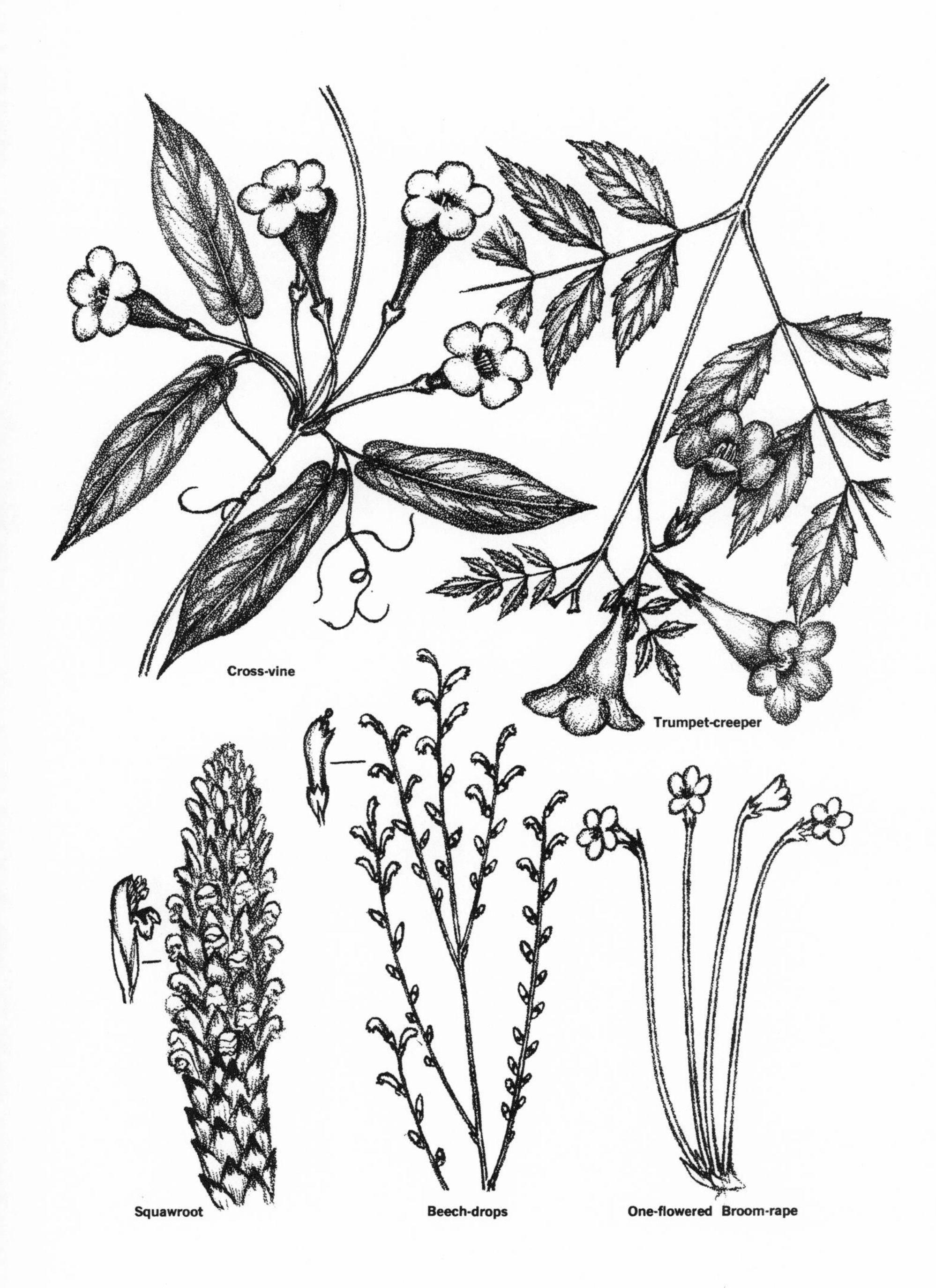
Cross-vine
Trumpet-creeper
Squawroot
Beech-drops
One-flowered Broom-rape

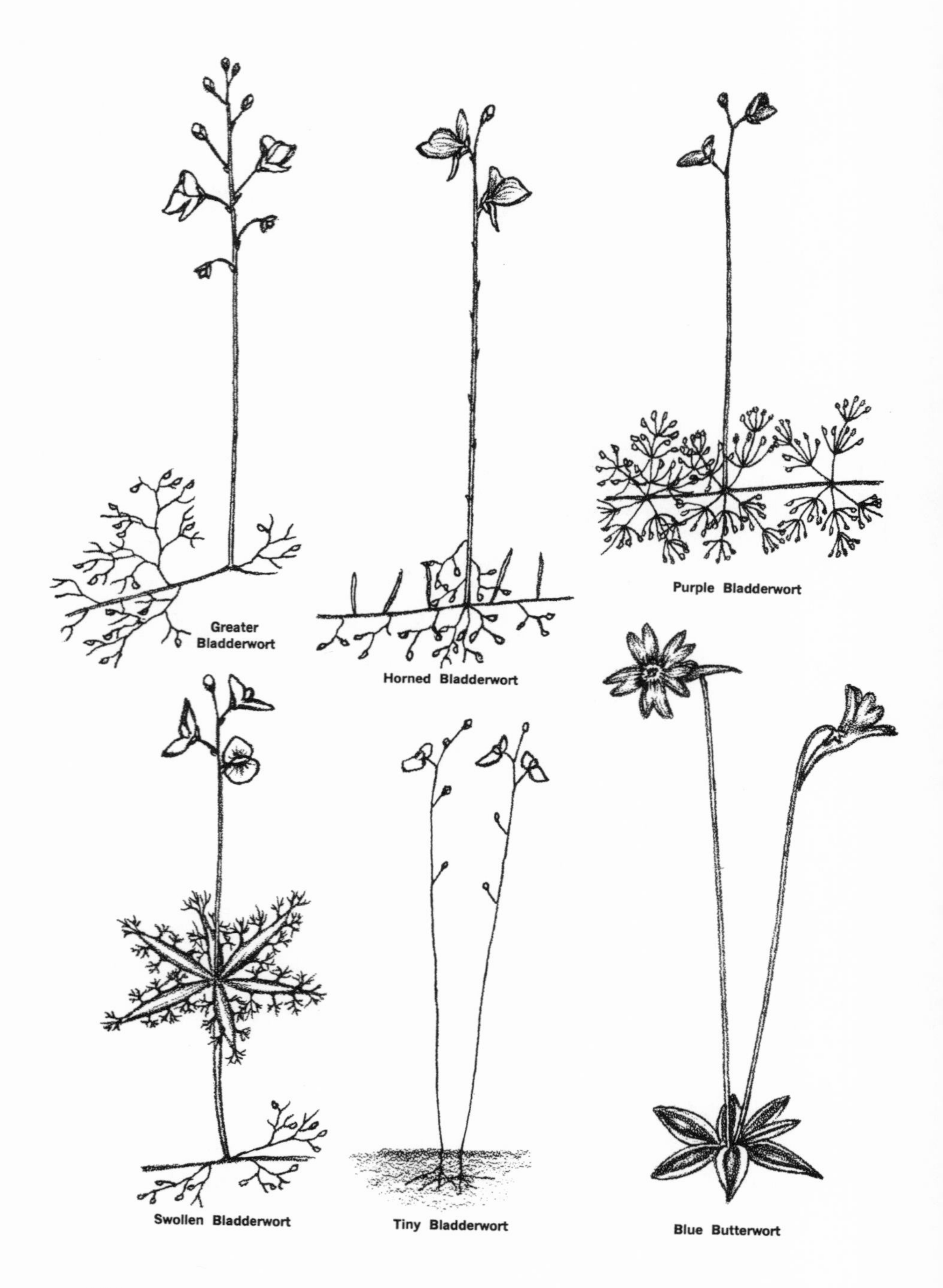
Greater Bladderwort
Horned Bladderwort
Purple Bladderwort
Swollen Bladderwort
Tiny Bladderwort
Blue Butterwort

BLADDERWORT FAMILY (Lentibulariaceae)

Members of the Bladderwort Family are aquatic plants, or plants of wet soils which get part of their nourishment by trapping insects or small aquatic animals. Their flowers have a bilateral symmetry, the corolla tube being spurred and very often 2-lipped; and they are borne on naked or minutely scaly stalks. There are always 2 stamens and a single 1-celled pistil.

The aquatic bladderworts have very finely dissected leaves, some or all of them bearing small bladders which serve as traps in catching tiny water animals. The older bladders often appear blackish because they are full of the indigestible remains of such animals. Besides the following, quite a few other species occur in the shallower waters of lakes, ponds, and slow-moving streams.

GREATER BLADDERWORT *(Utricularia macrorhiza)*

This species has stems 1 to 3 feet long which float horizontally beneath the surface of the water; sending up stout stalks 4 inches to 2 feet high which bear from 6 to 20 or more yellow flowers ¾ inch or more long, between May and September.

RANGE: Lab. to Alaska south to Va., Ohio, Mo. and Tex.

HORNED BLADDERWORT *(Utricularia cornuta)*

This bladderwort grows in bogs or on the wet sandy or muddy shores of lakes and ponds, the main portion of the plant being hidden. It sends up slender, brownish, wiry stalks 1 to 12 inches high; bearing from 1 to 5 yellow flowers about ¾ inch long between June and September. RANGE: Nfd. to Ont. and Minn. south to Fla. and Tex.

PURPLE BLADDERWORT *(Utricularia purpurea)*

The leaves of this bladderwort are in whorls along the 1- to 3-foot free-floating stems, and have bladders at the tips of the hair-like divisions. Between May and September it has erect stalks 2 to 6 inches high which bear from 2 to 5 deep pink to lavender-purple flowers. RANGE: N.S. to Que. and Wis. south to Fla. and La.

SWOLLEN BLADDERWORT *(Utricularia inflata)*

This species can be distinguished by the whorl of 3 to 10 float-like inflated leaves at the bases of its flower stalks. The latter are 5 to 12 inches high and bear from 3 to 5 or more yellow flowers ½ to ¾ inch across, between May and November.

RANGE: N.S. to N.Y. and Ind. south to Fla. and Tex.

TINY BLADDERWORT *(Utricularia subulata)*

This bladderwort grows in wet sandy places; sending up extremely fine, wiry, usually zig-zag and reddish-purple stems 3 to about 8 inches high. Between March and August, each one has from 1 to 12 yellow flowers about ⅓ inch across.

RANGE: N.S. south to Fla., west to Tex. and north to Ark.

BLUE BUTTERWORT *(Pinguicula caerulea)*

Butterworts got their name from the fact that the European species have been used as rennin in curdling milk. They have basal rosettes of shiny, sticky, pale green leaves. Insects stick to them as to flypaper, and the edges of the leaves roll up as their bodies are digested. This one has leaves ½ to 2 inches long. Its flowers are pale violet, almost an inch across, the corolla having 5 deeply notched lobes and a spur. They are solitary on sticky-hairy stalks 4 to 10 inches tall. It grows in wet coastal plain pinelands, blooming during April and May. RANGE: N.C. south to Fla.

YELLOW BUTTERWORT *(Pinguicula lutea)*

This species is similar to the preceding but it has golden-yellow flowers with the lobes shallowly 2- or 3-notched. It grows in similar situations from N.C. south to Fla. and west to La.; blooming during April and May. (Not illustrated)

SMALL BUTTERWORT *(Pinguicula pumila)*

This is a smaller species with white, pale blue, or rarely yellow flowers about ½ inch across; and leaves ½ to 1 inch long.. It grows in wet coastal plain pinelands from S.C. to Fla. and Tex.; blooming in April and May. (Not illustrated)

Members of this family have flowers with usually 5 partly united sepals; 5 more or less united petals forming a nearly regular to 2-lipped corolla; 2 or 4 stamens, in pairs of different length, and attached to the corolla tube; and a pistil consisting of 2 united carpels, the ovary being 2-celled and commonly with 2 ovules in each cell. The fruit is a capsule; the seeds being borne on small, hooked projections.

WATER-WILLOW *(Justicia americana)*

The Water-willow grows in slowly moving streams, the shallower waters of lakes and ponds, and more rarely in swamps. It is a smooth plant with numerous thick, cord-like runners; and it has slender, erect, grooved and angled stems 1 to 3 feet high. It leaves are paired, stalkless or nearly so, lance-shaped (often narrowly so) and 3 to 6 inches long. The attractive 2-lipped, pale violet flowers are about ¾ inch long; in dense and long-stalked clusters arising from the leaf axils. The flowering season is between June and October.

RANGE: Vt. and Que. to Ont., Wis. and Kan. south to Ga. and Tex.

HAIRY RUELLIA *(Ruellia caroliniensis* var. *cinerascens)*

The Hairy Ruellia is a somewhat variable plant. It has a more or less hairy, simple or branched stem 6 inches to 2½ feet tall. The egg-shaped, lance-shaped, or elliptic leaves have short stalks and are 1½ to 4 inches long ;usually being rather crowded toward the tips of the stems. The lavender or lilac-blue flowers are 1 to 2 inches long, the united sepals being prolonged into bristle-like and usually hairy-fringed tips. They are nearly stalkless and several usually occur together in the leaf axils. It grows in dry sandy woods and clearings, blooming between May and September.

RANGE: N.J. to Ind. south to Fla. and Tex.

SMOOTH RUELLIA *(Ruellia strepens)*

This is a smooth or sparingly hairy plant with a simple or branched stem 1 to 3 feet tall. The egg-shaped leaves are 3 to 6 inches long and taper into slender stalks. The flowers are pale blue-violet, 1½ to 2 inches long, and the calyx lobes are merely taper-pointed at the tip. It grows in rich, open woods and thickets; blooming between May and September.

RANGE: N.J. to Ill., Mo. and Kan. south to Fla. and Tex.

DYSCHORISTE *(Dyschoriste oblongifolia)*

Dyschoriste has a slender, erect, downy or hairy stem from 6 to 15 inches tall. On it are pairs of ascending or erect, oblong to oval or top-shaped, stalkless leaves ½ to 1¼ inches long. The flowers are blue, sometimes mottled with purple, about an inch long, and are usually solitary in the axils of the upper leaves. It grows in dry, sandy coastal plain pinelands or on sandhills; blooming during April and May.

RANGE: se. S.C. south to Fla.

Water-willow
Hairy Ruellia
Smooth Ruellia
Dyschoriste

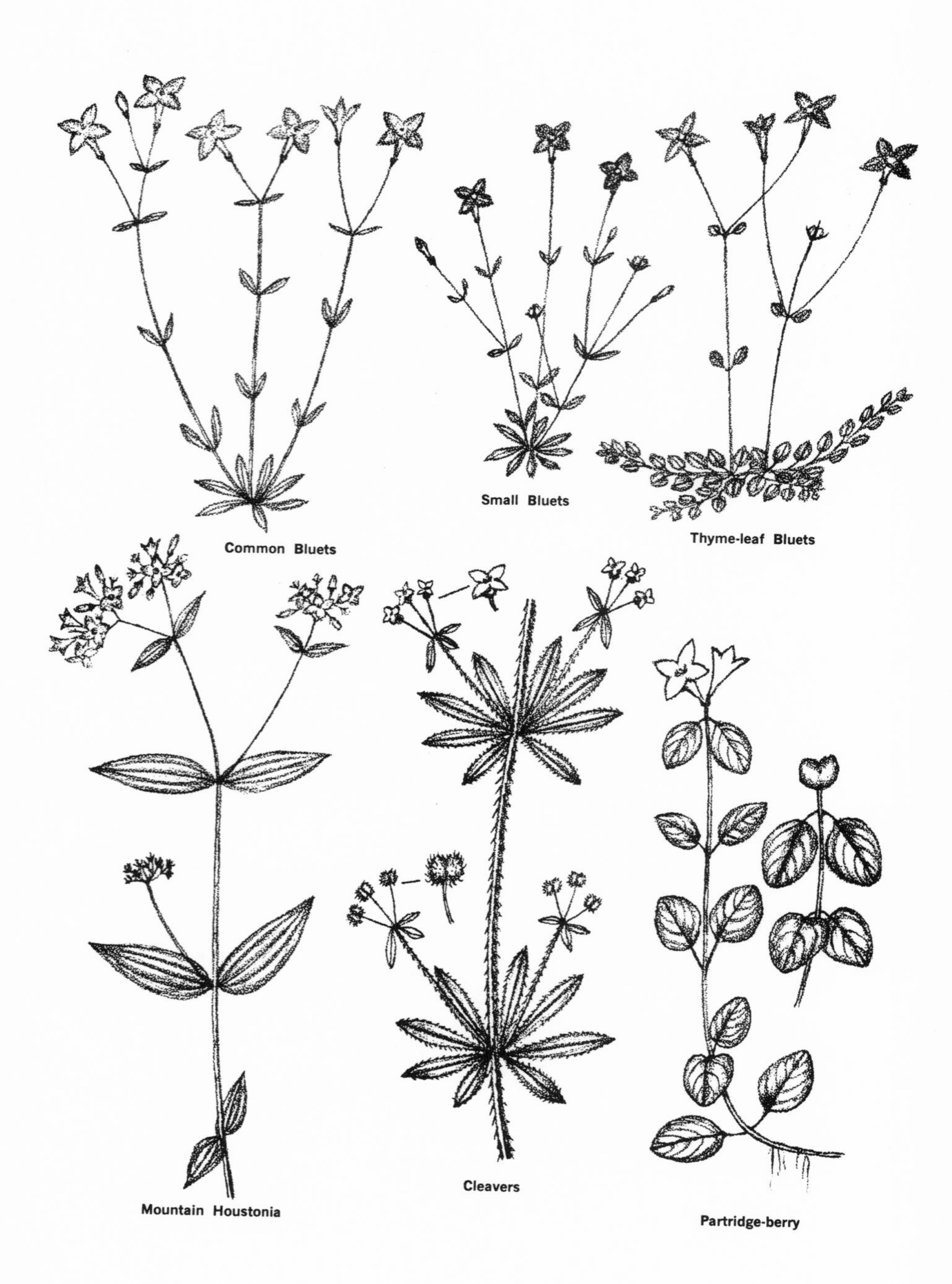
Common Bluets
Small Bluets
Thyme-leaf Bluets
Mountain Houstonia
Cleavers
Partridge-berry

This is a large family of plants with opposite, untoothed leaves; and flowers with usually 4 sepals more or less united with the ovary, 4 united petals, and 4 stamens which are attached to the corolla tube.

COMMON BLUETS *(Houstonia caerulea)*
Often called Innocence or Quakerladies, this is a smooth little plant with thread-like tufted stems 3 to 6 inches high. It has narrowly top-shaped leaves about ½ inch long in a basal rosette, and pairs of smaller ones on the flower stems. Its bright pale blue to whitish flowers have a yellow "eye". It is common in open woods, meadows and clearings; blooming between April and June.
RANGE: N.S. to Ont. and Wis. south to Ga., Ala. and Mo.

SMALL BLUETS *(Houstonia pusilla)*
This bluet has thread-like stems 1 to 4 inches high. The oval or egg-shaped basal leaves are less than ½ inch long; and the deep violet-blue or purplish flowers with a dark yellow "eye" are about ¼ inch across. Often called the Star-violet, it grows in sandy or rocky open places; blooming in March or April.
RANGE: Va. to Ill., Mo. and Okla. south to Fla. and Tex.

THYME-LEAF BLUETS *(Houstonia serpyllifolia)*
This is a delicate little plant with creeping, leafy stems 4 to 10 inches long. The deep blue flowers resemble those of the Common Bluet. It grows in wet places, often along mountain streams; blooming between April and July.
RANGE: Pa. and W.Va. south to n. Ga. and e. Tenn.

Houstonia procumbens is a coastal plain species which also has prostrate and creeping stems but its flowers are white. It grows on beach dunes and in sandy pinelands from se. S.C. south to Fla. and west to La.; blooming in March or April. (Not illustrated)

MOUNTAIN HOUSTONIA *(Houstonia purpurea)*
This plant has a simple or branched stem 4 to 18 inches tall; with pairs of egg-shaped to lance-shaped, 3- to 5-ribbed leaves ½ to 1 inch long. The small whitish to pale purple flowers are arranged in terminal clusters. It grows in open woods and on rocky slopes; blooming between May and July. Also called the Tall Houstonia.
RANGE: Del. to Iowa south to Fla., La. and Okla.

LONG-LEAF HOUSTONIA *(Houstonia longifolia)*
This species has narrow leaves ⅛ to ¼ inch wide; and calyx lobes longer than the corolla tube. Me. to Sask. south to Ga., Ky. and Okla. (Not illustrated)

CLEAVERS *(Galium aparine)*
Bedstraws or goose-grasses are slender plants with weak, mostly reclining, 4-sided stems; leaves arranged in whorls; and small, white to yellowish, 4-parted flowers in axillary or terminal clusters. A number of species occur in the eastern United States. This common one has narrow leaves 1 to 3 inches long, with rough margins and midribs, and arranged in whorls of from 6 to 8. The angles of its stems are bristly-prickly, the flowers white, and the fruits bristly. It grows in woods, meadows, and roadsides; blooming between April and July. RANGE: Nfd. to Alaska south to Fla. and Tex.

PARTRIDGE-BERRY *(Mitchella repens)*
Also known as Twinberry, this is a smooth plant with trailing slender stems 6 to 12 inches long; with pairs of lustrous, evergreen, roundish and stalked leaves usually ½ to ¾ inch long. The white or pinkish, fringed flowers are in pairs and have their bases united. They are followed by bright red, double berries about ⅓ inch across; which often persist until the next flowering time between May and July.
RANGE: Nfd. to Ont. and Minn. south to Fla. and Tex.

HONEYSUCKLE FAMILY (Caprifoliaceae)

Members of this family are mostly shrubs or woody vines with opposite leaves. Their flowers have 4 or 5 sepals which are united with the 3- to 5-celled ovary of the pistil; 4 or 5 more or less united petals which form a nearly radial to strongly bilateral or 2-lipped corolla; and 4 or 5 stamens which are attached to the corolla tube.

GLAUCOUS HONEYSUCKLE *(Lonicera dioica)*

This is a somewhat woody, climbing plant with smooth stems 3 to 10 feet long. The greenish-yellow and often purplish-tinged flowers are about ⅔ inch long. They are in a dense end cluster and the pair of leaves immediately below them are joined together at the base. The other leaves are oval or egg-shaped, stalkless or short-stalked, 1½ to 3½ inches long, and are pale or whitened beneath. The fruits are salmon-colored berries. It grows in rocky woods, thickets, and on slopes; blooming during May or June.

RANGE: Me. and Ont. to Man. south to Ga. and Mo.

TRUMPET HONEYSUCKLE *(Lonicera sempervirens)*

Also known as the Coral Honeysuckle, this native species is quite often cultivated. It is a climbing vine with pairs of oval to narrowly elliptic leaves 1½ to 3 inches long, which are dark green above and strongly whitened beneath. The flowers are slenderly trumpet-shaped with 5 short lobes at the summit, 1½ to 2 inches long, bright red on the outside and yellowish within. They are arranged in several whorls on a terminal stalk often several inches long, and bloom between April and September. It grows in the borders of woods, thickets, and in fencerows and is much more common southward.

RANGE: Me. to N.Y., Iowa and Neb. south to Fla. and Tex.

YELLOW HONEYSUCKLE *(Lonicera flava)*

This is one of the most attractive of our native climbing honeysuckles. It is a rather woody and twining plant somewhat like the Glaucous Honeysuckle, but its leaves are paler and grayish beneath. In April or May it has clusters of fragrant, bright yellow to orange-yellow flowers about an inch long, which are shaped somewhat like those of the familiar Japanese Honeysuckle. It grows naturally in rocky woods and on bluffs and is very desirable as a cultivated flowering vine.

RANGE: N.C. to Mo. south to Ga., Ala., Ark. and Okla.

JAPANESE HONEYSUCKLE *(Lonicera japonica)*

The Japanese Honeysuckle is a high-climbing and trailing vine with densely hairy stems and branchlets. It has pairs of egg-shaped to oval, short-stalked leaves 1 to 3 inches long which are downy beneath. The flowers are tubular and 2-lipped, the upper lip having 3 lobes and the lower one having 2. They are white at first but turn yellow, about 1¼ inches long, and are extremely fragrant. This Asiatic species has been widely cultivated in America and in many places—particularly in the South—it has escaped to become a very troublesome forest weed. It blooms between May and August and the flowers are followed by black berries.

TWINFLOWER *(Linnaea borealis* ssp. *longiflora)*

The Twinflower is a slender-stemmed, trailing or creeping plant which grows in cold woods and bogs in the northern portions of both the Old and the New Worlds. Between June and August it sends up 3- to 10-inch stalks bearing pairs of stalked, roundish or oval, obscurely toothed leaves ⅓ to ⅔ of an inch wide. At their tips they have a pair of nodding, bell-shaped, pink flowers about ⅓ of an inch long. The genus was named for the immortal Linnaeus who was especially fond of this dainty little plant.

RANGE: Lab. to Alaska south to Md., W.Va., the Great Lakes region, S.D., Colo., Utah and Calif.

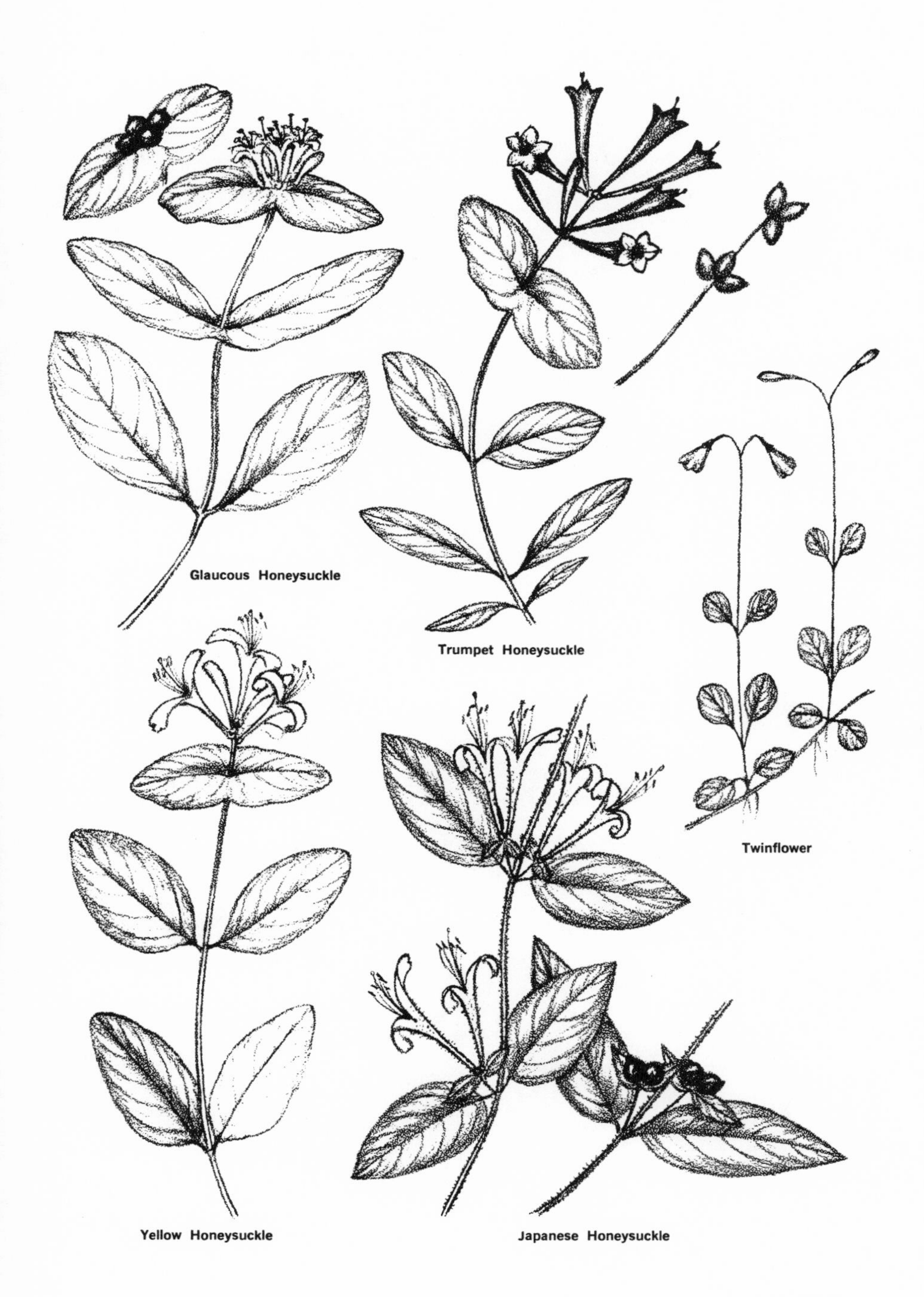
Glaucous Honeysuckle
Trumpet Honeysuckle
Twinflower
Yellow Honeysuckle
Japanese Honeysuckle

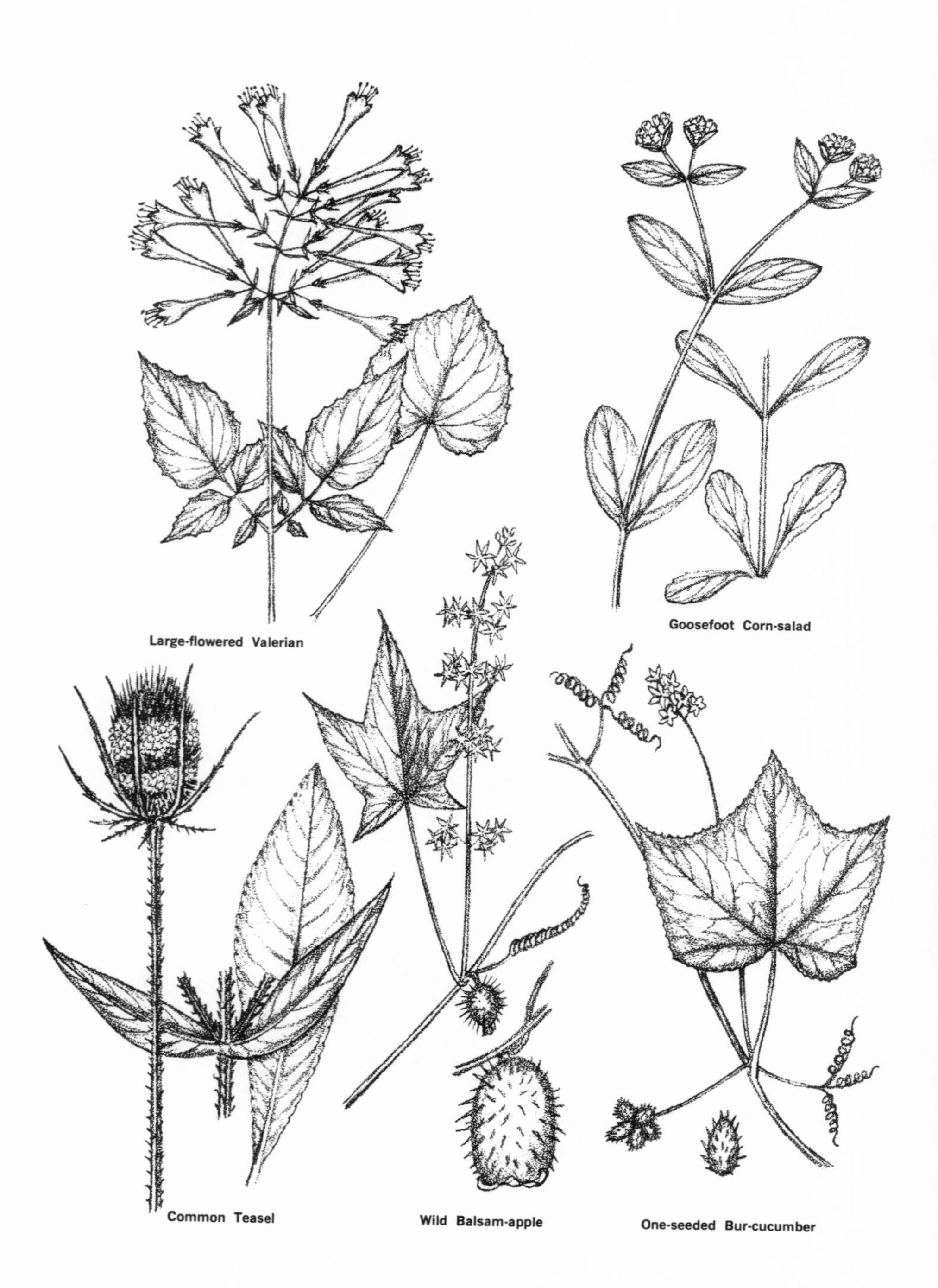
Large-flowered Valerian
Goosefoot Corn-salad
Common Teasel
Wild Balsam-apple
One-seeded Bur-cucumber

VALERIAN FAMILY (Valerianaceae)

Members of this family have a calyx united with the 1- to 3-chambered ovary of the pistil, the lobes often not evident or represented by plume-like bristles; a corolla of usually 5 united petals; and usually 3 stamens which are attached to the corolla tube. Each pistil usually produces only 1 seed.

LARGE-FLOWERED VALERIAN *(Valeriana pauciflora)*

Also known as the American Wild Valerian, this is a smooth plant with many spreading rootstocks and a slender stem 1 to 2½ feet high. The leaves on the stem are paired and divided into 3 to 7 thin leaflets, the terminal one being much the larger. The basal leaves are more or less heart-shaped, long-stalked, and sometimes have an additional pair of small leaflets. Its flowers have pale pink corolla tubes about ⅔ inch long, 5-lobed at the summit; and 5 stamens which extend well beyond the corolla. The calyx has feathery bristles which unroll as the fruits develop. It grows in rich moist woods and in stream bottoms, blooming in May or June. RANGE: Pa. to Ill. south to Va. and Tenn.

GOOSEFOOT CORN-SALAD *(Valerianella chenopodiifolia)*

This is a smooth plant from 1 to 2 feet tall, with forking branches. It has pairs of lance-shaped to oblong, stalkless leaves 1 to 3 inches long on the stems. The lower and basal ones are broader toward the tip and often somewhat wavy-toothed on the margin. Its small white flowers are densely crowded in head-like clusters about ⅔ inch broad, with leafy bracts at the base. It grows in moist open woods, meadows, and in stream bottoms, blooming in May or June. RANGE: N.Y. and Pa. to s. Ont. and Ind.

TEASEL FAMILY (Dipsacaceae)

Members of this family have flowers with a calyx united with the ovary of the pistil; a corolla of 2 to 5 united petals; and 2 to 4 stamens attached to the corolla tube. The fruits are small, dry, and 1-seeded.

COMMON TEASEL *(Dipsacus sylvestris)*

This native of Europe is widely naturalized in waste places, old fields, and along roadsides in eastern North America. It has a stout prickly stem 3 to 6 feet tall. The pairs of lance-shaped or oblong leaves are 6 to 12 inches long, the upper ones clasping the stem at the base. The flowers are very small, with a tubular lilac or pinkish-purple corolla; and are crowded in an egg-shaped or cylindrical head, intermixed with prickly bristles. It blooms between July and October.

GOURD FAMILY (Cucurbitaceae)

Members of this family are usually tendril-bearing vines with alternate leaves. The stamens and the pistils are always in separate flowers, sometimes on separate plants. The flowers have 4 to 6 usually partly united sepals; and a like number of more or less united petals. The 3 stamens are joined together by their contorted anthers. The pistil is united with the calyx tube.

WILD BALSAM-APPLE *(Echinocystis lobata)*

This tendril-climber grows in rich soils, usually along streams, climbing about over bushes and other vegetation. It has rather thin, roughish leaves which are deeply and sharply 5-lobed as well as toothed on the margin. The stamen-bearing flowers are greenish-white and grouped in long and narrow clusters. The pistil-bearing ones are solitary and develop into an egg-shaped, green, prickly fruit about 2 inches long. It blooms from June to October. Other names often given it are Wild Cucumber or Prickly Cucumber.
RANGE: N.B. to Sask. south to Fla. and Tex.

ONE-SEEDED BUR-CUCUMBER *(Sicyos angulatus)*

Stems of this tendril-climber are angled and sticky-hairy, and its tendrils are branched. Its leaves are 5-angled or rather shallowly 5-lobed, toothed on the margin, and rough on both sides. The stamen-bearing flowers are whitish and clustered at the ends of long stalks. The pistil-bearing ones are also clustered but are on much shorter stalks. They develop into yellowish, dry, egg-shaped, prickly, 1-seeded fruits about ½ inch long. It grows along streams and in low moist woods and thickets; blooming between July and September. RANGE: Me. to Que. and Minn. south to Fla. and Tex.

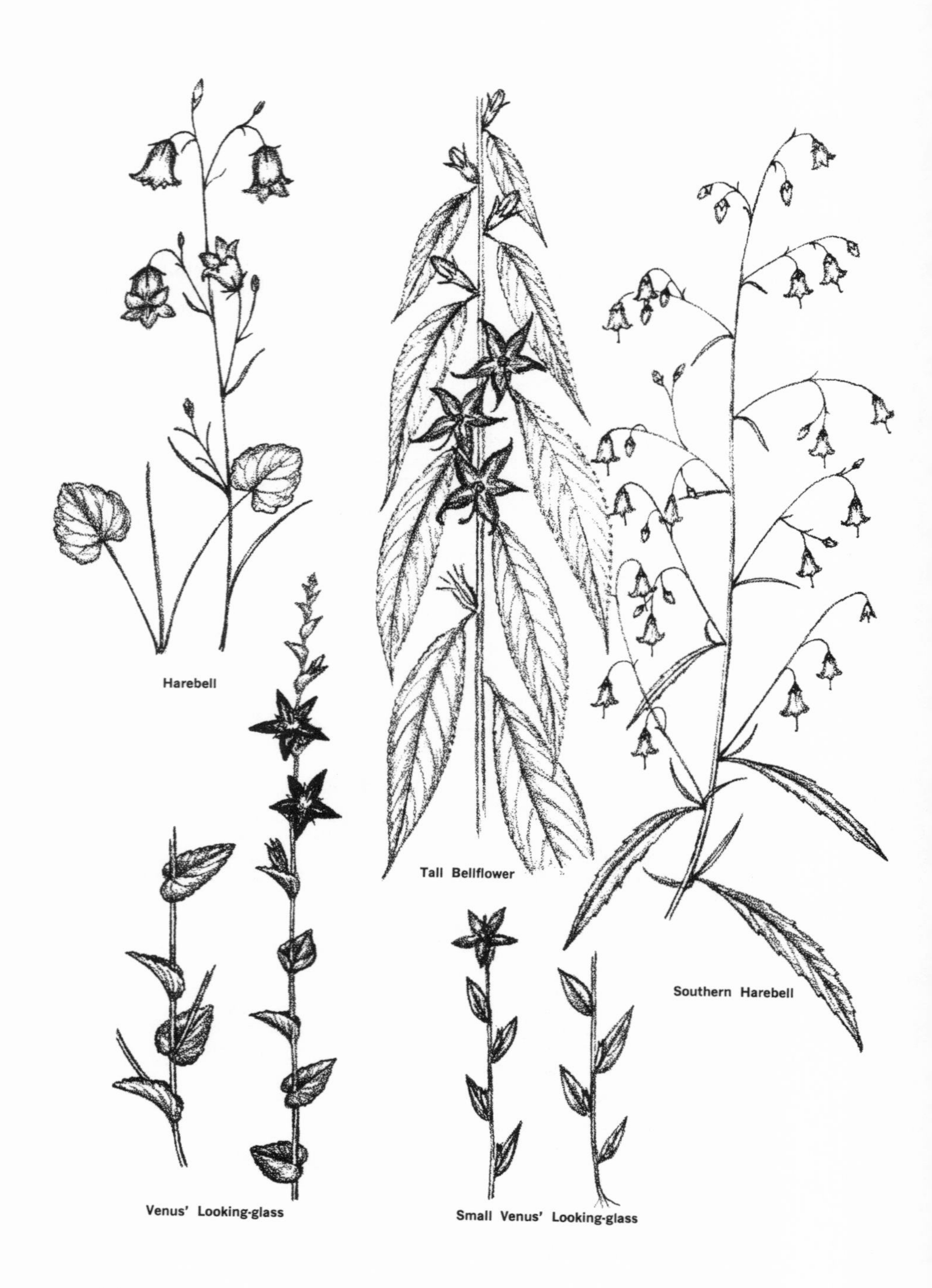
Harebell
Tall Bellflower
Southern Harebell
Venus' Looking-glass
Small Venus' Looking-glass

Members of the Bluebell Family have flowers with 5 sepals which are united with the ovary of the pistil; 5 partly united petals forming a 5-lobed corolla; and 5 stamens which are usually free from the corolla and all separate. The plants have a milky juice and alternate leaves.

HAREBELL *(Campanula rotundifolia)*

Also called the Bluebells-of-Scotland, the Harebell is found in the northern parts of both the Old and New World and is frequently cultivated. It has a slender, weak, and much-branched stem 6 to 18 inches high. The stem leaves are all very narrow but the basal ones are roundish, rather heart-shaped, ¼ to 1 inch wide, and long-stalked. Its hanging flowers are bell-shaped, purplish-blue, and about ¾ inch long; blooming between June and September. It grows on moist rocky slopes, cliffs, and meadows.

RANGE: Nfd. to Alaska south to N.J., W.Va., Ohio, Ill., Mo., Neb., Tex. and Calif.

TALL BELLFLOWER *(Campanula americana)*

The Tall or American Bellflower usually has a simple, more or less hairy stem, from 2 to about 6 feet tall; along which are scattered lance-shaped to narrowly egg-shaped, toothed leaves 3 to 6 inches long, most of them tapering at the base into short stalks. Its flowers are star-shaped rather than bell-like, with 5 long and pointed lobes. They are about an inch across, light violet-blue, and are in the axils of leaf-like bracts; forming a long and slender end cluster. It grows in rich moist woods and thickets, blooming between June and August. RANGE: Ont. to Minn. south to Fla., Ala. and Mo.

SOUTHERN HAREBELL *(Campanula divaricata)*

This is a smooth plant with a slender and much-branched stem 1 to 3 feet high, which grows on dry rocky slopes and in open woods in the mountains. The leaves are lance-shaped to narrowly egg-shaped, pointed at both ends, coarsely and sharply toothed, and usually 1 to 3 inches long. The light lavender-blue flowers are bell-shaped, about ¼ inch long, and nod on slender stalks. They are usually very numerous and arranged in a rather large but loose cluster, blooming between June and September. It is sometimes called the Panicled Bellflower. RANGE: Md., W.Va. and Ky. south to Ga. and Ala.

MARSH BELLFLOWER *(Campanula aparinoides)*

This species has a weak and somewhat 3-sided, branching stem which is minutely roughish-bristly on the angles. The narrow leaves are 1 to 2½ inches long and are roughish on the margins and midribs. Whitish to pale blue, open bell-shaped flowers ½ inch or less long are produced on slender stalks between June and August. The plant grows in wet meadows, swales, and on shores and stream banks. It is also known as the Slender or Bedstraw Bellflower. Me. to Minn. south to Ga., Ky., Iowa, Neb. and Colo. (Not illustrated)

VENUS' LOOKING-GLASS *(Triodanis perfoliata)*

The Venus' Looking-glass is a common plant of dry open woods, thickets, and fields. It is a more or less hairy plant with a simple stem, or sometimes branched toward the base, and 6 to 18 inches high. The numerous leaves are roundish or egg-shaped, toothed on the margin, ¼ to 1 inch wide, and clasp the stem by a heart-shaped base. The flowers in the axils of the lower leaves never open, remaining bud-like but producing seeds. The ones along the upper part of the stem are violet or violet-blue, about ½ inch across, and are star-shaped with 5 long corolla lobes; blooming between April and August. RANGE: Me. to Ont. and B.C. south to Fla. and Tex.

SMALL VENUS' LOOKING-GLASS *(Triodanis perfoliata* var. *biflora)*

Although quite similar to the preceding species, the leaves of this plant are untoothed or inconspicuously toothed and are stalkless but do not clasp the stems. It produces only 1 or 2 showy flowers toward the tip of the 6- to 18-inch stem, all of the others remaining closed and bud-like. It grows in dry open woods and fields, blooming between April and June.

RANGE: Va. to Ky., Kan. and Ore. south to Fla., Tex. and s. Calif.

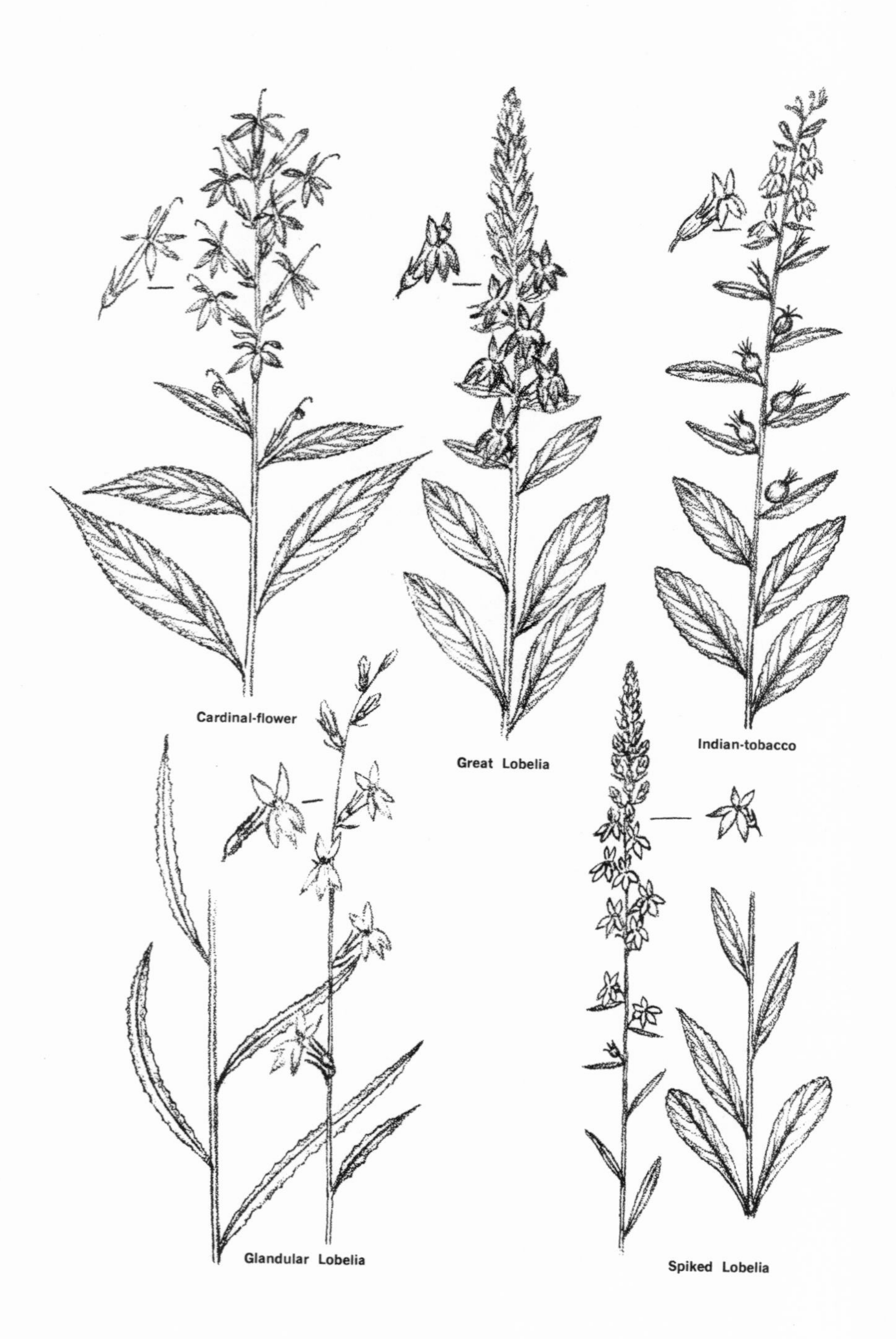
Cardinal-flower
Great Lobelia
Indian-tobacco
Glandular Lobelia
Spiked Lobelia

CARDINAL-FLOWER *(Lobelia cardinalis)*

Also known as the Red Lobelia, this plant always attracts one's attention by its terminal clusters of brilliant red flowers. Each flower is about 1½ inches long and the tube of stamens projects upward through the cleft in the corolla. It usually has a simple stem from 1 to 3 feet tall. The numerous leaves are 2 to 6 inches long, lance-shaped to narrowly egg-shaped, and are toothed on the margin. It grows in moist meadows, thickets, swamps, and along the banks of streams; blooming between July and October.

RANGE: N.B. to Ont. and Minn. south to Fla. and Tex.

GREAT LOBELIA *(Lobelia siphilitica)*

Sometimes the Great Lobelia is called the Blue Cardinal-flower. It has a nearly smooth, erect, rather stout, and simple stem from 1 to 3 feet tall. The many leaves are lance-shaped to egg-shaped, stalkless, untoothed or irregularly toothed on the margin, and from 2 to 6 inches long. Its bright blue flowers are about an inch long and often have white marks on the lower corolla lobes. They are borne in the axils of crowded leaf-like bracts toward the summit of the stem; blooming between July and October. It grows in wet or swampy places.

RANGE: Me. to Minn. south to Va., nw. S.C., n. Ala., La. and Tex.

DOWNY LOBELIA *(Lobelia puberula)*

This pretty lobelia is similar to the Great Lobelia but it has densely downy stems and leaves; and somewhat smaller bright blue flowers. It grows in wet woods and swampy places from N.J. to W.Va., Ill. and Okla. south to Fla. and Tex.; blooming between August and October. (Not illustrated)

INDIAN-TOBACCO *(Lobelia inflata)*

The Indian-tobacco is easily recognized by its small pale violet-blue to whitish flowers, the oval-shaped bases of which become swollen and bladder-like as the fruits form. It is a somewhat hairy and usually much-branched plant 1 to 3 feet high. The numerous leaves are lance-shaped, egg-shaped, or oval and are often broadest above the middle. They are stalkless, toothed on the margin, and from 1 to 2½ inches long. The flowers are numerous and in slender clusters at the tips of the branches; blooming between July and October. It grows in open woods, thickets, and fields.

RANGE: Lab. to Sask. south to Ga., Miss., Ark. and Kan.

GLANDULAR LOBELIA *(Lobelia glandulosa)*

Also known as the Swamp Lobelia, this plant grows in the wetter coastal plain pinelands and swamps in the Southeast. It is a smoothish plant with a slender stem 1 to 4 feet tall, which is chiefly leafy on the lower part. Its leaves are all very narrow, rather stiff and thickish, and are wavy-toothed with hard-tipped teeth on the margin. The lower ones are often 6 inches long but they are greatly reduced in size upward along the stem. The flowers are lavender with a white "eye", which is surrounded by an area of deeper violet-blue. They are about an inch long and the calyx teeth have prominent gland-tipped teeth on their margins. It blooms during September and October. RANGE: se. Va. south to Fla.

SPIKED LOBELIA *(Lobelia spicata)*

This lobelia usually has a simple stem 1 to 3 feet high which is densely downy toward the base but rather smooth above. The lower leaves are mostly top-shaped and 1 to 3½ inches long. Those upward along the stem are smaller and lance-shaped or narrower. Its pale bluish or whitish flowers are about ⅓ inch long and in a rather dense but long and narrow terminal cluster. It grows in open woods, thickets, fields and along roadsides; blooming between June and August. RANGE: N.B. to Minn. south to Ga., La. and Kan.

This is the largest of all plant families. The various members of the family have small flowers in a dense cluster or *head,* seated on the expanded end of the flower stalk which is called the *receptacle.* The group of flowers is surrounded by 1 or more series of *bracts,* collectively known as the *involucre.* In some composites, such as the asters, daisies, and sunflowers the heads contain two kinds of flowers. The ones in the central part, or *disk,* of the head are known as the *disk flowers.* These flowers have a radial symmetry. The calyx is united with the ovary of the pistil; the sepals often being represented by a tuft of hairs, bristles, or scales on its summit. Collectively they are called the *pappus.* The corolla is tubular, usually with 5 lobes at the summit and it is apparently situated on the top of the ovary. Within the corolla there are 5 stamens, their anthers united into a ring which surrounds the style of the pistil with the 2-branched stigma emerging from the top. The flowers around the border of the head, which look like petals, are the *ray flowers.* These flowers have a bilateral symmetry, the 5 petals all being united into a flattened or strap-shaped corolla.

The family was formerly called the Composites (Compositae). In some composites, such as the dandelions and the Chicory, all of the flowers in the heads are ray flowers. In others, such as the ironweeds and thistles, the heads are made up entirely of disk flowers. In many composites the disk flowers are mixed with what is called the *chaff.* It may be seen on the receptacle when some of the disk flowers are pulled off.

The fruits are dry, 1-celled, and 1-seeded bodies called *achenes* which are generally taken to be seeds. A sunflower "seed," for example, is such a fruit, or achene. In many composites the pappus remains on the mature fruits, as silky tufts of hairs, awns, etc.

KEY TO GENERA OF THE ASTER FAMILY

GROUP I. HEADS CONTAINING ONLY FLOWERS WITH STRAP-SHAPED COROLLAS (RAY FLOWERS).

1. Flowers blue or bluish.
    2. Flowers an inch or more across. — CHICORY *(Chicorium)*
    2. Flowers ½ inch or less across. — LETTUCES *(Lactuca)*
1. Flowers other than blue or bluish in color.
    3. Flowers pinkish to cream-colored or greenish-white; the heads borne in drooping clusters. — RATTLESNAKE-ROOTS *(Prenanthes)*
    3. Flowers yellow to orange-red.
4. Flower heads borne on leafless or nearly leafless stems, leaves mostly basal.
    5. Flower heads clustered at the summit of the stem. — HAWKWEEDS *(Hieracium)*
    5. Flower heads solitary at the summit of the stem or its few branches.
6. Involucre with a row of recurved bracts at the base. Pappus of fine white hairs on the stalk-like summit of the achene. — DANDELIONS *(Taraxacum)*
6. Involucre without recurved bracts at the base. Pappus of a few hairs and scales seated on summit of the achene. — DWARF DANDELIONS *(Krigia)*
    4. Flower heads borne on leafy stems.
7. Flower heads solitary on the stem branches; pappus of reddish-brown hairs. — FALSE-DANDELION *(Pyrrhopappus)*
7. Flower heads several to very numerous.
    8. Achenes with a beak or stalk with a pappus of fine hairs on the summit. — LETTUCES *(Lactuca)*
    8. Achenes with a tuft of bristly hairs seated directly on the summit. — HAWKWEEDS *(Hieracium)*

GROUP II. HEADS CONTAINING ONLY TUBULAR FLOWERS.

1. Plant climbing or twining. CLIMBING-HEMPWEED *(Mikania)*
1. Plants otherwise.
   2. Flowers white or whitish to cream-colored.
3. Flower heads in a cone-shaped end cluster. Stem white-woolly and with green wings running down from the leaf bases. BLACKROOT *(Pterocaulon)*
3. Heads not in cone-shaped end clusters. Stems not winged.
   4. Involucre with a single row of erect bracts with edges meeting, and sometimes with a row of minute ones at the base.
5. Involucre with 1 row of bracts. Flower heads nodding. INDIAN-PLANTAINS *(Cacalia)*
5. Involucre with a row of small bracts at the base. Flower heads held erect. FIREWEED *(Erechtites)*
   4. Involucre with 2 to several rows of overlapping bracts.
6. Involucral bracts wholly whitish and dry. Plants more or less woolly or cobwebby.
7. Leaves chiefly basal, or with a few small ones on the flower stalk. EVERLASTINGS *(Antennaria)*
7. Leaves scattered along the stems.
   8. Involucral bracts pearly-white and spreading. PEARLY EVERLASTING *(Anaphalis)*
   8. Involucral bracts tawny-white to brownish and close-fitting. CUDWEEDS *(Gnaphalium)*
6. Involucral bracts greenish or at least partly green.
   9. Leaves opposite. THOROUGHWORTS *(Eupatorium)*
   9. Leaves alternate. FALSE BONESET *(Brickellia)*
2. Flowers purple to lavender or pink.
   10. Leaves opposite or in whorls. JOE-PYE-WEEDS AND THOROUGHWORTS *(Eupatorium)*
   10. Leaves alternate.
11. Pappus of 5 or 6 nearly equal scales. BARBARA'S-BUTTONS *(Marshallia)*
11. Pappus of fine hairs or bristles.
   12. Heads in small groups surrounded by a few leaf-like bracts. ELEPHANT'S-FOOTS *(Elephantopus)*
   12. Heads otherwise.
13. Heads solitary at the tip of the stem or its branches. Leaves and often the involucral bracts prickly. THISTLES *(Carduus)*
13. Heads clustered. Plants not spiny or prickly.
   14. Involucral bracts nearly all of the same length. TRILISAS *(Carphephorus)*
   14. Involucral bracts in several rows, the outer ones being successively shorter.
15. Heads arranged in a long and narrow end cluster. Leaves usually very numerous and becoming smaller upward. BLAZING-STARS *(Liatris)*
15. Heads in a more or less flat-topped or open cluster.
   16. Plants with a rank camphor-like odor. MARSH-FLEABANES *(Pluchea)*
   16. Plants otherwise.
17. Leaves toothed or, if untoothed narrow and with the margin rolled inward on the lower side. Style branches long and slender. IRONWEEDS *(Vernonia)*
17. Leaves untoothed and narrow but flat. Style branches rather club-shaped. CARPHEPHORUS *(Carphephorus)*

GROUP III. HEADS WITH A CENTRAL DISK CONTAINING TUBULAR FLOWERS (DISK FLOWERS) AND MARGINAL FLOWERS WITH STRAP-SHAPED COROLLAS (RAY FLOWERS).

1. Stems with rather prominent wings extending downward from the leaf-bases.
    2. Receptacle without chaffy scales. SNEEZEWEEDS *(Helenium)*
    2. Receptacle with chaffy scales.
3. Involucral bracts few and nearly equal in length. Disk roundish, the flowers pointing in all directions. Rays yellow. WING-STEM *(Verbesina)*
3. Involucral bracts many and in 2 or more rows, the outer ones shorter. Rays yellow or white. CROWN-BEARDS *(Verbesina)*
    1. Stems not winged or but very slightly so.
4. Heads with yellow ray flowers.
    5. Leaves opposite or in whorls, or mostly opposite with some of the uppermost ones alternate.
6. Ray flowers sterile or without a pistil.
    7. Involucre with several rows of overlapping green bracts. Plants usually more or less roughish or hairy. SUNFLOWERS *(Helianthus)*
    7. Involucre with 2 rows of bracts; the outer ones green and spreading, the inner ones erect and usually not green.
8. Pappus of 2 or 4 barbed awns, the achenes maturing as "sticktights." Flowers sometimes without rays. BUR-MARIGOLDS *(Bidens)*
8. Pappus of 2 short teeth or none. TICKSEEDS *(Coreopsis)*
    6. Ray flowers fertile or with a pistil.
9. Disk flowers perfect but never producing achenes.
    10. Plant low, its stems more or less prostrate. GOLDEN-STAR *(Chrysogonum)*
    10. Plants 2 to several feet tall, the stems erect.
11. Achenes flattened and with a winged margin. Plants with a resinous juice. ROSINWEEDS *(Silphium)*
11. Achenes thick and scarcely flattened. Plants usually sticky-hairy. LEAF-CUPS *(Polymnia)*
    9. Disk flowers perfect and forming achenes.
12. Involucre with 4 large leaf-like outer bracts united to form a 4-angled cup. PINELAND-GINSENG *(Tetragonotheca)*
12. Involucre otherwise.
    13. Ray flowers persisting on the achenes and becoming papery. Herbaceous upland plant. OX-EYE *(Heliopsis)*
    13. Ray flowers not persisting. Shrubby plant of coastal salt and brackish marshes. SEA OX-EYE *(Borrichia)*
5. Leaves alternate.
    14. Pappus of hair-like bristles.

15. Heads 2 to 4 inches broad, the ray flowers very numerous and narrow. ELECAMPANE (*Inula*)
15. Heads smaller, mostly 1 inch or less broad.
    16. Involucral bracts apparently in 1 row. RAGWORTS *(Senecio)*
    16. Involucral bracts in several rows, the outer successively shorter.
17. Flower heads mostly less than ½ inch across, with short rays. GOLDENRODS *(Solidago)*
17. Flower heads ½ inch or more across, with rather long and showy rays.
    18. Plant with a rank camphor-like odor. CAMPHORWEED *(Heterotheca)*
    18. Plants otherwise. GOLDEN-ASTERS *(Heterotheca)*
14. Pappus, if any, of scales.
    19. Rays with a pistil. SNEEZEWEEDS *(Helenium)*
    19. Rays neutral or without a pistil.
20. Receptacle but slightly dome-shaped, deeply honeycombed. ENDORIMA *(Baludina)*
20. Receptacle conical to cylindrical, merely chaffy.
    21. Disk roundish or conical. CONEFLOWERS *(Rudbeckia)*
    21. Disk oblong or cylindrical. CONEFLOWERS *(Ratibida)*
4. Heads with white to bluish, violet, or pink rays.
    22. Heads solitary on leafless stalks. Leaves basal. SUN-BONNETS *(Chaptalia)*
    22. Heads clustered. Plants more or less leafy-stemmed or with some small leaves on the flower stalks.
23. Pappus of hair-like bristles.
    24. Involucral bracts in 1 or 2 rows and all of about the same length. FLEABANES *(Erigeron)*
    24. Involucral bracts in several rows, the outer successively shorter.
25. Rays short. GOLDENRODS *(Solidago)*
25. Rays rather long and showy.
    26. Rays about 5. Involucral bracts firm and with green tips. WHITE-TOPPED ASTER *(Aster)*
    26. Rays more numerous. Involucral bracts thin and usually green. ASTERS *(Aster)*
23. Pappus, if any, not of bristles.
    27. Heads 1½ inches or more broad, with 12 or more ray flowers.
28. Heads with a rather flat yellow disk and 20 to 30 bright white rays. OX-EYE DAISY *(Chrysanthemum)*
28. Heads with a cone-shapped dark disk and 12 to 20 purplish rays. PURPLE CONEFLOWER *(Echinacea)*
    27. Heads 1 inch or less across, with 4 to 8 white or pinkish rays.
29. Plant with finely dissected leaves and a strong odor. Heads ¼ inch or less across, with 4 to 6 small rays. YARROWS *(Achillea)*
29. Plant with narrow but not dissected leaves. Heads ½ to 1 inch across, with 4 to 8 rather long and narrow rays. TICKSEEDS *(Coreopsis)*

NEW YORK IRONWEED *(Vernonia noveboracensis)*

The New York Ironweed has a smoothish stem 3 to 6 feet tall; and scattered, lance-shaped, rather finely toothed leaves from 3 to 10 inches long. The numerous heads of deep purple, tubular flowers are in a big flat-topped cluster. Each head contains from 30 to 50 flowers which have a ring of purple or purplish-tinged bristles at the base of the corolla tube. The bracts of the involucre are also purplish and their tips are prolonged into slender and spreading "tails". It grows in moist thickets and fields, and along the banks of streams; blooming between July and September.

RANGE: Mass. to W.Va. and Ohio south to Ga. and Miss.

BROAD-LEAF IRONWEED *(Vernonia glauca)*

Like the preceding species, this one has slender and loose, tail-like tips on its involucral bracts; but the bristles accompanying its flowers are cream- to straw-colored. The stems are usually whitened with a bloom, and its broader leaves are more sharply toothed. It grows in rich woodlands from N.J. and Pa. south to Ga. and Ala.; blooming between late June and September. (Not illustrated)

TALL IRONWEED *(Vernonia gigantea)*

The Tall Ironweed differs from both of the preceding species in having involucral bracts which are merely pointed instead of tapering and loose at the tips. Its heads have but 15 to 30 flowers, and the leaves are broader than those of the New York Ironweed as well as downy beneath. The smoothish stems are from 3 to about 8 feet tall. It grows in rich moist thickets and stream bottoms, blooming between August and October.

RANGE: N.Y. to Ill. and Mo. south to Fla. and La.

MISSOURI IRONWEED *(Vernonia missurica)*

This species is similar to the Tall Ironweed but its stem and lower leaf surfaces are densely woolly-hairy. The heads contain from 35 to 55 flowers, and the purplish bracts of the involucre are cob-webby. It is usually only 3 to 5 feet tall and blooms between July and September. Rich low grounds and prairies from s. Ont. and Ohio to Iowa south to Ala., Miss., Ark., Okla. and Tex. (Not illustrated)

NARROW-LEAF IRONWEED *(Vernonia angustifolia)*

This little ironweed has a slender stem 1 to 3 feet tall with numerous and very narrow leaves 1½ to 5 inches long. The leaves are roughish above and the margins are rolled inward on the lower side. The heads of dark purple flowers have an involucre of small bracts with spreading green tips. It grows in sandy woods and pinelands in the coastal plain, blooming between late June and September.

RANGE: se. N.C. south to Fla. and west to Miss.

SMOOTH ELEPHANT'S-FOOT *(Elephantopus nudatus)*

This plant has a basal rosette of top-shaped leaves 2 to 10 inches long, which usually lie flat on the ground. It has a smooth or slightly hairy stem 1 to 2 feet tall which forks above but is leafless or nearly so. The branches end in clustered heads of tubular purple flowers which are surrounded by large leaf-like bracts. The tubular flowers are more deeply cleft on one side. It grows in sandy woods and openings; blooming between late July and September.

RANGE: Del. and Ark. south to Fla. and La.

WOOLLY ELEPHANT'S-FOOT *(Elephantopus tomentosus)*

This species very closely resembles the preceding one but it is densely hairy, its leaves being velvety to the touch. It grows in dry woodlands; blooming between late July and September. Md. to Ky. south to Fla. and Tex. (Not illustrated)

CAROLINA ELEPHANT'S-FOOT *(Elephantopus carolinianus)*

Unlike either of the preceding species, this one has leaves scattered along the stem rather than in a basal rosette. It has a somewhat roughish-hairy stem, sometimes nearly smooth above, which is 1 to 3 feet high. As in the other species, the heads of tubular flowers are in clusters surrounded by leaf-like bracts. It grows in dry open woods and thickets; blooming between August and October.

RANGE: N.J. to Ill. and Kan. south to Fla. and Tex.

New York Ironweed
Tall Ironweed
Narrow-leaf Ironweed
Smooth Elephant's-foot
Carolina Elephant's-foot

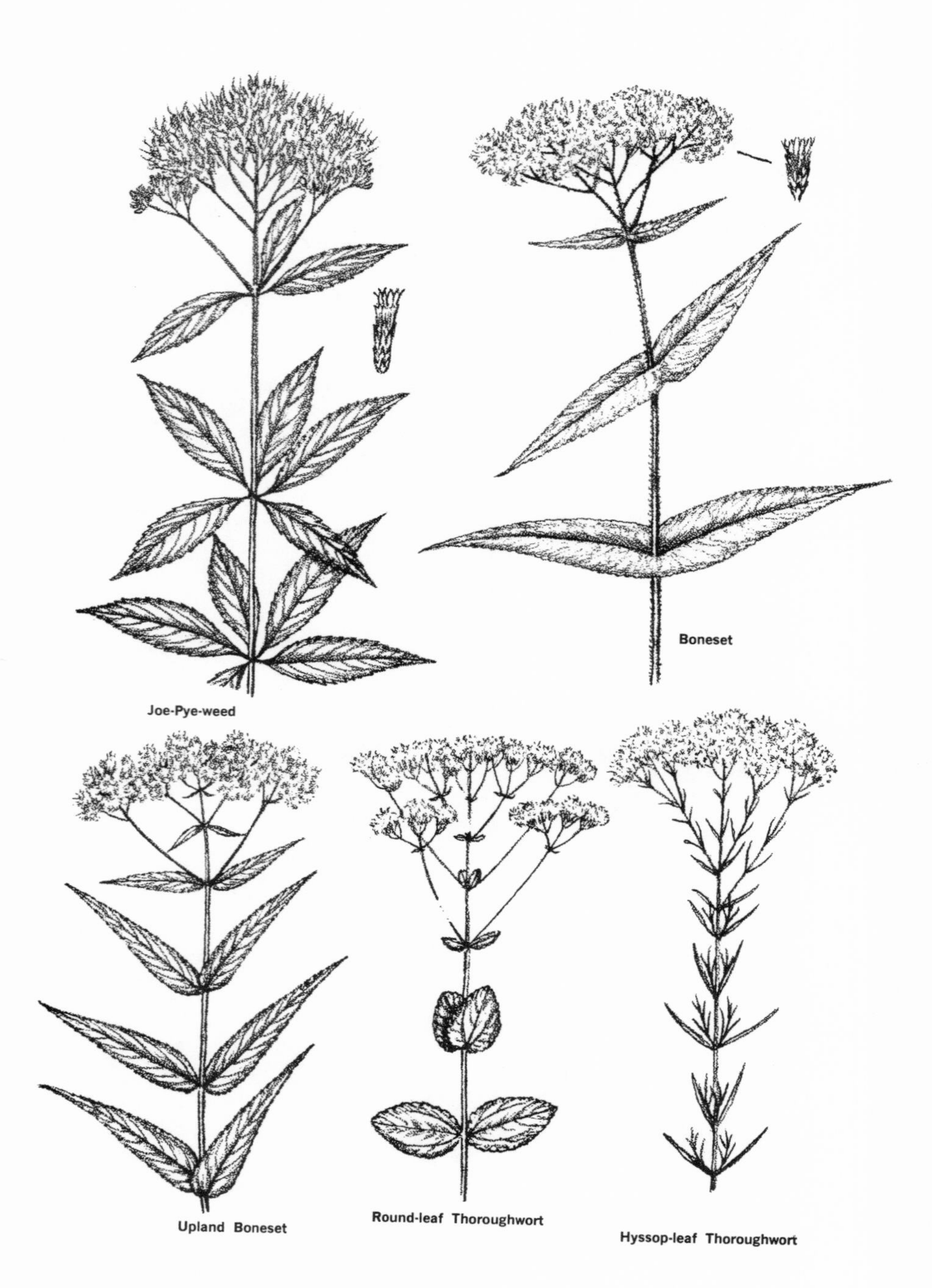
Boneset
Joe-Pye-weed
Upland Boneset
Round-leaf Thoroughwort
Hyssop-leaf Thoroughwort

JOE-PYE-WEED *(Eupatorium fistulosum)*

This species of Joe-Pye-Weed has a rather slender, smoothish, hollow, green to purplish-tinged and often somewhat whitened stem 3 to 10 feet tall. The lance-shaped leaves are in whorls of 4 to 7. They are bluntly toothed and 4 to 12 inches long. There are 5 to 8 tubular, purplish- to lilac-pink flowers in each head; and the heads are arranged in round-topped or dome-like clusters; blooming between July and September. It grows in moist meadows and thickets.

RANGE: Me. to Que., Ill., Iowa and Okla. south to Fla. and Tex.

JOE-PYE-WEED *(Eupatorium dubium)*

In general this species is similar to the preceding one; but the summit of the stem and the lower surfaces of the leaves are sprinkled with lustrous, sticky atoms. The leaves are usually in whorls of 3 or 4, often have 3 main veins, and are coarsely toothed. It grows in sandy swamps and thickets, chiefly in the coastal plain, from N.S. and sw. Me. to S.C. and Ala.; blooming between July and October.

GREEN-STEMMED JOE-PYE-WEED *(Eupatorium purpureum)*

This species has solid stems marked with purplish where the whorls of usually 3 or 4 leaves are attached. A vanilla-like odor is emitted when the plant is bruised. The heads have 3 to 7 creamy white to pale lilac flowers, and they are arranged in dome-like clusters; blooming between July and October. N.H. to Minn. and Neb. south to Fla., Tenn. and Okla. (Not illustrated)

SPOTTED JOE-PYE WEED *(Eupatorium maculatum)*

This species has a deep purple or purple-spotted stem often sticky-hairy above; with usually 4 or 5 sharply toothed leaves in each whorl. The heads have from 8 to 20 purplish flowers and they are arranged in a rather flat-topped cluster. It grows in moist places from Nfd. to B.C. south to nw. S.C., Ind., Neb., N.Mex. and Wash. (Not illustrated)

BONESET *(Eupatorium perfoliatum)*

The Boneset or Thoroughwort is a hairy plant with a stem 2 to 4 feet tall, usually branching above. The pairs of wrinkled-looking, lance-shaped, finely toothed leaves are 3 to 6 inches long; and are joined together at the base so that the stem appears to pass through them. The heads contain 10 to 20 small, white, tubular flowers; and they are arranged in flat-topped clusters. It grows in low moist thickets, open woods, and meadows; blooming between August and October. The plant was a favorite home remedy for colds and other afflictions. RANGE: Que. to Man. south to Fla. and Tex.

UPLAND BONESET *(Eupatorium sessilifolium)*

This is a rather smooth plant 2 to nearly 5 feet tall, usually branching above. It has pairs of lance-shaped, sharply toothed leaves 3 to 6 inches long, which are stalkless but not united at the base. The heads have about 5 tubular white flowers and are arranged in a flat-topped cluster. It grows in rather dry woods, thickets, and on hillsides; blooming between July and October. RANGE: Mass. to Minn. south to Ga., Ala. and Mo.

ROUND-LEAF THOROUGHWORT *(Eupatorium rotundifolium)*

The stem of this plant is hairy, rather slender, 1 to 3 feet tall, and usually branched above. The paired leaves are egg-shaped or roundish, stalkless, rather coarsely toothed, and from 1 to 2 inches long. They show 3 prominent veins from the base, are roughish above, and very veiny and downy beneath. The heads contain about 5 tubular white flowers and they are arranged in rather flat-topped clusters. It grows in dry to wet sandy or peaty soils in open woods, clearings, or savannahs; blooming between August and October. RANGE: se. N.Y., Tenn. and Ark. south to Fla. and Tex.

HYSSOP-LEAF THOROUGHWORT *(Eupatorium hyssopifolium)*

This is a more or less minutely rough-hairy plant with a slender stem 1 to 2 feet tall, which is rather bushy-branched above. It has narrow leaves ½ to 2 inches long which frequently have clusters of smaller leaves in their axils. The leaf margins are toothless or nearly so but often rolled inward on the lower side. The heads have about 5 tubular white flowers and they are arranged in a rather dense, flat-topped cluster. It grows in dry open woods, clearings, and old fields; blooming between July and October.

RANGE: Mass. to Ohio south to Fla. and Tex.

WHITE THOROUGHWORT *(Eupatorium album)*

The White Thoroughwort has a hairy and often roughish stem 1 to 3 feet high, which branches above. Its leaves are broadly lance-shaped to oblong, often broader above the middle, toothed on the margin, stalkless or nearly so, 1 to 4 inches long, and very veiny in appearance. The small tubular white flowers number 5 to 7 in each head, and the heads are arranged in flat-topped clusters which have opposite branches. It grows in dry and usually sandy open woods and clearings; blooming between late June and September.

RANGE: se. N.Y. to Md., Ky. and Ark. south to Fla. and La.

PINK THOROUGHWORT *(Eupatorium incarnatum)*

This plant has a downy, loosely branched, and often reclining or straggling stem 2 to 4 feet long. Its leaves are rather triangular in outline, long-pointed at the tip, broadly pointed to somewhat heart-shaped at the base, ¾ to 2¼ inches long, and have slender stalks. The pinkish to lavender tubular flowers number about 20 in each head; and the heads are in rather loose and open clusters, mostly on stalks arising from the axils of the leaves. It grows in open sandy woods and in swampy places; blooming between September and November.

RANGE: Va. to Ohio, Ill. and Mo. south to Fla. and Ariz.

MISTFLOWER *(Eupatorium coelestinum)*

The Mistflower is also known as the Blue Boneset, and it is often cultivated under the name Hardy Ageratum. It is a somewhat downy plant with a branching stem from 1 to 3 feet high; and pairs of rather triangular, stalked, bluntly toothed leaves 1½ to 3 inches long. The bell-shaped heads contain a number of bright violet-blue, tubular flowers; and they are arranged in quite dense, flat-topped clusters. Its heads differ from all other species of *Eupatorium* in that they have a cone-shaped rather than a flat receptacle, and some botanists place it in a different genus. In general appearance and color the flowers resemble the cultivated ageratums. It grows in moist woods, thickets, and along the banks of streams; blooming between July and October.

RANGE: N.J. to Ill. and Kan. south to Fla. and Tex.

DOG-FENNEL *(Eupatorium capillifolium)*

This is a bushy plant 3 to almost 10 feet tall; with crowded, mostly alternate, finely dissected and feathery leaves. The small 3- to 5-flowered heads are very numerous and arranged in large, leafy, pyramid-shaped end clusters; blooming between September and November. It is very common and conspicuous in old fields, along roadsides, and in the borders of woods and clearings, especially in the South.

RANGE: Mass. to N.J., Va .and Tenn. south to Fla. and Tex. (Not illustrated)

WHITE SNAKEROOT *(Ageratina altissima)*

The White Snakeroot usually has a smoothish and much-branched stem 1 to 4 feet high. The pairs of egg-shaped leaves are 3 to 6 inches long, sharply toothed on the margin, heart-shaped at the base, and have taper-pointed tips. They are rather thin in texture and have slender stalks usually 1 to 2½ inches long. The heads contain from 15 to 30 bright white, tubular flowers and they are arranged in rather dense flat-topped clusters; blooming between July and October. It grows in rich woods, thickets, and clearings. This plant is the principal cause of "milk sickness", which is transmitted to humans through the milk of cattle which have eaten it.

RANGE: N.B. to Sask. south to Va., n. Ga. and Tex.

SMALLER WHITE SNAKEROOT *(Ageratina aromatica)*

In a general way this plant resembles the preceding species. It is a somewhat more slender plant 1 to 2 feet tall, often with long ascending branches toward the summit. Its leaves are rather thickish, 1½ to 3 inches long, and they have stalks less than an inch in length. The tubular flowers are bright white but the flower heads are smaller. It grows in dry woods, thickets, pinelands, and clearings; blooming between late August and October.

RANGE: Mass. to N.J., W.Va. and Ohio south to Fla. and La.

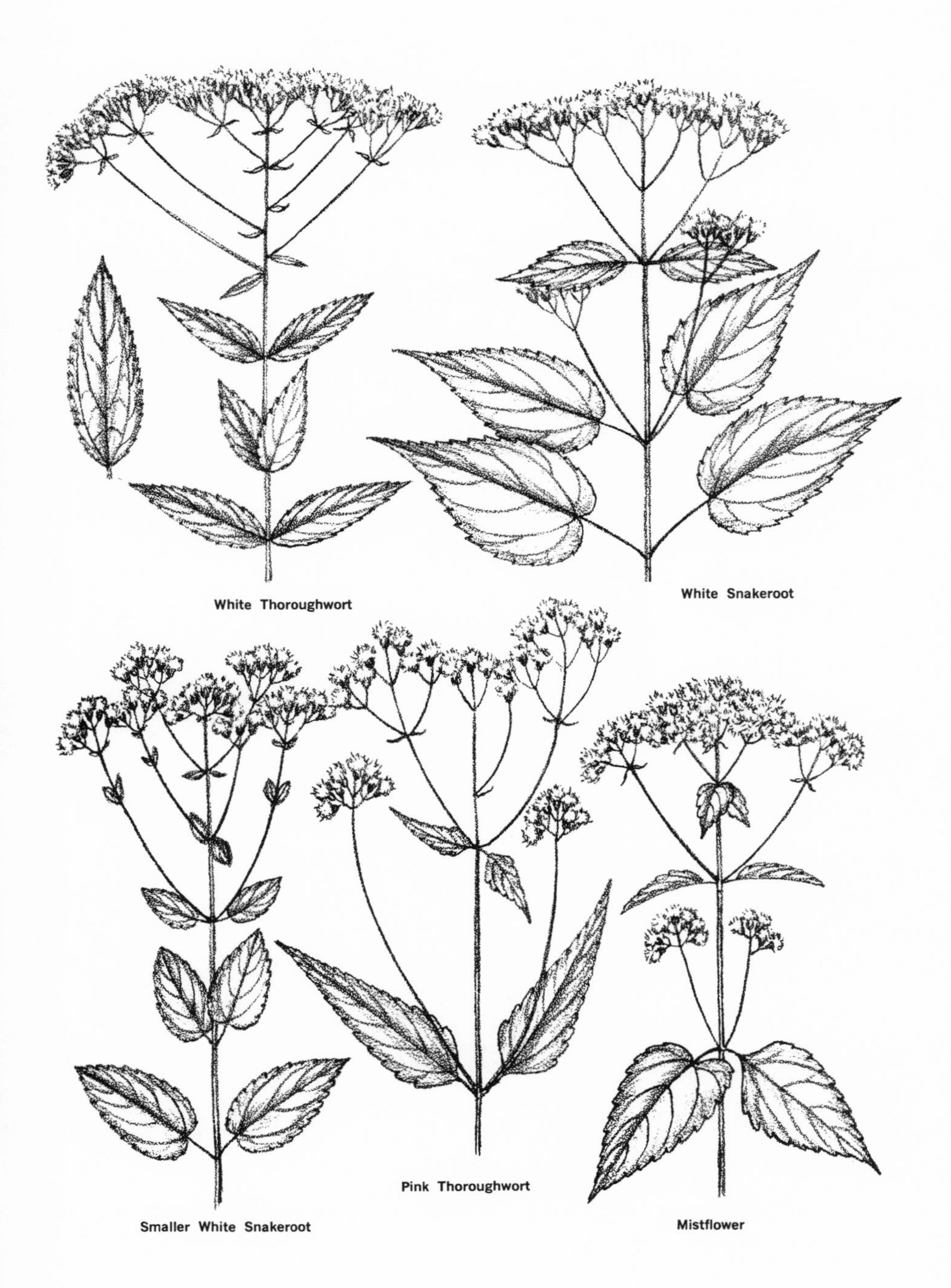
White Thoroughwort
White Snakeroot
Smaller White Snakeroot
Pink Thoroughwort
Mistflower

Climbing Hempweed
False Boneset
Hairy Trilisa
Vanilla-leaf
Hairy Carphephorus

CLIMBING HEMPWEED *(Mikania scandens)*

This plant is quite unusual among our composites as it is a climbing and twining vine, with smoothish stems from 5 to 15 feet long. Its leaves are somewhat triangular, heart-shaped at the base, taper-pointed at the tip, sometimes toothed, 2 to 4 inches long, and have long stalks. The whitish to pale purplish, tubular flowers number about 4 in each head; and they are clustered at the ends of long stalks arising from the leaf axils. It grows in swamps, moist thickets, and along the banks of streams; blooming between July and October.

RANGE: Me. to N.Y. and Ont. south to Fla. and Tex.

FALSE BONESET *(Brickellia eupatorioides)*

The False Boneset is a resin-dotted and more or less hairy plant from 1 to 3 feet high, which is usually branched above. It has scattered, narrow or lance-shaped leaves which are untoothed or sparingly toothed and usually 1½ to 4 inches long. The heads contain from 10 to 25 tubular, creamy-white to purplish flowers. They are about ⅓ inch long and arranged in a rather loose terminal cluster. It grows in dry open woods, thickets, clearings, and on rocky slopes; blooming between June and October.

RANGE: N.J. to Ill. and Mo. south to Fla. and Tex.

HAIRY TRILISA *(Carphephorus paniculatus)*

This is a common and conspicuous plant in the low, wet cosatal plain pinelands which blooms from August to October. It has a stiffly erect, sticky-hairy stem from 1 to 2 feet tall along which are numerous, scattered, small, lance-shaped leaves which tend to stand erect. The basal leaves are either lance-shaped or narrowly oblong and from 3 to 10 inches in length. Its numerous ¼ inch heads of tubular, rose purple flowers are arranged in a narrowly cylindrical cluster along the upper portion of the stem.

RANGE: N.C. south to Fla. and west to La.

VANILLA-LEAF *(Carphephorus odoratissimus)*

Unlike the preceding species, this is a smooth plant with an erect stem 2 to 3 feet tall, along which are scattered small leaves. The basal ones are smooth, thickish, usually broader above the middle but rather narrow, and from 4 to 10 inches long. When bruised they emit a characteristic vanilla-like odor. The numerous heads of small, tubular, rose-purple flowers are in a rather flat-topped terminal cluster. It grows in dry to wet coastal plain pinelands; blooming from August to October. Also known as Deer's-tongue, Hound's-tongue, and Carolina-vanilla; large quantities of its leaves are used in the flavoring of tobacco.

RANGE: N.C. south to Fla. and west to La.

HAIRY CARPHEPHORUS *(Carphephorus tomentosus)*

The Hairy Carphephorus has a simple, erect stem 1 to 2¼ feet tall which is quite hairy or woolly. On it are scattered, small, erect, lance-shaped leaves. The basal leaves are narrow and broadest toward the tip, and from 2 to about 6 inches long. Toward the summit of the stem it has several long-stalked heads, each one about ⅓ inch long and containing a number of rose-purple flowers. The heads are disposed in rather open but flat-topped terminal clusters. It grows in dry to moist coastal plain pinelands; blooming between August and October.

RANGE: se. Va. south to Fla.

CORYMBED CARPHEPHORUS *(Carphephorus corymbosus)*

This species also has rather downy stems 1 to 3 feet tall, but they are conspicuously leafy up to the flat-topped flower cluster which is rather dense. The bracts of the flower heads have blunt or roundish tips and thin, pale, dry margins. It grows in coastal plain pinelands from Ga. south to Fla. (Not illustrated)

Carphephorus bellidifolius often has clustered stems 1 to 2 feet tall which are smooth or nearly so, with scattered and spreading leaves. The ones on the lower part of the stem and at its base are narrowly top-shaped and 2 to about 6 inches long. The numerous flower heads have round-tipped bracts and are arranged in an open, slenderly branched cluster. It grows in dry coastal plain pinelands from se. Va. south to Fla.; blooming August to October. (Not illustrated)

Blazing-Stars

The Blazing-stars are also known as the Button-snakeroots and Gayfeathers. They are plants with usually simple, erect stems crowded with numerous and narrow leaves; and long and narrow clusters of stalkless or short-stalked heads of rose-purple, tubular flowers. The Blazing-stars are among the most gorgeous flowers of late summer and fall, blooming between July or August and September or October. In the coastal plain of the southeast, they occur in great abundance but some species are common in the mountain region. There are a number of species besides the ones included here, and they are often difficult to identify as to species.

LARGE BUTTON-SNAKEROOT *(Liatris scariosa)*

This beautiful blazing-star has flower heads which are bowl-shaped, up to an inch across, and contain from 25 to 60 flowers. Its stem is from 1 to 5 feet tall, finely downy at least above, and the leaves are hairy and often roughish. The basal ones vary from lance-shaped to slenderly top-shaped, up to a foot long, and ½ to 2 inches broad. It grows in dry woods and clearings chiefly in the mounain region.
RANGE: Pa. and W.Va. south to Ga. and Miss.

NORTHERN BUTTON-SNAKEROOT *(Liatris scariosa* var. *novae-angliae)*

This species resembles the preceding one. It has similar large heads but the involucral bracts are dark reddish, with narrow and almost petal-like margins. The leaves are smoother and the lance-shaped basal ones are seldom over an inch broad. It grows in dry woods, thickets, or clearings from sw. Me. to N.Y. south to N.J. and cent. Pa. (Not illustrated)

DENSE BUTTON-SNAKEROOT *(Liatris spicata)*

This species has smoothish stems 1 to 6 feet tall with crowded, very narrow, smooth leaves gradually becoming larger toward the base. The lower ones are still narrow, ¾ inch or less across, but often 5 to 15 inches long. The upper portion of the stem has crowded heads of flowers which bloom progressively downward from the tip. Each head is usually about ⅓ inch across, containing between 5 and 15 flowers. It grows in wet open woods and fields. RANGE: Del. to W.Va., Ohio and Neb. south to Fla. and Tex.

SCALY BLAZING-STAR *(Liatris squarrosa)*

The large heads of this species have spreading or recurving, thickish, and stiff bracts; and they usually contain from 25 to 40 rather large flowers. The plant has rather stout stems 1 to 3 feet high; and numerous narrow, rigid, resin-dotted leaves. It grows in dry open woods and fields.
RANGE: Del. to W.Va., Ohio. and Neb. south to Fla. and Tex.

SANDHILL BLAZING-STAR *(Liatris secunda)*

This species and the similar *Liatris pauciflora* are unique blazing-stars of dry coastal plain pinelands and sand-hills in the Southeast. Both have narrowly cylindrical heads about ¾ inch long which are all turned to the upper side of the arching stems; the heads having relatively few but large flowers. The plants are 1 to 2½ feet high and have narrow leaves. *Liatris secunda* has downy stems while those of *Liatris pauciflora* are smooth. The former ranges from N.C. south to Fla. and west to Ala.; the latter, from Ga. south to Fla.

HANDSOME BLAZING-STAR *(Liatris elegans)*

The flower heads of this species have enlarged, petal-like, rose-colored or sometimes white tips. Another distinctive feature is the bending downward of the leaves along the upper part of the stem. It grows in dry sandy woods and pinelands in the coastal plain.
RANGE: S.C. south to Fla. and west to Tex.

GRASS-LEAF BUTTON-SNAKEROOT *(Liatris graminifolia)*

This species usually has somewhat hairy stems 1 to 3 feet tall, with numerous and very narrow leaves becoming smaller upward. The lower leaves often have hairy-fringed margins or leaf stalks, the widest being less than ½ inch broad. The narrowly top-shaped heads contain 5 to 15 flowers and are usually arranged in a long and slender, but sometimes branched cluster. It grows in open woods and pinelands.
RANGE: N.J. south to Fla. and Ala.

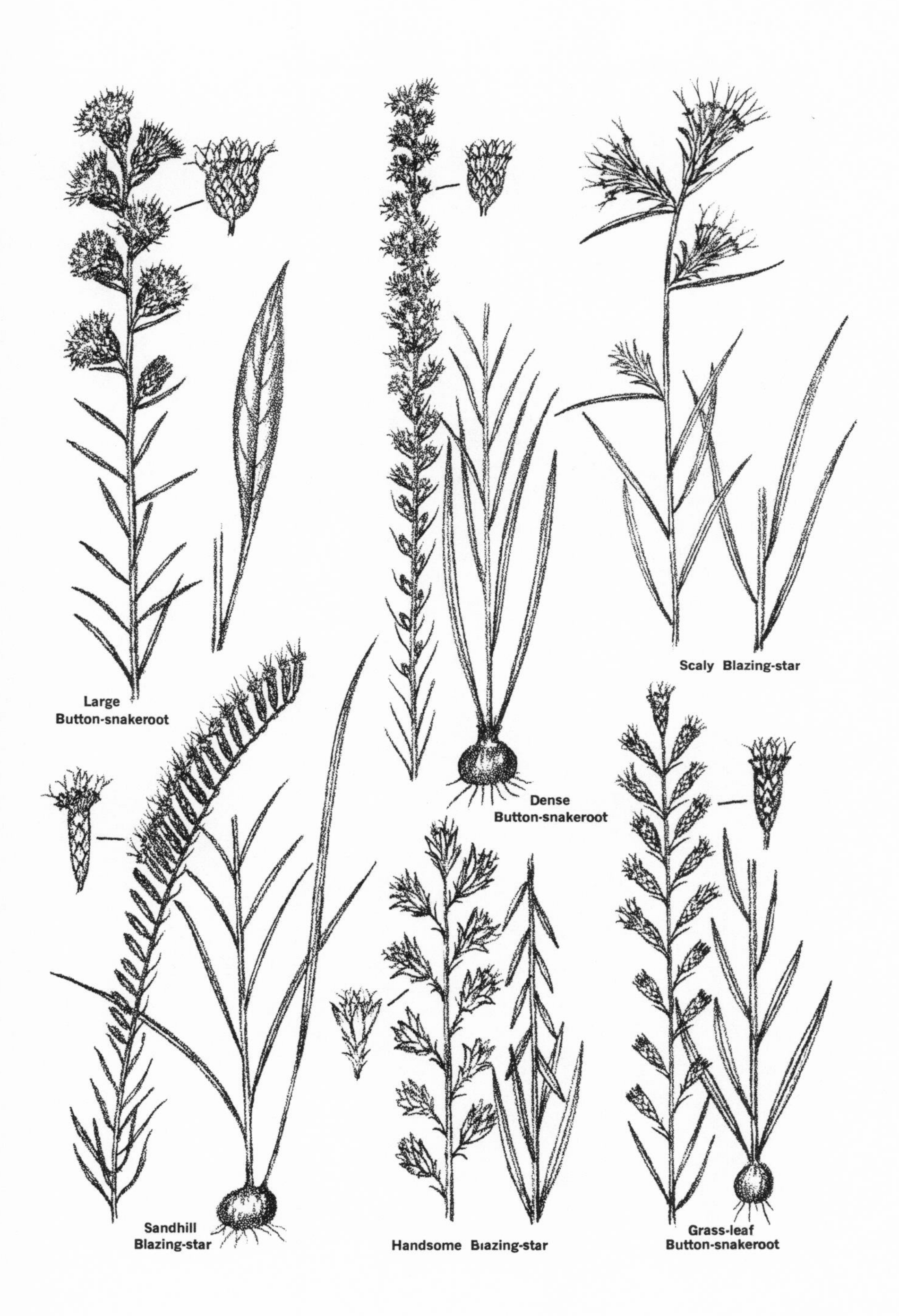
Large
Button-snakeroot
Scaly Blazing-star
Dense
Button-snakeroot
Sandhill
Blazing-star
Handsome Blazing-star
Grass-leaf
Button-snakeroot

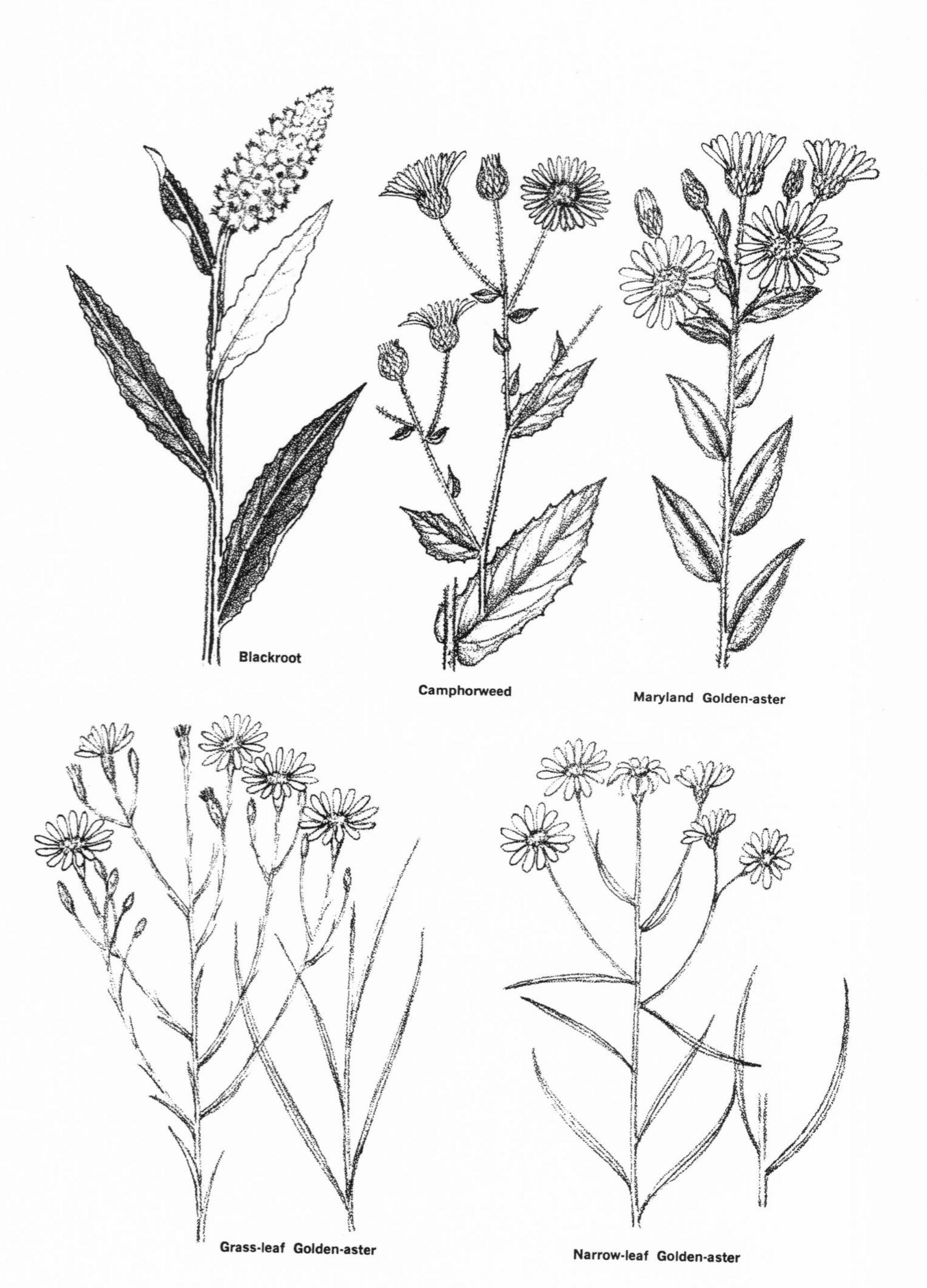
Blackroot
Camphorweed
Maryland Golden-aster
Grass-leaf Golden-aster
Narrow-leaf Golden-aster

BLACKROOT *(Pterocaulon virgatum)*

The Blackroot is one of the unique and striking plants found in the more moist parts of the coastal plain pinelands. Usually 1 to 2 feet tall, it has a simple stem with scattered, wavy-margined, narrowly elliptic or lance-shaped leaves from 2 to 4 inches long. It is terminated in a dense but narrowly cone-shaped cluster of creamy-white heads of tubular flowers. The undersides of the leaves, the stem, and the flower cluster are densely coated with a felted whitish wool; the upper surfaces of the leaves, and the narrow wings which run down the stem from their bases, being a bright green. It gets its name from the thick black root. The flowering season is in May and June.
RANGE: N.C. south to Fla. and west to Miss.

CAMPHORWEED *(Heterotheca subaxillaris)*

The original home of this plant was in the West; but in recent years it has become a common weed in fields and along roadsides, especially in the Southeast. It has a branching, hairy stem from 1 to 3 feet high; with scattered egg-shaped to oblong, sharply-toothed leaves that are mostly 2 to 3 inches long. Those along the stem and its branches have heart-shaped and clasping bases. The basal ones have stalks but they rarely persist until flowering time which is between July and October. The crushed foliage has a camphor-like odor. RANGE: Del., N.J. to Ill. and Kan. south to Fla. and Ariz.

MARYLAND GOLDEN-ASTER *(Chrysopsis mariana)*

This is a stout-stemmed, loosely hairy and often cobwebby plant from 1 to 2½ feet high; with scattered, oblong to lance-shaped, mostly bright green and stalkless leaves 1 to 2 inches long. The lower leaves are larger, broadest above the middle, and taper at the base into stalks. It has bright yellow, aster-like heads of flowers nearly an inch across; the bowl-shaped involucre having numerous but minute glands. It grows in open woods and fields, blooming between August and October.
RANGE: N.Y. to Ohio south to Fla. and Tex.

COTTONY GOLDEN-ASTER *(Chrysopsis gossypina)*

This is one of the most attractive species of golden-asters; and, in a general way, it resembles the preceding species. Its stems, leaves, and even the involucres of the flower heads are permanently coated with a whitish and rather cobwebby wool. The flower heads are an inch or a little more across and have a number of bright golden-yellow rays. It grows in the dry coastal plain pinelands and on sandhills from N.C. south to Fla. and Ala., blooming during September or October. (Not illustrated)

GRASS-LEAF GOLDEN-ASTER *(Pityopsis graminifolia)*

Also known as Silk-grass and Silver-grass, this is a silvery-silky plant with a slender stem 1 to 3 feet tall which branches above and has numerous, narrow, and erect leaves. Its basal leaves are narrow, soft, 4 to 12 inches long and grass-like. The numerous bright-yellow, aster-like flower heads are about ½ inch across and are on the ascending branches. They have a narrowly top-shaped involucre which is cobwebby, at least toward the base. It grows in dry, sandy or rocky, open woods and in fields; blooming between July and October. RANGE: Del. and W.Va., Ky., Ark. and Okla. south to Fla. and Tex.

NARROW-LEAF GOLDEN-ASTER *(Pityopsis falcata)*

This golden-aster usually has several stiff, cottony stems 4 to 12 inches high which branch above. Its numerous scattered leaves are narrow, stiff, spreading, often curved, and from 1 to 4 inches long. The relatively few aster-like flower heads are bright yellow, about ⅓ inch across, and tend to form a flat-topped cluster. It grows in dry sandy soils in the coastal region, blooming between July and October. RANGE: Mass. south to N.J.

Goldenrods

Anybody can tell a goldenrod from other kinds of flowers, but to distinguish the various species is quite another matter. That is a task which quite often perplexes even the trained botanist, for there are numerous similar species and they often hybridize. Only a few of the better marked ones are presented here. The heads contain both tubular (or disk) and strap-shaped (or ray) flowers.

BLUE-STEMMED GOLDENROD *(Solidago caesia)*

The Blue-stemmed or Wreath Goldenrod has slender, smooth, roundish stems 1 to 3 feet high, which are commonly purplish and usually coated with a whitish bloom. It has narrowly oblong or lance-shaped leaves which are stalkless, narrowed at both ends, sharply toothed, and 2 to 5 inches long. Its heads of flowers are in the axils of the leaves, blooming during September and October. It grows chiefly on wooded slopes or banks.
RANGE: N.S. to Ont. and Wis. south to Fla. and Tex.

BROAD-LEAF GOLDENROD *(Solidago flexicaulis)*

This is a well-marked goldenrod with a slender, zig-zag, somewhat angled green stem 1 to 3 feet high. The broad, egg-shaped, sharply toothed leaves are 2 to 7 inches long and are pointed at the tip and abruptly narrowed into winged stalks. Its heads of flowers are also clustered in the axils of the leaves, blooming between late August and October. It grows in rich woods and in ravines. Some know it as the Zig-zag Goldenrod.
RANGE: N.S. to N.D. south to Ga., Tenn. and Kan.

WHITE GOLDENROD *(Solidago bicolor)*

Silver-rod is another name often given to this pale-flowered goldenrod, the short outer rays of the heads being white or cream-colored. It is a grayish-downy plant up to 2 feet tall. The basal leaves are stalked, top-shaped, shallowly toothed, and 2 to 4 inches long. Those of the stem gradually decrease in size upward and are stalkless. It grows in dry open woods, thickets, and on slopes; blooming during September and October.
RANGE: N.B. to Ont. south to Ga. and Ark.

SLENDER GOLDENROD *(Solidago speciosa* var. *erecta)*

This species has a slender stem 2 or 3 feet tall which may be minutely hairy above. The lower leaves are usually broadest above the middle, shallowly toothed, and taper into winged stalks. The middle and upper ones are narrow, smaller, and stalkless. Its flower heads are arranged in a long narrow, end cluster; blooming between August and October. It may be found in dry to moist woods and thickets.
RANGE: N.J. to Ind. south to Ga. and Miss.

MOUNTAIN GOLDENROD *(Solidago roanensis)*

As its name indicates, this is a common goldenrod in the mountain woods and thickets. It has smoothish stems 1 to 3 feet tall. Its leaves are rather thin, 1 to 6 inches long, the lower ones being narrowly egg-shaped, sparingly toothed, and tapering into stalks; the upper ones lance-shaped, less toothed, and stalkless. The flower heads are in a slender but sometimes branched terminal cluster; blooming from August to October.
RANGE: Md. to Ky. south to n. Ga. and n. Ala.

CLUSTER GOLDENROD *(Solidago glomerata)*

This is the goldenrod whose large basal leaves are so conspicuous at high altitudes in the mts. of N.C. and Tenn. They are rather thin, long-stalked and sharply toothed. The flower heads are large, about ½ inch high, and clustered toward the summit of the 2- to 4-foot tall stems. Blooming August to October. (Not illustrated)

SEASIDE GOLDENROD *(Solidago sempervirens)*

True to its name, this goldenrod is found in coastal marshes and on the sea beaches. It has a smoothish, stout stem 2 to 8 feet tall; with numerous thickish, rather fleshy, smooth and untoothed leaves, the lowermost ones sometimes a foot long. The large flower heads are in a branched terminal cluster, blooming between August and November.
RANGE: Nfd. south to Fla.

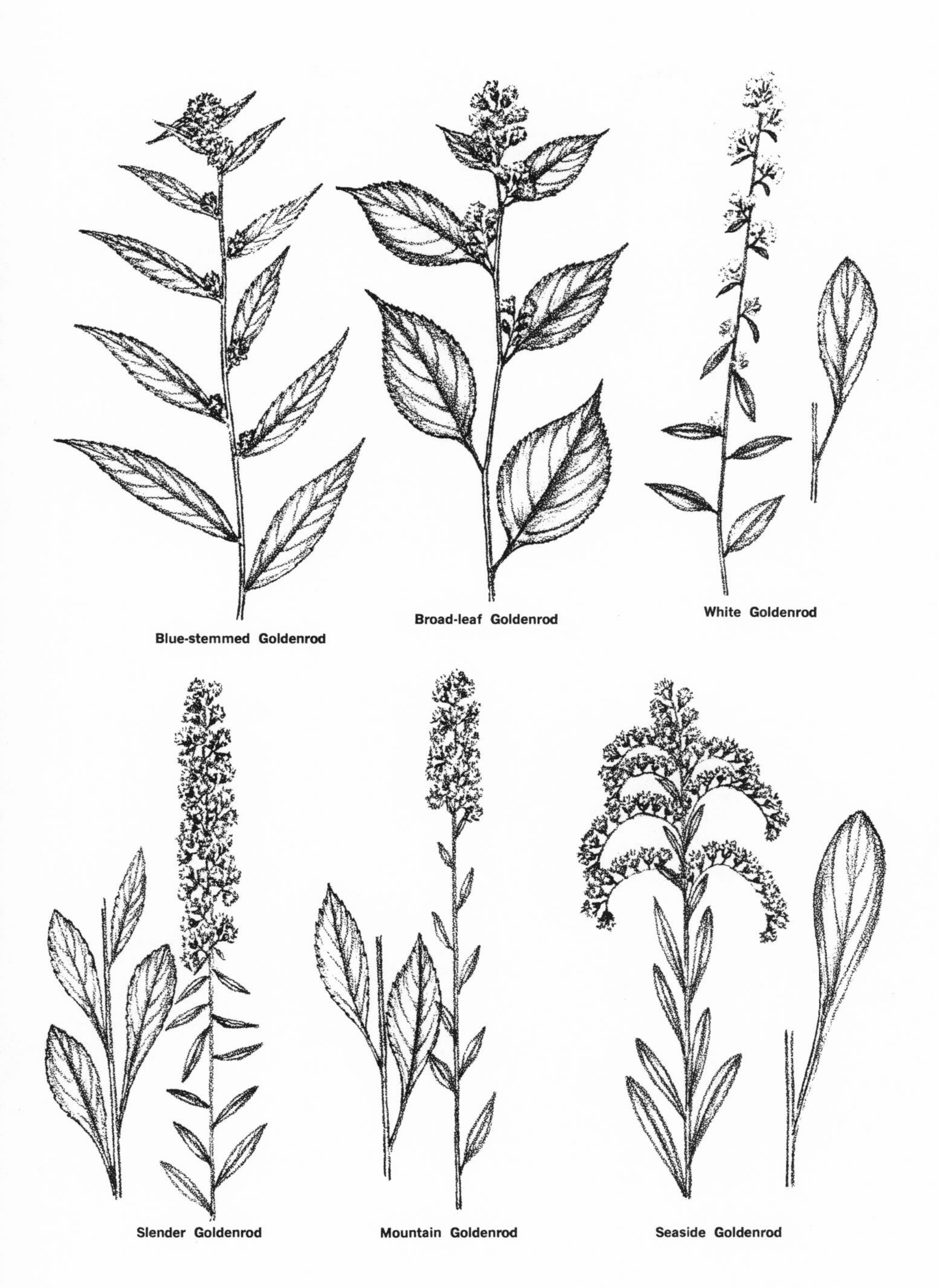
Blue-stemmed Goldenrod
Broad-leaf Goldenrod
White Goldenrod
Slender Goldenrod
Mountain Goldenrod
Seaside Goldenrod

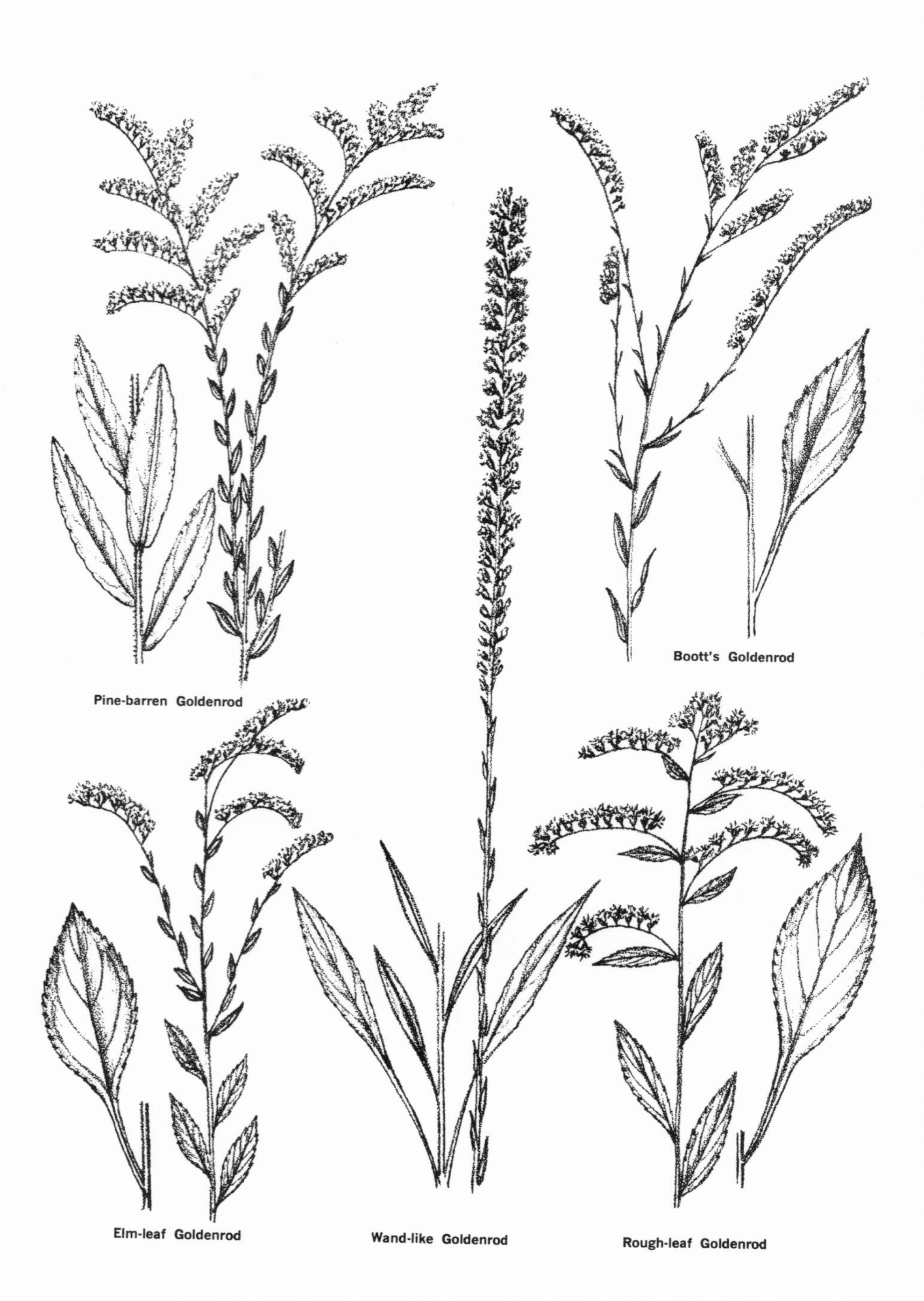
Pine-barren Goldenrod
Boott's Goldenrod
Elm-leaf Goldenrod
Wand-like Goldenrod
Rough-leaf Goldenrod

PINE-BARREN GOLDENROD *(Solidago fistulosa)*

This goldenrod has a rather stout, simple or branched stem 2 to 6 feet tall; with numerous, ascending, broadly lance-shaped or oblong leaves which are stalkless and have clasping bases. The lower ones are 1 to 4 inches long, broad at base, and sparingly toothed; the upper ones smaller and untoothed. Its small heads of flowers are in a terminal cluster which has several arching or spreading branches. It grows in low, moist, coastal plain pinelands; blooming between August and October or later.
RANGE: N.J. south to Fla.

BOOTT'S GOLDENROD *(Solidago arguta* var. *boottii)*

Boott's Goldenrod has a smoothish, slender stem 2 to 5 feet tall; usually with long ascending branches at the summit, along which many small flower heads are arranged in more or less 1-sided clusters. The leaves are rather firm; the lower ones 3 to 6 inches long, broadly lance-shaped or egg-shaped, sharply toothed, and tapered into hairy-fringed stalks. The upper ones are much smaller, stalkless, and often untoothed. It grows in dry sandy or rocky woods and fields; blooming between July and October.
RANGE: Va. to Ky. south to Fla. and Tex.

ELM-LEAF GOLDENROD *(Solidago ulmifolia)*

This species has a slender, usually somewhat downy stem 2 to 4 feet tall which branches toward the summit. The leaves are thinnish, elliptic or lance-shaped, coarsely and irregularly toothed, and hairy beneath. The lower ones are 3 to 5 inches long and abruptly narrowed into winged stalks; the upper ones much smaller, stalkless or nearly so, and almost untoothed. The flower cluster has a few arching or ascending branches. It grows in rocky woods and along streams, blooming between August and October.
RANGE: Mass. to Minn. south to Ga., Miss., Okla. and Tex.

WAND-LIKE GOLDENROD *(Solidago stricta)*

This unique goldenrod is a smooth plant with a simple, slender, wand-like stem 2 to 8 feet tall; along which are numerous small, narrow, erect leaves. The basal leaves are lance-shaped or broadest near the tip, inconspicuously toothed, stalked and 3 to 8 inches long. It grows in wet coastal plain pinelands, meadows and savannahs; blooming in September or October.
RANGE: N.J. south to Fla. and west to Tex.

ROUGH-LEAF GOLDENROD *(Solidago patula)*

This species has a smooth, reddish, 4-angled and usually narrowly winged stem 2 to 6 feet tall. Its leaves are harshly rough above but smooth and veiny beneath and have sharply toothed margins. The basal ones are 1 to 4 inches long, elliptic or oval, and narrowed into winged stalks. Those along the stem are much smaller, lance-shaped and stalkless. The flower cluster has several spreading or recurved branches. It grows in wet meadows, bogs, and swamps; blooming between August and October.
RANGE: Vt. to Ont. and Mich. south to Ga., La. and Mo.

DOWNY GOLDENROD *(Solidago puberula)*

This species has minutely downy, often purplish stems 1½ to 3 feet tall; with numerous narrow leaves, the lower ones 2 to 4 inches long and sharply toothed. The flower heads are grouped in a rather narrow end cluster, the lower ones on short branches from the axils of the upper leaves. It grows in sandy or rocky open woods and fields from Que. south to Ga. and Tenn.; blooming between September and November. (Not illustrated)

DOWNY RAGGED GOLDENROD *(Solidago petiolaris)*

Downy Ragged Goldenrod has a slender stem 1 to 3 feet tall which is downy or roughish above. The numerous leaves are stalkless or nearly so, rather firm, minutely rough-hairy on both surfaces, usually untoothed, and ½ to 3 inches long. The flower heads have bracts with pointed green tips and are in a narrow end cluster. It grows in dry woods and open places from N.C. to Mo. and Neb. south to Fla. and Tex. (Not illustrated)

EARLY GOLDENROD *(Solidago juncea)*

The Early Goldenrod is a quite smooth plant with a rather stout and light green stem from 1 to 4 feet high; branching toward the summit into several ascending, spreading, or recurved flower-bearing branches. Its leaves are quite firm; the lower and basal ones 4 to 12 inches long, broadly lance-shaped to oval, sharply toothed, and tapering into winged and usually hairy-fringed stalks. The upper ones are gradually reduced in size upwards becoming lance-shaped stalkless, and untoothed. It grows in dry and rocky open places, blooming from June to October. RANGE: N.B. to Sask. south to N.C., Tenn. and Mo.

LATE GOLDENROD *(Solidago gigantea)*

This species has a stout, smooth stem from 2 to 8 feet tall which is usually whitened with a bloom. The leaves are lance-shaped or narrowly oblong, stalkless, sharply toothed at least above the middle, 3 to 6 inches long, conspicuously 3-veined, and somewhat hairy on the veins beneath. The small flower heads are arranged along the spreading or recurved branches of the terminal flower cluster, blooming between July and October. It often forms extensive colonies in the borders of damp woods and thickets.
RANGE: N.B. to B.C. south to Fla. and Tex.

WRINKLED-LEAF GOLDENROD *(Solidago rugosa)*

This is a common but variable goldenrod with a usually stout stem 2 to 6 feet high, crowded with leaves and often with lines running down from the leaf bases. The leaves, like the stem, may be either smooth or hairy but they present a very wrinkled appearance above and are very veiny beneath. They are 1 to 4 inches long, lance-shaped to oval, sharply toothed, and are often rough above. The flower cluster has a number of spreading or recurved branches. It grows in dry to moist open woods, clearings, and thickets; blooming between August and October.
RANGE: Nfd. to Ont. and Mich. south to Fla. and Tex.

TALL GOLDENROD *(Solidago canadensis* var. *scabra)*

The Tall or Canada Goldenrod has a stout grayish-downy or sometimes roughish stem 2 to 8 feet tall; with numerous, crowded, lance-shaped, sharply toothed, stalkless or short-stalked leaves 3 to 6 inches long. They are thickish, 3-veined, and rough on the upper surface. Involucres of the flower heads vary from less than ⅛ inch to ¼ inch in height; and the numerous heads are arranged in a large pyramid-shaped terminal cluster, with many spreading or recurved branches. It grows in old fields, meadows and thickets; blooming in September and October. RANGE: Nfd. to Sask. south to Fla. and Tex.

SWEET GOLDENROD *(Solidago odora)*

The crushed leaves of this goldenrod have an anise-like odor. It has a slender smooth or slightly downy stem 2 to 4 feet tall; and numerous narrowly lance-shaped leaves 2 to 4 inches long which are stalkless, untoothed, and marked with tiny translucent dots. The flower heads are in a loose, 1-sided, plume-like cluster. It grows in dry open woods, fields, and thickets; blooming between July and October. Also called the Anise-scented Goldenrod. RANGE: N.H. to Ohio, Ky., Mo. and Okla. south to Fla. and Tex.

BUSHY GOLDENROD *(Euthamia graminifolia)*

This is a bushy-branched plant 2 to 4 feet tall, with numerous narrowly lance-shaped leaves 1 to 5 inches long which show 3 prominent veins. The small flower heads are in flat-topped end clusters. It grows in wet fields and stream bottoms, blooming between July and October. Also called Fragrant or Flat-topped Goldenrod.
RANGE: Nfd. to Sask. south to N.C. and Mo.

SLENDER FRAGRANT GOLDENROD *(Euthamia tenuifolia)*

This species has narrow leaves which show but 1 prominent vein and often has small leaves in the axils of the larger ones. It grows in sandy soils from N.S. to Mich. south to Tex., Ga. and Ind. (Not illustrated)

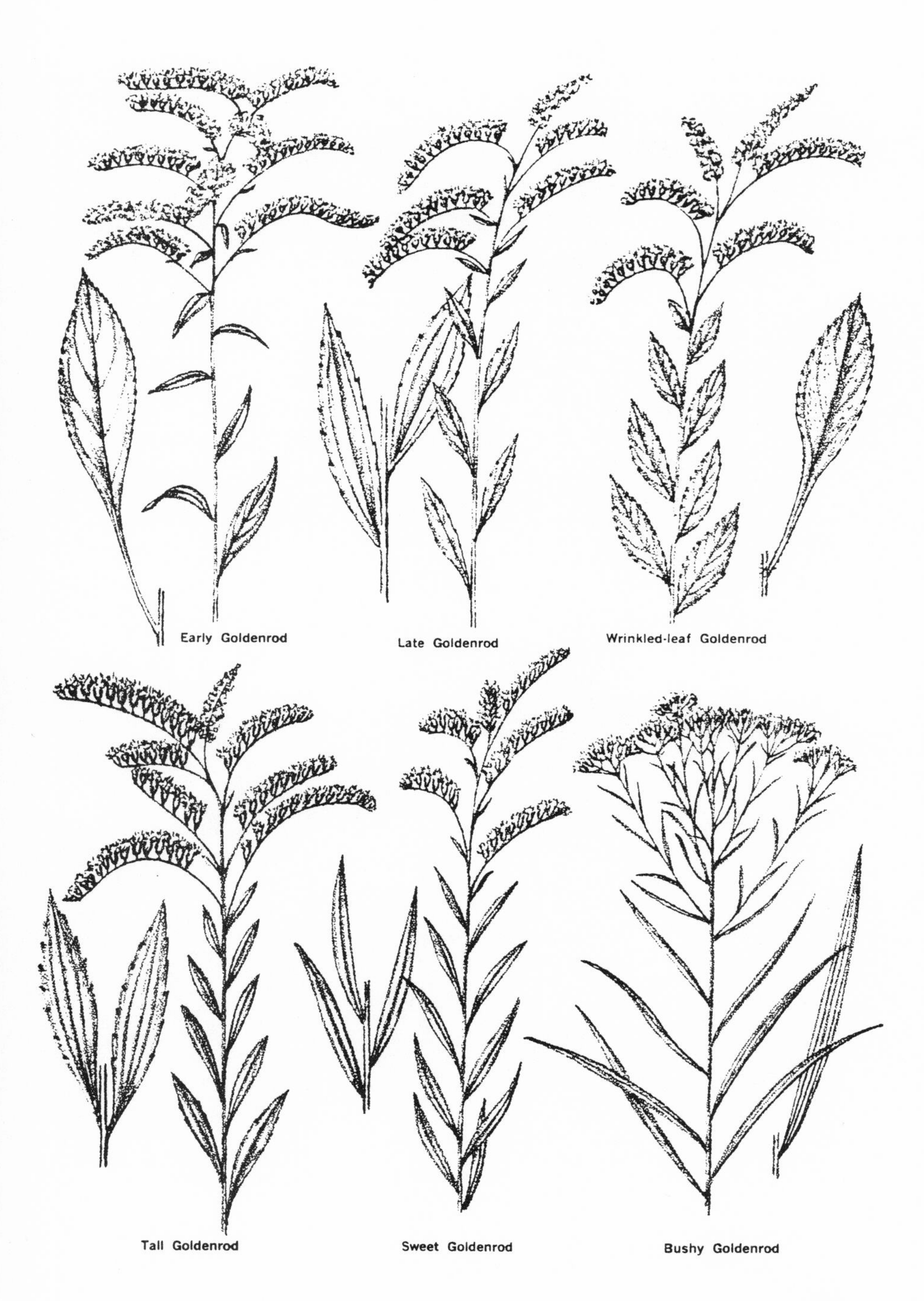
Early Goldenrod
Late Goldenrod
Wrinkled-leaf Goldenrod
Tall Goldenrod
Sweet Goldenrod
Bushy Goldenrod

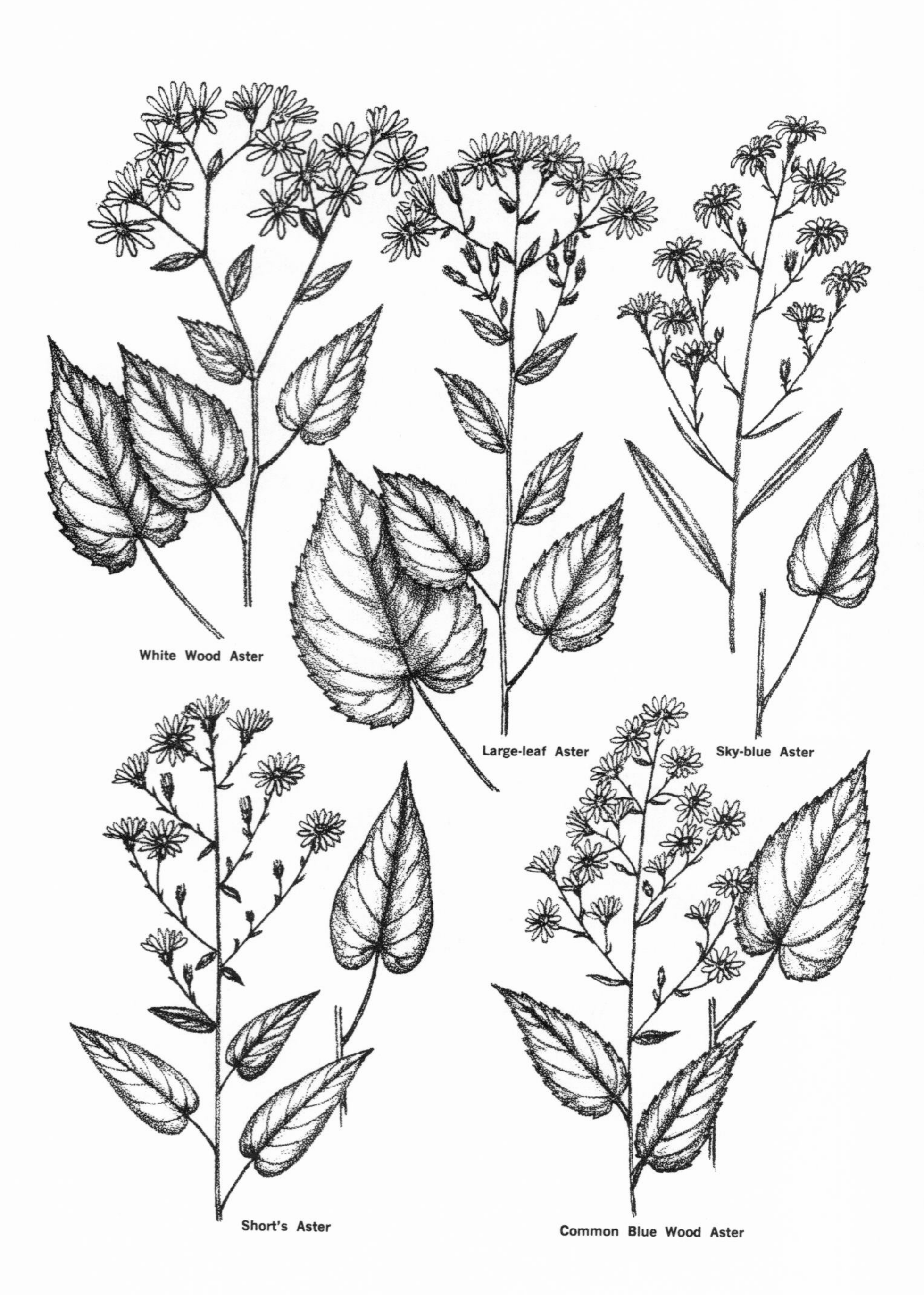
White Wood Aster
Large-leaf Aster
Sky-blue Aster
Short's Aster
Common Blue Wood Aster

Asters

Wild asters bloom chiefly in the late summer and fall, when they are among the most conspicuous wild flowers. The heads have a disk of tubular flowers in the center which are yellow, often changing to red or purple; and white, lavender, blue, pink or purple ray flowers about the margin. Some species of our native asters have contributed to the garden forms commonly called "hardy asters" or Michaelmas-daisies. There are more than 75 species in eastern North America and their identification is not always an easy matter.

WHITE WOOD ASTER *(Aster divaricatus)*

This aster grows in dry open woods and clearings, blooming from August to October. It has a slender, often zig-zag, smoothish stem 1½ to 2½ feet tall. The thin heart-shaped leaves are smooth or sparingly hairy, mostly stalked, and at least the lower ones are very coarsely but sharply toothed. The flower heads are almost an inch across. They have from 6 to 10 white rays and the disk flowers soon become brownish.

RANGE: Me. to Ohio south to Ga., Ala. and Tenn.

LARGE-LEAF ASTER *(Aster macrophyllus)*

Very large, thickish, rough, sharply toothed, and heart-shaped basal leaves characterize this aster of the open woodlands. The flower heads have white to lavender or bluish rays, and the branches of the flower cluster have small stalked glands. It blooms from July to September. RANGE: Que. to Minn. south to Md., n. Ga., n. Ala., Ohio and Ill.

SCHREBER'S ASTER *(Aster schreberi)*

This aster also has basal tufts of large heart-shaped leaves but they are rather thin and smooth or barely roughish. The flower heads have a narrowly cylindrical involucre and about 10 white rays. It grows in moist open woods or thickets from N.Y. to Ill. south to W. Va. and Ky; blooming between July and September. (Not illustrated)

SKY-BLUE ASTER *(Aster oolentangiensis)*

The leaves of this aster are thickish, rough on both surfaces, and often untoothed. The lower ones are egg-shaped to lance-shaped, heart-shaped at the base, 2 to 6 inches long, and have slender stalks; the upper ones being smaller, narrow, and almost stalkless. The 1- to 4-foot stems are stiffly erect, branched above, and also roughish. Its flower heads are about ½ inch across and they have from 10 to 20 deep blue or violet-blue rays; and they may be seen between August and October. It grows in dry open woods, thickets, and on prairies. RANGE: Que. to Minn. south to Ga. and Tex.

SHORT'S ASTER *(Aster shortii)*

This aster has a smooth or roughish slender stem 2 to 4 feet tall. On it are numerous broadly lance-shaped or egg-shaped leaves with heart-shaped bases. They are 2 to 6 inches long and all but the uppermost ones have slender stalks. The attractive flower heads are about ½ inch across, with from 10 to 15 violet pale rays. It grows in open woods and on rocky banks, blooming between August and October.

RANGE: Pa. to Wis. south to Ga., Ala. and Tenn.

COMMON BLUE WOOD ASTER *(Aster cordifolius)*

This is a common but quite variable woodland aster. It usually has smoothish and branching stems 1 to 5 feet high. Most or all of the lower leaves are heart-shaped, coarsely toothed, 2 to 5 inches long, and have slender stalks. The ones on the stem above are smaller, shorter stalked, less toothed, and often taper at the base. It usually has numerous flower heads about ⅓ inch across, with 10 to 20 whitish, pinkish, or pale blue-violet rays; and often a reddish disk. It flowers between August and October.

RANGE: N.S. to Que. and Wis. south to Ga., Ala., Mo. and Kan.

LOWRIE'S ASTER *(Aster cordifolius* var. *laevigatus)*

Lowrie's Aster is somewhat similar to the preceding species but its leaves are very smooth, in fact they have an almost greasy feeling. They are also less prominently toothed and those of the stem have broadly winged stalks. It grows in open woods and thickets from Conn. to Ont. south to Md., w. N.C. and n. Ga.; blooming between August and October. (Not illustrated)

ARROW-LEAF ASTER *(Aster cordifolius* var. *sagittifolius)*

This aster has a stiffly erect, smooth or sparingly hairy stem 2 to 5 feet tall, which branches only in the flower cluster. The lower leaves are egg-shaped with heart-shaped bases, sharply toothed, 3 to 6 inches long, and have broadly winged stalks. The upper ones are smaller, narrower, and stalkless or nearly so. Its heads of flowers are about ¾ inch across and have 10 to 15 pale blue, pinkish, or whitish rays. It grows in dry open woods, thickets, and open places; blooming between August and October.

RANGE: Vt. to Minn. south to Ga. and Tex.

WAVY-LEAF ASTER *(Aster undulatus)*

The Wavy-leaf Aster has a stiffly erect stem which is usually somewhat minutely roughish-hairy, branched above, and from 1 to 3½ feet tall. Its lower leaves are lance-shaped or egg-shaped, heart-shaped at the base, and 2 to 6 inches long. They are wavy on the margin, or sometimes with some low and blunt teeth, and have winged leaf-stalks which broaden at the clasping base. The ones upward along the stem are smaller and stalkless or nearly so. All of the leaves are roughish on the upper surface and downy underneath. The flower heads are about ½ inch across and have from 8 to 15 pale violet or bluish rays. It grows in dry open woods and thickets; blooming between August and November.

RANGE: N.S. to Ont. and Minn. south to Fla., La. and Ark.

LATE PURPLE ASTER *(Aster patens)*

The stems of this aster are slender, roughish-hairy, 1 to 3 feet tall, and are branched above. Its leaves are oblong or oval, mostly 1 to 2 inches long, untoothed, rather thickish and somewhat rigid, and roughish or hairy. They clasp the stem by their heart-shaped bases. The uppermost leaves in the flower cluster are small and often almost bract-like. It has beautiful flower heads about ⅔ inch across, with from 20 to 30 deep violet or bluish-purple rays, produced between August and October. This species also grows in dry open woods, thickets, and fields.

RANGE: Me. to Minn. south to Fla. and Tex.

CROOKED-STEMMED ASTER *(Aster prenanthoides)*

Usually this aster has a stout, somewhat zig-zag, branched stem 1 to 2 feet high, which may be smooth or have finely hairy lines. The leaves are lance-shaped to narrowly egg-shaped, taper-pointed at the tip, sharply toothed, and abruptly contracted into a broad and untoothed stalk-like portion expanding into 2 "ears" at the clasping base. The middle and lower ones are 3 to 8 inches long, usually roughish above but smooth beneath. Its flower heads are almost an inch across and have from 20 to 30 pale blue-violet rays. It grows in rich woods, damp thickets, and along streams; blooming between August and October.

RANGE: Mass. to Minn. south to Va., Ky. and Iowa.

PURPLE-STEMMED ASTER *(Aster puniceus)*

This is another common but variable aster. It usually has a stout stem which is reddish- or purplish-tinged, often rough-hairy, much-branched, and from 3 to 8 feet high. The leaves are more or less lance-shaped, sharply toothed, tapering gradually both to the tip and to the somewhat "eared" and clasping base. Those on the main part of the stem are 3 to 6 inches long, and they may be quite smooth or roughish-hairy. The flower heads are almost an inch across and have from 20 to 40 lilac, blue-violet, pinkish, or even whitish rays. It grows quite abundantly in moist meadows, thickets, and in swampy places; blooming between August and October. Also known as the Red-stalked Aster and Early Purple Aster.

RANGE: Nfd. to Man. south to Ga., Ala., Tenn. and Iowa.

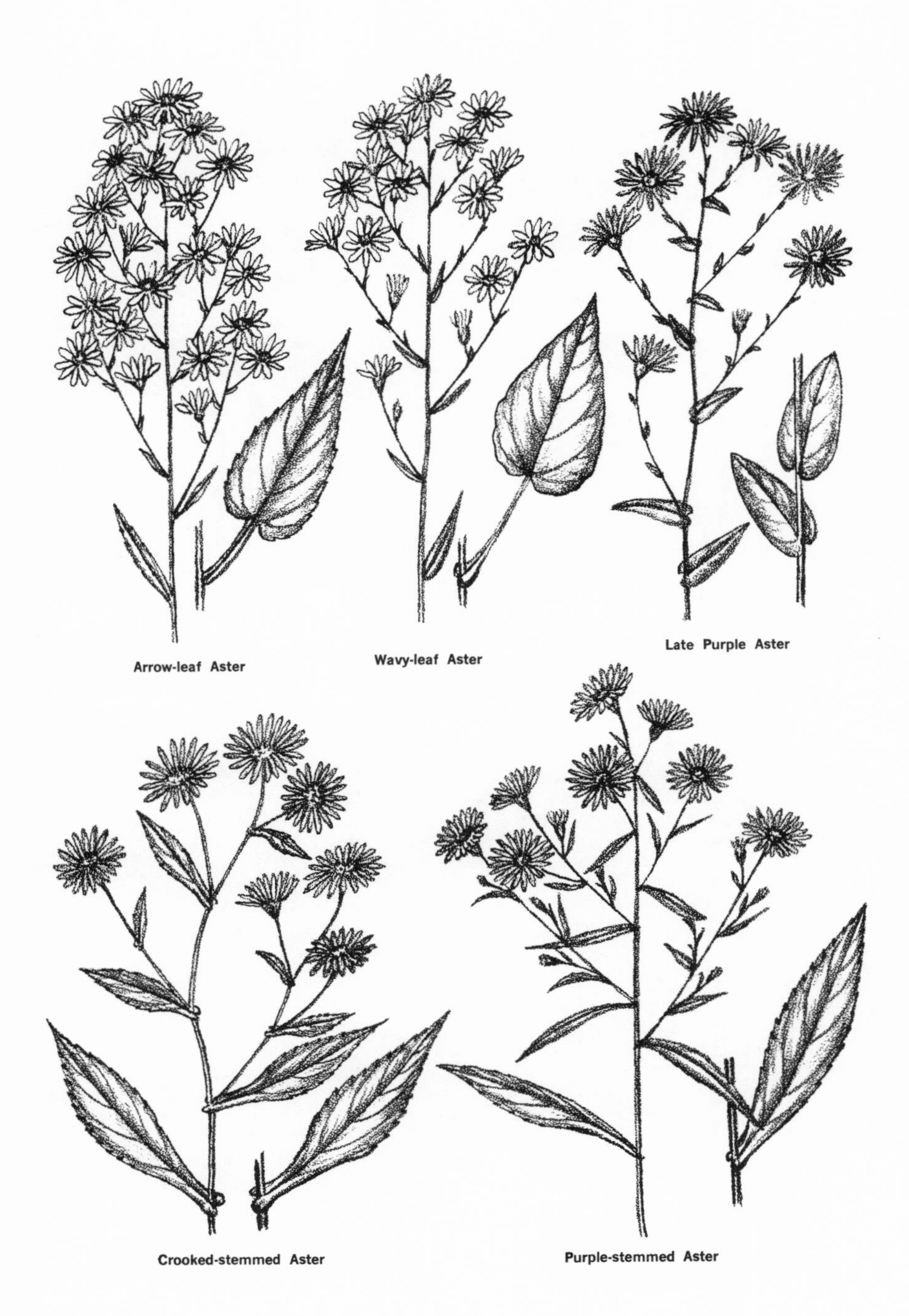
Arrow-leaf Aster
Wavy-leaf Aster
Late Purple Aster
Crooked-stemmed Aster
Purple-stemmed Aster

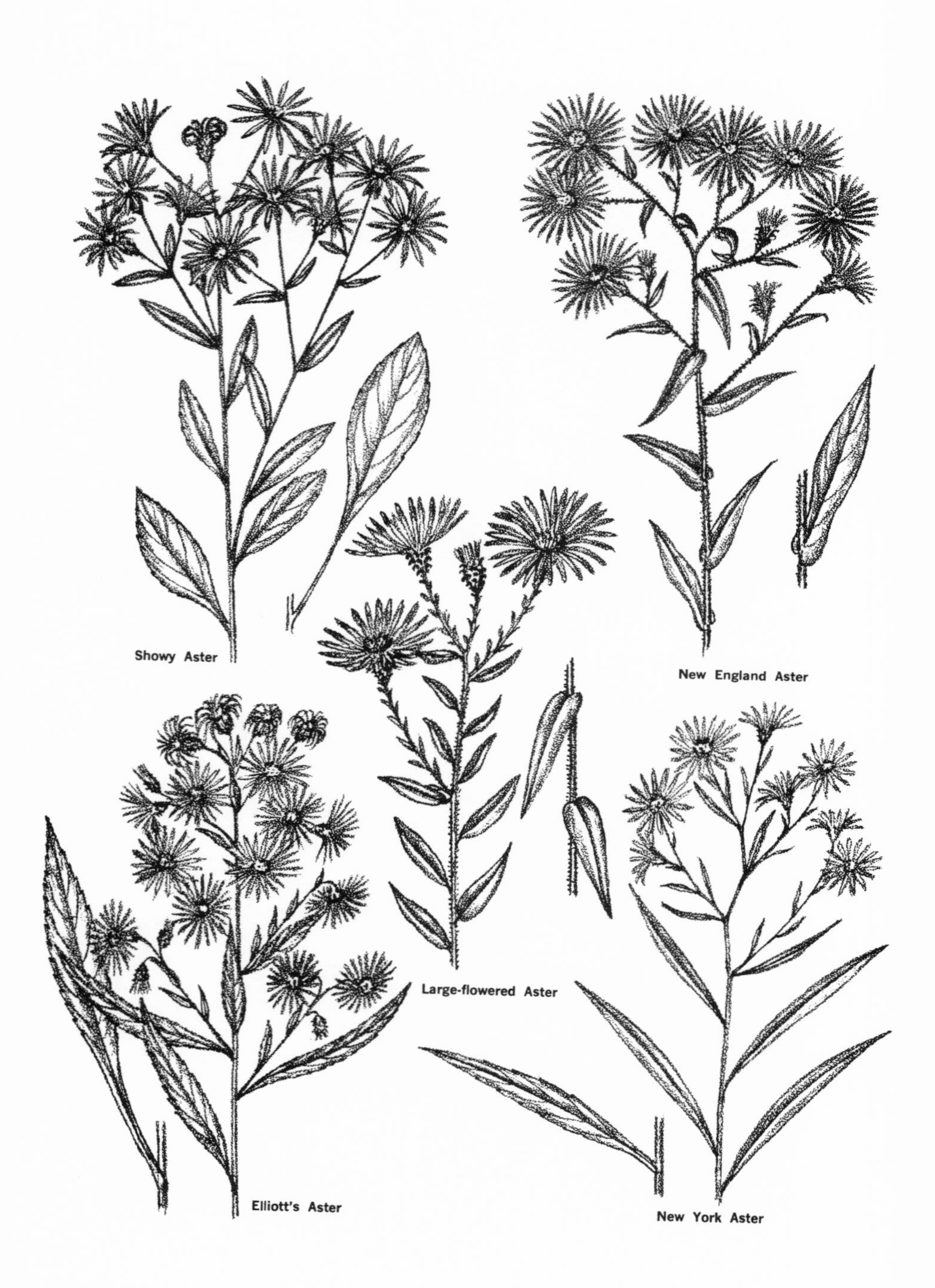
Showy Aster
New England Aster
Large-flowered Aster
Elliott's Aster
New York Aster

SHOWY ASTER *(Aster spectabilis)*

Although this aster is also known as the Seaside Purple Aster, and is usually found in sandy pinelands near the coast, it sometimes occurs in the mountains. As a rule it has several slightly roughish and somewhat glandular stems 1 to 2 feet high, which are sometimes branched above. Its leaves are thickish and firm; the lower and basal ones being oval, pointed, sparingly and inconspicuously toothed, 3 to 5 inches long, and tapering into slender stalks. Those along the stem become smaller and narrower upwards, and are stalkless and untoothed or nearly so. The flower heads are large, almost 1½ inches across, and have from 15 to 30 bright violet rays. It flowers between August and October.

RANGE: Mass. south to e. S.C.; also in mts. of w. N.C.

LARGE-FLOWERED ASTER *(Aster grandiflorus)*

This is the largest and showiest of our purple-flowered wild asters, with flower heads often 2 inches across which have numerous violet-purple rays. It is a rough-hairy plant 1 to 2½ feet high with long and slender, ascending, very leafy branches. The leaves are rigid, untoothed, narrow to oblong, stalkless and with somewhat clasping bases. On the main stem and branches they are up to 2 inches long and generally turned downward; but the uppermost ones are very small or bract-like, merging into the bracts of the cup-shaped involucre with their spreading green tips. It grows in dry oak woods and pinelands in the coastal plain and outer piedmont, blooming during September or October.

RANGE: Va. south to Fla.

NEW ENGLAND ASTER *(Aster novae-angliae)*

This is a very attractive wild aster which is sometimes cultivated. It is a hairy plant 2 to 8 feet high with a rather stout stem which is branched above and very leafy. The leaves are lance-shaped, untoothed, mostly 2 to 4 inches long, and clasp the stem by their heart-shaped or "eared" bases. The numerous flower heads are about 1½ inches across and have about 40 or 50 narrow violet-purple rays. Both the stalks of the heads and the involucres are covered with minute and sticky glands, and the involucral bracts have rather long, spreading tips. It grows in moist thickets, fields, swamps, and along roadsides; blooming between August and October.

RANGE: Que. to Alb. south to Md., w. N.C., e. Tenn., Kan. and Colo.

ELLIOTT'S ASTER *(Aster puniceus* var. *elliottii)*

Elliott's Aster grows in the coastal marshes of the Southeast, where it blooms from late September to November. It has a stout and stiffly erect stem which is smooth or sparingly hairy, 2 to 5 feet tall, and branches above. Its leaves are roughish on the upper surface but smooth beneath, the margins being sharply toothed. The upper ones are lance-shaped and stalkless but those on the lower portion of the stem are much larger, 4 to 8 inches long, more elliptic, and taper at the base into winged stalks. The flower heads are about an inch across, with numerous pale violet-purple rays and bracts with tapering and spreading tips.

RANGE: se. Va. south to Fla. and west to La.

NEW YORK ASTER *(Aster novi-belgii)*

The New York Aster is a rather smooth plant with a slender and branching stem from 1 to 3 feet high, sometimes being slightly downy above. Its leaves are all narrowly lance-shaped, often slightly fleshy, untoothed or with a few quite inconspicuous teeth on the margin, 2 to 6 inches long, and the upper ones more or less clasping at the base. The flower heads are about an inch across, usually numerous; with 15 to 25 blue-violet, or sometimes pinkish or whitish rays, and involucral bracts with slender and spreading tips. It grows in damp thickets, meadows, and on shores; blooming between August and November.

RANGE: Nfd. to Que. south to Ga.

EASTERN SILVERY ASTER *(Aster concolor)*

Also called the Lilac-flowered Aster, this species has a slender and usually minutely downy stem 1 to 2½ feet tall, which is simple or with a few almost erect branches. The numerous leaves are minutely silvery-silky, stalkless, and untoothed; the lower ones being elliptic to lance-shaped and 1½ to 2 inches long. Those upward on the stem are very much smaller or even bract-like. The pretty flower heads are about ¾ inch across, with from 10 to 15 lilac to violet-purple rays, and arranged in a long and narrow cluster. It grows in dry and sandy open woods and pinelands, chiefly in the coastal plain. Its flowering season is during September and October.

RANGE: Mass. and Ky. south to Fla. and La.

BUSHY ASTER *(Aster dumosus)*

This is a slender-stemmed and usually minutely downy plant 1 to 3 feet high, with spreading or ascending branches. Its heads of flowers are about ½ inch across, with 15 to 25 pale lavender or bluish rays. They terminate slender branchlets 2 to 4 inches long, on which are crowded, small, bract-like leaves. The larger leaves are quite narrow and from 1 to 3 inches long. Also known as the Rice-button Aster, it grows in dry to moist sandy fields and thickets; blooming between late August and October.

RANGE: Me. to Ont. and Mich. south to Fla. and Tex.

WALTER'S ASTER *(Aster walteri)*

This is a unique species of aster with a slender, almost smooth, loosely branched stem 1 to 2 feet tall. It has numerous rigid leaves which are stalkless, reflexed, and mostly less than ⅓ inch long. The upper ones, and those crowding the flower stalks, are very small and scale-like. The flower heads which terminate the branches are little more than ½ inch across and have from 10 to 15 lilac rays. It grows only in the dry coastal plain pinelands, blooming between August and October.

RANGE: N.C. south to Fla.

WHITE OLD-FIELD ASTER *(Aster pilosus)*

Sometimes this aster is pilose—meaning hairy—but just as often, it is smooth. It grows in dry thickets, clearings, and along roadsides; and it is often very abundant in abandoned fields. It grows from 2 to about 5 feet tall and has several slender branches which are ascending or stiffly spreading. Its leaves are all narrow or lance-shaped, stalkless, stiffish, slightly if at all toothed, and 1 to 3 inches long. The numerous flower heads are about ¾ inch across, with 15 to 25 white or sometimes pale purplish rays, and grow along the upper side of the branches; blooming between August and November.

RANGE: Me. to Minn. and Kan. south to Ga., Ala. and Ark.

CALICO ASTER *(Aster lateriflorus)*

The Calico or Starved Aster differs from the preceding species in having thinner and flexible leaves which are usually broader, 2 to 6 inches long, sharply toothed as a rule, and quite roughish. Its flower heads are about ½ inch across, with from 10 to 20 white rays and purplish disk flowers. The bracts of the involucre have a prominent green midrib. The heads are usually very numerous and borne in a branching flower cluster; blooming between August and November. It grows in dry to moist open woods, thickets, fields, and along roadsides.

RANGE: Que. to Minn. south to Ga. and Tex.

SMOOTH ASTER *(Aster laevis)*

This is a smooth plant with a rather stout and whitened stem 2 to 4 feet tall, which is often more or less branched. Its leaves are thick, lance-shaped to elliptic or egg-shaped, 1 to 3 inches long, and mostly untoothed. The ones on the upper part of the stem are stalkless but the lower and basal ones taper into winged stalks. Its flower heads are almost an inch across and have 15 to 30 violet or bluish rays; blooming between August and October. It grows on the borders of woods and in thickets or dry fields.

RANGE: Me. to Sask. south to Ga., La. and Kan.

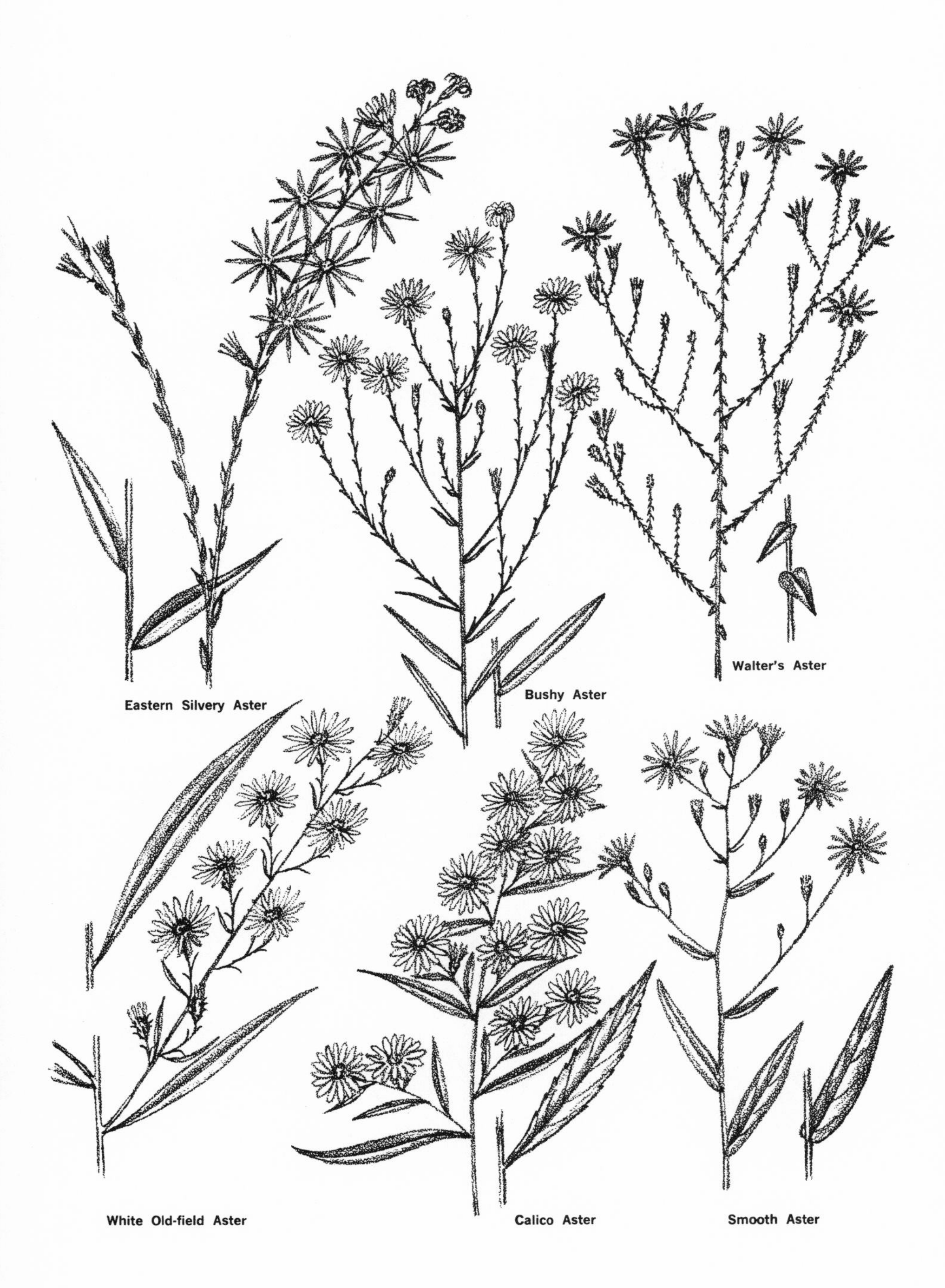
Eastern Silvery Aster
Bushy Aster
Walter's Aster
White Old-field Aster
Calico Aster
Smooth Aster

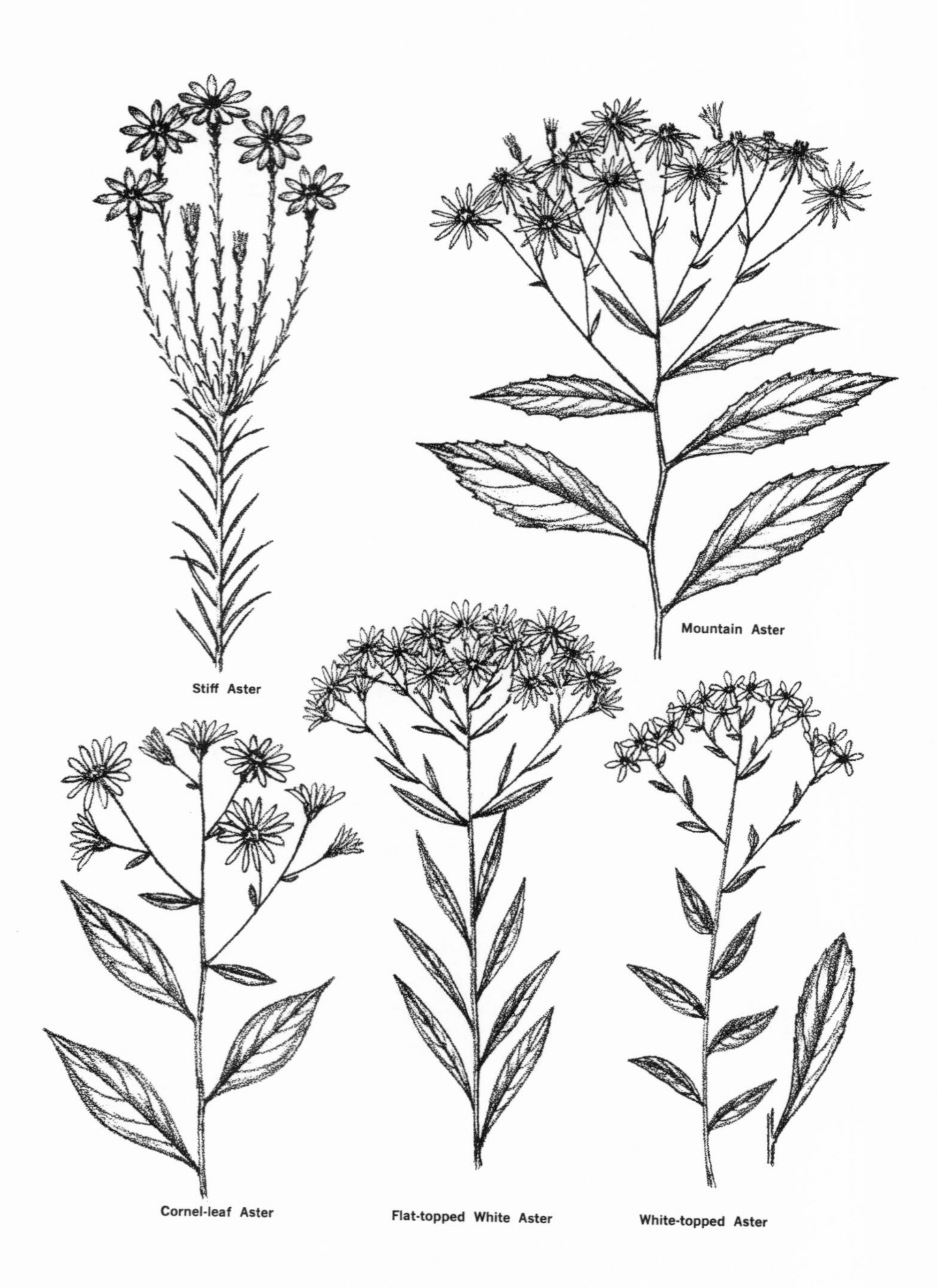
Stiff Aster
Mountain Aster
Cornel-leaf Aster
Flat-topped White Aster
White-topped Aster

STIFF ASTER *(Aster linariifolius)*

Savory-leaf Aster and Pine-starwort are other names given to this aster. It has a tuft of stiff, roughish, minutely hairy stems 6 inches to 2 feet high which are branched above. On them are numerous and rather closely crowded, very narrow, stiff, rough-margined leaves. The larger ones are ¾ to 1½ inches long, the upper ones being reduced to rigid bracts. The flower heads are solitary at the ends of the erect or ascending branches; each one about an inch across, with from 10 to 15 bright lavender rays. It grows in dry sandy or rocky open places, blooming between September and November.

RANGE: N.B. to Que. and Minn. south to Fla. and Tex.

MOUNTAIN ASTER *(Aster acuminatus)*

The Mountain or Whorled Aster has a somewhat zig-zag and minutely downy stem 1 to 3 feet high, branching toward the summit. Its leaves are thin, broadly lance-shaped or elliptic, 3 to 6 inches long, pointed at both ends, and rather coarsely toothed. Those on the upper part of the stem are often so close together that they appear to be whorled. The flower heads are about 1¼ inches across and have 12 to 18 narrow white or purplish-tinged rays. It grows in dry to moist woods and clearings, blooming between July and November.

RANGE: Nfd. to Que. south to N.J., Pa., n. Ga. and Tenn

CORNEL-LEAF ASTER *(Aster infirmus)*

This aster has a somewhat zig-zag, slender stem 1½ to 3 feet tall which is sparingly branched at the summit. Its leaves are elliptic or quite often broadest above the middle, untoothed, stalkless or nearly so, smooth above but slightly roughish-hairy on the veins beneath and from 2 to 5 inches long. The flower heads are about an inch across, with from 8 to 15 white or cream-colored rays, and arranged in an open terminal cluster. It grows in dry and usually rocky woods, thickets, and on slopes; blooming between late June and September.

RANGE: Mass. to Ohio south to Ga., Ala. and Tenn.

FLAT-TOPPED WHITE ASTER *(Aster umbellatus)*

This is a rather smooth plant with leafy stems from 1 to 8 feet tall. The leaves are lance-shaped to very narrowly egg-shaped, untoothed, pointed at both ends, up to about 6 inches long, and stalkless or very short-stalked. They are roughish on the upper surface and sometimes downy beneath. Its heads of flowers are ½ to about an inch across, with from 2 to 15 white rays; and are arranged in a broad, flat-topped cluster. It grows in both moist or dry open woodlands, thickets, and meadows; blooming between July and October.

RANGE: Nfd. to Ont. and Minn. south to Ga., Ky., Iowa and Neb.

CAROLINA ASTER *(Aster carolinianus)*

The Carolina or Climbing Aster is a unique species in that it has woody trailing, arching, or climbing stems. Its flower heads are solitary or sometimes clustered at the ends of conspicuously leafy branches. They are fairly large and have numerous pale purplish or pinkish ray flowers. The involucre is bowl-shaped and its bracts have prominent dark green and spreading tips. Its leaves are lance-shaped or elliptic, untoothed, minutely hairy, 1½ to 4 inches long, and have clasping bases. It grows in and about the coastal plain swamps from S.C. south to Fla.; blooming in late September and October or later. (Not illustrated)

WHITE-TOPPED ASTER *(Aster paternus)*

This plant usually has a somewhat downy and slightly angled stem from 1 to 2 feet high which is branched toward the summit. Its leaves are thin, veiny in appearance, and usually somewhat toothed. The lower ones are narrowly top-shaped, 2 to 4 inches long, and taper into winged stalks. The upper ones are smaller and stalkless or very nearly so. Its flower heads are about ½ inch across and have 5 white rays. They are arranged in a flat-topped cluster and bloom between June and September. It grows in dry woods, thickets, and clearings.

RANGE: N.H. to Ohio south to Ga. and La.

Fleabanes

Fleabanes differ from the asters in that their flower heads are on naked stalks; and the involucral bracts are narrow and in a single row, or sometimes in a row of long ones with very short ones at the base. As in the asters, the heads have a central disk of tubular flowers and a marginal row of strap-like ray flowers.

ROBIN'S-PLANTAIN *(Erigeron pulchellus)*

Between April and July, the attractive flowers of this fleabane may be seen on the wooded slopes and banks, or even along roadsides. The heads are an inch to 1½ inches broad and have about 50 narrow lavender or violet rays. They are usualiy arranged in a loose cluster of between 2 and 6 heads. The whole plant is quite hairy, with a simple stem from 1 to 2 feet high. The basal leaves are narrowly top-shaped or spoon-shaped, toothed, and 1 to 3 inches long. Those of the stem are few and widely spaced, smaller, lance-shaped, and untoothed or very nearly so.

RANGE: Que. to Ont. and Minn. south to Fla., La. and Kan.

PHILADELPHIA FLEABANE *(Erigeron philadelphicus)*

The Philadelphia Fleabane has a slender and downy stem 1 to 3 feet tall; along which are scattered, mostly untoothed, stalkless and somewhat clasping leaves. The lower and basal leaves are much larger, 1 to 3 inches long, narrowly top-shaped, toothed, and taper into winged stalks. The flower heads are ½ to an inch across and have 100 or more very narrow, rose-purple or pinkish rays. They may be relatively few or quite numerous. It grows in rich woods and fields or on springy slopes, blooming between April and July.

RANGE: Nfd. to B.C. south to Fla. and Tex.

OAK-LEAF FLEABANE *(Erigeron quercifolius)*

This fleabane resembles the preceding one but it is usually a smaller plant from 6 inches to 1½ feet tall. The lower and basal leaves are quite distinctly lobed, hence the name of Oak-leaf. Its flower heads are about ½ inch across and have a number of narrow bluish or violet rays. It grows in fields and along roadsides, generally in sandy soils of the coastal plain; blooming April to June.

RANGE: Va. and Tenn. south to Fla. and Tex.

EARLY FLEABANE *(Erigeron vernus)*

The Early Fleabane has a basal rosette of thickish, top-shaped, untoothed or indistinctly toothed leaves 2 to 4 inches long. Its slender stem is simple or loosely branched, usually quite smooth, 1 to 2½ feet tall, and has but a few very small and bract-like leaves. The flower heads are about ½ inch across and have from 20 to 30 lilac or whitish rays. There are usually from 2 to 12 together in a loose cluster. This fleabane grows in the wet coastal plain pinelands or in shallow ponds, blooming late March to June.

RANGE: se. Va. south to Florida and west to La.

NARROW-LEAF DAISY FLEABANE *(Erigeron strigosus)*

This is a branching plant from 1 to 3 feet high; the stems, and sometimes the leaves, with minute and closely pressed hairs. The leaves are firm and all but the lowest and basal ones are untoothed or nearly so. Those on the main part of the stem are quite narrow and rather scattered. The lowest and basal ones are narrowly top-shaped and taper at the base into stalks. Its flower heads are numerous, about ½ inch across, and have about 40 narrow, white or purplish-tinged rays. It grows in dry fields, pastures, waste places, and along roadsides; flowering between late April and October.

RANGE: N.S. to B.C. south to Fla., Tex. and Calif.

DAISY FLEABANE *(Erigeron annuus)*

Also known as the White-top and Sweet-scabious, this species is quite similar to the preceding one. It differs in having spreading hairs, leaves which are sharply toothed, the lower ones being broader and egg-shaped; and is also a somewhat taller plant, sometimes 4 or 5 feet high. Common everywhere in fields and waste places, it blooms between late March and August.

RANGE: N.S. to Man. south to Ga., Ala. and Miss.

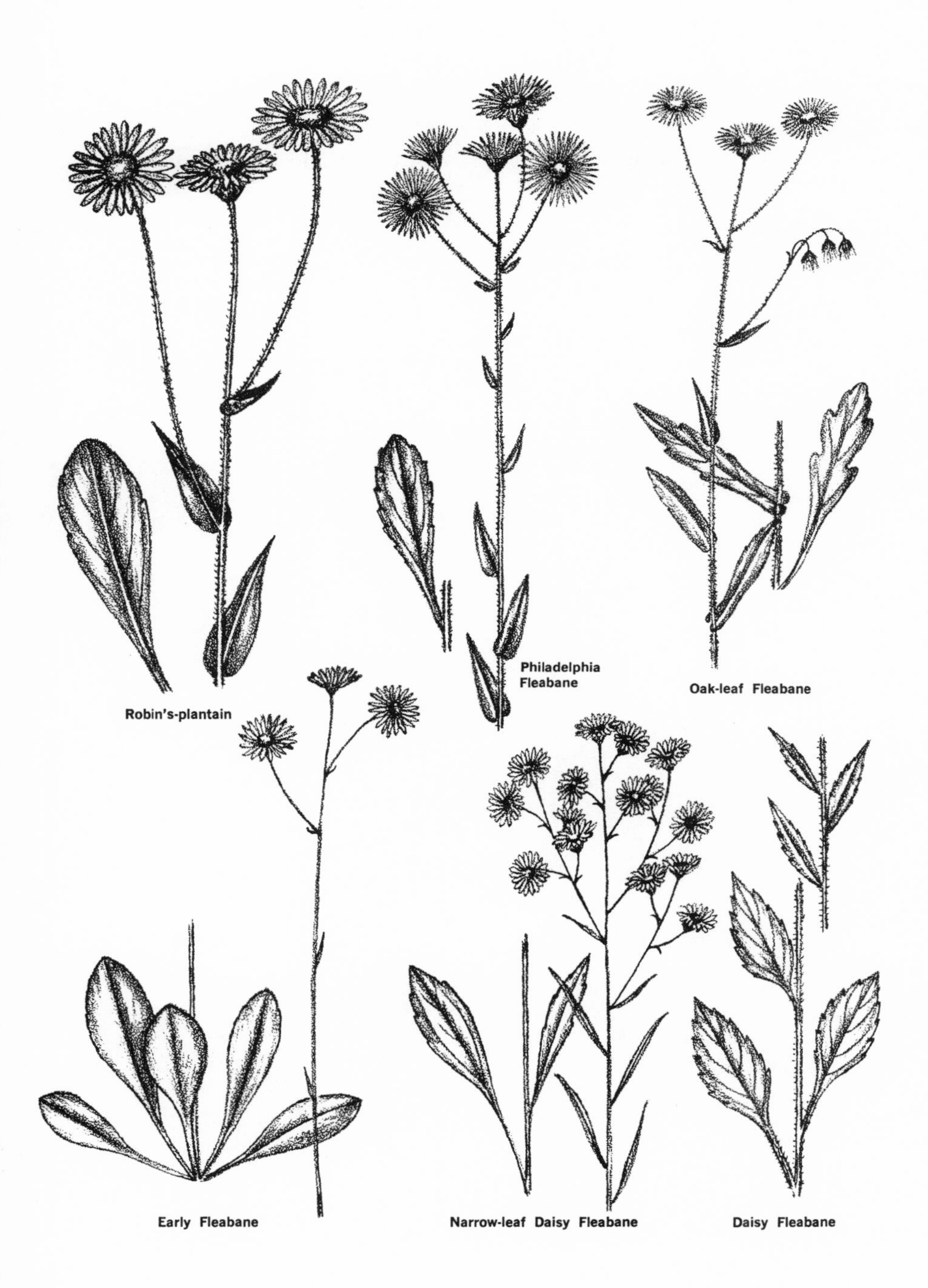
Robin's-plantain
Philadelphia
Fleabane
Oak-leaf Fleabane
Early Fleabane
Narrow-leaf Daisy Fleabane
Daisy Fleabane

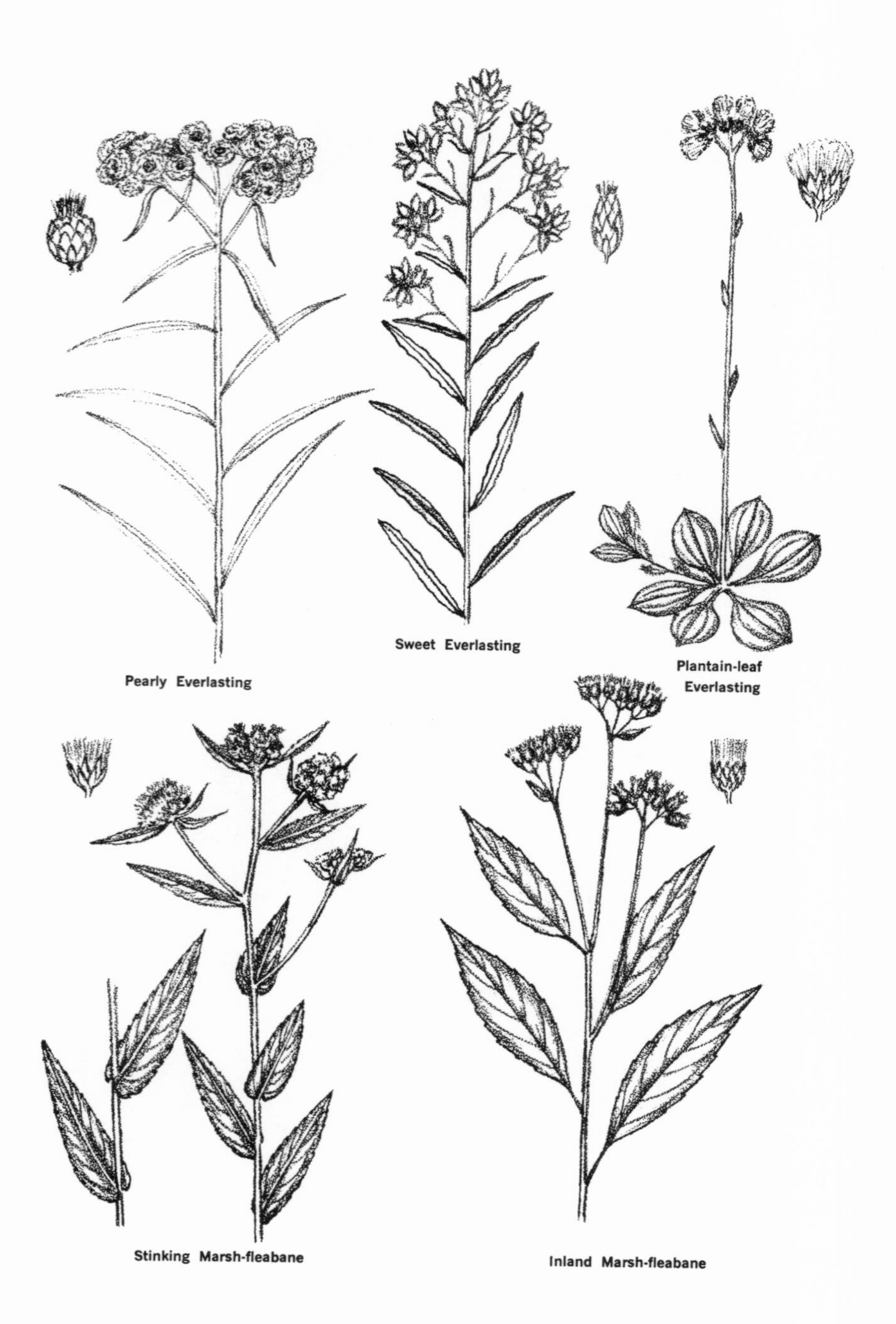
Pearly Everlasting
Sweet Everlasting
Plantain-leaf Everlasting
Stinking Marsh-fleabane
Inland Marsh-fleabane

PEARLY EVERLASTING *(Anaphalis margaritacea)*

This plant gets its common name from the pearly-white involucral bracts of its flower heads. It is a snowy-white, woolly plant with an erect leafy stem 1 to 3 feet tall which branches toward the summit. Its leaves are all very narrow, white-woolly beneath and often also above, 1 to 4 inches long, with the margins rolled inward beneath. The flower heads are close to ½ inch across and contain many tubular flowers. It grows in dry fields and pastures, blooming between July and September.

RANGE: Nfd. to Alaska south to W. Va., Ohio, Wis., S.D., Colo. and Calif.

SWEET EVERLASTING *(Gnaphalium obtusifolium)*

This plant has pleasantly fragrant, wavy-margined, narrow leaves which are smooth and green above but white-woolly beneath, and 1 to 3 inches long. It is a whitish-cobwebby plant from 1 to 3 feet tall, often with a branched stem. The bracts of the involucres are pale yellowish. It grows abundantly in dry fields, clearings, and along roadsides; blooming August to October.

RANGE: Nfd. to Alaska south to W. Va., Ohio, Wis., N. Mex., and Calif.

PURPLISH CUDWEED *(Gamochaeta purpurea)*

This species is similar to the preceding but its leaves are noticably broader toward their tips. The flower heads are stalkless or nearly so and form a more slender end cluster, and their involucral bracts are brownish or purplish. It grows in open places from Me. to Pa., Ky., and Kan. south to Fla. and Tex.; blooming from March to June or later. (Not illustrated)

PLANTAIN-LEAF EVERLASTING *(Antennaria plantaginifolia)*

Members of this genus are often called Pussy-toes or Cat's-foot. All are small, woolly or cobwebby plants with chiefly basal leaves; and stems 6 to 12 inches tall bearing the heads of flowers. The Plantain-leaf Everlasting is one of the more common of the several species found in eastern North America. It has roundish to top-shaped basal leaves with 3 or 5 prominent veins, cobwebby above, silvery-hairy beneath, and 1½ to 3 inches long. It is sometimes called Ladies'-tobacco and grows in dry open woods or thickets, blooming between late March and June.

RANGE: Me. to Minn. south to Fla., Ala. and Mo.

STINKING MARSH-FLEABANE *(Pluchea foetida)*

Marsh-fleabanes are often called Camphorweeds or Stinkweeds because of their strong camphor-like odor. This species has a minutely downy and sticky stem 1½ to 3 feet tall which may be branched at the summit. The oblong to lance-shaped leaves are sharply toothed, 2 to 4 inches long, and are stalkless with clasping bases. The flower heads are about ¼ inch high and contain a number of small creamy-white, tubular flowers. It grows in marshes, ditches, and savannahs of the coastal plain; blooming July to October.

RANGE: N.J. south to Fla. and west to Ala.

Pluchea rosea is similar to the preceding but it has heads of pinkish flowers. It is found in similar situations in the coastal plain from N.C. south to Fla. and west to Miss.; blooming in June or July. (Not illustrated)

INLAND MARSH-FLEABANE *(Pluchea camphorata)*

This is a rather smooth plant from 2½ to 4 feet tall; with narrowly egg-shaped to oval, sharply toothed leaves, 4 to 10 inches long, on slender stalks up to about an inch in length. The lateral branches terminated by flower clusters are shorter than the terminal one; and the heads of purplish flowers are about ¼ inch tall. It grows in wet woods, marshes, meadows and ditches; blooming between August and October.

RANGE: Del. to Ohio, Ill., Mo. and Kan. south to Fla. and Tex.

SALT-MARSH-FLEABANE *(Pluchea odorata)*

This is a species of salt or brackish marshes. It is a tall plant with rather short-stalked leaves which are lance-shaped to egg-shaped, sometims toothed, and quite firm or fleshy in texture. The lateral branches terminated by clusters of flowers are as long as the terminal one, giving a flat-topped appearance. The flowers are pink or purplish; blooming between August and October. S. Me. s. to Fla. west to Tex. and Mex. (Not illustrated)

ELCAMPANE *(Inula helenium)*

Elcampane is a stout plant which usually has several simple, densely woolly stems from 2 to 6 feet high. Its leaves are roughish-hairy above and densely woolly beneath, and toothed on the margin. The basal ones are from 10 to 20 inches long and from 4 to 8 inches wide, and are narrowed at the base into long stalks. Those on the stem above are much smaller, stalkless, and often clasp the stem by a heart-shaped base. The relatively few flower heads are 2 to 4 inches across and have numerous, narrow, yellow rays from 1 to 1½ inches long. This plant has been introduced from Europe but it is now widely naturalized; growing in fields and along roadsides and fencerows from N.S. to Ont. and Minn. south to N.C. and Mo. It blooms between July and September.

CUP-PLANT *(Silphium perfoliatum)*

Indian-cup is another name often given to this plant which usually has a smooth square stem 4 to 8 feet tall, with branch*es* toward the summit. It has pairs of egg-shaped or somewhat triangular leaves which are rough on both surfaces. The lower ones are abruptly contracted into a winged stalk and are coarsely toothed. The upper ones are untoothed and their bases are united, forming a cup about the stem. Its flower heads are numerous, 2 to 3 inches across, and have from 20 to 30 yellow rays. It grows in rich open woods, thickets, along riverbanks, and on prairies; blooming between July and September. RANGE: Ont. to S.D. south to Ga., Miss., Mo. and Okla.

ROSINWEED *(Silphium compositum)*

The Rosinweed has a smooth, whitened, wand-like stem from 3 to 9 feet tall; along which there are very small and widely scattered leaves. Its basal leaves are large, from 4 to 8 inches broad, long-stalked, and deeply cut-lobed or divided; or in the variety *reniforme,* simply roundish egg-shaped with a heart-shaped base and toothed margins. The flower heads are about 1½ inches across and have 10 or fewer yellow rays. They are quite numerous and arranged in a large but loosely forked terminal cluster; blooming June to September. It grows in dry sandy or rocky open woods and clearings.
RANGE: Va. to e. Tenn. south to Ga.

WHORLED ROSINWEED *(Silphium trifoliatum)*

This species of rosinweed has lance-shaped or narrowly egg-shaped, short-stalked leaves 3 to 7 inches long, which are almost always arranged in whorls of 3 or 4. They are roughish above and smooth to hairy beneath, toothed to nearly untoothed on the margin. The flower heads are 1½ to 2 inches across and have from 15 to 20 yellow rays. The plant has a smooth, sometimes whitened, slender stem from 3 to 7 feet tall which branches toward the summit. It grows in woods, thickets, and on prairies; blooming between June and September. RANGE: Pa. to Ind. south to S.C. and Ala.

TOOTHED ROSINWEED *(Silphium asteriscus* var. *laevicaule)*

The Toothed Rosinweed somewhat resembles the preceding species but its leaves are mostly opposite, or the upper ones may be alternate, and the margins are usually coarsely toothed. It grows in sandy woodlands and thickets from N.C. and Ky. south to Ga. and Ala.; blooming in August and September. (Not illustrated)

LARGE-FLOWERED LEAF-CUP *(Smallanthus uvedalia)*

Also known as the Bear's-foot, this is a roughish-hairy, very odorous, stout-stemmed and branching plant from 3 to 9 feet high. Its paired leaves are broadly egg-shaped, angled, and coarsely angular-toothed. The upper ones are stalkless or nearly so, but the lower ones are abruptly narrowed into a winged stalk. Its flower heads are 1½ to 3 inches across with from 10 to 15 bright yellow rays. The cup-shaped involucre has very large and hairy-fringed outer bracts. It grows in rich woods and thickets, blooming between July and October. RANGE: N.Y. to Ill. south to Fla. and Tex.

SMALL-FLOWERED LEAF-CUP *(Polymnia canadensis)*

This species is a more slender, somewhat sticky-hairy plant from 2 to 5 feet tall. The leaves are pinnately 3- to 5-lobed, stalked, and 4 to 10 inches long. Its flower heads are small and have 5 whitish or pale yellow rays which are shorter than the involucre, or they may even lack rays. It grows in moist woods and ravines from w. Vt. to Ont. south to Ga., Tenn., La. and Okla; blooming between July and October. (Not illustrated)

Elcampane
Cup-plant
Rosinweed
Whorled Rosinweed
Large-flowered Leaf-cup

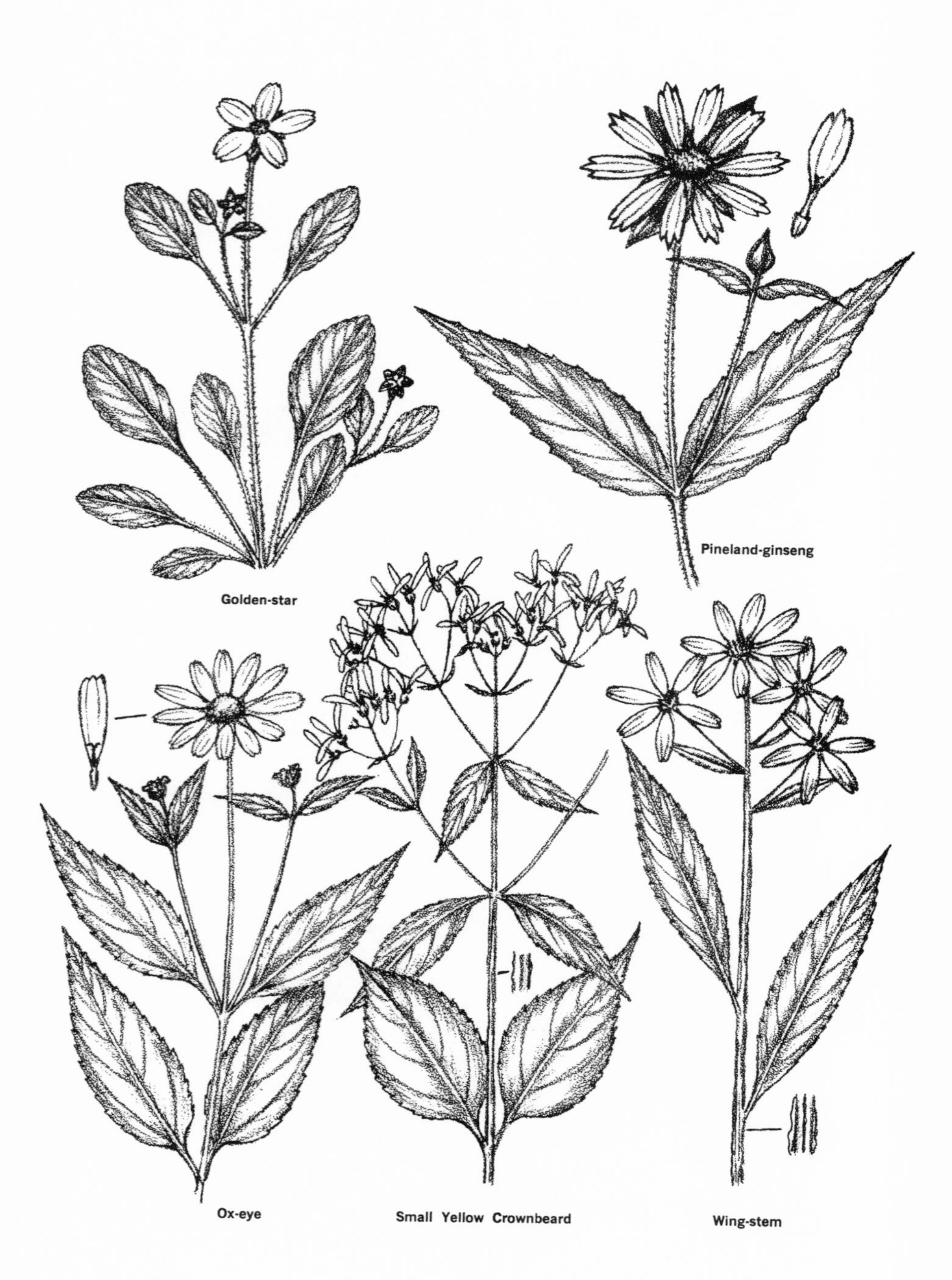
Golden-star
Pineland-ginseng
Ox-eye
Small Yellow Crownbeard
Wing-stem

GOLDEN-STAR *(Chrysogonum virginianum)*

In early spring, when the Golden-star begins to bloom, it appears to be a stemless, low, hairy plant; but it later becomes from 3 to 12 inches high through the growth of branches. It has pairs of oblong or egg-shaped leaves which are 1 to 3 inches long, bluntly toothed, and rather long-stalked. The attractive flower heads are about 1¼ inches across, with usually 5 broad and bright yellow rays. It grows in sandy or rocky woodlands and on banks, blooming between late March and June.
RANGE: Pa. and W. Va. south to Fla. and La.

PINELAND-GINSENG *(Tetragonotheca helianthoides)*

This is a sticky-hairy plant which usually has a simple stem from 1 to 2½ feet tall. It has pairs of egg-shaped or oblong leaves 2 to 6 inches long, which are toothed and have stalkless but narrowed and often somewhat clasping bases. The heads of flowers are 1½ to 3 inches across and have from 6 to 10 pale yellow rays. An unusual feature is its double involucre in which the 4 large and leaf-like outer bracts are united to form a 4-sided cup. It grows in sandy open woods and thickets; blooming from April to July and sometimes again in the fall. RANGE: N.C. south to Fla. and west to Miss.

OX-EYE *(Heliopsis helianthoides)*

The Ox-eye is a sunflower-like plant with a smoothish stem 3 to 5 feet tall which may be simple or branched above. Its leaves are paired, egg-shaped or broadly lance-shaped, sharply toothed, slender-stalked, often more or less rough on the upper surface, and from 3 to 6 inches long. The heads of flowers are about 2 inches across and have 10 yellow rays which, unlike the rays of the true sunflowers, have pistils. It grows in open woods, thickets, and on dry banks; blooming between May and October. Also called False Sunflower. RANGE: N.Y. to Ont. and Minn. south to Ga. and Tex.

SMALL YELLOW CROWNBEARD *(Verbesina occidentalis)*

This plant has a narrowly 4-winged and usually smooth stem from 3 to 7 feet tall, with branches above which are also winged and often downy. It has pairs of egg-shaped to lance-shaped leaves which are toothed, pointed at both ends, stalked, and 4 to 10 inches long. The numerous flower heads are about an inch across, with from 1 to 5 yellow rays. They are arranged in a large and rather flat-topped terminal cluster, blooming between August and October. It grows in rich woods, thickets, and openings.
RANGE: Pa to Ill. south to Fla. and Miss.

SMALL WHITE CROWNBEARD *(Verbesina virginica)*

This Crownbeard differs from the preceding species in having scattered rather than paired leaves, and smaller flowers with from 3 to 5 white rays. It grows in dry open woods, thickets and clearings from Pa. to Ky., Mo. and Kan. south to Fla. and Tex.; blooming July to October. (Not illustrated)

SUNFLOWER CROWNBEARD *(Verbesina helianthoides)*

This species has alternate leaves, and relatively few heads of flowers which are 2 to 3 inches across with a bowl-shaped involucre and from 8 to 15 yellow rays. It has a rough-hairy stem up to about 3 feet high which is rather broadly 4-winged. The leaves are egg-shaped to lance-shaped, stalkless, toothed, 2 to 4 inches long, rough above but softly hairy beneath. It grows in dry open woods and on prairies from Ohio to Ill. and Iowa south to Ga. and Tex.; blooming between June and October. (Not illustrated)

WING-STEM *(Verbesina alternifolia)*

The Wing-stem, or Yellow Ironweed, has a stem from 4 to 9 feet tall which is winged along the upper portion, usually branched above, and often somewhat hairy. The leaves are scattered, or the lower ones may be opposite or in whorls of 3. They are mostly broadly lance-shaped, toothed, pointed at both ends, 4 to 12 inches long, and stalkless or short-stalked. The flower heads are numerous, 1 to 2 inches across, and the 2 to 10 yellow rays often differ quite a bit in size. It grows in the borders of moist woods and thickets, blooming during August or September.
RANGE: N.Y. to Ont. and Iowa south to Fla. and La.

TALL CONEFLOWER *(Rudbeckia laciniata)*

The Tall Coneflower has a smooth, whitened, and branching stem from 2 to 9 feet high. Its lower leaves are stalked, usually divided into 5 to 7 deeply cut or 3-lobed leaflets, and often a foot wide. Those upward along the stem are often similar but stalkless; or sometimes 3- to 5-parted, or merely toothed. The long-stalked flower heads are quite numerous, 2½ to 4 inches across; with 6 to 10 bright yellow rays and a greenish-yellow, dome-shaped disk. It grows in rich, moist, open woods and thickets; blooming between July and October. A form with more numerous ray flowers is cultivated as the Golden-glow.

RANGE: Que. to Mont. south to Fla., Tex. and Ariz.

THIN-LEAF CONEFLOWER *(Rudbeckia triloba)*

This coneflower usually has a somewhat roughish-hairy and branching stem from 1 to 5 feet high. The leaves are thin and rough on both sides; the lower ones being 2 to 4 inches long, stalked, mostly 3-lobed or 3-parted, and sharply toothed. Those along the upper part of the stem are short-stalked or stalkless and may even be untoothed. The flower heads are often very numerous, almost 2 inches across, and have from 8 to 12 bright yellow or orange-yellow rays and an egg-shaped dark purplish-brown disk. It grows in open woods, thickets, fields, and on rocky slopes; blooming between July and October.

RANGE: N.Y. to Minn. south to Ga., Tenn., Ark. and Okla.

SHOWY CONEFLOWER *(Rudbeckia fulgida* var. *speciosa)*

The Showy Coneflower has a rather sparsely hairy stem from 1 to 3 feet tall which is loosely branched above. Its leaves are quite firm, 2 to 5 inches long, broadly lance-shaped, coarsely toothed or somewhat cut, and taper into winged stalks. The flower heads are about 2½ inches across; with from 12 to 20 bright yellow and usually orange-based rays and a somewhat flattened, globular, purple-brown disk. It grows in moist woods, bottomlands, and swamps; blooming between July and October.

RANGE: N.Y. to Mich. and Mo. south to Ga., Ala. and Ark.

BLACK-EYED SUSAN *(Rudbeckia hirta)*

Often called the Yellow Daisy, this is one of our best known wild flowers. It grows very commonly in fields, on banks, and along the roadsides; blooming between May and September. A roughish-hairy plant, it has simple or few-branched stems from 1 to 3 feet tall; and lance-shaped or narrowly top-shaped leaves which are 2 to 7 inches long, with untoothed or indistinctly toothed margins. The larger, lower ones taper into rather long stalks. Its flower heads are 2 to 4 inches broad and have from 10 to 20 bright orange-yellow rays; and an egg-shaped disk which is dark purplish-brown.

RANGE: Ont. to Man. south to Fla. and Tex.

GRAY-HEADED CONEFLOWER *(Ratibida pinnata)*

This is a minutely roughish-hairy, hoary plant with a simple or branched stem 1½ to about 4 feet high. The leaves are pinnately divided into from 3 to 7 segments which are sometimes toothed. Its flowers are quite large; with from 4 to 7 drooping yellow rays 1 to 3 inches long, and an oblong grayish disk which may become brown. It grows in dry open woods and on prairies from n. N.Y. and Ont. to Minn. and Neb. south to Ga. and Okla.; blooming between June and September. (Not illustrated)

PURPLE CONEFLOWER *(Echinacea purpurea)*

The Purple Coneflower has flower heads 2½ to 5 inches across; with from 12 to 20 purple and drooping rays and a cone-shaped, purplish disk. It has a simple or branched, often bristly hairy stem 3 to 5 feet tall. The leaves are roughish above and sharply toothed; the basal ones egg-shaped, 3 to 8 inches long, 5-veined, and stalked. The upper ones are lance-shaped, stalkless and much smaller. It grows in dry open woods and on prairies; blooming between June and September.

RANGE: Va. to Mich., Ill. and Iowa south to Ga. and La.

SMOOTH-LEAF PURPLE CONEFLOWER *(Echinacea laevigata)*

This species is similar to the preceding one but it has smooth and whitened leaves. It grows in woods and fields from Pa. south to Ga. (Not illustrated)

Tall Coneflower
Thin-leaf Coneflower
Showy Coneflower
Black-eyed Susan
Purple Coneflower

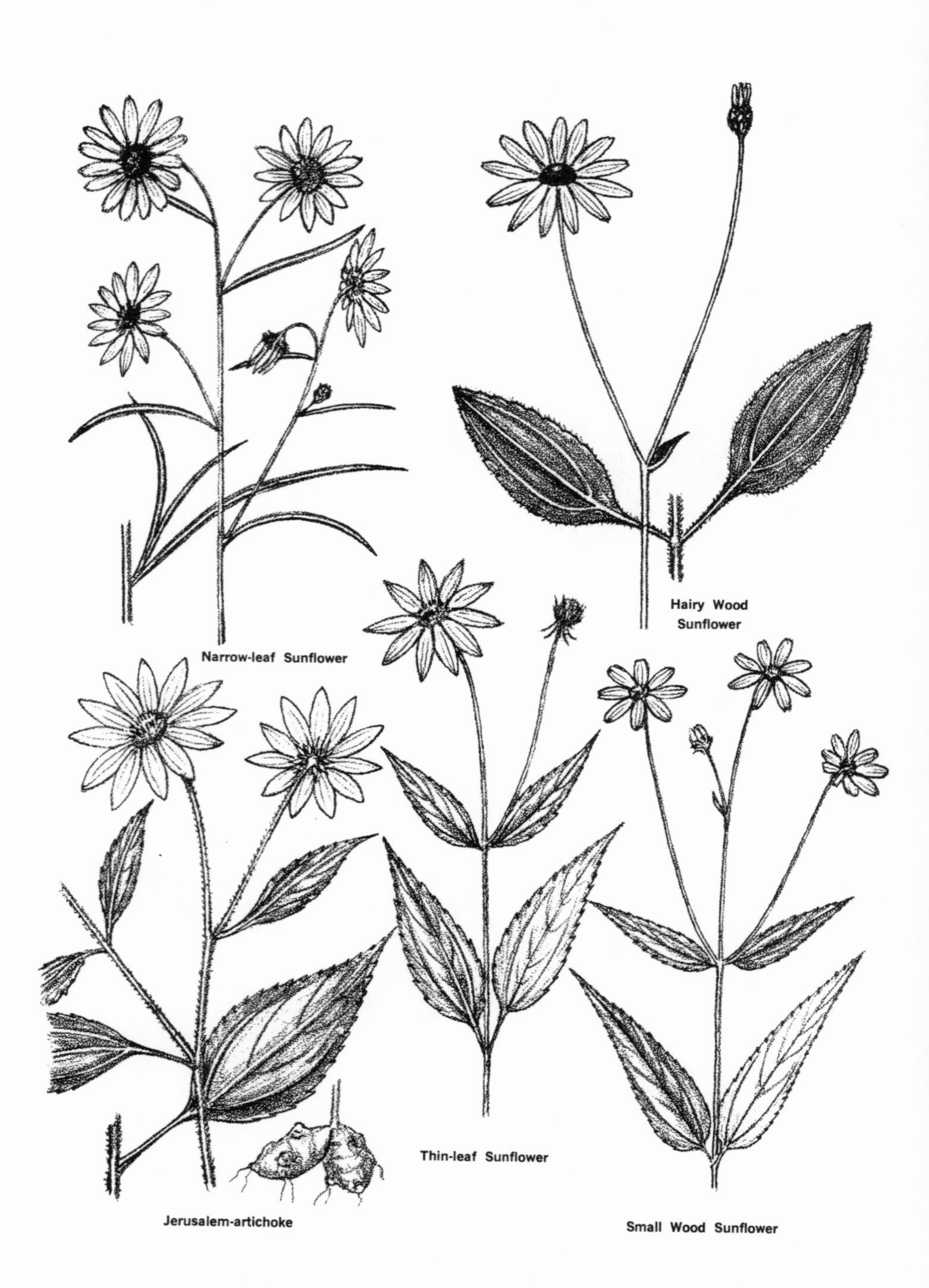
Narrow-leaf Sunflower
Hairy Wood Sunflower
Thin-leaf Sunflower
Jerusalem-artichoke
Small Wood Sunflower

Sunflowers

Sunflowers are mostly tall plants, usually with opposite leaves, and with showy flower heads with a prominent disk and marginal ray flowers. The latter are neutral, containing no sexual organs and producing no seeds. The involucre has several rows of overlapping bracts which are green and more or less leaf-like. Close to 40 species have been described from eastern North America.

NARROW-LEAF SUNFLOWER *(Helianthus angustifolius)*

This sunflower of the wet pinelands and swampy thickets has a slender stem 2 to 6 feet tall. It may be simple or branched above and is more or less rough, at least on the lower part. Its leaves are all very narrow, stiff, rough, untoothed, stalkless, 2 to 7 inches long, and often have clusters of smaller leaves in their axils. The flower heads are 2 to 3 inches across, with a dark purplish-brown disk and from 12 to 20 bright yellow rays. It blooms between July and October.
RANGE: se. N.Y. to s. Ind. and Mo. south to Fla. and Tex.

HAIRY WOOD SUNFLOWER *(Helianthus atrorubens)*

Often called the Purple-disked Sunflower, this species is widely distributed from the coastal plain pinelands to open woods and thickets in the mountains. Its stem is 2 to 5 feet tall and is densely hairy below, but less so in the branching portion above. The leaves are mostly on the lower part of the stem and are egg-shaped to oval, usually abruptly narrowed into a long and winged stalk. They are 4 to 10 inches long, rough above, more or less hairy beneath, and rather shallowly toothed. The flower heads are about 2 inches across, with from 10 to 20 bright yellow rays and a purplish-brown disk; blooming between late July and October.
RANGE: Va. to Ky. south to Fla. and La.

JERUSALEM-ARTICHOKE *(Helianthus tuberosus)*

In spite of its common name, this is a native American sunflower. Its thick and fleshy rootstocks bear edible tubers which were once much used by the Indians, and still sometimes cultivated. It has a very roughish-hairy stem 6 to 10 feet tall which is branched above. Its leaves are thick and hard, very rough-hairy above, downy beneath, and prominently 3-veined. Those along the main part of the stem are 4 to 8 inches long, egg-shaped, coarsely toothed, and contracted into broadly winged stalks. The flower heads are quite numerous, 2 to 3 inches across, with from 12 to 20 bright yellow rays and a yellowish disk. It grows in moist thickets and open places, blooming between July and October.
RANGE: Ont. to Sask. south to Ga., Tenn. and Ark.; and more widely naturalized.

THIN-LEAF SUNFLOWER *(Helianthus decapetalus)*

Also known as the Ten-petalled Sunflower, this species has a smoothish stem 1 to 5 feet tall which is branched and may be slightly roughish above. Its thin textured leaves are smooth or but slightly roughish above and are rather indistinctly 3-veined. The larger and lower ones are egg-shaped, 3 to 8 inches long, sharply toothed, and slender stalked. Those of the upper part of the stem are smaller, lance-shaped, and taper into short stalks. The flower heads are 2 to 3 inches across, with from 8 to 15 light yellow rays and a yellowish disk. It grows in open woods, thickets, and often along streams; blooming between August and October.
RANGE: Me. to Que., Minn. and Neb. south to Ga., Ky. and Mo.

SMALL WOOD SUNFLOWER *(Helianthus microcephalus)*

This woodland sunflower has a slender, smooth stem from 3 to 6 feet high which is branched above. Its leaves are lance-shaped to narrowly egg-shaped, thin, roughish above, pale and downy beneath, sharply toothed, stalked, and from 3 to 7 inches long. The flower heads are numerous, about an inch across, and have 5 to 10 yellow rays and a yellowish disk. It usually grows in moist places, blooming from August to October.
RANGE: Pa. to Ill. south to Fla., Miss. and Mo.

WOODLAND SUNFLOWER *(Helianthus divaricatus)*

The Woodland Sunflower is also known as the Rough Sunflower. It has a smooth, slender, often whitened stem from 2 to 7 feet tall which is usually branched above. Its leaves are lance-shaped to narrowly egg-shaped, broad at the base, stalkless or nearly so, toothed, 3 to 8 inches long, more or less hairy beneath, and rough-hairy on the upper surface. The flower heads are about 2 inches across, with 8 to 15 yellow rays and a yellowish disk. It grows in dry open woods, thickets, and clearings; blooming between June and September.

RANGE: Me. to Que. and Sask. south to Ga., Tenn. and Ark.

PALE-LEAF WOOD SUNFLOWER *(Helianthus strumosus)*

This species has a smooth or slightly roughish stem from 3 to 7 feet tall, which is usually branched above. Its leaves are egg-shaped or broadly lance-shaped, distinctly stalked, shallowly toothed or almost toothless, thick and firm, very rough above, pale to whitish and downy beneath, and from 3 to 8 inches long. The flower heads are 2½ to 4 inches across, with 5 to 15 yellow rays and a yellowish disk. It grows in dry open woods, thickets, and clearings; blooming between July and September.

RANGE: Que. to N.D. south to Ga., Ala., Ark. and Okla.

TALL SUNFLOWER *(Helianthus giganteus)*

The Tall Sunflower has a roughish stem which is usually 5 to 12 feet tall, and often much-branched above. Its leaves are mostly scattered, lance-shaped, stalkless or short-stalked, shallowly or inconspicuously toothed, green and roughish on both sides, and 2 to 6 inches long. The numerous flower heads are 1½ to 2½ inches across, with from 10 to 20 rather pale lemon-yellow rays and a relatively large yellowish disk. It grows in swamps and wet thickets or meadows, blooming between July and October.

RANGE: Que. to Sask. south to Fla., La. and Colo.

VARIED-LEAF SUNFLOWER *(Helianthus heterophyllus)*

This sunflower has a slender and roughish stem from 3 to 4 feet high, which is usually simple or very rarely with a few long and ascending branches. Its lower leaves have egg-shaped to elliptic or narrowly top-shaped blades which taper into stalks, and may be as much as 8 inches long. The ones on the stem above are very narrow, stalkless, and smaller. All are rather coarse in texture, rough-hairy, and untoothed. The long-stalked flower heads are 3 or 4 inches across, with about 20 bright yellow rays and a dark purplish-brown disk. It grows in wet coastal plain pinelands and the borders of swamps and marshes; blooming between August and October.

RANGE: N.C. south to Fla. and west to La.

HAIRY SUNFLOWER *(Helianthus mollis)*

The Hairy Sunflower has a stout, densely grayish- or whitish-hairy, and more or less roughish stem from 2 to 4 feet tall which is simple or has a few ascending branches above. The leaves are egg-shaped to broadly lance-shaped, stalkless, toothed, grayish and rough above, pale and downy beneath. They are 2 to 5 inches long and more or less clasp the stem by a rounded to somewhat heart-shaped base. The flower heads are 2 to 3 inches across, and have from 15 to 25 yellow rays and a yellowish disk. It grows in dry open woods and fields, blooming between July and September.

RANGE: Ohio and Mich. to Iowa south to Ga. and Tex.

STIFF-HAIRED SUNFLOWER *(Helianthus hirsutus)*

In a general way this sunflower resembles the Woodland Sunflower but its stem has stiffish and spreading hairs and is often roughish. The narrow leaves are very rough-hairy above and somewhat so beneath, usually prominently 3-veined, short-stalked, and sharply toothed. It grows in dry open woods and thickets from w. Pa. to Minn. south to Fla. and Tex.; blooming between July and October. (Not illustrated)

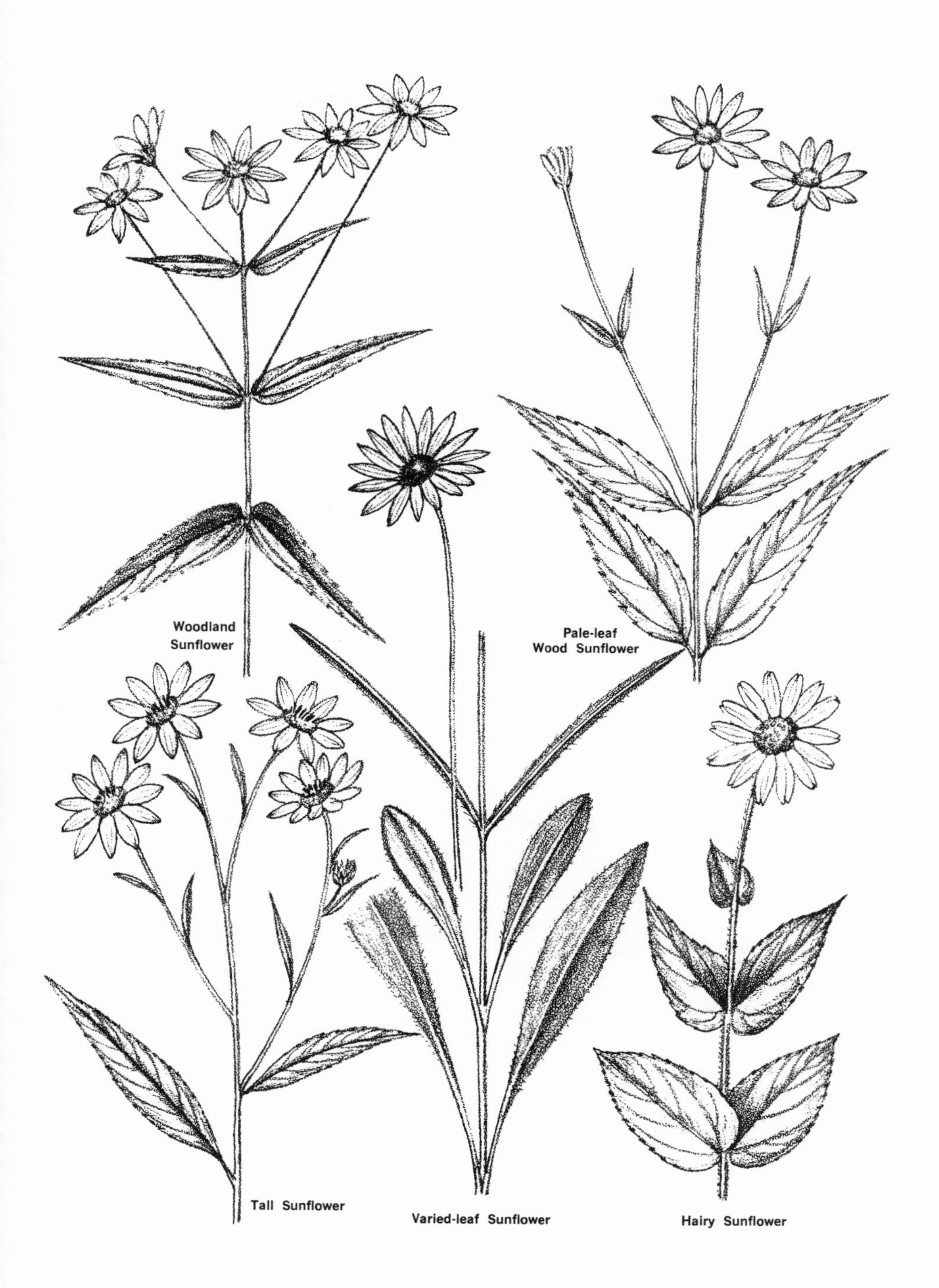
Woodland
Sunflower
Pale-leaf
Wood Sunflower
Tall Sunflower
Varied-leaf Sunflower
Hairy Sunflower

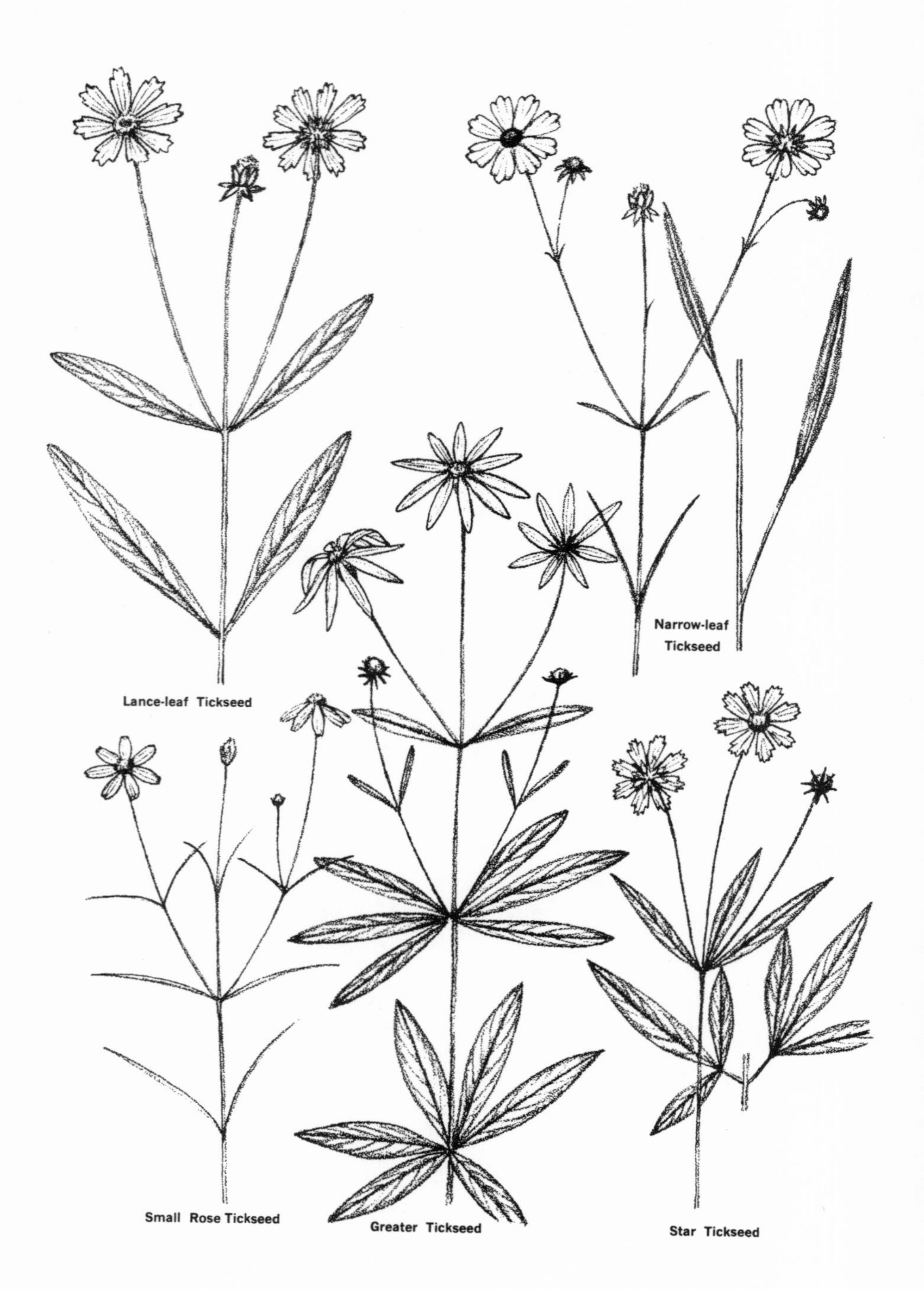
Lance-leaf Tickseed
Narrow-leaf Tickseed
Small Rose Tickseed
Greater Tickseed
Star Tickseed

Tickseeds

Tickseeds usually have opposite leaves and showy flower heads with a disk and about 8 marginal rays. The involucre is made up of two series of bracts: the outer ones green, leaf-like, and somewhat spreading; the inner ones broader, closely pressed, more or less erect, and usually not green in color.

LANCE-LEAF TICKSEED *(Coreopsis lanceolata)*

This plant has a slender, smoothish or hairy stem from 1 to 2 feet tall. Its leaves are lance-shaped to inversely lance-shaped, occasionally with 1 or 2 small lobes near the base. They are untoothed, 2 to 6 inches long, and the lower ones are stalked. The long-stalked flower heads are 1½ to 2½ inches across, with 6 to 10 bright yellow rays which are rather deeply lobed at the summit and a yellowish disk. The outer involucral bracts are lance-shaped, green and spreading; the inner ones egg-shaped and brownish or reddish tinged. It grows in dry, sandy or rocky, open woods, thickets, or clearings; blooming between April and July.

RANGE: Va. to Ont., Mich., Wis. and Mo. south to Fla. and N.Mex.

NARROW-LEAF TICKSEED *(Coreopsis gladiata)*

This species has a slender and smoothish stem 1½ to 2½ feet tall. Its lower leaves are scattered, inversely lance-shaped, untoothed, 3 to 5 inches long, and taper into stalks. The upper ones are much smaller, very narrow, and usually paired. The flower heads are 1½ to 2 inches across and have 6 to 10 deep yellow rays which are lobed at the summit, and a dark brown disk. The outer involucral bracts are egg-shaped and greenish; the inner ones larger and mostly red. It grows in low, wet, coastal plain pinelands; blooming between August and October.

RANGE: N.C. south to Fla. and west to Tex.

SMALL ROSE TICKSEED *(Coreopsis rosea)*

The flower heads of this tickseed are ½ to 1 inch across and have from 4 to 8 pink, rose-colored, or occasionally white rays and a yellowish disk. It has slender, branching, smooth stems 6 inches to 2 feet tall. The leaves are very narrow, untoothed, 1 to 2½ inches long, and often have tufts of smaller leaves in their axils. It grows in moist to wet open places along the Atlantic Coast; blooming between July and September. Also called Pink Tickseed.

RANGE: N.S. south to Ga.

GREATER TICKSEED *(Coreopsis major)*

The Greater, or Wood Tickseed has a slender and more or less downy stem 2 to 3 feet tall. Its leaves are stalkless and all except a few of the uppermost ones are divided into 3 lance-shaped or narrowly egg-shaped, untoothed segments 2 to 4 inches long and from ¼ to an inch broad, the plant thus appearing to have whorls of 6 leaves. The flower heads are slender-stalked, 1 to 2 inches across, with 6 to 10 yellow rays and a yellowish disk. The outer involucral bracts are narrow and about as long as the inner ones. It grows in open woods, thickets, and clearings; blooming between May and August.

RANGE: Va. to Ohio south to Fla. and Miss.

WHORLED TICKSEED *(Coreopsis verticillata)*

This species is quite similar to the preceding one but its leaves are divided into very narrow segments less than ⅛ inch broad, and they are often branched. It grows in dry open woods and clearings from Md. to Ky. and Ark. south to Fla. and Ala. (Not illustrated)

STAR TICKSEED *(Coreopsis pubescens)*

The Star Tickseed has a slender and downy stem 2 to 4 feet tall, which may have a few branches. The upper leaves are short-stalked or stalkless, lance-shaped, or sometimes cut into 3 to 5 segments, and 2 to 3 inches long. The lower ones are top-shaped and taper into slender stalks. Its flower heads are about 1¼ inches across, with 8 to 10 yellow rays which are lobed at the summit and a yellowish disk. The spreading narrow outer bracts of the involucre are star-like and about as long as the inner ones. It grows in dry open woods and on slopes or cliffs, blooming between July and September.

RANGE: Va. to s. Ill. and Mo. south to Fla. and La.

TALL TICKSEED *(Coreopsis tripteris)*

The Tall Tickseed has a smooth stem 4 to 8 feet tall which is branched above. The principal leaves are 2 to 5 inches long, definitely stalked, and divided into from 3 to 5 lance-shaped, untoothed segments. The uppermost leaves are not divided but are lance-shaped, smaller, and stalkless. Its slender-stalked flower heads are 1 to 1½ inches across, with 6 to 10 yellow rays and a yellowish disk. The outer involucral bracts are only ⅛ inch or less long and very much shorter than the inner ones. It grows in moist open woods and thickets, blooming between July and September. The heads have an anise-like odor.

RANGE: Ont. to Wis. south to Ga., La. and Kan.

Beggar-Ticks, Bur-Marigolds, and Tickseed Sunflowers

As in the tickseeds, the flower heads have involucral bracts in 2 series, the outer ones being quite large and usually leaf-like. The heads have many flowers, often with 3 to 8 rays, but some species have only disk flowers. The seed-like fruits have 2 to 4 barbed awns. They are usually called "beggar-ticks" or "stick-tights" as they commonly stick to one's clothing in trips afield in late summer and fall. The plants have opposite leaves.

SMALLER BUR-MARIGOLD *(Bidens cernua)*

This is a variable plant, sometimes with simple and slender stems only a few inches high, stalked and narrow leaves, and heads of flowers about ½ inch across. Again it may have a rather stout stem 3 feet tall, with lance-shaped or narrowly oblong leaves 3 to 6 inches long which are stalkless or nearly so, and with flower heads about an inch across. The flower heads commonly nod and they may have 6 to 10 yellow rays or none at all. It grows in swamps and other wet places, blooming between July and October. Also called the Nodding Bur-marigold.

RANGE: Que. and N.S. to B.C. south to Md., w.N.C., Tenn., Mo. and Colo.

LARGER BUR-MARIGOLD *(Bidens laevis)*

Also known as the Smooth Bur-marigold, this species is a smooth plant with a branching stem 1 to 3 feet high; and lance-shaped, toothed leaves 3 to 8 inches long. Its flower heads are much more showy than those of the preceding species, 1 to 1½ inches across, with 8 to 10 rather large and bright yellow rays. It grows in swamps, wet meadows, or along streams; blooming between September and November.

RANGE: N.H. to W.Va. and Ind. south to Fla. and Mex.

TICKSEED-SUNFLOWER *(Bidens coronata)*

This is a rather smooth plant with a much-branched stem from 1 to 5 feet high. Its leaves are mostly 3 to 5 inches long and pinnately divided into from 3 to 7 narrow or lance-shaped segments which are sharply cut or toothed; the lower ones being distinctly stalked. The flower heads are 1 to 2 inches across and have from 15 to 25 rather short yellow rays and a large, deeper yellow disk. The involucre has from 6 to 8 outer bracts. It grows in wet places and blooms between August and October.

RANGE: Conn. to Ont. and Neb. south to Fla. and Ala.

WESTERN TICKSEED-SUNFLOWER *(Bidens aristosa)*

This species rather closely resembles the preceding one. It is best distinguished by the 8 to 12 spreading outer involucral bracts which are no longer than the inner ones. It grows in low wet grounds from Me. to Minn. south to S.C., Ala., La. and Tex.; blooming between August and October. (Not illustrated)

SEA OX-EYE *(Borrichia frutescens)*

This is the shrubby and more or less fleshy or leathery composite, with yellow heads of flowers, that is so conspicuous on the borders of salt or brackish marshes in the Southeast. It has a sparingly branched stem 1 to 4 feet high; and pairs of lance-shaped to top-shaped leaves from 1 to 3 inches long, which sometimes have a few teeth toward the base. Its flower heads are about an inch across, with 15 to 25 rather short yellow rays and a large dusky-yellow disk. All parts of the plant are covered with minute silky-white hairs, giving it a rather dusty appearance. It blooms between May and September.

RANGE: se. Va. south to Fla. and west to Tex.

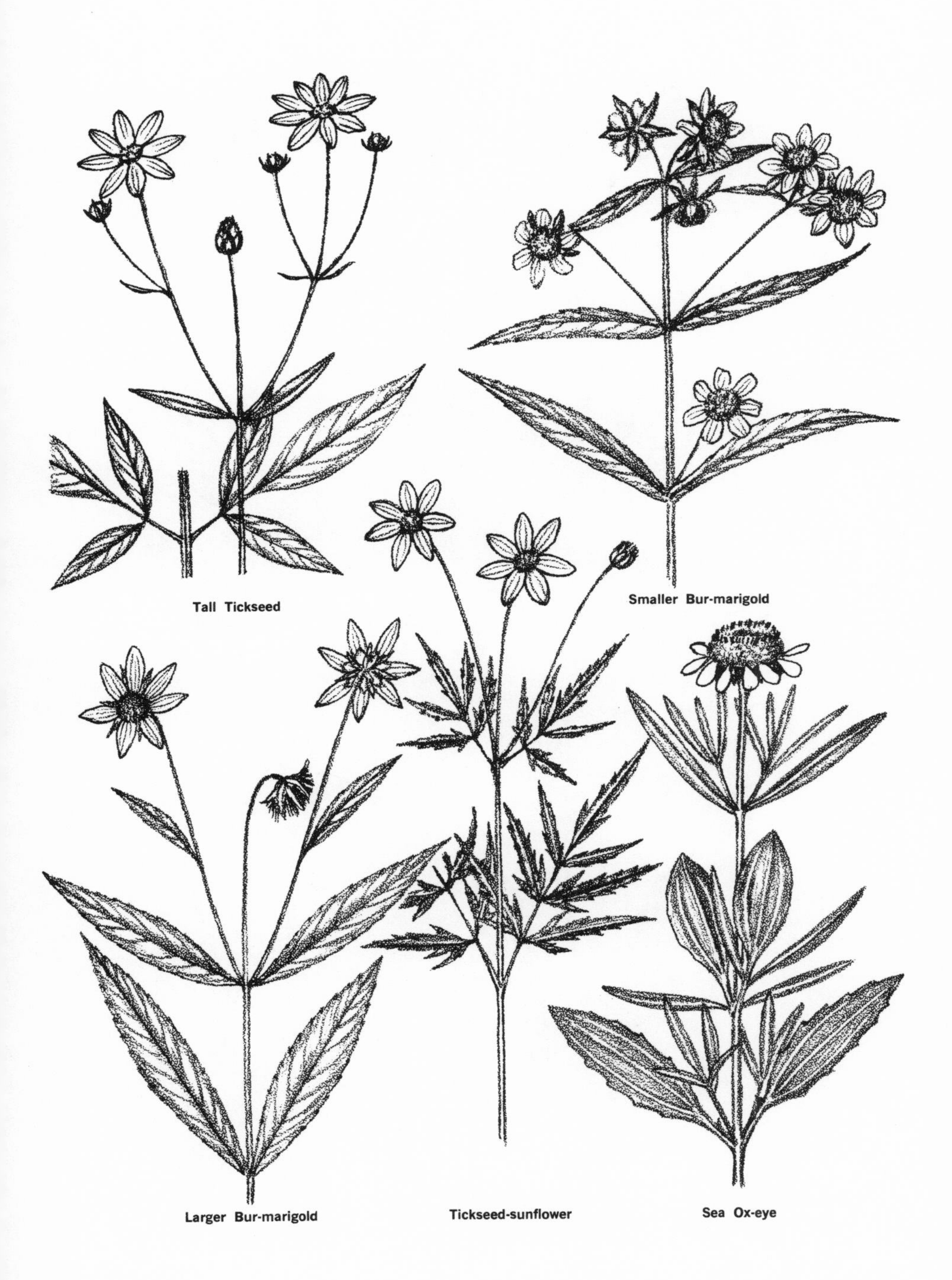
Tall Tickseed
Smaller Bur-marigold
Larger Bur-marigold
Tickseed-sunflower
Sea Ox-eye

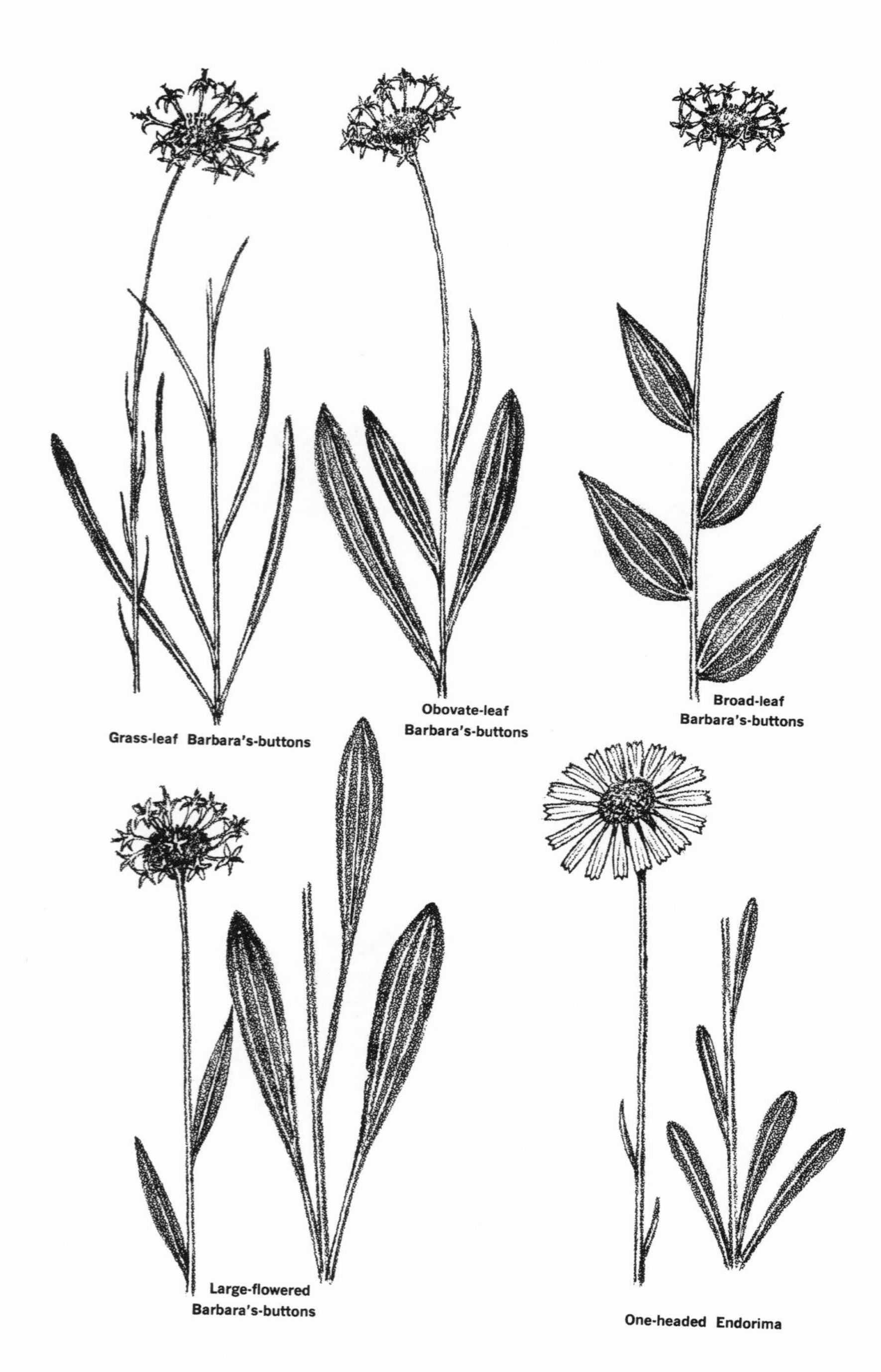
Grass-leaf Barbara's-buttons
Obovate-leaf
Barbara's-buttons
Broad-leaf
Barbara's-buttons
Large-flowered
Barbara's-buttons
One-headed Endorima

GRASS-LEAF BARBARA'S-BUTTONS *(Marshallia graminifolia)*

This is a smooth plant with a simple or sparingly branched and slender stem 1 to 2 feet tall; along which are scattered, narrow leaves. The leaves toward the base are larger, up to 4½ inches long, and mostly broadest near or somewhat above the middle. Its flower heads are long-stalked, about 1 inch across, and made up of a number of pink flowers having a slender tube with 5 flaring lobes at the summit and bluish anthers. This very attractive wildflower grows in moist to wet coastal plain pinelands and savannahs, blooming between July and September.

RANGE: N.C. south to Fla. and west to La.

OBOVATE-LEAF BARBARA'S-BUTTONS *(Marshallia obovata)*

Unlike the preceding species, this one grows in old fields and open woods chiefly in the piedmont region. It is also a smooth plant, usually with a slender and simple stem from 6 inches to about 2 feet tall, which is leafy chiefly below the middle or only toward the base. Its leaves are mostly narrowly top-shaped and from 1 to 4½ inches long. The flower heads are about an inch across; made up of a number of whitish to pale pink flowers with slender tubes. It blooms during April and May.

RANGE: N.C. and Mo. south to Fla. and Ala.

BROAD-LEAF BARBARA'S-BUTTONS *(Marshallia trinervia)*

This is a smooth plant with a simple or sparingly branched stem 1 to 2 feet tall, leafy to or somewhat beyond the middle. Its leaves are thin, egg-shaped to broadly lance-shaped, conspicuously 3-ribbed, stalkless, untoothed and 2 to 3 inches long. The flower heads are about ¾ inch across and made up of a number of purplish flowers with slender tubes. It grows in damp woods and pinelands, blooming in May and June.

RANGE: Va. and Tenn. south to Ga. and Miss.

LARGE-FLOWERED BARBARA'S-BUTTONS *(Marshallia grandiflora)*

While most Barbara's-buttons are plants of the lowlands, this species grows in boggy places in the mountains. It is a smooth plant with a simple slender stem 1 to 2 feet tall, leafy to or beyond the middle. The lower leaves are narrowly top-shaped, 2 to 6 inches long, thickish, and taper into stalks which are often as long as the blades. The upper ones are much smaller, lance-shaped, stalkless and more or less clasping at the base. Its flower heads are an inch or more across and made up of a number of slender-tubed purplish flowers about ⅔ inch long. It blooms in May and June.

RANGE: Pa. to Ky. south to w. N.C.

ONE-HEADED ENDORIMA *(Baludina uniflora)*

This plant usually has a simple, rather stout, minutely downy stem 1 to 3 feet tall. On it are just a few scattered leaves for most of the leaves are basal. The leaves are thickish, narrow, stalkless, untoothed and 1 to 2 inches long; the lower ones usually being somewhat broader toward the tip. The long-stalked flower head is 2 to 2½ inches across; with from 20 to 30 yellow rays which are lobed at the tip, and a large yellowish disk. The receptacle has a peculiar honey-combed surface. It grows in moist coastal plain pinelands, blooming between July and September.

RANGE: N.C. south to Fla. and west to La.

PURPLE-DISKED ENDORIMA *(Baludina atropurpurea)*

This species is very similar to the preceding but its flower heads have purple disks. It is less common in the coastal plain of S.C. and Ga. (Not illustrated)

COMMON SNEEZEWEED *(Helenium autumnale)*

The conspicuously wing-angled stem of this plant is 2 to 6 feet tall and branches toward the summit. Its narrowly oblong to lance-shaped leaves are stalkless, usually toothed, and 2 to 5 inches long. The usually numerous flower heads are 1 to 2 inches across; with from 10 to 18 drooping bright yellow rays which are 3-lobed at the summit, and a large ball-shaped yellowish disk. It grows in swamps, wet meadows and thickets; blooming in September and October.

RANGE: Que. to Minn. and Neb. south to Fla., Tex. and Ariz.

PURPLE-HEAD SNEEZEWEED *(Helenium flexuosum)*

This species has a narrowly wing-angled, more or less downy stem 1 to 3 feet tall, with stiffly ascending branches above. Its leaves are lance-shaped or narrower, mostly untoothed, 1½ to 3 inches long, and the lower ones taper into winged stalks. The flower heads are slightly more than an inch across; with from 10 to 15 drooping yellow rays 3-lobed at the summit, and a large ball-shaped purplish-brown disk. It grows in moist meadows, fields, or along roadsides; blooming between June and October.

RANGE: Ky. to Kan. south to Ga. and Tex.; spreading as a weed into the N.E.

BITTERWEED *(Helenium amarum)*

This yellow-flowered composite is common in fields and along roadsides in the Southeast and is gradually spreading northward. It is a smooth or minutely downy plant with a slender, much-branched stem 6 inches to 2 feet tall; with very numerous narrow leaves ½ to 1½ inches long, which often have clusters of small leaves in the axils of the larger ones. The numerous flower heads are about an inch across; with 4 to 8 drooping yellow rays 3-lobed at the summit, and a large ball-shaped yellowish disk. Also called Fine-leaf Sneezeweed or Yellow Dog-fennel. It blooms between June and November.

RANGE: Va. to s. Ill., Mo. and Kan. south to Fla., Tex. and Mex.

SAVANNAH SNEEZEWEED *(Helenium vernale)*

This species has a simple, erect stem 1 to 2 feet tall. The lower and basal leaves are long and narrow, from 2 to 6 inches in length, and have sparingly wavy-toothed margins; but on the stem they are rapidly reduced in size upward. The solitary flower head is about 2 inches across; with 15 to 25 spreading bright yellow rays 3-lobed at the summit, and a large yellowish disk which is flattened and dome-shaped. It grows in wet coastal plain pinelands and savannahs, blooming in April or May.

RANGE: S.C. south to Fla. west to Miss.

SHORT-LEAF SNEEZEWEED *(Helenium brevifolium)*

This species is distinguished from the preceding by the purplish-brown disks of its flower heads, and also by its leaves whose bases run for some distance down the stem. It grows in wet pinelands, chiefly in the coastal plain, from se. Va. south to Fla. and west to Miss.; blooming in May or June. (Not illustrated)

COMMON YARROW *(Achillea millefolium)*

Often called Milfoil, this European plant is now widely naturalized in old fields, waste places and along roadsides. It is a smoothish to cobwebby plant with a very strong and distinctive odor, with a simple or branched stem 1 to 2 feet tall. The narrowly oblong to lance-shaped leaves are finely dissected into numerous and very slender segments, the lower ones often as much as 10 inches long and tapered at the base into stalks. The numerous small flower heads are arranged in a dense, flat-topped, terminal cluster. Each head has 4 to 6 small white or occasionally pink or purplish rays. It blooms from late April to September or October.

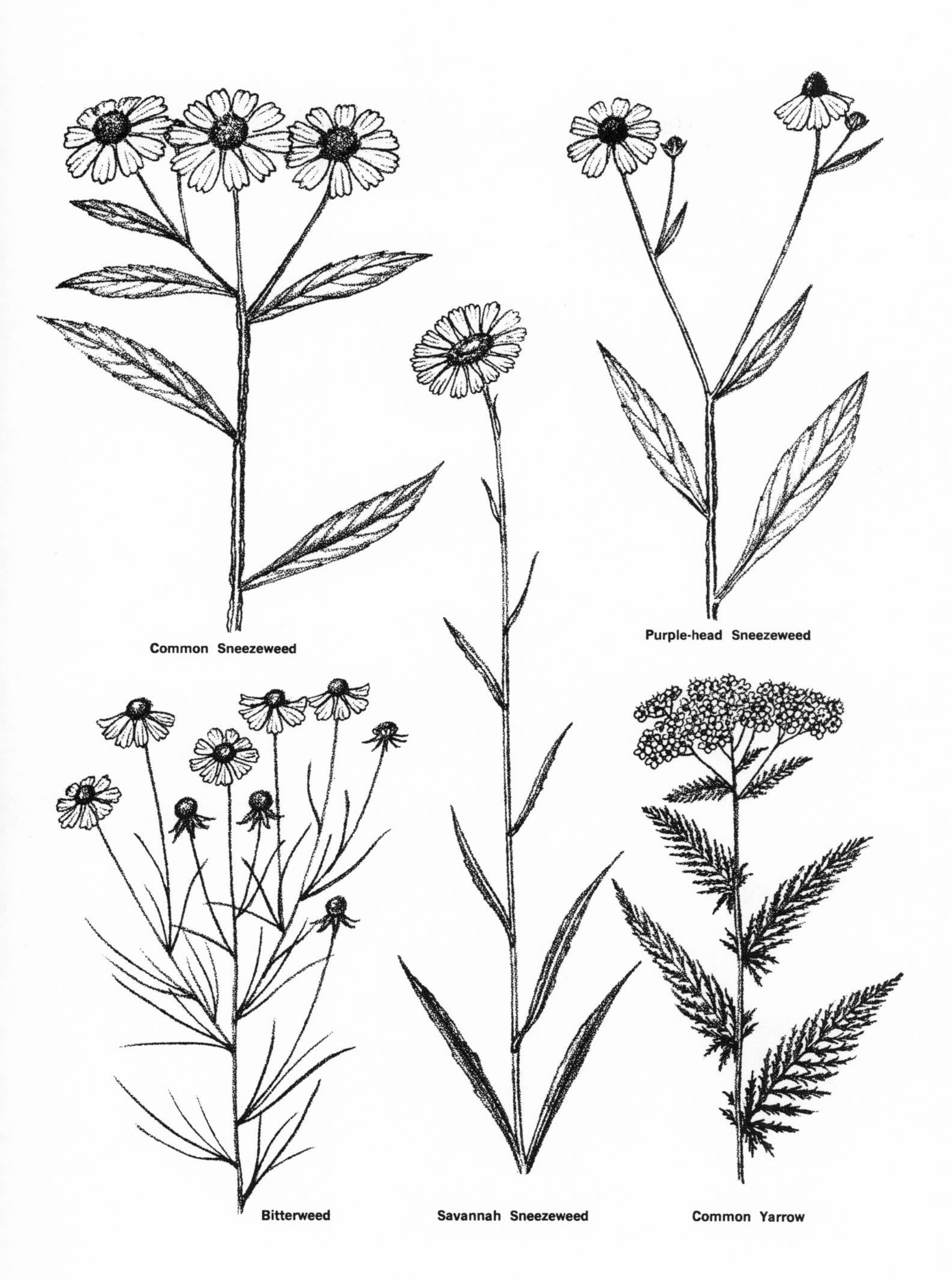
Common Sneezeweed
Purple-head Sneezeweed
Bitterweed
Savannah Sneezeweed
Common Yarrow

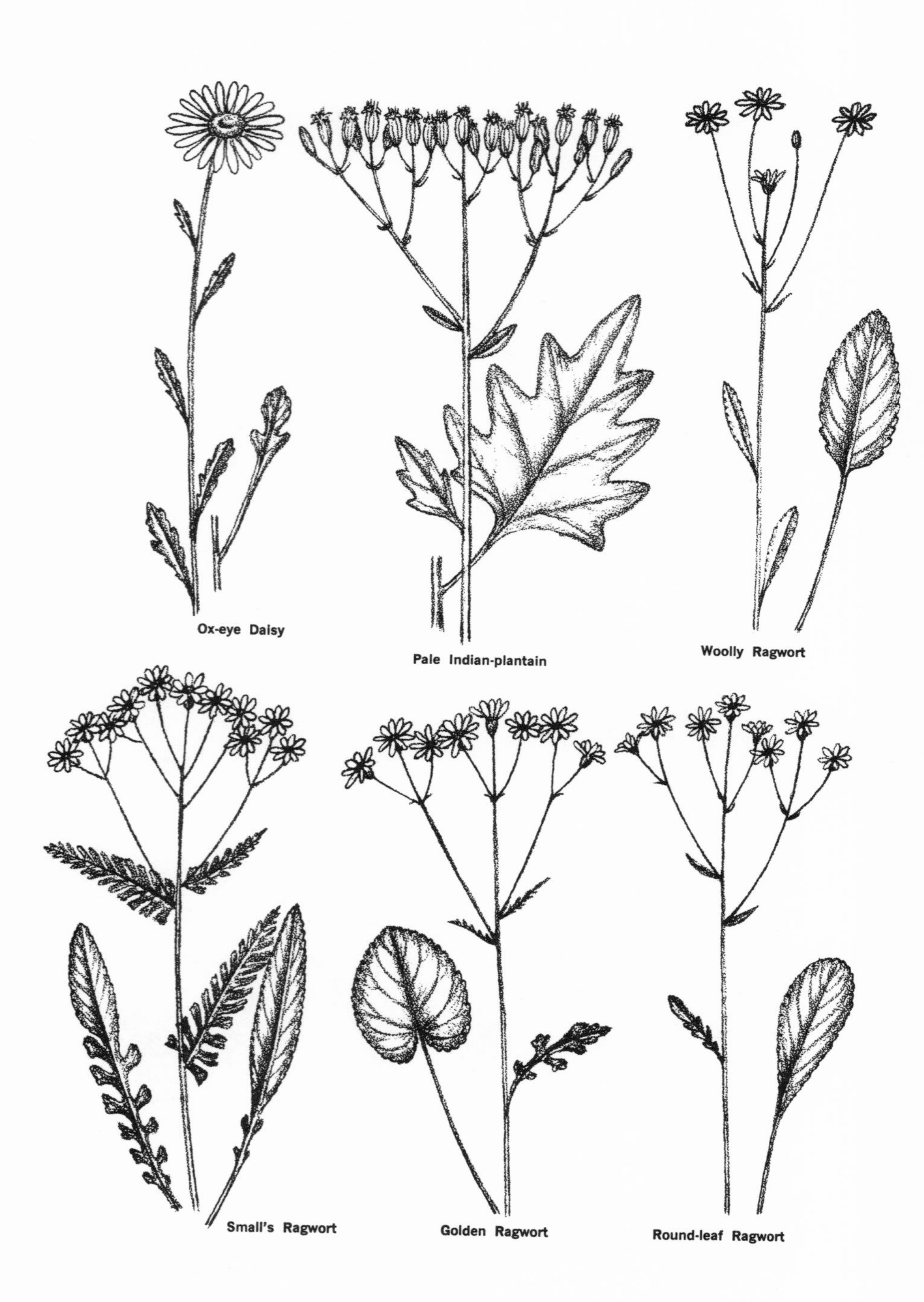
Ox-eye Daisy
Pale Indian-plantain
Woolly Ragwort
Small's Ragwort
Golden Ragwort
Round-leaf Ragwort

OX-EYE DAISY *(Leucanthemum vulgare)*

This common and well-known plant, often called the White or Field Daisy, is a native of Europe which is now widely naturalized in fields, meadows, waste places and along roadsides. It is a smooth or slightly hairy plant with a simple or sparingly branched stem 1 to 3 feet tall. The basal leaves are top-shaped or spoon-shaped, toothed or often lobed, and rather long-stalked. Those along the stem are narrow, stalkless, toothed or often cut-lobed, and 1 to 3 inches long. The flower heads are 1 to 2 inches across, with 20 to 30 bright white rays and a somewhat flattened yellow disk. It blooms from April to July or August.

PALE INDIAN-PLANTAIN *(Arnoglossum atriplicifolium)*

This striking plant has a smooth, round, whitened stem 3 to 6 feet tall. The leaves are somewhat triangular or fan-shaped with a broad base and angular lobes. They are thin, pale green above, whitened beneath, and up to 6 inches across. It numerous heads have a cylindrical involucre about ⅓ inch high and contain several tubular white flowers; and they are arranged in a large, loose, more or less flat-topped cluster. It grows in dry open woods and thickets, blooming between June and September.

RANGE: N.Y. to Minn. south to Ga., Tenn. and Okla.

Arnoglossum ovatum is a similar plant which grows in low wet pinelands and savannahs of the coastal plain from N.C. south to Fla. and west to La. It has narrowly egg-shaped or lance-shaped leaves which are untoothed or very sparingly toothed, and it blooms between August and October. (Not illustrated)

WOOLLY RAGWORT *(Senecio tomentosus)*

This ragwort usually has a tuft of several whitish-woolly stems 1 to 2½ feet high. The stem leaves are small, narrow, and scattered; but the basal ones are oblong to lance-shaped, stalked, toothed, white-woolly beneath and often cobwebby above, and 2 to 6 inches long. The flower heads are about ¾ inch across, with 10 to 15 yellow rays and a small yellow disk. It grows in moist sandy or rocky open woods and thickets, blooming between April and June. RANGE: N.J. to Ark. south to Fla. and La.

SMALL'S RAGWORT *(Senecio anonymus)*

This species has several stems 1½ to 2½ feet tall which are loosely woolly toward the base. The basal leaves are lance-shaped or narrowly oblong, bluntly toothed and 3 to 6 inches long. Those upward along the stem become progressively smaller and deeply cut into many segments. The numerous flower heads are about ½ inch across, with 8 to 10 bright yellow rays. It grows in open woods, fields and roadsides; blooming from May to early July. RANGE: N.J. to Pa. and Ky. south to Fla. and Ala.

GOLDEN RAGWORT *(Senecio aureus)*

Also called Squaw-weed, this is a rather smooth plant with usually several slender stems 6 inches to 2½ feet tall. The basal leaves are oblong-heart-shaped, bluntly-toothed, slender-stalked, often purplish beneath and 1 to 6 inches long. Those upward along the stem are small, narrow, and more or less deeply cut. The many flower heads are about ¾ inch across, with 8 to 12 bright yellow rays. It grows in rich moist woods, wet meadows, swamps and bogs; blooming late March to July.

RANGE: Nfd. to Que. and Wis. south to Fla., Tenn. and Mo.

ROUND-LEAF RAGWORT *(Senecio obovatus)*

This species has a rather smooth and slender stem 1 to 1½ feet tall. The basal leaves are usually more top-shaped than round, bluntly toothed, and 1 to 3½ inches long. Those upward along the stem are small, narrow, and more or less cut. The flower heads are about ⅔ inch across with 8 to 12 bright yellow rays. It grows on wooded slopes and in ravines, blooming between April and June.

RANGE: N.H. to Ont. and Mich. south to Fla. and Tex.

BUTTERWEED *(Senecio glabellus)*

This species is an annual plant with a succulent, hollow stem 1 to 3 feet tall. Its stem leaves have an end lobe but slightly larger than the others, and the basal ones are pinnately divided into more or less toothed segments. The flowers have 7 to 10 yellow rays. It grows in wet places from N.C. to sw. Ohio, s. Ill and Mo. south to Fla. and Tex.; blooming between March and June. (Not illustrated)

Thistles

The flower heads of thistles contain a large number of slender tubular flowers which have a deeply 5-lobed rim. There are no ray flowers. The feathery pappus becomes the "thistledown", a sort of parachute attached to the top of the seed-like fruits or achenes. The plants have alternate prickly leaves, and most species also have prickly involucral bracts. During the first year the plant forms a rosette of leaves. After blooming the second year, the plant dies.

YELLOW THISTLE *(Cirsium horridulum)*

This thistle has a very stout, leafy, branching stem 2 to 5 feet tall which is somewhat woolly when young. Its leaves are 2 to 5 inches long, green on both sides, and cut into triangular lobes which have many yellowish prickles. The flower heads are 2 to 4 inches across, with numerous pale yellow to purple flowers. The involucre is surrounded by narrow, spiny, bract-like leaves but the true bracts of the involucre have soft and unarmed tips. It grows in sandy fields, savannahs, roadsides and waste places; blooming between late March and June.

RANGE: Me. to Pa. south to Fla. and Tex.

COMMON THISTLE *(Cirsium vulgare)*

Often called the Bull Thistle, this native of Europe is now widely naturalized in fields, pastures, and along roadsides. It has a stout, branching, leafy stem 3 to 6 feet tall which is more or less wooly and has prickly wings running down from the leaf bases. Its leaves are green above, pale or with a web-like wool beneath, 3 to 6 inches long, and pinnately cut into a number of spiny segments. The flower heads are 1½ to 2 inches across, with numerous purple flowers; and all of the involucral bracts are tipped with a prickle. It blooms between June and October.

SWAMP THISTLE *(Cirsium muticum)*

This is a rather slender plant with a smoothish or somewhat woolly stem 3 to 8 feet tall. Its leaves are rather thin, green above, pale and more or less webby beneath, and pinnately cut into broad prickly segments. The flower heads are about 1½ inches across, with numerous purple flowers; and the involucral bracts lack prickles but are usually webby and stickly. It grows in wet woods or swamps, blooming between July and October.

RANGE: Nfd. to Sask. south to N.C., Tenn. and n. La.

TALL THISTLE *(Cirsium altissimum)*

Often called the Roadside Thistle, this species has a fairly stout, branching, woolly, and leafy stem 3 to 10 feet tall. Its leaves are undivided but have shallowly lobed or bristly-toothed margins, and they are densely felted with white wool beneath. The flower heads are about 2 inches across, with numerous rose-purple flowers; while the outer involucral bracts have a dark glandular spot and also a weak prickle. It grows in open woods, fields, pastures and thickets; blooming from August to late fall.

RANGE: N.Y. to Mich. and Minn. south to Fla. and Tex.

PASTURE THISTLE *(Cirsium pumilum)*

This species is also called the Fragrant or Bull Thistle. It has a stout stem 1 to 3 feet tall; with leaves which are pinnately cut into shallow, very spiny lobes and green on both sides. The flower heads are 2 to 3 inches across, with numerous purple flowers; but only the lower bracts of the broadly egg-shaped involucre are prickly. It grows in woods and fields from Me. to Ohio south to n. S.C., blooming between late May and August. (Not illustrated)

SANDHILL THISTLE *(Cirsium repandum)*

This small thistle grows in the dry coastal plain pinelands and sandhills. It has a simple or few-branched stem 6 inches to 1½ feet high, with narrow leaves 1½ to 3 inches long which are wavy-lobed and very prickly. Its flower heads are about an inch across and contain purple flowers. The outer involucral bracts are prickly and have a pale sticky band down the center. It blooms from May to July.

RANGE: N.C. south to Fla.

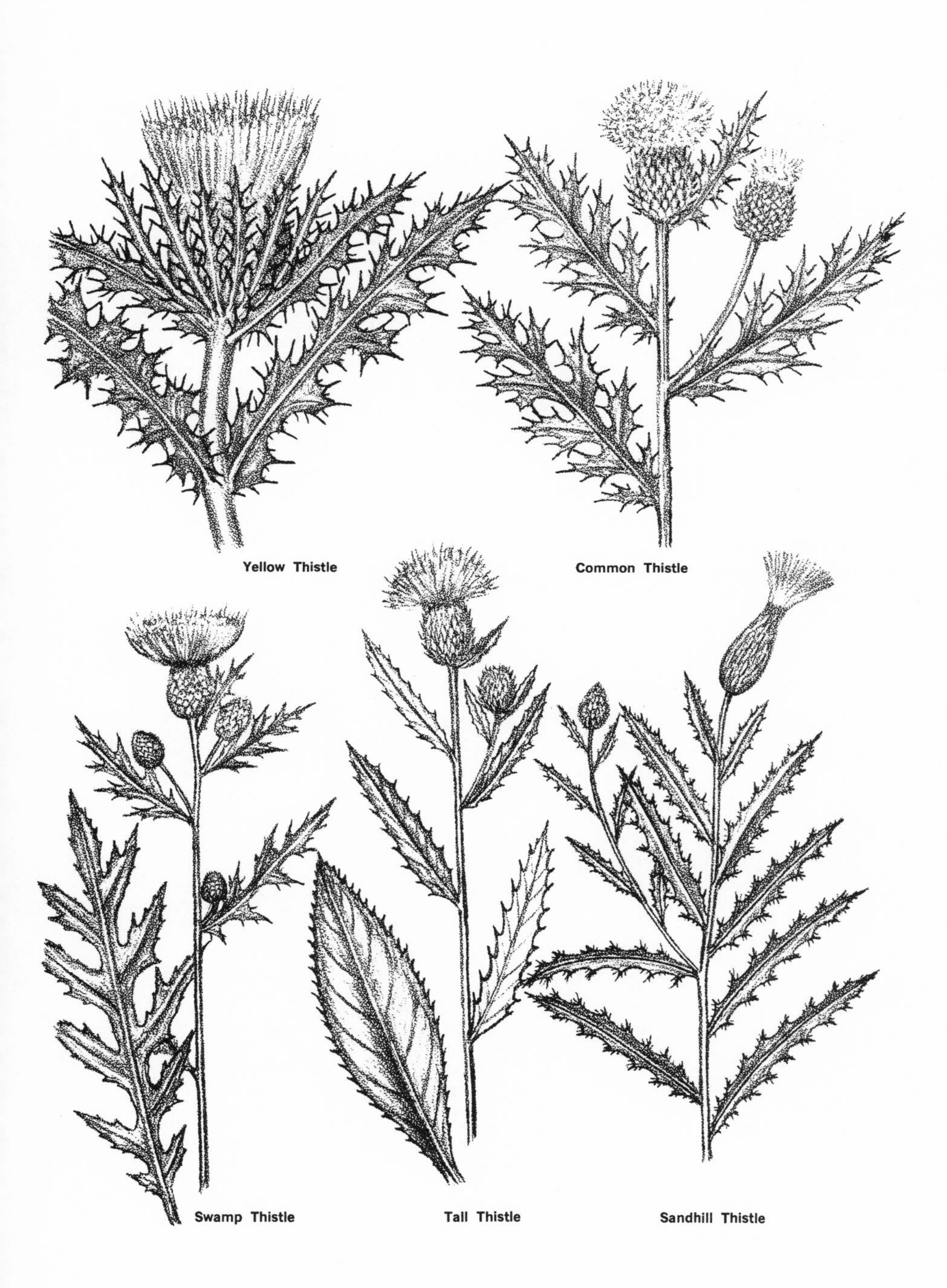
Yellow Thistle
Common Thistle
Swamp Thistle
Tall Thistle
Sandhill Thistle

Sun-bonnets
Fireweed
Dwarf Dandelion
Cynthia
Potato Dandelion

SUN-BONNETS *(Chaptalia tomentosa)*

This little daisy-like composite is one of the first flowers to appear in the spring in the coastal plain pinelands. The plant has a basal rosette of leaves which are bright green and smooth above but densely white-woolly beneath, the new ones not being fully grown at the time the plant blooms. Its flower heads are almost an inch across when fully expanded, and they have a number of marginal rays which are white or creamy-white on the upper surface and violet-tinged on the back. Each head nods at the summit of a slender, naked, woolly-coated stem from 4 to 12 inches tall. It grows in moist places, blooming between late February and May.

RANGE: N.C. south to Fla. and west to Tex.

FIREWEED *(Erechtites hieraciifolia)*

This is a smooth or somewhat hairy, rank-odored plant with a grooved and usually branched stem from 1 to 8 feet tall. Its leaves are thin, lance-shaped or oblong, with ragged teeth and sometimes deeply cut, and from 2 to 8 inches long. The numerous flower heads have a cylindrical involucre about ⅔ of an inch high and contain a number of tubular whitish flowers. The pappus is of silky whitish hairs. It grows in moist thickets, clearings, and open woods and it is often abundant on recently burned-over areas; blooming between July and October.

RANGE: Nfd. to Ont. and Minn. south to Fla. and Tex.

The Chicory Tribe

The remaining composites have heads made up entirely of flowers with strap-shaped corollas (rays), which are usually 5-toothed at the summit. Most of them have a bitter and milky juice. Some botanists consider them as belonging to a separate plant family, the Chicory Family.

DWARF DANDELION *(Krigia virginica)*

The Dwarf Dandelion has bright yellow heads of dandelion-like flowers about ½ inch across, on slender stalks from 2 to about 10 inches tall. Its leaves are almost always in a basal cluster or rosette. Most of them are between 2 and 4 inches long and they are variously toothed or lobed, or more rarely entire. It grows in dry open woods, fields, and along roadsides; blooming between late March and July.

RANGE: Me. to Ont., Wis. and Iowa south to Fla. and Tex.

CYNTHIA *(Krigia biflora)*

Cynthia is a smooth and somewhat whitened plant with a stem from 1 to 2 feet tall, usually with a single clasping leaf below the middle. The stem forks above the leaf and each branch ends in a head of orange-yellow flowers from 1 to 1½ inches across. Its clustered basal leaves are 2 to 6 inches long and may be either wavy-toothed or lobed, or quite entire. It grows in moist open woods, fields, and meadows; blooming between May and October. Also known as the Virginia Goat's-beard.

RANGE: Ont. to Man. south to Ga., Ky., Mo. and Colo.

MOUNTAIN CYNTHIA *(Krigia montana)*

This plant is also called the Blue Ridge Dandelion as it grows on rocky slopes and cliffs in the mountains from w. N.C. and e. Tenn. south to n. Ga. It has a branched stem from 5 to 15 inches high, usually with a reclining base, and bearing several scattered leaves. The leaves are quite narrow and vary from being untoothed to irregularly cut-toothed. Its flower heads are bright yellow, dandelion-like, almost an inch across, and bloom between May and September. (Not illustrated)

POTATO-DANDELION *(Krigia dandelion)*

The Potato-dandelion derives its name from tuber-bearing roots. It is a smooth and somewhat whitened plant with a cluster of basal leaves; and a slender and leafless stalk from 6 to 18 inches high which bears a single, bright yellow, dandelion-like flower head about an inch across. The leaves are 3 to 6 inches long and untoothed to sparingly toothed or lobed. It grows on the borders of woods, prairies, and in meadows; blooming between April and June.

RANGE: N.J. to Ky., Ill. and Kan. south to Fla. and Tex.

FALSE DANDELION *(Pyrrhopappus carolinianus)*

The False-dandelion is a smoothish plant with a leafy and branching stem 1 to 3 feet high. Its leaves are variable; the larger and lower ones 3 to 8 inches long and pinnately lobed, cleft or toothed. Those along the upper part of the stem are quite small, lance-shaped, and have clasping bases. Its flower heads are dandelion-like, about 1½ inches across, and usually a pale yellow; but a form which occurs locally in n. Fla. and Ala. has cream-colored rays which are tipped with rose. The pappus is very soft and tawny or reddish in color. It grows in dry fields, clearings, and along roadsides; blooming between April and June.

RANGE: Del. to s. Ind. and Kan. south to Fla. and Tex.

COMMON DANDELION *(Taraxacum officinale)*

The familiar and weedy dandelion occurs abundantly in open or grassy places throughout much of North America but, in spite of its widespread distribution, it is an immigrant from Europe. It is a stemless plant with a basal rosette of narrow leaves which are variously and pinnately lobed and cut-toothed. The bright golden-yellow heads of flowers are about 1½ inches across and are borne on hollow and leafless stalks from 2 to 10 inches long. The involucre has an outer circle of small bracts which curve outward and backward, and an inner circle of long ones. It blooms abundantly between late February and June and occasionally much later or in winter. Many people enjoy the leaves as spring greens.

CHICORY *(Cichorium intybus)*

The Chicory, or Succory, is a common but quite pretty weed which also came from Europe. It is now common in fields and along roadsides throughout much of eastern North America except in the Deep South. It is an erect plant often 3 feet or more high, and on its upright and rigid branches it produces a procession of flower heads a bit more than an inch across, their numerous rays being a most beautiful bright blue color. Its flowering season often extends from late May to October. The lower leaves are 3 to 6 inches long, variously toothed, cut, or lobed, and taper into long stalks. Those upward along the stem are smaller and have clasping bases. Some of its varieties are cultivated either as a leafy vegetable or for the roots which are used as a substitute or adulterant for coffee. Endive is a closely related species, *Cichorium endivia.*

GRASS-LEAF LETTUCE *(Lactuca graminifolia)*

This is a quite unique species of wild lettuce with a remotely leafy stem from 1 to 3 feet tall. Its leaves are all very long and narrow, the larger ones toward the base sometimes as much as 15 inches in length and often with a few coarse teeth or spreading lobes. The bright purplish-blue heads of flowers are about ½ inch across. It grows in dry, open, sandy woods and fields chiefly in the coastal plain; blooming between April and July.

RANGE: S.C. south to Fla. and west to Tex.

WILD LETTUCE *(Lactuca canadensis)*

This is a variable but smooth and somewhat whitened plant with a leafy stem from 3 to 10 feet tall, which branches above into a large cluster of innumerable flower heads. The leaves are 2 to 8 inches long and may be untoothed to deeply lobed; those of the stem being stalkless or having clasping bases. Its flower heads are little more than ¼ inch across and contain a number of yellow-rayed flowers. The Wild Lettuce is commonly seen in moist thickets, clearings, and the borders of woods; blooming between June and October. It is also known as the Tall Lettuce.

RANGE: N.S. to B.C. south to Fla. and Tex.

PRICKLY LETTUCE *(Lactuca serriola)*

The Prickly Lettuce is a widespread and common weed introduced from Europe. It is now found in fields, waste places and along roads quite generally in eastern North America. Although it quite closely resembles the preceding species, it may be readily distinguished by the row of bristly prickles along the midrib on the undersides of its leaves, which quite frequently are turned vertically. Its yellow flower heads are also small and produced between July and October. (Not illustrated)

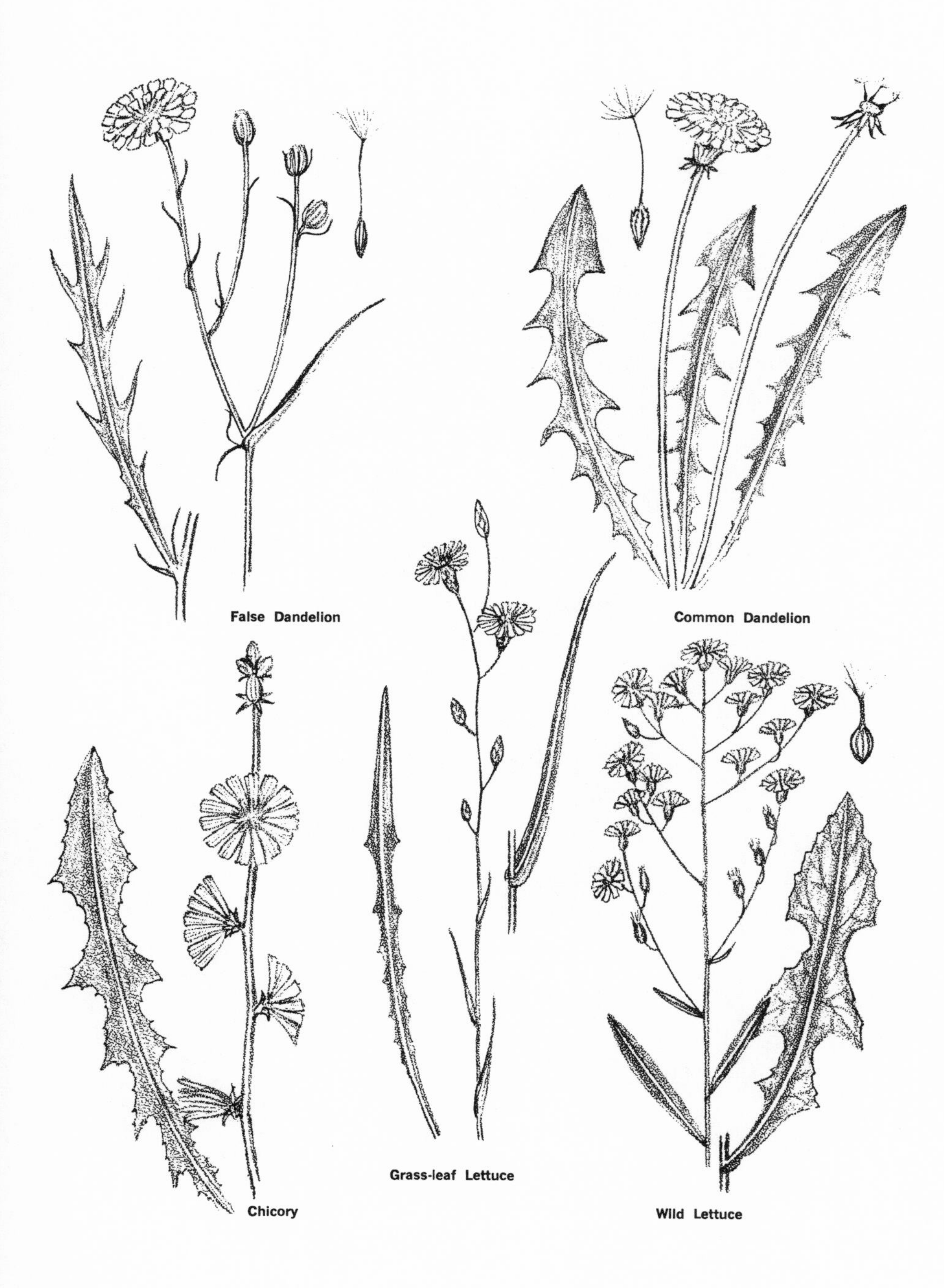
False Dandelion
Common Dandelion
Chicory
Grass-leaf Lettuce
Wild Lettuce

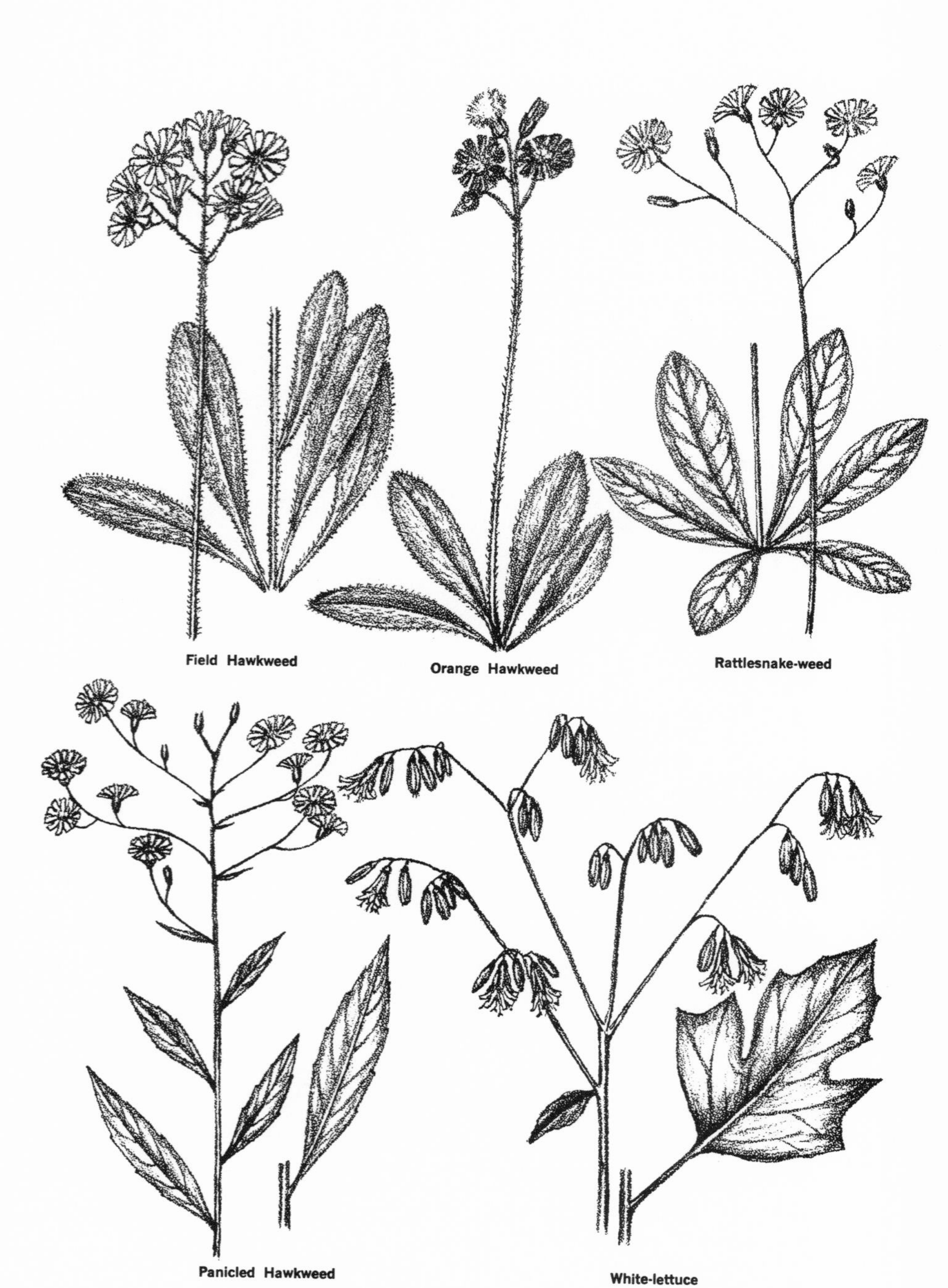
Field Hawkweed
Orange Hawkweed
Rattlesnake-weed
Panicled Hawkweed
White-lettuce

FIELD HAWKWEED *(Hieracium caespitosum)*

Our most common and weedy species of hawkweeds, such as this one, are naturalized immigrants from Europe. The Field Hawkweed, or King-devil, is a hairy plant with a cluster of basal leaves and a simple stem 1 to 2 feet high, which has but 1 or 2 small leaves. The basal leaves are oblong or narrowly top-shaped, 2 to 5 inches long, mostly untoothed, and tapered into winged stalks. The flower heads are borne in a more or less compact terminal cluster and contain a number of yellow-rayed flowers. It is often common in fields, clearings, and along roadsides as far south as w. N.C. and n. Ga.; blooming between May and August. Some of the hairs on the upper part of the stem are blackish.

ORANGE HAWKWEED *(Hieracium aurantiacum)*

Also known as the Devil's-paintbrush, this plant is quite similar to the preceding species, but it has heads of bright orange-red flowers. A very pretty but often extremely troublesome weed, it is also a native of Europe. It and the preceding species often occur together, blooming at the same time.

RATTLESNAKE-WEED *(Hieracium venosum)*

This native species is often common in dry open woods and clearings, and is sometimes called Poor Robin's-plantain. It is quite a smooth plant with a basal rosette of elliptic to narrowly top-shaped leaves from 1 to 4 inches long. They are beautifully veined or mottled with purple above and are pale to purplish and somewhat hairy beneath. The heads of bright yellow flowers are little more than ½ inch across and are in a loose cluster at the summit of a leafless or few-leaved stalk usually 1 to 2 feet high. It blooms between May and September. RANGE: Me. to Ont. south to Fla. and La.

PANICLED HAWKWEED *(Hieracium paniculatum)*

This hawkweed has a stem from 1 to 3 feet tall, smooth except toward the base, and leafy up to the open and slenderly-branched cluster of flower heads. The leaves are rather thin, mostly lance-shaped, sparingly toothed, whitish beneath, and from 2 to 6 inches long. Its numerous flower heads are about ½ inch across, containing a number of yellow-rayed flowers. It grows in dry open woods, blooming between July and October.
RANGE: N.S. to Ont. and Mich. south to Ga. and Tenn.

ROUGH HAWKWEED *(Hieracium scabrum)*

This species has a slender, roughish stem with leaves mostly below the middle; and yellow-rayed flower heads in a rather flat-topped cluster. It grows in dry open woods from Mass. to Mich. and Kan. south to Ga. and Tex.; blooming from July to October. (Not illustrated)

WHITE-LETTUCE *(Prenanthes alba)*

Also called Rattlesnake-root, this is a smooth whitened plant, usually with a purplish stem 2 to 5 feet tall. Its leaves are variable but commonly egg-shaped, heart-shaped or triangular; and coarsely toothed, lobed or divided. Most of them are stalked and the lower ones may be as much as 8 inches long. The usually numerous flower heads are in an open cluster. They contain 8 or more whitish flowers with a cinnamon-brown pappus, and the purplish involucre is whitened with a bloom. It grows on wooded slopes and roadbanks; blooming from August to October.
RANGE: Me. to Que. and Sask. south to N.C., Tenn. and Mo.

TALL WHITE-LETTUCE *(Prenanthes altissima)*

This species is similar but the flower heads have 5 or 6 greenish flowers with a creamy-white pappus. It grows in woods from Que. to Man. south to Ga. and Ala, blooming July to October. (Not illustrated)

LION'S-FOOT *(Prenanthes serpentaria)*

Lion's-foot usually has a purplish stem, and thickish leaves variously lobed. The flower heads have 8 to 12 pinkish or creamy flowers and a creamy pappus. Woods and roadbanks from Mass. to Ohio south to Fla. and Miss.; blooming Aug. to Oct. (Not illustrated)

SLENDER RATTLESNAKE-ROOT *(Prenanthes autumnalis)*

This plant has pink-rayed heads in a spike-like cluster. It grows in wet coastal plain pinelands and savannahs from N.J. south to Fla. and west to Miss.; blooming September to November. (Not illustrated)

Shrubs

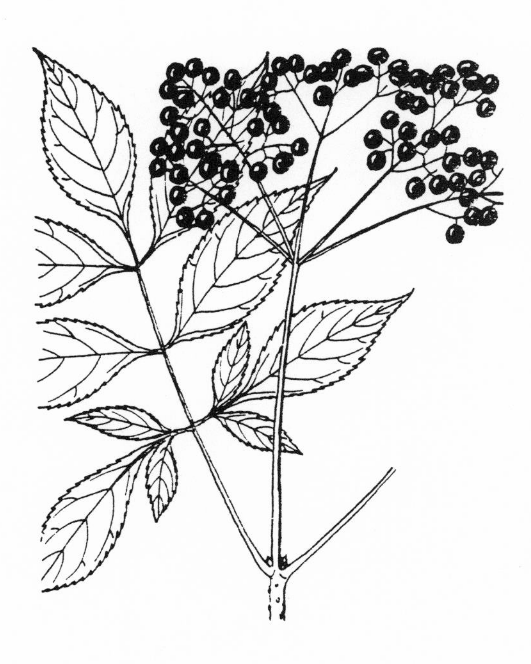

PAGE GUIDE TO SHRUB FAMILIES

USING THIS BOOK TO RECOGNIZE SHRUBS

To most people a shrub is a bush. It is a woody plant smaller than a tree, with several stems growing together in a clump. Sometimes an arbitrary height of twenty feet is set as the maximum height for a shrub. The distinction between shrubs and trees, however, is often very vague, and some woody plants can justifiably be considered as both trees and shrubs. Some of the larger shrubs often have a single stem or trunk a short distance above the ground. On the other hand, there are trees which commonly have several trunks growing together in a clump. Botanists also consider woody vines as being shrubs. Then, too, we often run into difficulty in distinguishing herbaceous plants from shrubs. Some plants are intermediate between the two. They have stems which are at least partly woody, especially toward the base, and they are often called "semiwoody" plants or "sub-shrubs." Woody vines are included in the present work, and so are some of these so-called "sub-shrubs."

Our native shrubs are useful in many ways. They help to prevent the erosion of the soil and of stream banks; they provide much needed cover for wildlife. A great many species also provide food for the wild creatures of our fields and woods. Some produce fruits which are enjoyed by human beings as well, or which have some medicinal value. The beauty of many of our native shrubs all too often is not appreciated. Most of the ornamental shrubs which we plant about our homes and in our gardens are shrubs from foreign lands. Very few native American species are ever listed in the nursery catalogs, though many are singularly attractive and well worth introducing into our home grounds and gardens. Some of them will thrive in places where most cultivated shrubs refuse to grow—on dry barren soils or in very wet environments.

One way to use a book such as this is to thumb through it until you find the name of the plant in question, and there is really nothing wrong about this method. Another way is to use a key; this usually turns out to be a shortcut to identification. Two sets of simple keys are provided in this book. One is a key to the genera of shrubs when leaves, and perhaps fruits, are present. The other is a key to the genera of leaf-losing shrubs in winter condition, when the leaves are absent. In addition, keys to most of the larger genera are provided in the Appendix.

To find a particular shrub, it is first necessary to determine which key should be used. To determine this, consult the information which immediately follows this Introduction. For instance, if the plant is a *woody vine,* you will be directed to use Key II; if it is a *trailing shrub,* or a dwarf one less than a foot high, you will be told to use Key I. The keys have two alternatives bearing the same number. The user should begin with the first step, which is number 1. At each step, or set of numbers, a choice must be made between each pair of characteristics. Of course it is necessary to make accurate observations and correct choices at each progressive step in the key until the genus name of the plant is finally determined. Sometimes it will be necessary to take many steps through a key before arriving at the genus name, while at other times the steps may be relatively few.

As this book is intended for the layman, technical terms have been avoided as much as possible. It should therefore be useful to anyone who has an interest in, and wishes to identify, native shrubs. A few of the particularly difficult genera, such as the blackberries (*Rubus*) and hawthorns (*Crataegus*), are treated here in only a very general sort of way. Even professionally trained botanists find them most exasperating. The average person can hardly be expected to do better than recognize them as blackberries or hawthorns.

While most of the characteristics used for identification are ones which can be easily seen with the unaided eye—or sometimes even detected by the nose—the use of such characteristics was not always possible. A good hand lens is therefore quite an asset. This is epecially true in trying to identify the leaf-losing shrubs in winter, when we often have to rely on characteristics that are not easily seen with the unaided eye. Usually it is necessary to use a hand lens to determine the number and arrangement of the bundle scars within the leaf scars, or other minute details.

For convenience, the text describing each shrub is placed on the page opposite the illustration. While the descriptions are necessarily brief, they will tell you what you need to know in order to identify a given shrub. Identification, as far as possible, is based upon leaf characteristics.

Times of flowering and fruiting are included. In the case of wide-ranging species, the earliest month mentioned is when the shrub usually blooms in the southern portion of its range; the latest month is when it usually finishes blooming in the north. The time of blooming is, of course, later at higher elevations in the mountains than in the lowlands in the same latitude.

The range of the shrub is also useful in arriving at an identification. If the shrub has a distinctly northern range, obviously it can be eliminated when you are trying to identify shrubs in the south; and strictly southern species need not be expected to occur in the north. Some shrubs are found only in the mountain regions; others in the coastal lowlands. Such facts are given in the text.

The beginner will undoubtedly find occasional shrubs which he will be unable to identify. If possible consult a local botanist or other person who is familiar with the local plants. Should such a person not be available, botanists at the various colleges and universities will be more than glad to help you make correct identifications. Fresh specimens may be sent to them by mail, wrapped in plastic wrap or waxed paper to prevent their withering before arrival. If dried specimens are sent, they should be carefully pressed while still fresh and the parts well arranged and flattened out. Specimens may be dried between blotters or folded newspapers, then securely wrapped and sent by parcel post. Address them to the Department of Botany at the institution of your choice. Your specimens may be a valuable addition to the collection of plants in the college or university herbarium, so always state when and where they were collected.

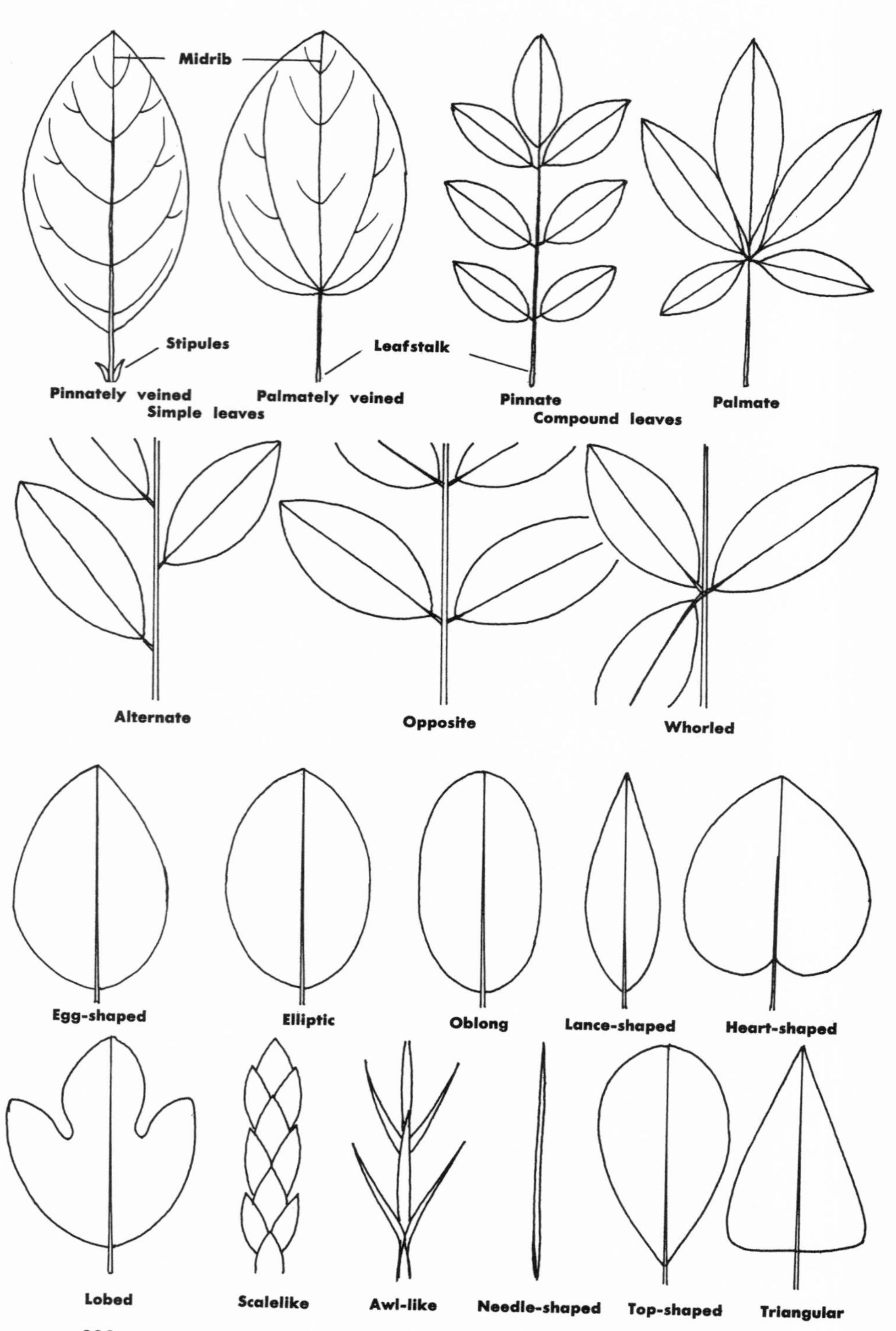
Midrib
Stipules
Leafstalk
Pinnately veined
Palmately veined
Simple leaves
Pinnate
Palmate
Compound leaves
Alternate
Opposite
Whorled
Egg-shaped
Elliptic
Oblong
Lance-shaped
Heart-shaped
Lobed
Scalelike
Awl-like
Needle-shaped
Top-shaped
Triangular

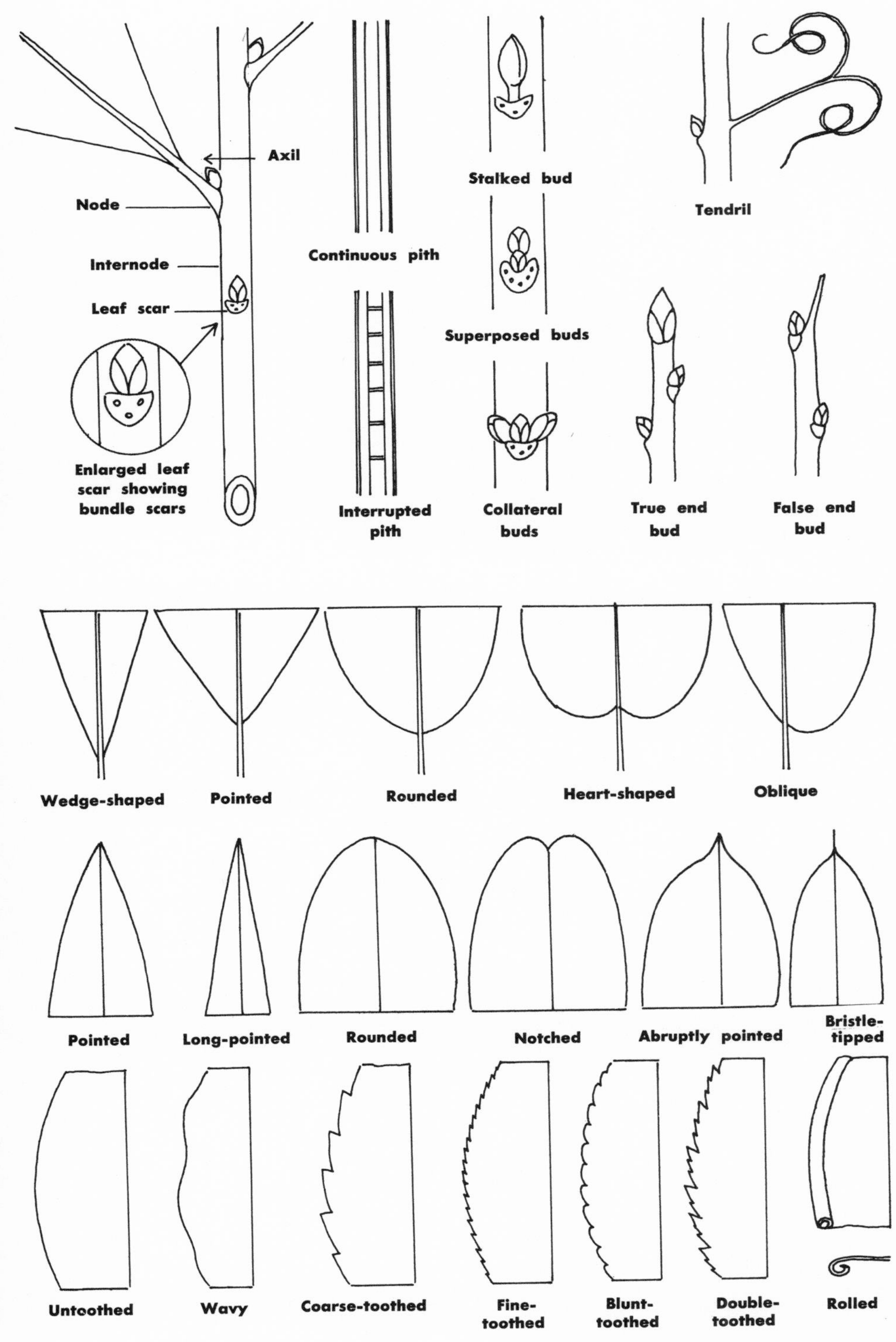

Axil
Node
Internode
Leaf scar
Enlarged leaf scar showing bundle scars
Continuous pith
Interrupted pith
Stalked bud
Superposed buds
Collateral buds
Tendril
True end bud
False end bud
Wedge-shaped
Pointed
Rounded
Heart-shaped
Oblique
Pointed
Long-pointed
Rounded
Notched
Abruptly pointed
Bristle-tipped
Untoothed
Wavy
Coarse-toothed
Fine-toothed
Blunt-toothed
Double-toothed
Rolled

KEYS TO THE GENERA OF SHRUBS

Shrubs with the leaves present	
Prostrate or trailing shrubs, or shrubs less than one foot high	KEY I
Climbing or scrambling shrubs (vines)	KEY II
Erect shrubs more than one foot high	
Leaves compound	KEY III
Leaves simple	
Leaves opposite or whorled	KEY IV
Leaves alternate	KEY V
Leaf-losing shrubs when the leaves are absent	KEY VI

KEY I

PROSTRATE, TRAILING, OR ERECT SHRUBS LESS THAN ONE FOOT HIGH

1 Stems fleshy and jointed, usually armed with slender spines; leaves small, scalelike and soon shed, bearing in their axils clusters of barbed hairs — PRICKLY-PEARS *(Opuntia)* 478
1 Stems and leaves otherwise—2

2 Leaves scalelike, awl-like, needle-like, or flattened but not over ⅛ inch wide—3
2 Leaves with broader blades, usually over ⅛ inch wide—10

3 Leaves scalelike or awl-like, the bases often overlapping—4
3 Leaves neither scalelike nor awl-like—5

4 Plants green and more or less hairy, or densely woolly and whitish; flowers yellow — HUDSONIAS *(Hudsonia)* 518
4 Plants smooth, mosslike, with leaves in 4 rows; flowers pink or white — MOSS-PLANT *(Cassiope)* 550

5 Leaves opposite—6
5 Leaves alternate—7

6 Leaves narrow and flat or almost needle-like, with clusters of smaller leaves in the axils; flowers yellow — ST. JOHN'S-WORTS *(Hypericum)* 518
6 Leaves never with axillary clusters of smaller leaves, not needle-like but with margins inrolled; flowers white or pink, bell-shaped — ALPINE-AZALEA *(Loiseleuria)* 538

7 Leaves flat, margins rough or minutely toothed — MOUNTAIN-HEATH *(Phyllodoce)* 542
7 Leaf margins more or less rolled inward beneath—8

8 Leaf margins untoothed — CROWBERRIES *(Empetrum)* 481
8 Leaf margins minutely toothed—9

9 Leaves under ¼ inch long, sides parallel — BROOM-CROWBERRY *(Corema)* 481
9 Leaves over ¼ inch long, broader toward tip — MOUNTAIN-LOVER *(Paxistima)* 494

10 Leaves compound—11
10 Leaves simple—18

11 Stem or leafstalks prickly or bristly—12
11 Stems and leafstalks entirely unarmed—14

12 Leaflets untoothed; pairs of short spines often present at the bases of the leafstalks; fruit a flattened pod — LOCUSTS *(Robinia)* 470
12 Leaflets toothed; fruits other than pods—13

13 Stipules joined to leafstalks half or more of their length — ROSES *(Rosa)* 461
13 Stipules not joined to the leafstalks or merely attached at the base — BLACKBERRIES *(Rubus)* 457

14 Leaflets 3—15
14 Leaflets 5 or more—17

15 Leaflets less than ½ inch wide

16 Stems erect, to 3 feet high, much branched; leaves pinnately compound with 3 to 7 (usually 5) leaflets; flowers bright yellow
SHRUBBY CINQUEFOIL *(Pentaphylloides)* 454

16 Stems ascending or trailing; leaves fan shaped with 3 leaflets, each toothed apically; flowers white
THREE-TOOTHED CINQUEFOIL *(Sibbaldiopsis)* 454

15 Leaflets broader POISON-IVY *(Toxicodendron)* 485

17 Interior of stems bright yellow; leaflets usually 5, coarsely cut-toothed and often deeply cleft
SHRUB YELLOWROOT *(Xanthorhiza)* 422

17 Interior of stems not bright yellow, leaflets 7 to 11, untoothed, elliptic or egg-shaped LOCUSTS *(Robinia)* 470

18 Leaves opposite or sometimes in 3's—19

18 Leaves alternate—25

19 Leaf margins toothed—20

19 Leaf margins not toothed—21

20 Plant with a mintlike odor; leaves shallowly toothed, usually less than ½ inch wide BASILS *(Calamintha)* 574

20 Plant not aromatic; leaves sharply and finely toothed, usually over ½ inch wide; branchlets green and 4-sided
STRAWBERRY-BUSHES *(Euonymus)* 494

21 Leaves with minute clear or black dots; flowers yellow—22

21 Leaves otherwise; flowers not yellow—23

22 Flowers 4-petalled; fruit a capsule enclosed by a pair of large heart-shaped sepals ST. PETER'S-WORTS *(Hypericum)* 510

22 Flowers 5-petalled; fruit a capsule with 5 sepals at the base
ST. JOHN'S-WORTS *(Hypericum)* 510

23 Leaves very white beneath; branchlets 2-edged LAURELS *(Kalmia)* 542

23 Leaves not noticeably whitened beneath; branchlets not 2-edged—24

24 Leaves oval or elliptic, leathery, very short-stalked; plant bushy or prostrate; fruit a capsule SANDMYRTLES *(Leiophyllum)* 530

24 Leaves roundish or heart-shaped, scarcely leathery, distinctly stalked; plant trailing; fruits red, berry-like PARTRIDGEBERRY *(Mitchella)* 578

25 Leaves lobed CURRANTS *(Ribes)* 438

25 Leaves not lobed—26

26 Largest leaves less than 1 inch long—27

26 Largest leaves more than 1 inch long—38

27 Leaf margins conspicuously toothed—28

27 Leaf margins untoothed or the teeth very inconspicuous—31

28 Leaves thick, leathery, evergreen; branchlets strongly angled
HUCKLEBERRIES *(Gaylussacia)* 553

28 Leaves and branchlets otherwise—29

29 Buds covered with a solitary hoodlike scale WILLOWS *(Salix)* 398

29 Buds with 1 or more pairs of visible scales—30

30 Plant low and trailing BEARBERRIES *(Arctostaphylos)* 553
30 Plant erect or bushy, over 6 inches high BLUEBERRIES *(Vaccinium)* 557

31 Leaves with minute black dots beneath
MOUNTAIN-CRANBERRY *(Vaccinium)* 565
31 Leaves not black-dotted beneath—32

32 Lower leaf surface with minute brown scales
RHODODENDRONS *(Rhododendron)* 530
32 Lower leaf surface otherwise—33

33 Shrubs upright or bushy, over 6 inches high—34
33 Shrubs trailing or with upright branches less than 6 inches high—35

34 Largest leaves less than ½ inch long; whole plant bristly-hairy; fruit a capsule LAURELS *(Kalmia)* 541
34 Largest leaves over ½ inch long; plants smooth or somewhat hairy; fruit a berry BLUEBERRIES AND BILBERRIES *(Vaccinium)* 557

35 Plant with a wintergreen odor; lower surfaces of the leaves with some stiff brown hairs; berries white, bristly
CREEPING-SNOWBERRY *(Gaylussacia)* 550
35 Plant not aromatic; leaves otherwise; berries not white—36

36 Leaves whitened beneath CRANBERRIES *(Vaccinium)* 565
36 Leaves merely paler green beneath—37

37 Leaf margins rolled inward on the lower side; plant of the southeastern coastal plain BLUEBERRIES *(Vaccinium)* 557
37 Leaf margins flat; plant of northeast and Canada
BEARBERRIES *(Arctostaphylos)* 550

38 Leaf margins conspicuously toothed—39
38 Leaf margins untoothed or the teeth very inconspicuous—42

39 Leave more or less clustered at tips of stems or almost in whorls—40
39 Leaves well spaced along the stem—41

40 Plant with a wintergreen odor; leaves oval and with a few bristly teeth on the margin; fruits berry-like, red TEABERRY *(Gaultheria)* 550
40 Plants not aromatic; leaves egg-shaped to lance-shaped or broadest above the middle; fruits otherwise PIPSISSEWAS *(Chimaphila)* 529

41 Leaves 1½ inches or more long, toothed above the middle; branchlets lustrous brown, the main stems prostrate
CHERRIES *(Prunus)* 465
41 Leaves less than 1½ inches long, the margin finely toothed throughout; branchlets otherwise; plants bushy
BILBERRIES AND BLUEBERRIES *(Vaccinium)* 557

42 Stems trailing or low climbing—43
42 Stems more or less erect—44

43 Leafstalks with a pair of tendrils at base; fruit a berry; plants woolly or smooth, sometimes bristly or prickly
GREENBRIERS *(Smilax)* 390

43 Leafstalks never with tendrils; stems with bristly brown hairs and leaf margins hairy-fringed; fruit a capsule
TRAILING-ARBUTUS *(Epigaea)* 549

44 Leaves with small yellow resin dots, at least beneath—45
44 Leaves never resin-dotted—46

45 Leaves spicy-aromatic when crushed, the margins usually toothed; fruit a wax-coated nutlet BAYBERRIES *(Myrica)* 397

45 Leaves not aromatic, margins untoothed; fruit a berry
HUCKLEBERRIES *(Gaylussacia)* 553

46 Leaves minutely black-dotted beneath STAGGERBUSH *(Lyonia)* 546
46 Leaves not black-dotted beneath—47

47 Branchlets and lower leaf surfaces strongly whitened; leaf margins rolled inward on lower side BOG ROSEMARY *(Andromeda)* 542

47 Branchlets and lower leaf surfaces very slightly if at all whitened; leaf margins flat or but slightly rolled—48

48 Branchlets velvety hairy or minutely warty-dotted
BLUEBERRIES *(Vaccinium)* 557

48 Branchlets otherwise—49

49 Leaves mostly rounded at tip, lustrous and wrinkled-veiny above, evergreen; fruit a whitish "plum" DEER-PLUM *(Licania)* 466

49 Leaves otherwise; fruit a capsule—50

50 Leaf margins slightly rolled inward on the lower side, the leaves finely hairy beneath; capsules globe-shaped LAURELS *(Kalmia)* 541

50 Leaf margins flat, hairy-fringed, the leaves with scattered stiff hairs along the midrib beneath; capsules oblong AZALEAS *(Rhododendron)* 533

KEY II

CLIMBING OR SCRAMBLING SHRUBS (VINES)

1 Climbing by means of tendrils or twining leafstalks—2
1 Climbing by other means—10

2 Climbing by twining tendril-like leafstalks CLEMATIS *(Clematis)* 421

2 Climbing by means of true tendrils—3

3 Leaves simple—4
3 Leaves compound—8

4 Tendrils in pairs on the bases of the leafstalks; stems usually green, often with prickles or stiff bristles GREENBRIERS *(Smilax)* 390
4 Tendrils not on the bases of the leafstalks – 5

5 Tendrils at tips of branchlets; leaves neither toothed nor lobed BRUNNICHIA *(Brunnichia)* 425
5 Tendrils on the stems opposite some of the leaves; leaves toothed and often lobed – 6

6 Tendrils branching GRAPES *(Vitis)* 505
6 Tendrils not branching – 7

7 Stems with a brownish pith GRAPES *(Vitis)* 505
7 Stems with a white pith HEARTLEAF AMPELOPSIS *(Ampelopsis)* 502

8 Leaves opposite; leaflets 2, with a tendril prolonged between them CROSS-VINE *(Bignonia)* 577
8 Leaves alternate; leaflets 5 or more; the tendrils on the stems opposite some of the leaves – 9

9 Leaflets 5, radiating from the summit of the long leafstalks, often with branches ending in an adhesive disk CREEPERS *(Parthenocissus)* 505
9 Leaflets usually more numerous and not radiating from the summit of the leafstalk PEPPER VINE *(Ampelopsis)* 502

10 Climbing by means of rootlets along the stems – 11
10 Climbing by twining stems or otherwise – 13

11 Leaves alternate, divided into 3 coarsely few-toothed leaflets POISON-IVY *(Toxicodendron)* 485
11 Leaves opposite – 12

12 Leaves simple, egg-shaped, thickish, toothed CLIMBING-HYDRANGEA *(Decumaria)* 433
12 Leaves compound, with 9 to 11 rather thin and sharply toothed leaflets TRUMPET-CREEPER *(Campsis)* 577

13 Leaves opposite – 14
13 Leaves alternate – 16

14 Leaves broadly pointed or blunt at tip, the upper ones united at base to form a disk; fruit a berry HONEYSUCKLES *(Lonicera)* 581
14 Leaves sharply pointed at tip, none united at the base – 15

15 Leaves usually less than 1 inch wide, somewhat leathery, evergreen; fruit a capsule less than 1 inch long YELLOW JESSAMINES *(Gelsemium)* 573
15 Leaves often over 1 inch wide, thin, not evergreen; fruit a slender pod 5 to 9 inches long TRACHELOSPERMUM *(Trachelospermum)* 574

16 Leaves compound – 17
16 Leaves simple – 18

17 Leaflets usually 3, sharply toothed on margin; stems usually with scattered prickles — ROSES *(Rosa)* — 461
17 Leaflets 9 to 15, the margins untoothed; stems never with prickles — WISTERIAS *(Wisteria)* — 474

18 Plant climbing beneath the bark of cypress trees, sending out branches at intervals — VINE-WICKY *(Pieris)* — 549
18 Plants with twining stems—19

19 Leaves with the main veins radiating from the summits of the leafstalks—20
19 Leaves with the main veins arising along a central midrib—23

20 Leafstalk attached to the blade a little inside of the margin; fruits blue, with a large disklike seed — MOONSEED *(Menispermum)* — 422
20 Leafstalks attached to the bases of the blades—21

21 Leaves heart-shaped, 3 to 8 inches wide — PIPEVINES *(Aristolochia)* — 418
21 Leaves more or less angled or lobed—22

22 Leaves less than 4 inches long, often triangular or shallowly lobed; fruits red — RED-BERRIED MOONSEED *(Cocculus)* — 422
22 Leaves over 4 inches long, deeply 3- to 5-lobed; fruits black — CUPSEED *(Calycocarpum)* — 422

23 Leaves on slender stalks at least ½ as long as the blades; plant spicy-aromatic when bruised — WILD SARSAPARILLA *(Schisandra)* — 426
23 Leaves with shorter leafstalks; plant not aromatic—24

24 Leaves 8 to 12 pairs of prominent straight veins, margin untoothed — SUPPLEJACK *(Berchemia)* — 498
24 Leaves with fewer pairs of branching veins, margin sharply toothed — AMERICAN BITTERSWEET *(Celastrus)* — 494

KEY III

ERECT SHRUBS ONE FOOT OR MORE HIGH WITH COMPOUND LEAVES

1 Leaves opposite—2
1 Leaves alternate—4

2 Leaflets 3, finely and sharply toothed — BLADDERNUT *(Staphylea)* — 494
2 Leaflets 5 or more—3

3 Leaflets radiating from the summit of a long leafstalk — BUCKEYES *(Aesculus)* — 497
3 Leaflets along an extension of the leafstalk — ELDERS *(Sambucus)* — 593

4 Branchlets or leafstalks prickly, thorny, or bristly—5
4 Branchlets and leafstalks otherwise—9

5 Leaves more than once compound — ARALIAS *(Aralia)* — 521
5 Leaves merely compound—6

6 Branchlets with a lemon- or limelike odor when bruised — PRICKLY-ASHES *(Zanthoxylum)* — 477
6 Branchlets not aromatic—7

7 Leaflets with untoothed margins — LOCUSTS *(Robinia)* — 470
7 Leaflets with toothed margins—8

8 Stipules joined to the leafstalks half or more of their length — ROSES *(Rosa)* — 461
8 Stipules otherwise — RASPBERRIES AND BLACKBERRIES *(Rubus)* — 457

9 Leaflets 3—10
9 Leaflets 5 or more—12

10 Stalk of the end leaflet distinctly longer than the other two, leaflets usually coarsely toothed or lobed — POISON-IVY AND POISON-OAK *(Toxicodendron)* — 485
10 Stalk of the end leaflet little if any longer than the other two—11

11 Leaflet margins coarsely toothed; branchlets aromatic if bruised or broken — SUMACS *(Rhus)* — 482
11 Leaflet margins untoothed or the teeth obscure; plant very ill-scented if bruised or broken — HOPTREE *(Ptelea)* — 477

12 Leaflets usually 5—13
12 Leaflets 7 or more—16

13 Leaflets radiating from the summit of the leafstalk — BLACKBERRIES *(Rubus)* — 457
13 Leaflets arranged along an extension of the leafstalk—14

14 Stems bright yellow inside; leaflets sharply cut-toothed and often deeply cleft — SHRUB YELLOWROOT *(Xanthorhiza)* — 422
14 Stems otherwise; leaflets untoothed, narrow, silky-hairy, and 1 inch or less long

15 Stems erect, to 3 feet high, much branched; leaves pinnately compound with 3 to 7 (usually 5) leaflets; flowers bright yellow — SHRUBBY CINQUEFOIL *(Pentaphylloides)* — 454
15 Stems ascending or trailing; leaves fan-shaped with 3 leaflets, each toothed apically; flowers white — THREE-TOOTHED CINQUEFOIL *(Sibbaldiopsis)* — 454

16 Extension of the leafstalk winged between the leaflets — SUMACS *(Rhus)* — 482
16 Extension of the leafstalks otherwise—17

17 Leaflet margins sharply toothed—18
17 Leaflet margins untoothed—19

18 Branchlets velvety-hairy, or smooth and whitened with a bloom, exuding a milky sap when cut; leaflets whitened or green and hairy beneath — SUMACS *(Rhus)* — 482
18 Branchlets and leaves otherwise; buds gummy — MOUNTAIN-ASH *(Sorbus)* — 450

19 Leaflets usually less than 2 inches long—20
19 Leaflets mostly over 2 inches long—21

20 Flowers and small 1- or 2-seeded pods in rather long and narrow end clusters; leaflets usually with minute clear dots or densely hairy INDIGOBUSHES *(Amorpha)* 469
20 Flowers and 2- to 3-inch flattened pods in axillary clusters; leaflets otherwise LOCUSTS *(Robinia)* 470

21 Leafstalks red or tinged with red; leaflets 2 to 3 times as long as broad; fruits small, whitish, in drooping axillary clusters POISON SUMAC *(Toxicodendron)* 485
21 Leafstalks otherwise; leaflets 4 or more times as long as broad; fruits rather large, yellowish, in more or less erect end clusters SOAPBERRIES *(Sapindus)* 498

KEY IV

SHRUBS MORE THAN ONE FOOT HIGH WITH OPPOSITE OR WHORLED SIMPLE LEAVES

1 Leaves awl-like, needle-like, or very narrow and less than ⅛ inch wide—2
1 Leaves with distinct blades more than ⅛ inch wide—4

2 Leaves arranged in 3's, awl-like, stiff, and sharp at the tip JUNIPERS *(Juniperus)* 386
2 Leaves opposite, very narrow or sometimes almost needle-like; with clusters of smaller leaves present in the axils—3

3 Leaves almost needle-like or flattened, smooth; flowers yellow; fruit a capsule with 5 sepals at the base ST. JOHN'S-WORTS *(Hypericum)* 510
3 Leaves club-shaped, grayish-downy; flowers pale bluish; fruit a seedlike nutlet, usually grouped in 4's and within the persistent calyx CONRADINA *(Conradina)* 574

4 Leaves lobed or with toothed margins—5
4 Leaves neither lobed nor with margins conspicuously toothed—21

5 Leaves lobed or both toothed and lobed—6
5 Leaves not lobed but the margins toothed—9

6 Main veins of the leaves arising along a midrib—7
6 Main veins of the leaves radiating from the summit of the leafstalk—8

7 Leaves usually less than 4 inches long, downy beneath, wavy-lobed; branchlets very slender. SNOWBERRIES *(Symphoricarpos)* 585
7 Leaves usually over 4 inches long, rather leathery, lobed like an oak leaf, densely tawny-woolly beneath HYDRANGEAS *(Hydrangea)* 433

8 Leafstalks usually with stipules at the base or glands at the summit, or the lower leaf surfaces minutely black-dotted or densely hairy — VIBURNUMS *(Viburnum)* 586

8 Leafstalks with neither stipules or glands, the lower surface smooth or minutely downy — MAPLES *(Acer)* 497

9 Leaf margins with large and coarse teeth averaging 5 or fewer to the inch of leaf margin — 10

9 Leaf margins with smaller and finer teeth — 14

10 Leaves with prominent and rather straight veins which end in the marginal teeth — VIBURNUMS *(Viburnum)* 586

10 Leaves with branching veins which do not end in the marginal teeth — 11

11 Leaf margins with a few bluntish or wavy teeth; leaves grayish green with silky hairs — SEA-OXEYE *(Borrichia)* 594

11 Leaf margins with rather sharply pointed teeth — 12

12 Leaves without evident stalks — MARSH ELDERS *(Iva)* 593

12 Leaves with evident stalks — 13

13 Leafstalks usually less than ½ inch long; marginal teeth usually widely spaced — MOCK-ORANGES *(Philadelphus)* 430

13 Leafstalks usually 2 or more inches long — HYDRANGEAS *(Hydrangea)* 433

14 Branchlets 4-sided and green — STRAWBERRY-BUSHES *(Euonymus)* 493

14 Branchlets otherwise — 15

15 Branchlets with hairy-lined ridges running down from a line between the leaf bases; leaves hairy-fringed; fruit a capsule — BUSH-HONEYSUCKLES *(Diervilla)* 578

15 Branchlets, leaves, and fruits otherwise — 16

16 Largest leaves less than 1 inch long — 17

16 Largest leaves more than 1 inch long — 18

17 Leaves with clusters of smaller leaves in the axils; plant usually less than 2 feet high and with a pleasant odor — BASILS *(Calamintha)* 574

17 Leaves otherwise; plant with straggling stems and often short spiny branchlets; not aromatic; fruits berry-like — SAGERETIA *(Sageretia)* 498

18 Leaves whitish-woolly beneath; fruits berry-like, magenta purple — BEAUTYBERRY *(Callicarpa)* 573

18 Leaves otherwise; fruits not berry-like, with a large stone — 19

19 Leaves roundish heart-shaped, 4 or more inches wide — VIBURNUMS *(Viburnum)* 586

19 Leaves smaller — 20

20 Leafstalks winged or grooved on the upper side; the leaf margins toothed quite to the base — VIBURNUMS *(Viburnum)* 586

20 Leafstalks otherwise; the leaf margins finely and often inconspicuously toothed above the middle — SWAMP-PRIVETS *(Foresteria)* 570

21 Plant parasitic, growing on the branches of deciduous trees
MISTLETOE *(Phoradendron)* 418

21 Plants rooted in the soil—22

22 Leaves with the main veins curving, tending to parallel the margin and to meet at the tip DOGWOODS *(Cornus)* 522

22 Leaves with the veins otherwise—23

23 Leaves leathery in texture—24

23 Leaves very slightly if at all leathery—26

24 Leaves usually less than ½ inch long SANDMYRTLES *(Lyonia)* 530

24 Leaves larger—25

25 Leaves mostly less than 2 inches long, often in whorls of 3; fruits globe-shaped and long-stalked capsules LAURELS *(Kalmia)* 541

25 Leaves 2 inches or more long, always opposite; fruit an olive-like drupe DEVILWOOD *(Osmanthus)* 573

26 Leaves commonly with clusters of smaller leaves in the axils—27

26 Leaves otherwise—29

27 Flowers red; fruit a seedlike nutlet usually in groups of 4 and enclosed within a persistent calyx BASILS *(Calamintha)* 574

27 Flowers yellow; fruit a capsule; leaves with minute clear or black dots—28

28 Flowers 4-petalled; capsule enclosed with a pair of large heart-shaped sepals ST. PETER'S-WORTS *(Hypericum)* 510

28 Flowers 5-petalled; capsules with 5 sepals at the base
ST. JOHN'S-WORTS *(Hypericum)* 510

29 Leaves and branchlets spicy-aromatic when bruised; fruit a large, leathery, pear-shaped pod containing large seeds
STRAWBERRY SHRUB *(Calycanthus)* 425

29 Leaves and branchlets not aromatic; fruits otherwise—30

30 Branchlets and lower leaf surfaces coated with silvery-white star-shaped hairs and rusty scales BUFFALOBERRY *(Shepherdia)* 521

30 Branchlets and lower leaf surfaces otherwise—31

31 Leaves often arranged in 3's—32

31 Leaves always opposite—33

32 Leaves lance-shaped; fruit an urn-shaped capsule in the leaf axils; plant woody chiefly toward the base
SWAMP LOOSESTRIFE *(Decodon)* 518

32 Leaves oval, elliptic, or egg-shaped; flowers and fruits in long-stalked dense, ball-shaped heads BUTTONBUSH *(Cephalanthus)* 577

33 Largest leaves usually more than 4 inches long—34
33 Largest leaves usually less than 4 inches long—35

34 Leaves rusty-hairy beneath; fruit a large roundish capsule
PINCKNEYA *(Pinckneya)* 577
34 Leaves at most pale downy beneath; fruit an olive-like drupe
FRINGETREE *(Chionanthus)* 573

35 Leaves stalkless or very nearly so, broadest at the tip and tapering to the base, sometimes obscurely toothed above the middle
VIBURNUMS *(Viburnum)* 586
35 Leaves with more or less evident stalks—36

36 Leafstalks grooved above or slightly winged; buds elongate, with one pair of rusty-scurfy scales VIBURNUMS *(Viburnum)* 586
36 Leafstalks neither grooved nor winged; buds otherwise—37

37 Leaves broadly pointed, blunt, or rounded at the tip; fruit a many-seeded berry—38
37 Leaves sharply pointed at the tip; fruits 1-seeded—39

38 Leaves but slightly longer than broad; berries in more or less dense axillary or end clusters
CORALBERRY AND SNOWBERRIES *(Symphoricarpos)* 585
38 Leaves usually distinctly longer than broad; berries on long axillary stalks, in distinct pairs or the pairs sometimes more or less united HONEYSUCKLES *(Lonicera)* 578

39 Leaves lance-shaped, broadly pointed or roundish at base; shrub more than 3 feet high, with greenish branchlets BUCKLEYA *(Buckleya)* 418
39 Leaves oval or egg-shaped, pointed at base; shrub less than 3 feet high, with purplish-brown branchlets NESTRONIA *(Nestronia)* 418

KEY V

ERECT SHRUBS ONE FOOT OR MORE HIGH WITH ALTERNATE SIMPLE LEAVES

1 Leaves 1 foot or more wide, fanlike and folded, and with long leafstalks; more or less clustered on a short, trunklike stem—2
1 Leaves less than 1 foot wide, neither fanlike nor folded—4

2 Leafstalks edged with sharp sawlike teeth
SAW-PALMETTO *(Serenoa)* 386
2 Leafstalks smooth-edged—3

3 Stems armed with long needle-like spines
NEEDLE-PALM *(Rhapidophyllum)* 386
3 Stems unarmed PALMETTOS *(Sabal)* 386

4 Leaves grasslike; or long and narrow, stiff, swordlike or dagger-like—5
4 Leaves otherwise—6

5 Plants bamboo-like; leaves grasslike; flowers not showy; fruits grainlike CANE *(Arundinaria)* 389

5 Plants with clusters of stiff and narrow or dagger-like leaves; flowers creamy white, large and showy; fruits capsules YUCCAS *(Yucca)* 389

6 Leaves very narrow, more or less stiff, less than ⅛ inch wide—7

6 Leaves with broader blades more than ⅛ inch wide—9

7 Leaves flat, with parallel margins and pointed tips, usually over ½ inch long YEWS *(Taxus)* 386

7 Leaves with a deep longitudinal groove on the lower side, usually less than ½ inch long, often nearly whorled—8

8 Leaves ¼ to ½ inch long; branches stiffly erect; plant aromatic SANDHILL ROSEMARY *(Ceratiola)* 481

8 Leaves ¼ inch or less long; branches spreading; plant not aromatic BROOM-CROWBERRY *(Corema)* 481

9 Leaves lobed or both toothed and lobed—10

9 Leaves not lobed, the margins either toothed or untoothed—18

10 Leaves aromatic when crushed—11

10 Leaves not especially aromatic when crushed—12

11 Leaves long and narrow, with deep rounded lobes on each side of the midrib SWEETFERN *(Comptonia)* 394

11 Leaves broad, mitten-shaped or 3-lobed SASSAFRAS *(Sassafras)* 430

12 Leaves large, 4 inches or more wide—13

12 Leaves smaller, less than 4 inches wide—14

13 Branches and veins on the lower surface of the leaves prickly DEVIL'S-CLUB *(Oplopanax)* 522

13 Branches and leafstalks covered with sticky-glandular and reddish hairs RASPBERRIES *(Rubus)* 457

14 Main veins of the leaves radiating from the summit of the leafstalk—15

14 Main veins of the leaves arising along a midrib—16

15 Leaves often clustered on lateral spurs; bark not freely peeling; fruit a berry CURRANTS AND GOOSEBERRIES *(Ribes)* 434

15 Leaves not clustered on lateral spurs; bark peeling in papery layers; fruits inflated papery pods NINEBARK *(Physocarpus)* 442

16 Branches without spines or thorns; fruit an acorn OAKS *(Quercus)* 414

16 Branches spine-tipped or with slender thorns; fruits apple-like—17

17 Branches often spine-tipped or spurlike; fruits greenish about 1 inch in diameter CRAB APPLES *(Malus)* 450

17 Branches with slender thorns; fruits smaller, usually red or yellowish HAWTHORNS *(Crataegus)* 450

18 Leaf margins toothed—19

18 Leaf margins untoothed—67

19 Leaves often paired or clustered on lateral spurlike branches—20
19 Leaves not in such pairs or clusters—22

20 Leaves in pairs; catkins often present; fruits small nutlets borne in conelike structures BIRCHES *(Betula)* 409
20 Leaves in clusters; catkins absent; fruits berry-like—21

21 Branchlets with 3-parted spines, the interior yellow; fruits oval-shaped, pulpy berries in drooping clusters BARBERRIES *(Berberis)* 425
21 Branchlets unarmed, the interior not yellow; fruits with large, bony, seedlike nutlets HOLLIES *(Ilex)* 486

22 Leaves with minute yellow resin dots at least on the lower surface, usually aromatic when crushed
SWEETGALE AND WAX MYRTLES *(Myrica)* 394
22 Leaves not resin-dotted—23

23 Leaves with 3 prominent veins from near the base—24
23 Leaves with one prominent midrib from which smaller veins branch—25

24 Leaf margins finely toothed, the leaf bases quite symmetrical
REDROOTS *(Ceanothus)* 502
24 Leaf margins coarsely toothed, the bases not symmetrical
HACKBERRIES *(Celtis)* 417

25 Branches spine-tipped or with slender thorns—26
25 Branches not armed—28

26 Fruits with a large stone surrounded by flesh PLUMS *(Prunus)* 462
26 Fruits apple-like—27

27 Fruits greenish, about 1 inch in diameter; branches often spine-tipped
CRAB APPLES *(Malus)* 450
27 Fruits usually red or yellowish; branches with slender thorns
HAWTHORNS *(Crataegus)* 450

28 Leaves leathery in texture; usually evergreen—29
28 Leaves not leathery; mostly deciduous—38

29 Leaves coarsely wavy-toothed; fruit an acorn OAKS *(Quercus)* 414
29 Leaves with small or obscure teeth; fruit not an acorn—30

30 Leaves spine-tipped or with a few bristly or spiny teeth above the middle HOLLIES *(Ilex)* 486
30 Leaves otherwise—31

31 Leaves with minute black dots on the lower surface
FETTERBUSHES *(Pieris)* 549
31 Leaves not black-dotted beneath—32

32 Leaves usually broadest toward the rounded or indented tip; fruit a whitish "plum" DEER-PLUM *(Licania)* 466
32 Leaves broadest at or below the middle, not rounded at tip—33

33 Largest leaves more than 3 inches long—34
33 Largest leaves less than 3 inches long—35

34 Sprawling or arching shrubs; fruit a capsule
LEUCOTHOËS *(Leucothoe)* 545
34 Erect shrub or small tree; fruit not a capsule but rather dry and 1-seeded SWEETLEAF *(Symplocos)* 570

35 Leaves with minute rusty and silvery scales
LEATHERLEAF *(Chamaedaphne)* 549
35 Leaves otherwise, usually quite smooth—36

36 Leaf margins obscurely toothed; leaves often whitened beneath; fruit a capsule ZENOBIA *(Zenobia)* 542
36 Leaf margins with low or bluntish teeth; leaves not whitened beneath; fruits berry-like—37

37 Colony-forming shrubs less than 1½ feet high; branchlets angled; fruit a bluish berry HUCKLEBERRIES *(Gaylussacia)* 553
37 Taller erect shrub; branchlets not angled, rather stiff and spiky; fruits usually red, with large seedlike nutlets HOLLIES *(Ilex)* 486

38 Leaf margins toothed mainly above the middle—39
38 Leaf margins toothed quite to the base—42

39 Marginal teeth rather large and coarse, irregular, or wavy—40
39 Marginal teeth rather small and sharp—41

40 Leaves wedge-shaped at base; fruits small achenes with long silky hairs GROUNDSEL BUSHES *(Baccharis)* 594
40 Leaves broadly pointed to roundish at base, the base often not symmetrical; fruit a 2-beaked woody pod
WITCH-ALDERS *(Fothergilla)* 441

41 Branchlets and lower leaf surfaces with starry-branched hairs; tall shrubs SNOWBELLS *(Styrax)* 569
41 Branchlets and lower leaf surfaces smooth or nearly so; low shrubs; the leaf margins often double-toothed SPIRAEAS *(Spiraea)* 442

42 Leaf margins with coarse teeth, averaging 5 or fewer to an inch—43
42 Leaf margins more finely and sometimes double-toothed—45

43 Leaves with veins ending in sharp marginal teeth
CHINQUAPINS *(Castanea)* 413
43 Leaf margins with bluntish, rounded, or wavy teeth—44

44 Leaf bases not symmetrical; fruit a woody pod
WITCH-HAZEL *(Hamamelis)* 441
44 Leaf bases quite symmetrical; fruit an acorn OAKS *(Quercus)* 414

45 Aquatic shrub of cypress ponds with a spindle-shaped stem branching above; leaves rather crowded, narrow, minutely toothed, and yellowish green STILLINGIA *(Stillingia)* 478
45 Shrubs otherwise—46

46 Buds with a solitary hoodlike scale; fruits small capsules filled with cottony seeds WILLOWS *(Salix)* 398

46 Buds without evident scales or with 2 or more scales—47

47 Branchlets usually bearing catkins or spikelike clusters of flower buds—48

47 Branchlets never with catkins or spikelike clusters of flower buds—50

48 Spikelike clusters of flower buds often present at tips of branchlets; leaves narrowly elliptic; fruit a capsule LEUCOTHOËS *(Leucothoe)* 545

48 True catkins often present; leaves broader; fruits not capsules—49

49 Branchlets bristly-hairy; catkins along the branchlets and all alike; fruit a nut enclosed within leaflike bracts HAZELNUTS *(Corylus)* 413

49 Branchlets not bristly hairy; catkins of 2 sizes at branch tips; fruits small nutlets borne in woody conelike structures ALDERS *(Alnus)* 410

50 Upper leaf surfaces with dark glands along the midrib CHOKEBERRIES *(Aronia)* 445

50 Upper leaf surfaces without such glands—51

51 Leafstalks with small glands at the summit; leaves often with an almond-like or rank odor if crushed PLUMS AND CHERRIES *(Prunus)* 462

51 Leafstalks not with glands at the summit- 52

52 Branchlets 4-angled or 4-sided BILBERRIES *(Vaccinium)* 557

52 Branchlets otherwise—53

53 Leaf margins doubly toothed—54

53 Leaf margins regularly and simply toothed, or the teeth rather obscure—55

54 Leaves broadest at or below the middle, roundish or broadly pointed at base; branchlets with a sharp ridge running down below the leafstalks or leaf scars ALABAMA SNOW-WREATH *(Neviusia)* 442

54 Leaves broadest above the middle, wedge-shaped at base; branchlets not ridged PEPPERBUSHES *(Clethra)* 526

55 Leaves more or less clustered near the tips of the branchlets; leaf margins both minutely toothed and hairy-fringed AZALEAS *(Rhododendron)* 530

55 Leaves well spaced along the branchlets—56

56 Branchlets showing partitions in the pith when cut lengthwise—57

56 Branchlets showing a continuous pith when cut lengthwise—58

57 Leaves broadly pointed to roundish at base, with some starry-branched hairs on the lower surface; fruits winged SILVERBELLS *(Halesia)* 569

57 Leaves wedge-shaped at base, smooth or nearly so; fruits oblong, 2-grooved capsules about ¼ inch long VIRGINIA WILLOW *(Itea)* 434

58 Largest leaves 3 or more inches long—59
58 Largest leaves usually less than 3 inches long—61

59 Branchlets and lower leaf surfaces with small starry-branched hairs PEPPERBUSHES *(Clethra)* 526
59 Branchlets and lower leaf surfaces with simple hairs, if any—60

60 Leaf margins hairy-fringed as well as toothed; fruit a capsule STEWARTIAS *(Stewartia)* 510
60 Leaf margins toothed but not hairy-fringed; fruits berry-like BUCKTHORNS *(Rhamnus)* 498

61 Leafstalks slender, usually ¼ inch or more long—62
61 Leafstalks moderate or stout, less than ¼ inch long—63

62 Fruits juicy, with about 10 small seeds; buds narrow, several times as long as wide SERVICEBERRIES *(Amelanchier)* 446
62 Fruits rather dry, with large seedlike nutlets; buds short, not much longer than wide HOLLIES *(Ilex)* 486

63 Branchlets minutely warty-dotted, green or reddish, smooth or hairy BLUEBERRIES *(Vaccinium)* 557
63 Branchlets not warty-dotted—64

64 Largest leaves less than 2 inches long BILBERRIES *(Vaccinium)* 557
64 Largest leaves usually 2 or more inches long—65

65 Leaf margins with small and rather obscure teeth; fruit a small roundish capsule MALEBERRY *(Lyonia)* 546
65 Leaf margins with sharp and quite conspicuous teeth—66

66 Leaves broadest near the often roundish base; fruit a juicy berry SOUTHERN MOUNTAIN-CRANBERRY *(Vaccinium)* 562
66 Leaves broadest near or above the middle, pointed at base; fruits berry-like but with large seedlike nutlets HOLLIES *(Ilex)* 486

67 Shrubs with thorns on the branchlets and with a milky sap BUMELIAS *(Bumelia)* 566
67 Shrubs not thorny—68

68 Leaves or branchlets aromatic when bruised—69
68 Leaves and branchlets odorless or nearly so, or with an unpleasant odor—75

69 Sap gummy; leaves broadest above the middle and often indented at tip SMOKETREE *(Cotinus)* 481
69 Sap not gummy; leaves usually with pointed tips—70

70 Leaves with small yellowish resin dots, at least on the lower surface WAX MYRTLES *(Myrica)* 397
70 Leaves not resin dotted—71

71 Leaves more or less leathery—72
71 Leaves not leathery—74

72 Leaves less than 2 inches long; branchlets very slender; spicy-aromatic PONDSPICE *(Litsea)* 430
72 Leaves more than 2 inches long; branchlets moderate—73

73 Leaves whitened beneath; spicy-aromatic MAGNOLIAS *(Magnolia)* 425
73 Leaves not whitened beneath; odor anise-like PURPLE-ANISE *(Ilicium)* 426

74 Branchlets slender; leaves sharply pointed at tip; fruits red SPICEBUSHES *(Lindera)* 429
74 Branchlets moderate; leaves broadly pointed or blunt at tip, some usually mitten-shaped or 3-lobed; fruits dark blue SASSAFRAS *(Sassafras)* 430

75 Leaves heart-shaped; leafstalks swollen at summit; fruit a flattened pea-like pod REDBUD *(Cercis)* 469
75 Leaves not heart-shaped and leafstalks not swollen at summit; fruits not pealike pods—76

76 Young branchlets and lower leaf surfaces densely coated with minute silvery scales ALABAMA CROTON *(Croton)* 478
76 Young branchlets and leaves otherwise—77

77 Leaves usually with clusters of smaller leaves in the axils—78
77 Leaves without axillary clusters of smaller leaves—79

78 Largest leaves scarcely ¼ inch long; fruits 3-lobed, 3-seeded capsules REDROOTS *(Ceanothus)* 502
78 Largest leaves often 1½ inches long, broadest near the blunt or rounded summit; fruit a berry MATRIMONY-VINES *(Lycium)* 566

79 Largest leaves less than 1 inch long—80
79 Largest leaves usually more than 1 inch long—81

80 Leaves less than ½ inch long, stalkless or nearly so, rather crowded; both leaves and branchlets with bristly hairs; fruit a capsule LAURELS *(Kalmia)* 541
80 Largest leaves over ½ inch long, short-stalked, often downy beneath; branchlets minutely warty-dotted; fruit a berry BLUEBERRIES *(Vaccinium)* 557

81 Leaves more or less leathery—82
81 Leaves not leathery—101

82 Branchlets ill-scented if broken; buds hairy and without visible scales; fruit a large pulpy berry with big seeds PAWPAWS *(Asimina)* 426
82 Branchlets nearly or quite odorless if broken—83

83 Leaves densely rusty-woolly beneath, the margin strongly rolled inward on the lower side LABRADOR-TEA *(Ledum)* 530
83 Leaves not rusty-woolly beneath—84

84 Leaves densely coated with rust-colored scales or brown dots on the lower surface—85
84 Leaves otherwise—87

85 Leaves 2 to 4 inches long; branchlets with large end buds
RHODODENDRONS *(Rhododendron)* 530

85 Leaves smaller; branchlets not with large end buds—86

86 Leaves with minute silvery scales on the upper surface; flowers and fruits in the axils of leaflike bracts toward the ends of the branchlets LEATHERLEAF *(Chamaedaphne)* 549

86 Leaves smooth above; flowers and fruits in clusters on growth of the previous year LYONIAS *(Lyonia)* 546

87 Branchlets 3-sided; leaves with a conspicuous vein paralleling the leaf margin, glandular-dotted beneath; shrub with spreading and arching branches FETTERBUSHES *(Lyonia)* 546

87 Branchlets and leaves otherwise—88

88 Leaves either spine-tipped or minutely black-dotted beneath; fruits berry-like but with large bony nutlets HOLLIES *(Ilex)* 486

88 Leaves neither spine-tipped nor black-dotted beneath—89

89 Leaves very white on the lower surface, the leaf margins strongly rolled inward on the lower side BOG ROSEMARY *(Andromeda)* 542

89 Leaves slightly if at all whitened on the lower surface, and leaf margins slightly if at all rolled inward on the lower side—90

90 Leaves usually less than 3 inches long—91

90 Leaves usually 3 or more inches long—97

91 Leaf blades more than 4 times as long as broad, tipped with a small bristle; fruit an acorn OAKS *(Quercus)* 414

91 Leaves usually less than 4 times as long as broad; fruits otherwise—92

92 Branchlets with large end buds; leaves often whitened and with small rusty hairs beneath AZALEAS *(Rhododendron)* 530

92 Branchlets without large end buds—93

93 Leaves sharply pointed at tip and more or less twisted, standing rather upright along hairy branchlets TAR-FLOWER *(Befaria)* 529

93 Leaves otherwise—94

94 Branchlets minutely warty-dotted or hairy; fruit a berry
BLUEBERRIES *(Vaccinium)* 557

94 Branchlets neither warty-dotted nor hairy; fruit a capsule—95

95 Leaves broadest about the middle and broadly pointed at both ends, often whitened beneath ZENOBIA *(Zenobia)* 542

95 Leaves wedge-shaped at base, broadly or bluntly pointed at tip—96

96 Shrub less than 3 feet high; leaves deciduous; capsules globe-shaped and long-stalked LAURELS *(Kalmia)* 541

96 Shrub taller or small tree; leaves evergreen; capsules 2- to 4-winged, arranged in narrow clusters
BUCKWHEAT-TREE *(Cliftonia)* 486

97 Leaves more or less clustered near the ends of the branchlets—98

97 Leaves all well spaced along the branchlets—100

98 Branchlets with large end buds; fruit an oblong capsule ½ inch or more long RHODODENDRONS *(Rhododendron)* 530

98 Branchlets without large end buds; capsules smaller, globe-shaped or egg-shaped — 99

99 Leaves usually broadest near the middle and pointed at both ends; capsules globe-shaped and long-stalked LAURELS *(Kalmia)* 541

99 Leaves usually broadest toward the blunt or rounded tip, wedge-shaped at base; capsules small, egg-shaped, in long and narrow clusters SWAMP CYRILLA *(Cyrilla)* 485

100 Leaves broadest toward the roundish or indented tip, very veiny and lustrous above; fruit a whitish "plum" DEER-PLUM *(Licania)* 466

100 Leaves broadest about the middle, pointed at both ends, not lustrous or especially veiny above; fruit brownish, dry, 1-seeded SWEETLEAF *(Symplocos)* 570

101 Leaves 6 to 12 inches long, broadest toward the tip and tapering to the base; buds hairy and without visible scales; branchlets ill-scented when broken; fruit a large pulpy berry with big seeds PAWPAWS *(Asimina)* 426

101 Leaves usually less than 6 inches long — 102

102 Leaves more or less clustered toward the tips of the branchlets — 103

102 Leaves all well spaced along the branchlets, or clustered on spurs — 105

103 Leaves mostly oval-shaped, the main veins curving and tending to parallel the margin and meet at the tip DOGWOODS *(Cornus)* 522

103 Leaves with veins otherwise, the margins usually hairy fringed; branchlets with large end buds — 104

104 Leaves rough-hairy above, usually with chaffy scales on the veins beneath; branchlets rusty-hairy; fruit an egg-shaped capsule less than ¼ inch long MENZIESIA *(Menziesia)* 538

104 Leaves usually smooth above, smooth or downy and with some stiff hairs along the midrib beneath; branchlets smooth, downy, or with scattered bristly hairs; fruits oblong capsules ⅜ inch or more long AZALEAS *(Rhododendron)* 530

105 Branchlets and lower leaf surfaces with starry-branched hairs; fruits pea-sized, dry, 1-seeded, grayish-downy SNOWBELLS *(Styrax)* 569

105 Branchlets and lower leaf surfaces smooth or with simple unbranched hairs — 106

106 Largest leaves more than 4 inches long — 107

106 Largest leaves less than 4 inches long — 109

107 Branchlets with large catkin-like flower buds toward the tip; fruits dry, wrinkled, 1-seeded CORKWOOD *(Leitneria)* 394

107 Branchlets not with catkin-like flower buds — 108

108 Leaves conspicuously veiny, clear dots visible when leaf is held up to light; fruits about 1 inch long, pear-shaped, leathery, 1-seeded, and very oily BUFFALONUT *(Pyrularia)* 417

108 Leaves otherwise; fruit a capsule about ⅜ inch in diameter SOUTHERN-PLUME *(Elliottia)* 529

109 Branchlets sharply 4-angled BILBERRIES *(Vaccinium)* 558

109 Branchlets round or but slightly angled—110

110 Leaves often clustered on short lateral spurs, blunt or with a small bristle-point at tip; leafstalks slender, ¼ inch or more long MOUNTAIN HOLLY *(Nemopanthus)* 486

110 Leaves never clustered on lateral spurs; leafstalks moderate to stout, mostly less than ¼ inch long—111

111 Bases of the stout leafstalks hollow and covering the buds; leaves broadest above the middle and blunt at tip; branchlets enlarged upwards LEATHERWOOD *(Dirca)* 521

111 Bases of the leafstalks otherwise—112

112 Buds covered with a solitary hoodlike scale; fruits small capsules filled with cottony seeds WILLOWS *(Salix)* 398

112 Buds with 2 or more evident scales; fruits otherwise—113

113 Leaves 5 or more times as long as broad, less than ½ inch wide; chiefly plants of salt marshes—114

113 Leaves 4 times or less as long as broad, usually ½ inch or more wide—115

114 Plants resinous GROUNDSEL BUSHES *(Baccharis)* 594

114 Plants not resinous MARSH ELDERS *(Iva)* 593

115 Leaves with glistening yellow resin dots, at least on the lower surface; fruit a berry HUCKLEBERRIES *(Gaylussacia)* 553

115 Leaves not resin-dotted—116

116 Leaves with minute black dots on the lower surface; fruit a vase-shaped capsule STAGGERBUSH *(Lyonia)* 546

116 Leaves not black-dotted beneath—117

117 Branchlets exuding a milky sap when cut or broken; fruit a 3-lobed capsule SEBASTIANBUSH *(Sebastiania)* 477

117 Branchlets not with a milky sap; fruits otherwise—118

118 Branchlets minutely warty-dotted, otherwise smooth or hairy, usually greenish or reddish; fruit a berry BLUEBERRIES *(Vaccinium)* 557

118 Branchlets not warty-dotted, smooth or minutely downy—119

119 Branchlets purplish; lower leaf surfaces whitened; fruit a berry, borne in dangling leafy-bracted clusters DEERBERRY *(Vaccinium)* 557

119 Branchlets yellowish brown to ashy gray; lower leaf surfaces not whitened; fruit a small globe-shaped capsule MALEBERRY *(Lyonia)* 546

KEY VI

WINTER KEY TO LEAF-LOSING SHRUBS

1 Leaf scars and buds opposite or whorled — 2
1 Leaf scars and buds alternate — 39

2 Climbing or twining vines — 3
2 Stems otherwise — 8

3 Climbing by means of rootlets along the stems — 4
3 Not with rootlets along the stems — 5

4 Leaf scars half-round, with a solitary C-shaped bundle scar TRUMPET CREEPER *(Campsis)* 577
4 Leaf scars horseshoe-shaped, with 3 bundle scars CLIMBING-HYDRANGEA *(Decumaria)* 433

5 Climbing by means of tendrils — 6
5 Scrambling or with twining stems — 7

6 Stems soft-woody, with longitudinal ridges; leaf scars hidden by persisting bases of the leafstalks VIRGIN'S-BOWERS *(Clematis)* 421
6 Stems and leaf scars otherwise; the leaf scars with a solitary bundle scar CROSS-VINE *(Bignonia)* 577

7 Stems hollow inside between the nodes HONEYSUCKLES *(Lonicera)* 581
7 Stems with pith inside between the nodes TRACHELOSPERMUM *(Trachelospermum)* 574

8 Stems prostrate or trailing — 9
8 Stems more or less erect — 10

9 Twigs 2-edged or 2-winged ST. PETER'S-WORTS *(Hypericum)* 510
9 Twigs more or less 4-sided or square STRAWBERRY-BUSHES *(Euonymus)* 493

10 Buds naked, or without visible scales — 11
10 Buds with visible scales, or the buds not evident — 13

11 Leaf scars with a solitary bundle scar; twigs roughish BEAUTYBERRY *(Callicarpa)* 573
11 Leaf scars with 3 or more bundle scars — 12

12 Leaf scars with 3 bundle scars VIBURNUMS *(Viburnum)* 586
12 Leaf scars with 5 or 7 bundle scars HYDRANGEAS *(Hydrangea)* 433

13 Twigs often spiny-tipped SAGERETIA *(Sageretia)* 498
13 Twigs never spiny-tipped — 14

14 Twigs spicy-aromatic when broken STRAWBERRY-SHRUBS *(Calycanthus)* 425
14 Twigs not aromatic — 15

15 Twigs with a true end or terminal bud — 16
15 Twigs without a true end or terminal bud — 25

16 Leaf scars with a solitary bundle scar — 17
16 Leaf scars with 3 or more bundle scars — 20

17 Twigs rather stout, reddish brown and usually hairy; end bud much larger than the lateral ones PINCKNEYA *(Pinckneya)* 577
17 Twigs moderate or slender; end bud little if any larger than lateral ones—18

18 Twigs 4-lined or 4-sided, usually green STRAWBERRY-BUSHES *(Euonymus)* 493
18 Twigs otherwise—19

19 Twigs more or less coated with small rusty scales; buds stalked and with 2 visible scales BUFFALOBERRY *(Shepherdia)* 521
19 Twigs smooth except for conspicuous lenticels; buds with about 6 visible scales FRINGETREE *(Chionanthus)* 573

20 Leaf scars large, shield-shaped or triangular, usually with 5 or more bundle scars but sometimes in 3 groups BUCKEYES *(Aesculus)* 497
20 Leaf scars small or narrow, always with 3 bundle scars—21

21 Buds with 4 or more visible scales—22
21 Buds with only 2 visible scales—23

22 Buds with but 2 pairs of visible scales VIBURNUMS *(Viburnum)* 586
22 Buds with more than 2 pairs of visible scales MAPLES *(Acer)* 497

23 End bud flanked by a pair of persistent leafstalk bases DOGWOODS *(Cornus)* 522
23 End bud otherwise—24

24 Tips of the opposing leaf scars meeting in a point MAPLES *(Acer)* 497
24 Tips of the opposing leaf scars not meeting but sometimes connected with a transverse line VIBURNUMS *(Viburnum)* 586

25 Leaf scars indistinct, raised or covered by the persistent bases of the leafstalks—26
25 Leaf scars rather distinct, sometimes slightly raised—27

26 Twigs very fine and usually downy; buds about 1/16 inch long; red berries often persisting CORALBERRY *(Symphoricarpos)* 585
26 Twigs slender, smooth or nearly so; buds about ⅛ inch long HONEYSUCKLES *(Lonicera)* 581

27 Leaf scars with a solitary bundle scar—28
27 Leaf scars with 3 or more bundle scars—33

28 Leaf scars often 3 at a node; buds buried in the bark and not visible BUTTONBUSH *(Cephalanthus)* 577
28 Leaf scars always opposite; buds evident—29

29 Twigs prominently 2-angled or 2-winged below the leaf scars—30
29 Twigs otherwise—31

30 Twigs 2-winged below the leaf scars ST. PETER'S-WORTS *(Hypericum)* 510
30 Twigs 2-edged below the leaf scars ST. JOHN'S-WORTS *(Hypericum)* 510

31 Buds roundish and usually superposed SWAMP-PRIVETS *(Foresteria)* 570
31 Buds otherwise—32

32 Twigs green or grayish olive — BUCKLEYA *(Buckleya)* — 418
32 Twigs dark purplish brown — NESTRONIA *(Nestronia)* — 418

33 Buds not visible above the leaf scars—34
33 Buds plainly visible above the leaf scars—35

34 Buds imbedded in the bark; twigs with fine longitudinal ridges; shrubs of coastal marshes — MARSH ELDERS *(Iva)* — 593
34 Buds concealed beneath the leaf scars; bark often soon peeling from the branchlets; upland shrubs — MOCK-ORANGES *(Philadelphus)* — 430

35 Leaf scars with 5 or more bundle scars — ELDERS *(Sambucus)* — 593
35 Leaf scars with 3 bundle scars—36

36 Twigs with raised hairy lines running down from a transverse line connecting the leaf scars — BUSH-HONEYSUCKLES *(Diervilla)* — 578
36 Twigs otherwise—37

37 Scales of the previous year's buds persisting at the twig bases; buds with 4 to 6 exposed scales — HYDRANGEAS *(Hydrangea)* — 433
37 Scales of the previous year's buds not so persisting; buds with 2 to 4 exposed scales—38

38 Twigs smooth; buds with 2 to 4 smooth scales; leaf scars not connected by a transverse line; bark close — BLADDERNUT *(Staphylea)* — 494
38 Twigs downy; buds with 2 downy scales; leaf scars connected by a transverse line; bark soon peeling — MOCK-ORANGES *(Philadelphus)* — 430

39 Climbing or twining vines—40
39 Stems otherwise—54

40 Climbing by means of rootlets along the stems — POISON-IVY *(Toxicodendron)* — 485
40 Not with rootlets along the stems—41

41 Climbing by means of tendrils—42
41 Scrambling or with twining stems—46

42 Tendrils on the persistent bases of the leafstalks; stems usually green and often prickly or bristly — GREENBRIERS *(Smilax)* — 390
42 Tendrils on the stems or branchlets—43

43 Tendrils at the tips of the branchlets — BRUNNICHIA *(Brunnichia)* — 425
43 Tendrils on the stems opposite some of the leaf scars—44

44 Pith brownish, usually with woody partitions at the nodes — GRAPES *(Vitis)* — 505
44 Pith white, never with woody partitions at the nodes—45

45 Pith becoming somewhat chambered — AMPELOPSIS *(Ampelopsis)* — 502
45 Pith continuous; tendrils often ending in adhesive disks — CREEPERS *(Parthenocissus)* — 505

46 Stems aromatic when bruised or broken
SCHISANDRA *(Schisandra)* 426

46 Not aromatic—47

47 Buds small, superposed on a silky area within the U-shaped leaf scars
PIPEVINES *(Aristolochia)* 418

47 Buds and leaf scars otherwise—48

48 Leaf scars very narrow and but slightly curved; stems usually with scattered prickles ROSES *(Rosa)* 461

48 Leaf scars otherwise; stems never with prickles—49

49 Leaf scars flanked with prominent knobs; buds silky-hairy
WISTERIAS *(Wisteria)* 474

49 Leaf scars and buds otherwise—50

50 Buds small or sunken in depressions above the leaf scars—51

50 Buds of moderate size and never sunken—53

51 Twigs downy; fruits red RED-BERRIED MOONSEED *(Cocculus)* 422

51 Twigs smooth or nearly so; fruits blue or black—52

52 Stems slender; fruits blue COMMON MOONSEED *(Menispermum)* 422

52 Stems moderate or rather stout; fruits black
CUPSEED *(Calycocarpum)* 422

53 Buds oblong and appressed; leaf scars raised SUPPLEJACK *(Berchemia)* 498

53 Buds roundish and pointed outward; leaf scars not raised
AMERICAN BITTERSWEET *(Celastrus)* 494

54 Stems prostrate or creeping—55

54 Stems more or less erect—60

55 Buds small, often more than one above a leaf scar; twigs lustrous reddish brown CHERRIES *(Prunus)* 462

55 Buds moderate, always solitary; twigs otherwise—56

56 Leaf scars covered by persistent leafstalk bases; stems prickly or bristly BLACKBERRIES *(Rubus)* 457

56 Leaf scars and stems otherwise—57

57 Buds densely hairy and without visible scales; leaf scars with 5 or more bundle scars POISON-IVY *(Toxicodendron)* 485

57 Buds with 1 or more visible scales; leaf scars with 3 bundle scars—58

58 Buds covered with a solitary hoodlike scale WILLOWS *(Salix)* 398

58 Buds with 2 or more visible scales—59

59 Buds usually with 3 visible scales; twigs glandular-warty or densely downy, with a very small green pith BIRCHES *(Betula)* 409

59 Buds with about 6 visible scales; twigs with a large pale pith and sometimes with a skunklike odor when broken CURRANTS *(Ribes)* 434

60 Plant with a solitary spindle-shaped stem and more or less erect branches above — STILLINGIA *(Stillingia)* 478

60 Plants usually with several stems growing in a clump from the base, or with wide-spreading branches—61

61 Stems or twigs spine-tipped or with thorns, prickles, or prickly bristles—62

61 Stems or twigs never armed—75

62 Leaf scars indistinct or hidden by persistent leafstalk bases—63

62 Leaf scars quite distinct—64

63 Stems unbranched, usually arching, bristly or with broad-based prickles — RASPBERRIES AND BLACKBERRIES *(Rubus)* 457

63 Stems branched, yellow inside, with 3-branched spines at the nodes — BARBERRIES *(Berberis)* 425

64 Branches often with sharp spinelike tips—65

64 Branches otherwise—66

65 Leaf scars nearly half-round; twigs usually with a bitter almond-like taste — PLUMS *(Prunus)* 462

65 Leaf scars narrowly crescent-shaped; twigs otherwise — CRAB APPLES *(Malus)* 450

66 Buds not evident, buried in the bark above the leaf scars; twigs bristly or with pairs of spines at the nodes — LOCUSTS *(Robinia)* 470

66 Buds more or less evident above the leaf scars—67

67 Stems woody chiefly toward the base, very bristly; twigs nearly encircled by narrow leaf scars — BRISTLY-SARSAPARILLA *(Aralia)* 521

67 Stems more or less woody throughout—68

68 Twigs exuding a milky sap when cut — BUMELIAS *(Bumelia)* 566

68 Twigs otherwise—69

69 Twigs very stout, ½ inch or more in diameter—70

69 Twigs more moderate or slender—71

70 Twigs with broad-based stout prickles; fruits black — HERCULES'-CLUB *(Aralia)* 521

70 Twigs with slender prickles; fruits red — DEVIL'S-CLUB *(Oplopanax)* 522

71 Twigs 5-angled, pale, slender; leaf scars with a solitary bundle scar — MATRIMONY-VINES *(Lycium)* 566

71 Twigs and leaf scars otherwise—72

72 Twigs with a lemon- or limelike odor when bruised or broken; buds hairy and without visible scales — PRICKLY-ASHES *(Zanthoxylum)* 477

72 Twigs not aromatic; buds with visible scales—73

73 Buds with rather fleshy red scales; twigs with sharp thorns ½ inch or more long — HAWTHORNS *(Crataegus)* 450

73 Buds otherwise; twigs bristly or with shorter thorns or prickles—74

74 Leaf scars narrow and but slightly curved; twigs round
ROSES *(Rosa)* 461

74 Leaf scars crescent-shaped to U-shaped; twigs with lines or ridges running down below each leaf scar
CURRANTS AND GOOSEBERRIES *(Ribes)* 434

75 Leaf scars hidden by persistent leafstalk bases—76
75 Leaf scars otherwise—78

76 Bases of the leafstalks and the persistent stipules forming a sheath about the twig—77
76 Bases of the leafstalks otherwise RASPBERRIES *(Rubus)* 457

77 Stems erect, to 3 feet high, much branched; leaves pinnately compound with 3 to 7 (usually 5) leaflets; flowers bright yellow
SHRUBBY CINQUEFOIL *(Pentaphylloides)* 454

77 Stems ascending or trailing; leaves fan-shaped with 3 leaflets, each toothed apically; flowers white
THREE-TOOTHED CINQUEFOIL *(Sibbaldiopsis)* 454

78 Leaf scars with a solitary bundle scar—79
78 Leaf scars with 3 or more bundle scars—104

79 Twigs with a spicy-aromatic odor when bruised or broken—80
79 Twigs not spicy-aromatic—81

80 Twigs very slender; terminal bud little if any larger than the lateral ones and often flanked by long-stalked flower buds
PONDSPICE *(Litsea)* 430

80 Twigs moderate; terminal bud much larger than the lateral ones
SASSAFRAS *(Sassafras)* 430

81 Twigs ending in a true end or terminal bud—82
81 Twigs without a true end bud, the tip dying back to a lateral bud—89

82 End bud usually much larger than the others, lateral buds solitary—83
82 End bud little if any larger than the others, lateral buds sometimes superposed—85

83 Outer bud scales about as long as the bud but falling off early, the inner scales downy or silky-hairy PEPPERBUSHES *(Clethra)* 526

83 Outer bud scales much shorter than the bud, the scales all persisting—84

84 Bark soon shredding; fruit a capsule splitting into 4 parts
MENZIESIA *(Menziesia)* 538

84 Bark not soon shredding; fruit a capsule splitting into 5 parts
AZALEAS *(Rhododendron)* 530

85 Twigs showing interruptions in the pith when cut lengthwise—86
85 Twigs showing a continuous pith when cut lengthwise—87

86 Buds egg-shaped, reddish, and pointed SILVERBELLS *(Halesia)* 569
86 Buds roundish, greenish to brown, blunt
SWEETLEAF *(Symplocos)* 570

87 Leaf scars shield shaped, distinctly longer than broad, flat with the twig SOUTHERN-PLUME *(Elliottia)* 529
87 Leaf scars half-round or triangular, about as long as broad and usually slightly raised—88

88 Leaf scars with a persistent stipule or stipule scar at each end; buds often superposed, the upper one larger, usually blunt
HOLLIES *(Ilex)* 486
88 Leaf scars otherwise; buds solitary, cone-shaped, usually pointed, and with but 2 visible scales MOUNTAIN HOLLY *(Nemopanthus)* 486

89 Twigs exuding a milky sap when cut; buds almost hidden by the raised leaf scars which have prominent stipules at each end
SEBASTIANBUSH *(Sebastiania)* 477
89 Twigs not with milky sap; buds and leaf scars otherwise—90

90 Twigs 4-sided or 4-angled BILBERRIES *(Vaccinium)* 557
90 Twigs otherwise—91

91 Buds densely hairy and without distinct scales—92
91 Buds otherwise—93

92 Buds superposed, densely coated with minute whitish and rusty scalelike hairs; leaf scars with a central bundle scar
SNOWBELLS *(Styrax)* 569
92 Buds solitary, silvery-silky; leaf scars with the bundle scar near the top edge STEWARTIAS *(Stewartia)* 510

93 Buds about ¼ inch long, narrowly egg-shaped, pointed, greenish, and sometimes in groups BUFFALONUT *(Pyrularia)* 417
93 Buds smaller—94

94 Flower buds often present in a long spikelike end cluster
LEUCOTHOËS *(Leucothoe)* 545
94 Flower buds, if present, not in a spikelike end cluster—95

95 Twigs usually with buds of 2 sizes, the larger flower buds with several scales, the smaller ones with usually 2 visible scales—96
95 Twigs with buds all more or less alike—97

96 Scales of the smaller buds with long-pointed tips; twigs usually warty-dotted and sometimes hairy BLUEBERRIES *(Vaccinium)* 557
96 Scales of the smaller buds not long-pointed; twigs often downy and sometimes with minute yellow resin dots
HUCKLEBERRIES *(Gaylussacia)* 553

97 Buds with but 2 visible scales and closely pressed against the twig—98
97 Buds with 3 or more visible scales, not closely pressed against the twig—100

98 Buds minute; fruits long-stalked capsules in end clusters
LAURELS *(Kalmia)* 541

98 Buds 1/16 to 1/8 inch long—99

99 Buds about 1/8 inch long, pointed at tip; fruits small globe-shaped capsules in branched clusters MALEBERRY *(Lyonia)* 546

99 Buds usually smaller, blunt at tip; fruits not persisting
SOUTHERN MOUNTAIN-CRANBERRY *(Vaccinium)* 562

100 Leaf scars distinctly raised—101

100 Leaf scars but slightly if at all raised—102

101 Leaf scars with minute narrow stipule scars at each side; silvery-lined, cup-shaped fruit bases often persisting REDROOTS *(Ceanothus)* 502

101 Leaf scars without such stipule scars; fruits small pods in groups of 3 to 5, in flat-topped or cone-shaped end clusters SPIRAEAS *(Spiraea)* 442

102 Leaf scars half-round; twigs purplish red, usually downy or whitened with a bloom DEERBERRY *(Vaccinium)* 557

102 Leaf scars shield shaped or triangular; twigs smooth—103

103 Buds roundish or egg-shaped; twigs often with minute black dots; fruit an urn-shaped capsule STAGGERBUSH *(Lyonia)* 546

103 Buds cone-shaped; twigs sometimes whitened with a bloom; fruit a globe-shaped capsule ZENOBIA *(Zenobia)* 542

104 Leaf scars with 3 bundle scars—105

104 Leaf scars with 5 or more bundle scars—137

105 Twigs usually with catkins or catkin-like flower buds—106

105 Twigs never with catkins or catkin-like flower buds—111

106 Twigs rather stout, about 1/4 inch in diameter; large many-scaled flower buds present toward the tips of the twigs
CORKWOOD *(Leitneria)* 394

106 Twigs much more slender—107

107 Twigs spicy-aromatic when broken—108

107 Twigs not spicy-aromatic—109

108 Twigs smooth; buds pointed SWEETGALE *(Myrica)* 394

108 Twigs hairy; buds roundish SWEETFERN *(Comptonia)* 394

109 Catkins of 2 sizes, the smaller ones pistil-bearing; conelike fruiting bodies of the previous year often present ALDERS *(Alnus)* 410

109 Catkins all more or less alike—110

110 Buds with 2 or 3 visible scales; twigs densely downy or glandular-warty, some of them short and spurlike BIRCHES *(Betula)* 409

110 Buds with 4 or more visible scales, the flower-bearing ones usually larger; twigs more or less bristly-hairy HAZELNUTS *(Corylus)* 413

111 Buds naked or without evident scales, rather hairy—112

111 Buds with one or more visible scales—115

112 Buds stalked; fruits urn-shaped woody capsules — 113
112 Buds not stalked; fruits if persisting not woody capsules — 114

113 Capsules 2-beaked and in elongate clusters
WITCH-ALDERS *(Fothergilla)* 441
113 Capsules with 4 short beaks, grouped in short clusters
WITCH-HAZEL *(Hamamelis)* 441

114 Twigs ill-scented if broken; buds silvery-silky, almost surrounded by the U-shaped leaf scars HOPTREE *(Ptelea)* 477
114 Twigs not ill-scented; buds otherwise BUCKTHORNS *(Rhamnus)* 498

115 Buds covered with a solitary hoodlike scale WILLOWS *(Salix)* 398
115 Buds with 2 or more visible scales — 116

116 Twigs with a true end or terminal bud — 117
116 Twigs without a true end bud, the tips dying back to a lateral bud — 127

117 Twigs spicy-aromatic when broken — 118
117 Twigs not spicy-aromatic — 120

118 Flower buds usually clustered at the nodes SPICEBUSHES *(Lindera)* 429
118 Buds always solitary above the leaf scars — 119

119 Twigs exuding a gummy sap when broken; buds pointed, usually with 2 visible scales SMOKETREE *(Cotinus)* 481
119 Twigs not exuding a gummy sap; buds blunt, with more than 2 visible scales BAYBERRY *(Myrica)* 397

120 Buds superposed, pointed outward from the twig; twigs showing interruptions in the pith when cut lengthwise
VIRGINIA WILLOW *(Itea)* 434
120 Buds not superposed; pith continuous — 121

121 Buds often present on each side of the usual axillary bud
CHERRIES *(Prunus)* 466
121 Buds always solitary above the leaf scars — 122

122 Twigs with prominent lines or ridges running downward below the leaf scars — 123
122 Twigs otherwise — 124

123 Leaf scars half-round or somewhat 3-lobed; bark soon peeling from the stems NINEBARK *(Physocarpus)* 442
123 Leaf scars rather narrowly crescent-shaped; bark not readily peeling
CURRANTS AND GOOSEBERRIES *(Ribes)* 434

124 Twigs with a rank or disagreeable odor when broken
CHERRIES *(Prunus)* 462
124 Twigs otherwise — 125

125 Buds usually with 2 visible scales, the end bud much larger — 126
125 Buds with several visible scales, the end one not much larger — 127

126 Buds smooth, usually more or less clustered toward the tips of the twigs; twigs greenish, the bark becoming streaked with white
DOGWOODS *(Cornus)* 522

126 Buds densely downy; twigs and bark otherwise
WITCH-ALDERS *(Fothergilla)* 441

127 Bud scales dark-tipped, more or less twisted, the second scale usually less than half as long as the bud SHADBUSHES *(Amelanchier)* 446

127 Bud scales otherwise, the second scale half or more as long as the bud; small apple-like fruits often persisting CHOKEBERRIES *(Aronia)* 445

128 Buds superposed, or with buds along side of the usual axillary ones—129
128 Buds always solitary above the leaf scars—132

129 Buds present at the sides of the usual axillary bud; twigs with a bitter or almond-like taste PLUMS *(Prunus)* 462
129 Buds superposed—130

130 Twigs stout; leaf scars large and triangular or 3-lobed
SOAPBERRIES *(Sapindus)* 498

130 Twigs moderate or slender, the leaf scars rather small—131

131 Leaf scars fringed at the top, stipule scars not present; larger buds with 6 or more overlapping scales REDBUD *(Cercis)* 469
131 Leaf scars not fringed at the top, small stipule scars evident; larger buds with 2 or 3 scales INDIGOBUSHES *(Amorpha)* 469

132 Twigs greenish or grayish green and with fine longitudinal grooves; buds resin-coated GROUNDSEL BUSHES *(Baccharis)* 594
132 Twigs and buds otherwise—133

133 Twigs showing interruptions in the pith (at least at the nodes) when cut lengthwise; buds closely pressed against the twigs
HACKBERRIES *(Celtis)* 417

133 Twigs with a continuous pith; buds not closely pressed against the twigs—134

134 Buds with 2 or 3 exposed scales INDIGOBUSHES *(Amorpha)* 469
134 Buds with more numerous exposed scales—135

135 Twigs with a prominent sharp ridge running down below each leaf scar; leaf scars somewhat heart-shaped
ALABAMA SNOW-WREATH *(Neviusia)* 442
135 Twigs and leaf scars otherwise—136

136 Leaf scars raised; twigs often with a spiny tip and usually with a bitter or almond-like taste PLUMS *(Prunus)* 462
136 Leaf scars raised; twigs not spine-tipped or with an almond-like taste BUCKTHORNS *(Rhamnus)* 498

137 Twigs with a true end or terminal bud—138
137 Twigs without a true end bud, the tip dying back to a lateral bud—144

138 Twigs encircled at the nodes with linelike stipule scars, pleasantly aromatic when bruised or broken MAGNOLIAS *(Magnolia)* 425

138 Twigs otherwise—139

139 Leaf scars crescent-shaped and almost encircling the twigs; twigs bright yellow inside SHRUB-YELLOWROOT *(Xanthorhiza)* 422

139 Leaf scars and twigs otherwise—140

140 Buds naked or without evident scales—141

140 Buds with evident scales—142

141 End bud very much larger than the lateral ones; twigs ill-scented if bruised or broken PAWPAWS *(Asimina)* 426

141 End bud little if any longer than the lateral ones; waxy white berry-like fruits often persisting POISON-IVY AND POISON-OAK *(Toxicodendron)* 485

142 Buds tending to be clustered at the tips of the twigs OAKS *(Quercus)* 414

142 Buds well distributed along the twigs, the end one much larger than the lateral buds—143

143 Buds smooth; leaf scars broadly crescent-shaped or nearly triangular; fruits waxy white, in drooping axillary clusters POISON SUMAC *(Toxicodendron)* 485

143 Buds gummy and often with matted hairs; leaf scars narrowly crescent-shaped; fruits like small orange-red apples and in flat-topped clusters MOUNTAIN-ASH *(Sorbus)* 450

144 Buds with 2 visible scales, placed a little to one side of the center of the half-round leaf scars CHINQUAPINS *(Castanea)* 413

144 Buds without distinct scales, usually very hairy—145

145 Buds hidden behind the raised circular leaf scars; catkin-like flower buds often present; twigs aromatic if broken SUMACS *(Rhus)* 482

145 Buds not so hidden, often almost encircled by the leaf scars—146

146 Twigs slender and enlarged at the nodes, very tough and pliable LEATHERWOOD *(Dirca)* 521

146 Twigs stout, with a large pith and rather brittle; clusters of hairy red fruits often present SUMACS *(Rhus)* 482

HOW PLANTS ARE NAMED

Our modern method of naming plants was devised by the Swedish botanist Linnaeus in 1753. The first part of the scientific name is that of the genus to which a plant belongs, and it is always written with a capital letter. The second part is the name of the particular species, and it is customary to write this in lower case, even though it may be a geographical name or name of a person. Sometimes varieties, forms, or subspecies are recognized, and these names follow the species name. In this book only some of the more outstanding or distinct varieties or forms are recognized. Following the name of the plant it is customary to give the name of the author, or person who described the plant. In most cases these names are abbreviated.

Scientific names are derived from various sources. Sometimes the generic names are simply the ancient Greek or Latin names for groups of plants: *Rosa* for the roses, *Vitis* for grapes, *Prunus* for the plums and cherries, etc. In quite a few cases the names signify some characteristic of the plant: *Xanthorhiza,* for example, means "yellow root." Again, the name may honor some person, as *Lyonia* for John Lyon; or some character from mythology, such as *Andromeda*. Specific names, too, often tell us of some distinctive characteristic: *acerifolium* means "with leaves like a maple," *macrocarpon* means "large-fruited," and *pubescens* tells us that perhaps the leaves, at least, are downy.

In the belief that many of you who will use this book will be interested in knowing the meanings of the scientific names, the following lists of generic and specific names have been included.

Acer—Latin name of the maples, from the Celtic, meaning hard.

Aesculus—Ancient Latin name of an oak or other mast-bearing tree.

Alnus—Ancient Latin name of the alders.

Amelanchier—French name of a related plant.

Amorpha—From a Greek word meaning deformed, alluding to the absence of four of the petals.

Ampelopsis—From two Greek words meaning grape and likeness.

Andromeda—Named for Andromeda, beautiful Ethiopian princess in Greek mythology.

Aralia—From the French-Canadian name *araile.*

Arctostaphylos—From two Greek words meaning a bear and a bunch of grapes.

Aristolochia—From two Greek words meaning best and delivery, in allusion to the supposed value in aiding childbirth.

Arundinaria—From the Latin word for a reed or cane.

Asimina—From the American Indian name *assimin.*

Baccharis—Name of some shrub apparently dedicated long ago to the wine god Bacchus and transferred to this genus by Linnaeus.

Befaria—Apparently named for José Bejar, professor of botany at Cadiz, Spain; but mispelled Befar by Linnaeus.

Berberis—Name latinized from the Arabic name for the fruits.

Berchemia—Named for Berthout van Berchem, an 18th-century Dutch botanist.

Bignonia—Named for Abbé Jean-Paul Bignon, once court librarian at Paris.

Borrichia—Named for Ole Borrich, 15th-century Danish botanist.

Brunnichia—Named for M. T. Brünnich, 18th-century Norwegian naturalist.

Buckleya—Named for Samuel B. Buckley, 19th-century American botanist.

Calamintha—From two Greek words meaning beautiful and mint.

Callicarpa—From two Greek words meaning beauty and fruit.

Calycanthus—From two Greek words meaning a calyx, or cup, and flower.

Calycocarpum—From two Greek words meaning a cup and fruit.

Campsis—From the Greek meaning curvature, alluding to the curved stamens.

Cassiope—Named for Cassiope, wife of Cepheus and mother of Andromeda.

Castanea—Ancient Latin name from the Greek *castana,* a chestnut.

Ceanothus—An obscure name used by Theophrastus.

Celastrus—Ancient Greek name for some evergreen tree.

Celtis—Classical Latin name of a species of lotus.

Cephalanthus—From two Greek words meaning a head and a flower.

Ceratiola—From the Greek word meaning a horn.
Chamaedaphne—From two Greek words meaning on the ground and laurel.
Chimaphila—From two Greek words meaning winter and to love.
Clematis—Name used by Dioscorides for a climbing plant with long, lithe branches.
Clethra—Ancient Greek name of the alder.
Cliftonia—Named for Francis Clifton, an early English physician.
Cocculus—An old name meaning a small berry.
Comptonia—Named for Henry Compton, Bishop of London and patron of botany.
Conradina—Named for Solomon W. Conrad, early Philadelphia botanist.
Corema—From the Greek meaning a broom.
Cornus—From the Latin meaning a horn, alluding to the hard wood of some species.
Corylus—From the Greek meaning a helmet, alluding to the leafy involucres surrounding the fruits.
Cotinus—Ancient Greek name for the wild olive.
Crataegus—From the Greek meaning strength.
Croton—From the Greek meaning a tick, from the similarity of the seed to a tick.
Cyrilla—Named for Domenico Cirillo, 18th-century professor of medicine at Naples.

Decodon—From two Greek words meaning ten and tooth, referring to the calyx.
Decumaria—From the Latin meaning ten, alluding to the number of flower parts.
Diervilla—Named for Dr. N. Dierville, an early French traveler.
Dirca—Named for Dirce, wife of Lycus, in Greek mythology.

Elliottia—Named for Stephen Elliott, early South Carolina botanist.
Empetrum—From two Greek words meaning upon and a rock.
Epigaea—From two Greek words meaning upon and the earth.
Euonymus—From two Greek words meaning good name.

Forestieria—Named for Charles LeForestier, French physician and naturalist of Saint-Quentin.
Fothergilla—Named for Dr. John Fothergill, 18th-century physician and botanist of London.

Gaultheria—Named for Jean-Francois Gaulthier, early French physician and naturalist.
Gaylussacia—Named for Louis Joseph Gay-Lussac, early French chemist.
Gelsemium—From the Italian name of the true jasmine.

Halesia—Named for Stephen Hale, author of *Vegetable Staticks*.

Hamamelis—An ancient Greek name, probably of some plant producing flowers and fruits at the same time.

Hudsonia—Named for William Hudson, 18th-century English botanist.

Hydrangea—From two Greek words meaning water and vessel, alluding to the urn-shaped fruits.

Hypericum—Ancient Greek name of obscure meaning.

Ilex—Ancient Latin name of the Holly Oak.

Ilicium—From the Latin meaning allurement, alluding to the aromatic odor.

Itea—Greek name of the willow.

Iva—Old name of a medicinal plant.

Juniperus—Classical Latin name of the junipers.

Kalmia—Named for Peter Kalm, early Swedish botanist.

Ledum—From the Greek *ledon,* a plant yielding an aromatic resin.

Leiophyllum—From two Greek words meaning smooth and leaf.

Leitneria—Named for E. F. Leitner, German botanist killed in Florida during the Seminole War.

Leucothoe—Named for Leucothoë, daughter of Orchamus, King of Babylon.

Licania—A name of uncertain origin.

Lindera—Named for Johann Linder, early Swedish botanist.

Litsea—A name of Chinese origin.

Loiseleuria—Named for Jean Auguste Loiseleur-Delongchamps, early French physician and botanist.

Lonicera—Named for Adam Lonitzer, early German herbalist.

Lycium—Ancient Greek name of a prickly shrub growing in Lycia.

Lyonia—Named for John Lyon, early American botanist and explorer.

Magnolia—Named for Pierre Magnol, early professor of botany at Montpellier.

Menispermum—From two Greek words meaning moon and seed.

Menziesia—Named for Archibald Menzies, early English surgeon and naturalist.

Mitchella—Named for Dr. John Mitchell, early American botanist.

Myrica—Latin name for the tamarisk, transferred to this genus by Linnaeus.

Nemopanthus—From the Greek, meaning a flower with a threadlike stalk.

Nestronia—Said to be derived from a Greek word for Daphne.

Neviusia—Named for the Rev. R. D. Nevius of Alabama, who discovered the shrub.

Oplopanax—From two Greek words meaning a weapon and all-healing.

Opuntia—Ancient Greek name of a plant growing near Opous in Boeotia.

Osmanthus—From two Greek words meaning odor and flower.

Paxstima—From two Greek words meaning thick and stigma.

Parthenocissus—From two Greek words meaning virgin and ivy.

Pentaphylloides—From two Greek words meaning five and leaf.

Philadelphus—Named for King Ptolemy Philadelphus, King of Egypt in the 3rd century B.C.

Phoradendron—From two Greek words meaning a thief and tree, alluding to its parasitic habit.

Phyllodoce—Name of a sea-nymph mentioned by Virgil.

Physocarpus—From two Greek words meaning bellows and fruit, alluding to the inflated pods.

Pieris—Named for Pieris, a muse in Greek mythology.

Pinckneya—Named for Charles Coatesworth Pinckney, early South Carolina statesman, Revolutionary War general, and botanist.

Prunus—Ancient Latin name of the plum tree.

Ptelea—Ancient Greek name for the elm.

Pyrularia—A diminutive of *Pyrus,* a pear, alluding to the shape of the fruits.

Quercus—Classical Latin name of the oak tree.

Rhamnus—Ancient Greek name of the buckthorns.

Rhapidophyllum—From two Greek words meaning a needle and leaf.

Rhododendron—From two Greek words meaning rose and tree.

Rhus—From ancient Greek and Latin names for the sumacs.

Ribes—Ancient name of uncertain derivation but probably Arabic.

Robinia—Named for Jean Robin and his son Vespasian Robin, who first cultivated the locust tree in Europe.

Rosa—Ancient Latin name for a rose.

Rubus—Roman name for a bramble and meaning red.

Sabal—Name possibly of American Indian origin.

Sageretia—Named for Auguste Sageret, early French botanist.

Salix—Classical Latin name of the willows.

Sambucus—Ancient Latin name of the elders.

Sapindus—From two Latin words meaning soap and Indian.

Schisandra—From two Greek words meaning to cleave and anther.

Sebastiania—Possibly named for Don Sebastian, King of Portugal, 1554-1578.

Serenoa—Named for Sereno Watson, 19th-century American botanist.

Shepherdia—Named for John Shepherd, early English botanist.

Sibbaldiopsis—From Greek, indicating likeness to the plant genus *Sibbaldia.*

Sideroxylon—Greek, literally "iron wood."

Smilax—Ancient Greek name of an evergreen oak.
Spiraea—From the Greek *spira,* meaning a wreath.
Staphylea—From the Greek meaning a bunch of grapes.
Stewartia—Named for John Stuart, third Earl of Bute and patron of botany.
Stillingia—Named for Dr. Benjamin Stillingfleet, 18th-century English naturalist.
Styrax—Ancient Greek name of the tree producing storax.
Symphoricarpos—From two Greek words meaning to bear together and fruit.
Symplocos—From a Greek word meaning connected, alluding to the union of the stamens.

Taxus—Classical Greek name for the yew tree.
Toxicodendron—From two Greek words meaning poison and tree.
Trachelospermum—From two Greek words meaning a neck and seed.

Vaccinium—Classical Latin name of the Old World species called cowberry, alluding to the fondness of cattle for the fruit.
Viburnum—Classical Latin name of these plants.
Vitis—Classical Latin name of the grapes.

Wisteria—Named for Caspar Wister, distinguished Philadelphia anatomist who died in 1818.

Xanthorhiza—From two Greek words meaning yellow and root.

Yucca—A name of Haitian origin.

Zanthoxylum—From two Greek words meaning yellow and wood.
Zenobia—Named for Zenobia, a queen of Palmyra.

SPECIFIC NAMES

acerifolium—with maple-like leaves.
acicularis—with needle-like prickles.
acuminata—with a narrowly pointed tip.
aestivalis—flowering in summer.
alabamense, *albamensis*—of Alabama.
alba—white.
albovestita—coated with white hairs.
alleghaniensis, *allegheniensis*—of the Allegheny Mountains.
alnifolia, *alnifolium*—with alder-like leaves.
aloifolia—with aloe-like leaves.
alpina, *alpinum*—Alpine.
alternifolia—with alternate leaves.

altissima—very high.
ambigua—doubtful.
amelanchier—with leaves like a shadbush.
americana
americanum —of America.
americanus
amomum—Latin name of some shrub.
amygdaloides—like the peach.
angustifolia
angustifolium —with narrow leaves.
aquatica—aquatic, growing in water.
arborea
arborescens —Treelike.
arboreum
arbutifolia—with leaves like arbutus.
argentifolia—with silvery leaves.
argutus—with sharp-toothed leaves.
argyrocarpa—with silvery fruits.
aromatica—with a fragrant or spicy odor.
ashei—named for William W. Ashe, American forester and botanist.
asperifolia—with rough leaves.
atlanticum—of the Atlantic Coast.
atrococcum—with black fruit.
atropurpureum
atropurpureus —dark purple.
atrox—cruel; with formidable thorns or spines.
auriculata—with earlike lobes.
axillaris—with flowers in the axils of the leaves.

baccata—with berries.
baileyanus
baileyi —named for Liberty Hyde Bailey, American horticulturist and botanist.
barbara—of Barbary, the plant mistakenly thought to be African.
bartramiana—named for William Bartram, early Philadelphia botanist.
bebbiana—named for Michael S. Bebb, noted student of the willows.
benzoin—named for the official benzoin because of a similar odor.
betulifolia
betulifolius —with leaves like a birch.
blanda—smooth; without thorns.
bona-nox—good night.
borealis—northern.
boyntonii—named for Frank E. Boynton, American botanist.
brachycera—short-horned.
buckleyi—named for Samuel B. Buckley, American botanist.
buxifolium—with leaves like the box.

caespitosum—forming tufts.
calendulaceum—colored like Calendula, the marigold.
calyculata—with bracts simulating an outer calyx.
canadense
canadensis }—of Canada.
candida—whitened or hoary.
canescens—grayish-downy or hoary.
capreolata—twining.
carolina
carolinensis }—of the Carolinas.
caroliniana
cassine—older name for *Ilex vomitoria,* now applied to this species.
cassinoides—resembling *Ilex cassine.*
catawbiense—named from the Catawba River, North Carolina.
cerifera—bearing wax.
cirrhosa—having tendrils.
cisatlantica—on this side of the Atlantic.
cistifolium—with leaves like *Cistus,* the Old World rock-rose.
clava-herculis—Hercules' club.
coactilis—with feltlike hairs.
coccinea—scarlet.
communis—growing together; in a clump.
compressa—compressed or flattened.
conradii—named for Solomon W. Conrad, its discoverer.
copallina—exuding a copal-like gum.
cordata—heart-shaped.
coriacea—leathery.
cornuta—horned.
corymbosa
corymbosum }—in flat-topped clusters, or corymbs.
crassifolium—with thick or leathery leaves.
crinitum—with long hairs.
crispa—crisped, from the wavy margin.
croceolanata—with yellowish wool.
crus-galli—meaning a cock's spur.
cumberlandense—of the Cumberland Plateau.
cuneata—wedge-shaped.
cuneifolius—with wedge-shaped leaf bases.
curvatum—curved or arched.
cynosbati—the dogberry.

decidua—deciduous, leaf-losing.
densiflorum—with flowers in dense clusters.
dentatum—dentate, with sharp outwardly pointed teeth.
depressa—lying down flat.
difforme—of two forms; with dissimilar leaves.

dioica—with stamen-bearing and pistil-bearing flowers on separate plants.
diptera—two winged.
discolor—partly colored; with two or more colors.
distichophylla—with leaves in two ranks.
dolabriforme—shaped like an ax.
drummondii—named for Thomas Drummond, early American botanist.
dumosa—bushy.
durior—tougher.

echinellum—prickly.
edule—edible.
elliottii—named for Stephen Elliott, early South Carolina botanist.
ericoides—heath-like.
eriocephala—cottony-headed.
erythrocarpum—with red fruits.

fasciculatum—in clusters; with clustered leaves.
filamentosa—bearing slender threads, referring to the leaves.
flagellaris—whip-like, referring to the stems.
flava—yellow.
flavescens—yellowish.
floribunda—full of flowers.
floridus—flowering.
fontanesiana—named for René Louiche Desfontaines, early French botanist.
frondosa, *frondosum*, *frondosus*—leafy; with leaflike sepals.
fruticosa—shrubby.

gale—old generic name of the plant.
galioides—like Galium, the bedstraws.
gardeni—named for Alexander Garden, early Charleston, S. C., physician and correspondent of Linnaeus.
georgiana—of Georgia.
gigantea—very large.
glabra—without hairs.
glandulosa, *glandulosum*—glandular, bearing glands.
glauca, *glaucescens*—whitened with a bloom.
glaucophylla—with bluish-green leaves.
glomeruliflora—with clustered flowers or flower heads.
gloriosa—glorious.
gracilis—slender.
grandifolia—with large leaves.

gravesii—named for Charles Burr Graves, its discoverer.
groenlandicum—of Greenland.

halimifolia—with leaves resembling those of *Halimus.*
hartwegii—named for K. G. Hartwig, who discovered it among cultivated plants in Germany.
herbacea—herbaceous.
heterophylla—with variable leaves.
hirsuta
hirsutum —roughish, with stiff or bristly hairs.
hirsutus
hirtella
hirtellum —bristly.
hispida
hispidula —with stiff hairs or bristles.
hispidus
horizontalis—lying flat or horizonal.
horridus—very prickly.
humilis—low-growing.
hypericoides—like *Hypericum,* the St. John's-worts.
hypnoides—resembling *Hypnum,* a genus of mosses.
hypolasium—hairy beneath, refering to the leaves.
hystrix—a hedgehog, in reference to the needle-like spines.

idaeus—of Mt. Ida, ancient name of Mt. Psiloriti in Crete.
ilicifolia—with holly-like leaves.
imbricata—overlapping.
inodora
inodorus —without an odor.
inserta—inserted, referring to some mode of attachment.
interior
interius —inland, pertaining to the interior.
intermedia—intermediate.
intonsa—having hairs.
involucrata—with an involucre.

kalmianum—named for Peter Kalm, early Swedish botanist.
kelseyi—named for Harlan P. Kelsey, American horticulturist and nurseryman.

labrusca—old Latin name for the grape vine.
lacustre—of the lakes.
laevifolium—with smooth leaves.
laevigata
laevigatus —smooth.
laevis
lanceolata
lancifolia —with lance-shaped leaves.

lanuginosa—woolly.
lapponicum—of Lapland.
latifolia—with broad leaves.
laurifolia—with laurel-like leaves.
lentago—an old name for some shrub, meaning flexible.
leucoderme—with whitish skin or bark.
leurophylla—with smooth leaves.
ligustrina—resembling *Ligustrum,* the privets.
lincecumii—named for Gideon Lincecum, its discoverer.
lloydii—named for Dr. A. J. Lloyd, American botanist.
longipes—with a long stalk.
lonicera—resembling *Lonicera,* the honeysuckles.
lucida—lustrous or shining.
lucidulum—somewhat shining.
lycioides—resembling Lycium, the matrimony-vines.
lyoni—named for John Lyon, early English botanical explorer.

macrocarpon—large-fruited.
macrosperma—large-seeded.
macrostachya—large-spiked; with a large flower cluster.
maculata—spotted or variegated.
major—larger.
malachodendron—older generic name of the plant.
margaretta—named for Margaret H. Wilcox, later Mrs. W. W. Ashe.
marginatus—margined.
mariana—of Maryland.
maritima—bordering on the sea.
marshallii—named for Humphrey Marshall, early American botanist.
maximum—largest.
melanocarpa, *melanocarpum*—with black or dark-colored fruits.
melissaefolium—with leaves like *Melissa.*
michauxii—named for André Michaux, early French botanist.
microphylla, *microphyllus*—with small leaves.
minima—smallest.
minor, *minus*—smaller.
missouriense—of Missouri.
molle, *mollis*—soft; with soft hairs.
monophylla—one-leaved; with a simple leaf.
montana—of the mountains.
mucronata—tipped with a mucro or short point.
multiflora—with many flowers.
myrsinites—resembling a shrub of the myrtle family.

myrtifolia
myrtifolium }—with myrtle-like leaves.

myrtilioides—resembling *Vaccinium myrtillus,* the European whortleberry.

nana—dwarf.

nigra
nigrum }—black.

nitens
nitida
nitidum }—shining.

novae-angliae—of New England.

nudiflorum—naked-flowered; flowering before the leaves appear.

nudum—naked.

obliqua—oblique, uneven.

oblonga—oblong, longer than broad.

oblongifolia
oblongifolius }—with oblong leaves.

obovalis
obovatum
obovatus }—inversely egg-shaped or obovate.

occidentalis—of the western hemisphere.

odoratum
odoratus }—fragrant.

opulifolius—with leaves like *Viburnum opulus,* the snowball-bush.

ovalifolium—with oval leaves.

ovata
ovatus }—egg-shaped or ovate.

oxycanthoides—resembling *Crataegus oxycantha,* an Old World hawthorn.

oxycoccus—old generic name meaning a sour berry.

pallidum—pale.

palmata—with palmate leaves, with veins radiating from summit of leaf-stalk.

palustris—of swamps.

parviflora—with small flowers.

parvifolia—with small leaves.

pavia—named for Peter Paaw of Leyden.

pedicellaris
pedicellata }—having flower stalks or pedicels.

pellita—meaning clad in skins.

pensylvanica—of Pennsylvania

peregrina—foreign or strange.

perrostrata—very long-beaked, referring to the fruits.

phillyreifolia—with leaves like *Phillyrea,* an evergreen shrub of the Mediterranean region.
pilosa—having long, soft hairs.
planifolia—with flat leaves.
polifolia—with leaves like *Polium.*
prinoides—resembling *Quercus prinus,* the chestnut oak.
procumbens—lying flat on the ground.
prolifera—reproducing freely by vegetative means.
prostratum—lying flat.
prunifolia, *prunifolium*—with leaves like a plum.
pubera, *pubens*, *pubescens*—downy; with short, soft hairs.
pulverulenta—powdery; with very fine down.
pumila—dwarf.
punctata—dotted.
pusilla, *pusillum*—very small.
pyrifolia—with leaves like the pear.

quercifolia—with leaves like an oak.
quinquefolia—five-leaved; a compound leaf with five leaflets.

racemiflora—with flowers in racemes.
racemosa—in racemes.
radiata—having rays or showy sterile flowers.
radicans—rooting.
rafinesquianum—named for C. S. Rafinesque-Schmaltz, early French botanist.
rankinii—named for H. A. Rankin, who originally discovered the plant.
recurva—recurved; bent backwards.
reductum—reduced; smaller.
repens—creeping and rooting.
rigida—stiff.
rigidiuscula—somewhat stiff.
riparia—pertaining to the banks of streams.
rivularis—of rills.
roseum—rosy.
rotundifolia, *rotundifolium*—with round leaves.
rufidulum—reddish.
rugosa—wrinkled.
rupestris—of rocky places.

sanguinea—blood-red.
saxatilis—of rocks.

scandens—climbing.
schwerinii—apparently named for someone by the name of Schwerin.
sempervirens—evergreen.
sericea—silky.
serrissima—late-fruiting.
serrulata—finely saw-toothed.
sessilifolia—with stalkless or sessile leaves.
setigera—bearing bristles.
setosum—bristly.
simplicissima—simple or unbranched.
smallii—named for John Kunkle Small, American botanist.
spathulatum—shaped like a spatula.
speciosa, *speciosum*—showy.
spicata, *spicatum*—with flowers in spikes.
spinosa—with spines.
stamineum—with prominent stamens.
stans—standing upright.
stolonifera—with stolons or runners.
stragalum—forming a mat or carpet.
stricta—upright or straight.
strigosus—covered with bristles.
suffruticosum—very low and partly woody.
susquehanae—of the Susquehanna River.
sylvatica—of the woods.
syrticola—growing on sand dunes.

tenax—tough.
tenellum—slender.
tenuifolia—with slender leaves.
textoris—of the basket-maker.
tinctoria—used as a dye.
tomentosa—wooly; covered with matted hairs.
toxicodendron—an old name meaning poison tree.
tridentata—three-toothed.
trifolia, *trifoliata*—three-leaved; a compound leaf with three leaflets.
triloba, *trilobum*—three-lobed.
triste—dull-colored; gray.
trivialis—ordinary.
typhina—resembling *Typha,* the cattails.

uliginosum—growing in swamps.
umbellata—having flowers in umbels.
umbellula—with flowers in small umbels.

uniflora—one-flowered.
ursina—pertaining to a bear.
uva-ursi—an old generic name meaning bear's grape.

vacillans—vacillating; unsteady.
vaseyi—named for George R. Vasey, American botanist.
vernix—varnish.
verticillaris
verticillatus } —whorled.
villosa—soft-hairy.
viorna—an old generic name.
virgata—wandlike.
virginiana
virginica } —of Virginia.
viscosa
viscosum } —sticky.
vitis-idea—name meaning grape of Mt. Ida.
vomitoria—inducing vomiting.
vulgaris—common.
vulpina—pertaining to a fox.

walteri—named for Thomas Walter, 18th-century English botanist and author of *Flora Caroliniana*.

CYPRESS FAMILY (Cupressaceae)

COMMON JUNIPER *Juniperus communis*

FIELD MARKS. An evergreen shrub or small tree; growing in exposed rocky or sandy places. *Leaves* awl-like, sharply-pointed, arranged in 3's, grooved and whitened above, ¼ to ¾ inch long. *Fruits* berry-like, roundish to egg-shaped, aromatic, about ¼ inch in diameter, bluish black and whitened with a bloom. The typical variety has an erect or columnar form. The following two varieties in our range are low, spreading, or mat-forming shrubs: var. *montana* (Mountain Juniper) has relatively short, broad, curved leaves up to ⅛ inch long and is northern in distribution; var. *depressa* Pursh (Oldfield or Prostrate Juniper) has almost straight leaves ⅜ to ¾ inch long and is the most common and widespread variety.

RANGE. Newfoundland to Alberta; south to eastern Virginia, the region of the Great Lakes, and in the mountains to South Carolina and Georgia.

Sometimes used in ornamental planting. The fruits are used medicinally and are eaten by a number of species of wild birds.

CREEPING JUNIPER *Juniperus horizontalis*

FIELD MARKS. A prostrate or creeping evergreen shrub, usually with long trailing branches and numerous short branchlets; growing on the sandy or rocky borders of swamps and bogs. *Leaves* bluish green, small and scalelike, with pointed tips; or awl-like on young specimens or vigorous growth. *Fruits* berry-like, light blue, ¼ to ⅜ inch in diameter, on backward curving stalks.

RANGE. Newfoundland to Alaska; south to northern portions of New England States, northwestern New York, and the region of the Great Lakes.

YEW FAMILY (Taxaceae)

AMERICAN YEW *Taxus canadensis*

FIELD MARKS. A sprawling evergreen shrub 1 to 3 feet high; growing in cool, moist, usually coniferous woods. *Leaves* narrow, flattened, rigid, abruptly pointed, lustrous dark green above, paler and yellowish green beneath, ⅜ to about 1 inch long; with short stalks running slightly down the branchlets. *Fruits* waxy-looking, orange red, fleshy, about ½ inch across, partly enclosing a large bony seed.

RANGE. Newfoundland to Manitoba; south to western Virginia, Kentucky, and Iowa. Also called Ground-hemlock.

PALM FAMILY (Arecaceae)

SAW-PALMETTO *Serenoa repens*

FIELD MARKS. A dwarf palm with stout, creeping underground stems; growing in hammocks and sandy coastal plain pinelands. *Leaves* fan-shaped, nearly circular, deeply cleft into many radiating divisions, green or yellowish green, 1 to 3 feet broad; leafstalks slender, armed with numerous, small, very sharp spines. *Flowers* small, creamy white, in a large branched cluster; blooming May to July. *Fruits* oval-shaped, black, 1-seeded, ½ to about 1 inch long; ripening October or November.

RANGE. Southeastern South Carolina south to Florida, west to Louisiana. Fruits edible and used medicinally. Important as a honey plant.

BLUE-STEM PALMETTO *Sabal minor*

FIELD MARKS. Similar to the preceding but leafstalks are smooth and fairly heavy, and leaves have a bluish-green cast. Fruits are smaller and roundish. Grows in low woods or swamps, often along streams, in the coastal plain from North Carolina south to Florida, west to Arkansas and Texas. (Not illustrated)

NEEDLE-PALM *Rhapidophyllum hystrix*

FIELD MARKS. Distinguished by its short, thick, erect or reclining trunklike stems covered with loose fibers and numerous long black spines. Mature fruits are red. Grows in coastal plain hammocks and swamps from Florida west to Mississippi. (Not illustrated)

Common Juniper

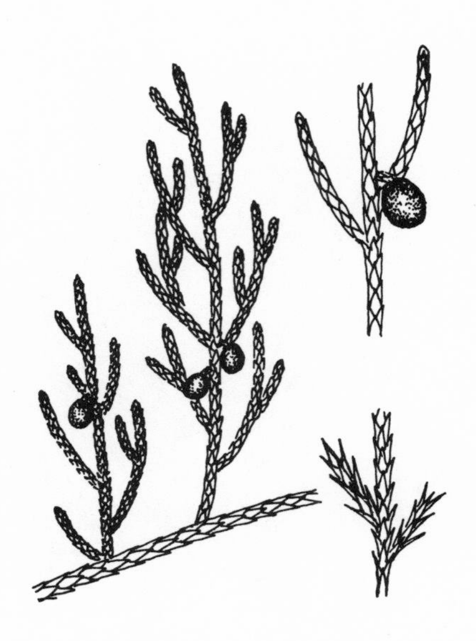

Creeping Juniper

American Yew

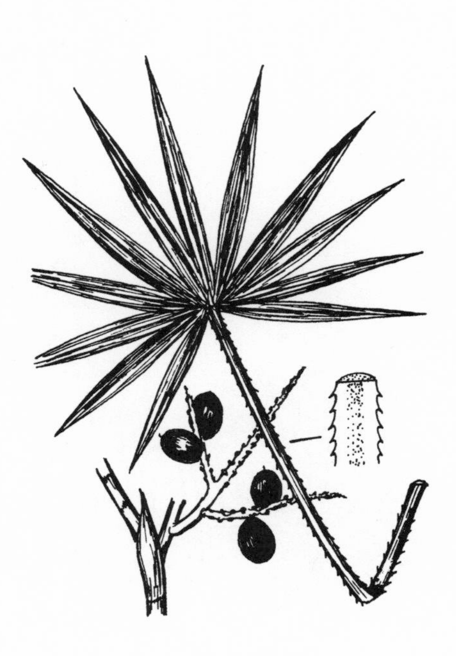

Saw-palmetto

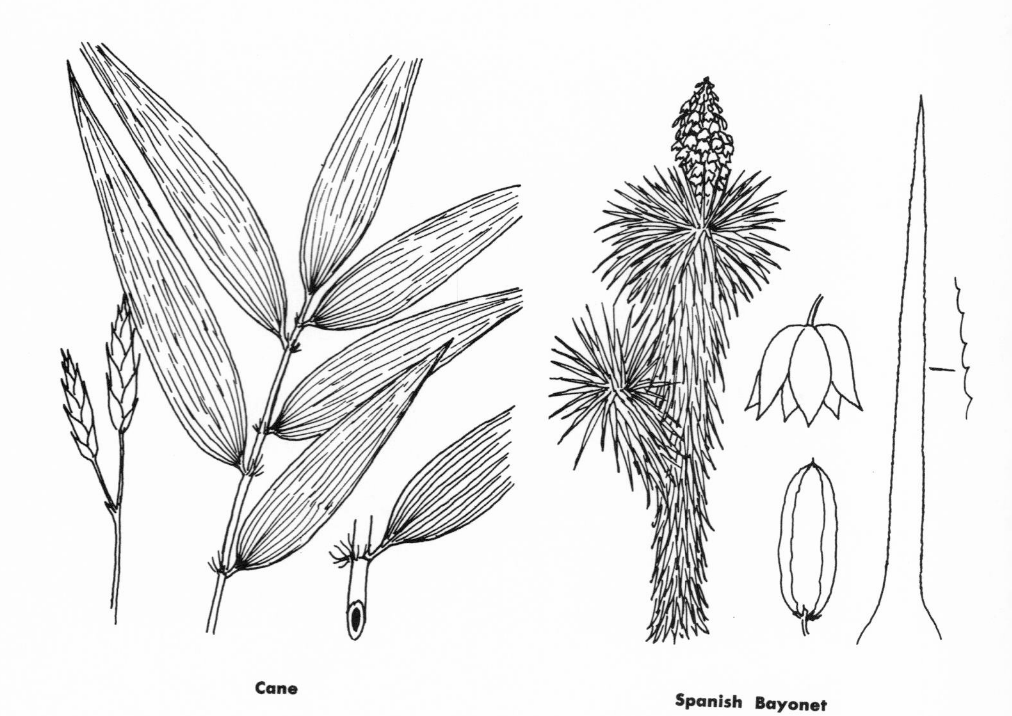

Cane

Spanish Bayonet

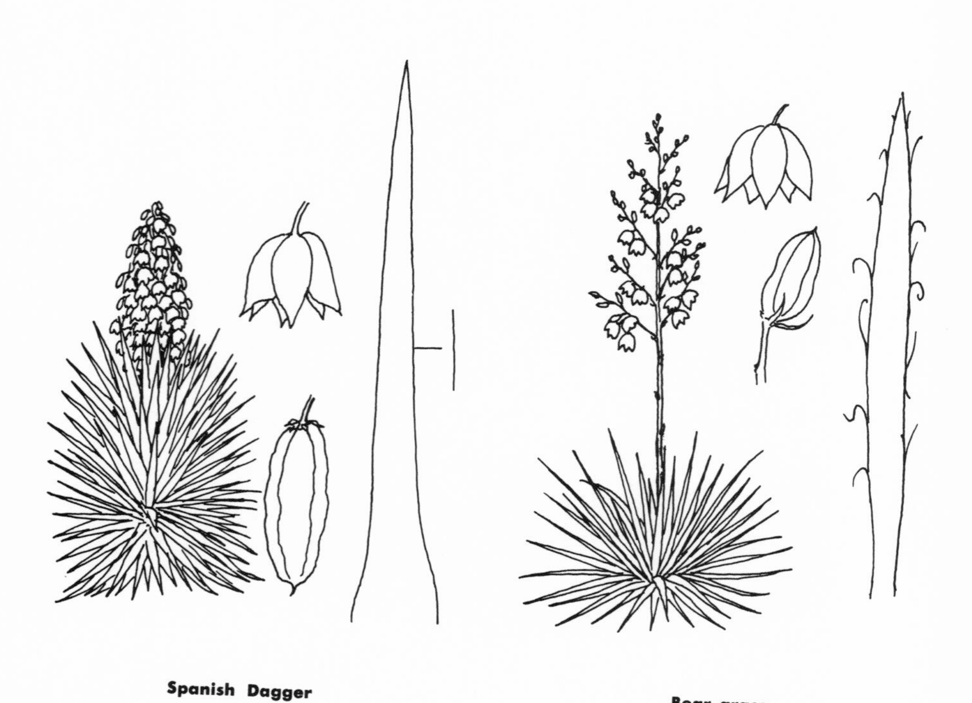

Spanish Dagger

Bear-grass

GRASS FAMILY (Poaceae)

CANE *Arundinaria gigantea*

FIELD MARKS. A woody-stemmed, more or less evergreen, bamboo-like grass commonly 3 to 15 feet or higher; growing in low wet woods, stream bottoms, savannahs, and bogs. *Leaves* grasslike with flat blades 3 to 18 inches long and ⅜ to 1½ inches wide. *Flowers* small, greenish, in 5- to 15-flowered spikelets, either on short leafless basal shoots or on the leafy branches of the older stems; blooming irregularly in April or May.

RANGE. New Jersey and southern Delaware to southern Ohio, Illinois, Missouri, and Oklahoma; south to Florida and Texas.

In primitive America, canebreaks were extensive from Virginia and Missouri southward, with stems 3 or more inches in diameter. Stems used for fishing poles and other purposes. The starchy seeds were used as grain by the Indians and early settlers, and the young shoots as a potherb.

CENTURY-PLANT FAMILY (Agavaceae)

YUCCAS (Yucca)

Yuccas are somewhat woody-stemmed, evergreen plants with alternate, crowded, swordlike or dagger-like leaves. The showy white or creamy-white flowers have a 6-parted, bell-shaped perianth and 6 stamens. They are pollinated by a small moth, the larvae of which feed upon the seeds. The fruits are elliptical, more or less 6-sided capsules containing a number of seeds. They were eaten to some extent by the Indians. Yuccas are popular ornamental plants.

SPANISH BAYONET *Yucca aloifolia*

FIELD MARKS. A plant with a stem to about 15 feet high, commonly covered with downward-pointing old leaves, and with a dense cluster of spreading ones at the summit or ends of branches; growing in sandy woods and among coastal sand dunes. *Leaves* rigid, dagger-like, sharply pointed at tip, minutely saw-toothed on margin, 1½ to 2½ feet long. *Flowers* 1 to 2 inches long, white or creamy white, in a large and dense end cluster; blooming May or June. *Fruits* 2½ to 4 inches long, drooping, containing plump, marginless seeds.

RANGE. Coastal plain; North Carolina south to Florida, west to Alabama.

SPANISH DAGGER *Yucca gloriosa*

FIELD MARKS. A plant usually with a simple stem 2 to 8 feet high, densely covered with spreading leaves except near the base; growing in sandy woods and among coastal sand dunes. *Leaves* rigid, dagger-like, sharply pointed at tip, smooth on margin, 1 to 2½ feet long. *Flowers* 2 to 4 inches long, white or creamy white, in a large and dense end cluster; blooming May or June. Fruits 2 to 3 inches long, drooping, prominently 6-ridged, containing flattened, thin-margined seeds.

RANGE. Coastal plain; North Carolina south to Florida.

BEAR-GRASS *Yucca filamentosa*

FIELD MARKS. A plant with a short stem and cluster of spreading leaves close to the ground; growing in dry sandy open woods, clearings, and old fields. *Leaves* somewhat flexible, spine-pointed at tip, margin fraying into loose threads, 1 to 2 feet long. *Flowers* about 2 inches long, white or creamy white, in a large but rather open end cluster on a stalk 3 to 5 feet high; blooming May or June, or later northward. *Fruits* about 2 inches long, erect, containing elongate seeds.

RANGE. Southern New Jersey to West Virginia and Tennessee; south to Florida and Louisiana.

Also called Adam's-needle. Very popular as a garden plant.

(Key Appendix A)

Greenbriers are green-stemmed and often prickly vines which climb by means of paired tendrils on the bases of the leafstalks, these remaining after the leaves fall. The alternate leaves have from 3 to 7 prominent parallel veins and a network of smaller veins. The flowers are small, yellowish green, and borne in stalked umbels in the axils of the leaves. The fruits are small, usually roundish, 1- to 3-seeded berries.

The greenbriers often form impenetrable thickets. The starchy, tuberous roots of some species were used by the Indians as food. Rabbits and deer often eat the stems; the berries are eaten by many birds, including the ruffed grouse, wild turkey, and ring-necked pheasant. In the South, the greenbriers are popularly known as "bamboos."

COMMON GREENBRIER *Smilax rotundifolia*

FIELD MARKS. A scrambling or climbing leaf-losing vine with round or sometimes 4-angled stems and branchlets; armed with scattered, stout, broad-based prickles. *Leaves* egg-shaped to broadly egg-shaped or nearly round, often with a heart-shaped base, green and lustrous on both surfaces, 5-veined, 2 to 6 inches long. *Fruits* bluish black, coated with a whitish bloom, about ¼ inch in diameter; the cluster on a stalk hardly as long as the leafstalk.

RANGE. Nova Scotia to Minnesota, south to Florida and Texas.

Widely distributed in moist woods and thickets. The most common species of greenbrier northward as well as southward along the mountains. Also known as Horsebrier and Round-leaf Brier.

BRISTLY GREENBRIER *Smilax hispida*

FIELD MARKS. A high-climbing, leaf-losing vine with stems thickly beset with weak bristly or needle-like blackish prickles (at least toward the base). *Leaves* thin, egg-shaped or broadly egg-shaped, lustrous green on both surfaces, 5- to 7-veined, 2 to 5 inches long, the margins roughish with minute bristle-tipped teeth. *Fruits* black, about ¼ inch in diameter, usually with a single shiny reddish-brown seed; the cluster on a stalk usually much longer than the leafstalk.

RANGE. New York to southern Ontario and Minnesota; south to Georgia and Mississippi.

A common greenbrier, especially in low moist thickets and woodlands and along the banks of streams.

GLAUCOUS GREENBRIER *Smilax glauca*

FIELD MARKS. A leaf-losing, or sometimes partly evergreen, scrambling or high-climbing vine; its round stems whitened with a bloom and usually armed with numerous stout prickles. *Leaves* egg-shaped to broadly egg-shaped, green above but conspicuously whitened and sometimes downy beneath, 3- to 5-veined, 2 to 5 inches long. *Fruits* bluish black, whitened with a bloom, about ¼ inch in diameter; the cluster on a stalk much longer than the leafstalk.

RANGE. Massachusetts to Illinois, south to Florida and Texas.

Also known as Catbrier and Sawbrier. Often common in dry to moist woods, thickets, and clearings, and spreading by underground runners.

CHINABRIER *Smilax bona-nox*

FIELD MARKS. A variable species which is sometimes high-climbing or partly evergreen; the 4-angled stems with usually some rigid prickles and starry-branched, scalelike hairs toward the base. *Leaves* more or less thick and leathery, commonly triangular or fiddle-shaped, the margins thickened and usually prickly, lustrous green on both surfaces or sometimes mottled with white, 5- to 7-veined, 2 to 5 inches long. *Fruits* bluish black, whitened with a bloom, about ¼ inch in diameter.

RANGE. Massachusetts to Illinois, south to Florida and Texas.

Also known as Bullbrier and Sawbrier. Often common southward in deciduous woods, old fields, and on sand dunes, but not present in the mountains.

Common Greenbrier

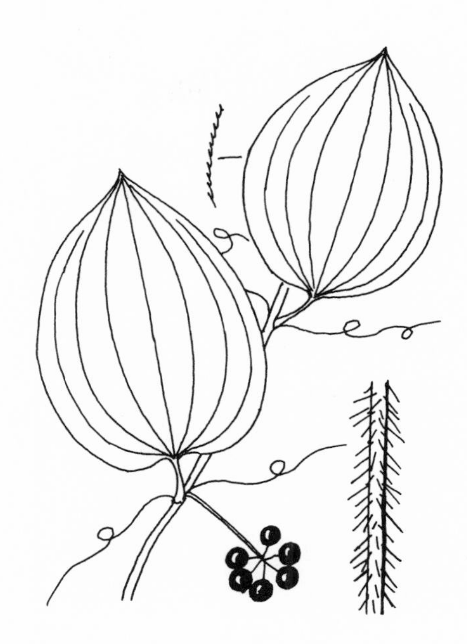

Bristly Greenbrier

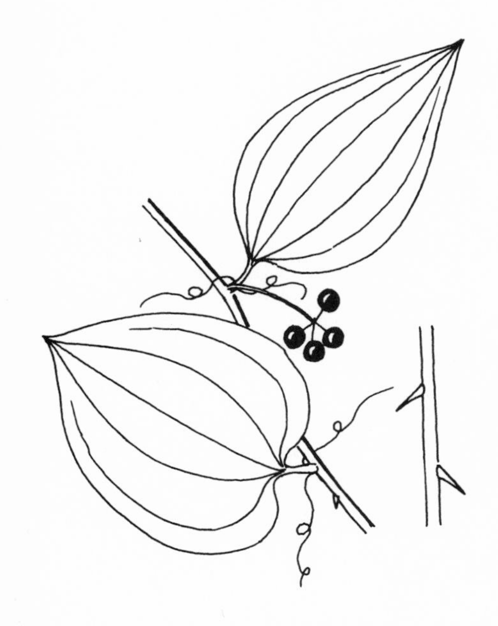

Glaucous Greenbrier

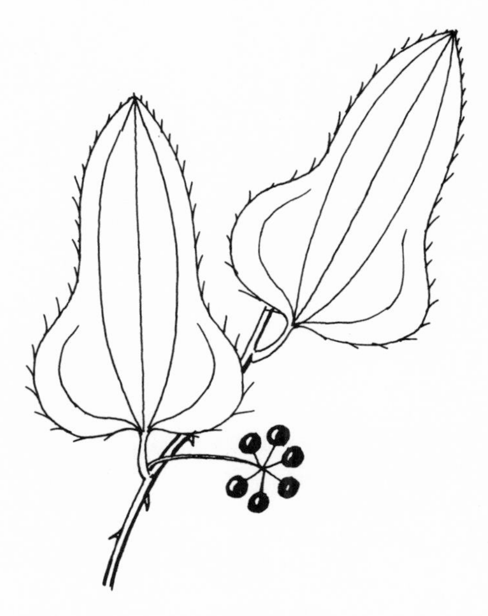

Chinabrier

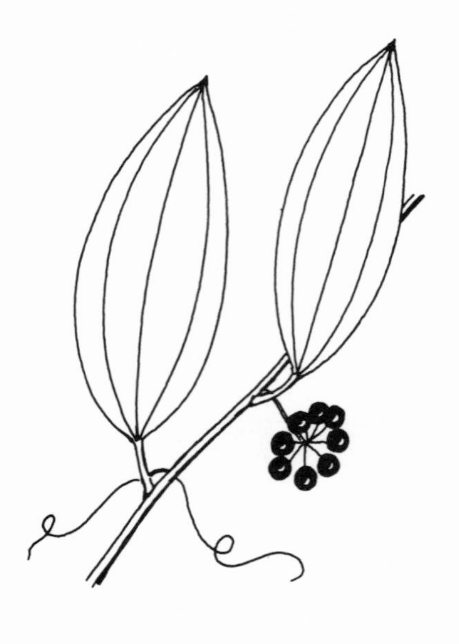

Laurel-leaf Greenbrier

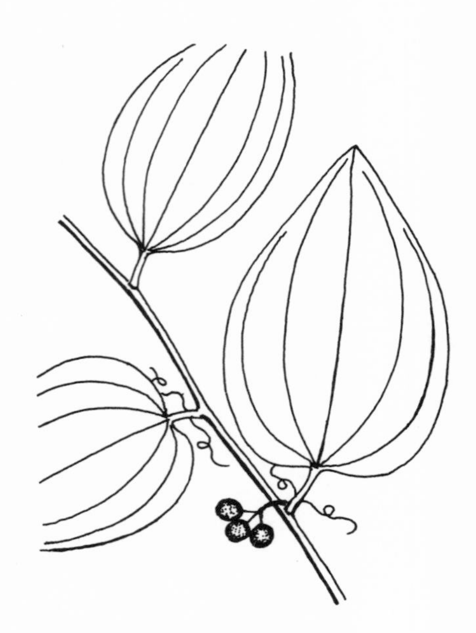

Red-berried Greenbrier

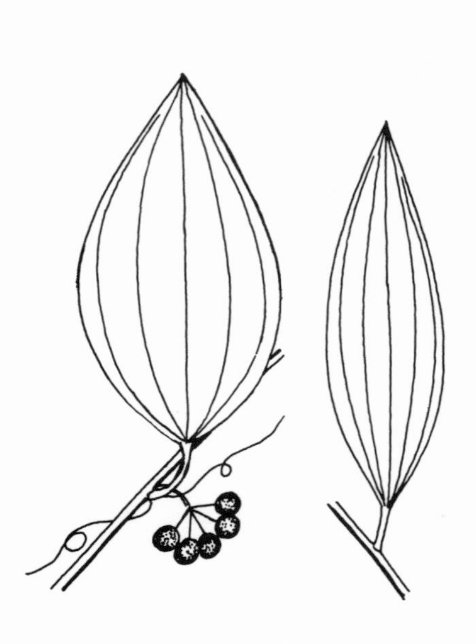

Lanceleaf Greenbrier

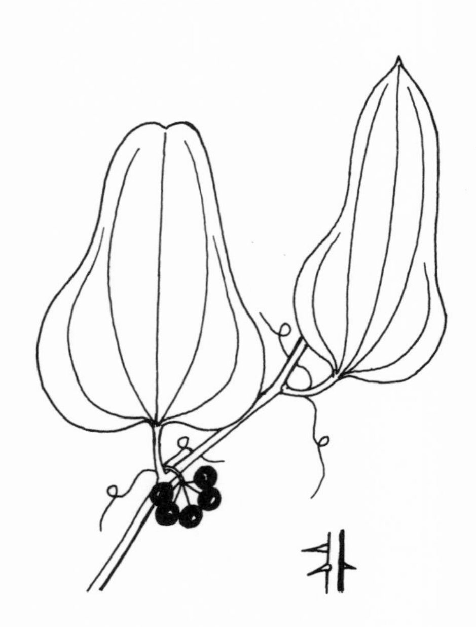

Wild Bamboo

LAUREL-LEAF GREENBRIER *Smilax laurifolia*

FIELD MARKS. A high-climbing evergreen vine of low grounds and swamps; the round stems chiefly prickly toward the base and sometimes on the more vigorous shoots. *Leaves* thickish and leathery, pointed at both ends, dark green and lustrous above, paler and sometimes slightly whitened beneath, 3-veined, 2½ to 5 inches long. *Fruits* black, about ¼ inch in diameter.

RANGE. Coastal plain and piedmont; New Jersey south to Florida, west to Arkansas and Texas.

Unique among our greenbriers in having fruits which do not ripen until the second year; thus both ripe fruits and smaller green ones are commonly present. Also known as the Bamboo-vine or Blaspheme-vine.

RED-BERRIED GREENBRIER *Smilax walteri*

FIELD MARKS. A slender-stemmed, leaf-losing vine with angled branchlets; usually scrambling over bushes about the borders of swamps and in low wet pinelands. The stems usually have scattered and slender prickles only toward the base. *Leaves* rather thin, egg-shaped to broadly egg-shaped, green on both surfaces, 5- to 7-veined, 2 to 4 inches long. *Fruits* bright coral red, about ¼ inch in diameter, persistent throughout the winter.

RANGE. Chiefly coastal plain; New Jersey south to Florida, west to Louisiana.

Easily recognized by its bright-red berries. Also known as the Coral Greenbrier.

LANCELEAF GREENBRIER *Smilax smallii*

FIELD MARKS. A vigorous evergreen vine, the stems often whitened with a bloom and either unarmed or with a few stout prickles toward the base; growing in low wet woodlands, hammocks, and wayside thickets. *Leaves* rather thin but firm in texture, lance-shaped to egg-shaped, more or less pointed at the base and tapering gradually to the pointed tip, dark green and shiny above, pale green and dull or slightly whitened beneath, 5- to 7-veined, 2 to 4½ inches long. *Fruits* dull red, about ¼ inch in diameter.

RANGE. Chiefly coastal plain; North Carolina south to Florida, west to Texas and Arkansas.

Also known as Jacksonbrier. Often gathered and sold to florists where abundant.

WILD BAMBOO *Smilax auriculata* Walt.

FIELD MARKS. A scrambling or sometimes high-climbing evergreen vine with more or less 4-angled stems and noticeably zigzag branchlets, sometimes armed with very small prickles. *Leaves* variable but commonly fiddle-shaped, margins thickened but not spiny, notched or abruptly pointed at the tip, usually 5-veined, 1 to 4½ inches long. *Fruits* black, often with a whitish bloom, ¼ to ½ inch in diameter.

RANGE. Outer coastal plain; North Carolina south to Florida, west to Mississippi.

Often grows among the coastal sand dunes.

WOOLLY GREENBRIER *Smilax pumila*

FIELD MARKS. A low-climbing or often trailing evergreen vine which is easily recognized by its woolly and unarmed stems. *Leaves* broadly lance-shaped or egg-shaped, heart-shaped at the base, woolly on the lower surface, 3-veined, 2 to 4 inches long. *Fruits* red, egg-shaped, about ¼ inch long.

RANGE. Coastal plain; South Carolina south to Florida, west to Texas.

This greenbrier is unique in that it blooms in September or October and matures its berries the following spring. Also known as Sarsaparilla-vine.

CORKWOOD FAMILY (Leitneriaceae)

CORKWOOD *Leitneria floridana*

FIELD MARKS. A leaf-losing shrub or small tree which occurs rather rarely and locally in swamps along tidewater rivers. *Branchlets* rather stout, often with catkin-like flower buds clustered near the tip, and much smaller buds with 3 exposed scales. *Leaves* alternate, elliptic or lance-shaped, pointed at both ends, untoothed on the margin, bright green and smooth above, paler and downy beneath, 4 to 6 inches long. *Fruits* about ¾ inch long, dry, brown, wrinkled, 1-seeded.

RANGE. Coastal plain; southeastern Georgia and northern Florida west to Texas; also in portions of Arkansas and Missouri.

The wood is lighter than cork, weighing but 13 pounds per cubic foot, and is used locally to make floats for fishing nets. This is the only known member of its family.

BAYBERRY FAMILY (Myricaceae)

(Key Appendix B)

SWEETFERN *Comptonia peregrina*

FIELD MARKS. An aromatic, much-branched, leaf-losing shrub from 1 to about 3 feet high; growing on dry rocky or sandy, and usually sterile, soils. *Branchlets* slender, downy, resin-dotted when young, fragrant when broken, often with clusters of catkins at the tips. *Leaves* alternate, fernlike, and deeply cut into numerous lobes, dark green above, paler and downy beneath, with numerous small resin dots on both surfaces, 1½ to 4 inches long, pleasantly fragrant when crushed. *Fruits* small olive-brown nutlets which are surrounded by pointed bracts and borne in little burrlike heads.

RANGE. Nova Scotia to Manitoba, south to interior North Carolina, and along the mountains to northern Georgia.

The leaves have astringent and tonic properties and were once used in home remedies and as a substitute for tea. Often heavily browsed by deer.

SWEETGALE *Myrica gale*

FIELD MARKS. An aromatic leaf-losing shrub 1 to 4 feet high; growing in cool swamps and the boggy borders of streams and ponds. *Branchlets* often have large pointed flower buds toward the tips, and they are fragrant when broken. *Leaves* alternate, broadest toward the tip and tapering to the base, toothed above the middle, resin-dotted and sometimes with scattered hairs on the lower surface, 1 to 2 inches long, pleasantly fragrant when crushed. *Fruits* small, 2-winged, resin-dotted nutlets borne in conelike clusters.

RANGE. Newfoundland to Alaska; south to northern New Jersey, the region of the Great Lakes, Oregon, and in the Appalachians to western North Carolina.

The leaves have been used in clothes closets to repel moths, as a vermifuge, and as a substitute for tea.

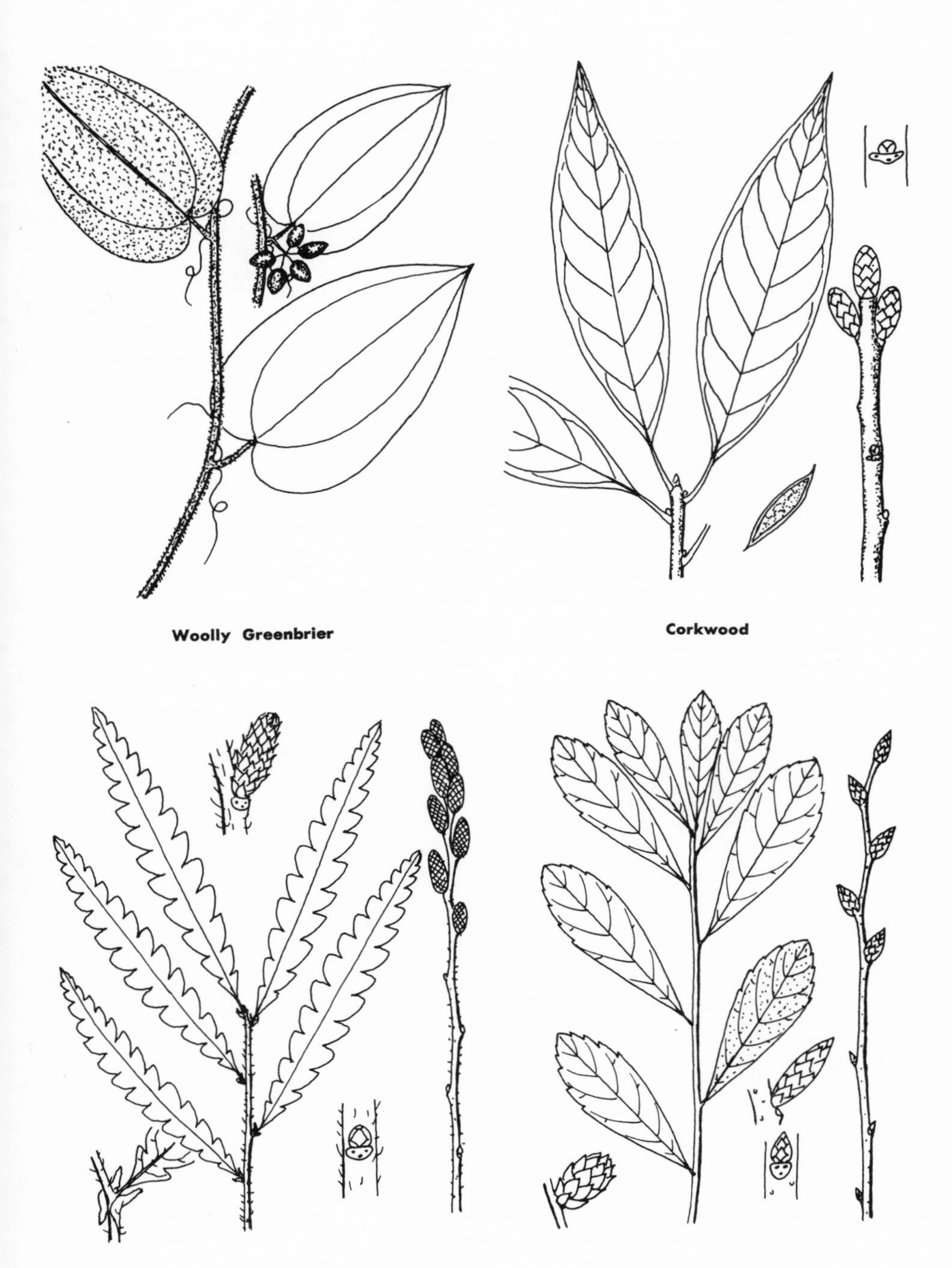

Woolly Greenbrier

Corkwood

Sweetfern

Sweetgale

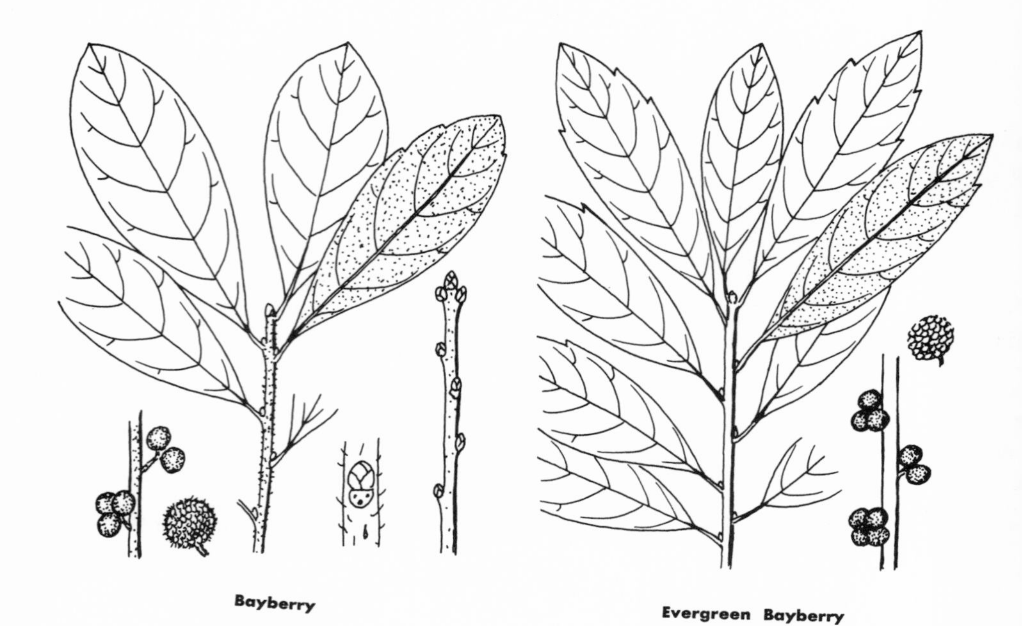

Bayberry

Evergreen Bayberry

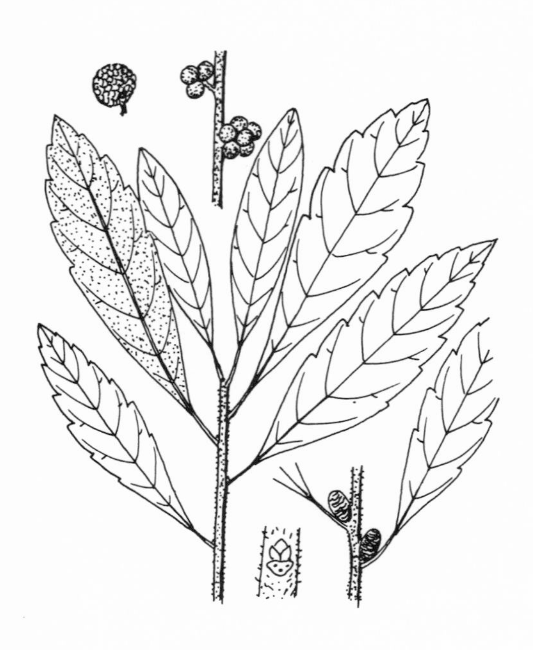

Wax Myrtle

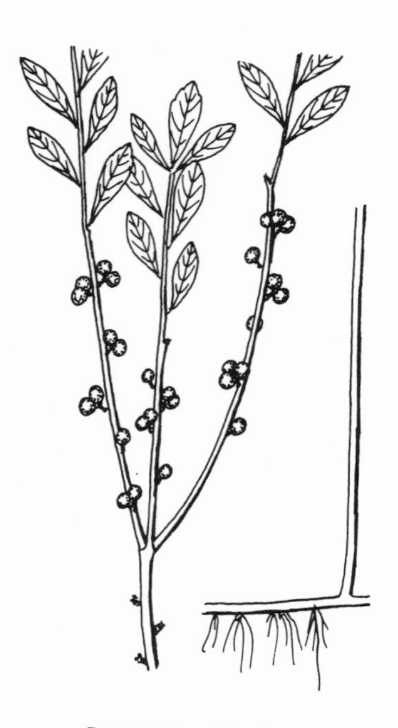

Dwarf Wax Myrtle

BAYBERRY *Myrica pensylvanica*

FIELD MARKS. An aromatic leaf-losing shrub 3 to about 6 feet high; usually growing in dry sandy soils or on sand dunes. *Branchlets* grayish to ashy brown, often slightly hairy and resin-dotted, pleasantly fragrant when broken. *Leaves* alternate, elliptic or slightly broader above the middle, untoothed or sometimes with a few teeth toward the blunt tip, dull green and sometimes a little hairy above, paler and downy and resin-dotted beneath, 1½ to 3½ inches long, pleasantly fragrant when crushed. *Fruits* are bony nutlets which are minutely hairy and coated with whitish wax, about 3/16 inch in diameter.

RANGE. Coastal plain from Newfoundland south to North Carolina, inland near the Great Lakes to northern Ohio.

Wax of the fruits is used to make bayberry candles. The fruits are eaten by many wild birds. Leaves have been used as a substitute for bay leaves in seasoning. Rather attractive as an ornamental shrub and very useful in soil conservation.

EVERGREEN BAYBERRY *Myrica heterophylla*

FIELD MARKS. A large aromatic shrub which grows in wet woods and sandy bogs. *Branchlets* dark brown or blackish and sometimes hairy, pleasantly fragrant when broken. *Leaves* alternate, rather leathery, elliptic or broadest above the middle, usually with several distinct teeth, more or less sharply pointed at the tip, lustrous green above, paler and dull and resin-dotted beneath, 1½ to 3½ inches long, pleasantly fragrant when crushed. *Fruits* are bony nutlets which are wax-coated but not hairy, about ⅛ inch in diameter.

RANGE. Coastal plain; southern New Jersey south to Florida, west to Louisiana.

Wax of the fruits is used to make bayberry candles.

WAX MYRTLE *Myrica cerifera*

FIELD MARKS. An aromatic evergreen shrub or small tree which grows in wet sandy pinelands and bogs. *Branchlets* smooth or sparsely hairy, resin-dotted, pleasantly fragrant when broken. *Leaves* alternate, narrow, broadest toward the pointed tip and gradually tapered to the pointed base, untoothed or more often sharply toothed on the margin, resin-dotted on both surfaces, 2 to 4 inches long, pleasantly fragrant when crushed. *Fruits* are bony nutlets coated with whitish wax, about ⅛ inch in diameter.

RANGE. Coastal plain; southern New Jersey south to Florida, west to Texas, and northward to Arkansas and Oklahoma.

Wax of the fruits used to make bayberry candles. Fruits eaten by many birds, including the bobwhite quail and wild turkey. Makes a rather attractive ornamental shrub for wet places.

DWARF WAX MYRTLE *Myrica cerifera* var. *pumila*

FIELD MARKS. An aromatic evergreen shrub 1 to (rarely) 3 feet high, forming colonies from its underground stems. *Leaves* similar to those of the Wax Myrtle but only ½ to 1½ inches long. It grows in low woods and sandy pinelands.

RANGE. Coastal plain; southern Delaware south to Florida, west to Texas and north to Arkansas.

ODORLESS WAX MYRTLE *Myrica inodora*

FIELD MARKS. A shrub similar to the Wax Myrtle, but the leaves are usually untoothed, 2 to 5 inches long, and not aromatic when crushed; and the wax-coated fruits are somewhat larger.

RANGE. Coastal plain; northern Florida west to Louisiana. (Not illustrated)

WILLOW FAMILY (Salicaceae)

WILLOWS (Salix)

(Key Appendix C)

Willows are leaf-losing shrubs or trees with bitter bark and usually slender branchlets. They have alternate simple leaves, and the buds are covered with a solitary hoodlike scale. The flowers are in catkins which appear either before or with the leaves, the stamen-bearing and pistil-bearing ones being on separate plants. Willow fruits are capsules which split down the middle at maturity, releasing the silky-hairy seeds. Many willows are useful in controlling the erosion of stream banks, or for ornamental planting. Twigs and bark are eaten by deer, moose, beaver, muskrats, and rabbits; the buds or young leaves are eaten by ptarmigan and grouse. The bark contains tannin and salicin which are used medicinally. Identification is often difficult and is complicated by the fact that hybrids often occur.

SLENDER WILLOW *Salix gracilis*

FIELD MARKS. A shrub 3 to 10 feet high; growing in low wet meadows and swales. *Branchlets* smooth or nearly so, often clustered at the ends of the branches. *Leaves* narrowly lance-shaped, long-pointed at tip, pointed at base, sharply toothed on margin but untoothed toward base, lustrous above, whitened and sometimes with silky hairs beneath, 2 to 4 inches long; leafstalks slender; stipules small and soon shed.

RANGE. Quebec to Manitoba; south to northern New Jersey, Pennsylvania, the region of the Great Lakes, Iowa, and Nebraska.

AUTUMN WILLOW *Salix serissima*

FIELD MARKS. A shrub 3 to 12 feet high; growing in swamps or bogs, usually in marl or limestone regions. *Branchlets* smooth, lustrous, yellow brown to olive brown. *Leaves* narrowly elliptic or lance-shaped, pointed at tip, rounded to broadly pointed at base, finely and sharply toothed on margin, firm, lustrous above, smooth and pale or whitened beneath, 2 to 4 inches long; slender leafstalks usually have a pair of glands at the summit. *Flowers* blooming in June or July. *Fruits* maturing August or September.

RANGE. Newfoundland to Alberta; south to northern New Jersey, Pennsylvania, the region of the Great Lakes, North Dakota, and Colorado.

SILKY WILLOW *Salix sericea*

FIELD MARKS. A shrub 4 to 12 feet high; common and widely distributed along stream banks and other wet places. *Branchlets* brittle at base, purplish brown, smooth or minutely silky. *Leaves* lance-shaped, long-pointed at tip, pointed to rounded at base, margin finely and sharply toothed, smooth above, pale and with silvery-silky hairs beneath, 2 to 4 inches long; stipules small and soon shed.

RANGE. Nova Scotia to Wisconsin; south to Georgia, Tennessee, and Missouri.

SANDBAR WILLOW *Salix interior*

FIELD MARKS. A shrub 3 to 15 feet high; forming thickets on sand or gravel deposits along streams and places subject to frequent flooding. *Branchlets* reddish brown, smooth or nearly so. *Leaves narrowly* lance-shaped, pointed at both ends, margin with shallow and widely spaced teeth, green and smooth on both surfaces or sometimes silvery-silky, 2 to 5 inches long; leafstalks very short; stipules very small if present.

RANGE. New Brunswick to Quebec and Alaska; south to Maryland, Kentucky, Louisiana, and New Mexico.

Slender Willow

Autumn Willow

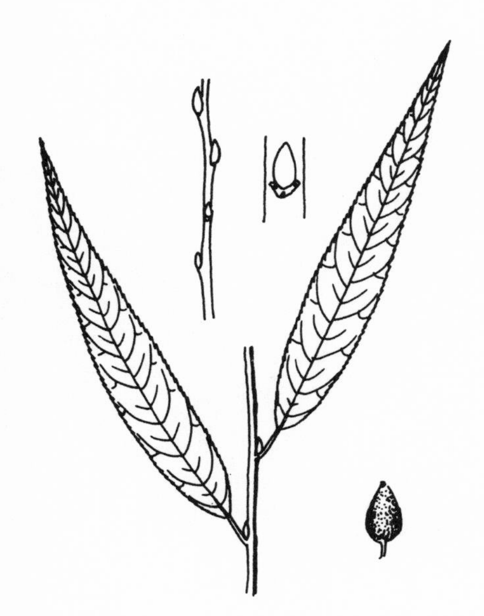

Silky Willow

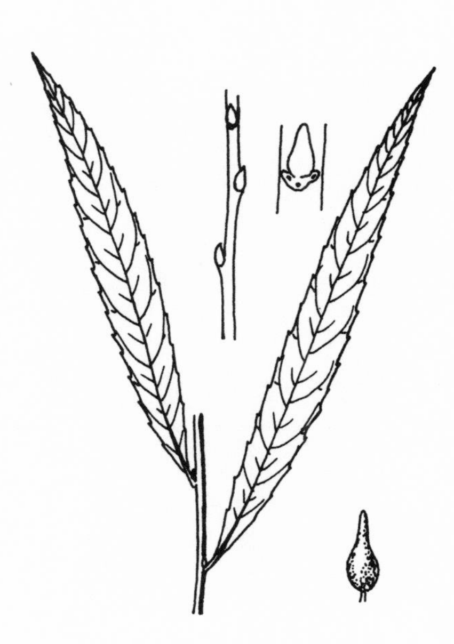

Sandbar Willow

Ward Willow

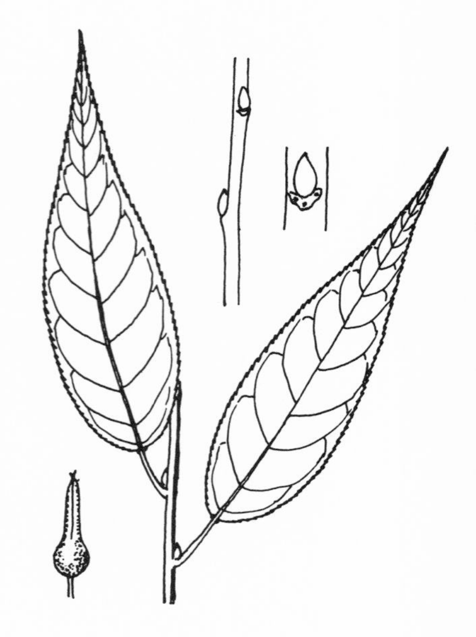

Shining Willow

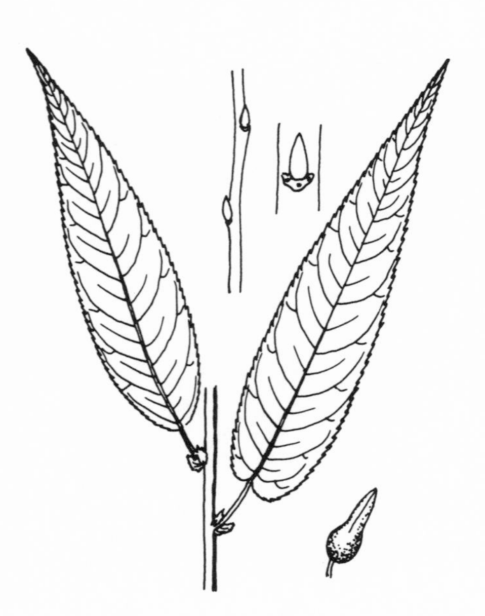

Missouri River Willow

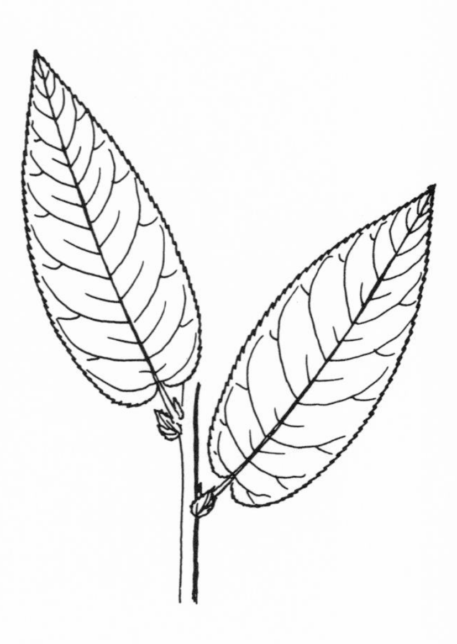

Heartleaf Willow

WARD WILLOW *Salix caroliniana*

FIELD MARKS. A shrub or small tree of low moist places and stream banks. *Branchlets* smooth or somewhat downy, yellowish brown to reddish brown, sometimes brittle at base. *Leaves* narrowly lance-shaped, long-pointed at tip, pointed to rounded at base, finely and sharply toothed on margin, smooth or nearly so on both surfaces, dark green above, whitened beneath, 2½ to 6 inches long. Large stipules are often present at bases of the leafstalks.

RANGE. Maryland to southern Illinois and Missouri, south to Florida and Texas.

PEACHLEAF WILLOW *Salix amygdaloides*

FIELD MARKS. A large shrub or small tree of stream banks, shores, and low wet places. *Branchlets* somewhat drooping, yellowish to orange brown, lustrous, flexible. *Leaves* lance-shaped or broadly lance-shaped, long-pointed at tip, pointed to rounded at base, finely and sharply toothed on margin, pale green and lustrous above, smooth and whitened beneath, 2 to 6 inches long. Stipules very small or absent.

RANGE. Vermont and southern Quebec to southeastern British Columbia; south to Massachusetts, New York, Michigan, Kentucky, Kansas, western Texas, and Arizona.

Also known as the Almondleaf Willow. (Not illustrated)

SHINING WILLOW *Salix lucida*

FIELD MARKS. A shrub 4 to 10 feet high, or occasionally a small tree; growing in swampy places and along the banks of streams. *Branchlets* lustrous, smooth or hairy. *Leaves* broadly lance-shaped, long-pointed at tip, broadly pointed to rounded at base, finely and sharply toothed on margin, dark green and very lustrous above, slightly paler green and smooth or hairy beneath, 2½ to 5 inches long. The leafstalks usually have large stipules at the base.

RANGE. Labrador to Manitoba; south to Delaware, Maryland, the region of the Great Lakes, and South Dakota.

MISSOURI RIVER WILLOW *Salix eriocephala*

FIELD MARKS. A shrub or small tree of lakeshores, borders of swamps, and stream banks. *Branchlets* smooth or somewhat downy, not brittle at base. *Leaves* lance-shaped to broadly lance-shaped, long-pointed at tip, pointed to rounded or heart-shaped at base, finely and sharply toothed on margin, dark green and smooth above, paler green or whitened to silvery-silky beneath, 2½ to 6 inches long. The leafstalks are slender, either smooth or downy, and usually have large stipules at the base.

RANGE. Newfoundland to Saskatchewan and Minnesota; south to Virginia, Kentucky, western Tennessee, Missouri, Kansas, and Nebraska.

HEARTLEAF WILLOW *Salix cordata*

FIELD MARKS. A shrub 1 to 10 feet high; growing on lakeshores, beaches, borders of swamps, and banks of streams. *Branchlets* densely gray-hairy. *Leaves* broadly lance-shaped to egg-shaped, pointed at tip, rounded to heart-shaped at base, finely and sharply toothed on margin, often gray-hairy on both surfaces, 1½ to 5 inches long. The leafstalks are hairy and usually have large and sometimes glandular-toothed stipules at the base.

RANGE. Labrador to Ontario; south to Massachusetts, northern New York, Michigan, Illinois, and Wisconsin.

PUSSY WILLOW *Salix discolor*

FIELD MARKS. A large shrub usually 6 to 12 feet high, or occasionally a small tree; growing in swamps and other wet places. *Branchlets* dark purplish red, smooth or softly hairy. *Leaves* elliptic to lance-shaped, pointed at both ends, occasionally untoothed but usually with irregular and somewhat wavy teeth mostly above the middle, bright green and with a wrinkled-veiny appearance above, whitened and sometimes with rusty hairs beneath, 2 to 4 inches long. Large stipules are often present on vigorous shoots.

RANGE. Nova Scotia to Manitoba; south to Delaware, Maryland, West Virginia, Indiana, northeastern Missouri, and Nebraska.

The large furry catkins appear before the leaves and are a familiar harbinger of spring.

BEBB WILLOW *Salix bebbiana*

FIELD MARKS. A large shrub usually 5 to 15 feet high, or occasionally a small tree; growing in either wet or dry places. *Branchlets* gray-downy. *Leaves* rather broad, elliptic or often broadest above the middle, pointed at tip, pointed to rounded at base, wavy-toothed or (more rarely) untoothed on the margin, thick and firm in texture; dull green, minutely downy, and wrinkled-veiny above; densely whitish- or grayish-woolly beneath; 1½ to 3 inches long. The slender, downy leafstalks are often reddish in color, and stipules are small or lacking.

RANGE. Newfoundland and Labrador to Alaska; south to New Jersey, Maryland, the region of the Great Lakes, Nebraska, New Mexico, and California.

Also known as the Beaked Willow.

PRAIRIE WILLOW *Salix humilis*

FIELD MARKS. A shrub 2 to 10 feet high which is often common on dry and barren soils. *Branchlets* wandlike, erect, usually coated with a dirty-grayish down. *Leaves* narrowly elliptic, often broadest above the middle, pointed at both ends, margin slightly rolled, wavy, usually untoothed but sometimes sparingly toothed, bright green to dull grayish green and wrinkled-veiny above, pale and usually grayish-woolly beneath, 2 to 5 inches long. The leafstalks are stout and often have stipules at the base.

RANGE. Newfoundland and Labrador to Ontario and Minnesota; south to Florida and eastern Texas.

DWARF GRAY WILLOW *Salix humilis* var. *microphylla*

FIELD MARKS. A tufted shrub 1 to (rarely) 3 feet high; growing in dry to moist sandy or stony soils. *Branchlets* usually grayish-downy. *Leaves* rather crowded, narrow, usually broadest near the tip, wedge-shaped at base, margin rolled and untoothed, dark green and often somewhat downy above, densely white-woolly beneath, ½ to 2 inches long. The leafstalks are very short.

RANGE. Southeastern Maine to southern Minnesota; south to Virginia, northwestern Florida, Louisiana, and Oklahoma.

Also known as the Dwarf Prairie Willow, Dwarf Pussy Willow, and Sage Willow.

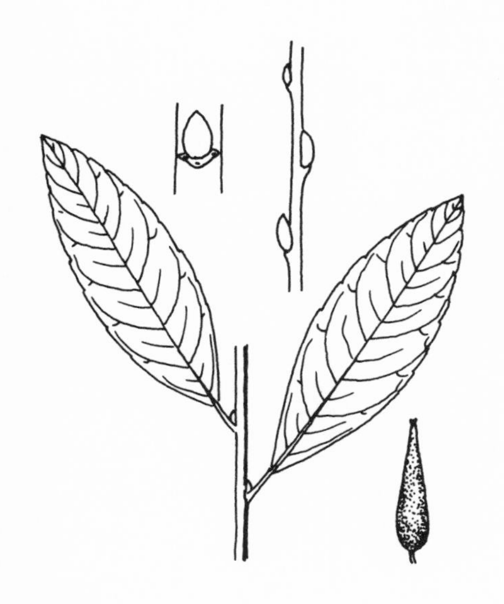

Pussy Willow

Bebb Willow

Prairie Willow

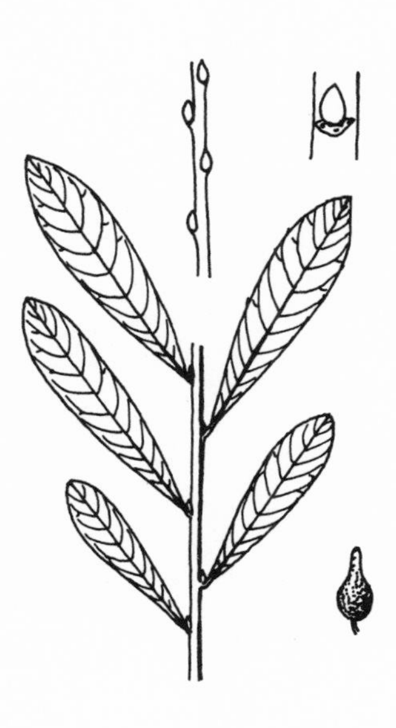

Dwarf Gray Willow

Blueleaf Willow

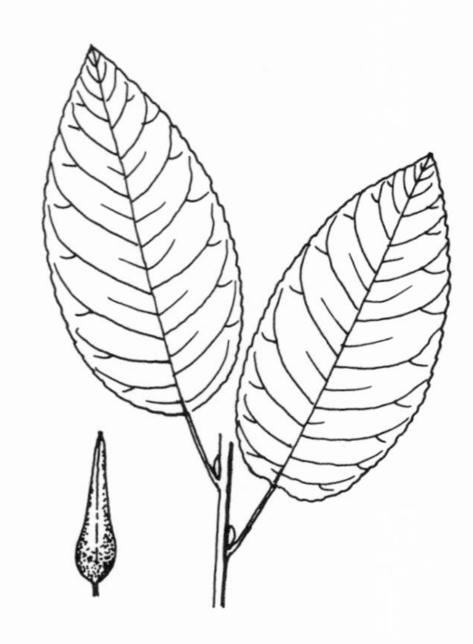

Balsam Willow

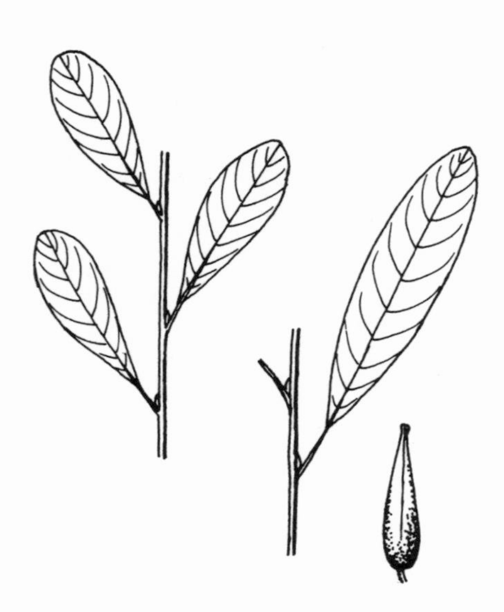

Bog Willow

BLUELEAF WILLOW *Salix glaucophylloides*

FIELD MARKS. A shrub 2 to 15 feet high, or (rarely) a small tree, of gravelly shores, beaches, and sand dunes. *Branchlets* smooth. *Leaves* egg-shaped to broadly lance-shaped, pointed at tip, rounded to somewhat heart-shaped at base, finely and sharply toothed on the margin, thickish and firm in texture, dark green and lustrous above, whitened and smooth beneath, 2 to 4½ inches long. The stout leafstalks are smooth or downy.

RANGE. Newfoundland and northern Ontario to British Columbia; south to northern Maine and the region of the Great Lakes.

BALSAM WILLOW *Salix pyrifolia*

FIELD MARKS. A shrub 3 to 10 feet high, or (rarely) a small tree; growing in swamps, low thickets, and the borders of woods. *Branchlets* smooth, lustrous, reddish brown. *Leaves* with a pleasant balsamic or resinous odor when crushed, elliptic to egg-shaped, pointed at tip, broadly rounded to heart-shaped at base, slightly glandular-toothed on margin, thin in texture, dark green and smooth above, whitened and prominently veiny beneath, 1½ to 4 inches long. The leafstalks are smooth and slender.

RANGE. Newfoundland and Labrador to British Columbia; south to Maine, northern New York, Michigan, Minnesota, and Saskatchewan.

BOG WILLOW *Salix pedicellaris*

FIELD MARKS. A loosely branched and creeping shrub 1 to 3 feet high; growing in cold northern bogs. *Branchlets* long, rather erect, flexible, brown, and smooth. *Leaves* narrowly elliptic, often broadest well above the middle, bluntly pointed or rounded at tip, wedge-shaped at base, margin untoothed and rolled, more or less leathery in texture, smooth and green on both surfaces, ½ to 2 inches long. The leafstalks are smooth and slender.

RANGE. Labrador and Newfoundland to British Columbia; south to northern New Jersey, the region of the Great Lakes, northern Iowa, Idaho, and Oregon.

HOARY WILLOW *Salix candida*

FIELD MARKS. A much-branched shrub 1 to (rarely) 5 feet high; growing in cold northern bogs. *Branchlets* densely coated with loose white wool. *Leaves* narrowly oblong or lance-shaped, bluntly pointed at tip, narrowed at the base, untoothed or with inconspicuous wavy teeth on the rolled margin, dull dark green above, densely coated with snow-white wool beneath, firm in texture, 1½ to 4 inches long. Leafstalks white-woolly. The narrow stipules are sometimes as long as the leafstalks.

RANGE. Labrador and Newfoundland to British Columbia; south to northern New Jersey, northeastern Pennsylvania, the region of the Great Lakes, northern Iowa, South Dakota, and Colorado.

SATINY WILLOW *Salix pellita*

FIELD MARKS. A shrub 3 to 8 feet high, or sometimes a small tree; growing along stream banks and in swampy places. *Branchlets* reddish brown to olive brown, smooth, usually coated with a whitish bloom. *Leaves* lance-shaped and often narrowly so, sometimes broadest above the middle, pointed at both ends, untoothed or obscurely toothed on the margin, thick and firm in texture, bright green and smooth above, pale or whitened and sometimes silky-haired or even velvety beneath, 1½ to 4½ inches long.

RANGE. Labrador and Newfoundland to northern Ontario; south to Maine, Vermont, and Michigan.

TEALEAF WILLOW *Salix planifolia*

FIELD MARKS. A much-branched shrub 8 inches to 10 feet high; growing in swampy places and along stream banks. *Branchlets* smooth, often whitened with a bloom. *Leaves* elliptic to oblong, pointed or sometimes blunt at both ends, untoothed or nearly so on margin, dark green and lustrous above, smooth and whitened beneath, 1 to 3 inches long.

RANGE. Labrador to Alberta; south to Newfoundland, Quebec, and the higher mountains of northern New England.

SILVER WILLOW *Salix argyrocarpa*

FIELD MARKS. A shrub 8 inches to about 5 feet high; growing on tundras and southward in high mountain meadows or on wet rocks. *Branchlets* dark green, lustrous. *Leaves* narrowly elliptic to narrowly lance-shaped, often broadest well above the middle, pointed or bluntly pointed at tip, pointed at base, untoothed to obscurely wavy-toothed on the slightly rolled margin, bright green and smooth above, silvery-silky with minute hairs beneath, ¾ to 2½ inches long.

RANGE. Arctic Region south to Labrador, Quebec, and mountains of New Hampshire.

Hoary Willow

Satiny Willow

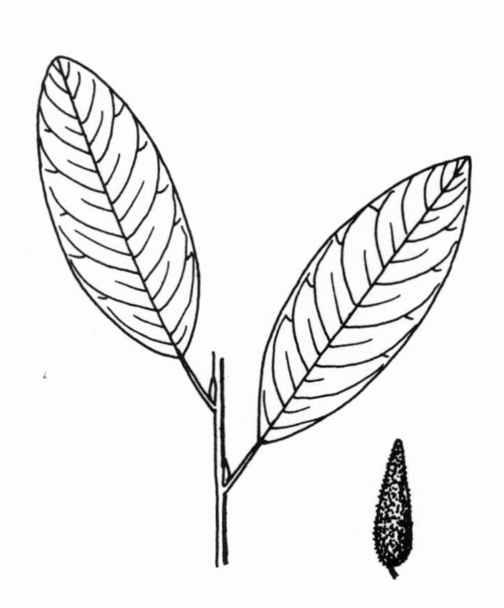

Tealeaf Willow

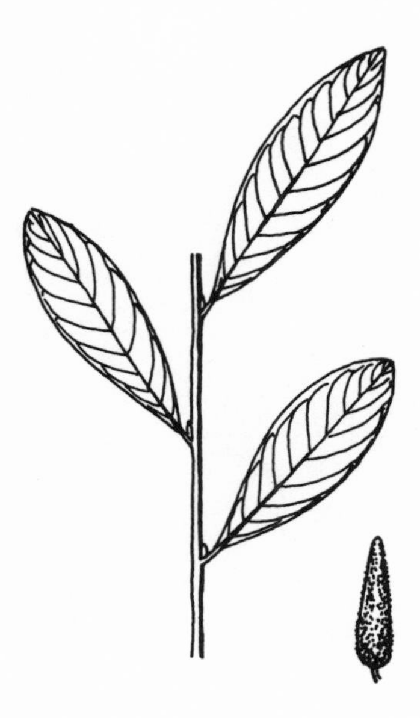

Silver Willow

Bearberry Willow

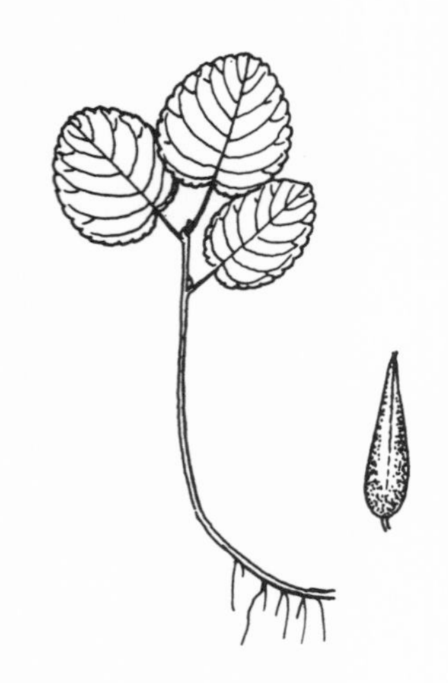

Dwarf Willow

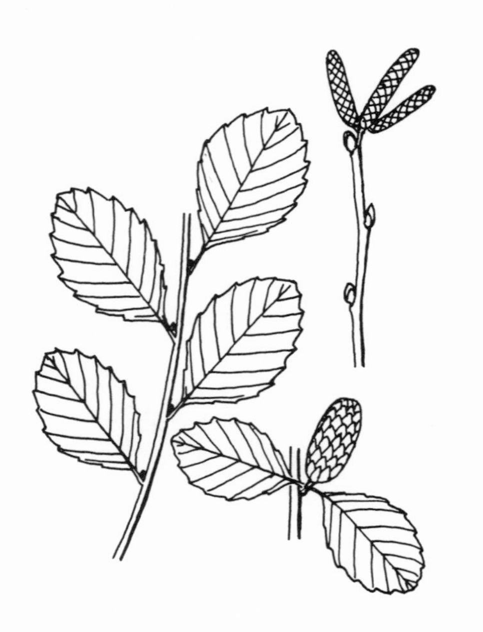

Low Birch

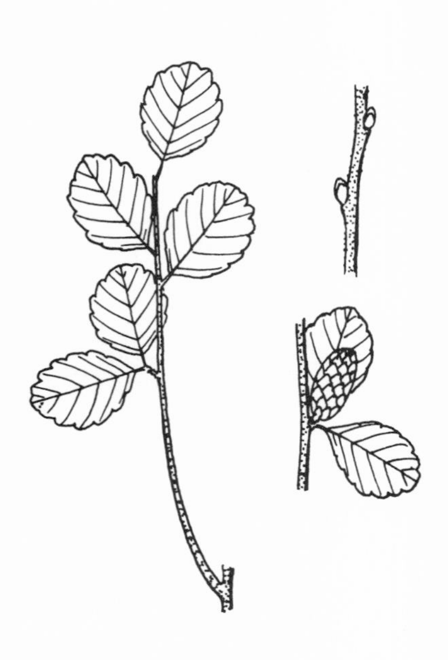

Dwarf Birch

BEARBERRY WILLOW *Salix uva-ursi*

FIELD MARKS. A prostrate, trailing, mat-forming shrub of tundras and alpine mountain summits. *Branches* brown, 6 to 12 inches long, not rooting. *Leaves* elliptic or broadest above the middle, pointed to blunt at tip, wedge-shaped at base, margin finely and sharply toothed, firm, prominently veined, lustrous and smooth or (rarely) soft-hairy above, pale and usually smooth beneath, ¼ to 1 inch long; leafstalks very short.
RANGE. Greenland and Arctic Region south to Newfoundland, Nova Scotia, New England, northern New York, and Quebec.

DWARF WILLOW *Salix herbacea*

FIELD MARKS. A creeping or mat-forming shrub of tundras and alpine mountain summits. *Branches* mostly below the surface and rooting at the nodes. *Branchlets* above ground 1 to 6 inches long, very slender, angled. *Leaves* roundish, heart-shaped at base, margin with blunt or rounded teeth, smooth and bright green and lustrous on both surfaces, ½ to 1¼ inches across; leafstalks slender.
RANGE. Arctic region south to Newfoundland, Quebec, northern portions of Maine, New Hampshire, and New York.

BIRCH FAMILY (Betulaceae)

LOW BIRCH *Betula pumila*

FIELD MARKS. A leaf-losing, often low and mat-forming, shrub 1 to 6 feet high; growing in northern bogs and swamps. *Branchlets* smooth, hairy, or glandular-warty. *Leaves* oval to roundish or broadest above the middle, rounded or blunt at tip, rounded to pointed at base, margin coarsely toothed, dull above, pale or whitened and smooth to hairy or glandular-dotted beneath, ½ to 1½ inches long; leafstalks very short.
RANGE. Newfoundland and Labrador to British Columbia; south to northern New Jersey, New York, and the region of the Great Lakes.
Also known as Swamp Birch.

DWARF BIRCH *Betula nana*

FIELD MARKS. A leaf-losing shrub 1 to 6 feet high, often low and mat-forming; growing on tundras and south chiefly on alpine mountain summits. *Branchlets* thickly warty-dotted. *Leaves* oval, roundish, or broadest above the middle, rounded at tip, round to pointed at base, coarsely toothed on margin, leathery, green on both surfaces but glandular-dotted beneath, ¼ to 1 inch long; leafstalks short.
RANGE. Newfoundland to Alaska; south to Maine, New Hampshire, New York, Michigan, Minnesota, Colorado, and California.

DWARF WHITE BIRCH *Betula minor*

FIELD MARKS. Spreading or upright leaf-losing shrub 1 to 6 feet high. *Branchlets* warty-dotted. *Leaves* egg-shaped, double-toothed on margin, ½ to about 2 inches long.
RANGE. Labrador and Newfoundland to mountains of New England, northwestern New York, and Ontario. (Not illustrated)

NORTHERN BIRCH *Betula borealis*

FIELD MARKS. Spreading or erect leaf-losing shrub or small tree. *Branchlets* densely whitish-hairy. *Leaves* elliptic to broadly egg-shaped, double-toothed on margin, ½ to 2¼ inches long.
RANGE. Labrador and Ungava south to Newfoundland, Northern Maine and Vermont, and Quebec. (Not illustrated)

Alders are leaf-losing shrubs with alternate, simple leaves. The stamen-bearing flowers are in drooping catkins; the pistil-bearing ones are in much smaller catkins. Both are fully developed before the leaves are shed in the fall. The fruits are small, slightly winged nutlets, borne in the axils of woody scales of conelike structures which are peculiar to the alders. Alders are useful in controlling stream bank erosion. The bark and twigs are eaten by beavers, deer, and rabbits; many small birds feed on the seeds.

COMMON ALDER *Alnus serrulata*

FIELD MARKS. A thicket-forming shrub 5 to 15 feet high; growing along streams and in swampy places. *Bark* with scattered, small, dotlike lenticels. *Leaves* oval or broadest above the middle, broadly pointed to rounded at tip, pointed at base, margin finely toothed with simple teeth, smooth above, paler green and smooth or somewhat rusty-downy beneath, 1 to 6 inches long. *Flowers* blooming late February to April, the smaller catkins at branch tips and erect. *Buds* 2-scaled and noticeably stalked.

RANGE. Nova Scotia and Maine to the region of the Great Lakes, Missouri, and Oklahoma; south to northern Florida and Louisiana.

Also called Smooth Alder.

SPECKLED ALDER *Alnus rugosa*

FIELD MARKS. A shrub very much like the preceding. *Bark* with numerous whitish, horizontally elongated lenticels. *Leaves* oval or broadest above the middle, short-pointed at tip, rounded to slightly heart-shaped at base, margin double-toothed, smooth above, paler or more often whitened and sometimes pale- to rusty-downy beneath, 2 to 5 inches long. *Flowers* blooming March or April, the smaller catkins drooping and not at branch tips. *Buds* 2-scaled and noticeably stalked.

RANGE. Labrador to Saskatchewan; south to Maine, Maryland, West Virginia, the region of the Great Lakes, and northeastern Iowa.

Also called Hoary Alder.

GREEN ALDER *Alnus crispa*

FIELD MARKS. A northern shrub 2 to 10 feet high growing on rocky shores, slopes, and southward on mountain balds. *Leaves* roundish to egg-shaped or somewhat heart-shaped, finely and sharply but irregularly toothed on the often puckered margin, somewhat sticky, green and smooth on both sides or sometimes downy or velvety beneath, 1 to 3 inches long. *Flowers* blooming in May or June as the leaves expand. *Buds* have 3 or 4 visible scales and are not stalked.

RANGE. Newfoundland to Alaska; south to Massachusetts, northern New York, Michigan, and British Columbia; south in the high mountains to western North Carolina and eastern Tennessee.

Also called Mountain Alder.

SEASIDE ALDER *Alnus maritima*

FIELD MARKS. A shrub or small tree growing about ponds and on stream banks in the coastal plain of Delaware and Maryland. *Leaves* oval to egg-shaped or broadest above the middle, pointed at tip, wedge-shaped at base, margin with small but sharp and widely spaced teeth, lustrous above, paler green and smooth or nearly so beneath, 2 to 4 inches long. *Flowers* blooming in August or September. *Fruits* not maturing until the following year. *Buds* pointed, hairy, stalked, and several-scaled.

RANGE. Near coast; Delaware and Maryland.

Common Alder

Speckled Alder

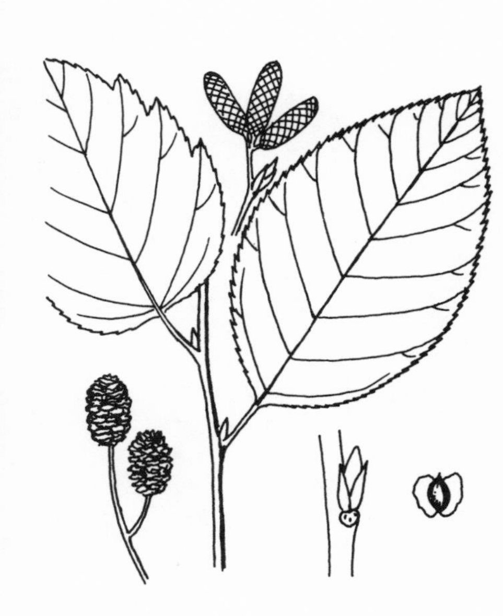

Green Alder

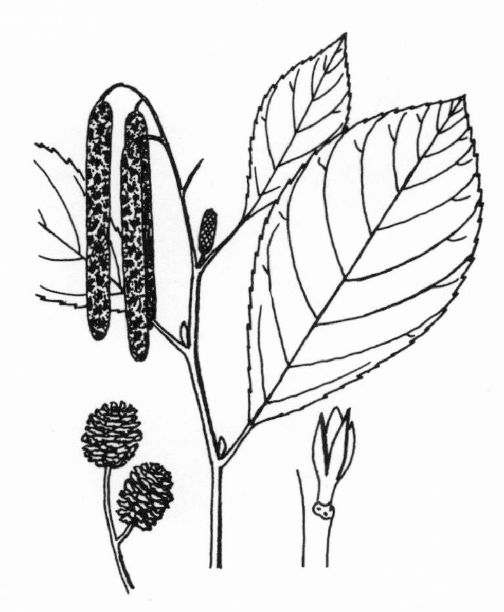

Seaside Alder

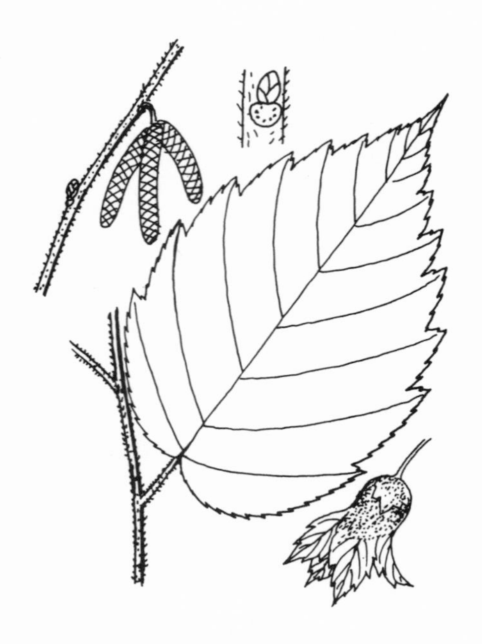

American Hazelnut

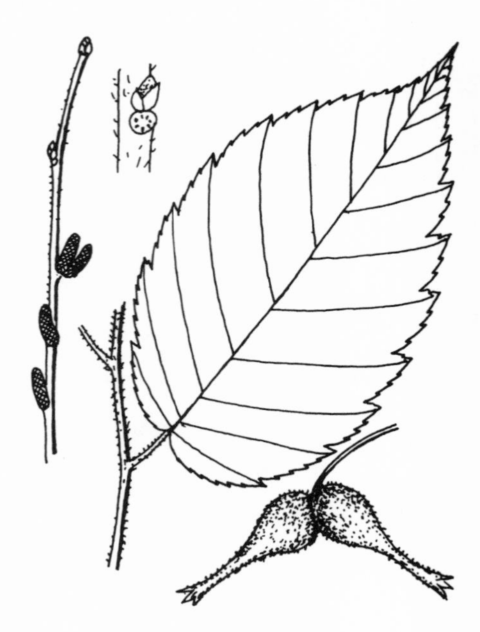

Beaked Hazelnut

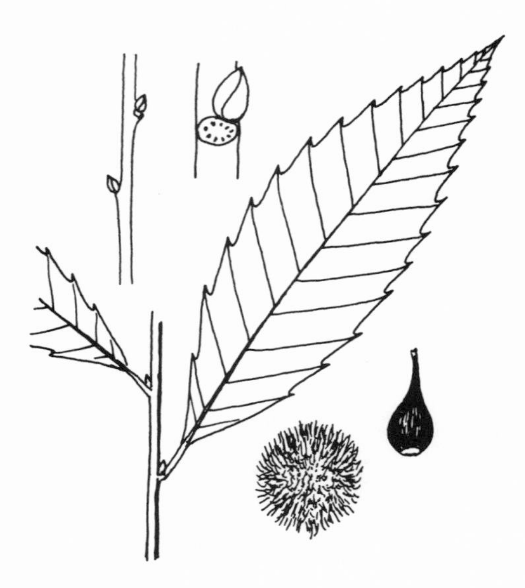

Common Chinquapin

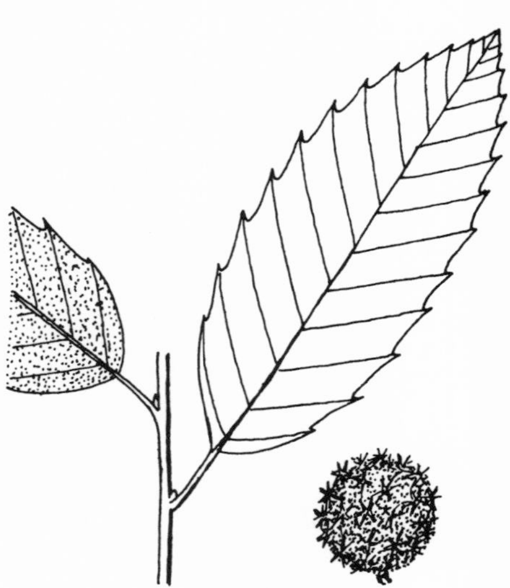

Coastal Chinquapin

HAZELNUTS (Corylus)

Hazelnuts are leaf-losing shrubs with simple, alternate leaves. The branchlets are rather slender and somewhat zigzag. Along them are well-developed catkins of the stamen-bearing flowers. They bloom in early spring before the leaves appear; the catkins elongating and shedding their pollen while the bright-red stigmas of the pistil-bearing flowers protrude from buds along the branchlets. The fruits are light-brown, roundish nuts which are enclosed by pairs of leafy bracts.

The nuts are sweet and edible. They are eaten by deer, squirrels, chipmunks, blue jays, and a number of other kinds of wildlife.

AMERICAN HAZELNUT *Corylus americana*

FIELD MARKS. A shrub 3 to 10 feet high growing in moist to dry thickets, fence rows, roadsides, and borders of woods. *Branchlets* bristly-hairy; *catkins* stalked and often 1 inch long. *Leaves* egg-shaped to roundish, somewhat heart-shaped at base, abruptly pointed at tip, double-toothed on margin, smooth or nearly so above, downy on the veins beneath, 2½ to 6 inches long; leafstalks bristly-hairy. *Fruits* enclosed by a pair of ragged-edged leafy bracts.

RANGE. Maine to Saskatchewan; south to northern Florida, Missouri, and Oklahoma.

BEAKED HAZELNUT *Corylus cornuta*

FIELD MARKS. A shrub quite similar to the American Hazelnut and found in similar situations. *Branchlets* sparingly if at all bristly-hairy; *catkins* stalkless and less than ½ inch long. *Fruits* enclosed by a pair of bracts which are united and form a prolonged, tubelike, bristly-hairy beak.

RANGE. Newfoundland to British Columbia; south principally in the mountains to Georgia, Missouri, and Colorado.

BEECH FAMILY (Fagaceae)

CHINQUAPINS (Castanea)

Chinquapins are leaf-losing shrubs or small trees with alternate, simple leaves. They bloom in the late spring after the leaves are developed, and the flower clusters are quite showy. The stamen-bearing flowers are whitish and arranged in slender clusters. Those bearing the pistils are in small, prickly involucres at the base of the stamen-bearing clusters or in the axils of the leaves. The fruits are edible nuts which are enclosed in a prickly bur.

COMMON CHINQUAPIN *Castanea pumila*

FIELD MARKS. A shrub or small tree to about 30 feet high; growing in dry woods and thickets. *Leaves* narrowly elliptic, pointed at both ends or occasionally rounded at base, coarsely and sharply toothed on margin, smooth above, paler and whitish-downy beneath, 3 to 5 inches long. *Fruits* shiny brown, silky-hairy nuts ¼ to ½ inch across; usually solitary in a densely prickly bur.

RANGE. Massachusetts and New Jersey to Tennessee and Arkansas; south to Florida and eastern Texas.

COASTAL CHINQUAPIN *Castanea pumila* var. *ashei*

FIELD MARKS. A shrub or small tree; growing in dry sandy soils of the coastal plain. *Leaves* somewhat shorter, broader, and of a firmer texture than those of the preceding, bluntly pointed or rounded at tip, and densely white-woolly beneath. *Fruits* similar but the burs with widely spaced clusters of spines.

RANGE. Virginia south to Florida, west to eastern Texas and Arkansas.

RUNNING CHINQUAPIN *Castanea alnifolia*

FIELD MARKS. A leaf-losing shrub 1 to (rarely) 3 feet high, with creeping underground stems; growing in dry sandy pinelands and sandhills. *Leaves* elliptic or broadest above the middle, rounded to somewhat pointed at the tip, usually pointed at the base, becoming nearly smooth beneath at maturity, 2 to 5 inches long. *Fruits* similar to those of the Coastal Chinquapin but with even more widely spaced clusters of spines.
RANGE. Coastal plain; North Carolina south to Florida, west to Louisiana.

OAKS (Quercus)

Oaks are trees, or sometimes shrubs, with simple, alternate leaves which tend to be more or less clustered toward the tips of the branchlets. The flowers appear in the spring along with the developing leaves; they are greenish, yellowish, or reddish. Stamen-bearing ones are in slender and drooping catkins. The pistil-bearing ones are small and inconspicuous, being in the axils of the new leaves. Oak fruits are called acorns. They are nuts which are seated in scaly-bracted cups.

The tree oaks are valuable shade and timber trees. Shrubby species are useful for the control of soil erosion. Acorns were a staple food of the American Indians. They are eaten by many kinds of wildlife, including deer, bears, squirrels, wild turkeys, wild ducks, blue jays, and some woodpeckers.

SCRUB OAK *Quercus ilicifolia*

FIELD MARKS. A leaf-losing shrub or small tree 3 or more feet high; growing on dry and barren uplands and slopes. *Leaves* oblong-oval or often broadest above the middle, with 3 to 7 (usually 5) short, broadly triangular, bristle-tipped, and sparingly bristle-toothed lobes, dark green and lustrous above, whitish-downy beneath, 2 to 4 inches long. *Fruits* egg-shaped acorns about ⅜ inch long which are about half covered by a bowl-shaped cup.
RANGE. Maine to New York and western Pennsylvania; south to Virginia, West Virginia, and western North Carolina.
Also known as the Bear Oak.

DWARF CHESTNUT OAK *Quercus prinoides*

FIELD MARKS. A leaf-losing shrub or small tree 2 to 12 feet high; growing on dry and barren uplands and slopes. *Leaves* oblong or broadest above the middle, blunt-pointed at tip, pointed at base, margin with 3 to 7 pairs of bluntish or pointed teeth, bright green and smooth above, grayish-downy beneath, 3 to 5 inches long. *Fruits* lustrous egg-shaped acorns ½ to ¾ inch long, about half covered by a bowl-shaped cup.
RANGE. Maine to Minnesota and Nebraska; south to Virginia, western North Carolina, northern Alabama, and Texas.
Also known as Scrub Chestnut Oak and Chinquapin Oak.

SAND POST OAK *Quercus margarettiae*

FIELD MARKS. A leaf-losing shrub or small tree of dry sandy pinelands and sandhills. *Leaves* broadest and with usually 3 (seldom 5) short and irregularly rounded lobes above the middle, pointed at base, smooth or nearly so above, somewhat downy to nearly smooth beneath, 2 to 4 inches long. *Fruits* egg-shaped or oval acorns about ½ inch long, about half covered by a deeply bowl-shaped cup.
RANGE. Coastal plain; southeastern Virginia south to Florida, west to Texas, and north to Missouri and Oklahoma.
Also known as Dwarf Post Oak and Scrub Post Oak.

Running Chinquapin

Scrub Oak

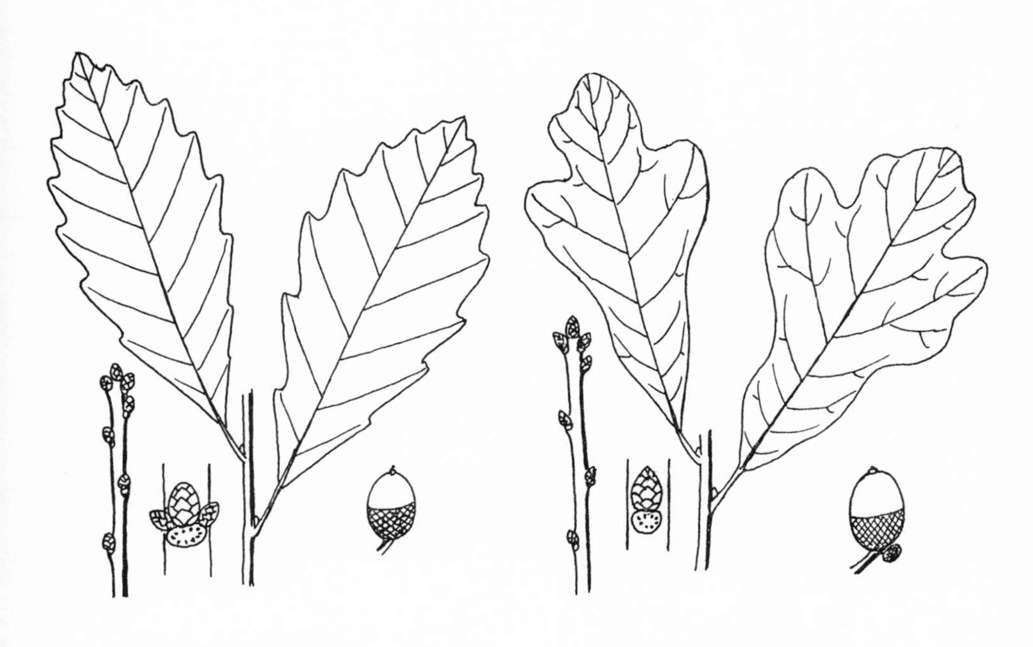

Dwarf Chestnut Oak

Sand Post Oak

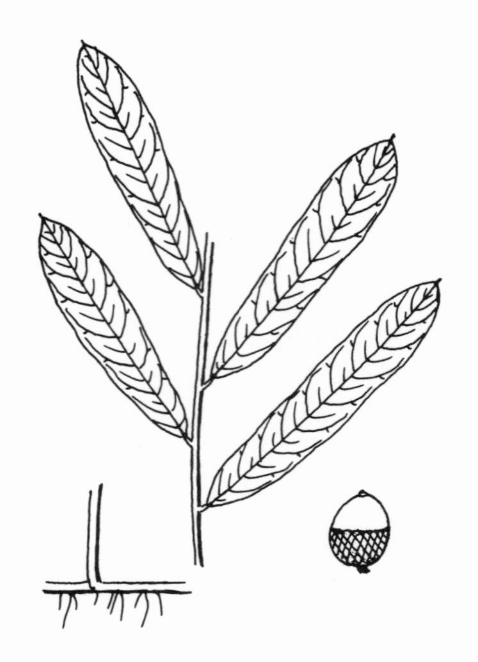

Running Oak

Creeping Live Oak

Dwarf Hackberry

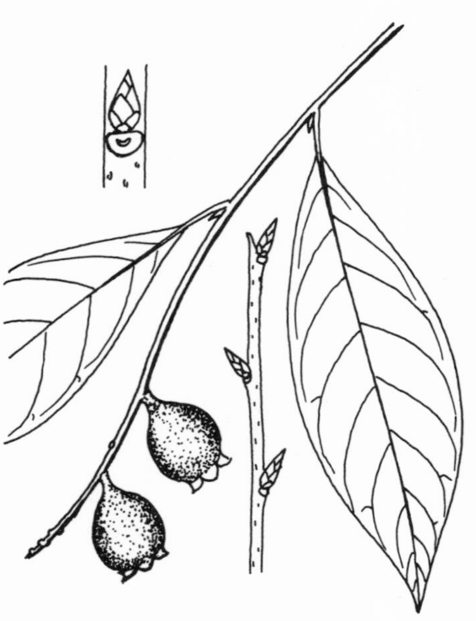

Buffalonut

RUNNING OAK *Quercus pumila*

FIELD MARKS. A tardily leaf-losing shrub with running underground stems and erect branches 1 to 3 feet high; growing in flat and sandy pinelands. *Leaves* narrowly elliptic, lance-shaped or broadest near the tip, rounded to broadly pointed at both ends but with a bristle-point at tip, margin wavy but untoothed, bright green and smooth above, paler and downy beneath, leathery, 1½ to 3 inches long; leafstalks very short. *Fruits* egg-shaped acorns about ½ inch long, about half covered by a deep bowl-like cup.

RANGE. Coastal plain; North Carolina south to Florida, west to Mississippi.

CREEPING LIVE OAK *Quercus minima*

FIELD MARKS. An evergreen shrub with running underground stems and erect branches 1 to 3 feet high; growing on sandhills and in dry sandy pinelands. *Leaves* thick and leathery, elliptic to narrowly elliptic or often broadest above the middle, usually wavy-toothed on margin, smooth above, smooth or downy beneath, 1¼ to 4 inches long. *Fruits* narrowly egg-shaped acorns about ⅝ inch long, about half covered by deep bowl-like cups.

RANGE. Coastal plain; Georgia, Florida, and southern Alabama.

ELM FAMILY (Ulmaceae)

DWARF HACKBERRY *Celtis tenuifolia*

FIELD MARKS. A straggling, leaf-losing shrub or small tree 3 to about 20 feet high; growing on dry uplands and slopes. *Leaves* have 3 prominent veins from near the base, egg-shaped, unevenly rounded or heart-shaped at base, pointed at tip, untoothed or sparingly toothed on margin, thin, often roughish above, 1 to 3 inches long. *Flowers* small, greenish, blooming in April or May. *Fruits* roundish, orange brown to purplish, 1-seeded, flesh thin and sweet, about ¼ inch in diameter.

RANGE. Pennsylvania to Indiana and Missouri; south to northern Florida and Louisiana.

SANDALWOOD FAMILY (Santalaceae)

BUFFALONUT *Pyrularia pubera*

FIELD MARKS. An upright, but often straggling, leaf-losing shrub 3 to about 12 feet high; growing on moist, often rocky, wooded slopes. *Leaves* alternate, elliptic or broadest above the middle, pointed at both ends, margin untoothed, soft and very veiny in appearance, smooth or nearly so at maturity, with minute clear dots which are visible when the leaf is held up to light, 2 to about 6 inches long. *Flowers* small, pale green, in short clusters; blooming in April or May. *Fruits* more or less pear-shaped, leathery, yellowish green, about 1 inch long; containing a solitary large oily seed; maturing July to October.

RANGE. Chiefly mountains; Pennsylvania south to Georgia and Alabama.

A root parasite of other shrubs and trees. All parts of the plant, but particularly the seeds, contain a bitterly pungent and poisonous oil. Also known as Oilnut.

NESTRONIA *Nestronia umbellula*

FIELD MARKS. A leaf-losing and often colonial shrub 1 to (rarely) 3 feet high; growing in dry woodlands. *Branchlets* slender, dark purplish brown. *Leaves* opposite, oval to egg-shaped or often broadest above the middle, usually pointed at both ends, margin untoothed, bright yellowish green and smooth or nearly so, 1 to 2½ inches long; leaf-stalks short. *Flowers* small, greenish, the stamen-bearing ones in long-stalked clusters in the leaf axils, the pistil-bearing ones on separate plants; blooming in April or May. *Fruits* berry-like, egg-shaped, 1-seeded, yellowish green, almost ½ inch across; ripening in July.

RANGE. Western Virginia south to northern Georgia and Alabama.

A rather rare shrub of local occurrence, parasitic on the roots of various broad-leaved trees and shrubs.

BUCKLEYA *Buckleya distichophylla*

FIELD MARKS. A leaf-losing shrub 6 to 12 feet high; growing on cliffs or bluffs along streams in the southern Appalachians. *Branchlets* green or grayish green, slender, with prominent pale lenticels. *Leaves* opposite or nearly so, lance-shaped to narrowly egg-shaped, taper-pointed at tip, roundish or broadly pointed at base, margin untoothed, smooth or nearly so and bright green on both surfaces, 1 to 3 inches long; stalkless or short-stalked. *Flowers* small, greenish, the stamen-bearing ones clustered, the pistil-bearing ones solitary at the tips of branches; blooming in April or May. *Fruits* berry-like, ellipsoid, 1-seeded, yellowish green to dull orange, about ½ inch long; ripening in August.

RANGE. Western Virginia south to western North Carolina and eastern Tennessee.

A rare shrub of local occurrence, parasitic on the roots of hemlock and possibly other trees.

MISTLETOE FAMILY (Viscaceae)

AMERICAN MISTLETOE *Phoradendron leucarpum*

FIELD MARKS. An evergreen, yellowish-green, much-branched shrubby plant, parasitic on the branches of oaks and other broad-leaved trees. *Branchlets* jointed, brittle at base. *Leaves* opposite, oblong or narrowly oblong and often broadest above the middle, rounded at tip, pointed at base, untoothed on margin, thickish and leathery in texture, smooth on both sides, ¾ to 2 inches long. *Flowers* small and inconspicuous, yellowish green; blooming in October or November. *Fruits* maturing the following fall, berry-like, round, waxy white, 1-seeded, about ⅛ inch across, with a sticky-gummy pulp.

RANGE. New Jersey to southern Illinois and southeastern Kansas, south to Florida and Texas.

Well known as a Christmas decoration. The fruits are eaten by birds, and the seeds are placed on tree branches when birds wipe the sticky pulp from their bills.

BIRTHWORT FAMILY (Aristolochiaceae)

DUTCHMAN'S-PIPE *Aristolochia macrophylla*

FIELD MARKS. A twining and high-climbing, leaf-losing, woody vine; growing in rich woods and along streams. *Leaves* alternate, heart-shaped, untoothed on margin, thin in texture, dark green and smooth above, paler and smooth or nearly so beneath, 3 to 8 inches wide. *Flowers* with a yellowish-green, U-shaped and pipelike calyx tube, spreading at summit into a 3-lobed and brownish-purple border; blooming in May and June. *Fruits* cylindrical, 6-ribbed capsules, 2 to 3 inches long, and containing many seeds.

RANGE. Appalachian region; western Pennsylvania and West Virginia south to Georgia and Alabama.

Cultivated and grown on trellises and porches; often listed in nursery catalogs as *Aristolochia sipho*.

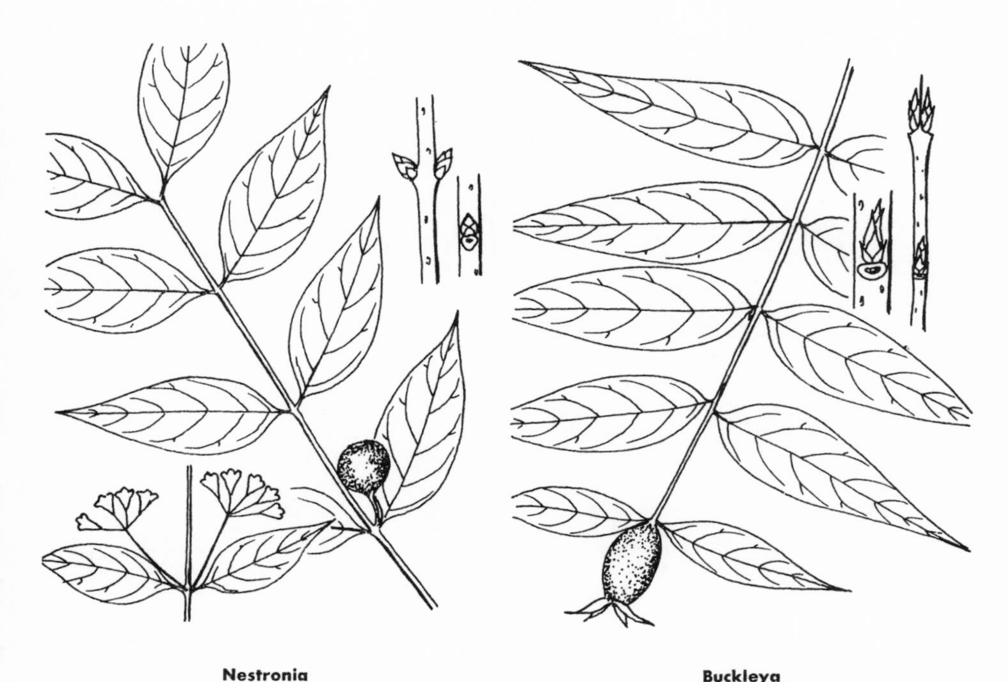

Nestronia

Buckleya

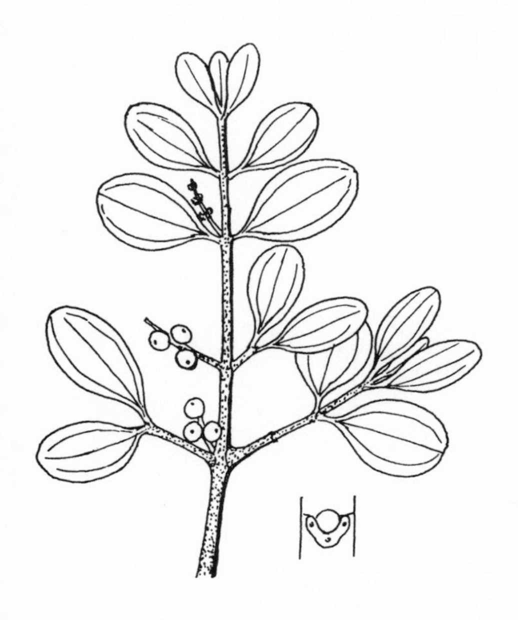

American Mistletoe

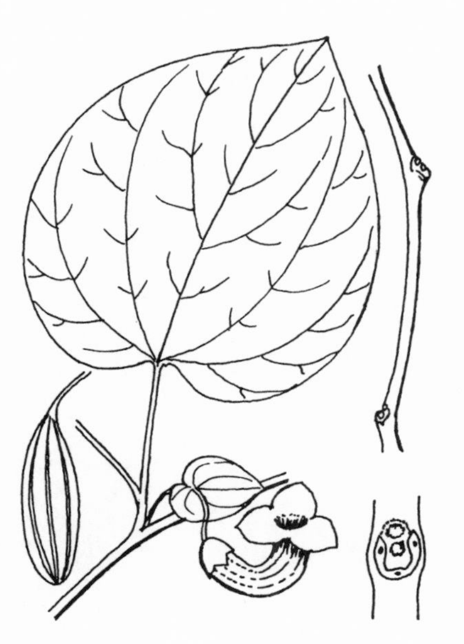

Dutchman's-Pipe

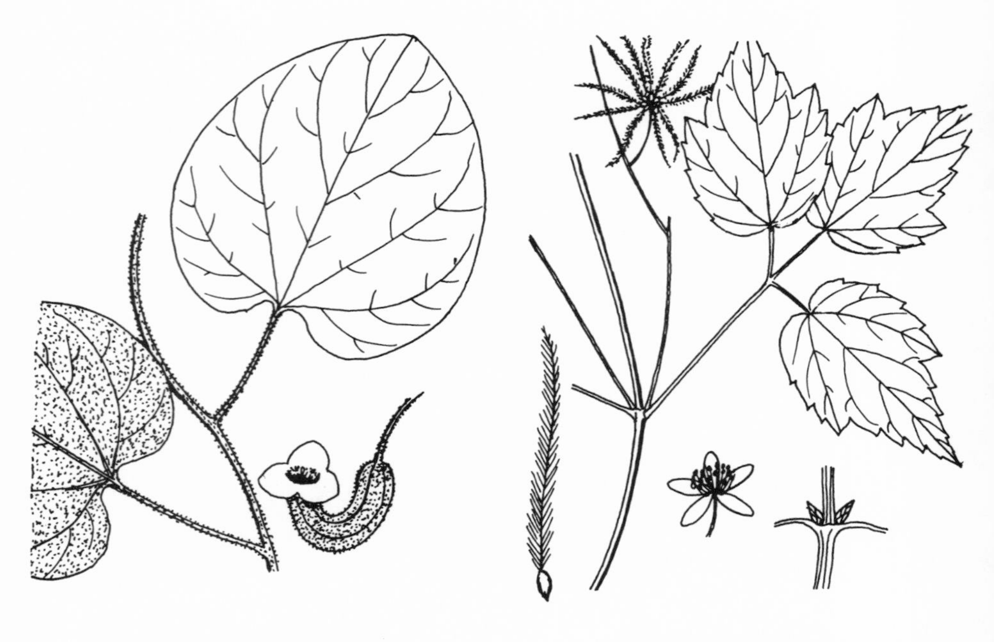

Woolly Pipevine

Common Virgin's-bower

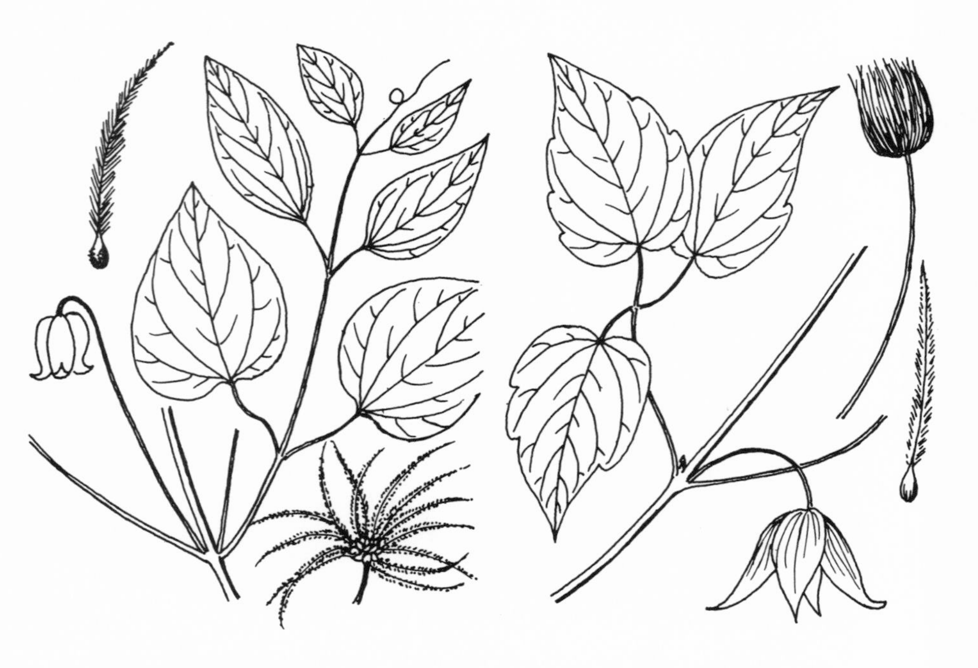

Leatherflower

Purple Virgin's-bower

WOOLLY PIPEVINE *Aristolochia tomentosa*

FIELD MARKS. A twining, leaf-losing vine usually found in bottomlands or along the banks of streams. *Branchlets* densely white-woolly. *Leaves* alternate, roundish heart-shaped, untoothed on margin, smooth or nearly so above, densely white-woolly beneath, 3 to 6 inches broad; leafstalks white-woolly beneath, 3 to 6 inches broad; leafstalks white-woolly. *Flowers* similar to those of the Dutchman's Pipe but the yellowish calyx tube downy. *Fruits* similar to those of Dutchman's Pipe but more or less woolly.

RANGE. North Carolina to southwestern Indiana, southern Illinois, Missouri, and Kansas; south to Florida and eastern Texas.

CROWFOOT (OR BUTTERCUP) FAMILY (Ranunculaceae)

VIRGIN'S-BOWERS (Clematis)

Virgin's-bowers are more or less soft-woody climbing vines or sometimes erect plants. They have opposite, compound leaves which sometimes end in a tendril. The leafstalks of the climbing species bend and twine about supporting objects. Their flowers are usually showy and have 4 petal-like parts which are really sepals. The fruits are achenes on which the style remains as a long, silky or plumelike tail; and they are grouped in fluffy and showy clusters.

COMMON VIRGIN'S-BOWER *Clematis virginiana*

FIELD MARKS. A climbing, leaf-losing vine with stems to 20 feet long; growing over bushes along streams, borders of swamps and woods, and in wayside thickets. *Branchlets* furrowed, downy. *Leaves* usually with 3 leaflets which are egg-shaped, pointed at tip, rounded to somewhat heart-shaped at base, coarsely toothed on margin, smooth or nearly so on both surfaces, 1½ to 4 inches long. *Flowers* white, about ¾ inch across, in leafy-bracted clusters on stalks from the axils of leaves; blooming in June and July. *Fruits* silver gray, in dense clusters about 2 inches across.

RANGE. Nova Scotia to Manitoba; south to Georgia, Louisiana, and eastern Kansas.

LEATHERFLOWER *Clematis viorna*

FIELD MARKS. A slightly woody, leaf-losing vine with slender stems 8 to 12 feet long; growing in thickets on rich soils. *Branchlets* furrowed, finely downy. *Leaves* with 3 to 7 leaflets which are egg-shaped, pointed at tip, bluntly pointed to rounded or heart-shaped at base, untoothed on margin, firm in texture, smooth or nearly so on both surfaces, 1¼ to 3 inches long. *Flowers* solitary on a long stalk from leaf axils, bell-shaped, nodding, with thick, leathery, reddish-purple sepals; blooming between May and August. *Fruits* with brownish plumelike tails 1½ or more inches long.

RANGE. Southern Pennsylvania to southern Illinois and southeastern Iowa, south to Georgia and Texas.

Also called Vasevine.

PURPLE VIRGIN'S-BOWER *Clematis verticillaris*

FIELD MARKS. A climbing, woody, leaf-losing vine with smooth, 6-sided, brown stems to 15 feet long; growing in rich, rocky woods. *Leaves* with 3 leaflets which are egg-shaped to heart-shaped, untoothed or sparingly toothed on margin, thin, smooth or nearly so on both surfaces, 1¼ to 3 inches long. *Flowers* solitary on long stems from the leaf axils, purple or bluish purple, 2 to 3 inches across; blooming in May or June. *Fruits* with grayish to brownish plumelike tails about 2 inches long.

RANGE. Eastern Quebec to Manitoba; south to Delaware, Maryland, West Virginia, Ohio, Illinois, and northeastern Iowa.

SHRUB YELLOWROOT *Xanthorhiza simplicissima*

FIELD MARKS. A sparingly-branched, leaf-losing shrub 8 inches to 2 feet high; growing in cool, moist woods and along the banks of streams. *Wood* of both stems and roots is very bright yellow. *Leaves* alternate, more or less clustered toward the summits of the short branchlets, usually divided into 5 leaflets which are deeply cut-toothed and often cleft, thin, bright green and lustrous on both surfaces. *Flowers* small, brownish purple, in drooping clusters; blooming in April or May and occasionally in the fall. *Fruits* small, light yellow, inflated, 1-seeded capsules, 4 to 8 grouped together.

RANGE. Southwestern New York southward, chiefly in the mountains, to Alabama and northwestern Florida.

Roots are used medicinally and to make a yellow dye.

MOONSEED FAMILY (Menispermaceae)

COMMON MOONSEED *Menispermum canadanse*

FIELD MARKS. A soft-wooded, leaf-losing vine with slender, faintly grooved, twining stems; growing in woods and along streams. *Leaves* alternate, broadly egg-shaped or roundish, untoothed but usually 3- to 7-lobed or angled, dark green and smooth above, paler and often slightly downy beneath, attached to the long leafstalks a little within the margin of the leaf blades, 3 to 8 inches wide. *Flowers* small, greenish white, blooming in June and July. *Fruits* roundish, bluish black with a whitish bloom, about ⅜ inch in diameter; resembling small grapes but with a single flattened and crescent-shaped seed.

RANGE. Western New England and Quebec to Manitoba; south to Georgia, Alabama, Arkansas, and Oklahoma.

Roots have been used medicinally. The fruits are said to be poisonous.

RED-BERRIED MOONSEED *Cocculus carolinus*

FIELD MARKS. A scrambling, leaf-losing vine with slender and usually downy stems; growing in rich woods and thickets, usually along streams. *Leaves* alternate, heart-shaped, egg-shaped, or triangular, margin untoothed but sometimes lobed, smooth or nearly so above, downy beneath, 2 to 4 inches long; leafstalks slender, ¾ to 1½ inches long. *Flowers* small, greenish; blooming June to August. *Fruits* roundish, red, about ¼ inch in diameter, with a single flattened, crescent-shaped seed; ripe from August to October.

RANGE. Southeastern Virginia to Kentucky, southern Illinois, Missouri, and southeastern Kansas; south to Florida and Louisiana.

Also called Carolina Moonseed.

CUPSEED *Calycocarpum lyoni*

FIELD MARKS. A rather smooth, high-climbing, leaf-losing vine of rich, usually bottomland woods. *Leaves* alternate, heart-shaped with 3 to 5 rather long lobes which are pointed at the tips and sometimes have a few coarse wavy teeth, thin in texture, smooth above, often somewhat downy beneath, 5 to 8 inches long; leafstalks 2 to 6 inches long. *Flowers* small, greenish, blooming in May or June. *Fruits* roundish to oval, black, nearly 1 inch long, with a single flattened and dish-shaped seed; ripe from August to October.

RANGE. Kentucky to southern Illinois, Missouri, and eastern Kansas; south to Florida and Louisiana.

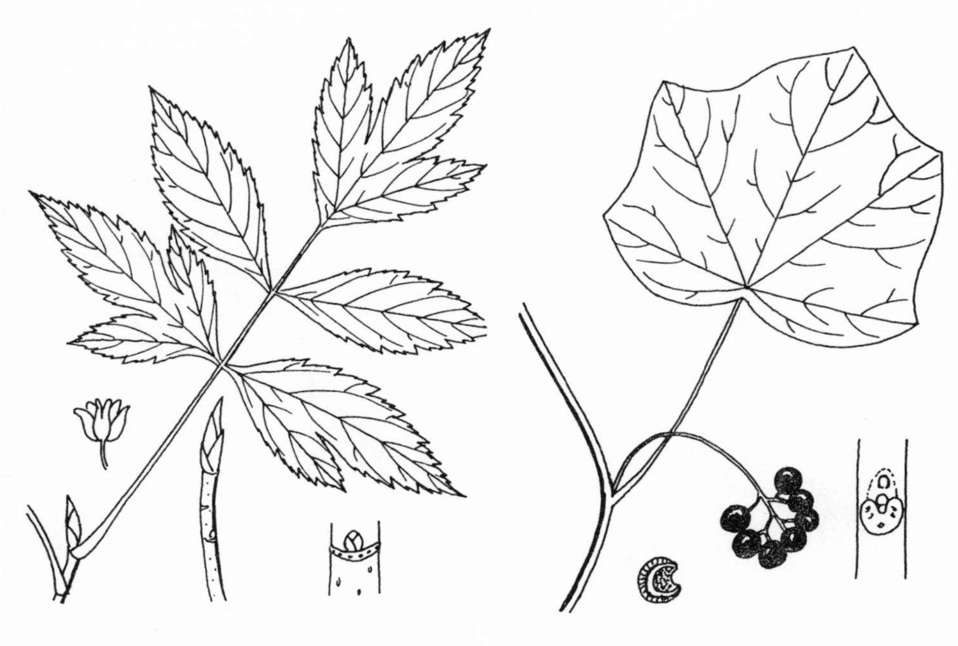

Shrub Yellowroot

Common Moonseed

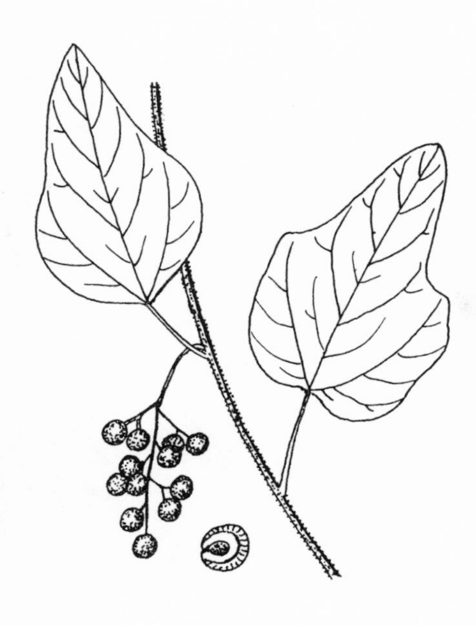

Red-berried Moonseed

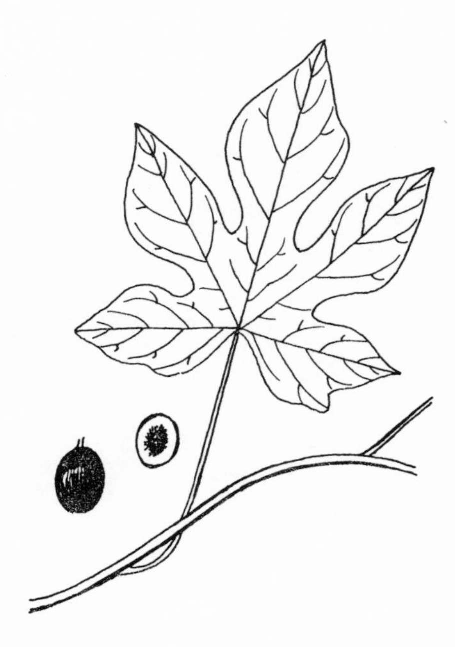

Cupseed

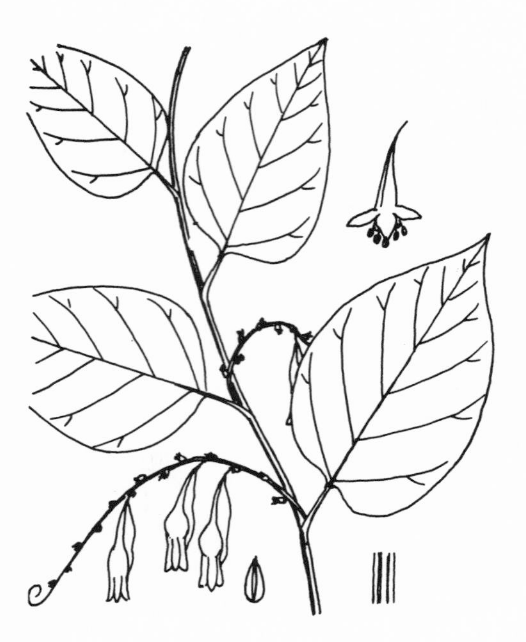

Brunnichia

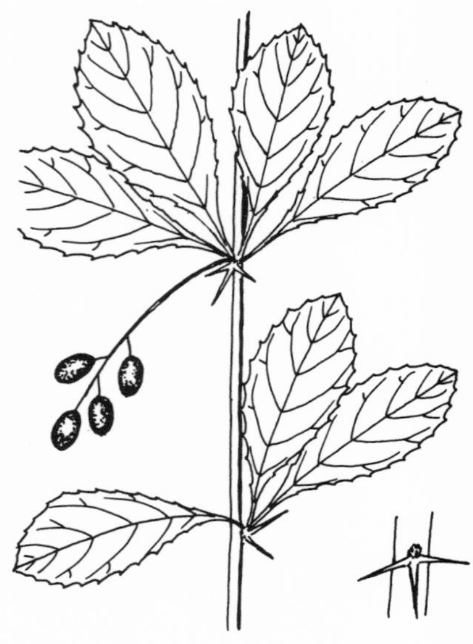

American Barberry

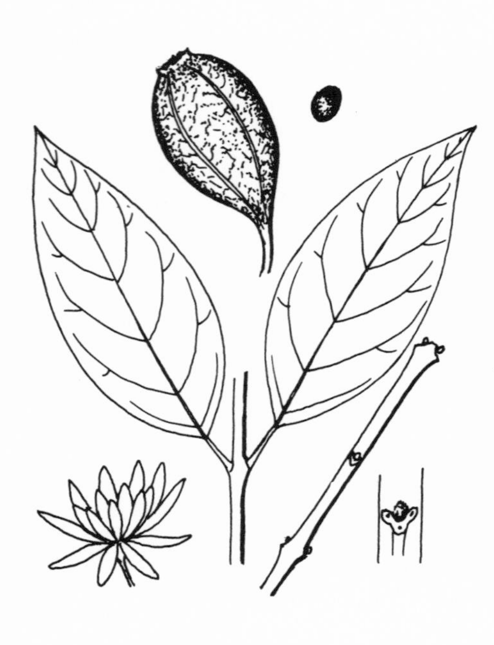

Strawberry-shrub

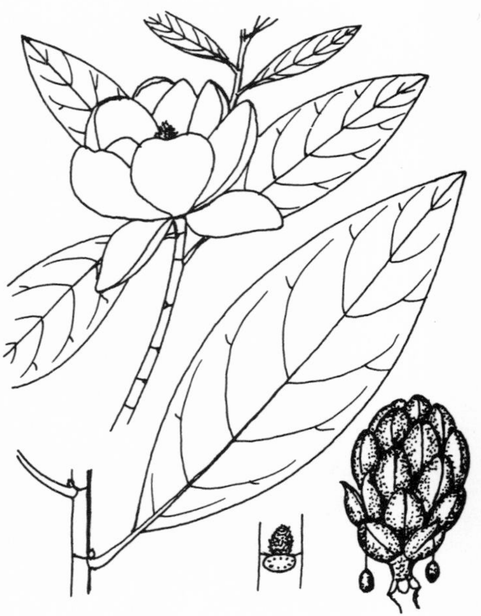

Sweetbay Magnolia

BUCKWHEAT FAMILY (Polygonaceae)

BRUNNICHIA *Brunnichia cirrhosa*

FIELD MARKS. A leaf-losing vine with grooved stems, climbing by means of tendrils at tips of branches; growing in swamps and along banks of streams. *Leaves* alternate, egg-shaped, pointed at tip, sometimes heart-shaped at the broad base, untoothed on margin, often slightly downy beneath, 1 to 6 inches long. *Flowers* greenish, 2 to 5 in a cluster. *Fruits* dry, 1-seeded, 3-angled, brown, enclosed by a persistent calyx which develops a broad wing.

RANGE. Chiefly coastal plain; South Carolina south to Florida, west to Texas; north in Mississippi Valley to southern Illinois and southeastern Missouri.

BARBERRY FAMILY (Berberidaceae)

AMERICAN BARBERRY *Berberis canadensis*

FIELD MARKS. A leaf-losing shrub 3 to 6 feet high, armed with 3-pronged spines at nodes; growing chiefly in woods of the southern Appalachians. *Branchlets* slender, dark brown, minutely warty, yellow within. *Leaves* alternate or in lateral clusters, broadest near or above the middle, usually rounded at tip, wedge-shaped at base, sharply toothed on margin, smooth or nearly so, ¾ to 2 inches long. *Flowers* yellow, in 5- to 15-flowered clusters; blooming May or June. *Fruits* oval-shaped, bright-red berries about ⅜ inch long, in drooping clusters; ripening September or October.

RANGE. South-central Pennsylvania south to Georgia; also in Missouri.

Serves as an alternate host of black stem rust of wheat.

CALYCANTHUS FAMILY (Calycanthaceae)

STRAWBERRY-SHRUB *Calycanthus floridus*

FIELD MARKS. A leaf-losing, aromatic shrub 3 to 9 feet high; growing in rich woods and along streams. *Branchlets* dark brown, somewhat enlarged and flattened at nodes, spicy-aromatic when bruised. *Leaves* opposite, elliptic or egg-shaped, pointed at tip, pointed to roundish at base, untoothed on margin, often roughish or rather lustrous above, smooth to somewhat downy beneath, 2 to 6 inches long, spicy-aromatic when crushed (or in var. *glaucus* lustrous green and smooth or whitened beneath). *Flowers* greenish brown to dark reddish brown or maroon, 1 to 2 inches across, almost odorless or with a decided strawberry-like fragrance; blooming April to August. *Fruits* urn-shaped capsules 2 to 2½ inches long; containing many large hard-coated seeds.

RANGE. South-central Pennsylvania and southern Ohio south, chiefly in mountains and upper piedmont, to Florida and Mississippi.

Also called Sweetshrub or Carolina-allspice. Commonly cultivated.

MAGNOLIA FAMILY (Magnoliaceae)

SWEETBAY MAGNOLIA *Magnolia virginiana*

FIELD MARKS. A large shrub to medium-sized tree, leaf-losing or evergreen southward; growing in swampy places and along streams. *Branchlets* green, ringed at the nodes with stipule scars, spicy-fragrant when bruised. *Leaves* alternate, oval to narrowly elliptic, usually pointed at both ends, untoothed on margin, lustrous above, whitened beneath, 3 to 5 inches long, spicy-aromatic when crushed. *Flowers* cup-shaped, creamy white, 2 to 3 inches across, very fragrant; blooming April to June. *Fruits* borne in conelike clusters; the individual fruits podlike, splitting down the outside and releasing the scarlet-coated seeds on silklike threads; ripening July to October.

RANGE. Chiefly coastal plain; southeastern Massachusetts and southern Pennsylvania south to Florida, west to Texas, north in Mississippi Valley to eastern Tennessee and southern Arkansas.

PURPLE-ANISE *Illicium floridanum*

FIELD MARKS. An aromatic evergreen shrub 6 to 18 feet high; growing in stream bottoms, swamps, or low hammocks. *Leaves* alternate, more or less crowded toward tips of the branchlets, elliptic, pointed at both ends, margin untoothed, smooth, dark green above, paler beneath, leathery, veins except the midrib not prominent, with an anise-like odor when crushed, 2½ to 6 inches long. *Flowers* dark red, ill-scented, about 1½ inches across, with many strap-shaped petals. *Fruits* 1-seeded greenish capsules in a wheel-shaped cluster about 1¼ inches across.

RANGE. Coastal plain; northern Florida west to Louisiana.

Also called Anise-tree, Starbush, and Stinkbush.

WILD-SARSAPARILLA *Schisandra glabra*

FIELD MARKS. A leaf-losing, aromatic, twining, woody vine; growing in low, rich woods. *Leaves* alternate, sometimes clustered on short lateral spurs, elliptic to oval or egg-shaped, pointed at both ends, sometimes with a few marginal teeth, bright green and smooth on both surfaces, 1½ to 5 inches long; leafstalks long and slender. *Flowers* crimson, about ½ inch across, 5- or 6-petalled; blooming May or June. *Fruits* berry-like, bright red, about ⅜ inch long, arranged along a drooping stem.

RANGE. Coastal plain; North Carolina south to Florida, west to Louisiana; north in Mississippi Valley to western Tennessee.

Also called Star-vine. The stamens and pistils are in separate flowers.

CUSTARD-APPLE FAMILY (Annonaceae)

PAWPAWS (Asimina)

Pawpaws are leaf-losing shrubs or small trees with alternate, simple leaves. The flowers have 3 sepals, 3 petals, numerous stamens in a globe-shaped cluster, and several pistils. The fruits are rather large, cylindrical berries which have several big seeds and a soft, pulpy, edible flesh.

DWARF PAWPAW *Asimina parviflora*

FIELD MARKS. A shrub 2 to 5 feet high, growing in dry oak and pine woods. *Branchlets* slender, densely rusty-hairy, ill-scented when bruised. *Leaves* broadest above the middle, blunt, short-pointed, or notched at tip, wedge-shaped at base, margin untoothed, thickish in texture, smooth or nearly so above, rusty-hairy beneath, 2 to 5 inches long. *Flowers* reddish brown to purplish brown, about ¾ inch across; blooming April or May, before the leaves appear. *Fruits* greenish yellow becoming brown, 1 to 2 inches long, with oval-shaped and slightly flattened seeds; ripening July to September.

RANGE. Southeastern Virginia to northwestern South Carolina, south to Florida and Louisiana.

Also called Possum-simmon.

COMMON PAWPAW *Asimina triloba*

FIELD MARKS. A shrub or small tree (rarely) to 40 feet high; growing in low woods and along streams. *Branchlets* slender, rusty-hairy to smooth, ill-scented when bruised. *Leaves* broadest toward tip and tapering to a wedge-shaped base, abruptly short-pointed at tip, margin untoothed, thin and veiny in appearance, smooth or nearly so at maturity, paler beneath, 6 to 12 inches long. *Flowers* purplish brown to greenish brown, 1 to 1½ inches across; blooming March to May, before the leaves appear. *Fruits* greenish yellow becoming brown, 2½ to 5 inches long, with elongate and slightly flattened seeds; ripening August to October.

RANGE. New Jersey to Michigan, south to Florida and Texas.

The tropical appearance of the foliage makes this plant a handsome ornamental.

Purple-Anise

Wild-sarsaparilla

Dwarf Pawpaw

Common Pawpaw

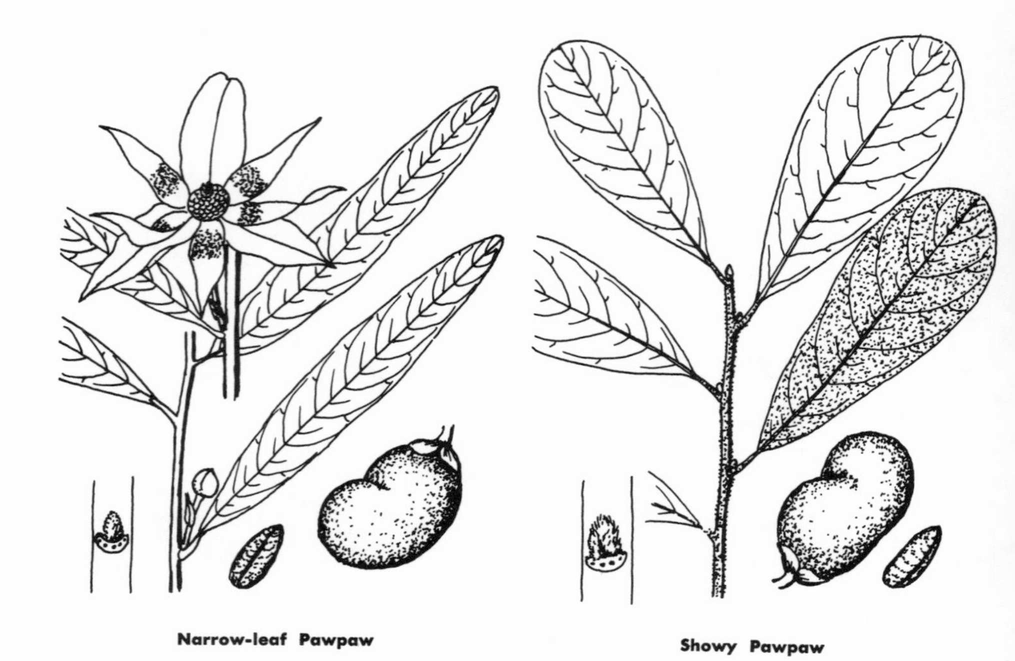

Narrow-leaf Pawpaw

Showy Pawpaw

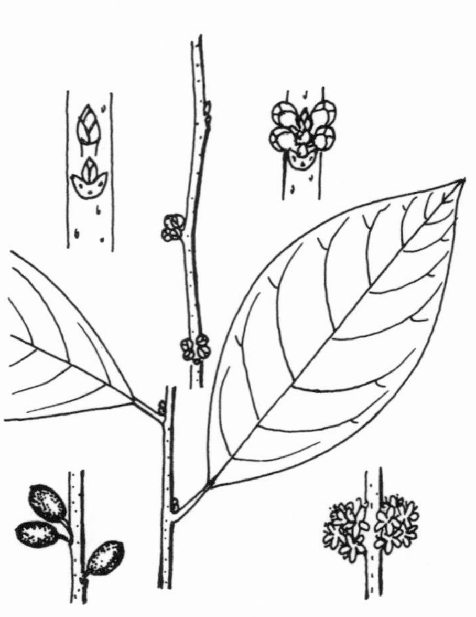

Common Spicebush

Hairy Spicebush

NARROW-LEAF PAWPAW *Asimina angustifolia*

FIELD MARKS. A shrub 2 to 4 feet high; growing in dry sandy pinelands. *Branchlets* minutely downy to smooth, ill-scented when bruised. *Leaves* narrow, blunt or somewhat pointed at tip, pointed at base, untoothed on margin, somewhat leathery in texture, smooth and rather lustrous above, paler and often tawny-downy beneath, 2½ to about 4 inches long, ½ to ¾ inch wide. *Flowers* creamy white, the inner petals often reddish purple at the base, 2 to 3 inches across, borne in the axils of the new leaves; blooming April or May. *Fruits* greenish yellow becoming brown, usually more or less lopsided, 1½ to 2½ inches long, the seeds elongate.

RANGE. Coastal plain; southern Georgia, northern Florida, and southeastern Alabama.

SHOWY PAWPAW *Asimina speciosa*

FIELD MARKS. A shrub 2 to about 4 feet high; growing in sandy pinelands. *Branchlets* densely tawny-woolly, ill-scented when bruised. *Leaves* narrowly top-shaped, broadest toward the rounded tip and gradually narrowed to the base, untoothed on margin, somewhat leathery in texture, smooth or nearly so above, velvety with tawny wool beneath, 1½ to about 3 inches long, often 1½ inches wide. *Flowers* white or creamy white, 3 to 4 inches across, developing from buds formed the previous season; blooming April or May. Fruits greenish yellow becoming brown, somewhat lopsided, 1½ to 2½ inches long, the seeds elongate.

RANGE. Coastal plain; southeastern Georgia and northeastern Florida.

This and the preceding species are often called Flag Pawpaw or Dog-apple.

LAUREL FAMILY (Lauraceae)

COMMON SPICEBUSH *Lindera benzoin*

FIELD MARKS. A leaf-losing, aromatic shrub 3 to 15 feet high; growing in moist woods and along streams. *Branchlets* slender, smooth, brittle, greenish brown to olive brown, spicy-fragrant when broken. *Leaves* alternate, elliptic or broadest above the middle, pointed at both ends, untoothed on margin, smooth on both sides or sparingly downy beneath, bright green above, paler beneath, spicy-fragrant when crushed, 2 to 6 inches long. *Flowers* small, honey-yellow, fragrant, clustered at the nodes; blooming in March or April, before the leaves appear. *Fruits* oval-shaped, bright red, 1-seeded, about ⅜ inch long, very spicy-aromatic when crushed; ripening July to September.

RANGE. Southeastern Maine to southern Ontario, southern Michigan, Iowa, and southeastern Kansas; south to Florida and Texas.

Also called Benjamin-bush and Wild-allspice. The leaves have been used to make a tea, and the dried powdered fruits are used as a substitute for allspice. The fruits are eaten by many wild birds, and the twigs are often eaten by deer and rabbits.

HAIRY SPICEBUSH *Lindera melissifolia*

FIELD MARKS. A leaf-losing, aromatic shrub 2 to about 6 feet high; growing in swamps and about the borders of ponds. *Branchlets* slender, downy, spicy-fragrant when broken. *Leaves* alternate, lance-shaped to elliptic, pointed at tip, rounded to slightly heart-shaped at base, smooth or nearly so above, densely downy beneath, spicy-fragrant when crushed, 2 to 6 inches long. *Flowers* similar to those of preceding species; blooming March or April. *Fruits* similar to those of the preceding species but often larger and inversely egg-shaped; ripening August or September.

RANGE. Chiefly coastal plain; North Carolina south to Florida, west to Alabama; north in Mississippi Valley to southern Missouri and southern Illinois.

Uncommon and local in distribution. Also known as Jove's-fruit.

PONDSPICE *Litsea aestivalis*

FIELD MARKS. A leaf-losing, aromatic shrub 6 to 10 feet high; growing in low wet woods and pond and swamp margins. *Branchlets* slender, zigzag, frequently forked, spicy-aromatic when broken. *Leaves* alternate, elliptic, pointed at both ends, untoothed on margin, smooth, somewhat leathery in texture, dark green above, paler beneath, spicy-aromatic when crushed, ¾ to 2 inches long. *Flowers* small, honey yellow, in clusters of 2 to 4; blooming February to April, before the leaves. *Fruits* roundish, red, 1-seeded, spicy-aromatic, about ¼ inch in diameter; ripening in May or June.

RANGE. Coastal plain; North Carolina south to Florida.

Rare and local in occurrence.

SASSAFRAS *Sassafras albidum*

FIELD MARKS. A leaf-losing, aromatic shrub or tree, sometimes 20 to 40 feet high; growing in borders of woods, old fields, and along fence rows. *Branchlets* yellowish green to reddish, brittle, smooth or finely downy, with a spicy-aromatic odor when broken. *Leaves* alternate, oval or elliptic and often with 1 or 2 lobes, bluntly pointed at tip, pointed at base, smooth or nearly so above but sometimes slightly downy beneath, spicy-aromatic when crushed, 3 to 6 inches long. *Flowers* greenish yellow, clustered; blooming March to May, before the leaves. *Fruits* oval-shaped, dark blue, 1-seeded, borne on club-shaped red stalks; ripening June to October.

RANGE. Southwestern Maine to Michigan and southeastern Iowa, south to Florida and Texas.

The bark of the roots is used to make sassafras tea and is the source of an oil used as a flavoring in candies and medicines and to perfume soaps. The fruits are eaten by many kinds of wild birds, including the bobwhite quail and wild turkey, and the twigs are eaten by deer and rabbits.

HYDRANGEA FAMILY (Hydrangeaceae)

MOCK-ORANGES (Philadelphus)

The mock-oranges are leaf-losing shrubs with simple, opposite leaves. The leaves have 3 to 5 prominent veins from near the base, and the leafstalks are short. They have showy flowers with 4 white petals and numerous stamens. The fruits are cup-shaped capsules on which the 4 calyx lobes persist at the summit, and they contain numerous minute seeds. The branchlets are often somewhat 6-sided or lined and have a large white pith. Excellent for ornamental planting.

COMMON MOCK-ORANGE *Philadelphus inodorus*

FIELD MARKS. A shrub 3 to 10 feet high, growing on rocky slopes and along streams. *Branchlets* smooth, with reddish-brown bark that is soon shed in thin papery flakes; the buds concealed by the leaf scars. *Leaves* egg-shaped, sharply pointed at tip, somewhat pointed or rounded at base, margin usually with some widely spaced but sharply pointed teeth, usually smooth on both surfaces but sometimes roughish above or sparingly hairy beneath, 2 to 4 inches long. *Flowers* 1 to 4 at tips of short branchlets, 1½ to 2 inches across, odorless or nearly so, with distinctly separate stigmas; blooming in May or June. *Fruits* usually smooth and with ascending calyx lobes.

RANGE. Virginia and Tennessee south to Florida and Alabama.

HAIRY MOCK-ORANGE *Philadelphus hirsutus*

FIELD MARKS. A shrub 2 to 5 feet high; growing on rocky slopes and along the banks of streams. *Branchlets* appressed-hairy, the brown bark soon shed in thin papery flakes; buds not hidden by the leaf scars. *Leaves* egg-shaped, pointed at tip, pointed to rounded at base, sharply toothed on margin, smooth or nearly so above, densely grayish-downy beneath, 1½ to 3 inches long. *Flowers* usually in clusters of 3 at the tips of short branchlets, about 1 inch across, odorless, with united stigmas; blooming in May or June. *Fruits* hairy, the calyx lobes spreading.

RANGE. North Carolina and Kentucky south to Georgia and Alabama.

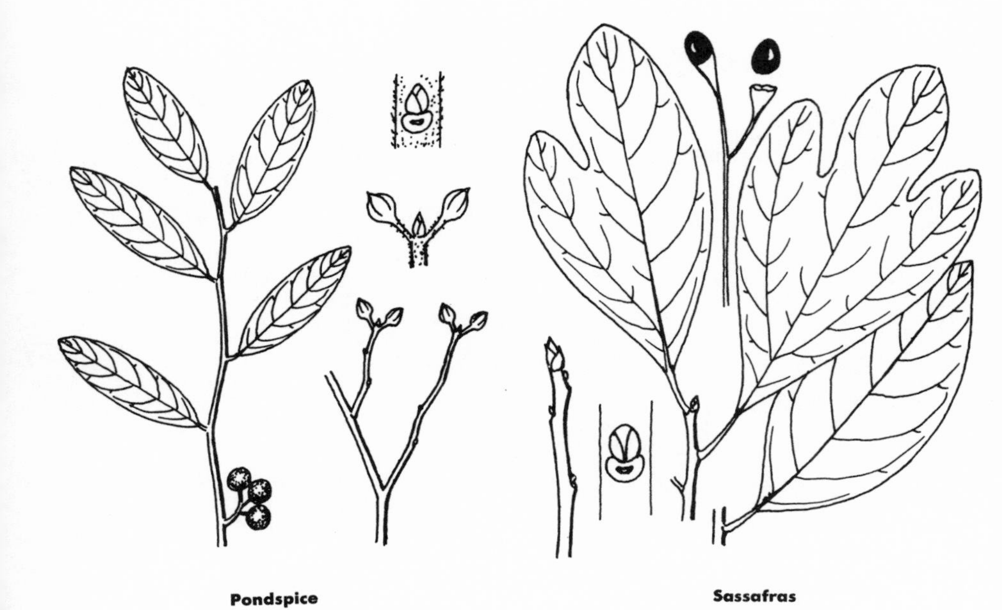

Pondspice

Sassafras

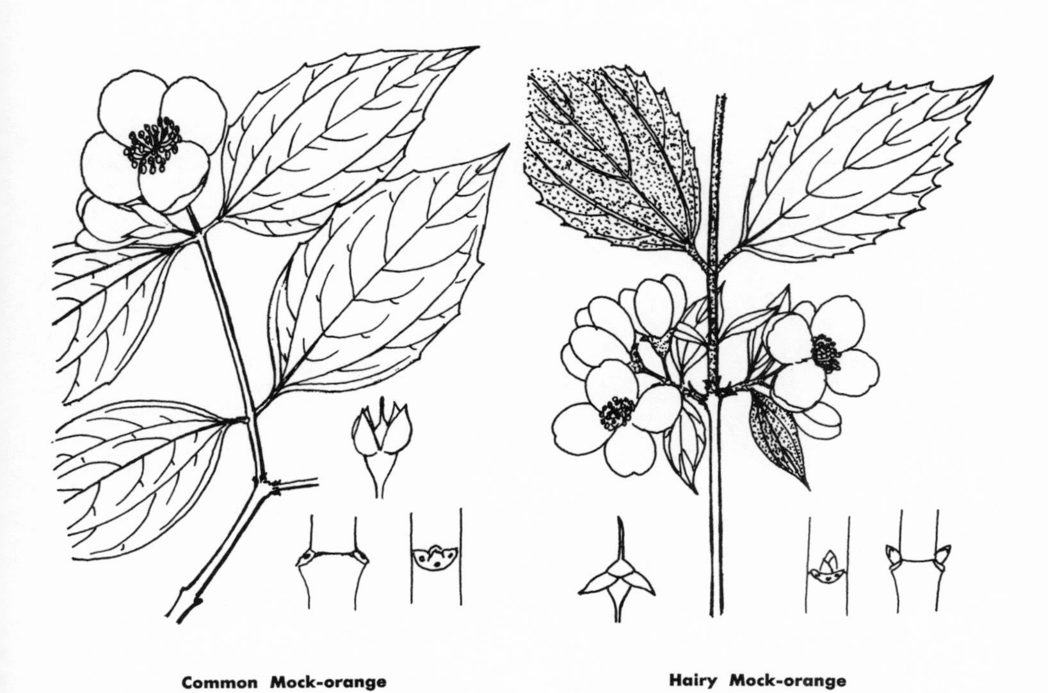

Common Mock-orange

Hairy Mock-orange

Downy Mock-orange

Climbing-Hydrangea

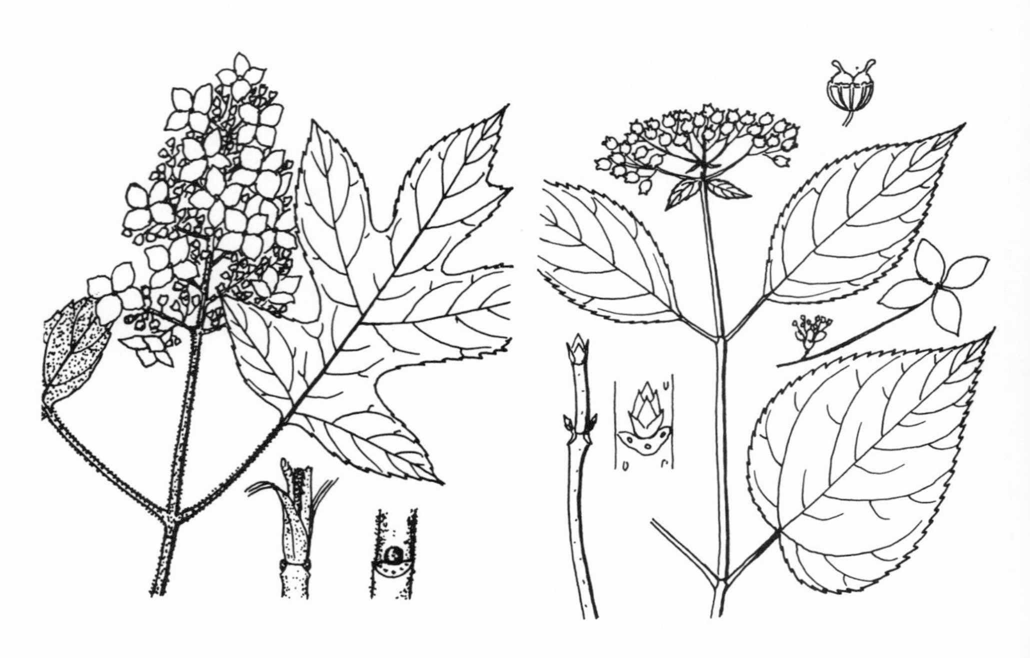

Oakleaf Hydrangea

Wild Hydrangea

DOWNY MOCK-ORANGE *Philadelphus pubescens*

FIELD MARKS. A shrub 3 to 10 feet high; growing on rocky bluffs and the banks of streams. *Branchlets* gray or sometimes straw-colored, the bark not freely peeling; buds hidden by the leaf scars. *Leaves* egg-shaped or elliptic, sharply pointed at tip, broadly pointed or rounded at base, usually with some sharp teeth on margin, dark green and smooth or nearly so above, densely grayish-downy beneath, 1½ to 4 inches long. *Flowers* in end clusters of 5 to 9, 1 to 1½ inches across, odorless or nearly so; blooming in May or June. *Fruits* downy, the calyx lobes spreading or reflexed.

RANGE. Tennessee to southern Illinois, south to Alabama and Arkansas.

CLIMBING-HYDRANGEA *Decumaria barbara*

FIELD MARKS. A semi-evergreen vine climbing trunks of trees by means of aerial rootlets on its stems; growing in rich, moist woods and swamps. *Leaves* opposite, egg-shaped or elliptic, broadly pointed to roundish at base, usually somewhat pointed at tip, margin usually with low teeth above the middle, bright green and lustrous above, paler and sometimes slightly downy beneath, thickish in texture, 1½ to 4 inches long. *Flowers* small, white, fragrant, in stalked flat-topped clusters at ends of branchlets; blooming April to June. *Fruits* small, urn-shaped, prominently ribbed capsules containing numerous small seeds.

RANGE. Southeastern Virginia to northwestern South Carolina and Tennessee, south to Florida and Texas.

HYDRANGEAS (Hydrangea L.)

Hydrangeas are leaf-losing shrubs with opposite, long-stalked, simple leaves; soft-wooded, mostly rather stout branchlets with a large whitish pith. The flowers are in dense and usually flat-topped clusters and are of two types: small ones which have stamens and pistils and produce fruits and seeds; and much larger, showy, but sterile ones which usually have 4 petal-like rays. The fruits are small urn-shaped capsules which contain a large number of very small seeds.

OAKLEAF HYDRANGEA *Hydrangea quercifolia*

FIELD MARKS. A shrub 3 to 7 feet high; growing on bluffs and stream banks. *Branchlets* rusty-woolly. *Leaves* egg-shaped to roundish, deeply 3- to 7-lobed and also sharply toothed on margin, thickish and somewhat leathery in texture, dark green and smooth above, whitish- to tawny-woolly beneath, 3 to 8 inches long; leafstalks rusty-woolly. *Flowers* in cone-shaped clusters 4 to 12 inches long, the numerous sterile ones about 1 inch across and turning purple.

RANGE. Chiefly coastal plain and adjacent piedmont; Georgia to northern Florida and west to Louisiana.

Often cultivated as an ornamental and sometimes escaping.

WILD HYDRANGEA *Hydrangea arborescens*

FIELD MARKS. A loosely branched shrub 3 to 6 feet high; growing on shaded banks and along streams. *Branchlets* pale brown, smooth or nearly so. *Leaves* roundish to egg-shaped, elliptic, or broadly lance-shaped, round to heart-shaped or broadly pointed at base, sharply pointed at tip, sharply toothed on margin, smooth or nearly so on both surfaces, dark green above, paler green beneath, 3 to 6 inches long. *Flowers* in flat-topped end clusters, creamy white, usually with several sterile flowers about ⅝ inch across; blooming May to July.

RANGE. New York to Illinois, Missouri, and Oklahoma; south to Florida and Louisiana.

The cultivated Hills-of-snow Hydrangea is a form with only sterile flowers.

ASHY HYDRANGEA *Hydrangea cinerea*

Differs from the preceding in having leaves which are grayish beneath. West Virginia to Indiana, Illinois, and Missouri; south to Georgia and Alabama. (Not illustrated)

SNOWY HYDRANGEA *Hydrangea radiata*

Leaves are snowy white with felted down on the lower surfaces. Grows in rocky woods from western North Carolina and eastern Tennessee to northern Georgia. (Not illustrated)

CURRANT AND GOOSEBERRY FAMILY (Grossulariaceae)

(Key Appendix D)

Currants and gooseberries are leaf-losing shrubs with alternate leaves and leaves clustered on short lateral spurs. The flowers are small, bell-shaped, and borne singly or in small clusters. The fruits are pulpy berries tipped with the remains of the calyx. These shrubs are alternate hosts of the blister rust which affects white pine trees. The fruits of most species are edible and used to make jelly, preserves, or pies.

VIRGINIA WILLOW *Itea virginica*

FIELD MARKS. A leaf-losing shrub 3 to 8 feet high with slender and wandlike branches; growing in swamps and along streams. *Branchlets* slender, green or reddish-tinged, often minutely downy, the white pith showing interruptions when cut lengthwise. *Leaves* alternate, elliptic or sometimes broadest above the middle, pointed at both ends, finely and sharply toothed on margin, smooth on both surfaces or sparingly hairy beneath, 2 to 4 inches long. *Flowers* rather small, 5-petalled, white, in narrow upright end clusters 3 to 5 inches long; blooming April to June. *Fruits* narrowly cone-shaped, 2-grooved, downy capsules about 1/4 inch long; containing numerous small flattened seeds.
RANGE. New Jersey and southeastern Pennsylvania south to Florida; southern illinois, Kentucky, and Missouri south to Louisiana and Texas.
Also called Sweet-spires and Tassel-white.

FLORIDA GOOSEBERRY *Ribes echinellum*

FIELD MARKS. A shrub 1½ to 3 feet high; growing in rich woods. *Branchlets grayish,* armed with slender nodal spines. *Leaves* roundish or broader than long, somewhat pointed to rather square at base, with 3 to 5 coarsely toothed lobes, smooth above, sparingly hairy beneath, ½ to 1 inch wide; leafstalks slender. *Flowers* whitish; blooming March or April. *Fruits* about ½ inch in diameter, greenish, densely covered with slender gland-tipped spines; ripening June to September.
RANGE. South-central South Carolina and north-central Florida.

ROUNDLEAF GOOSEBERRY *Ribes rotundifolium*

FIELD MARKS. A shrub 2 to 3 feet high; growing in cool rocky mountain woods. *Branchlets* pale brown or grayish, unarmed or with short nodal spines. *Leaves* roundish, often heart-shaped at base, usually with 3 bluntly toothed lobes, smooth or minutely downy above, sometimes sparingly hairy beneath, 3/4 to 3 inches wide. *Flowers* greenish purple, solitary or 2 or 3 in a cluster; blooming April to July. *Fruits* smooth, purplish, 1/4 to 3/8 inch in diameter, sweet; ripening June to September.
RANGE. Massachusetts to New York and Kentucky; south to western North Carolina and eastern Tennessee.
Also called Mountain Gooseberry.

PRICKLY GOOSEBERRY *Ribes cynosbati*

FIELD MARKS. A shrub 1½ to 4 feet high; growing in rocky woods and clearings. *Branchlets* pale brown or grayish, with slender nodal spines and often scattered prickles. *Leaves* roundish, often heart-shaped at base, 3- to 5-lobed and bluntly toothed, soft-hairy on both surfaces or nearly smooth, 1½ to 2½ inches wide. *Flowers* greenish, solitary or 2 or 3 in a cluster; blooming April to June. *Fruits* prickly, reddish purple, 1/3 to ½ inch in diameter, sweet; ripening July to September.
RANGE. New Brunswick to Manitoba; south to Georgia, Alabama, and Missouri.
Also called Pasture Gooseberry and Dogberry.

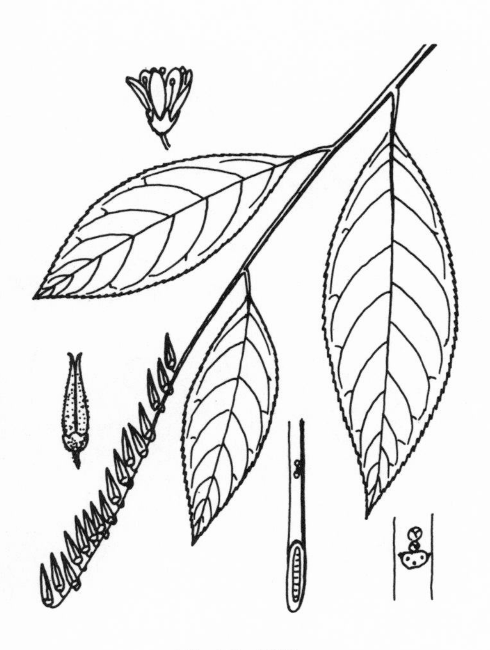

Virginia Willow

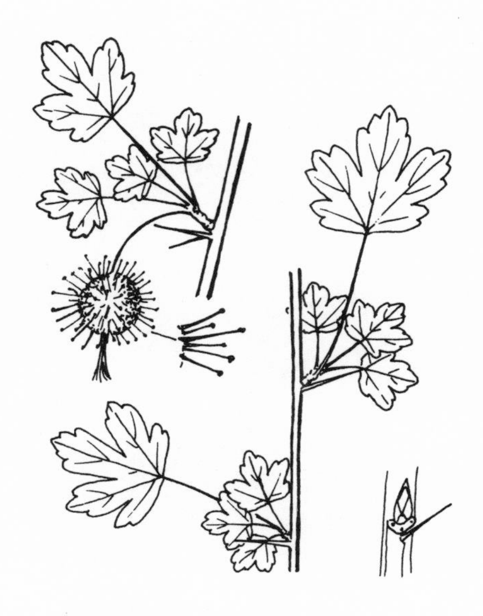

Florida Gooseberry

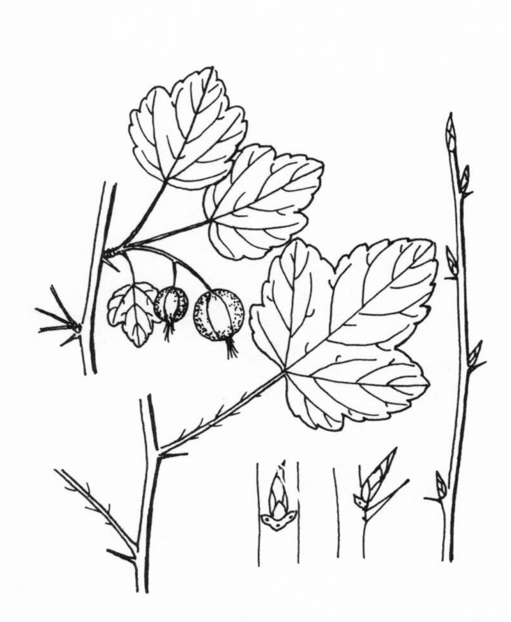

Roundleaf Gooseberry

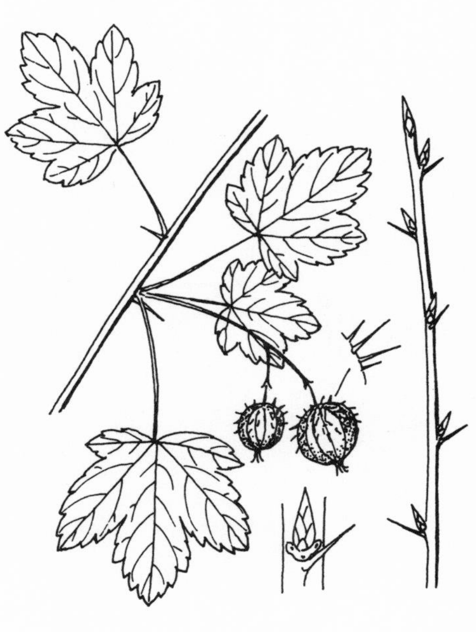

Prickly Gooseberry

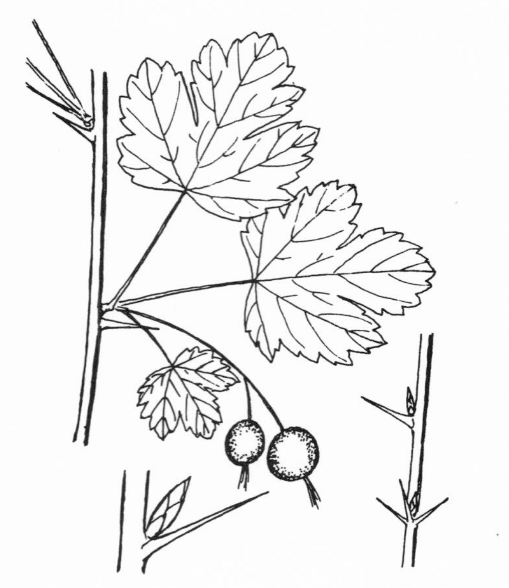

Missouri Gooseberry

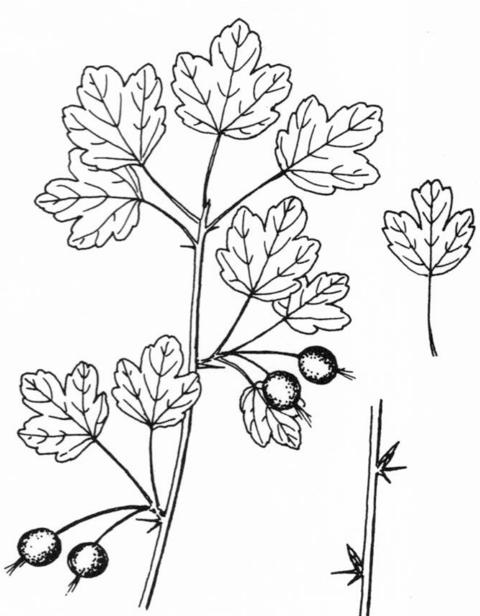

Granite Gooseberry

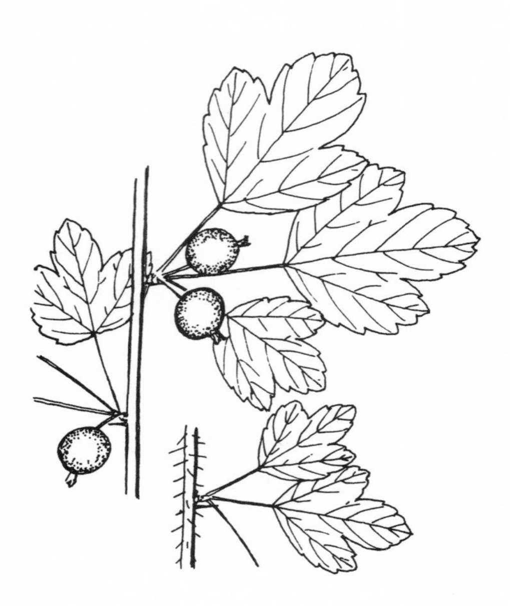

Northern Gooseberry

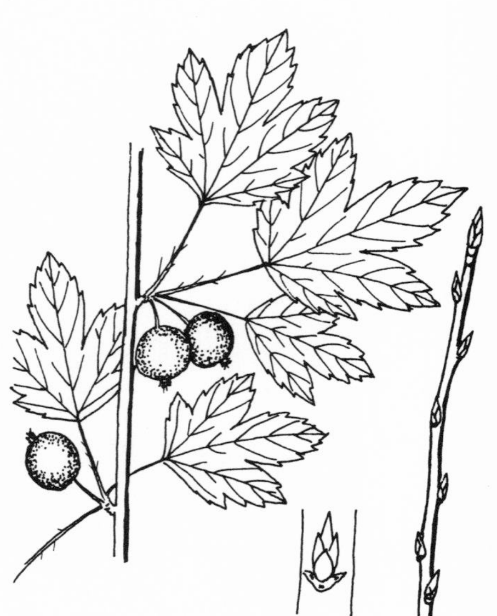

Smooth Gooseberry

MISSOURI GOOSEBERRY *Ribes missouriense*

FIELD MARKS. An erect shrub 3 to 6 feet high; growing in dry woods, ravines, and on stream banks. *Branchlets* whitish or gray, with stout red nodal spines to ⅝ inch long and often with scattered prickles. *Leaves* flattened to rounded, or heart-shaped at base, 3- to 5-lobed and toothed on margin, smooth or nearly so above, somewhat hairy beneath, ¾ to 1½ inches wide. *Flowers* white or greenish white, solitary or 2 to 3 on a slender stalk longer than the leafstalks; blooming April and May. *Fruits* smooth, purplish, about ½ inch in diameter, sweet; ripening July to September.

RANGE. Connecticut to Michigan, Minnesota, and South Dakota; south to Tennessee, Arkansas, and Kansas. Rare eastward.

GRANITE GOOSEBERRY *Ribes curvatum*

FIELD MARKS. A diffusely branched shrub 2 to 4 feet high, with recurving or drooping branches; growing in rocky woods and slopes. *Branchlets* reddish brown to purplish brown with short, slender, red nodal spines. *Leaves* roundish, broadly pointed to somewhat heart-shaped at base, 3- or sometimes 5-lobed and with few bluntish teeth on margin, sparingly hairy or nearly smooth, ⅜ to 1 inch wide. *Flowers* white, solitary or 2 to 5 on a slender stalk; blooming April or May. *Fruits* greenish, ¼ to ⅜ inch in diameter; ripening June to August.

RANGE. Northern Georgia and Alabama and southeastern Tennessee.

NORTHERN GOOSEBERRY *Ribes oxyacanthoides*

FIELD MARKS. A shrub 1½ to 3 feet high; growing in cool, moist, northern woods. *Branchlets* with short nodal spines and often bristly between. *Leaves* roundish, flattened to somewhat heart-shaped at base, 3- to 5-lobed and bluntly toothed on margin, somewhat hairy and also glandular beneath, ¾ to 1½ inches wide. *Flowers* greenish white or purplish, 1 or 2 on a stalk shorter than the leafstalks; blooming May to July. *Fruits* smooth, reddish purple, about ⅜ inch in diameter, sweet; ripening July and August.

RANGE. Northern Ontario to Yukon; south to Michigan, Minnesota, South Dakota, and Montana.

BRISTLY GOOSEBERRY *Ribes oxyacanthoides* ssp. *setosum*

FIELD MARKS. Differs slightly from the preceding. *Branchlets* reddish brown, unarmed or with nodal spines up to ¾ inch long and usually bristly. *Leaves* roundish, deeply 3- to 5-lobed and sharply toothed, usually somewhat hairy on both surfaces, ½ to 1½ inches wide. *Flowers* greenish white, 1 to 4 on a stalk; blooming in May. *Fruits* red or black, smooth or sparingly bristly; ripening July and August.

RANGE. North-central Ontario to Manitoba; south to northern Michigan, Wisconsin, and Nebraska. (Not illustrated)

SMOOTH GOOSEBERRY *Ribes hirtellum*

FIELD MARKS. An erect shrub 1 to 3 feet high; growing in swamps and cool, moist, rocky woods. *Branchlets* grayish, unarmed or with very small nodal spines, sometimes bristly between. *Leaves* roundish, usually more or less pointed but sometimes roundish to heart-shaped at base, deeply 3- to 5-lobed and sharply toothed on margin, smooth or nearly so on both surfaces, ¾ to 2½ inches wide. *Flowers* greenish or purplish, 1 to 3 on a stalk shorter than the leafstalks; blooming April to July. *Fruits* smooth, purplish to black, about ⅜ inch in diameter, sweet; ripening June to September.

RANGE. Southern Labrador to eastern Manitoba; south to Maryland, West Virginia, and the region of the Great Lakes.

WILD BLACK CURRANT *Ribes americanum*

FIELD MARKS. An erect unarmed shrub 1½ to 4 feet high, with spreading branches; growing in rich, moist, often rocky woods. *Branchlets* dotted with small yellow resin-glands. *Leaves* roundish, squarish to somewhat heart-shaped at base, sharply 3- to 5-lobed and double-toothed on margin, smooth or nearly so above, more or less downy and dotted with yellow resin-glands beneath, 1½ to 3 inches wide. *Flowers* rather large, yellow and whitish, several in a drooping cluster; blooming April to June. *Fruits* smooth, black, about 5/16 inch in diameter; ripening July to September.

RANGE. Nova Scotia to Alberta; south to Delaware, West Virginia, the region of the Great Lakes, Missouri, and Oklahoma.

SWAMP BLACK CURRANT *Ribes lacustre*

FIELD MARKS. An erect shrub 1 to 3 feet high, with prickly and bristly stems; growing in cool, moist woods and swamps. *Branchlets* with 1 to 3 long nodal spines and prickly bristles between. *Leaves* roundish, heart-shaped at base, deeply 3- to 5-lobed and toothed on margin, smooth or nearly so on both surfaces, 1 to 3 inches wide. *Flowers* greenish or purplish, several in a drooping cluster; blooming May to July. *Fruits* glandular-bristly, purplish black, about 5/16 inch in diameter, unpleasant to the taste; ripening July to September.

RANGE. Newfoundland to Alaska; south to western Massachusetts, New York, southwestern Pennsylvania, the region of the Great Lakes, Colorado, Utah, and California.

SKUNK CURRANT *Ribes glandulosum*

FIELD MARKS. A low, prostrate, sprawling or reclining unarmed shrub, all parts giving off a skunklike odor when crushed; growing in cold, damp, rocky woods. *Leaves* broader than long, deeply 5- to 7-lobed and doubly toothed on margin, heart-shaped at base, smooth above, sometimes slightly downy beneath, 1½ to 3 inches wide. *Flowers* yellowish green or purplish, several in a slender-stalked cluster; blooming May or June. *Fruits* glandular-bristly, coral red, unpleasant to the taste; ripening July to September.

RANGE. Newfoundland to British Columbia; south to New England, New York, northern Ohio, Michigan, Minnesota, and in the Appalachian Mountains to western North Carolina and northwestern South Carolina.

SWAMP RED CURRANT *Ribes triste*

FIELD MARKS. A low sprawling or reclining unarmed shrub, with stems often rooting; growing in cold mossy woods and boggy places. It does not have a skunklike or other disagreeable odor when crushed. *Leaves* roundish or somewhat broader than long, flattened to somewhat heart-shaped at base, 3- to 5-lobed and irregularly toothed on margin, light green and smooth above, paler and somewhat white-woolly beneath, 2 to 4 inches wide. *Flowers* brownish or purplish, several in a drooping cluster; blooming May to July. *Fruits* smooth, bright red, about ¼ inch in diameter; ripening July to September.

RANGE. Labrador to Alaska; south to northern New Jersey, West Virginia, Michigan, South Dakota, and Oregon.

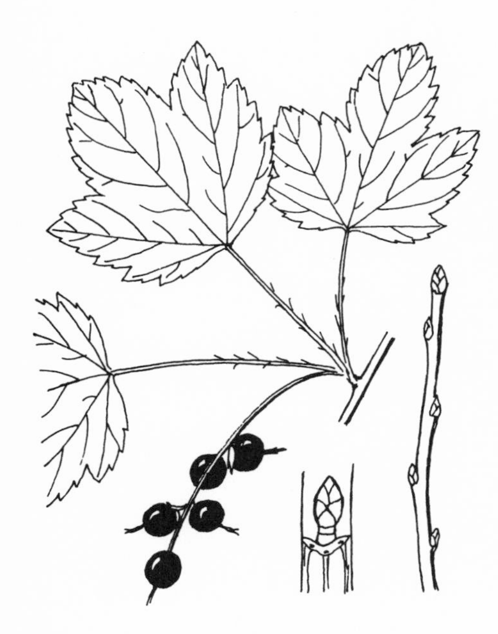

Wild Black Currant

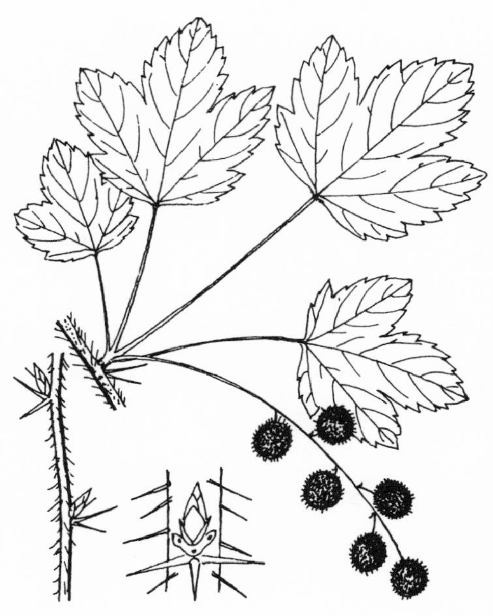

Swamp Black Currant

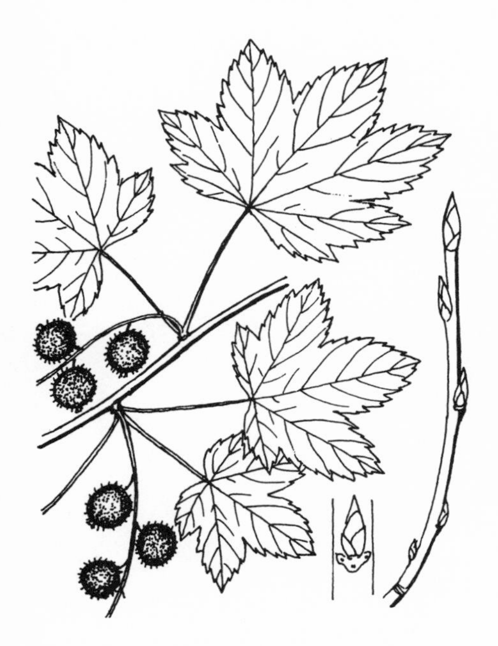

Skunk Currant

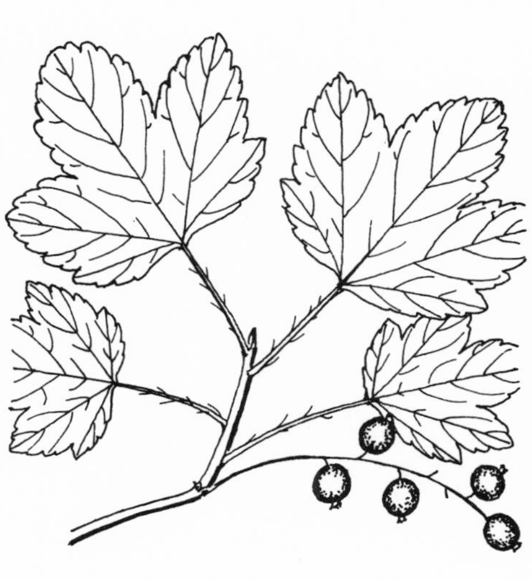

Swamp Red Currant

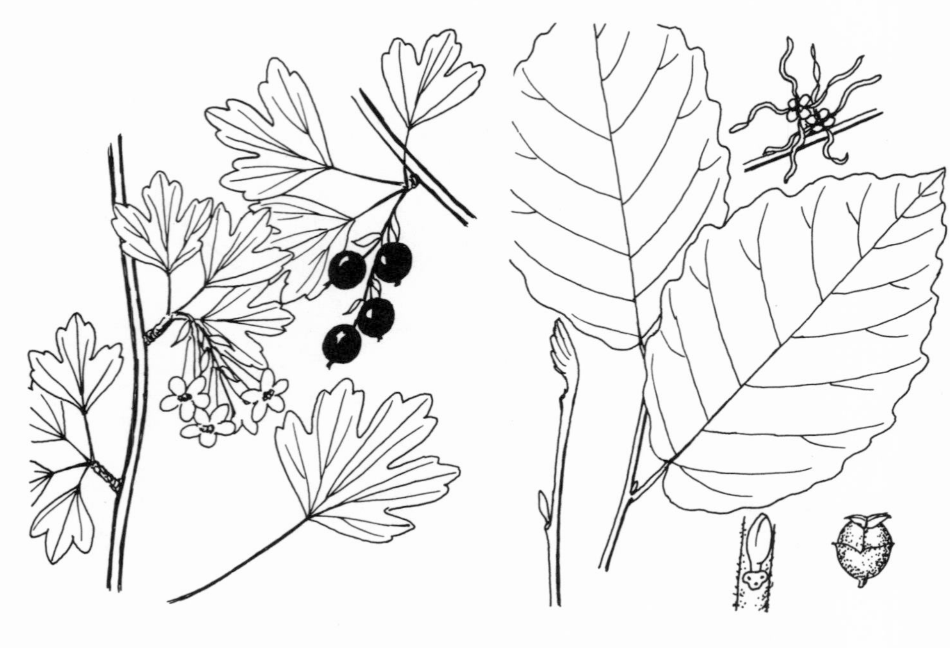

Buffalo Currant

Witch-Hazel

Mountain Witch-alder

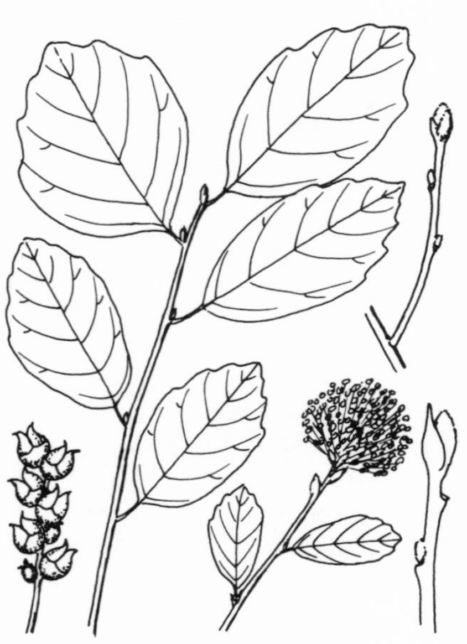

Dwarf Witch-alder

BUFFALO CURRANT *Ribes odoratum*

FIELD MARKS. An unarmed shrub 4 to 6 feet high; growing on rocky bluffs, slopes, and along streams. *Branchlets* more or less minutely grayish-downy. *Leaves* roundish, deeply 3- to 5-lobed and usually coarsely toothed, broadly pointed to rather flat at base, smooth or nearly so above, sometimes more or less downy beneath, 1 to 3 inches long; the slender leafstalks usually somewhat hairy. *Flowers* bright yellow, spicy-fragrant, 5 to 10 in nodding and leafy-bracted axillary clusters; blooming April to June. *Fruits* black or (rarely) yellow, smooth, about ⅜ inch in diameter; ripening June to August.

RANGE. Minnesota and South Dakota south to Missouri, Arkansas, and Texas. Commonly cultivated and often escaping eastward.

Also called Golden or Missouri Currant.

WITCH-HAZEL FAMILY (Hamamelidaceae)

WITCH-HAZEL *Hamamelis virginiana*

FIELD MARKS. A leaf-losing shrub or small tree 5 to 20 feet high; growing in rich woods, thickets, and clearings. *Branchlets* with stalked, naked buds; both roughish with tawny or rusty hairs. *Leaves* alternate, oval or broadest above the middle, unevenly rounded or heart-shaped at base, pointed or rounded at tip, irregularly wavy-toothed on margin, smooth above, paler to rusty-hairy or smooth to nearly so beneath, 1½ to 6 inches long. *Flowers* bright yellow, with 4 ribbon-like petals; blooming September to November or later, usually after the leaves have fallen. *Fruits* urn-shaped, grayish-downy, woody capsules containing 2 shiny black seeds which are forcibly ejected in October or November.

RANGE. Nova Scotia to southern Ontario, central Michigan, and southeastern Minnesota; south to Florida and Texas.

The plant has astringent properties, and commercial witch-hazel is distilled from the bark. Twigs or bark eaten by deer and rabbits; seeds by ruffed grouse and bobwhite quail.

MOUNTAIN WITCH-ALDER *Fothergilla major*

FIELD MARKS. A leaf-losing shrub 3 to 10 feet high; growing in woods of the southern Appalachians. *Branchlets* and buds downy. *Leaves* oval to roundish and often broadest above the middle, heart-shaped to broadly pointed at base, coarsely wavy-toothed on margin above the middle, smooth or nearly so above, paler or whitened and more or less hairy beneath, 2 to 4½ inches long. *Flowers* without petals but with numerous creamy-white stamens, in dense end clusters 1 to 2 inches long; blooming April or May. *Fruits* egg-shaped, downy, woody, 2-beaked capsules about ½ inch long, containing two lustrous brown seeds; maturing September or October and forcibly ejecting the seeds.

RANGE. Western North Carolina to northern Georgia and Alabama.

A handsome shrub suitable for ornamental planting.

DWARF WITCH-ALDER *Fothergilla gardenii*

FIELD MARKS. A leaf-losing shrub 1 to 3 feet high, sometimes with creeping underground stems; growing in sandy coastal plain bogs and savannahs. *Branchlets* and buds densely downy. *Leaves* elliptic to oval or oblong and often broadest above the middle, pointed to roundish or heart-shaped at base, margin usually with some wavy teeth above the middle, usually with scattered starry-branched hairs above, paler and more densely coated with such hairs beneath, ¾ to 2¼ inches long. *Flowers* white, similar to those of preceding species, in dense clusters ¾ to 1¼ inches long; blooming April or May. *Fruits* similar to those of the preceding species but somewhat smaller.

RANGE. Virginia south to Florida, west to Mississippi; north in Mississippi Valley to western Tennessee.

COMMON NINEBARK *Physocarpus opulifolius*

FIELD MARKS. A leaf-losing shrub 3 to 10 feet high, with arching branches and bark peeling off in thin papery layers; growing on rocky slopes, cliffs, and along streams. *Branchlets* with prominent ridges below the leaves or leaf scars, smooth to densely hairy with starry-branched hairs. *Leaves* alternate, roundish to egg-shaped, heart-shaped to roundish or broadly pointed at base, pointed to roundish at tip, commonly more or less 3-lobed and double-toothed on margin, smooth or nearly so above, smooth to downy or coated with starry-branched hairs beneath, ¾ to about 3 inches long. *Flowers* small, white or pinkish, 5-petalled, in often dense umbrella-shaped end clusters; blooming May to July. *Fruits* small, papery, red or purplish, smooth or hairy, inflated pods, 3 to 5 grouped on each stalk in the cluster; maturing July to September and persisting.

RANGE. Quebec and northern Ontario to Minnesota; south to Georgia, Tennessee, and Arkansas.

The hairy southern forms have been described as distinct species by some botanists but are not considered valid by others.

ALABAMA SNOW-WREATH *Neviusia alabamensis*

FIELD MARKS. A leaf-losing shrub 4 to 6 feet high, with recurving branches; growing on shady cliffs and bluffs in a limited area of northwestern Alabama. *Branchlets* yellowish brown, downy, with ridges below the leaves or leaf scars. *Leaves* alternate, egg-shaped to somewhat rhombic, roundish to somewhat pointed at base, pointed to blunt at tip, sharply and doubly toothed on margin, smooth or nearly so on both surfaces, ¾ to 2½ inches long. *Flowers* white, petals none, sepals 5 (pale green or whitish and toothed), stamens white and numerous in a tassel-like cluster; blooming March to May. *Fruits* somewhat fleshy-coated achenes about 3/16 inch long, 2 to 4 in a group.

A very attractive ornamental shrub deserving much wider use in ornamental planting; hardy as far north as Massachusetts.

SPIRAEAS (Spiraea)

The spiraeas are rather low, leaf-losing shrubs with alternate, simple, short-stalked leaves and slender, often wandlike branches. They have small 5-petalled white or pink flowers in rather dense, narrow or flat-topped clusters. The fruits are small dry pods, usually in groups of 5, and they persist for some time after maturing.

NARROW-LEAF MEADOWSWEET *Spiraea alba*

FIELD MARKS. An erect shrub 2 to 5 feet high; growing in wet open places, swamp thickets, and along streams. *Branchlets* yellowish brown, more or less angled, often slightly downy. *Leaves* numerous, crowded, narrowly elliptic or broadest above the middle, pointed at both ends, finely and sharply toothed on margin, smooth on both surfaces or slightly downy beneath, 2 to 3 inches long. *Flowers* white, in a narrow and somewhat downy inflorescence from 2 to 5 inches in length; blooming June to September.

RANGE. Northwestern Vermont to southwestern Quebec and Saskatchewan; south to Delaware, Ohio, Illinois, Missouri, North Dakota, and in the Appalachian Mountains to North Carolina.

BROADLEAF MEADOWSWEET *Spiraea alba* var. *latifolia*

Differs from the preceding in having reddish brown to purplish brown branches that are more or less angled, smooth or nearly so, and with smooth rather than hairy branchlets of the inflorescence.

RANGE. Newfoundland to northern Quebec and Michigan; south to New York and the mountains of western North Carolina.

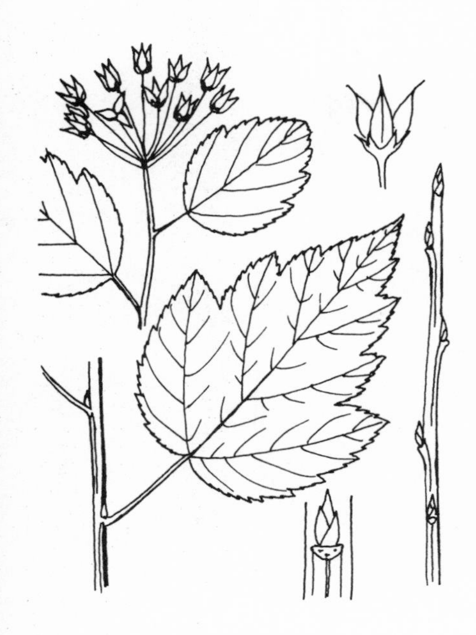

Common Ninebark

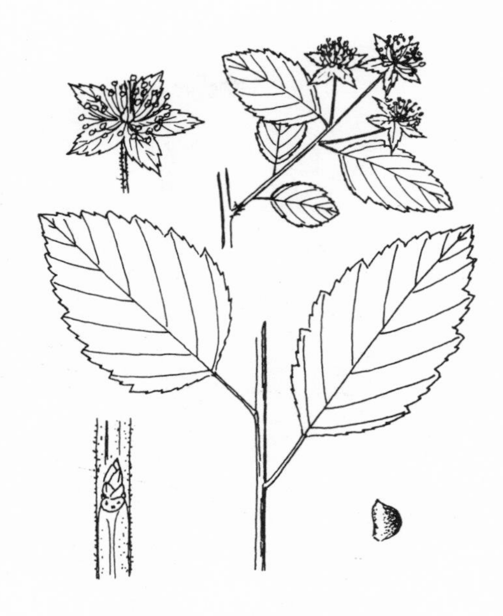

Alabama Snow-wreath

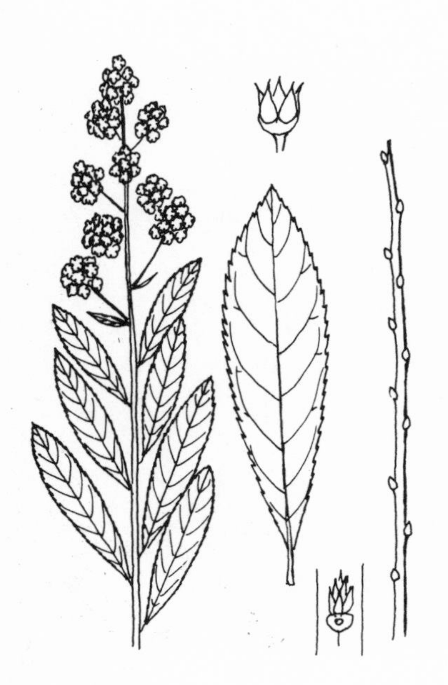

Narrow-leaf Meadowsweet

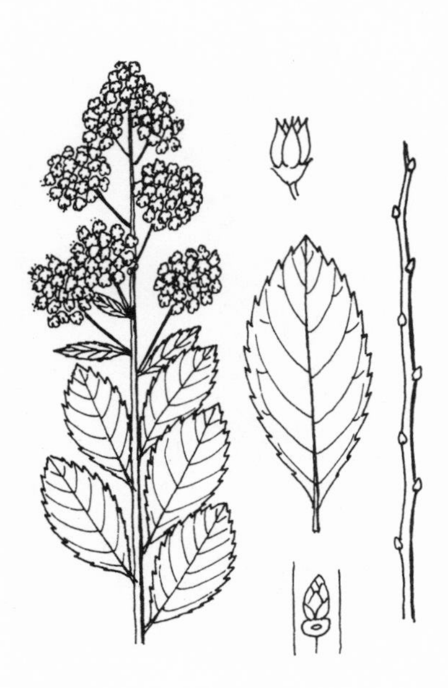

Broadleaf Meadowsweet

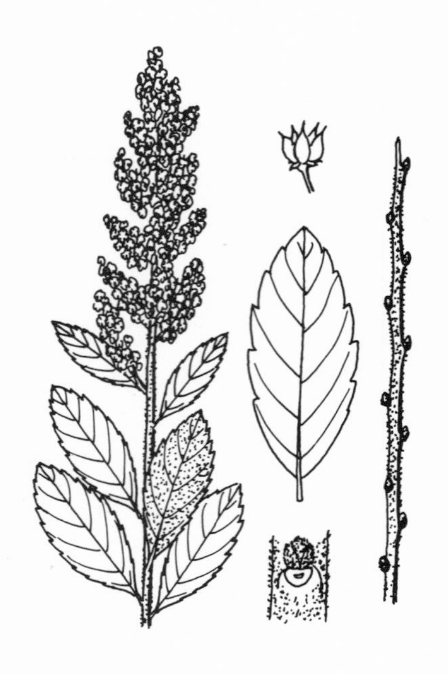

Hardhack

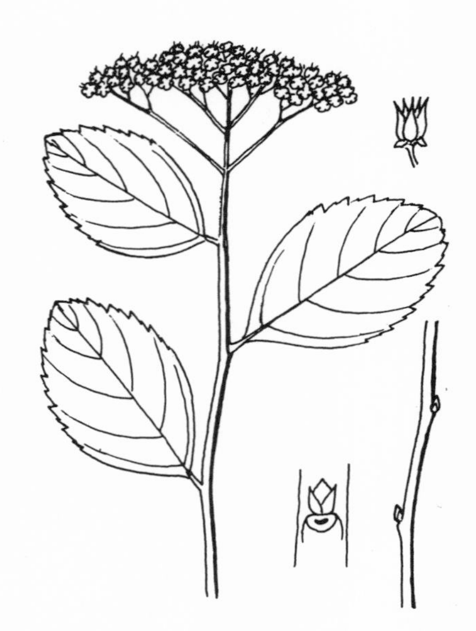

Dwarf Spiraea

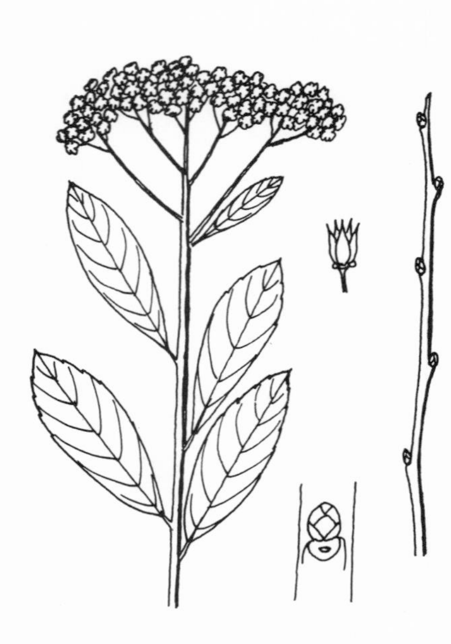

Virginia Spiraea

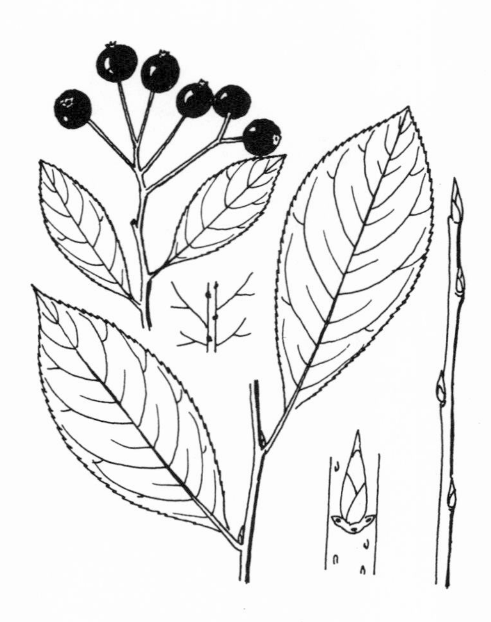

Black Chokeberry

HARDHACK *Spiraea tomentosa*

FIELD MARKS. A simple or sparingly branched shrub 1 to (rarely) 4 feet high; growing in wet meadows and bogs. *Stems* angular, purplish brown, at first densely coated with tawny or rusty wool. *Leaves* numerous, egg-shaped to elliptic, pointed at base, pointed to blunt at tip, sharply toothed on margin, bright green and smooth above, densely tawny- to rusty-woolly beneath, 1 to 3 inches long. *Flowers* pink or rose-colored (rarely white), in a dense spirelike end cluster 4 to 7 inches long; blooming June to September.

RANGE. Nova Scotia to Manitoba; south to Virginia, northern Georgia, Tennessee, and Arkansas.

Also known as Steeplebush.

DWARF SPIRAEA *Spiraea betulifolia*

FIELD MARKS. A simple or sparingly branched shrub 1 to (rarely) 3 feet high; growing on rocky slopes or stream banks in the Appalachian Region. *Stems* round, purplish brown to reddish brown, smooth. *Leaves* oval to broadly egg-shaped, rounded to broadly pointed at base, rounded or bluntly pointed at tip, margin coarsely and usually doubly toothed above the middle, firm in texture, smooth or nearly so on both surfaces, paler beneath, 1 to 3 inches long. *Flowers* white, in a flat-topped and finely downy end cluster 2 to 4 inches broad; blooming May to July.

RANGE. New Jersey, Pennsylvania, and Kentucky; south to northern Georgia.

VIRGINIA SPIRAEA *Spiraea virginiana*

FIELD MARKS. A slender, branching shrub 2 to 5 feet high; growing on damp rocky banks and slopes of the Appalachian Region. *Branchlets* at first downy but becoming smooth. *Leaves* narrowly oblong or broadest toward the tip, usually pointed at both ends and bristle-tipped, margin untoothed or with a few low teeth above the middle, thin, smooth above, whitened and sometimes finely downy beneath, 1 to 2 inches long. *Flowers* white, in downy flat-topped end clusters 1¾ to 2 inches across; blooming June or July.

RANGE. West Virginia south to western North Carolina and eastern Tennessee.

CHOKEBERRIES (Aronia)

The chokeberries are leaf-losing shrubs with alternate, simple leaves which have small, dark-colored glands along the midrib on the upper side. They have white or purplish-tinged 5-petalled flowers in flat-topped clusters, blooming in spring when the new leaves appear. The fruits are small, round, and apple-like; they ripen in the fall.

BLACK CHOKEBERRY *Aronia melanocarpa*

FIELD MARKS. A shrub 2 to 6 feet high, growing in swamps and moist to fairly dry thickets and woods. *Branchlets* smooth. *Leaves* elliptic or more often broadest above the middle, pointed at base, pointed to blunt at tip, finely and sharply toothed on margin, smooth on both surfaces, paler beneath, 1 to 3 inches long. *Flowers* blooming in June. *Fruits* black, about ¼ inch in diameter; ripening September or October.

RANGE. Newfoundland to northwestern Ontario and Minnesota, south to Pennsylvania and along the mountains to northern Georgia.

RED CHOKEBERRY *Aronia arbutifolia*

FIELD MARKS. A shrub 3 to 8 feet high; growing in wet or boggy places. *Branchlets* more or less woolly. *Leaves* elliptic or broadest above the middle, wedge-shaped to broadly pointed at base, pointed at tip, finely toothed on margin, smooth above, grayish-woolly beneath, 1½ to 4 inches long. *Flowers* white or purplish-tinged; blooming March to June. *Fruits* red, about ¼ inch in diameter; ripening August to November and persisting.

RANGE. Nova Scotia to Ontario, Michigan, and Missouri; south to Florida and Texas.

PURPLE CHOKEBERRY *Aronia × prunifolia*

FIELD MARKS. A hybrid shrub with characteristics intermediate between the two preceding parents. *Branchlets* at first woolly but becoming smooth. *Leaves* downy beneath or becoming smooth. *Fruits* deep reddish purple to blackish purple.

RANGE. Newfoundland to Ontario; south to North Carolina, Ohio, and Illinois. (Not illustrated)

JUNEBERRIES (Amelanchier)

(Key Appendix E)

The Juneberries are leaf-losing shrubs or small trees with alternate, simple leaves. They have showy white 5-petalled flowers, usually in narrow clusters and blooming in early spring. The small apple-like fruits are usually soft, sweet, and juicy. They are edible and can be made into jelly, preserves, or pies; they are eaten by many wild birds and mammals, including deer, bears, raccoons, wild turkeys, and ruffed grouse. The Juneberries are also known as serviceberries, sarvisberries, and shadbushes. Although readily recognizable as a group, the species are often difficult to identify.

LOW JUNEBERRY *Amelanchier humilis*

FIELD MARKS. A shrub 1 to 3 feet high, with creeping underground stems and colony-forming; growing in dry rocky or sandy places. Leaves oval to oblong, rounded to heart-shaped at base, round or bluntly pointed at tip, margin with rather large teeth to slightly below the middle, smooth or nearly so on both surfaces, 1 to 2 inches long. *Fruits* black but with a whitish bloom, sweet and juicy; ripening July or August.

RANGE. Quebec and Ontario; south to Vermont, West Virginia, the region of the Great Lakes, and South Dakota.

ROUNDLEAF JUNEBERRY *Amelanchier sanguinea*

FIELD MARKS. A somewhat straggling shrub 3 to 8 feet high, with a solitary stem or a few stems in a clump; growing in dry, upland, rocky woods. *Leaves* roundish to oblong oval, blunt or rounded at tip, often heart-shaped at base, margin coarsely and sharply toothed nearly to the base, 1 to 2½ inches long. *Fruits* purplish black with a whitish bloom, sweet and juicy; ripening July or August.

RANGE. Quebec and Ontario; south to New York, western North Carolina, Michigan, Wisconsin, and Iowa.

RUNNING JUNEBERRY *Amelanchier stolonifera*

FIELD MARKS. A shrub 1 to 4 feet high, with creeping underground stems and thicket-forming; growing in sandy or rocky woodlands. *Leaves* oval or roundish, rounded at base, bluntly pointed to roundish at tip, margin finely and sharply toothed to a little below the middle, smooth above, sometimes slightly downy beneath, ¾ to 2 inches long. *Fruits* purplish black, sweet and juicy; ripening July or August.

RANGE. Newfoundland to Ontario; south to South Carolina, Michigan, and Minnesota.

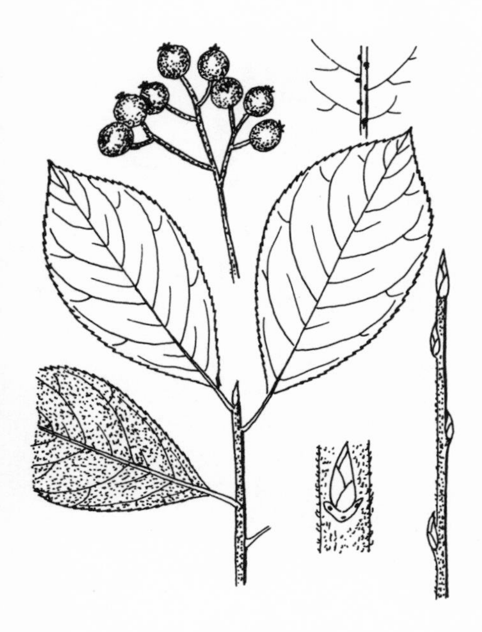

Red Chokeberry

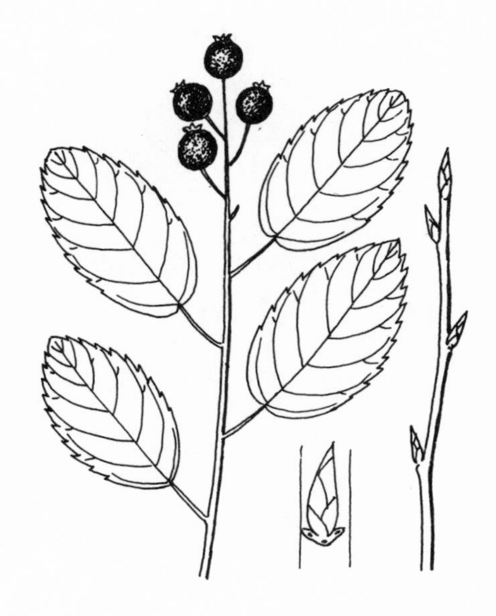

Low Juneberry

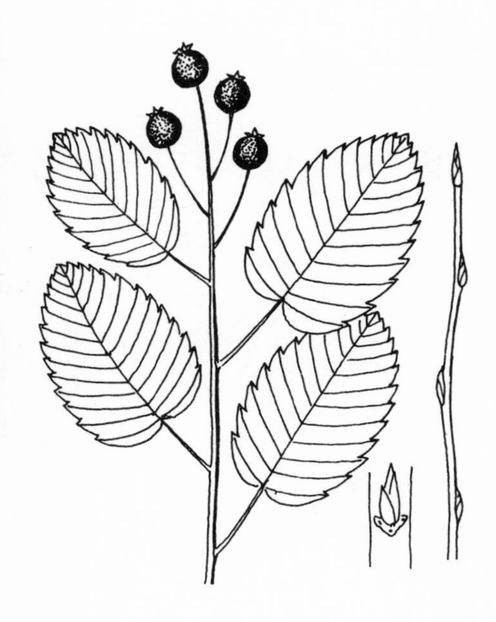

Roundleaf Juneberry

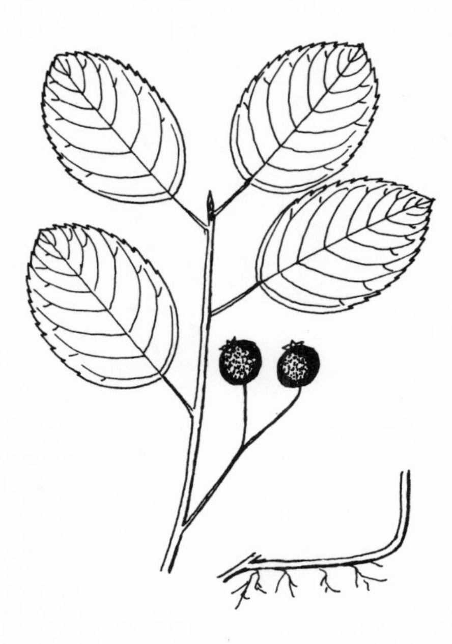

Running Juneberry

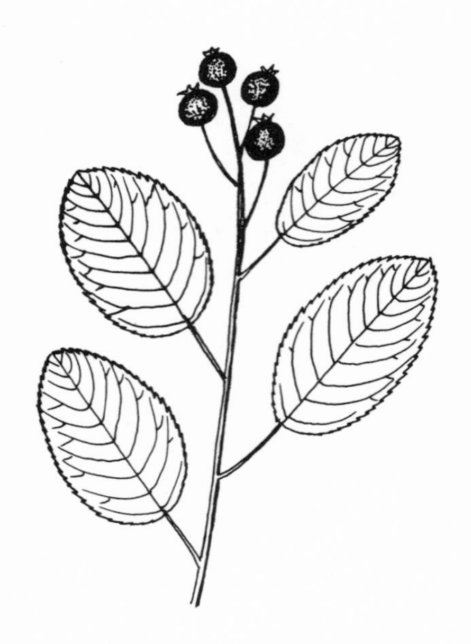

Oblong-leaf Juneberry

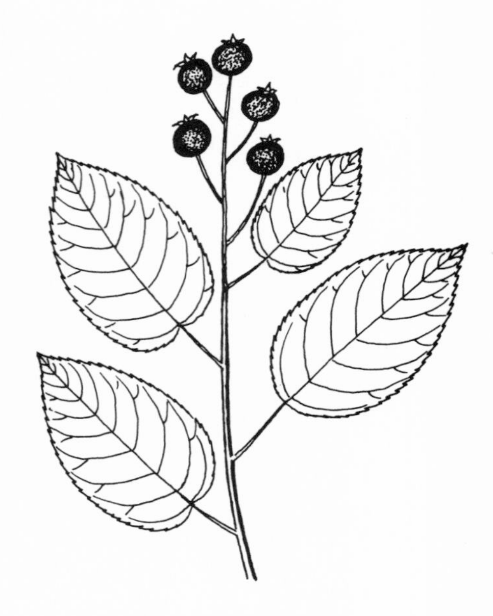

Swamp Juneberry

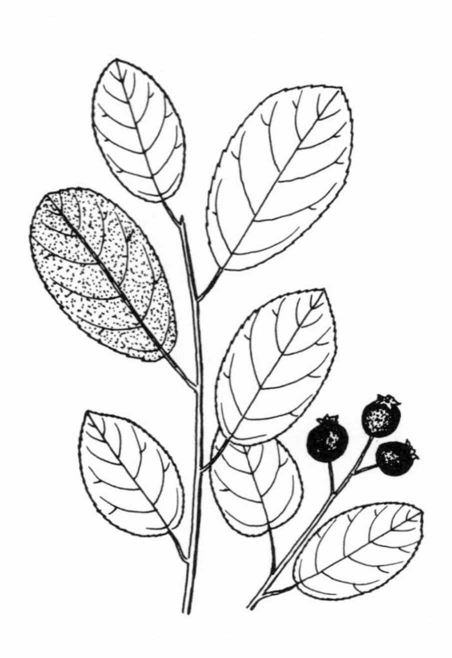

Coastal Juneberry

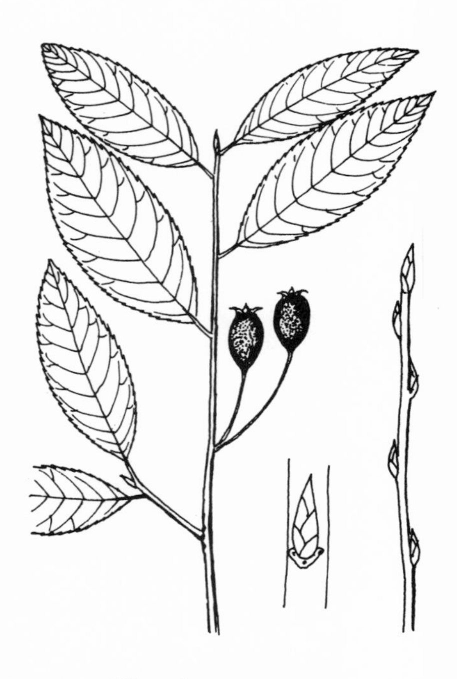

Oblong-fruited Juneberry

OBLONG-LEAF JUNEBERRY *Amelanchier canadensis*

FIELD MARKS. A shrub 5 to about 20 feet high with several slender stems in a clump; growing in low, moist places and swamps. *Leaves* oblong to oblong-elliptic, usually rounded at both ends, margin finely and sharply toothed nearly to the base, smooth above and smooth or nearly so beneath, 1 to 2½ inches long. *Fruits* blackish, about ¼ inch in diameter, sweet and juicy; ripening June and July.

RANGE. Maine to New York and southwestern Quebec, south to Georgia.

SWAMP JUNEBERRY *Amelanchier × intermedia*

FIELD MARKS. A hybrid shrub 5 to about 20 feet high with characteristics intermediate between *A. canadensis* and *A. arborea*, growing in swamps and bogs. *Leaves* oblong-elliptic or sometimes broadest above the middle, rounded to heart-shaped at base, pointed at tip, margin finely and sharply toothed nearly to the base, smooth or nearly so on both surfaces or sometimes slightly hairy beneath, 1 to 2½ inches long. *Fruits* dark purplish, sweet and juicy; ripening July and August.

RANGE. Newfoundland to northern Minnesota; south to western Virginia, Michigan, and southern Minnesota.

Also known as the Swamp Sugar-pear.

COASTAL JUNEBERRY *Amelanchier obovalis*

FIELD MARKS. A slender thicket-forming shrub 1 to 4 feet high, spreading by creeping underground stems; growing in low, moist, sandy woods and pinelands. *Leaves* oblong-elliptic, egg-shaped, or more often broadest above the middle, rounded at base, blunt or pointed at tip, margin finely and sharply toothed nearly to the base, thin in texture, dull pale green and smooth above, usually somewhat downy beneath, ¾ to 2 inches long. *Fruits* purplish black, sweet and juicy; ripening May or June.

RANGE. Chiefly coastal plain; southern New Jersey, eastern Maryland, and southeastern Pennsylvania; south to central Virginia and the interior of South Carolina.

OBLONG-FRUITED JUNEBERRY *Amelanchier bartramiana*

FIELD MARKS. A shrub 1½ to about 8 feet high, usually with several stems in a clump; growing in cold, wet, rocky woods or swamps and bogs. *Leaves* elliptic or narrowly so, more or less pointed at both ends, finely and sharply toothed on margin to below the middle, thin in texture, smooth on both surfaces, bright green above, paler and often somewhat whitened beneath, 1 to 2½ inches long; leafstalks short. *Flowers* 1 to 3 clustered together in the axils of the leaves; blooming May to August. *Fruits* oval or pear-shaped, about ½ inch long, dark purplish with a whitish bloom, sweet and juicy; ripening July to September.

RANGE. Labrador to western Ontario; south to northeastern Pennsylvania, West Virginia, northern Michigan, northern Wisconsin, and Minnesota.

AMERICAN MOUNTAIN-ASH *Sorbus americana*

FIELD MARKS. A tall shrub or small tree to 20 or 25 feet; growing in cold swamps or bogs and on rocky ridges at high elevations. *Branchlets* rather stout and with large gummy buds. *Leaves* alternate, compound, 6 to 10 inches long; leaflets 11 to 17, lance-shaped, rounded to pointed at base, long-pointed at tip, sharply toothed on margin, smooth on both surfaces, 2 to 3 inches long. *Flowers* small, white, 5-petalled, many in a broad flat-topped end cluster; blooming May or June. *Fruits* apple-like, bright orange red, shiny, about ¼ inch in diameter; ripening August to October.

RANGE. Newfoundland to Manitoba; south to New Jersey, northern Illinois, and along the mountains to northern Georgia.

The fruits are eaten by the ruffed grouse and many other wild birds; the twigs are browsed by deer.

AMERICAN CRAB APPLE *Malus coronaria*

FIELD MARKS. A bushy shrub or small tree to about 15 or 20 feet; often forming thickets along fence rows, roadsides, and abandoned fields. *Branchlets* smooth, often spurlike and ending in a sharp point. *Leaves* alternate, egg-shaped to oval (or lance-shaped in var. *dasycalyx*), rounded to pointed at base, pointed at tip, sharply and irregularly toothed on margin, thin-textured, smooth on both surfaces, 2 to 4 inches long. *Flowers* white to pink, 5-petalled, almost 1 inch across, very fragrant, 5 or 6 in clusters; blooming March to May. *Fruits* yellowish green, waxy or greasy, fragrant apples about 1 inch in diameter; ripening October or November.

RANGE. Western New York to Indiana and Missouri, south in the uplands to northern Georgia.

The fruits can be used to make jelly and vinegar.

SOUTHERN CRAB APPLE *Malus angustifolia*

FIELD MARKS. A shrub or small tree similar to the preceding. *Leaves* elliptic, pointed at base, rounded or blunt at tip, sparingly or bluntly toothed on margin, thick-textured, smooth on both surfaces, 1 to 2½ inches long. *Flowers* similar to those of preceding species. *Fruits* also similar but somewhat more flattened, 1 to 1½ inches in diameter.

RANGE. Chiefly coastal plain; Maryland to northern Florida, west to Louisiana and north in Mississippi Valley to southern Illinois.

Also known as Narrow-leaf Crab apple.

HAWTHORNS (Crataegus)

Hawthorns (also called haws or thorns) are shrubs or small trees with alternate, simple leaves; branches armed with large unbranched thorns or spines. They have showy 5-petalled white flowers about ½ inch across, which are usually in small, flat, end clusters. The *fruits* resemble small apples. They are tipped with the conspicuous remains of the calyx and have from 1 to 5 large, bony seeds.

As a group the hawthorns are quite distinctive, but the identity of the various species is extremely difficult even for trained botanists. Only a few species are presented here, either because they are usually easily recognized or because they are common. Most of us will be content if we recognize them as hawthorns.

MAY HAW *Crataegus aestivalis*

FIELD MARKS. A rather distinctive shrub or small tree; growing on wet soils, chiefly in coastal regions. *Leaves* broadest above the middle, wedge-shaped at base, rounded to pointed at tip, margin sharply toothed to or below the middle, sometimes 3-lobed, smooth or nearly so on both surfaces or somewhat rusty-hairy beneath, 1 to 2½ inches long; leafstalks short. *Fruits* red, dotted, about ½ inch in diameter, with 3 to 5 bony seeds; ripening in May.

RANGE. North Carolina south to Florida and west to Texas.

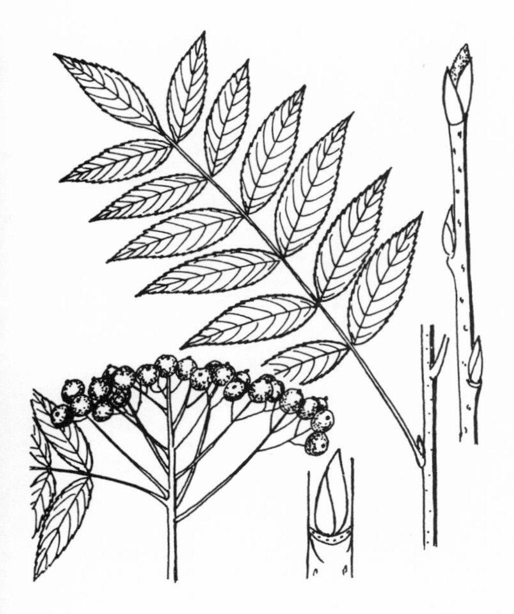

American Mountain-ash

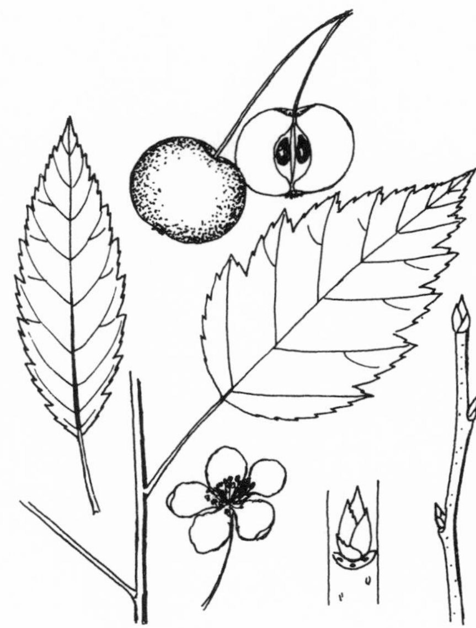

American Crab Apple

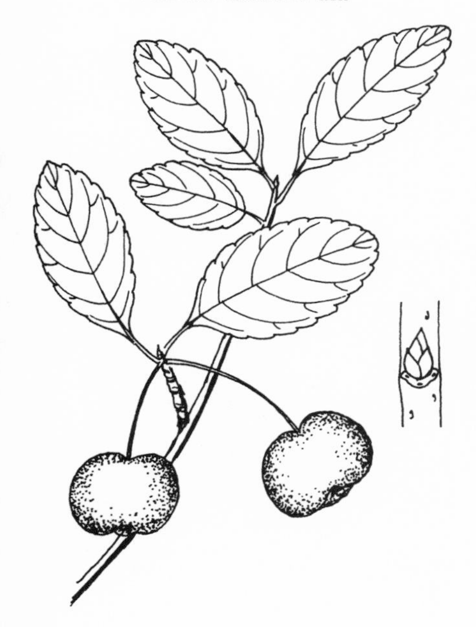

Southern Crab Apple

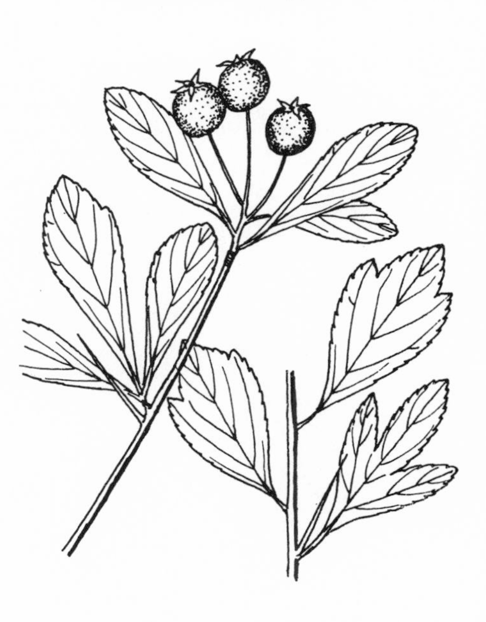

May Haw

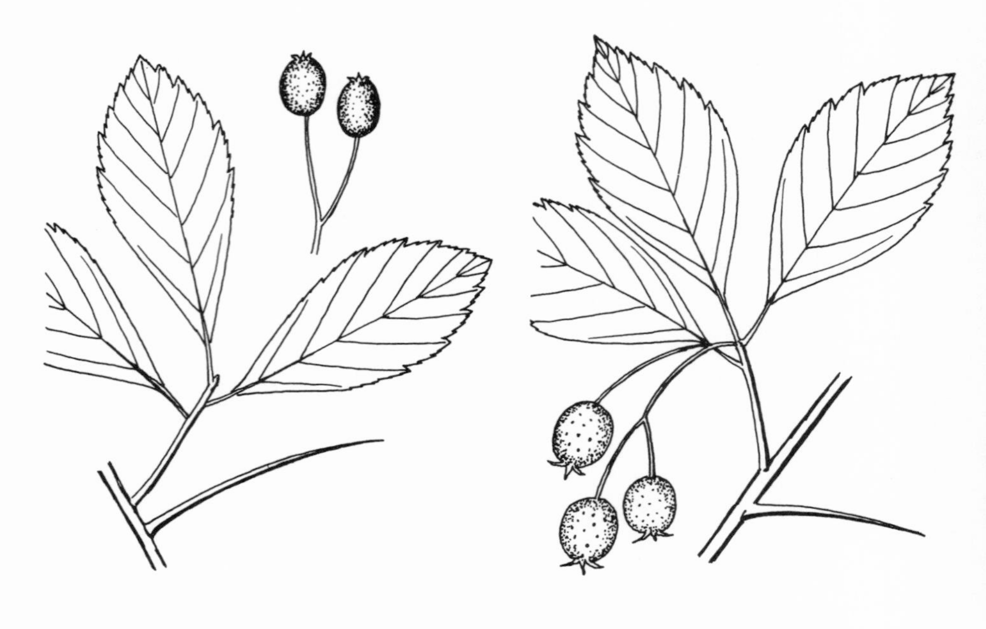

Cockspur Thorn

Dotted Hawthorn

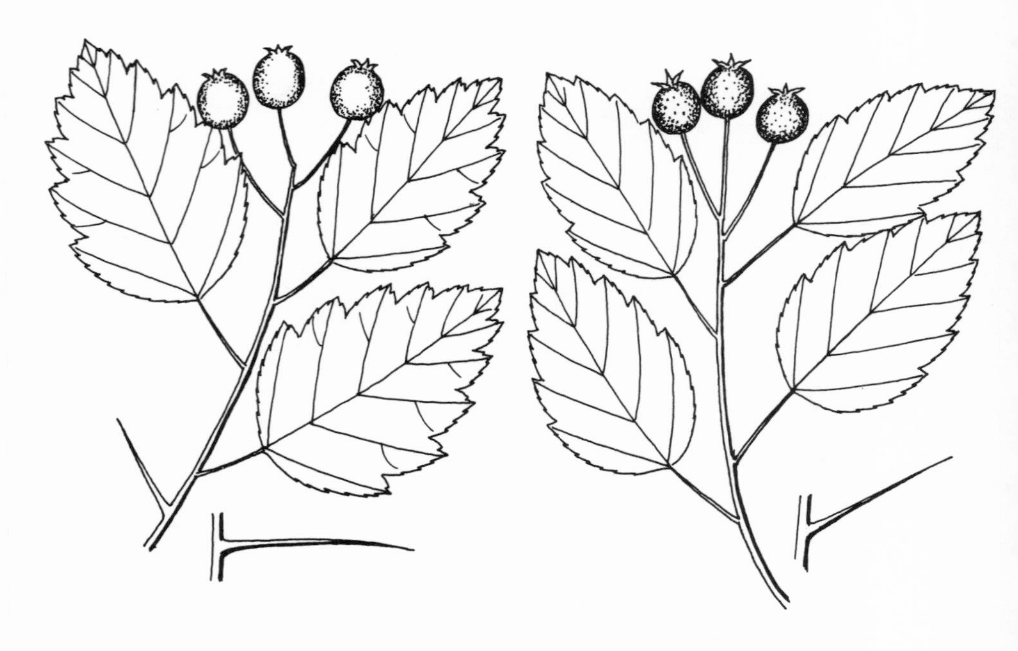

Variable Thorn

Waxy-fruited Thorn

COCKSPUR THORN *Crataegus crus-galli*

FIELD MARKS. A bushy shrub or small tree; growing in old pastures, borders of woods, thickets, etc. *Branches* with many slender, straight or slightly curved thorns 2 to 4 inches long. *Leaves* broadest above the middle, wedge-shaped at base, rounded to pointed at tip, margin sharply and irregularly toothed above the middle, thickish and somewhat leathery in texture, dark green and lustrous above, paler and smooth or nearly so beneath, 2 to 4 inches long. *Flowers* blooming April to June. *Fruits* greenish to dull red, hard, about ⅜ inch in diameter, usually 2-seeded; ripening September or October.
RANGE. Quebec to Ontario and Minnesota; south to Georgia, Kansas, and eastern Texas.

DOTTED HAWTHORN *Crataegus punctata*

FIELD MARKS. A rather flat-topped shrub or small tree; growing in stream bottoms and on adjoining slopes. *Branches* with rather scattered, stout, mostly straight thorns 2 to 3 inches long. *Leaves* broadest above the middle, wedge-shaped at base, pointed to rounded at tip, margin sharply and doubly toothed above the middle, firm in texture, dull grayish green and smooth but with sunken veins above, paler and often somewhat downy beneath, 2 to 3 inches long. *Flowers* blooming April to June. *Fruits* yellow to red, dark-dotted, ½ to ¾ inch in diameter, 3- to 5-seeded; ripening September or October.
RANGE. Quebec and Ontario south to North Carolina and Kentucky. Also known as the Large-fruited Thorn.

VARIABLE THORN *Crataegus macrosperma*

FIELD MARKS. A bushy shrub or small tree of stony woods and thickets. *Branches* with many slender thorns 1¼ to 2½ inches long. *Leaves* egg-shaped or broadly egg-shaped, rounded to broadly pointed or flattened at base, pointed at tip, margin with about 5 pairs of broadly triangular and sharply toothed lobes, thin in texture, dark yellowish green above, slightly paler beneath, smooth or nearly so on both surfaces, 1 to 3 inches long. *Flowers* blooming April or May. *Fruits* bright red, about ½ inch in diameter, 3- to 5-seeded; ripening August or September.
RANGE. Southeastern Canada and New England to northern Illinois and Wisconsin; south in the mountains to northern Georgia.

WAXY-FRUITED THORN *Crataegus pruniosa*

FIELD MARKS. A bushy shrub or small tree growing in stony woods and thickets. *Branches* with many slender thorns 1¼ to 2½ inches long. *Leaves* egg-shaped or broadly egg-shaped, rounded to pointed at base, pointed at tip, margin with 3 or 4 pairs of broad and shallow lobes and sharply and irregularly toothed, firm in texture, dark green or yellowish green above, paler beneath, smooth on both surfaces, 1 to 2½ inches long. *Flowers* blooming April or May. *Fruits* greenish to dull red or purplish, dark-dotted, waxy-coated, about ½ inch in diameter; ripening October or November.
RANGE. Newfoundland and southeastern Canada to Michigan and Wisconsin; south to North Carolina, Kentucky, and Arkansas.

PARSLEY HAWTHORN *Crataegus marshallii*

FIELD MARKS. A rather distinctive shrub or small tree; growing along streams, the borders of swamps, and on slopes. *Branches* unarmed or with occasional short thorns. *Leaves* broadly egg-shaped to roundish, deeply cleft between the 5 to 7 lobes, which are sharply and irregularly toothed, bright green and lustrous above, paler beneath, smooth or nearly so on both surfaces, 1½ to 2½ inches long. *Flowers* blooming April or May. *Fruits* bright red, about ⅓ inch long, 1- to 3-seeded; ripening about October.

RANGE. Virginia, Missouri, and Oklahoma south to Florida and Texas.

DWARF THORN *Crataegus uniflora*

FIELD MARKS. An irregularly branched and sprawling shrub 2 to 5 feet high; growing on sandy soils or rocky banks. *Branches* long, slender, often quite flexible, with thorns ½ to 1 inch long. *Leaves* elliptic or broadest above the middle, wedge-shaped at base, rounded or bluntly pointed at tip, margin bluntly toothed and sometimes obscurely lobed above the middle, thick-textured, dark green and lustrous but sometimes roughish-hairy above and with sunken veins, paler and more or less downy beneath, ½ to 1½ inches long; leafstalks short. *Flowers* usually solitary, blooming April or May. *Fruits* greenish yellow to dull red, 3- to 5-seeded; ripening in October.

RANGE. Southeastern New York and southeastern Pennsylvania south to Florida, west to Texas, and north in Mississippi Valley to southern Missouri.

SHRUBBY CINQUEFOIL *Pentaphylloides floribunda*

FIELD MARKS. A leaf-losing shrub 1 to 3 feet high, with erect or ascending branches and shreddy bark; growing in cold, moist, rocky places or in northern bogs. *Leaves* alternate, compound, with prominent pointed stipules at the bases of the leafstalks; leaflets usually 5 (rarely 3 or 7), stalkless, lance-shaped, pointed at both ends, margin untoothed but often slightly rolled, silvery-silky on both surfaces, ½ to about 1 inch long. *Flowers* bright yellow, 5-petalled, about ¾ inch across; blooming June to September. *Fruits* greenish becoming brown, small, dry, densely hairy, borne in heads.

RANGE. Newfoundland and southern Labrador to Alaska; south to northern New Jersey, northeastern Pennsylvania, the region of the Great Lakes, northern Iowa, South Dakota, New Mexico, and California.

Found in the northern portions of Europe and Asia, as well as North America, and cultivated in many varieties as an ornamental shrub.

THREE-TOOTHED CINQUEFOIL *Sibbaldiopsis tridentata*

FIELD MARKS. A small evergreen tufted and creeping plant 1½ to (rarely) 10 inches high, woody at the base; growing on dry, open, rocky or peaty soils, southward only at high elevations in the mountains. *Leaves* alternate, compound, long-stalked; leaflets 3, broadest at the 3-toothed tip and tapering to the base, stalkless, leathery-textured, dark green and lustrous above, paler beneath, smooth on both surfaces or somewhat hairy beneath, ½ to 2 inches long. *Flowers* white, 5-petalled, about ⅜ inch across, 1 to 6 in an end cluster; blooming late May to September. *Fruits* greenish becoming brown, small, dry, densely hairy, borne in heads.

RANGE. Labrador to Manitoba; south to New England, New York, southern Ontario, Michigan, Wisconsin, northeastern Iowa, North Dakota, and along the Appalachians to northern Georgia.

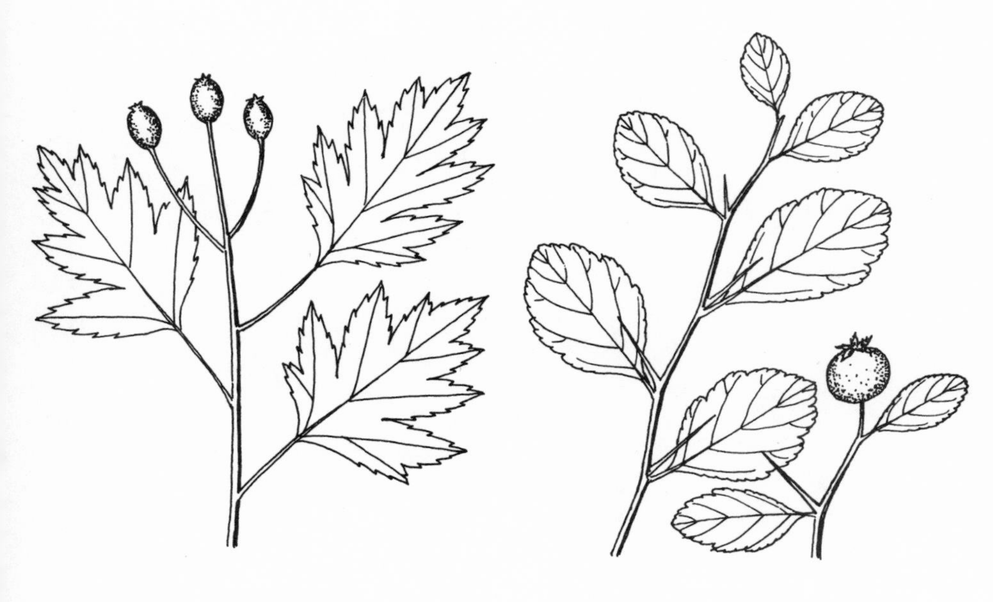

Parsley Hawthorn

Dwarf Thorn

Shrubby Cinquefoil

Three-toothed Cinquefoil

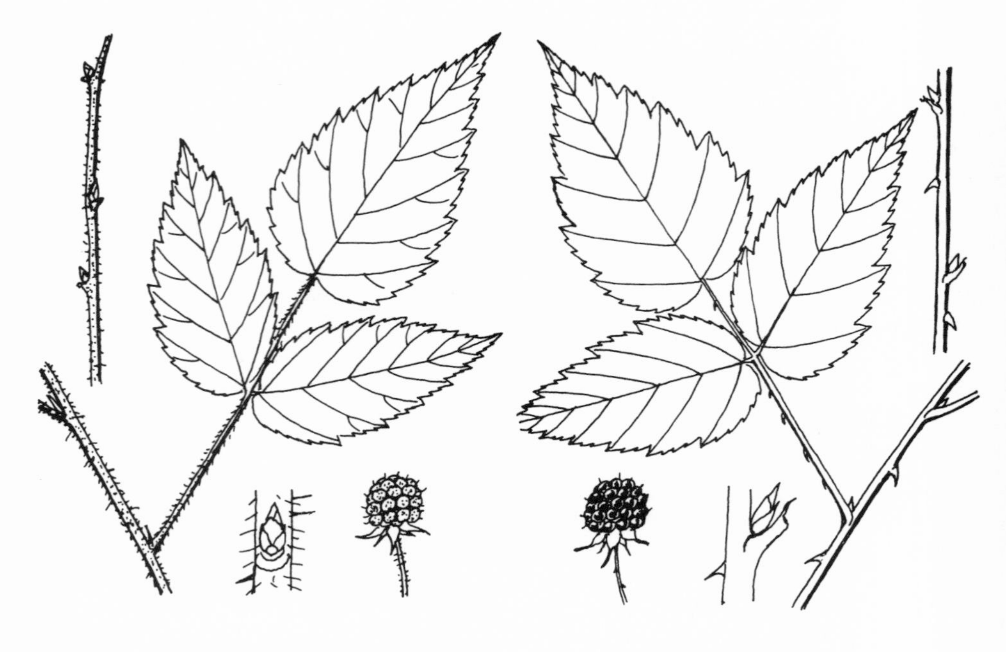

Red Raspberry

Black Raspberry

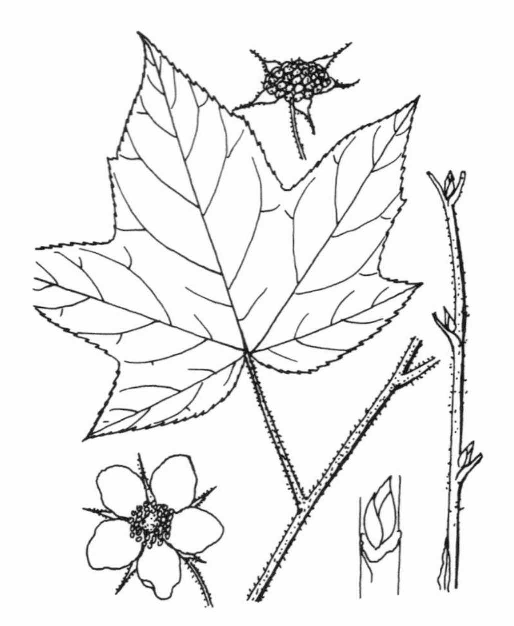

Purple-flowering Raspberry

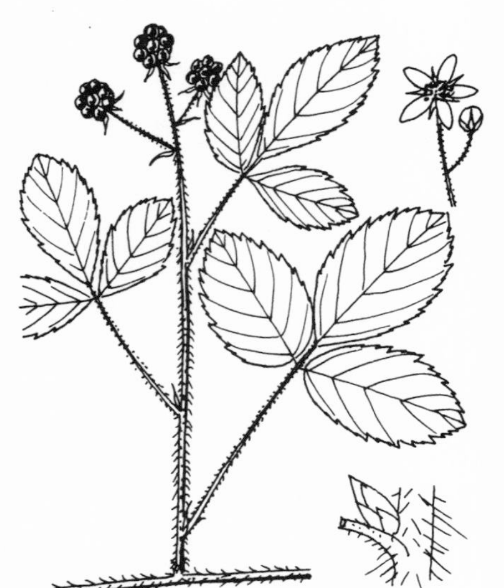

Swamp Dewberry

BLACKBERRIES AND RASPBERRIES (Rubus)

These are soft-wooded, short-lived, leaf-losing or (rarely) half-evergreen plants which are usually bristly or prickly. They have simple stems called "canes," which produce flowers and fruits the second year and then die. The leaves are alternate, long-stalked, and usually compound. The flowers are perfect, 5-petalled, and usually showy. The fruits are not true berries but compact clusters of small 1-seeded drupelets. Those of the raspberries separate readily from the receptacle and appear hollow inside when picked. Those of blackberries are firmly attached to the receptacle which is picked along with the cluster of drupelets.

As a group these plants are commonly called "briers" or "brambles." They are common and well known, but the various species are very difficult to identify. Only a few of the more common or distinctive ones are included here. The plants provide food and cover for wildlife; the fruits are often gathered to be eaten raw or made into pies, jelly, preserves, or wine.

RED RASPBERRY *Rubus idaeus* var. *strigosus*

FIELD MARKS. A shrub with arching, round canes 2 to 6 feet high, sometimes slightly whitened when young and usually bristly but with few or no prickles; growing on rocky slopes and in clearings. *Leaves* with 3 to 5 leaflets which are egg-shaped, rounded to somewhat heart-shaped at base, pointed at tip, double-toothed on margin, whitened and downy beneath, 2 to 3 inches long; leafstalks bristly. *Flowers* white, petals small; blooming May to July. *Fruits* red, in half-round clusters; ripening July to September.

RANGE. Newfoundland to British Columbia; south to western North Carolina, Ohio, Indiana, Nebraska, and Wyoming.

BLACK RASPBERRY *Rubus occidentalis*

FIELD MARKS. Different from the preceding in having strongly whitened canes with scattered, hooked prickles and leafstalks with smaller but similar prickles. *Fruits* purplish black, in half-round clusters; ripening June or July.

RANGE. New Brunswick to Minnesota, south to Georgia and Colorado.

PURPLE-FLOWERING RASPBERRY *Rubus odoratus*

FIELD MARKS. A straggling, unarmed shrub 2 to 5 feet high; growing in moist rocky woods and ravines. *Branchlets,* leafstalks, and flower stalks are covered with sticky, reddish, bristly hairs; the bark peels freely off the canes. *Leaves* maple-like, 3- to 5-lobed and sharply toothed on margin, bright green and more or less downy on both surfaces, 4 to 7 inches wide. *Flowers* 1½ to 2 inches across, roselike, rose purple; blooming June to August. *Fruits* dull red, rather insipid, in somewhat flattened dome-shaped clusters ½ to ¾ inch across; ripening July to September.

RANGE. Nova Scotia and southern Ontario to southern Michigan; south to New York, northern Georgia, and Tennessee.

SWAMP DEWBERRY *Rubus hispidus*

FIELD MARKS. A semi-evergreen plant with slender, trailing, bristly-hairy stems; growing in cool moist woods and bogs. *Leaves* usually with 3 leaflets which are elliptic to roundish or broadest above the middle, rounded to pointed at both ends, sharply toothed on margin, dark green and lustrous above, paler and often slightly downy beneath, ¾ to 2 inches long. *Flowers* small, white; blooming June to September. *Fruits* small, purplish to black, few in a cluster; ripening August to October.

RANGE. Nova Scotia to Wisconsin; south to Maryland, the region of the Great Lakes, and in the mountains to northern Georgia.

NORTHERN DEWBERRY *Rubus flagellaris*

FIELD MARKS. A leaf-losing plant with long, trailing stems which are armed with scattered, broad-based, often slightly curved prickles; growing in dry open places, often in poor soils. *Leaves* with 3 or sometimes 5 leaflets which are oval to egg-shaped or roundish, rounded to broadly pointed at base and tip, sharply and doubly toothed on margin, thin in texture, smooth or nearly so above, slightly paler and sometimes slightly downy beneath, 1½ to 4 inches long. *Flowers* white, often 1 inch across; blooming April to June. *Fruits* black, in roundish or somewhat elongate clusters; ripening May to August.

RANGE. Maine to Minnesota, south to northern Georgia and Missouri.

SOUTHERN DEWBERRY *Rubus trivialis*

FIELD MARKS. A somewhat evergreen plant with long trailing stems which are bristly and armed with rather small prickles; growing in open places. *Leaves* with 3 or sometimes 5 leaflets which are elliptic or lance-shaped, pointed to rounded at base, pointed or blunt at tip, sharply and doubly toothed on margin, rather leathery in texture, lustrous above, paler beneath, smooth on both surfaces, 1 to 4 inches long. *Flowers* usually solitary, white or pinkish, often 1¼ inches across; blooming March to May. *Fruits* black, in roundish or elongate clusters; ripening late April to June.

RANGE. Coastal plain and lower piedmont; Maryland south to Florida, west to Texas, north to Missouri and Oklahoma.

HIGHBUSH BLACKBERRY *Rubus allegheniensis*

FIELD MARKS. A shrub with erect or arching, often stout, angled, purplish-red canes 3 to 6 feet high; well armed with stout, straight, broad-based prickles; growing in woods borders, clearings, and wayside thickets. *Leaves* with prickly leafstalks and 3 or 5 leaflets which are egg-shaped, rounded to somewhat heart-shaped at base, pointed at tip, sharply and doubly toothed on margin, smooth or nearly so above, densely downy beneath, 2 to 4½ inches long. *Flowers* white, about 1 inch across, clustered; blooming April to June. The younger branchlets and the flower stalks are quite densely covered with gland-tipped hairs. *Fruits* black, in roundish to thimble-shaped clusters; ripening July or August.

RANGE. New Brunswick to Minnesota; south to Maryland, Missouri, and in the Appalachians to nothern Georgia.

TALL BLACKBERRY *Rubus argutus*

FIELD MARKS. Similar to the Highbush Blackberry, from which it can best be distinguished by the absence of gland-tipped hairs on the young growth and flower stalks.

RANGE. Massachusetts to southern Illinois, south to Florida and Mississippi. (Not illustrated)

MOUNTAIN BLACKBERRY *Rubus canadensis*

FIELD MARKS. Similar to Highbush Blackberry but stems unarmed or with only an occasional weak prickle. *Leaves* smooth and green on both surfaces. Young growth and flower stalks smooth.

RANGE. Newfoundland to Ontario and Minnesota, south along the higher Appalachians to northern Georgia.

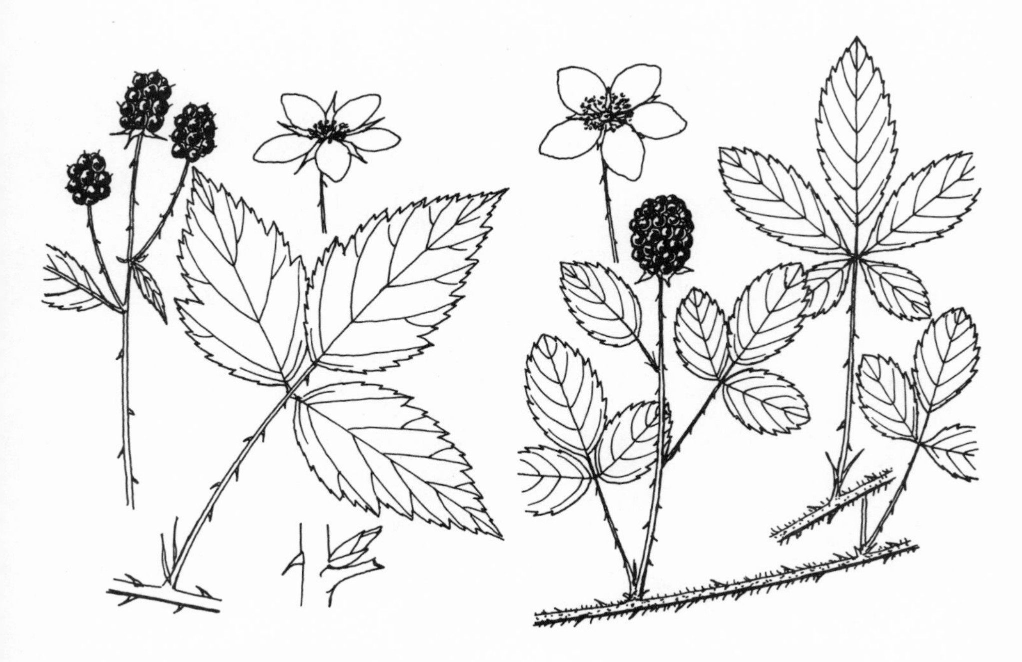

Northern Dewberry

Southern Dewberry

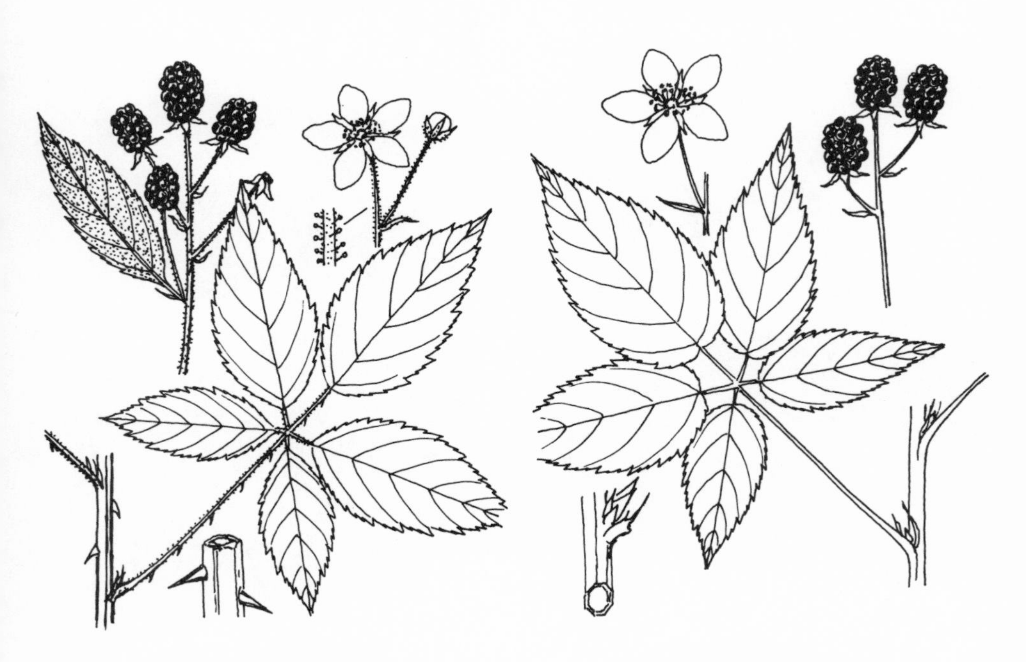

Highbush Blackberry

Mountain Blackberry

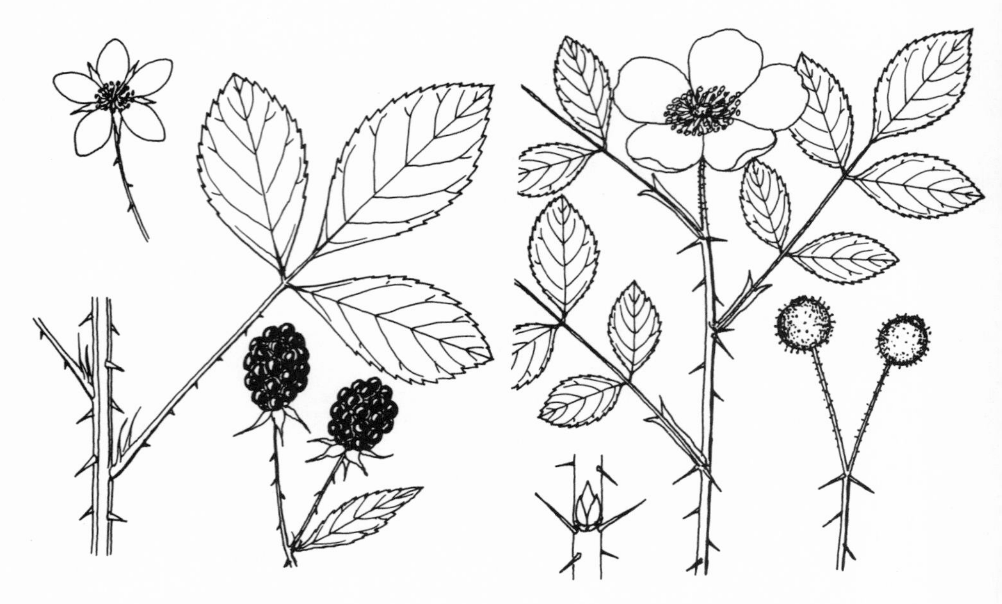

Sand Blackberry

Pasture Rose

Swamp Rose

Shining Rose

SAND BLACKBERRY *Rubus cuneifolius*

FIELD MARKS. A shrub with erect or slightly arching, angled, and densely prickly canes 1 to 3 feet high. *Leaves* with usually 3 (sometimes 5) leaflets which are broadest above the middle, tapered to the base, rounded or broadly pointed at tip, sharply toothed on margin to or slightly below the middle, thickish, smooth or nearly so above, densely white-woolly beneath, ¾ to 2½ inches long. *Flowers* white or (rarely) pink, about 1 inch across, clustered; blooming April to June. *Fruits* black, in short cylindrical clusters ½ to 1 inch long; ripening June or July.

RANGE. Chiefly coastal plain; Connecticut to southeastern Pennsylvania, south to Florida.

ROSES (Rosa)

(Key Appendix F)

Roses are usually prickly leaf-losing shrubs with alternate and pinnately compound leaves. A pair of stipules is united with the base of the leafstalk, often for half or more of the length of the leafstalk. The showy flowers have 5 sepals, 5 petals, numerous stamens, and several pistils within the calyx cup. The fruits are called "hips" and consist of the cuplike or urn-shaped receptacle on which the sepals sometimes persist; but within the fleshy receptacle are the small, hard, dry true fruits which are achenes.

The native roses are not always easy to identify, for numerous hybrids occur. Thus in many cases the amateur is justified in simply calling them "wild roses." Besides the native species described here, there are several introduced species which are now widely naturalized.

PASTURE ROSE *Rosa carolina*

FIELD MARKS. An erect shrub ½ to 2½ feet high, the stems armed with needle-like prickles; growing in dry open woods, thickets, and pastures. *Leaves* with usually 5 leaflets which are elliptic or egg-shaped, pointed to rounded at base, pointed at tip, sharply toothed on margin, usually dull above, paler and sometimes slightly downy beneath, ¾ to 1¼ inches long; the stipules narrow, flat, and untoothed. *Flowers* usually solitary, pink, about 2 inches broad; blooming May to July. *Fruits* roundish, red, about ½ inch in diameter, and with gland-tipped hairs.

RANGE. Nova Scotia to Minnesota and Nebraska, south to Florida and Texas. Also called Low or Carolina Rose.

SWAMP ROSE *Rosa palustris*

FIELD MARKS. An erect shrub 2½ to 6 feet high, the stems armed with broad-based and commonly hooked prickles; growing in swamps and other wet places. *Leaves* with 5 to 9 (usually 7) leaflets which are oval or elliptic, usually pointed at both ends, sharply toothed on margin, smooth above, paler and smooth or slightly downy beneath, ¾ to 2 inches long; the stipules often broadened upward or with slightly rolled margins. *Flowers* solitary or few in a cluster, pink, 2 to 2½ inches broad; blooming May to August. *Fruits* roundish, red, about ½ inch in diameter, smooth to rather densely glandular-bristly.

RANGE. Newfoundland and Nova Scotia to Ontario and Minnesota; south to Florida, Tennessee, and Arkansas.

SHINING ROSE *Rosa nitida*

FIELD MARKS. An erect shrub 8 inches to 2 feet high, the stems densely bristly and prickly; growing in bogs and wet thickets. *Leaves* usually with 7 or 9 leaflets which are narrowly elliptic or oblong, pointed at both ends, finely and sharply toothed on margin, lustrous above, paler and usually smooth beneath, ½ to (rarely) 2 inches long; the stipules broadened upward. *Flowers* solitary or few in a cluster, pink, 1½ to 2¼ inches broad; blooming June to September. *Fruits* roundish, red, bristly-hairy, about ⅜ inch in diameter, the erect sepals soon shed.

RANGE. Newfoundland to Quebec, south to Massachusetts and Connecticut.

SMOOTH ROSE *Rosa blanda*

FIELD MARKS. An erect shrub 2 to 5 feet high, unarmed or with a few weak or bristly prickles; growing in moist rocky places, along streams, or on lakeshores. *Leaves* with 5 or 7 leaflets which are elliptic or broadest above the middle, blunt or pointed at tip, pointed at base, sharply toothed on margin, smooth and dull above, sometimes downy beneath, ½ to 1½ inches long; stipules somewhat broadened upward. *Flowers* solitary or few, pink or white, 2 to 2½ inches broad; blooming June to August. *Fruits* roundish, red, smooth, about ½ inch in diameter, with persistent sepals.
RANGE. Newfoundland to Manitoba; south to New England, Pennsylvania, the region of the Great Lakes, Missouri, and Nebraska.

PRICKLY WILD ROSE *Rosa acicularis*

FIELD MARKS. Shrub 1 to 3 feet high, the stems densely covered with needle-like prickles; growing in thickets and on rocky slopes. *Leaves* with 3 to 7 leaflets which are blunt at tip, roundish at base, often hairy-resinous, 1 to 2 inches long; leafstalks glandular-hairy. *Flowers* solitary, pink, 1 to 2 inches broad; blooming June or July. *Fruits* roundish or egg-shaped, ⅝ to 1 inch long, usually smooth, sepals persistent.
RANGE. New England and Quebec to Yukon and British Columbia; south to Michigan, Wisconsin, South Dakota, and Colorado. (Not illustrated)

PRAIRIE ROSE *Rosa setigera*

FIELD MARKS. Stems 5 to 15 feet long, trailing, arching, or climbing; armed with scattered, broad-based, often hooked prickles; growing in open woods, clearings, or along streams. *Leaves* with 3 (rarely 5) egg-shaped leaflets which are pointed at tip, rounded at base, sharply toothed, usually lustrous above, often downy beneath, 1 to 3 inches long. *Flowers* in small clusters, usually pink, 2 to 3 inches broad; blooming May to August. *Fruits* roundish, red, usually bristly-hairy, about ⅜ inch in diameter.
RANGE. New York to Ontario, Missouri, and Nebraska; south to Florida and Texas.

PLUMS AND CHERRIES (Prunus)

These are shrubs or trees with alternate, simple leaves; often with glands at the summits of the leakstalks or with narrow stipules at the base. Bark of the younger stems and branches usually marked with horizontally elongated lenticels. The flowers are 5-petalled, often showy; the fruits are drupes with a bony pit enclosing the seed.

BEACH PLUM *Prunus maritima*

FIELD MARKS. A straggling, unarmed, leaf-losing shrub 1 to 6 feet high; growing on dunes and sands along the coast. *Branchlets* velvety, gradually becoming smooth. *Leaves* elliptic to egg-shaped or broadest above the middle, rounded or broadly pointed at base, pointed or blunt at tip, finely and sharply toothed on margin, smooth or nearly so above, downy beneath, 1¼ to 2¼ inches long. *Flowers* white; blooming April to June. *Fruits* roundish, purple with a white bloom, ½ to 1 inch in diameter; ripening August to October.
RANGE. Coastal plain; Maine south to Delaware and southeastern Pennsylvania.

ALLEGHENY PLUM *Prunus alleghaniensis*

FIELD MARKS. A straggling, leaf-losing shrub or small tree 3 to 15 feet high, sometimes with spiny-tipped branches; growing on mountain slopes or ridges and often thicket-forming. *Leaves* lance-shaped to egg-shaped, pointed at base, often long-pointed at tip, finely and sharply toothed on margin, smooth or nearly so above, paler and often somewhat downy beneath, 2 to 3½ inches long. *Flowers* white, ½ inch across; blooming April or May. *Fruits* roundish or oval, dark purple with whitish bloom, ½ inch in diameter; ripening August or September.
RANGE. Connecticut and Pennsylvania south to western Virginia. Also called Porter Plum or Sloe.

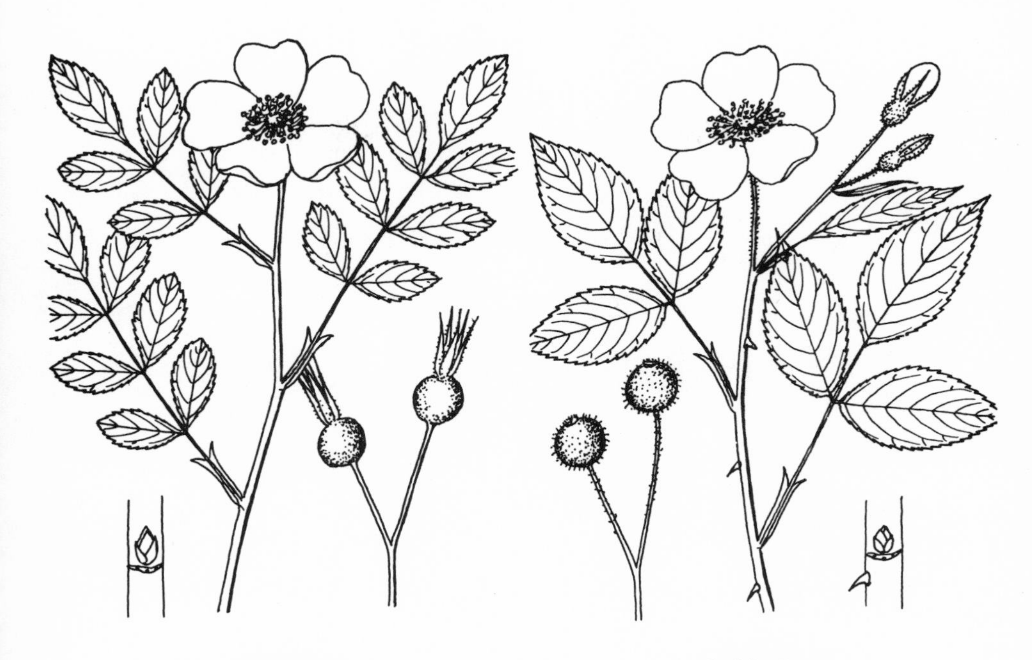

Smooth Rose **Prairie Rose**

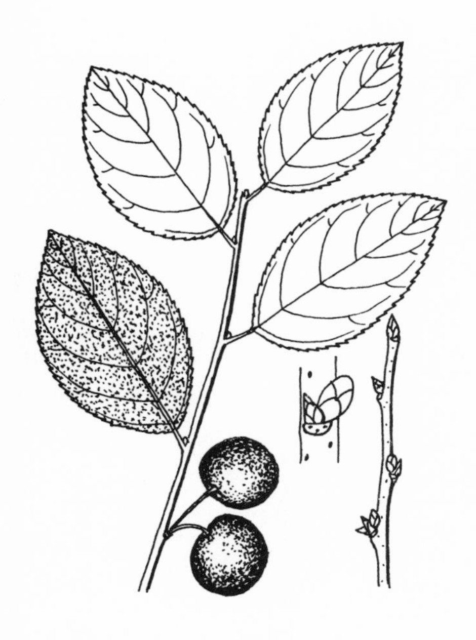

Beach Plum

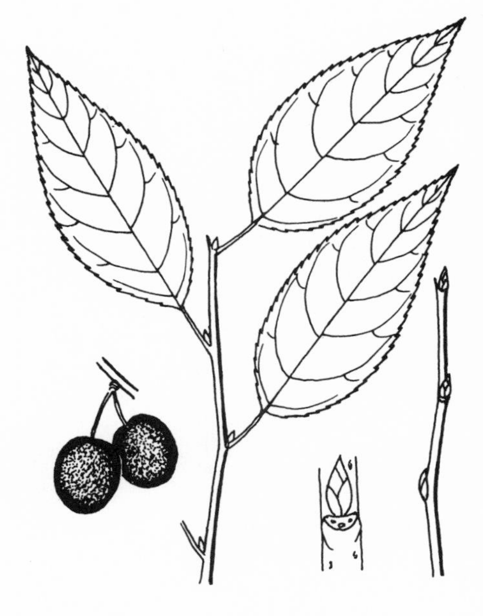

Allegheny Plum

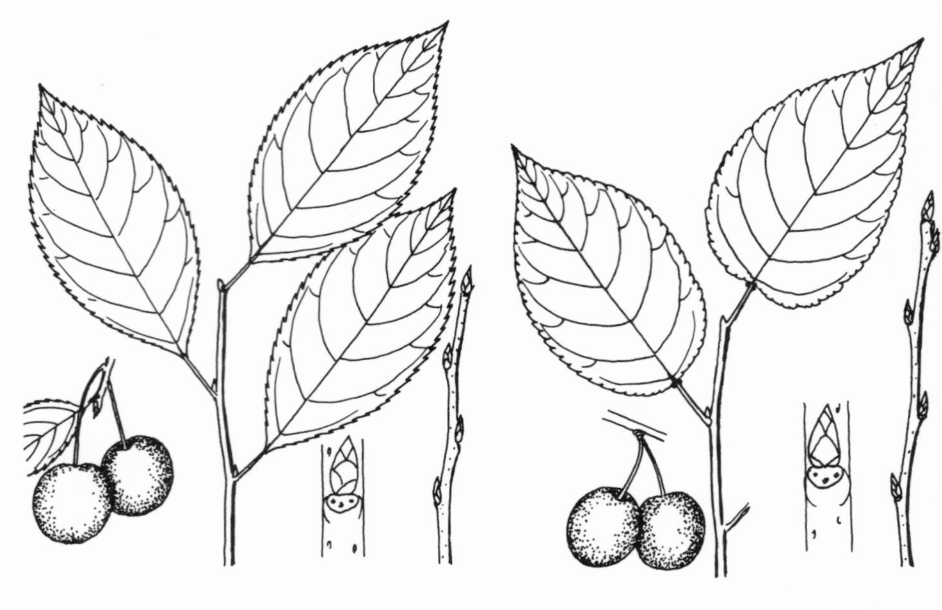

American Plum

Canada Plum

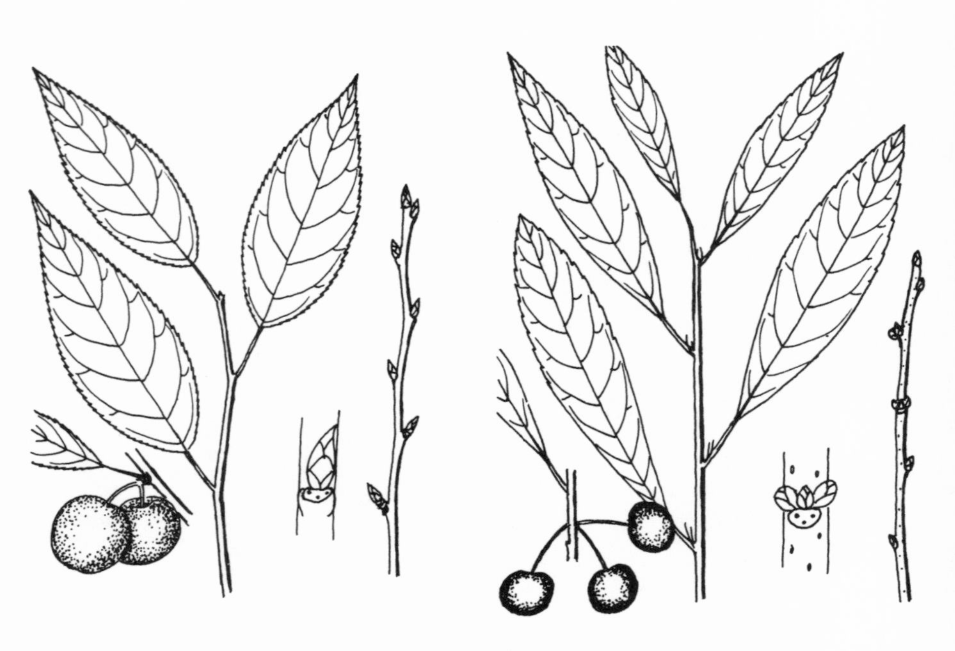

Chickasaw Plum

Sand Cherry

GRAVES PLUM *Prunus gravesii = Prunus maritima* var. *gravesii*

FIELD MARKS. A shrub similar to the Beach Plum but smaller, with roundish leaves, blunt or abruptly short-pointed at tip, ¾ to 1½ inches long. *Flowers* often 1 inch across. *Fruits* almost black, about ½ inch in diameter.

RANGE. Southeastern Connecticut. (Not illustrated)

AMERICAN PLUM *Prunus americana*

FIELD MARKS. A leaf-losing shrub or small tree with spine-tipped branches; often forming thickets in woods borders, or along fence rows and streams. *Leaves* oval or broadest above the middle, mostly broadly pointed at base and abruptly pointed at tip, finely and often doubly toothed on margin, smooth or nearly so on both surfaces, 2 to 4 inches long; leafstalks usually without glands. *Flowers* white, ¾ inch across, 2 to 5 in a cluster; blooming April to June. *Fruits* roundish, yellow or red, ¾ to 1 inch in diameter; ripening August to October.

RANGE. Massachusetts to Manitoba; south to Florida, Louisiana, and New Mexico.

CANADA PLUM *Prunus nigra*

FIELD MARKS. A shrub or small tree similar to the preceding. Leaves usually egg-shaped, rounded or somewhat heart-shaped at base, margin with more bluntish teeth; leafstalks with glands at summit. The flowers and fruits are similar to those of the American Plum.

RANGE. New Brunswick to southern Manitoba; south to New England, New York, Ohio, northern Illinois, and Iowa.

CHICKASAW PLUM *Prunus angustifolia*

FIELD MARKS. A leaf-losing shrub or small tree with spiny-tipped branches; often forming thickets along fence rows, abandoned fields, and woods borders. *Leaves* lance-shaped, pointed at base, long-pointed at tip, finely toothed on margin, smooth, 1 to 2½ inches long; leafstalks with glands at summit. *Flowers* white, about ⅓ inch across, 2 to 4 in a cluster; blooming March or April. *Fruits* roundish, red or yellow with a whitish bloom, ¾ to 1 inch in diameter; ripening June or July and often quite sweet.

RANGE. New Jersey to southern Illinois and Nebraska; south to Florida and Texas.

FLATWOODS PLUM *Prunus umbellata*

FIELD MARKS. A leaf-losing shrub or small tree with spiny-tipped branches; growing in dry sandy woods or along stream banks. *Leaves* elliptic, pointed at tip, very finely toothed on margin, smooth or sometimes slightly downy beneath, 1 to 2½ inches long. *Flowers* white, about ⅜ inch across, 2 to 4 in a cluster; blooming late February to April. *Fruits* roundish, reddish purple, ½ inch or less in diameter, sour or bitter; ripening June or July.

RANGE. Southeastern North Carolina south to Florida, west to Texas, north in Mississippi Valley to Arkansas.

Also called Hog Plum or Black Sloe. (Not illustrated)

SAND CHERRY *Prunus pumila*

FIELD MARKS. An erect leaf-losing shrub 1 to about 6 feet high; growing on sandy soils, dunes, and rocky shores. *Leaves* narrowly elliptic or broadest above the middle, wedge-shaped at base, pointed at tip, margin toothed with low and bluntish teeth above the middle, somewhat leathery in texture, dark green and lustrous above, paler and smooth beneath, 1½ to 3 inches long. *Flowers* white, about ½ inch across, few in a cluster; blooming May or June. *Fruits* roundish, purplish black, about ⅜ inch in diameter; ripening July to September.

RANGE. Ontario south to St. Lawrence Basin and the Great Lakes region from New York to Minnesota.

APPALACHIAN CHERRY *Prunus pumila* var. *susquehanae*

FIELD MARKS. Differs from the preceding in having more oblong leaves, sometimes broadest above the middle, wedge-shaped at base, blunt or broadly pointed at tip, margin toothed with low and bluntish teeth above the middle, thin but firm in texture, light green above, paler beneath, smooth on both surfaces, 1¼ to 2½ inches long. Grows in sandy or rocky woods, thickets, and clearings.

RANGE. Southwestern Maine and Quebec to southeastern Minnesota; south to Long Island, Virginia, central and western North Carolina, Ohio, Illinois, and Minnesota.

PROSTRATE SAND CHERRY *Prunus pumila* var. *depressa*

FIELD MARKS. Differs from the preceding two expressions by being a depressed or prostrate, spreading, leaf-losing shrub with more or less upright branchlets 8 inches to 2 feet high; growing on sandy or gravelly shores or along streams. *Leaves* narrowly elliptic or rather narrow but broader toward the tip, wedge-shaped at base, pointed or blunt at tip, margin toothed with low and bluntish teeth above the middle, light green above, pale or somewhat whitened beneath, smooth on both surfaces, 1½ to 4 inches long. *Flowers* white, about ½ inch across, 2 to 4 in a cluster; blooming May to July. *Fruits* roundish, reddish purple or blackish purple, about ⅜ inch in diameter; ripening July to September.

RANGE. Quebec to Ontario; south to western Massachusetts, New York, Pennsylvania, and Wisconsin.

CHOKECHERRY *Prunus virginiana*

FIELD MARKS. A large leaf-losing shrub or small tree; growing along woods borders, fence rows, and wayside thickets. *Branchlets* have a very disagreeable odor when bruised or broken. *Leaves* oval or broadest above the middle, rounded to broadly pointed at base, abruptly pointed at tip, finely and sharply toothed on margin, dull green above, paler beneath, smooth on both surfaces, 2 to 4 inches long; leafstalks with glands at summit. *Flowers* white, about ⅜ inch across, strong-scented, numerous, and in a dense and elongate cluster; blooming April to June. *Fruits* roundish, dark red to purplish black, about ⅜ inch in diameter, very astringent; ripening July to September.

RANGE. Newfoundland to British Columbia; south to Maryland, northern Georgia, eastern Kentucky, Illinois, Kansas, New Mexico and California.

COCOA-PLUM FAMILY (Chrysobalanaceae)

DEER-PLUM *Licania michauxii*

FIELD MARKS. An evergreen shrub ½ to 2 feet high, with creeping underground stems; growing in dry sandy woods or on sandhills. *Branchlets* smooth, dark purplish brown. *Leaves* alternate, narrowly elliptic or broadest above the middle, wedge-shaped at base, blunt to indented or sometimes with a minute bristle at tip, margin indistinctly toothed, veiny and lustrous above, slightly paler beneath, smooth, 1¼ to 4½ inches long. *Flowers* small, yellowish white to greenish white, in end clusters; blooming May or June. *Fruits* ivory white tinged with red or purple, oval-shaped, about 1 inch long, not edible; ripening September or October.

RANGE. Coastal plain; South Carolina south to Florida, west to Mississippi.

Also called Gopher-apple or Ground-oak.

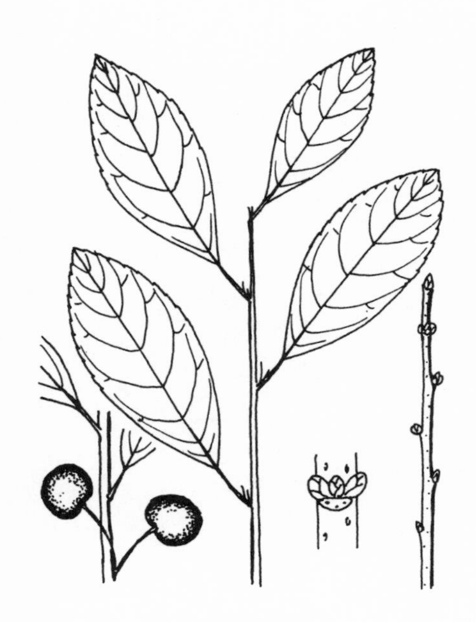

Appalachian Cherry

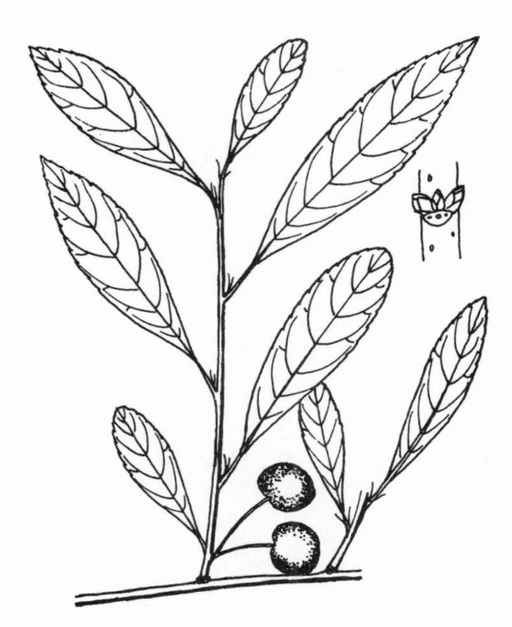

Prostrate Sand Cherry

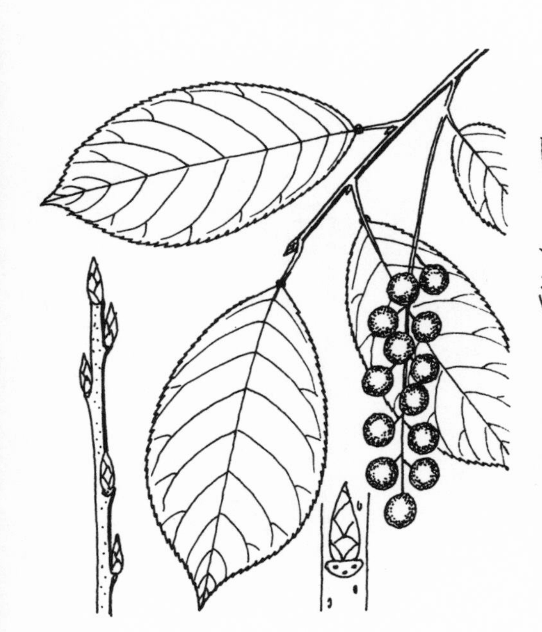

Chokecherry

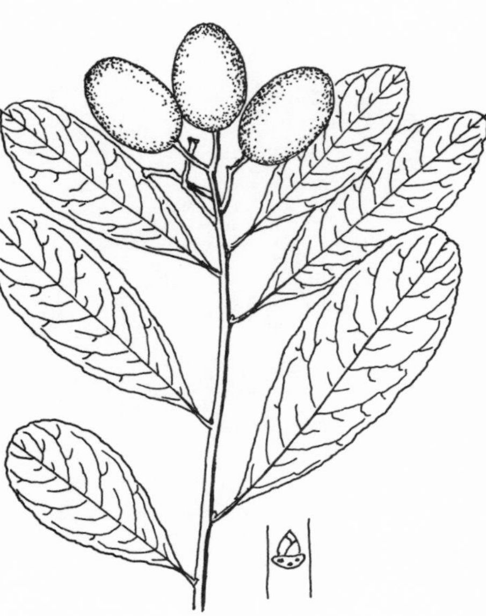

Deer-Plum

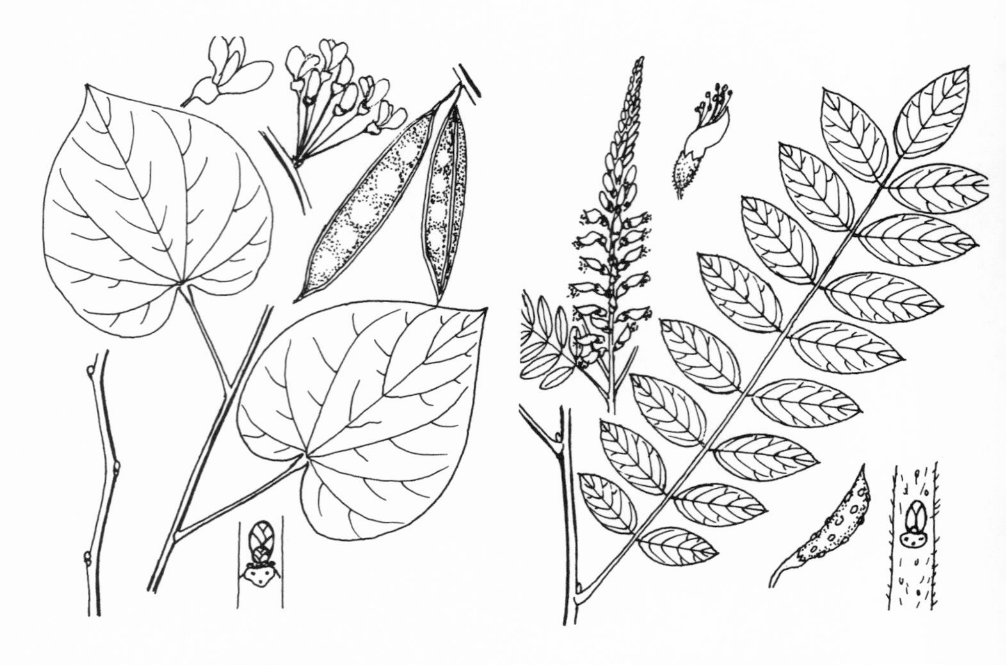

Redbud

Common Indigobush

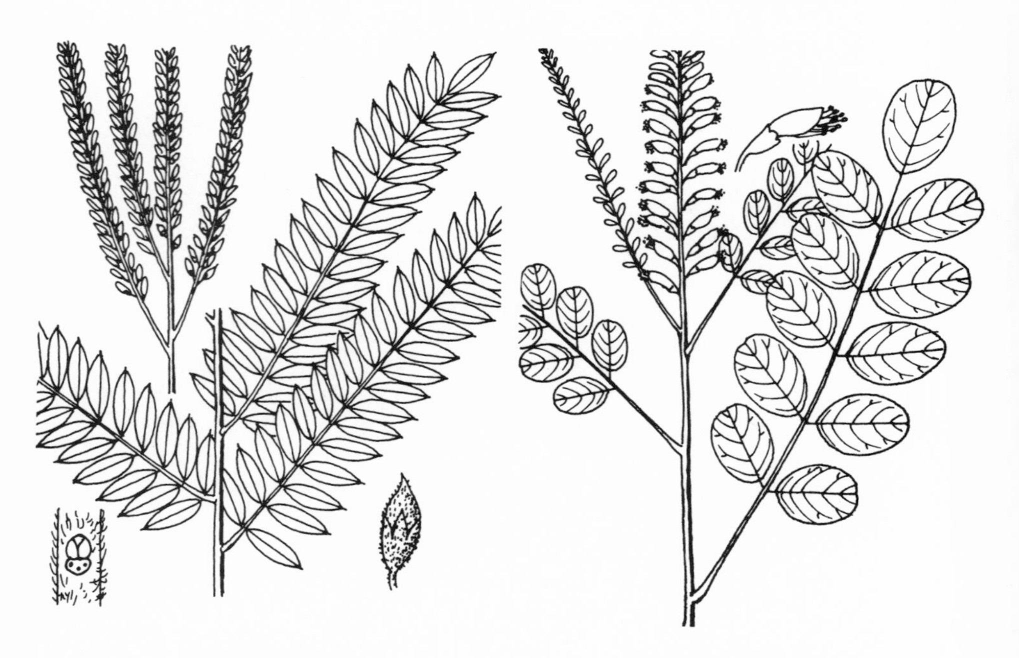

Leadplant

Mountain Indigobush

PEA FAMILY (Fabaceae)

REDBUD *Cercis canadensis*

FIELD MARKS. A leaf-losing shrub or small tree to 30 feet high; growing in rocky woods, ravines, or along streams. *Leaves* alternate, heart-shaped, untoothed, smooth or nearly so, 2 to 4 inches wide; leafstalks swollen at summit. *Flowers* somewhat pealike, rose purple or pink; blooming late March to May, before the leaves appear. *Fruits* light brown, flattened pods, 2 to 3 inches long; maturing July or August and persisting.
RANGE. Connecticut to southern Ontario, Wisconsin, and Nebraska; south to Florida and western Texas.
Very popular as an ornamental. Also called Judas-tree.

THE INDIGOBUSHES (Amorpha)

(Key Appendix G)

The indigobushes are leaf-losing shrubs with alternate, pinnately compound leaves; the numerous leaflets have untoothed margins. The small flowers have only a standard petal and are borne in long, dense, narrow end clusters. They are followed by small, usually 1-seeded pods, which are usually glandular-dotted.

COMMON INDIGOBUSH *Amorpha fruticosa*

FIELD MARKS. A shrub 4 to about 10 feet high; growing in open woods, along streams, or borders of swamps. *Leaves* 6 to 12 inches long; the 11 to 35 oval or elliptic leaflets ½ to 2 inches long, dull green, minutely glandular-dotted, usually with some short grayish hairs (but with long tawny hairs in the variety *croceolanata* [P. W. Wats.] Schneid.), usually rounded at both ends but with an abrupt little point at tip. *Flowers* violet purple, ¼ to ⅜ inch long, usually in several clusters 3 to 6 inches long; blooming April to June. *Fruits* curved, about ⅜ inch long, dotted with large raised glands; maturing June to August.
RANGE. Southern Pennsylvania to southern Michigan, Wisconsin, and Kansas; south to Florida and Texas.

LEADPLANT *Amorpha canescens*

FIELD MARKS. A shrub 1 to 3 feet high; growing on dry hillsides and prairies. *Branchlets* densely grayish-hairy. *Leaves* short-stalked, 2 to 5 inches long; the 21 to 51 leaflets elliptic or lance-shaped, abruptly short-pointed at tip, sometimes slightly hairy above, densely grayish-hairy beneath, ¼ to about ½ inch long. *Flowers* purplish blue, in several end clusters 2 to 6 inches long; blooming June to August. *Fruits* about 3/16 inch long, densely hairy; maturing August or September.
RANGE. Southern Michigan to Saskatchewan; south to northern Indiana, Arkansas, Texas, and New Mexico.

MOUNTAIN INDIGOBUSH *Amorpha glabra*

FIELD MARKS. A shrub 3 to 6 feet high; growing on wooded slopes of the southern Applachians. *Leaves* long-stalked, 3 to 6 inches long; the 9 to 19 leaflets quite broadly egg-shaped or oval, rounded or notched at tip, smooth on both sides, ¾ to 2 inches long. *Flowers* deep purple, about ⅜ inch long, usually in several end clusters 2½ to 6 inches long; blooming April to June. *Fruits* about 5/16 inch long, sparingly glandular-dotted; maturing July or August.
RANGE. North Carolina and Tennessee south to Georgia and Alabama.

PLUME-LOCUST *Amorpha herbacea*

FIELD MARKS. A shrub 1 to 3 feet high; growing in open sandy woods and pinelands. *Leaves* very short-stalked; the 11 to 37 downy leaflets elliptic to oblong or egg-shaped, rounded at both ends but with an abrupt little bristle-point at tip, ⅜ to 1 inch long. *Flowers* violet purple to white, the calyx with amber-colored dotlike glands, usually in several downy end clusters 4 to 12 inches long; blooming May to July. *Fruits* about 3/16 inch long, downy and glandular-dotted.

RANGE. Chiefly coastal plain; North Carolina south to Florida.

GEORGIA INDIGOBUSH *Amorpha georgiana*

FIELD MARKS. A shrub 1 to about 3 feet high, with creeping underground stems and few branches; growing in sandy open woods or woods borders. *Leaves* very short-stalked; the 11 to 25 leaflets oval-shaped, rounded at both ends but with an abrupt little point at tip, smooth or nearly so but conspicuously glandular-dotted on the lower surface, ⅜ to about 1 inch long. *Flowers* blue, the calyx smooth or nearly so, usually in a solitary end cluster 2 to 6 inches long; blooming June and July. *Fruits* about 3/16 inch long, smooth or nearly so.

RANGE. Coastal plain; North Carolina south to Georgia.

SCHWERIN INDIGOBUSH *Amorpha schwerinii*

FIELD MARKS. A shrub 3 to 6 feet; growing in rocky woods or on river bluffs. *Branchlets* brownish-hairy. *Leaves* rather long-stalked; the 11 to 23 leaflets elliptic or narrowly egg-shaped, broadly pointed and often with a small bristle at tip, rounded at base, quite densely brownish-hairy beneath, ⅜ to 1¼ inches long. *Flowers* purplish, about ⅜ inch long, in 1 or 2 hairy end clusters 1½ to 3 inches long; blooming April to June. *Fruits* about 3/16 inch long, hairy.

RANGE. Chiefly piedmont: North Carolina south to Georgia.

SHINING INDIGOBUSH *Amorpha nitens*

FIELD MARKS. A smooth shrub 3 to about 9 feet; growing in swamps or along streams. Both the branchlets and the upper surfaces of the 9 to 19 oblong egg-shaped leaflets are lustrous. The leaves are long-stalked, and the purple flowers are usually in a solitary end cluster 4½ to 10 inches long. Fruit pods are curved and quite smooth.

RANGE. Southern Illinois south to Arkansas and Georgia. (Not illustrated)

THE LOCUSTS (Robinia)

(Key Appendix H)

The locusts are leaf-losing trees or shrubs with alternate, pinnately compound leaves. The branchlets usually have pairs of nodal spines which are modified stipules; the leaflets are untoothed. Locusts have large, showy, pealike white or pink flowers in slender-stalked, usually drooping, axillary clusters. The fruits are flattened pealike pods with several hard seeds.

CLAMMY LOCUST *Robinia viscosa*

FIELD MARKS. A shrub or small tree to 20 or more feet high; growing in open woods or on wooded slopes in the southern Appalachians. *Branchlets* have small nodal thorns and are densely covered with sticky stalkless or with occasional short-stalked glands in the inflorescence. *Leaves* 6 to 12 inches long; the 11 to 25 leaflets elliptic to egg-shaped, broadly pointed to rounded at base, usually rounded but with an abrupt bristle-point at tip, 1 to 2 inches long; leafstalks glandular-sticky. *Flowers* pink or lavender pink, not fragrant, the calyx and stalks of the flower cluster sticky-glandular; blooming May or June. *Fruits* 2 to 3 inches long, at first sticky-glandular, later dry.

RANGE. Western Virginia south to northern Georgia and Alabama. Often found as an escape from cultivation northward.

Plume-locust

Georgia Indigobush

Schwerin Indigobush

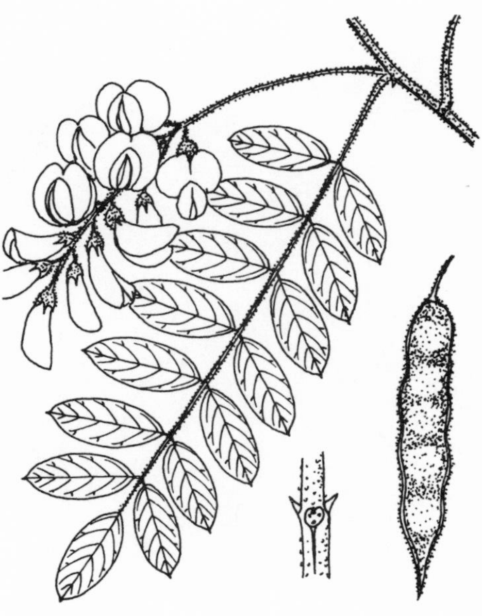

Clammy Locust

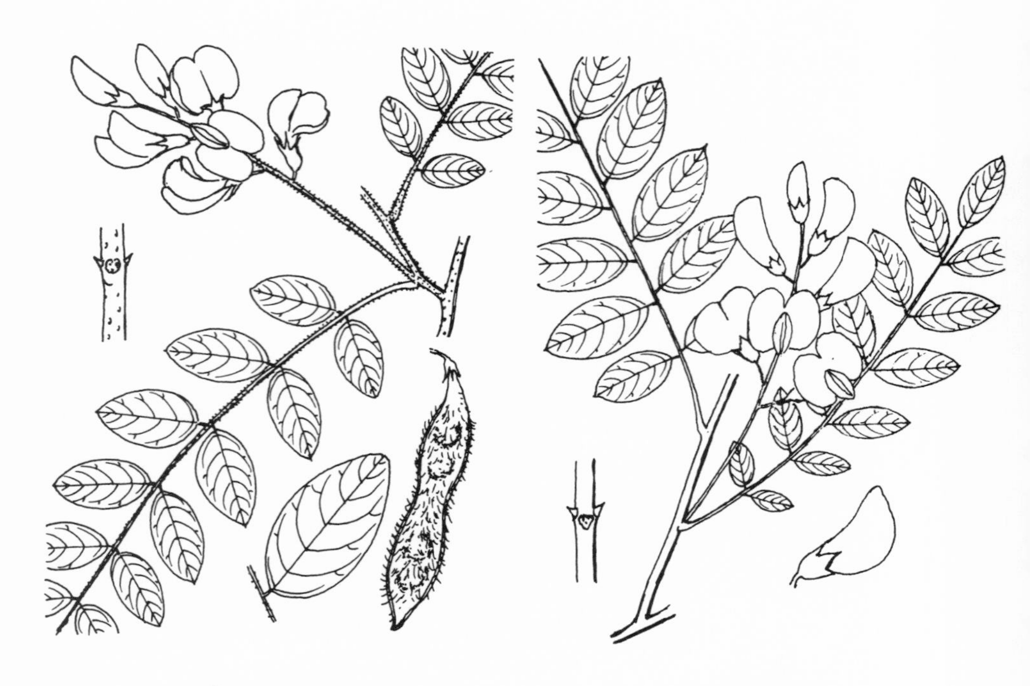

Hartweg Locust

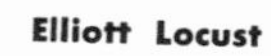

Elliott Locust

Bristly Locust

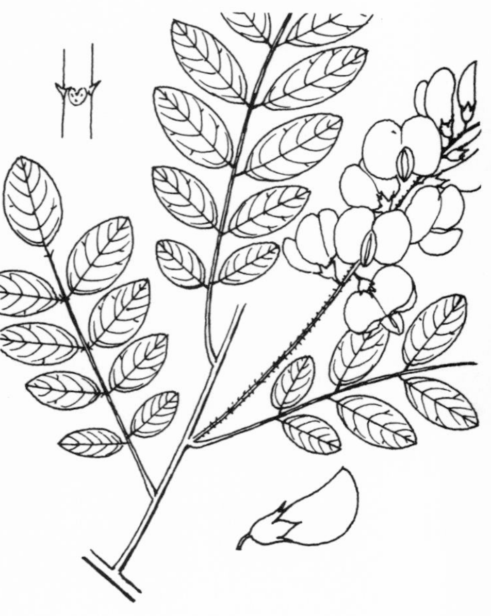

Boynton Locust

HARTWEG LOCUST *Robinia viscosa* var. *hartwegii*

FIELD MARKS. A shrub 3 to about 6 feet high, similar to the preceding but having the young branches and inflorescence covered with stout, short-stalked glands. *Leaves* with 13 to 23 leaflets which are elliptic to oval or egg-shaped, bluntish but often with a small bristle-point at tip, roundish at base, smooth or nearly so above, more or less downy beneath, ¾ to about 2 inches long. *Flowers* pale pink to rose purple, 10 to 20 in a downy and glandular-dotted axillary cluster; blooming June or July. *Fruits* downy and glandular-bristly, 2 to 3½ inches long; maturing August or September.

RANGE. Western North Carolina and eastern Tennessee south to northern Georgia and Alabama.

BRISTLY LOCUST *Robinia hispida*

FIELD MARKS. A shrub 2½ to about 10 feet high with stems, branchlets, leafstalks and stalks of the flower clusters densely covered with reddish-brown bristles; growing in open woods and on slopes and ridges. *Leaves* 4 to 8 inches long, with 9 to 15 leaflets which are egg-shaped, rounded at base, prominently bristle-pointed at tip, ¾ to about 2 inches long. *Flowers* large, rose pink, in clusters of 3 to 8; blooming May and June. *Fruits* at first sticky-glandular, later dry and bristly, 1½ to 2½ inches long.

RANGE. Virginia, West Virginia, and Kentucky south to North Carolina; and northern Georgia and Alabama.

Also called Rose-acacia. Often planted as an ornamental and escaping north of its natural range.

ELLIOTT LOCUST *Robinia hispida* var. *nana*

FIELD MARKS. A shrub 1 to about 3 feet high, similar to the preceding but being armed with short but stout nodal spines. *Leaves* with 9 to 15 leaflets which are oval or elliptic, blunt or somewhat pointed at both ends, more or less downy beneath, ¾ to 2 inches long. *Flowers* rose pink or purplish, 5 to 10 in a downy-stalked axillary cluster; the calyx teeth triangular and narrowly pointed; blooming April and May. *Fruits* narrow, bristly-hairy; maturing July to September.

RANGE. Chiefly coastal plain; North Carolina south to Georgia, west to Alabama.

BOYNTON LOCUST *Robinia hispida* var. *rosea*

FIELD MARKS. A shrub 3 to about 6 feet high, similar to the bristly locust but frequently unarmed and somewhat taller than the Elliott locust. It is commonly found growing on wooded slopes and ridges in the southern Appalachians. *Branchlets* smooth or nearly so; unarmed or with very short nodal spines. *Leaves* with 7 to 13 leaflets which are elliptic or narrowly egg-shaped, roundish to pointed at base, blunt but often with a minute bristle at tip, smooth on both surfaces, ¾ to 1½ inches long. *Flowers* pink or rose purple, 8 to 10 on a sparingly glandular-bristly stalk from the leaf axils; blooming April to June. *Fruits* seldom present.

RANGE. Western North Carolina and eastern Tennessee south to northern Georgia and Alabama.

KELSEY LOCUST *Robinia hispida* var. *kelseyi*

FIELD MARKS. An uncommon shrub 3 to about 10 feet high; reported to grow on wooded slopes and ridges in the southern Appalachians. *Branchlets* smooth; armed with slender nodal spines. *Leaves* with 9 to 13 leaflets which are narrowly elliptic or lance-shaped, pointed at tip, pointed or roundish at base, downy on both surfaces at flowering time but smooth or nearly so when mature, ¾ to about 2 inches long. *Flowers* rose pink, 5 to 8 in a cluster on a more or less glandular-bristly axillary stalk; calyx teeth narrow, long-pointed, longer than the often glandular-hairy tube; blooming April to June. *Fruits* brown, densely glandular-bristly, 1½ to 2½ inches long; maturing June to September.

RANGE. Western North Carolina.

This variety is similar to var. *fertilis*, which differs by being more densely prickly-hispid.

AMERICAN WISTERIA *Wisteria frutescens*

FIELD MARKS. A climbing, leaf-losing vine; growing on the borders of swamps and low woods or along streams. *Leaves* alternate, compound, 4 to 9 inches long; the 9 to 15 leaflets elliptic or egg-shaped, rounded to pointed at base, pointed at tip, untoothed on margin, smooth above, often sparingly hairy beneath, ¾ to 2¼ inches long. *Flowers* showy, lilac purple to white, in dense end clusters 2 to 4½ inches long; blooming April to June. *Fruits* knobby beanlike pods 2 to 4 inches long; maturing June to September.

RANGE. Chiefly coastal plain; eastern Maryland and Virginia south to Florida, west to Louisiana and Texas.

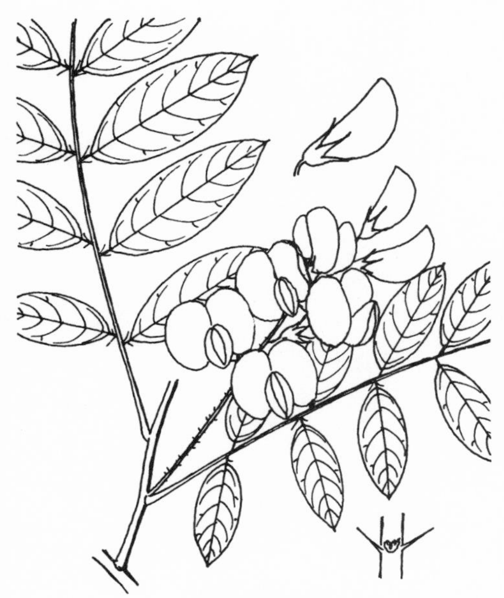

Kelsey Locust

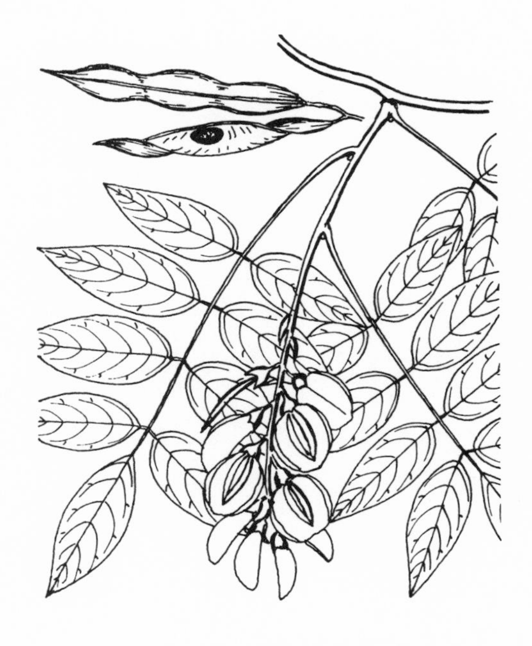

American Wisteria

Northern Prickly-ash

Southern Prickly-ash

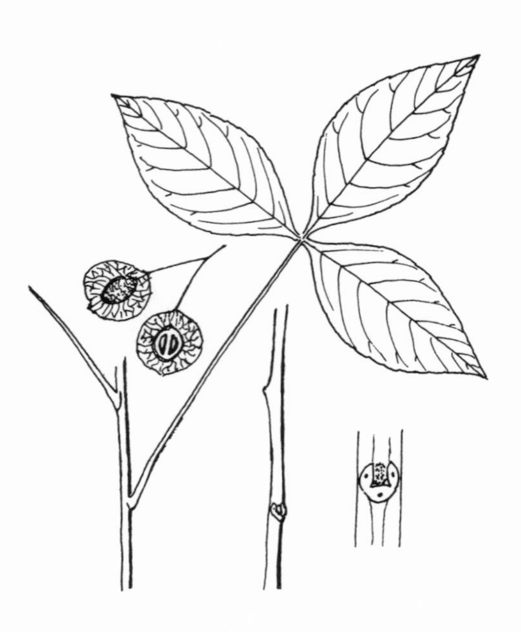

Hoptree

Sebastianbush

RUE FAMILY (Rutaceae)

NORTHERN PRICKLY-ASH *Zanthoxylum americanum*

FIELD MARKS. A leaf-losing shrub 4 to about 15 feet high; growing in rocky woods or along streams. *Branchlets* with pairs of broad-based nodal prickles, with a lemon-like odor when bruised; buds hairy, rusty-red. *Leaves* alternate, compound, 3 to 10 inches long; the 5 to 11 leaflets almost stalkless, egg-shaped, rounded to broadly pointed at base, pointed at tip, untoothed on margin, smooth or nearly so above, often slightly downy beneath, glandular-dotted, with lemon-like odor when crushed, 1 to 2½ inches long. *Flowers* small, greenish, clustered; blooming in April or May. *Fruits* dry, reddish-brown pods about 3/16 inch long, containing 1 or 2 shiny black seeds; maturing August or September.

RANGE. Quebec and Ontario to South Dakota; south to northern Georgia, Alabama, and northeastern Oklahoma.

Also called Toothache-tree; once used as a remedy for toothache, rheumatism, ulcers, colic, etc.

SOUTHERN PRICKLY-ASH *Zanthoxylum clava-herculis*

FIELD MARKS. A large shrub or small somewhat evergreen tree; growing in dry sandy woods or on sandhills. *Branchlets* with paired nodal prickles and some scattered ones, on larger stems developing into corky knobs with a stout prickle at tip; lime-like odor when bruised. *Leaves* alternate, compound, 5 to 8 inches long; the 7 to 19 leaflets almost stalkless, often rather sickle-shaped, unevenly rounded or pointed at base, pointed at tip, margin bluntly toothed, somewhat leathery in texture, lustrous above, paler and usually smooth beneath, 1 to 2½ inches long, with limelike odor when crushed. *Flowers* small, greenish, clustered; blooming April to June. *Fruits* dry, reddish-brown pods about 3/16 inch long, containing 1 or 2 shiny black seeds; maturing July to September.

RANGE. Coastal plain; southeastern Virginia south to Florida, west to Texas; north in Mississippi Valley to Arkansas and southeastern Oklahoma.

HOPTREE *Ptelea trifoliata*

FIELD MARKS. A leaf-losing shrub or small tree to 15 feet high, all parts of which are ill-scented when crushed; growing on sandy shores or rocky banks of streams. *Leaves* alternate, long-stalked, compound; the 3 leaflets egg-shaped, pointed at both ends, sometimes obscurely toothed on margin, smooth above, paler and sometimes downy beneath, minutely clear- or black-dotted, 2 to 6 inches long. *Flowers* small, greenish white, clustered; blooming April to July. *Fruits* wafer-like, roundish, thin, papery, with 2 seeds in a central chamber; maturing June to September.

RANGE. Southwestern Quebec to Ontario, south to Florida and Texas.

Also called Wafer-ash and Stinking-ash.

SPURGE FAMILY (Euphorbiaceae)

SEBASTIANBUSH *Sebastiania ligustrina*

FIELD MARKS. An erect half-evergreen shrub 3 to 12 feet high; growing in wet woods, swamps, or along streams. *Branchlets* slender, exuding a milky sap when broken. *Leaves* alternate, elliptic to oval, pointed at both ends or long-pointed at tip, margin untoothed, smooth or nearly so on both surfaces, slightly paler beneath, ¾ to 3 inches long; stipules small but quite evident. *Flowers* small, greenish, in short clusters; blooming May or June. *Fruits* 3-lobed, yellowish-green capsules about ¼ inch in diameter; maturing July or August, persisting until about October.

RANGE. Coastal plain; North Carolina south to Florida, west to Louisiana.

ALABAMA CROTON *Croton alabamensis*

FIELD MARKS. A rare evergreen or semi-evergreen shrub 5 to 10 feet high; growing on shale or limestone river bluffs in Tuscaloosa and Bibb counties, Alabama. *Branchlets* pale brownish gray, minutely scaly, with a strong odor when broken. *Leaves* elliptic to broadly lance-shaped, roundish at base, blunt or rounded at tip, untoothed on margin, bright green and smooth above, silvery white with minute scales beneath, 2 to about 4 inches long; leafstalks slender. *Flowers* small, arranged in short and narrow end clusters; blooming February or March. *Fruits* 3-parted, silvery grayish, minutely scaly capsules ¼ to ⅜ inch long.

STILLINGIA *Stillingia aquatica*

FIELD MARKS. A leaf-losing shrub 2 to about 6 feet high, with a solitary spindle-shaped stem which branches above; growing in shallow pineland ponds. *Branchlets* slender, smooth. *Leaves* alternate, rather crowded, narrow and willow-like, pointed at both ends, short-stalked, very finely but sharply toothed on margin, yellowish green and smooth on both surfaces, 1 to about 2¼ inches long. *Flowers* small, yellowish green, in a rather short and narrow end cluster; blooming May to September. *Fruits* 3-parted capsules about ⅜ inch long.

RANGE. Coastal plain; southeastern South Carolina south to Florida, west to Mississippi.

CACTUS FAMILY (Cactaceae)

EASTERN PRICKLY-PEAR *Opuntia humifusa*

FIELD MARKS. A more or less prostrate and sometimes mat-forming plant; growing in sandy or rocky, open places. *Stems* made up of firmly attached, flattened, thick and fleshy joints which are inversely egg-shaped to elliptic or roundish in outline, pale green, 2 to 4 inches long, and commonly with 1 or a few slender spines an inch or less in length. *Leaves* thick, scalelike, spirally arranged, about 3/16 inch long, and soon shed; bearing in their axils little clusters of prickly bristles. *Flowers* bright yellow, often with a red center, about 2 inches across; blooming May or June. *Fruits* inversely egg-shaped, red or purplish berries 1 to about 2 inches long; ripening August to October.

RANGE. Coastal plain; Massachusetts south to Florida, west to Alabama; inland to Tennessee and the mountains of West Virginia.

Also known as Indian-fig, as the fruits are edible.

DRUMMOND PRICKLY-PEAR *Opuntia pusilla*

FIELD MARKS. A small prostrate or mat-forming plant growing on sand dunes and in sandy pinelands near the coast. *Stems* made up of loosely attached fleshy joints which are pale green or partly purplish, usually ¾ to (seldom) 2 inches long, and viciously armed with slender spines which are often 1 inch or more long and commonly in groups of 2 to 4. *Leaves* thick, scalelike, and soon shed. *Flowers* similar to those of the preceding species; blooming in May or June. *Fruits* also similar but usually smaller; ripening August to October.

RANGE. North Carolina south to Florida, west to Mississippi.

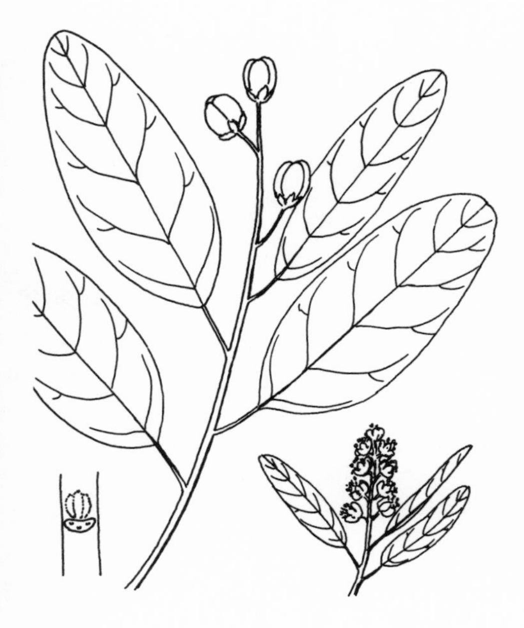

Alabama Croton

Stillingia

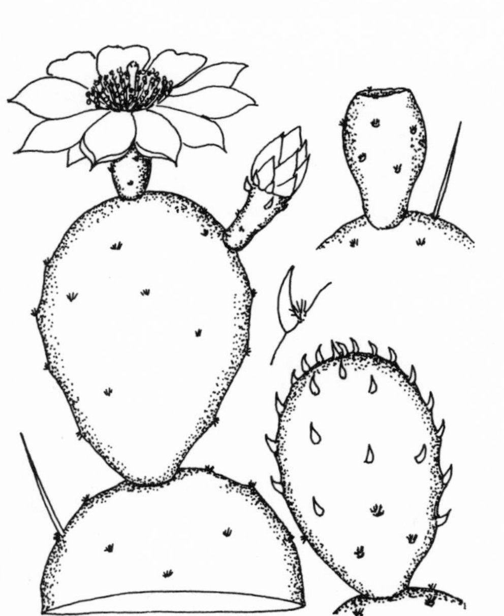

Eastern Prickly-pear

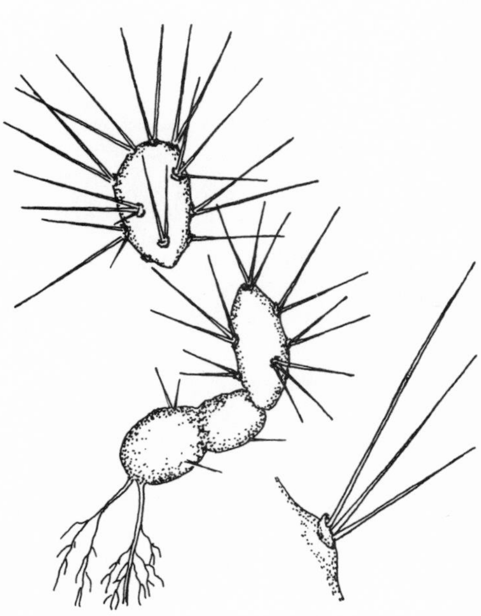

Drummond Prickly-pear

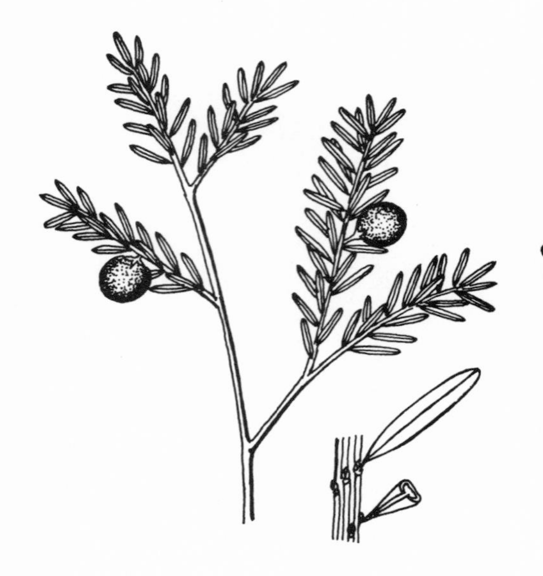

Black Crowberry

Broom-crowberry

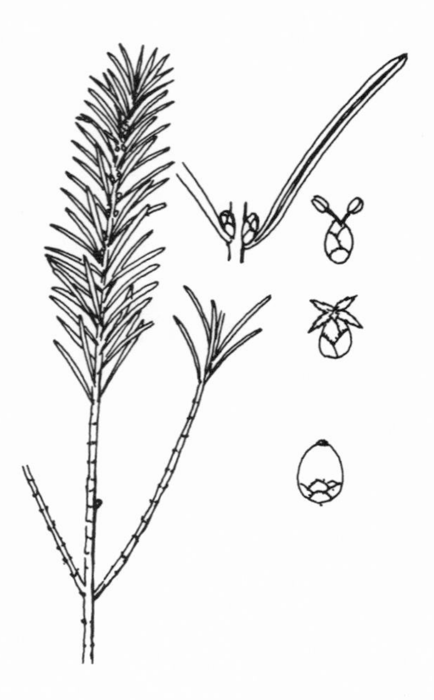

Sandhill-rosemary

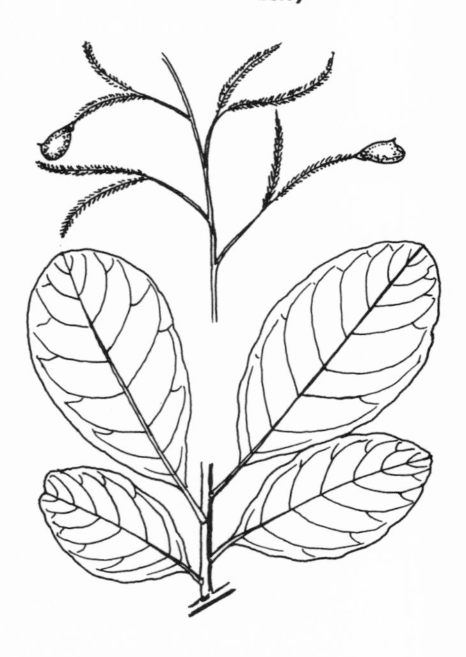

American Smoketree

CROWBERRY FAMILY (Empetraceae)

BLACK CROWBERRY *Empetrum nigrum*

FIELD MARKS. A prostrate and spreading evergreen shrub; growing in cold, sandy, or rocky places. *Branchlets* smooth or sometimes sticky-hairy. *Leaves* alternate or sometimes whorled, very narrow or narrowly elliptic, blunt at tip, margin rolled and roughish,thick, dark green, smooth or nearly so, ⅛ to ¼ inch long. *Flowers* small, purplish, solitary in the axils of the upper leaves; blooming June or July. *Fruits* round, berry-like, black or purple, juicy, 6- to 9-seeded, slightly less than ¼ inch in diameter; ripening July to November.

RANGE. Arctic region south to New England, northern New York, northern Michigan, northern Minnesota, southern Alberta, and northern California; also on eastern Long Island.

PURPLE CROWBERRY *Empetrum eamesii* ssp. *atropurpureum*

FIELD MARKS. A shrub similar to the preceding but with trailing branchlets and leaves white-wooly when young. *Fruits* red to purplish-black.

RANGE. Labrador to Quebec, south to Nova Scotia and the mountains of northern New England. (Not illustrated)

BROOM-CROWBERRY *Corema conradii*

FIELD MARKS. A spreading evergreen shrub 6 inches to about 2 feet high; growing in sandy or rocky places. *Leaves* alternate or in 3's, very narrow, blunt at tip, bright green and smooth when mature, ⅛ to 3/16 inch long. *Flowers* small, purplish, in heads at tips of the branchlets; blooming March to June. *Fruits* round, brownish, berry-like, usually 3-seeded, almost dry, less than ⅛ inch in diameter; ripening July to September.

RANGE. Newfoundland south near the coast to New Jersey; also in Shawangunk Mountains, New York.

SANDHILL-ROSEMARY *Ceratiola ericoides*

FIELD MARKS. A much-branched, evergreen, aromatic shrub 1 to about 5 feet high; growing in dry and sandy coastal-plain pinelands and on sandhills. *Branchlets* slender and stiffly erect. *Leaves* alternate but closely crowded, spreading, very narrow, the margins so inrolled that the leaves appear to be needle-like or tubular, ¼ to about ½ inch long. *Flowers* small and inconspicuous, red or yellowish, 2 or 3 together in the axils of the upper leaves; stamen-bearing and pistil-bearing flowers usually on separate plants; blooming spring or summer, or all year southward. *Fruits* roundish, berry-like, rather dry, yellow or red, scarcely ⅛ inch in diameter.

RANGE. Coastal plain; South Carolina south to Florida, west to Mississippi.

SUMAC FAMILY (Anacardiaceae)

AMERICAN SMOKETREE *Cotinus obovatus*

FIELD MARKS. A leaf-losing shrub or small tree 6 to sometimes 30 feet high; growing on rocky ridges and bluffs. *Branchlets* aromatic and exuding a gummy sap when broken. *Leaves* alternate, oval or broadest above the middle, somewhat pointed at base, rounded or notched at tip, untoothed and with a slightly rolled margin, dark green and smooth above, paler and often somewhat downy beneath, 2 to 5 inches long. *Flowers* small, yellowish green, a number in a much-branched pyramidal end cluster 6 to 8 inches long; blooming April or May. *Fruits* small, podlike, lopsided, dry; maturing June to September.

RANGE. Eastern Tennessee to Missouri, Arkansas, and Oklahoma; south to Alabama and Texas.

The hairy stalks of the flower clusters resemble puffs of smoke at a distance. Occasionally cultivated for its brilliant red and orange fall foilage.

SMOOTH SUMAC *Rhus glabra*

FIELD MARKS. A leaf-losing shrub or small tree 2 to about 15 feet high; growing in old fields, along fence rows, etc. *Branchlets* stout, smooth, whitened with a bloom, exuding milky sap when cut. *Leaves* alternate, compound, 12 to 20 inches long; the 11 to 31 leaflets stalkless, oblong lance-shaped, usually roundish at base, pointed at tip, sharply toothed on margin, smooth, whitened beneath, 2 to 5 inches long. *Flowers* small, yellowish green, in dense erect end clusters; blooming June or July. *Fruits* roundish, red, sticky-hairy, 1-seeded, about ⅛ inch in diameter, in compact end clusters; maturing August or September and persisting.

RANGE. Maine and Quebec to British Columbia, south to Florida and California.

Young fruits of this and other red-fruited sumacs have been used to make a pleasantly acid, lemonade-like drink. The fruits are eaten by many birds, but usually only in an emergency. Twigs and bark are often eaten by rabbits and deer.

STAGHORN SUMAC *Rhus hirta*

FIELD MARKS. Similar to the preceding but usually larger, to about 30 feet high. *Branchlets* are velvety-hairy. Leaves are usually both whitened and downy beneath, and the leafstalks are downy. The flowers and fruits are similar to those of Smooth Sumac.

RANGE. Nova Scotia to Quebec and Minnesota; south to northern Georgia, Tennessee, Illinois, and Iowa.

DWARF SUMAC *Rhus copallinum*

FIELD MARKS. A leaf-losing shrub or small tree 2 to about 10 feet high; growing in open woods, thickets, fence rows, and old fields. *Branchlets* moderate, smooth or finely downy, exuding a clear sap when cut. *Leaves* alternate, compound, 6 to 12 inches long, with conspicuous wings between the leaflets; the 9 to 21 leaflets stalkless, egg-shaped to lance-shaped, pointed or unevenly rounded at base, pointed at tip, untoothed or nearly so on margin, usually lustrous above, paler and often downy beneath, 1½ to 3 inches long. *Flowers* small, yellowish green, numerous, in large end clusters; blooming July to September. *Fruits* roundish, red, hairy, about ⅛ inch in diameter, in large and often somewhat drooping end clusters; maturing September or October.

RANGE. Southern Maine to New York, Michigan, central Wisconsin, and eastern Kansas; south to Florida and Texas.

Also called Shining Sumac. Rich in tannin used in tanning hides.

MICHAUX SUMAC *Rhus michauxii*

FIELD MARKS. A rather rare shrub. *Branchlets* moderate, densely brownish-hairy. *Leaves* with brownish-hairy leafstalks and 9 to 15 leaflets which are coarsely toothed on margin, brownish-hairy and green beneath. *Fruits* red and densely hairy.

RANGE. Coastal plain and lower piedmont; North Carolina south to Georgia. (Not illustrated)

FRAGRANT SUMAC *Rhus aromatica*

FIELD MARKS. A leaf-losing, aromatic shrub 2 to 6 feet high; growing on dry rocky banks and hillsides. *Branchlets* pleasantly fragrant when bruised, often with catkin-like flower buds. *Leaves* alternate, rather long-stalked, compound, 4 to 6 inches long; leaflets 3, the end one short-stalked and commonly diamond-shaped, the side ones egg-shaped and almost stalkless, all usually pointed at base, pointed or blunt at tip, coarsely and irregularly toothed on margin, somewhat downy, 1 to 3 inches long. *Flowers* small, yellowish green, in catkin-like clusters; blooming March or April. *Fruits* roundish, red, densely hairy, about 1¼ inch in diameter; maturing July or August.

RANGE. Vermont and southwestern Quebec to Indiana, Kansas, Nebraska, and Oklahoma; south to Florida and Texas.

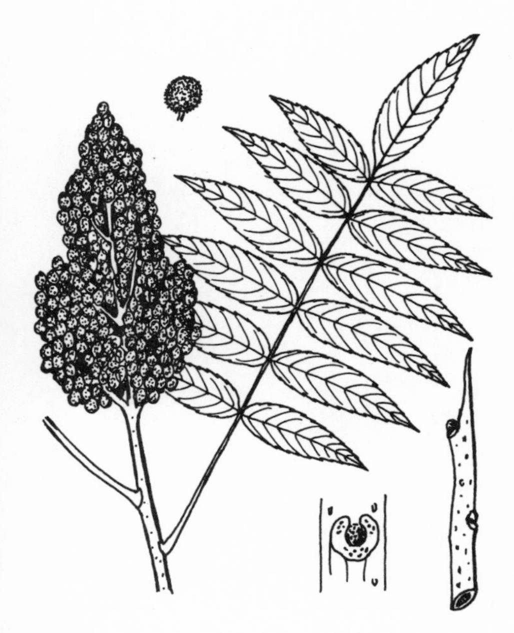

Smooth Sumac

Staghorn Sumac

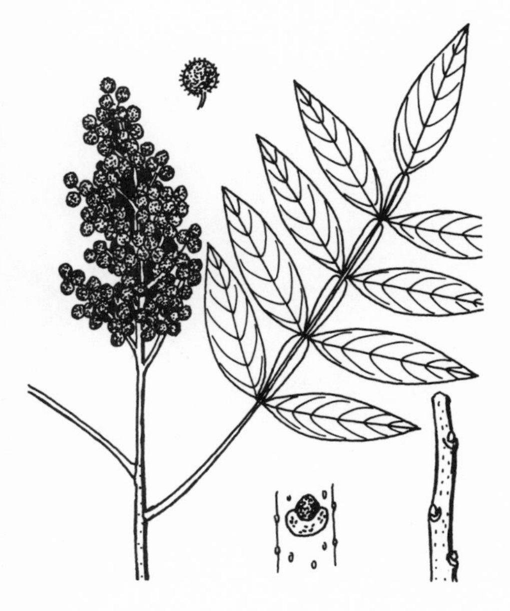

Dwarf Sumac

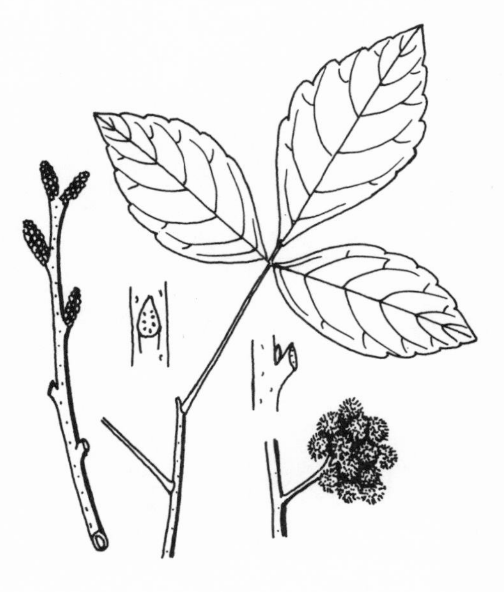

Fragrant Sumac

Poison Sumac

Poison-ivy

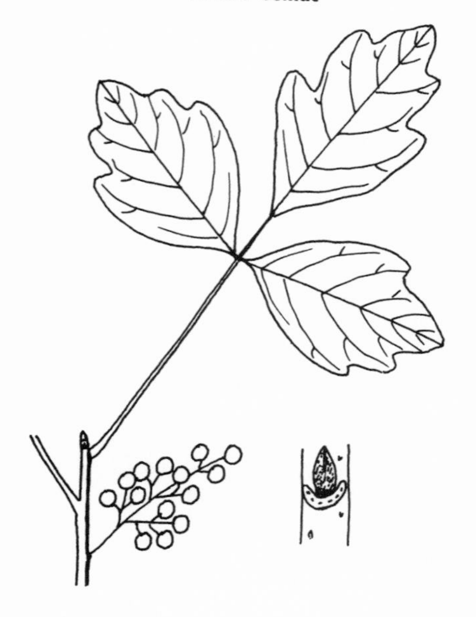

Poison-oak

Swamp Cyrilla

POISON SUMAC *Toxicodendron vernix*

CAUTION: All parts of this plant contain a dangerous skin irritant.

FIELD MARKS. A shrub or small tree 4 to about 15 feet high; growing in swamps, bogs, and other wet places. *Branchlets* moderately stout, smooth, the end bud present. *Leaves* alternate, compound, 6 to 12 inches long; the 7 to 13 leaflets short-stalked, elliptic or egg-shaped, broadly pointed at base, pointed at tip, margin untoothed, rather lustrous above, paler beneath, smooth, 2 to 4 inches long; leafstalks usually reddish. *Flowers* small, greenish, in axillary clusters; blooming May to July. *Fruits* roundish, smooth, waxy white, about 3/16 inch in diameter, in rather loose and drooping clusters; ripening August or September and persisting.

RANGE. Southwestern Maine to Ontario and Minnesota, south to Florida and Texas.

POISON-IVY *Toxicodendron radicans*

CAUTION: All parts of this plant contain a dangerous skin irritant.

FIELD MARKS. An erect or trailing leaf-losing shrub, or a woody vine climbing by means of aerial rootlets on the stems; growing in wooded areas, thickets, clearings, or along fence rows and roadsides. *Leaves* alternate, long-stalked, compound, 4 to 12 inches long; the 3 leaflets oval or egg-shaped, rounded or broadly pointed at base, pointed at tip, usually with a few coarse teeth on margin, often lustrous above, paler and slightly downy beneath, 1½ to 8 inches long; the end leaflet rather long-stalked, the side ones almost stalkless. *Flowers* small, yellowish green, in axillary clusters; blooming May to July. *Fruits* roundish, waxy white, about 3/16 inch in diameter; maturing August to October and persisting.

RANGE. Nova Scotia to British Columbia; south to Florida, Texas, and Arizona.

POISON-OAK *Toxicodendron pubescens*

CAUTION: All parts of this plant contain a dangerous skin irritant.

FIELD MARKS. A stiffly erect, simple or sparingly branched, leaf-losing shrub 1 to 2½ feet high; growing in dry sandy pine and oak woods and clearings. *Leaves* alternate, long-stalked, compound, 3 to 8 inches long; the 3 leaflets often broadly egg-shaped, pointed at base, blunt at tip, with 3 to 7 often deep lobes, somewhat downy above, more densely so and paler beneath, 2 to 5 inches long; the end leaflet rather long-stalked, the side ones almost stalkless; leafstalks downy. *Flowers* and *fruits* similar to those of Poison-ivy but usually more downy.

RANGE. Chiefly coastal plain; New Jersey and Maryland south to Florida; Tennessee to eastern Oklahoma south to Alabama and Texas.

CYRILLA FAMILY (Cyrillaceae)

SWAMP CYRILLA *Cyrilla racemiflora*

FIELD MARKS. A shrub or small tree to 15 (rarely 30) feet high; forming thickets on borders of swamps and ponds. *Leaves* alternate, narrowly elliptic or broadest above the middle, wedge-shaped at base, rounded or blunt-pointed at tip, untoothed on margin, thin but somewhat leathery, lustrous above, paler beneath, smooth, 2 to 4 inches long; leafstalks short. *Flowers* small, white; in long, narrow, drooping end clusters; blooming late April to July. *Fruits* small, yellowish-brown, 4-seeded capsules about ⅛ inch long; maturing August or September and persisting.

RANGE. Coastal plain; Virginia south to Florida, west to Texas.

Also called White-titi. Flowers are important source of honey. Leaves usually turn red in the fall but persist most of the winter.

BUCKWHEAT-TREE *Cliftonia monophylla*

FIELD MARKS. A large evergreen shrub or small tree usually 5 to 15 (sometimes to 30) feet high, often forming dense thickets on the borders of swamps or along streams. *Leaves* alternate, narrowly elliptic or broadest above the middle, wedge-shaped at base, rounded or bluntly pointed at tip, margin untoothed, lustrous green above, paler and dull beneath, smooth, 1 to 2 inches long; the leafstalks very short. *Flowers* small, white or pinkish, fragrant, in narrow end clusters; blooming March or April. *Fruits* 2- to 4-winged, dry, light-brown capsules about ¼ inch long.

RANGE. Coastal plain; Georgia and Florida west to Louisiana.

An important honey plant. Also called Black-titi.

HOLLY FAMILY (Aquifoliaceae)

MOUNTAIN-HOLLY *Nemopanthus mucronatus*

FIELD MARKS. An erect leaf-losing shrub 3 to 12 feet high; growing in cool, moist, rocky woods, lakeshores, and bogs. *Leaves* alternate, often clustered on short spurs, elliptic, pointed to roundish at base, blunt but with an abrupt little point at tip, (rarely) with a few teeth on margin, thin in texture, smooth, slightly paler beneath, ¾ to 2 inches long; leafstalks smooth and very slender. *Flowers* small, greenish white, on long and slender stalks; blooming about May. *Fruits* berry-like, 3- to 5-seeded, roundish, dull red, about ¼ inch in diameter, on long and slender stalks; ripening July to September.

RANGE. Newfoundland to Minnesota; south to West Virginia, Indiana, and northern Illinois.

HOLLIES (Ilex)

(Key Appendix I)

The hollies are evergreen or leaf-losing shrubs or trees with alternate and simple leaves. They have small greenish-white flowers which bloom in the spring. As a rule the stamens and pistils are in separate flowers, and they are found on separate plants. The fruits are roundish, berry-like, with 4 to 6 large bony nutlets and a mealy or pulpy flesh. Hollies are important as ornamental and honey plants; their fruits are eaten by many kinds of wild birds. The American Holly (*Ilex opaca* Ait.) is a tree commonly 20 to 40 feet or more high, the berried branches of which are widely used for Christmas decorations.

WINTERBERRY *Ilex verticillata*

FIELD MARKS. A leaf-losing shrub 3 to 15 feet high; growing along streams and in low wet woods and swamps. *Leaves* oval to lance-shaped or broadest above the middle, pointed at both ends, sharply and rather coarsely toothed on margin, sometimes slightly leathery in texture, dull green above, slightly paler and sometimes slightly downy beneath, 1½ to 3 inches long; remaining green until frost, then turning black. *Flowers* bearing the stamens and those bearing the pistils both short-stalked; blooming April to June. *Fruits* bright red (rarely yellow), short-stalked, about ¼ inch in diameter; ripening September and October and persisting after the leaves fall.

RANGE. Newfoundland to Minnesota; south to Georgia, southeastern Louisiana, and Missouri.

Also called Black-alder.

SMOOTH WINTERBERRY *Ilex laevigata*

FIELD MARKS. A leaf-losing shrub similar to the preceding and growing in similar swampy situations. *Leaves* oval to narrowly elliptic, pointed at both ends, with rather low but sharp teeth on margin, thin in texture, lustrous green above, paler and smooth or nearly so on both surfaces, 1½ to 3 inches long. *Flowers* bearing the stamens much longer than those bearing pistils; blooming April to June. *Fruits* orange red (rarely yellow), short stalked, about ¼ inch in diameter; ripening August to October, usually persisting after the leaves fall.

RANGE. Chiefly coastal plain; southern Maine and New York south to Georgia.

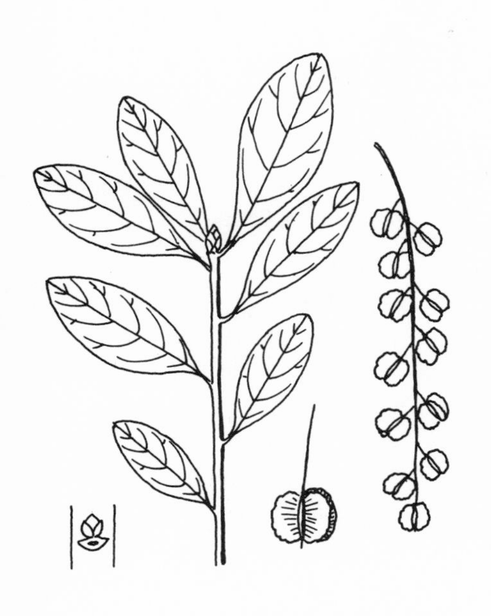

Buckwheat-tree

Mountain-holly

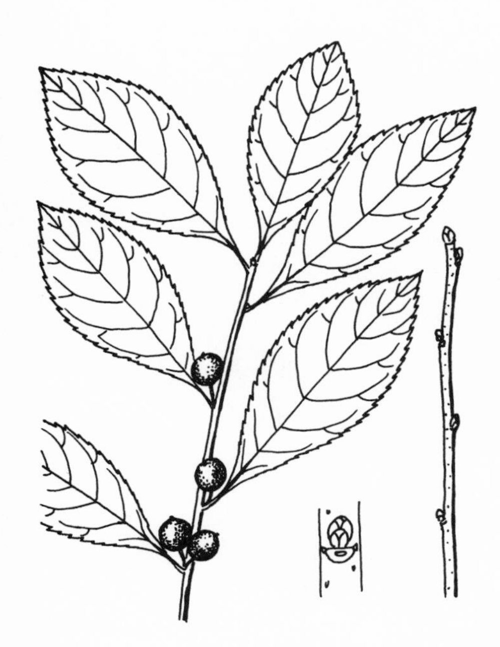

Winterberry

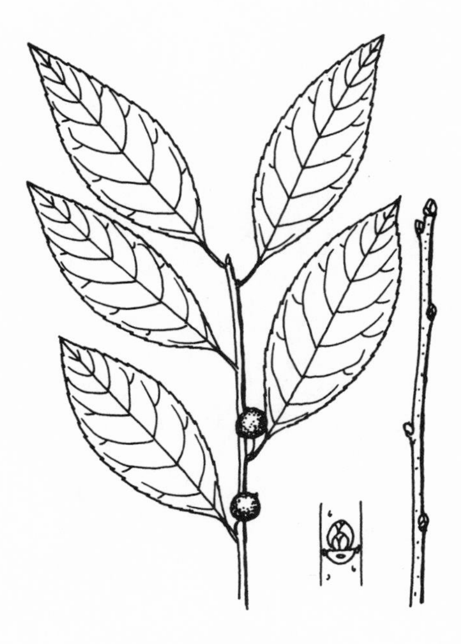

Smooth Winterberry

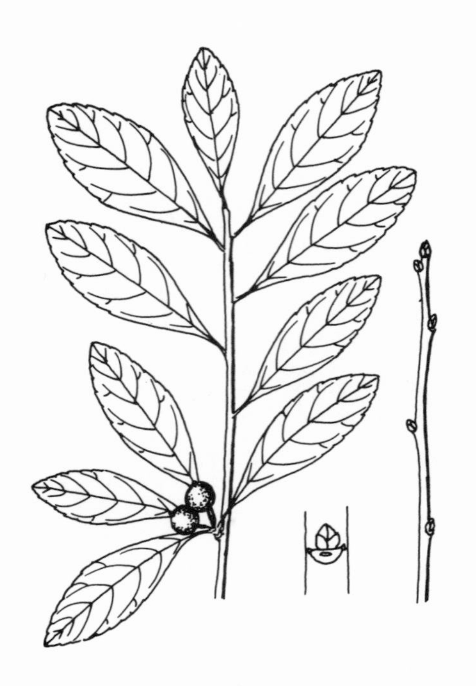

Deciduous Holly

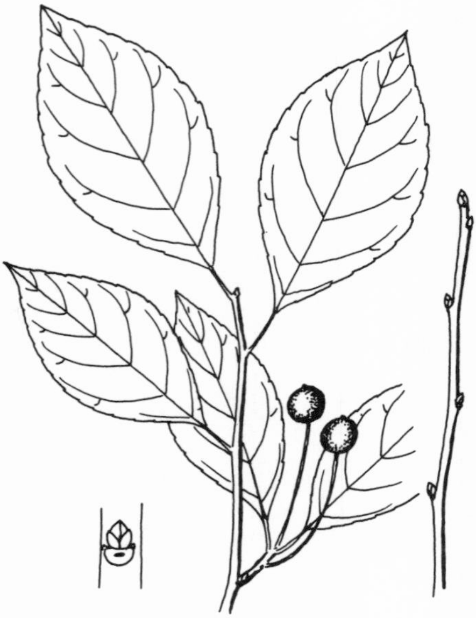

Long-stalked Holly

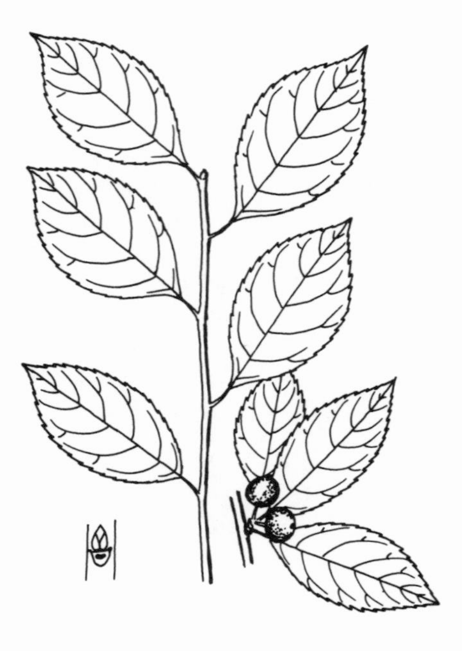

Carolina Holly

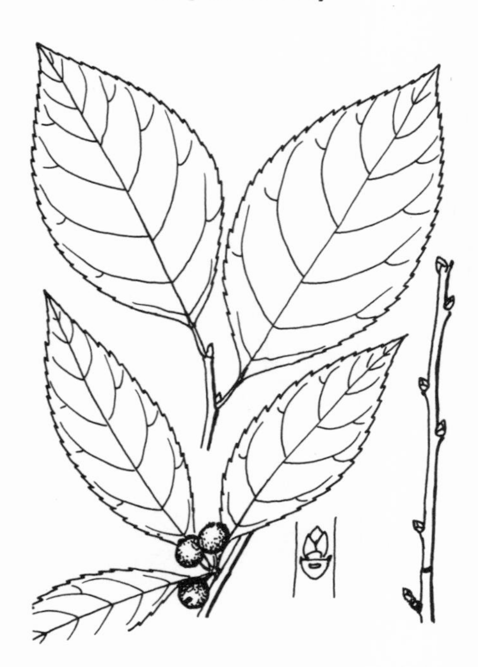

Largeleaf Holly

DECIDUOUS HOLLY *Ilex decidua*

FIELD MARKS. A leaf-losing shrub or small tree 5 to 25 feet high; growing in upland forests, about swamps, and along streams. *Leaves* sometimes clustered on short spurlike branches, narrowly elliptic or broadest above the middle, wedge-shaped at the base, broadly pointed or blunt at tip, margin with low and bluntish teeth above the middle, light green above, paler beneath, smooth or nearly so on both surfaces, 1½ to 3 inches long; leafstalks short and sometimes downy. *Flowers* blooming in April or May. *Fruits* bright red, short-stalked, about ¼ inch in diameter; ripening September and October and usually persisting.

RANGE. Virginia to southern Illinois, Missouri, and Oklahoma; south to Florida and south central Texas.

Also called Possum-haw.

LONG-STALKED HOLLY *Ilex longipes*

FIELD MARKS. A leaf-losing shrub 6 to 12 feet high; growing in upland woods and thickets and on cool rocky slopes. *Leaves* broadly lance-shaped to broadly elliptic, pointed at both ends, margin with rather widely spaced and bluntish teeth, thin in texture, smooth on both surfaces or sometimes sparingly hairy beneath, 2 to 4½ inches long. *Flowers* blooming in April or May. *Fruits* red or purplish (rarely yellow), about ¼ inch in diameter on stalks ½ to 1¼ inches long; ripening September or October.

RANGE. Virginia, West Virginia, and Tennessee; south to Florida and Alabama.

Also called Georgia Holly.

CAROLINA HOLLY *Ilex ambigua*

FIELD MARKS. A leaf-losing shrub or small tree usually 5 to 15 feet high; growing in sandy or rocky upland woods. *Leaves* often clustered on short lateral spurlike branches, elliptic to oval, roundish or broadly pointed at base, pointed at tip, sharply toothed on margin often nearly to base, smooth on both surfaces or somewhat downy beneath, 1 to about 2½ inches long. *Flowers* blooming April to June. *Fruits* bright red, short-stalked, about ¼ inch in diameter; ripening August or September but scarcely persisting.

RANGE. Coastal plain; North Carolina south to Florida, west to Texas, and north into Arkansas.

LARGELEAF HOLLY *Ilex montana*

FIELD MARKS. A leaf-losing shrub or small tree 5 to 20 feet high; growing in cool, moist, wooded areas. *Leaves* often clustered on short lateral spurlike branches, elliptic to egg-shaped or lance-shaped, roundish to pointed at base, pointed at tip, sharply toothed on margin, thin in texture, bright green and smooth above, sometimes with scattered soft hairs beneath, 2½ to 5 inches long. *Flowers* blooming April to June. *Fruits* bright red, short-stalked, about ⅜ inch in diameter; ripening August to October but scarcely persisting.

RANGE. Southwestern Massachusetts to western New York; south along the mountains to northern Georgia.

SARVIS HOLLY *Ilex amelanchier*

FIELD MARKS. A rather rare leaf-losing shrub 3 to 6 feet high; growing along streams and in sandy swamps. *Leaves* elliptic or oblong and sometimes broadest above the middle, pointed to roundish at both ends, margin with small and inconspicious teeth, veiny, smooth above, paler and somewhat downy beneath, 1½ to 3 inches long; the leafstalks slender. *Flowers* blooming in April or May. *Fruits* dull red, about ⅜ inch in diameter, on slender stalks ¼ to ⅜ inch long; ripening in October or November.

RANGE. Coastal plain; North Carolina south to Georgia, also in Louisiana.

INKBERRY *Ilex glabra*

FIELD MARKS. An evergreen shrub 1 to about 4 feet high; growing in low sandy woods and about the borders of swamps or bogs. *Leaves* narrowly elliptic to oval and often broadest above the middle, wedge-shaped at base, broadly pointed at tip, margin with some low and bluntish teeth above the middle, thickish and leathery in texture, dark green and lustrous above, paler and smooth and with minute black dots beneath, ¾ to 2 inches long. *Flowers* blooming May to August. *Fruits* black, about ¼ inch in diameter, on stalks about ⅜ inch long; ripening September to November and persisting.

RANGE. Coastal plain; Nova Scotia south to Florida, west to Louisiana.

Also called Gallberry. An important honey plant in the southeast. The fruits are eaten by many birds, including the bobwhite quail and wild turkey.

LARGE GALLBERRY *Ilex coriacea*

FIELD MARKS. An evergreen shrub 4 to about 10 feet high; growing in wet sandy woods and swampy places. *Leaves* narrowly elliptic to oval and often broadest above the middle, wedge-shaped at base, usually pointed at tip, margin untoothed or with a few spiny-tipped teeth above the middle, thickish and leathery in texture, dark green and lustrous above, paler and smooth and with minute black dots beneath, 1 to 3½ inches long. *Flowers* blooming in April or May. *Fruits* black, about ⅜ inch in diameter, on stalks ¼ to ⅜ inch long; ripening in September or October and persisting.

RANGE. Coastal plain; southeastern Virginia south to Florida, west to Louisiana.

YAUPON HOLLY *Ilex vomitoria*

FIELD MARKS. An evergreen shrub or small tree 4 to about 20 feet high; growing in sandy woods and on dunes. *Branchlets* stiff and spiky. *Leaves* oval to elliptic or egg-shaped, rounded to broadly pointed at the base, bluntly pointed at tip, margin with rounded or bluntish teeth, thickish and leathery in texture, lustrous green above, slightly paler beneath, smooth on both surfaces, ½ to 1½ inches long; the leafstalks very short. *Flowers* blooming March to May. *Fruits* bright red (rarely yellow) short-stalked, about ¼ inch in diameter; ripening in October or November and persisting.

RANGE. Coastal plain; southeastern Virginia south to Florida, west to south central Texas; north in the Mississippi Valley to Arkansas and Oklahoma.

Also called Cassine and Christmas-berry. The leaves have been used as a substitute for tea and were formerly used by the Indians to make their ceremonial "black drink." Cultivated as an ornamental shrub.

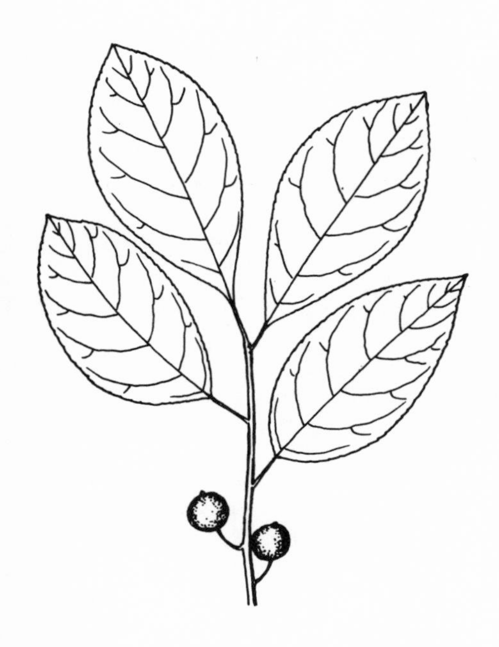

Sarvis Holly

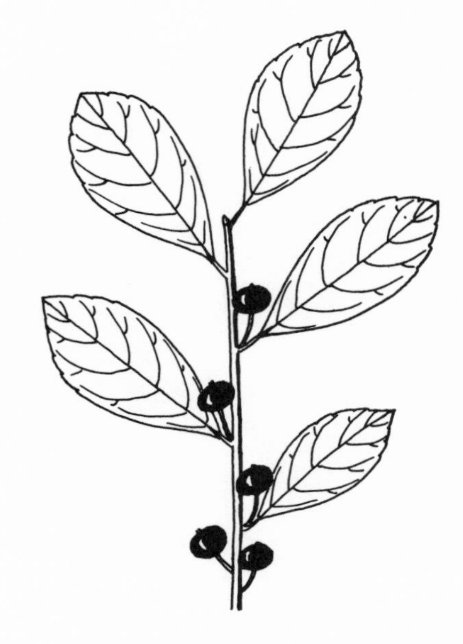

Inkberry

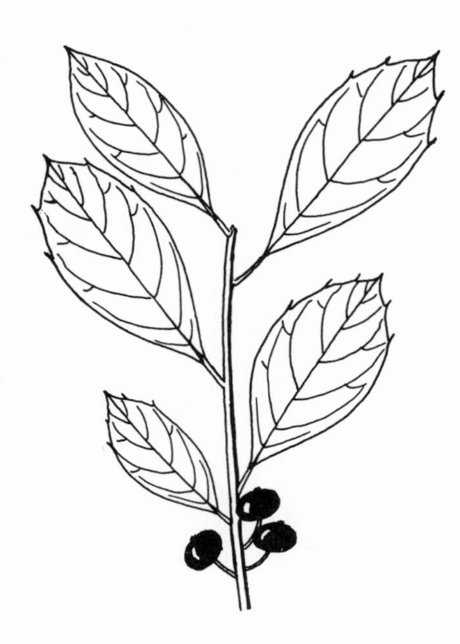

Large Gallberry

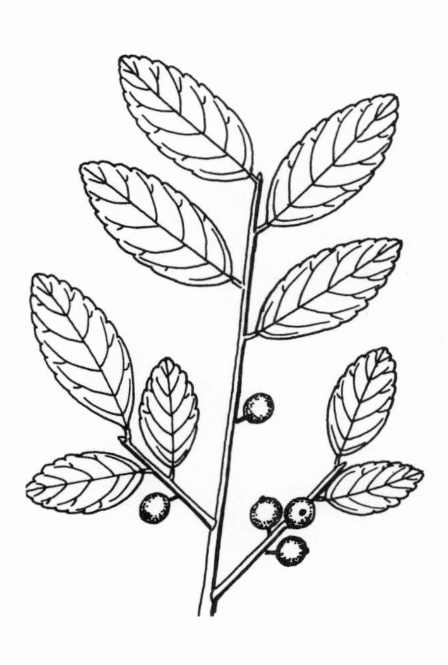

Yaupon Holly

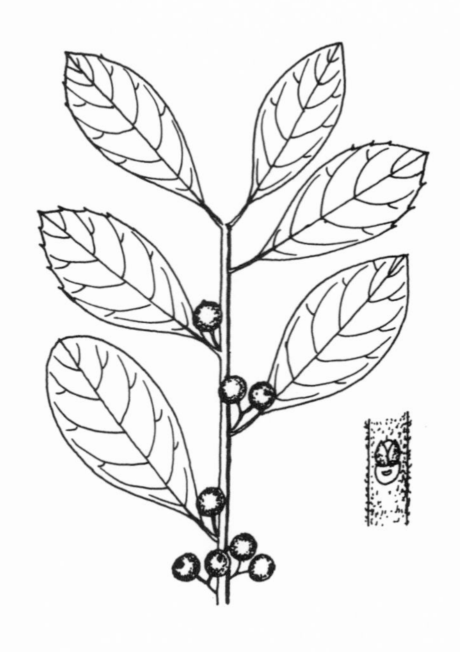

Dahoon Holly

Myrtle-leaf Holly

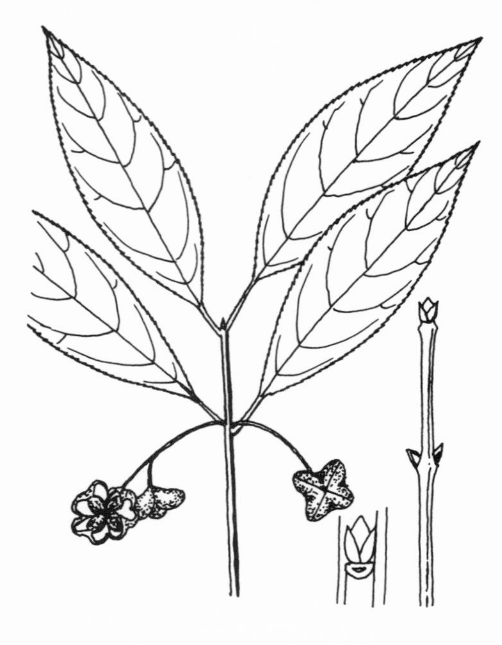

Burning-Bush

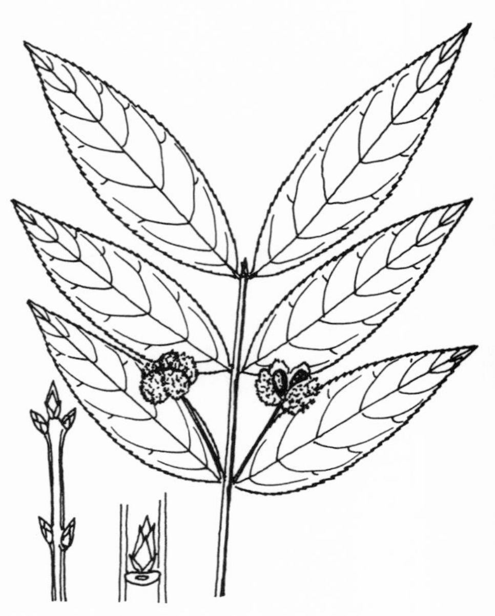

Strawberry-Bush

DAHOON HOLLY *Ilex cassine*

FIELD MARKS. An evergreen shrub or small tree 5 to 25 feet high; growing about the borders of swamps and cypress ponds. *Leaves* elliptic or oblong egg-shaped, pointed at base, pointed or blunt at tip, margin untoothed or with a few spiny teeth above the middle, thickish and leathery in texture, dark green and smooth above, usually with minute pale hairs along midrib beneath, 1½ to 3 inches long. *Flowers* blooming in May or June. *Fruits* red (rarely yellow), short-stalked, about ¼ inch in diameter; ripening October or November and persisting.

RANGE. Coastal plain; North Carolina south to Florida, west to Louisiana.

MYRTLE-LEAF HOLLY *Ilex myrtifolia*

FIELD MARKS. An evergreen shrub or small tree; growing in swampy woods and shallow pineland ponds. *Branchlets* straight, stiff, and spiky. *Leaves* narrowly elliptic, roundish or broadly pointed at base, abruptly tipped with a small spine, margin rolled and untoothed or with some small and inconspicuous teeth above the middle, stiff and leathery in texture, dark green above, paler beneath, smooth, ½ to 1½ inches long; the leafstalks very short. *Flowers* blooming in May or June. *Fruits* bright red (rarely yellow), short-stalked, about ¼ inch in diameter; ripening October or November and persisting.

RANGE. Coastal plain; North Carolina south to Florida, west to Louisiana.

BITTERSWEET FAMILY (Celastraceae)

SPINDLETREES (Euonymus)

Members of this genus are shrubs, or sometimes small trees, with opposite and simple leaves. The branchlets are characteristically green and 4-sided or 4-lined. They have small greenish-white to greenish-yellow or purplish flowers which have 4 or 5 petals. The fruits are 3- to 5-lobed capsules which split open at maturity and expose the seeds, which have fleshy bright-red coats. Many species, including the following native ones, are cultivated as ornamental shrubs.

BURNING-BUSH *Euonymus atropurpurea*

FIELD MARKS. An erect leaf-losing shrub 6 to about 15 feet high; growing in rich, moist woods, ravines, and stream bottoms. *Branchlets* roundish and 4-lined. *Leaves* oblong egg-shaped to broadly lance-shaped, usually pointed at base and sharply pointed at tip, finely and sharply toothed on margin, thin-textured, dark green and smooth above, paler and usually somewhat downy beneath, 2 to 5 inches long; leafstalks slender, ¼ to ¾ inch long. *Flowers* purplish brown, about ¼ inch across, in clusters of 5 to 15 from the leaf axils; blooming in June. *Fruits* smooth, purplish pink, about ¾ inch across before bursting; ripening September or October.

RANGE. Western New York and southern Ontario to southern Michigan, central Minnesota, and Montana; south to eastern Virginia, northern Georgia and Alabama, Arkansas, and Oklahoma.

Also known as Wahoo.

STRAWBERRY-BUSH *Euonymus americana*

FIELD MARKS. An erect or straggling leaf-losing shrub 2 to 6 feet high; growing in rich woods, ravines, and along stream banks. *Branchlets* 4-sided. *Leaves* egg-shaped to broadly lance-shaped, usually pointed at base and sharply pointed at tip, finely and sharply toothed on margin, bright green and smooth above, slightly paler and smooth or nearly so beneath, 1 to 3½ inches long; almost stalkless. *Flowers* greenish purple, about ¼ inch across, solitary or 2 or 3 in cluster from leaf axils; blooming in May or June. *Fruits* rough-warty, crimson, ½ to ¾ inch across before bursting; ripening September or October.

RANGE. Southeastern New York and Pennsylvania to southern Illinois, Missouri, and Oklahoma; south to Florida and Texas.

Also known as Bursting-heart.

RUNNING STRAWBERRY-BUSH *Euonymus obovata*

FIELD MARKS. A trailing leaf-losing shrub, rooting at the nodes, with ascending branchlets 6 to 12 inches high; growing in cool moist woods. *Branchlets* 4-sided. *Leaves* elliptical or broadest above the middle, pointed at base, broadly or abruptly pointed at tip, finely and sharply toothed on margin, thin in texture, dull light green above, slightly paler beneath, smooth or nearly so on both surfaces, ¾ to 2 inches long, very short-stalked. *Flowers* yellowish green, about ¼ inch across, 1 to 3 on stalks from leaf axils; blooming May or June. *Fruits* rough-warty, pale orange red, about ½ inch across before bursting; ripening September or October.

RANGE. Western New York to southern Ontario and southern Michigan; south to northern Georgia, Tennessee, and Missouri.

MOUNTAIN LOVER *Paxistima canbyi*

FIELD MARKS. A low, trailing, evergreen, mat-forming shrub with upright 4-sided branchlets 4 to 12 inches high; growing on rocky slopes in the mid-southern Appalachians. *Leaves* opposite, narrowly oblong or sometimes broader above the middle, pointed at base, blunt at tip, margin rolled and finely toothed, leathery in texture, green and smooth on both surfaces, ¼ to 1 inch long, almost stalkless. *Flowers* small, greenish or brownish, 1 to 3 on stalks from leaf axils; blooming April or May. *Fruits* small capsules about ⅛ inch in diameter, splitting at maturity in August or September.

RANGE. South central Pennsylvania and southeastern Ohio south to West Virginia, western Virginia, and eastern Kentucky.

Used in gardening as a ground cover or rock garden plant.

AMERICAN BITTERSWEET *Celastrus scandens*

FIELD MARKS. A twining, leaf-losing, woody vine climbing to a height of 20 or more feet or scrambling over low vegetation and sometimes trailing on the ground; growing in moist thickets and along the banks of streams. *Leaves* alternate, oval or egg-shaped, rounded to broadly pointed at base, abruptly pointed at tip, finely toothed on margin, thin in texture, dull dark green above, paler beneath, smooth or nearly so on both surfaces, 2 to 4 inches long. *Flowers* yellowish-green, small, many in an elongate end cluster; blooming May or June. *Fruits* ball-shaped, dull orange capsules about ⅜ inch in diameter; splitting when mature in September or October and exposing the bright-scarlet, fleshy-coated seeds.

RANGE. Southern Quebec to southern Manitoba; south to Georgia, Mississippi, Arkansas, and Oklahoma.

Also called Climbing Bittersweet and Waxwort. Often cultivated as an ornamental vine. The fruits are eaten by several species of wild birds.

BLADDERNUT FAMILY (Staphyleaceae)

AMERICAN BLADDERNUT *Staphylea trifolia*

FIELD MARKS. A shrub 3 to about 12 feet high; growing in moist thickets, on hillsides, and banks of streams. *Leaves* opposite, compound, long-stalked; the 3 leaflets egg-shaped or elliptic, pointed to roundish at base, abruptly pointed at tip, finely toothed on margin, dark green and smooth above, paler and sometimes downy beneath, end leaflet long-stalked, side ones short-stalked, 2 to 5 inches long. *Flowers* white or creamy white, bell-shaped, about ⅜ inch long, many in elongate and drooping clusters; blooming April to June. *Fruits* 3-sided, inflated, papery, baglike capsules 1 to 3 inches long which contain 3 to 5 light-brown and bony seeds which rattle when capsule is shaken; maturing August to October.

RANGE. Massachusetts to southwestern Quebec, southern Ontario, northern Michigan, and southern Minnesota; south to Georgia, Alabama, southeastern Oklahoma, and southeastern Nebraska.

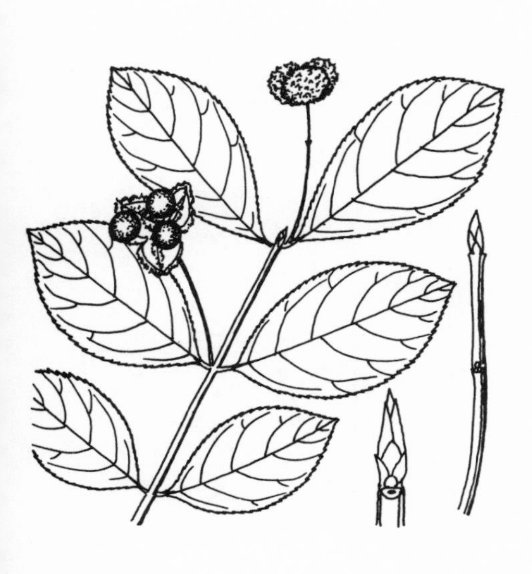

Running Strawberry-Bush

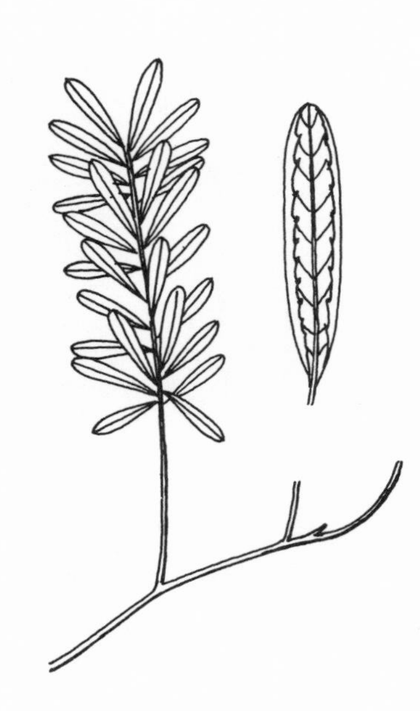

Mountain Lover

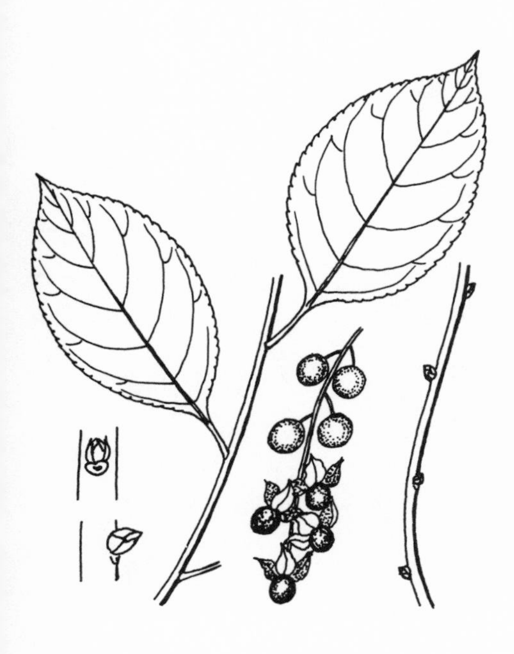

American Bittersweet

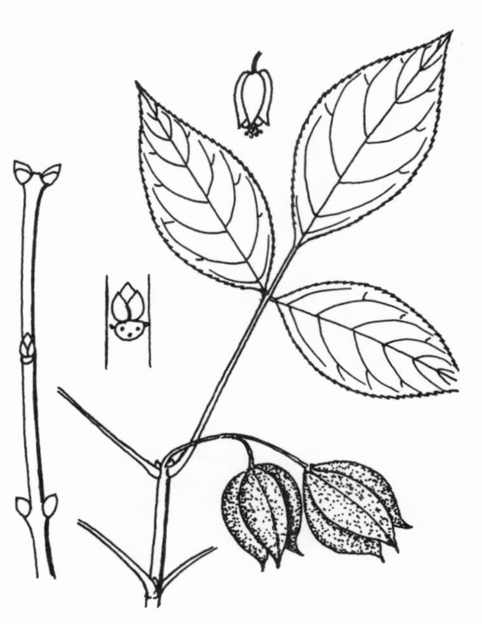

American Bladdernut

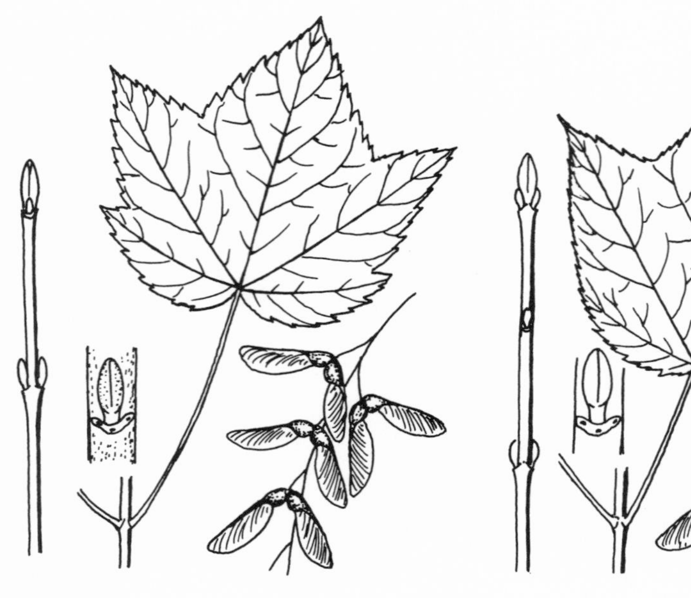

Mountain Maple

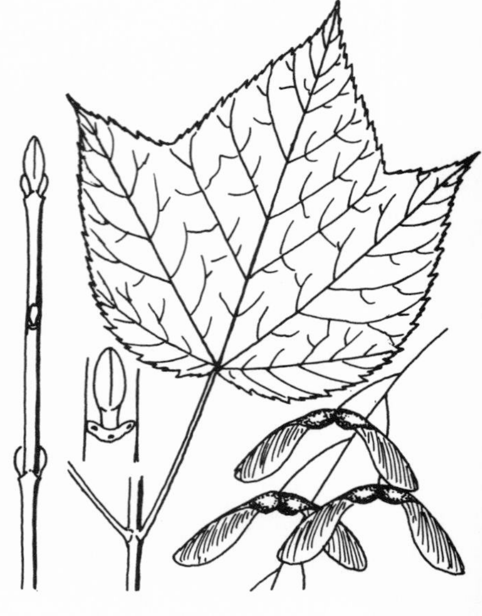

Striped Maple

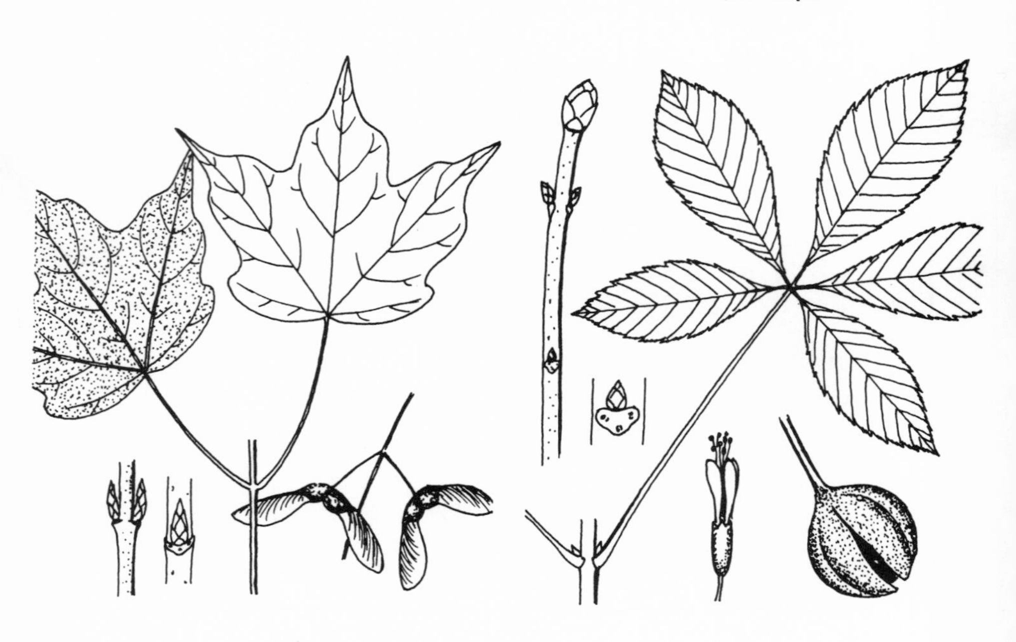

Chalk Maple

Red Buckeye

MAPLE FAMILY (Acer)

Maples are usually leaf-losing trees, sometimes shrubs, with opposite and usually simple 3- to 5-lobed leaves; with the main veins radiating from the summit of a long leafstalk. The flowers are rather small and seldom perfect. Maple fruits, called "keys," have a seed-bearing portion tipped with a broad flat wing. They are borne in pairs on slender stalks.

MOUNTAIN MAPLE *Acer spicatum*

FIELD MARKS. A shrub or small tree 5 to 20 feet high; growing in cool moist, usually rocky woods. *Leaves* usually heart-shaped at base, 3-lobed or sometimes 5-lobed above the middle, the lobes short but broad and pointed, margin coarsely toothed, smooth above, slightly paler and usually downy beneath, 3 to 5 inches broad. *Flowers* greenish yellow, in elongate upright clusters; blooming May or June. *Fruits* with wings slightly spread, about ½ inch long; maturing August or September.

RANGE. Newfoundland to Saskatchewan; south to New York, the region of the Great Lakes, northeastern Iowa, and in the mountains to northern Georgia.

STRIPED MAPLE *Acer pensylvanicum*

FIELD MARKS. A shrub or small tree to about 30 feet high; growing in cool, moist, rocky woods. *Bark* of the larger stems greenish and with conspicuous whitish streaks. *Leaves* rounded or heart-shaped at base, with 3 rather short but broad and taper-pointed lobes, margin finely toothed, smooth above, paler and smooth or nearly so beneath, 4 to 8 inches broad. *Flowers* greenish yellow, in loose drooping clusters; blooming May or June. *Fruits* with widely spread wings, about ¾ inch long; maturing August to October.

RANGE. Nova Scotia to Manitoba; south to New England, the region of the Great Lakes, and along the mountains to northern Georgia.

Also called Goosefoot Maple and Moosewood.

CHALK MAPLE *Acer leucoderme*

FIELD MARKS. A shrub or small tree 10 to 25 feet high; growing along streams and on rocky wooded banks. *Bark* light gray or whitish. *Leaves* with 3 to 5 long-pointed, broad, sparingly blunt-toothed lobes, usually heart-shaped at base, smooth above, paler yellowish green and downy beneath, 2 to 3 inches broad. *Flowers* yellowish green, on slender stalks in end clusters; blooming in April. *Fruits* with spreading wings about ½ inch long; maturing September or October.

RANGE. North Carolina south to Florida, west to Louisiana, north in Mississippi Valley to Tennessee and Arkansas.

HORSECHESTNUT FAMILY (Hippocastanaceae)

BUCKEYES (Aesculus)

Buckeyes are leaf-losing trees or shrubs with opposite, compound leaves; with 5 to 7 leaflets radiating from the summit of a long leafstalk. They have showy flowers in large, upright end clusters. The fruits are leathery pods with 1 to 3 large shiny-brown seeds having a conspicuous lighter-colored scar. The seeds are poisonous if eaten.

RED BUCKEYE *Aesculus pavia*

FIELD MARKS. A shrub or small tree usually 3 to 10 (rarely 30) feet high, spreading by underground runners; growing in hammocks and pinelands near the coast. *Leaves* with 5 (rarely 7) leaflets which are elliptic or broadest above the middle, wedge-shaped at base, pointed at tip, finely toothed on margin, paler yellowish green and smooth to downy beneath, 4 to 6 inches long. *Flowers* bright red, in erect clusters 5 or 6 inches long; blooming April or May. *Fruits* smooth, 1 to 2 inches across; maturing July or August.

RANGE. Coastal plain; Virginia south to Florida, west to Louisiana, north in Mississippi Valley to southern Illinois and Oklahoma.

PAINTED BUCKEYE *Aesculus sylvatica*

FIELD MARKS A shrub or small tree usually 2 to 10 (rarely 30) feet high, spreading by underground runners; growing in rich woods or along streams. *Leaves* usually with 5 leaflets which are lance-shaped or broadest above the middle, wedge-shaped at base, pointed at tip, finely toothed on margin, smooth above, paler and smooth to downy beneath, 3 to 8 inches long. *Flowers* pale yellow or greenish yellow, often tinged with red or sometimes wholly red, in erect clusters 4 to 8 inches long; blooming April or May. *Fruits* smooth, 1 to 1½ inches across, usually 1-seeded; maturing July or August.

RANGE. Chiefly piedmont; Virginia southwest to Georgia and Alabama; also in Tennessee.

BOTTLE-BRUSH BUCKEYE *Aesculus parviflora*

FIELD MARKS. A suckering shrub 4 to 10 feet high; growing in rich woods. *Leaves* with 5 to 7 leaflets which are almost stalkless, elliptic or broadest above the middle, wedge-shaped at base, finely toothed on margin, smooth above, paler and grayish-downy beneath, 2 to 7 inches long. *Flowers* small, white or pinkish, in slender upright clusters 6 to 12 inches long. *Fruits* smooth, 1 to 1½ inches across.

RANGE. Coastal plain and piedmont; Georgia, Alabama, and northern Florida.

A handsome ornamental often cultivated much farther north. (Not illustrated)

SOAPBERRY FAMILY (Sapindaceae)

FLORIDA SOAPBERRY *Sapindus marginatus*

FIELD MARKS. A large leaf-losing shrub or small tree to 30 feet high; growing in coastal hammocks. *Leaves* alternate, compound, 8 to 24 inches long; the 7 to 13 leaflets almost stalkless, lance-shaped, long-pointed at top, margin untoothed, smooth, lustrous above, paler beneath, 2 to 6 inches long; leafstalks with a narrow ridge on each side. *Flowers* small, reddish, in erect end clusters; blooming in spring. *Fruits* roundish, 3-lobed, smooth, pale yellowish berries about ¾ inch in diameter.

RANGE. Coastal plain; southeastern South Carolina (Small) south to Florida.

WESTERN SOAPBERRY *Sapindus saponaria* var. *drummondii*

FIELD MARKS. Usually a small tree. *Leaves* 6 to 12 inches long; the 9 to 18 leaflets short-stalked, often more or less sickle-shaped, long-pointed at tip, untoothed on margin, 2 to 3 inches long. *Flowers* small, whitish, in dense end clusters. *Fruits* roundish, yellow, about ½ inch in diameter.

RANGE. Louisiana to Arizona; north to southwestern Missouri, Kansas, and Colorado. (Not illustrated)

BUCKTHORN FAMILY (Rhamnaceae)

SUPPLEJACK *Berchemia scandens*

FIELD MARKS. A high-climbing, leaf-losing vine; growing in swamps, sandy woods, or along streams. *Leaves* alternate, elliptic or lance-shaped, roundish to broadly pointed at base, pointed at tip, slightly paler beneath, smooth, with 9 to 12 pairs of prominent straight veins, 1½ to 3 inches long. *Flowers* small, greenish white, in end clusters; blooming April to June. *Fruits* oval-shaped, bluish black, 1-seeded, about ¼ inch long; maturing August to October.

RANGE. Chiefly coastal plain; Virginia south to Florida, west to Texas; north in Mississippi Valley to Kentucky and Missouri.

SAGERETIA *Sageretia minutiflora*

FIELD MARKS. A straggling leaf-losing shrub to 10 feet high, with long weak branches and often spine-tipped branchlets; growing on sand or shell dunes and in coastal hammocks. *Leaves* opposite, roundish to egg-shaped or lance-shaped, roundish at base, pointed at tip, finely and inconspicuously toothed on margin, lustrous above, often somewhat downy beneath, short-stalked, ½ to a bit over 1 inch long. *Flowers* small, creamy white, fragrant, clustered. *Fruits* roundish, dark purple, about ¼ inch in diameter, containing 3 leathery nutlets.

RANGE. Coastal plain; South Carolina south to Florida, west to Mississippi.

Sometimes called Buckthorn.

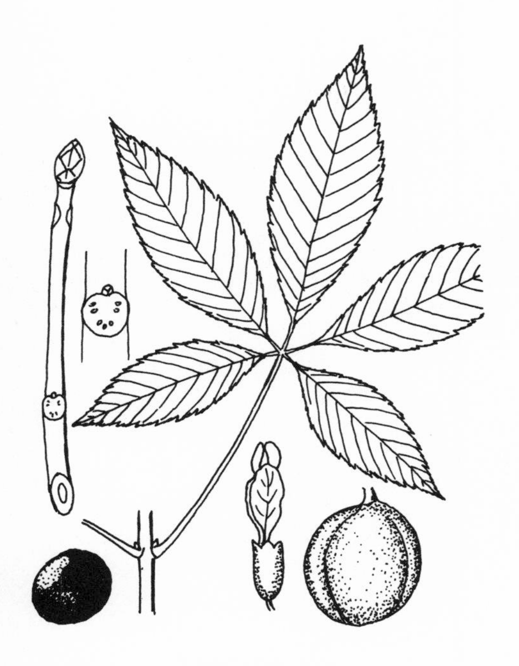

Painted Buckeye

Florida Soapberry

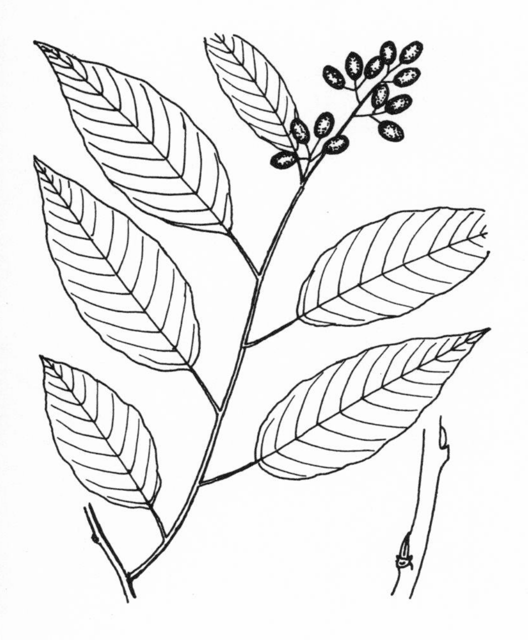

Supplejack

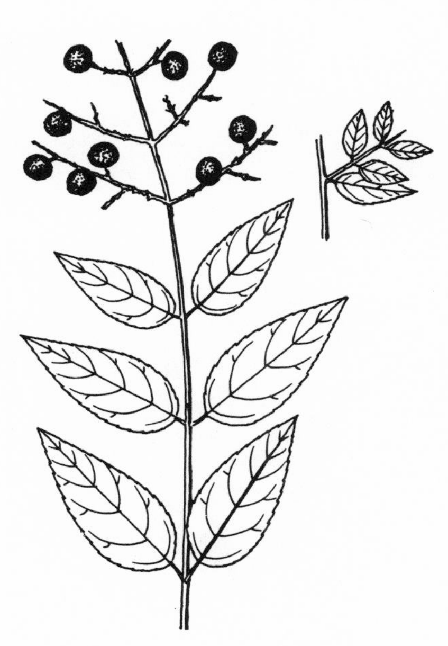

Sageretia

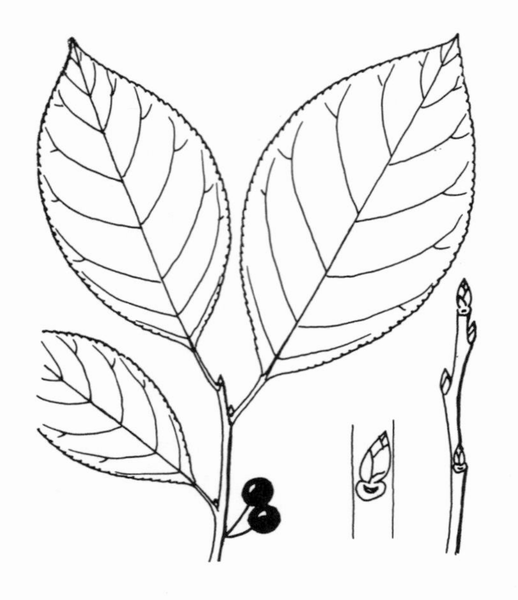

Alderleaf Buckthorn

Lanceleaf Buckthorn

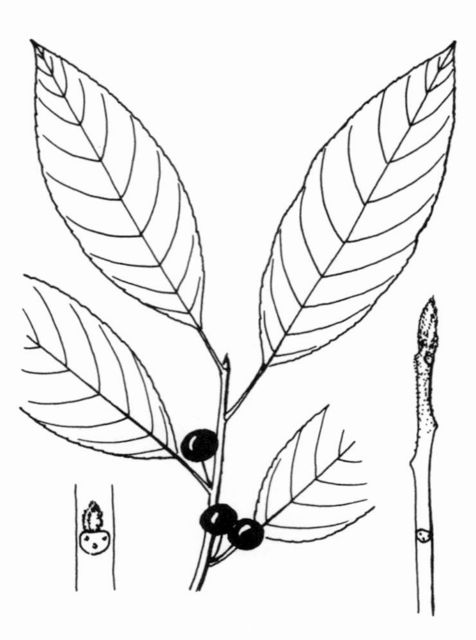

Carolina Buckthorn

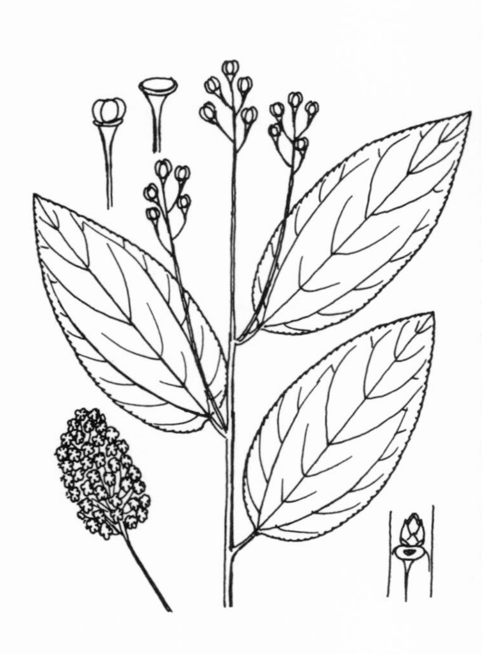

New Jersey-Tea

ALDERLEAF BUCKTHORN *Rhamnus alnifolia*

FIELD MARKS. A leaf-losing shrub 1½ to about 3 feet high; growing in cold swamps and bogs. *Branchlets* reddish brown, smooth or nearly so. *Leaves* alternate, oval or elliptic, broadly pointed at base, pointed at tip, finely and bluntly toothed on margin, lustrous and with sunken veins above, paler and usually downy on the veins beneath, 2 to 4½ inches long. *Flowers* small, greenish yellow, fragrant, 1 to 3 in leaf axils; blooming May or June. *Fruits* roundish or slightly egg-shaped, black, 1- to 3-seeded, about ¼ inch in diameter; ripening July or August.

RANGE. Newfoundland to British Columbia; south to northern New Jersey, West Virginia, the region of the Great Lakes, Nebraska, Wyoming, and California.

LANCELEAF BUCKTHORN *Rhamnus lanceolata*

FIELD MARKS. A leaf-losing shrub 4 to 8 feet high; growing on hillsides and stream banks. *Branchlets* ashy gray, often downy. *Leaves* alternate, broadly lance-shaped or elliptic, rounded or broadly pointed at base, often abruptly pointed at tip, finely and bluntly toothed on margin, smooth above, paler and sometimes slightly downy beneath, 1¼ to 3½ inches long. *Flowers* small, greenish yellow, fragrant, 1 to 6 in leaf axils; blooming in May. *Fruits* roundish, black, 2-seeded, about ¼ inch in diameter; ripening August or September.

RANGE. Southeastern Pennsylvania, West Virginia, southern Ohio, Wisconsin, and Nebraska; south to Alabama and Texas.

CAROLINA BUCKTHORN *Rhamnus caroliniana*

FIELD MARKS. A leaf-losing shrub or small tree to 30 feet high; growing along streams and on wooded hillsides. *Branchlets* reddish brown to ashy brown; with an elongate, naked, hairy end bud. *Leaves* alternate, elliptic, rounded to broadly pointed at base, pointed at tip, obscurely toothed on margin, lustrous above, paler and sometimes downy beneath, prominently veined, 2 to 6 inches long. *Flowers* small, yellowish green, 3 to 5 clustered in leaf axils; blooming May or June. *Fruits* roundish, black, 2- to 4-seeded, about ⅓ inch in diameter; ripening September or October.

RANGE. Southwestern Virginia, West Virginia, the Ohio Valley, and Nebraska; south to Florida and Texas.

Also called Indian-cherry.

NEW JERSEY-TEA *Ceanothus americanus*

FIELD MARKS. A leaf-losing shrub 1 to 3 feet high; growing on dry, rocky, wooded slopes and in clearings. *Leaves* alternate, egg-shaped, rounded or somewhat heart-shaped at base, pointed at tip, finely toothed on margin, more or less hairy above, paler and downy beneath, 1 to 3 inches long, with 3 prominent veins from near summit of the short leafstalks. *Flowers* small, white, in dense, cylindrical, long-stalked clusters; blooming May to Auugst. *Fruits* small, 3-lobed capsules containing 3 pale-brown seeds; maturing August to October; the silvery-lined, cup-shaped bases persisting into the winter.

RANGE. Maine to southern Quebec and Manitoba; south to Florida and Texas.

The leaves were used as a substitute for tea during the American Revolution; and the large red roots yield a dye.

REDROOT *Ceanothus herbaceus*

FIELD MARKS. A leaf-losing shrub 1 to 2 feet high; growing on dry rocky or sandy soils. *Leaves* alternate, elliptic or lance-shaped, rounded to broadly pointed at base, bluntly pointed at tip, finely toothed on margin, smooth above, sometimes slightly downy on the veins beneath, ¾ to 2 inches long; with 3 prominent veins arising near the summit of the short leafstalk. *Flowers* small, white, in rather short and somewhat flat-topped end clusters which have stalks shorter than the leaves; blooming April to July. *Fruits* similar to those of New Jersey-Tea but with dark-brown seeds.

RANGE. Western Maine to Quebec and Manitoba, south to western Georgia and Texas.

SMALL-LEAF REDROOT *Ceanothus microphyllus*

FIELD MARKS. A diffusely branched evergreen shrub 1 to 2 feet high; growing in dry sandy pineland, pine-oak woods, or on sandhills. *Branchlets* yellowish brown, very slender, smooth, ascending. *Leaves* alternate, roundish to oval or elliptic, untoothed on margin, rather fleshy in texture, ⅛ to ¼ inch long, commonly with clusters of smaller leaves in the axils. *Flowers* small, white, in flat-topped, stalkless end clusters; blooming April or May. *Fruits* similar to those of New Jersey-Tea but smaller.

RANGE. Coastal plain; southern Georgia and Alabama south into Florida.

GRAPE FAMILY (Vitaceae)

PEPPER VINE *Ampelopsis arborea*

FIELD MARKS. A leaf-losing, somewhat bushy to high-climbing vine; growing in rich moist woods and thickets. Tendrils may be present opposite some of the leaves. *Leaves* alternate, twice or thrice compound, 2 to about 8 inches long; the leaflets more or less egg-shaped, rounded to pointed at base, pointed at tip, sharply and coarsely toothed on margin, smooth on both surfaces or slightly downy along the veins beneath, ½ to 1½ inches long. *Flowers* small, greenish, in loose, long-stalked clusters opposite some of the leaves; blooming June or July. *Fruits* roundish or slightly flattened, dark purple to black, 1- to 3-seeded berries about ¼ inch in diameter, bitter and inedible; ripening August to October.

RANGE. Eastern Maryland, West Virginia, southern Illinois, Missouri, and Oklahoma; south to Florida and Texas.

HEARTLEAF AMPELOPSIS *Ampelopsis cordata*

FIELD MARKS. A leaf-losing climbing vine; growing in swamps and stream bottoms. Tendrils may be present opposite some of the leaves. *Branchlets* with a white pith. *Leaves* alternate, egg-shaped or broadly so, heart-shaped to flattened at base, pointed at tip, coarsely and irregularly toothed on margin and sometimes 3-lobed, smooth above, paler and sometimes slightly downy beneath, 2 to 5 inches long. *Flowers* small, greenish, in long-stalked, forking clusters opposite some of the leaves; blooming May to July. *Fruits* roundish or slightly flattened, bluish, usually 2-seeded berries about ¼ inch in diameter, inedible; ripening August to October.

RANGE. Virginia to southern Ohio and Illinois, southeastern Missouri, and Oklahoma; south to Florida and Texas.

Often confused with Muscadine Grape, the branchlets of which have a brownish pith.

Redroot

Small-leaf Redroot

Pepper Vine

Heartleaf Ampelopsis

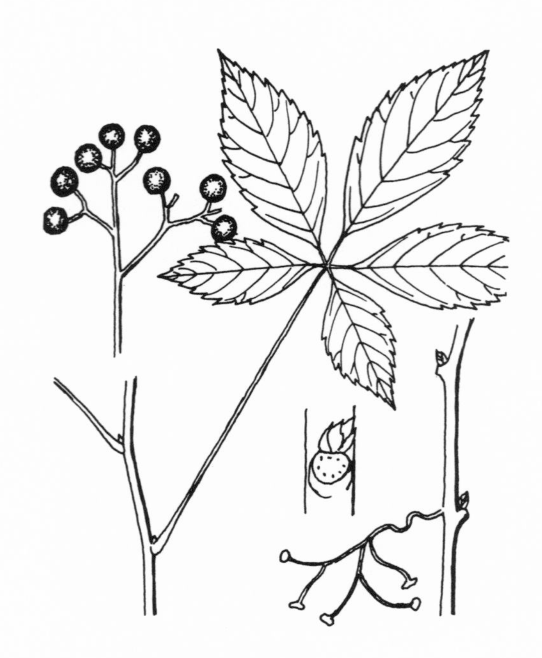

Virginia Creeper

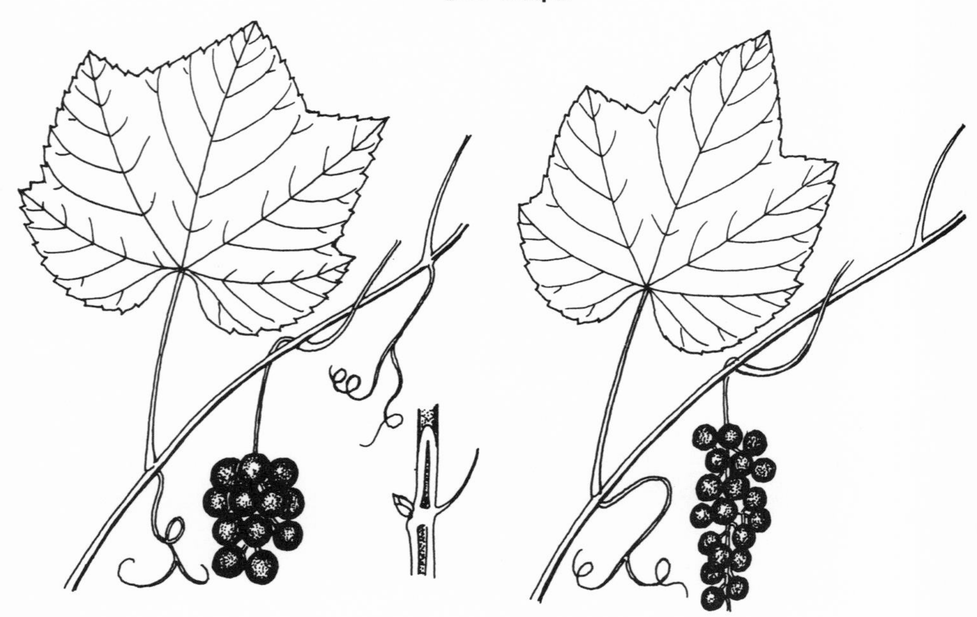

Fox Grape

Summer Grape

VIRGINIA CREEPER *Parthenocissus quinquefolia*

FIELD MARKS. A climbing or sometimes trailing leaf-losing vine; growing in moist woods and thickets. Tendrils opposite some of the leaves are 5- to 12-branched and end in small adhesive disks. *Leaves* alternate, compound, with usually 5 leaflets radiating from the summit of a long leafstalk; leaflets egg-shaped or elliptic, wedge-shaped at base, pointed or abruptly pointed at tip, margin sharply and coarsely toothed to or slightly below the middle, dull above, paler and sometimes downy beneath, 2 to 8 inches long. *Flowers* small, greenish, in branched end clusters; blooming June to August. *Fruits* roundish, dark blue, 2- to 3-seeded berries about ¼ inch in diameter; ripening August to October.

RANGE. Maine and Quebec to Minnesota, south to Florida and Texas.

Often confused with Poison-ivy. Cultivated as an ornamental.

GRAPES (Vitis)

(Key Appendix J)

Grapes are usually high-climbing vines with shredding bark, alternate simple leaves, and forked tendrils. The branchlets usually have a brownish pith which is interrupted by harder partitions (diaphragms) at the nodes. The leaves have main veins radiating from the summit of a long leafstalk, are coarsely toothed on the margin, and are often lobed. The flowers are small, greenish yellow, fragrant, and borne in compact clusters opposite the leaves; blooming in spring as the new leaves begin to expand. The fruits are few-seeded, round, pulpy berries ripening in late summer or fall. They are an important food for wildlife. Although most of our wild grapes are too tart to be eaten raw, they make very good jelly and preserves.

FOX GRAPE *Vitis labrusca*

FIELD MARKS. A high-climbing vine; growing in rich woods, thickets, or along streams. *Branchlets* more or less rusty-woolly, with a tendril or flower cluster opposite each leaf. *Leaves* heart-shaped, shallowly or sometimes deeply 3-lobed, margin coarsely but shallowly toothed, somewhat leathery, smooth above, felted with tawny or rusty wool beneath, 4 to 8 inches long. *Fruits* purplish black, brownish purple, or (rarely) amber-colored, 3- to 6-seeded, about ½ inch across, usually sweet and somewhat musky.

RANGE. Southern Maine to Michigan, south to Florida and Mississippi.

Parent of Concord, Catawba, Niagra, and other cultivated grapes.

SUMMER GRAPE *Vitis aestivalis*

FIELD MARKS. A high-climbing vine; growing in woods, thickets, or on stream banks. *Branchlets* usually somewhat downy, no tendril or flower cluster opposite every third leaf. *Leaves* heart-shaped, unlobed or shallowly to deeply 3- to 5-lobed, basal sinus narrowly to broadly U-shaped, smooth above, whitish and with some loose tawny or rusty wool beneath, 3 to 8 inches long. *Fruits* black with a whitish bloom, 2- to 3-seeded, about ⅜ inch across, very tart.

RANGE. Massachusetts to Michigan and Wisconsin, south to Georgia and Texas.

BLUELEAF GRAPE *Vitis aestivalis* var. *bicolor*

FIELD MARKS. This variety is readily distinguished by its smooth branchlets, which are whitened with a powdery bloom, and by its leaves, which are smooth or nearly so and very white beneath. (Not illustrated)

POST OAK GRAPE *Vitis aestivalis* var. *lincecumii*

FIELD MARKS. This variety is distinguished by the loose, rusty wool on lower surface of the leaves. *Fruits* purplish black with a slight white bloom, ⅜ to 1 inch in diameter, pleasant-tasting.

RANGE. Southern Indiana and Missouri, south to Mississippi and Texas. (Not illustrated)

RIVERBANK GRAPE *Vitis riparia*

FIELD MARKS. A trailing or high-climbing vine; growing along streams and in alluvial bottomlands. *Branchlets* smooth or nearly so, no tendril or flower cluster opposite every third leaf, pith with thin partitions at the nodes. *Leaves* heart-shaped, usually 3-lobed, margin with large and sharply pointed teeth, sinus at base V-shaped but broad and open, bright green and lustrous on both surfaces, sometimes slightly hairy on larger veins beneath, 2½ to 5 inches long. *Fruits* bluish black and with a white bloom, 2- to 4-seeded, about ⅜ inch in diameter, usually sour.
RANGE. New Brunswick to Manitoba; south to Virginia, Tennessee, Missouri, and Texas.

FROST GRAPE *Vitis vulpina*

FIELD MARKS. A high-climbing vine; growing in low rich woods or thickets and along streams. *Branchlets* smooth or nearly so, no tendril or flower cluster opposite every third leaf, pith with thick partitions at the nodes. *Leaves* heart-shaped, unlobed or occasionally slightly 3-lobed, margin coarsely and sharply toothed, sinus at base deep and narrowly V-shaped, bright green on both surfaces, lustrous above, sometimes sparingly hairy along the veins beneath, 3 to 5 inches long. *Fruits* lustrous black, 2- to 3-seeded, about ⅜ inch in diameter, often sweet after frost.
RANGE. Southeastern New York and Pennsylvania to Illinois and eastern Kansas; south to Florida and Texas.

NEW ENGLAND GRAPE *Vitis* × *novae-angliae*

FIELD MARKS. A hybrid high-climbing grape from *V. labrusca* and *V. riparia*; growing in dry woods and thickets. *Branchlets* reddish-hairy, a tendril or flower cluster opposite each leaf. *Leaves* heart-shaped, unlobed or sometimes with a short pointed lobe on each side above the middle, margin with broad and sharply pointed teeth, sinus at base broadly V-shaped or the base almost flattened, lustrous above, smooth or nearly so on both surfaces at maturity, rusty-hairy when young, 2 to 4 inches long. *Fruits* black, usually with a white bloom, about ⅜ inch in diameter, sour.
RANGE. New England south to eastern New York, New Jersey, and central Pennsylvania.

WINTER GRAPE *Vitis cinerea*

FIELD MARKS. A high-climbing vine; growing in dry woods and thickets. *Branchlets* angled, grayish-hairy, no tendril or flower cluster opposite every third leaf. *Leaves* heart-shaped, unlobed or with a short lobe on each side above the middle, sinus at base narrowly or broadly V-shaped, margin with broad and pointed teeth, smooth and dull green above, more or less whitish-wooly beneath, 2 to 6 inches long; leafstalks grayish-hairy. *Fruits* black, usually about a bloom, 1- to 3-seeded, about ⅜ inch in diameter, usually sweet when ripe.
RANGE. Southeastern Virginia to southern Ohio, Illinois, Iowa, and Nebraska; south to Florida and Texas.

PIGEON GRAPE *Vitis cinerea* var. *floridana*

A variety of the Winter Grape, with rusty or reddish-brown hairs on the branchlets, leafstalks, and lower leaf surfaces. More common than the typical variety eastward. Eastern Virginia and Arkansas south to Florida and Texas. (Not illustrated)

Riverbank Grape

Frost Grape

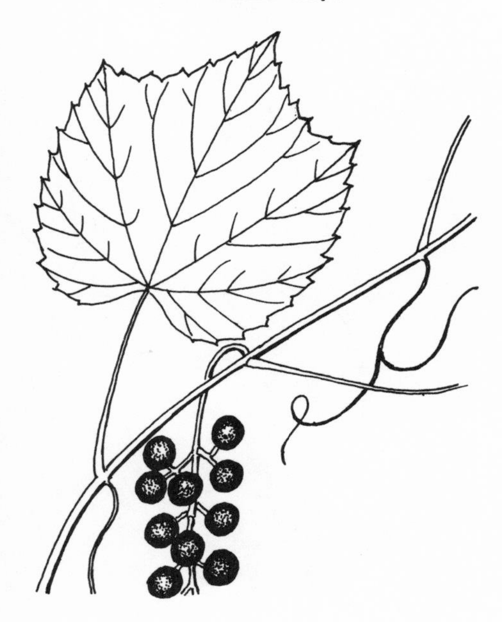

New England Grape

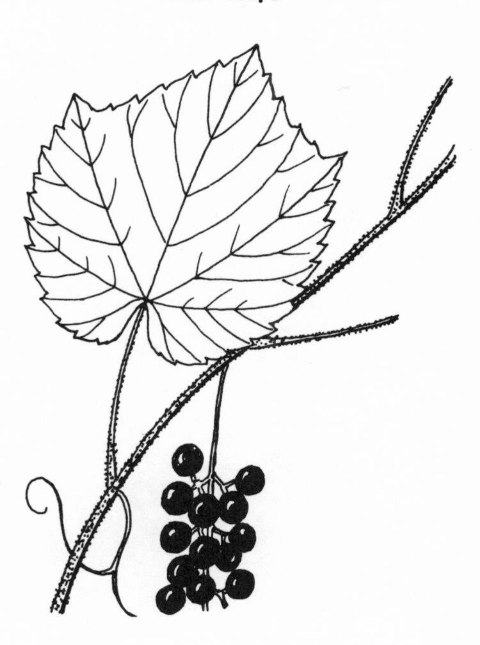

Winter Grape

Possum Grape

Red Grape

Bush Grape

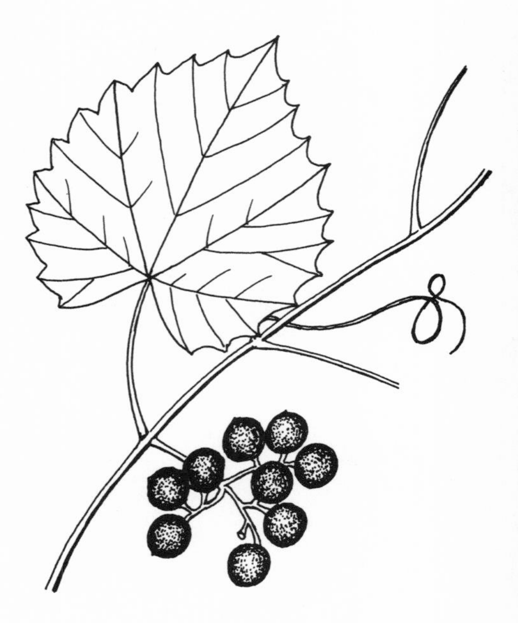

Muscadine Grape

POSSUM GRAPE *Vitis cinerea* var. *baileyana*

FIELD MARKS. Another variety of winter grape, distinguished by its nearly smooth, angled branches (hairy only when young).

RANGE. Eastern Virginia to Kentucky and Missouri south to Georgia and Arkansas.

RED GRAPE *Vitis palmata*

FIELD MARKS. A slender, high-climbing vine; growing in low, moist woods, borders of ponds, sloughs, etc. *Branchlets* at first bright red, later turning reddish brown, smooth or nearly so, no tendril or flower cluster opposite every third leaf. *Leaves* egg-shaped, sinus at base broadly U-shaped, deeply 3- to 5-lobed with rounded sinuses between the long-pointed lobes, margin coarsely and irregularly toothed, dark green and smooth above, bright green and sometimes downy on the veins (which are commonly red) beneath, 2 to 5 inches long. *Fruits* black, without a bloom, 1- or 2-seeded, about ⅜ inch across.

RANGE. Southern Indiana and Illinois to southeastern Iowa, south to Louisiana and Texas.

Also called Cat or Catbird Grape.

BUSH GRAPE *Vitis rupestris*

FIELD MARKS. A low, bushy, or slightly climbing shrub; growing on sandy banks and rocky slopes. *Branchlets* only occasionally with tendrils. *Leaves* roundish to broadly heart-shaped, the sinus at the base very broadly V-shaped, unlobed or sometimes slightly 3-lobed, margin with coarse and broad but sharply pointed teeth, lustrous above, slightly paler and smooth or nearly so beneath, 2 to 4 inches long. *Fruits* black, usually with a white bloom, 2- to 4-seeded, about ⅜ inch in diameter, sweet; in small but compact clusters.

RANGE. Maryland and southern Pennsylvania to Missouri; south to western Virginia, Tennessee, Arkansas, and Texas.

Also called Sand Grape and Sugar Grape.

MUSCADINE GRAPE *Vitis rotundifolia*

FIELD MARKS. A high-climbing vine; growing in woods, thickets, sandhills, and swamps. Differs from all other grapes in having smooth bark dotted with paler lenticels, no woody partitions in the pith at the nodes, and tendrils which are not branched. *Leaves* roundish or broadly egg-shaped, the sinus at the base broadly V-shaped and shallow, margin with large and triangular teeth, lustrous above, yellowish green and sometimes slightly downy on the veins beneath, 2 to 4 inches wide. *Fruits* purplish black to bronze, without a bloom, with very tough skin, sweet with musky flavor, ½ to 1 inch in diameter.

RANGE. Southern Delaware and Virginia to southern Indiana, southeastern Missouri, and Oklahoma; south to Florida and Texas.

TEA FAMILY (Theaceae)

SILKY-CAMELLIA *Stewartia malachodendron*

FIELD MARKS. A leaf-losing shrub or small tree 5 to about 20 feet high; growing in rich deciduous woods, usually near streams. *Leaves* alternate, elliptic, wedge-shaped at base, pointed or abruptly pointed at tip, minutely toothed and hairy-fringed on margin, dark green and smooth above, paler and downy at least on the veins beneath, 2½ to 4 inches long. *Flowers* very showy, 3 to 4 inches across, the 5 creamy-white petals crimped and ragged on margin, stamens purple; styles united; blooming May or June. *Fruits* roundish capsules ½ inch or slightly more in diameter; seeds shiny, not winged.

RANGE. Chiefly coastal plain; Virginia south to Florida, west to Louisiana, north in Mississippi Valley to Tennessee and eastern Arkansas.

MOUNTAIN-CAMELLIA *Stewartia ovata*

FIELD MARKS. A leaf-losing shrub or small tree similar to the preceding; growing in rich woods, usually along streams, chiefly in the mountains. *Leaves* alternate, oblong-elliptic or egg-shaped, mostly rounded at base, sharply pointed at tip, minutely toothed and hairy-fringed on margin, dark green and smooth above, grayish green and slightly hairy beneath, 2 to 5 inches long. *Flowers* similar to those of the Silky-camellia, but the stamens usually with yellow (rarely purple) anthers and the styles distinct; blooming June to August. *Fruits* egg-shaped, pointed, sharply 5-angled capsules about ⅝ inch long; seeds dull and winged.

RANGE. Virginia and southeastern Kentucky; south to central North Carolina, and northern Georgia and Alabama.

ST. JOHN'S-WORT FAMILY (Clusiaceae)

ST. PETER'S-WORTS AND ST. JOHN'S-WORTS (Hypericum)

(Key Appendix K)

These are herbaceous or shrubby plants, with opposite simple leaves with untoothed margins and usually with minute clear or blackish dots. The branchlets are commonly more or less 2-edged below the leaves, and often there are clusters of smaller leaves in the axils of the larger ones.

ST. PETER'S-WORT *Hypericum crux-andreae*

FIELD MARKS. A shrub with a simple or sparingly branched, more or less erect stem 1 to 3 feet high; growing in sandy woods and fields. *Branchlets* prominently 2-edged or 2-winged. *Leaves* oblong-oval, stalkless, rounded to somewhat heart-shaped and often clasping the stem at the base, rounded at tip, thickish and firm in texture, smooth, ½ to 1½ inches long. *Flowers* ¾ to 1¼ inches across, with 4 bright yellow petals, 2 outer sepals which are large and heart-shaped and 2 inner ones very much smaller and narrow, pistil with 3 or 4 styles; blooming July to September. *Fruits* egg-shaped capsules about ⅜ inch long.

RANGE. Southeastern Massachusetts to southeastern Pennsylvania and Kentucky, south to Florida and Texas.

DWARF ST. PETER'S-WORT *Hypericum suffruticosum*

FIELD MARKS. A shrub resembling the preceding but smaller and with prostrate stems often forming mats, the erect flowering branches only a few inches high; growing in sandy coastal-plain pinelands. *Leaves* stalkless, oval to narrowly elliptic, ⅛ to about 5/16 inch long. *Flowers* yellow, about ½ inch across, the 2 outer sepals very small or absent. *Fruits* about 3/16 inch long.

RANGE. North Carolina south to Florida, west to Mississippi.

Silky-camellia

Mountain-camellia

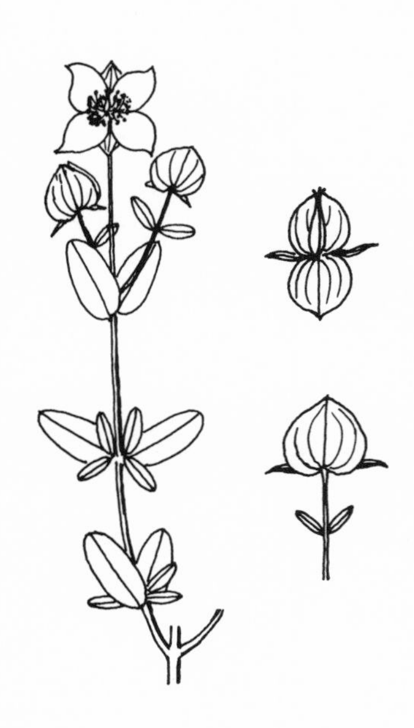

St. Peter's-wort

Dwarf St. Peter's-wort

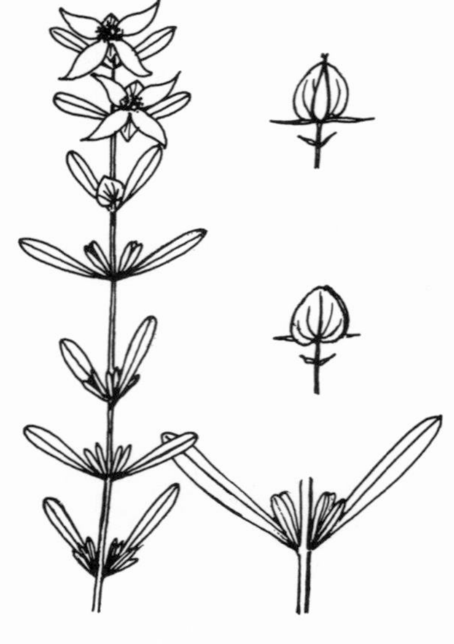
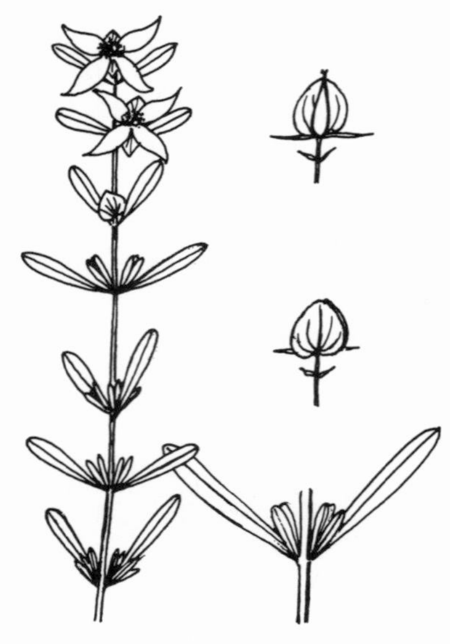

St. Andrew's-cross

Reclining St. Andrew's-cross

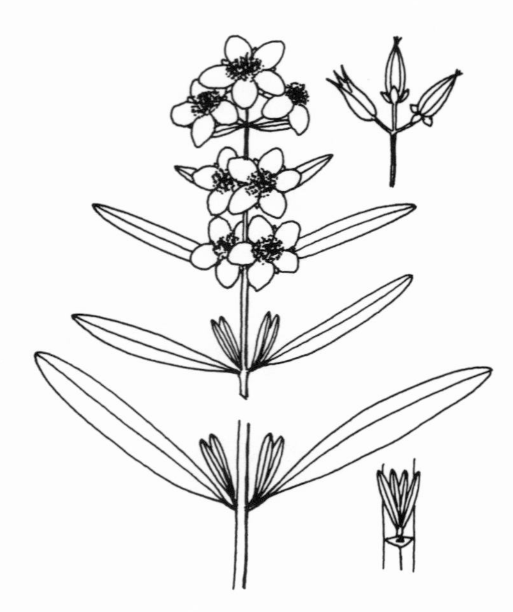

Shrubby St. John's-wort

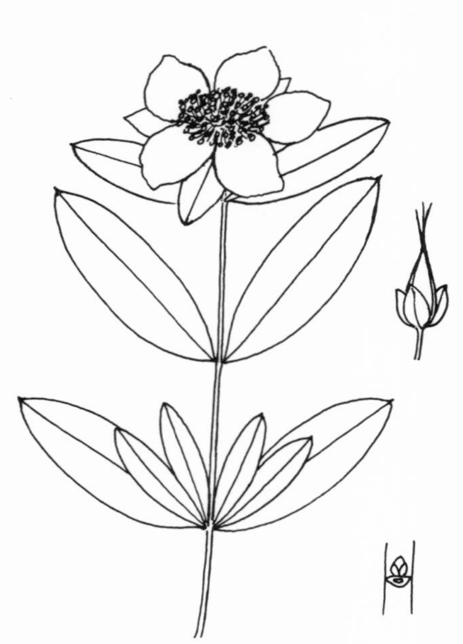

Golden St. John's-wort

ST. ANDREW'S-CROSS *Hypericum hypericoides*

FIELD MARKS. A more or less erect and branching shrub to about 3 feet high; growing in dry sandy or rocky fields and open woods. *Branchlets* flattened and 2-edged. *Leaves* stalkless, ¼ to about 1 inch long and mostly less than ⅛ inch wide, broadest toward the tip, wedge-shaped at base, blunt at tip, smooth. *Flowers* about ½ inch across, with 4 bright-yellow petals, 2 outer sepals which are large and egg-shaped or heart-shaped and 2 inner ones very much smaller and narrow, pistil with 2 styles; blooming May to September. *Fruits* narrowly egg-shaped and somewhat flattened 2-celled capsules about ¼ inch long.

RANGE. Virginia and Tennessee south to Florida and Texas.

RECLINING ST. ANDREW'S-CROSS *Hypericum stragalum*

FIELD MARKS. A shrub similar to the preceding but with reclining stems and numerous erect branchlets 5 inches to about 1 foot high; growing in dry sandy or rocky woods and clearings. *Leaves* similar to those of the preceding species but usually shorter and broader. *Flowers* and *fruits* also similar.

RANGE. Southeastern Massachusetts to southern Pennsylvania, Illinois, and Kansas; south to Georgia and Texas.

SHRUBBY ST. JOHN'S-WORT *Hypericum spathulatum*

FIELD MARKS. An erect, bushy-branched, leaf-losing shrub 2 to about 4 feet high; usually growing in moist sandy or rocky open woods, fields, and slopes. *Leaves* very short-stalked, narrowly oblong or sometimes broadest toward the tip, pointed at base, blunt at tip, light green above, pale beneath, smooth on both surfaces, ¾ to 3 inches long. *Flowers* bright yellow, about ¾ inch across, 5-petalled, clustered in the axils of the upper leaves; blooming June to September. *Fruits* narrowly egg-shaped, 3-celled capsules about ½ inch long.

RANGE. Southeastern New York to Ontario and Minnesota; south to Georgia, Mississippi, and Arkansas.

One of our most showy species of St. John's-worts; sometimes cultivated and locally escaping as far north as Massachusetts.

GOLDEN ST. JOHN'S-WORT *Hypericum frondosum*

FIELD MARKS. A widely branched, often semi-evergreen shrub 2 to about 6 feet high; growing on limestone or shale bluffs in the southern Appalachians. *Leaves* stalkless or nearly so, oblong to narrowly oblong, narrowed at base, roundish or blunt at tip but often with an abrupt little point, bluish green above, pale or somewhat whitened beneath, smooth on both surfaces, 1 to 3 inches long. *Flowers* bright golden yellow, 1 to 2 inches across, 5-petalled, stamens very numerous, sepals large and leaflike but not all of the same size; blooming June to August. *Fruits* narrowly egg-shaped, somewhat 3-celled capsules ½ to ¾ inch long, more or less enclosed by the sepals.

RANGE. Kentucky and southern Indiana; south to Georgia, Alabama, and Texas.

Sometimes cultivated for its large showy flowers.

KALM ST. JOHN'S-WORT *Hypericum kalmianum*

FIELD MARKS. A widely branching leaf-losing shrub 1 to 2½ feet high; growing in sandy or rocky open woods and clearings. *Leaves* stalkless, narrowly oblong or broadest toward the tip, pointed at base, blunt to pointed at tip, smooth on both surfaces, whitened beneath, 1 to 2 inches long. *Flowers* bright yellow, ¾ to 1 inch across, 5-petalled and with 5 distinct styles; blooming July to September. *Fruits* narrowly egg-shaped, usually 5-celled capsules about ⅜ inch long.

RANGE. Region of the Great Lakes from western New York and western Quebec to Illinois and western Ontario.

BUSHY ST. JOHN'S-WORT *Hypericum densiflorum*

FIELD MARKS. A much-branched, bushy, leaf-losing shrub 1 to about 6 feet high; growing in swamps, swales, boggy places or sometimes in dry woods. *Leaves* stalkless or almost so, very narrow or narrowly oblong, pointed at base, blunt or somewhat pointed at tip, margin slightly rolled beneath, light green above, slightly paler beneath, smooth on both surfaces, ¾ to 2 inches long. *Flowers* bright yellow, about ⅜ inch across, 5-petalled, in rather crowded, forking, flat-topped end clusters; blooming June to September. *Fruits* narrowly egg-shaped, 3-celled capsules usually less than ¼ inch long.

RANGE. Southeastern New York to southern Pennsylvania and Kentucky; south to Georgia and Alabama.

STRAGGLING ST. JOHN'S-WORT *Hypericum dolabriforme*

FIELD MARKS. A straggling shrub, prostrate at the base and with ascending or erect branches ½ to about 1½ feet high; growing in rocky woods or on balds and barrens. *Branchlets* slightly angled or quite round. *Leaves* narrow or narrowly lance-shaped, tapered at base, pointed or blunt at tip, smooth on both surfaces, ¾ to about 2 inches long. *Flowers* bright yellow, nearly 1 inch across, the sepals almost as long as the 5 oblique petals, in a forking, few-flowered, leafy-bracted end cluster; blooming June to August. *Fruits* egg-shaped, pointed, 1-celled capsules about ¼ inch long.

RANGE. Kentucky south to Georgia, west to southeast Missouri.

MYRTLE-LEAF ST. JOHN'S-WORT *Hypericum myrtifolium*

FIELD MARKS. A simple or sparingly-branched evergreen shrub 1 to 3 feet high; growing in low wet pinelands or shallow ponds. Leaves stalkless or nearly so, egg-shaped or elliptic, rounded at base, blunt at tip, veiny, smooth on both surfaces, paler beneath, more or less coated with a waxy bloom, ½ to 1¼ inches long. *Flowers* bright yellow, about ¾ inch across, 5-petalled, with large and leaflike sepals; blooming May to August. *Fruits* egg-shaped capsules about ¼ inch long.

RANGE. Coastal plain; Georgia and Florida west to Mississippi.

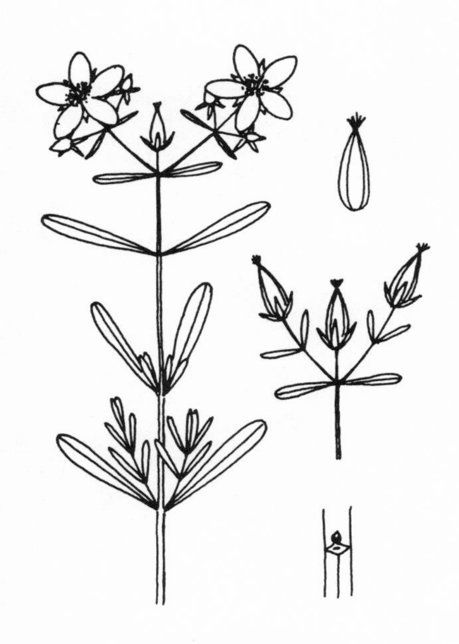

Kalm St. John's-wort

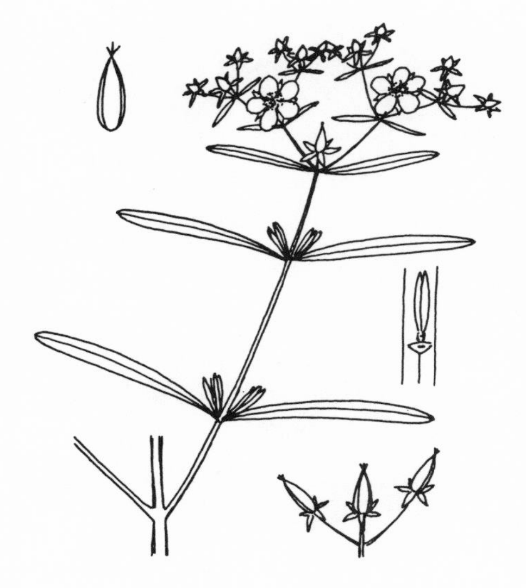

Bushy St. John's-wort

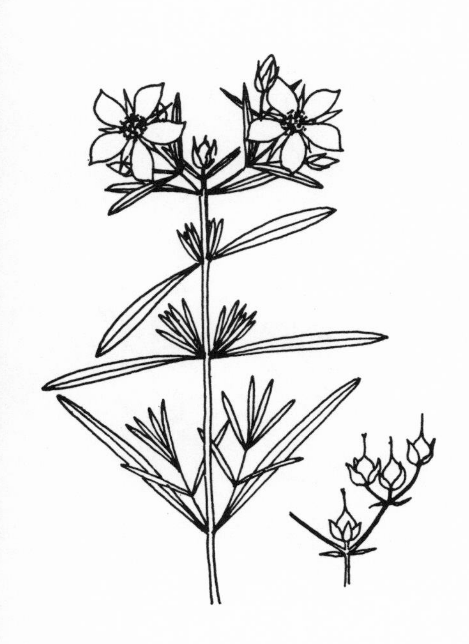

Straggling St. John's-wort

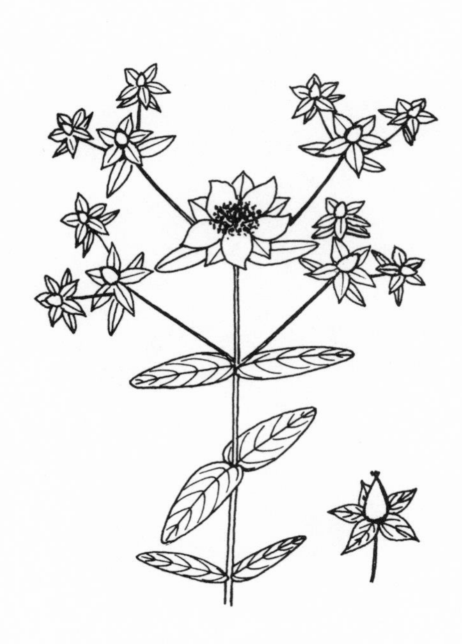

Myrtle-leaf St. John's-wort

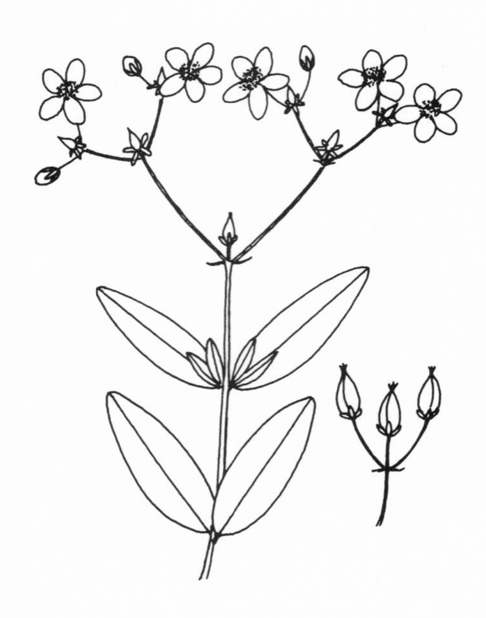
Naked-flowered St. John's-wort

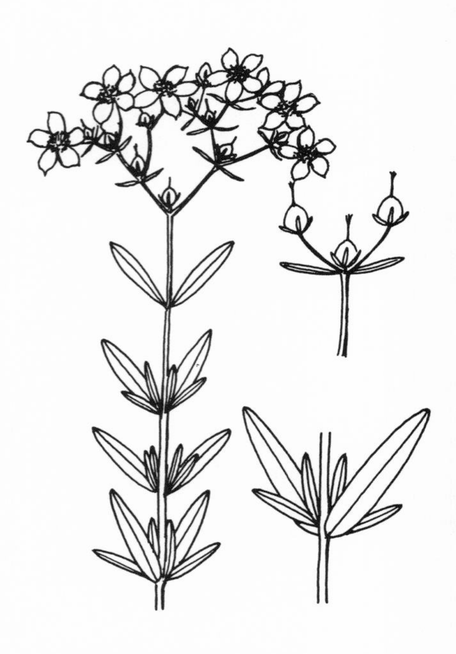
Round-podded St. John's-wort

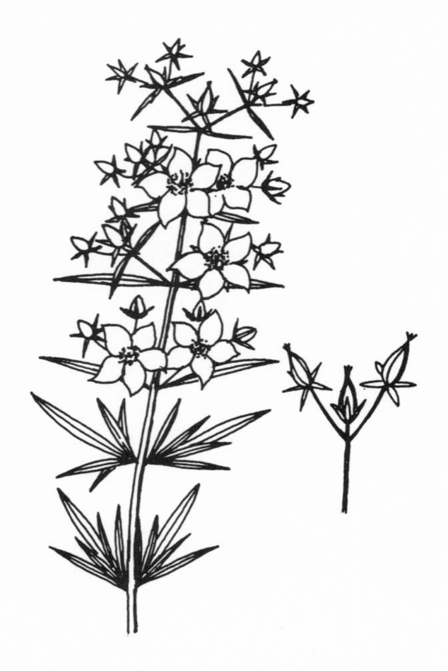
Bedstraw St. John's-wort

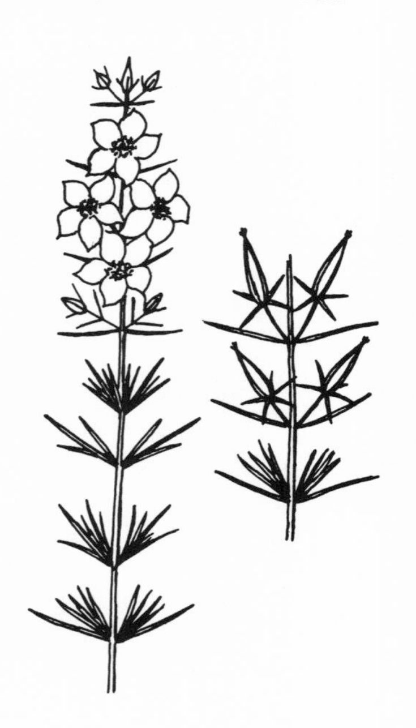
Sandweed

NAKED-FLOWERED ST. JOHN'S-WORT *Hypericum nudiflorum*

FIELD MARKS. A shrub 1 to 3 feet high, often more or less herbaceous above; growing in low woods, swamps, or along the banks of streams. *Branchlets* 2-edged, brown. *Leaves* elliptic or narrowly egg-shaped, pointed or roundish at base, blunt or bluntly pointed at tip, thin in texture, smooth, ¾ to 2¾ inches long. *Flowers* bright yellow, ½ inch or slightly more across, 5-petalled, in a forking and rather flat-topped end cluster; blooming May to September. *Fruits* narrowly egg-shaped, more or less 1-celled capsules about ¼ inch long.

RANGE. Virginia and Tennessee south to Florida and Mississippi.

ROUND-PODDED ST. JOHN'S-WORT *Hypericum cistifolium*

FIELD MARKS. A simple or sparingly-branched shrub 1 to 2½ feet high; growing in low pinelands and savannahs. *Branchlets* more or less erect, dark purplish brown, 4-angled. *Leaves* narrowly oblong or narrowly elliptic, stalkless or nearly so, narrowed or roundish at base, blunt or broadly pointed at tip, smooth on both surfaces, paler below, ¾ to about 2 inches long. *Flowers* bright yellow, about ½ inch across, 5-petalled, in a forking and rather flat-topped end cluster; blooming June to August. *Fruits* roundish egg-shaped, 1-celled capsules about 3/16 inch long.

RANGE. Coastal plain; North Carolina south to Florida, west to Mississippi.

BEDSTRAW ST. JOHN'S-WORT *Hypericum galioides*

FIELD MARKS. An evergreen shrub 1 to about 4 feet high; growing in low wet pinelands and swamps. *Branchlets* slender, nearly round. *Leaves* stalkless, flat, narrow but usually slightly broadened upward, pointed at tip, tapering to base, thickish and firm in texture, dark green above, slightly paler beneath, smooth on both surfaces, ½ to about 2 inches long. *Flowers* bright yellow, about ½ inch across, 5-petalled, the sepals narrow and similar to the leaves, in rather narrow and elongate end clusters; blooming June to August. *Fruits* conical egg-shaped, pointed capsules about ¼ inch long.

RANGE. Coastal plain; North Carolina south to Florida, west to Louisiana.

SANDWEED *Hypericum fasciculatum*

FIELD MARKS. An evergreen shrub 1 to about 4 feet high; growing in low, wet, sandy pinelands and the borders of shallow ponds. *Branchlets* slender, 4-angled. *Leaves* stalkless, narrow and almost needle-like, leathery in texture, bright green, ⅜ to ¾ inch long; those of the axillary clusters often about as long; with a hand lens 2 longitudinal groves may be seen on the lower leaf surface. *Flowers* bright yellow, about ½ inch across, 5-petalled, the sepals narrow like the leaves, in rather narrow and elongate end clusters; blooming May to September. *Fruits* narrowly egg-shaped capsules about ¼ inch long.

RANGE. Coastal plain; South Carolina south to Florida, west to Texas.

SANDWEED *Hypericum nitidum*

This species is similar to the preceding but often taller (up to 10 feet high), and the leaves are not distinctly grooved beneath. It grows in wet places in the coastal plain from North Carolina south to Florida, west to Alabama. (Not illustrated)

SANDWEED *Hypericum reductum*

Similar to *H. fasciculatum* but growing in moist to dry sandy woods in the coastal plain from southeastern North Carolina south to Georgia and Alabama. The largest leaves are less than ½ inch long. (Not illustrated)

SANDWEED *Hypericum lloydii*

Similar to *H. fasciculatum* but with the main stems reclining and with upright branches 1 foot or less high. It grows in dry sandy or rocky woodlands in the coastal plain from North Carolina south to Florida. (Not illustrated)

MOUNTAIN ST. JOHN'S-WORT *Hypericum buckleyi*

FIELD MARKS. A low, creeping, somewhat woody, often matted, leaf-losing plant with ascending branches 4 to 12 inches high; growing in rock crevices and on seepage slopes in the southern Appalachians. *Branchlets* slender, 4-angled. *Leaves* short-stalked, elliptic or broadest above the middle, wedge-shaped at base, rounded at tip, smooth on both surfaces, paler beneath, ¼ to ¾ inch long. *Flowers* bright yellow, ¾ to 1 inch across, 5-petalled, with spoon-shaped sepals; blooming June to August. *Fruits* egg-shaped, pointed, 3-celled capsules ¼ to ⅜ inch long.
RANGE. North Carolina south to Georgia, in mountains above 3,500 feet.

ROCKROSE FAMILY (Cistaceae)

BEACH-HEATHER *Hudsonia tomentosa*

FIELD MARKS. An intricately branched and matted evergreen shrub 4 to 8 inches high, densely coated with soft white wooly hair; growing in dry sands and on sand dunes. *Leaves* scalelike, usually less than ⅛ inch long, overlapping and closely pressed to the branchlets. *Flowers* bright yellow, 5-petalled, about ¼ inch across, stalkless or nearly so, and solitary at the tips of the branchlets; blooming May to July. *Fruits* small capsules enclosed by the calyx.
RANGE. Labrador south along the coast to North Carolina, inland chiefly about the Great Lakes to Minnesota and Saskatchewan; also in Pendelton County, West Virginia.

GOLDEN-HEATHER *Hudsonia ericoides*

FIELD MARKS. A bushy-branched, tufted, evergreen shrub 4 to about 7 inches high, softly hairy but green throughout; growing in dry sandy or rocky places. *Leaves* slender, awl-shaped, more or less spreading, about ¼ inch long; densely crowded and with bases overlapping on the younger branchlets, more widely scattered on the older ones. *Flowers* bright yellow (rarely white), 5-petalled, about ⅜ inch across, on slender hairy stalks ¼ to ½ inch long, and solitary at the tips of the branchlets; blooming May to July. *Fruits* small capsules enclosed by the calyx.
RANGE. Newfoundland south chiefly along the coast to Virginia, and in the mountains of central New Hampshire.

MOUNTAIN-HEATHER *Hudsonia montana*

This species is very similar to the preceding one. It occurs on the summits of a few mountains in the North Carolina Blue Ridge. (Not illustrated)

LOOSESTRIFE FAMILY (Lythraceae)

SWAMP LOOSESTRIFE *Decodon verticillatus*

FIELD MARKS. A soft-woody plant with angled, recurved stems 3 to 9 feet long, which are thickened and spongy at the base; growing in shallow waters of lakes, ponds, streams, and marshes. *Leaves* opposite or in 3's, lance-shaped, pointed at both ends, untoothed on margin, bright green and smooth above, paler and often downy beneath, 2 to 5 inches long. *Flowers* purplish, about 1 inch across, 5-petalled, in axillary clusters; blooming July to September. *Fruits* urn-shaped capsules about ¼ inch across, with remnants of the sepals at the summit.
RANGE. Maine to southern Ontario south to Florida and Louisiana.
Also called Water-willow.

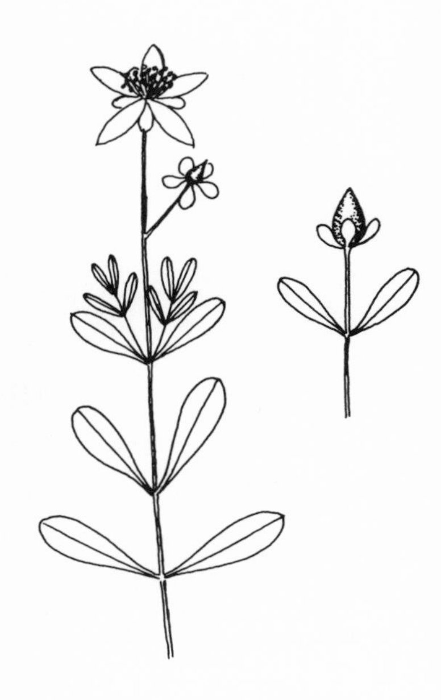

Mountain St. John's-wort

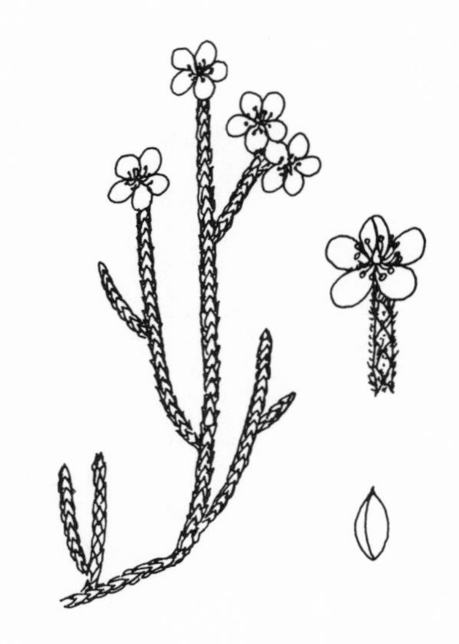

Beach-heather

Golden-heather

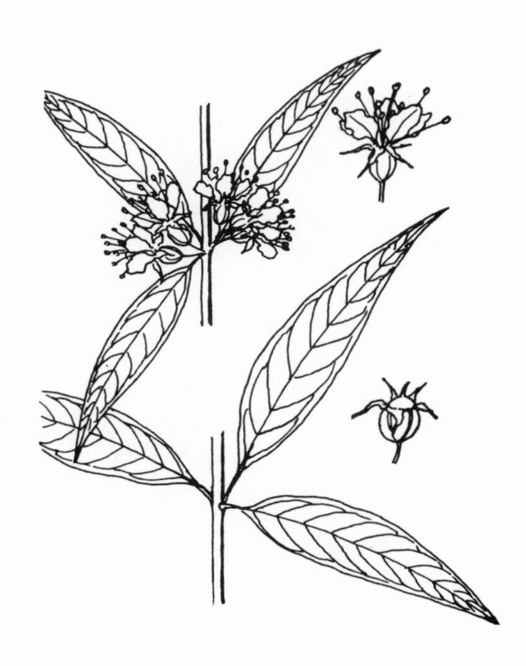

Swamp Loosestrife

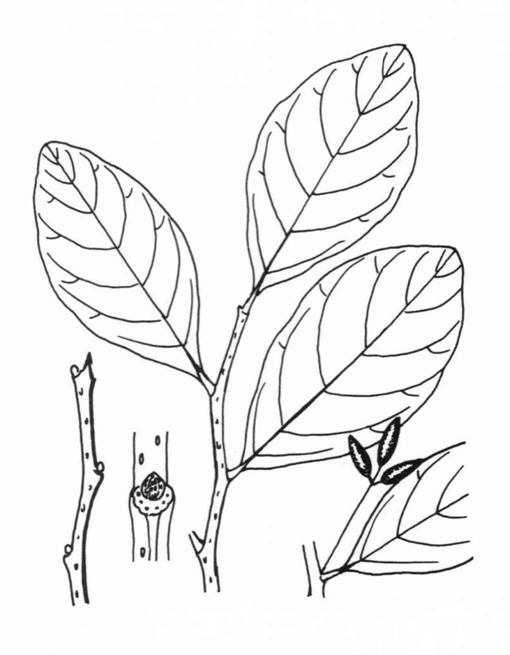

Leatherwood

Canada Buffaloberry

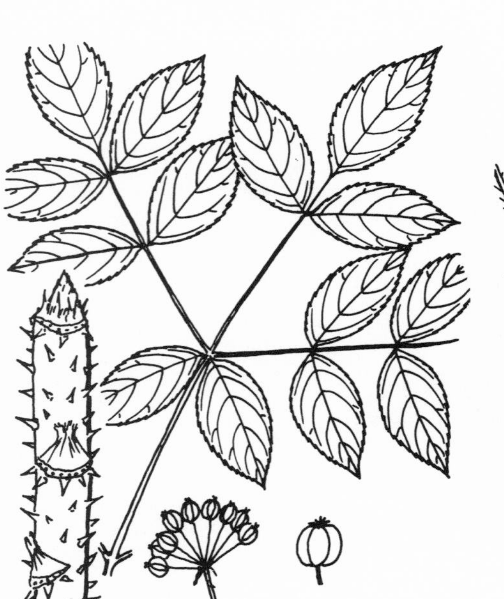

Hercules'-club

Bristly Sarsaparilla

MEZEREUM FAMILY (Thymelaeaceae)

LEATHERWOOD *Dirca palustris*

FIELD MARKS. A leaf-losing shrub 2 to 8 feet high, often with a solitary trunklike stem; growing in rich moist woods. *Branchlets* tough, pliable, enlarged at nodes and at tip, and appearing jointed. *Leaves* alternate, short-stalked, oval to elliptic or broadest above the middle, pointed to roundish at base, blunt or broadly pointed at tip, margin untoothed, smooth or nearly so, 1½ to 3½ inches long. *Flowers* small, pale yellow, 2 to 4 in clusters; blooming April or May, before the leaves appear. *Fruits* berry-like, oval, red, 1-seeded, about ⅜ inch long; ripening May or June.

RANGE. New Brunswick to Ontario and Minnesota, south to Florida and Mississippi.

Bark was used by Indians for bow strings, cordage, and baskets. Taken internally it induces vomiting and purging; externally it may irritate the skin.

OLEASTER FAMILY (Elaeagnaceae)

CANADA BUFFALOBERRY *Shepherdia canadensis*

FIELD MARKS. A leaf-losing shrub 3 to 5 feet high; growing in sandy or rocky woods and along streams. *Branchlets* slender; densely coated with rusty, branlike scales. *Leaves* opposite, short-stalked, elliptic to egg-shaped, roundish to broadly pointed at base, blunt at tip, margin untoothed, dull green above, coated with silvery starry-branched hairs and rusty scales beneath, ¾ to 1½ inches long. *Flowers* small, yellowish green, bell-shaped, clustered; blooming April or May, before the leaves appear. *Fruits* berry-like, egg-shaped, reddish or yellowish, 1-seeded, about ¼ inch in diameter; ripening June to August.

RANGE. Newfoundland to Alaska; south to Maine, Vermont, the region of the Great Lakes, South Dakota, and New Mexico.

GINSENG FAMILY (Araliaceae)

HERCULES'-CLUB *Aralia spinosa*

FIELD MARKS. A leaf-losing shrub or small tree 5 to 15 or more feet high; growing in woods, clearings, or along streams. *Branchlets* very stout and armed with prickles. *Leaves* alternate, 2 to 3 feet long, doubly or triply compound; the numerous leaflets egg-shaped, rounded or broadly pointed at base, pointed at tip, sharply toothed on margin, smooth above, paler and sometimes downy beneath, 1½ to 3½ inches long. *Flowers* small, white, in numerous clusters (umbels) and in a large pyramid-shaped end group; blooming June to September. *Fruits* berry-like, egg-shaped, black, about ¼ inch long; ripening August to October.

RANGE. Southern New England and central New York to Michigan and Iowa, south to Florida and Texas.

BRISTLY SARSAPARILLA *Aralia hispida*

FIELD MARKS. A woody-based, leaf-losing plant 1½ to about 10 feet high; growing in rocky or sandy open woods and clearings. *Branchlets* moderate; with numerous weak, needle-like or bristly prickles. *Leaves* alternate, 4 to 12 inches long, usually twice compound; the leaflets egg-shaped to lance-shaped, rounded to pointed at base, pointed at tip, sharply toothed on margin, paler beneath and sometimes bristly along the veins, 1 to 2 inches long. *Flowers* small, white, in 2 to about 7 clusters (umbels) at stem tip; blooming June to August. *Fruits* berry-like, roundish, purplish black, about ¼ inch in diameter; ripening August or September.

RANGE. Newfoundland to Manitoba; south to western Virginia, West Virginia, and the region of the Great Lakes.

DEVIL'S-CLUB *Oplopanax horridus* (Sm.) Miq.

FIELD MARKS. A leaf-losing shrub 6 to 10 feet high; growing in rocky places. *Branchlets* stout, covered with slender prickles. *Leaves* alternate, roundish, heart-shaped at base, margin sharply and irregularly toothed and with 3 to 11 pointed lobes, 4 to 12 inches wide; the long leafstalks and veins of the lower leaf surfaces prickly. *Flowers* small, greenish white, the many clusters (umbels) in an elongate end group; blooming June to August. *Fruits* berry-like, red, roundish, about ¼ inch in diameter; ripening August to October.

RANGE. Western Ontario and Isle Royale (Michigan) to Alaska, south to Montana and California.

DOGWOOD FAMILY (Cornaceae)

DOGWOODS (Cornus)

(Key Appendix L)

Dogwoods are leaf-losing shrubs or small trees with opposite (rarely alternate), simple leaves which are untoothed on the margin and have veins which curve and tend to parallel the margin or to meet at the tip. The flowers are small, perfect, greenish yellow or whitish, and borne in open or compact end clusters. The fruits have a large stone enclosing the seed surrounded by a fleshy portion. The Flowering Dogwood (*Cornus florida* L.) is a well-known small tree, with showy bracts surrounding the flower clusters.

RED-OSIER DOGWOOD *Cornus sericea*

FIELD MARKS. A shrub 3 to 9 feet high, often with stems partly prostrate and rooting; growing in wet or swampy places. *Branchlets* purplish red to blood-red, with a large white pith. *Leaves* elliptic to egg-shaped, roundish or broadly pointed at base, pointed at tip, veins 4 to 6 pairs, smooth or minutely hairy on both surfaces, whitened beneath, 2 to 4 inches long. Flowers whitish, in flat-topped clusters; blooming May to July. *Fruits* roundish, white, about ¼ inch in diameter; ripening July to September.

RANGE. Newfoundland to Yukon; south to western Maryland, West Virginia, the region of the Great Lakes, Iowa, New Mexico, and California.

SILKY DOGWOOD *Cornus amomum*

FIELD MARKS. A shrub 4 to about 10 feet high; growing in wet places and along streams. *Branchlets* purplish red, more or less covered with minute closely pressed hairs; pith large and brownish. *Leaves* egg-shaped to broadly elliptic, rounded at base, pointed at tip, veins 3 to 5 pairs, smooth or nearly so above, pale and usually with some small reddish hairs beneath, 2 to 4 inches long. *Flowers* creamy white, in flat-topped to slightly convex clusters; blooming May to July. *Fruits* roundish, dull blue or partly white, about ¼ inch in diameter; ripening August to October.

RANGE. Southern Maine to Illinois, south to Georgia and Alabama.

Also called Kinnikinnik.

PALE DOGWOOD *Cornus amomum* var. *obliqua*

FIELD MARKS. A shrub very similar to the preceding but differing in having lance-shaped or narrowly egg-shaped leaves which are pointed at both ends, and with whitish, closely pressed hairs beneath.

RANGE. New Brunswick to North Dakota; south to New Jersey, West Virginia, Kentucky, Arkansas, and Oklahoma.

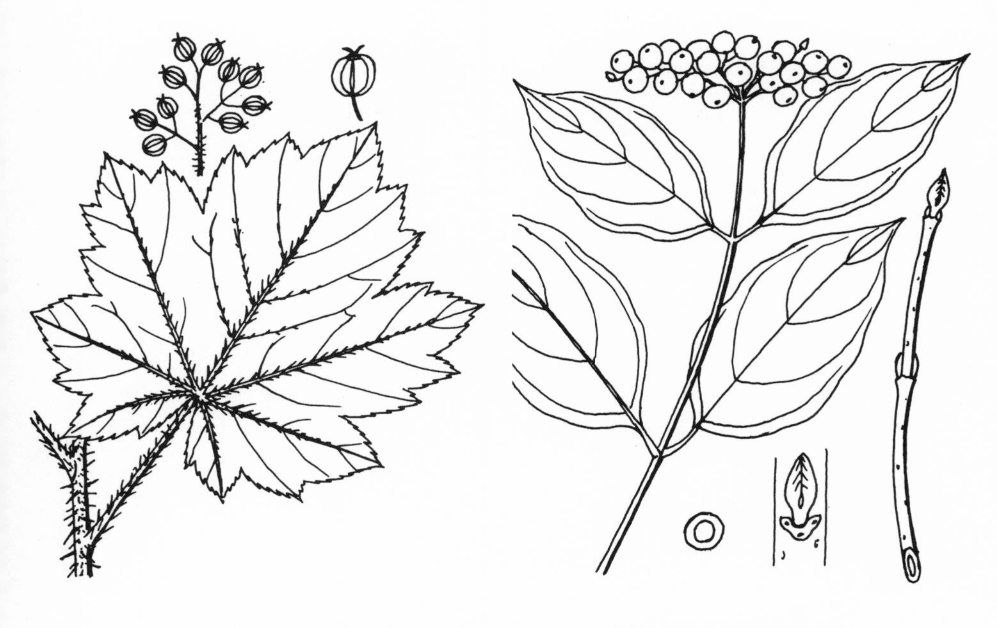

Devil's-club

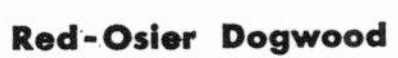

Red-Osier Dogwood

Silky Dogwood

Pale Dogwood

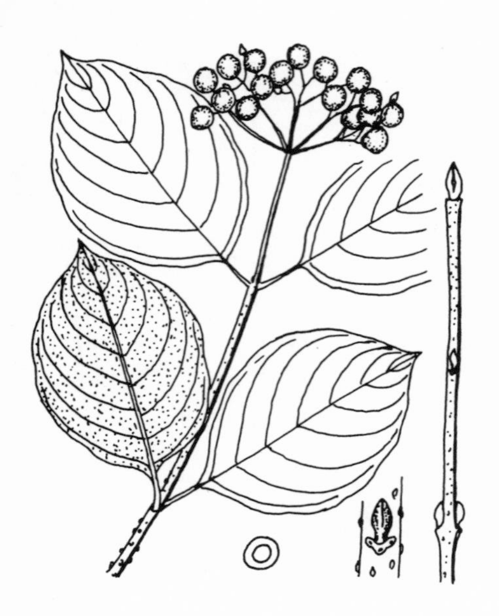

Roundleaf Dogwood

Gray-stemmed Dogwood

Rough-leaf Dogwood

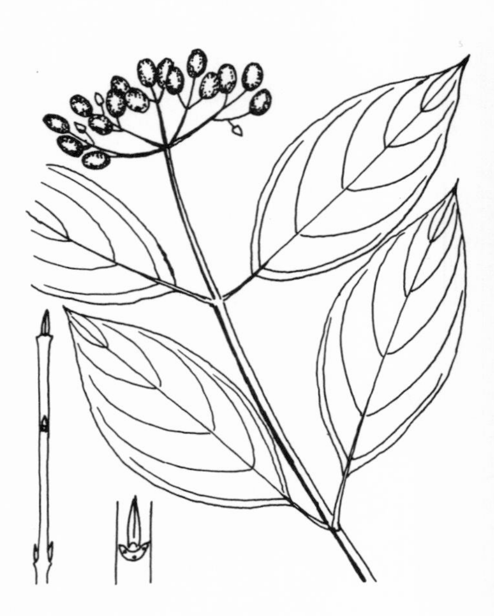

Stiff Dogwood

ROUNDLEAF DOGWOOD *Cornus rugosa*

FIELD MARKS. An erect, rather losely branched shrub 3 to 8 feet high; growing in dry to well-drained, rocky woods. *Branchlets* greenish usually spotted with purple, warty-dotted, and with white pith. *Leaves* broadly egg-shaped to roundish, rounded or broadly pointed at base, pointed at tip, usually with 7 to 9 pairs of veins, often with scattered close-pressed hairs on upper surface, pale and woolly or downy beneath, 2½ to 5 inches long. *Flowers* white, in flat-topped clusters; blooming May to July. *Fruits* lead-colored or pale blue, roundish, about ¼ inch in diameter; ripening August and September.

RANGE. Nova Scotia and eastern Quebec to Manitoba; south to New England, western Virginia, West Virginia, the Great Lakes region, and northeastern Iowa.

GRAY-STEMMED DOGWOOD *Cornus racemosa*

FIELD MARKS. A much-branched, very twiggy, gray-stemmed shrub 3 to 8 feet high; growing in dry to moist thickets, borders of woods, fence rows, etc. *Branchlets* very slender, light brown, and with pale brown or whitish pith. *Leaves* narrowly egg-shaped to lance-shaped, usually pointed at base, long-pointed at tip, with 3 to 5 pairs of veins, smooth or with minute close-pressed hairs on both surfaces, pale or whitened beneath, 1½ to 3 inches long. *Flowers* creamy white, in cone-shaped clusters; blooming June and July. *Fruits* white, roundish, about 3/16 inch in diameter, on bright-red stalks; ripening August or September.

RANGE. Central Maine to southern Ontario and Minnesota; south to Delaware, West Virginia, North Carolina, Kentucky, and Oklahoma.

Also called Panicled Dogwood.

ROUGH-LEAF DOGWOOD *Cornus drummondii*

FIELD MARKS. A shrub or small tree 4 to about 15 feet high; growing in moist woods, along streams, and on shores of lakes and ponds. *Branchlets* brownish or reddish brown, rough-hairy, with brownish (rarely whitish) pith. *Leaves* elliptic to broadly egg-shaped, usually rounded at base, pointed at tip, with 3 to 5 pairs of veins, rough-hairy above, downy beneath, 1½ to 5 inches long. *Flowers* creamy white, in flat or slightly convex clusters; blooming May and June. *Fruits* white, roundish, ⅛ to ¼ inch in diameter; ripening August or September.

RANGE. Southern Ontario and Michigan to Iowa and Nebraska; south to Alabama, Mississippi, and eastern Texas.

STIFF DOGWOOD *Cornus foemina*

FIELD MARKS. A shrub or small tree 5 to 15 feet high; growing in wet woods, swamps, and along streams. *Branchlets* slender, smooth, reddish or partly green and becoming gray, with a white pith. *Leaves* elliptic to egg-shaped or broadly lance-shaped, usually pointed at base, pointed at tip, smooth or nearly so on both surfaces, dark dull green above, slightly paler beneath, with 4 or 5 pairs of veins, 2 to 5 inches long. *Flowers* creamy white, in a flat-topped cluster; blooming April to June. *Fruits* pale blue, roundish, about ¼ inch in diameter; ripening July to September.

RANGE. Eastern Virginia to southern Indiana and southeastern Missouri; south to Florida and eastern Texas.

SOUTHERN ROUGH-LEAF DOGWOOD *Cornus asperifolia*

FIELD MARKS. A leaf-losing shrub 5 or more feet high; growing in low woods and swamps. *Branchlets* slender, brownish, with a white pith. *Leaves* elliptic, pointed to roundish at base, rather long-pointed at tip, thin in texture, roughish above, scarcely paler and rather hairy beneath, 1½ to about 3 inches long. *Flowers* creamy white, in rather small flat-topped clusters; blooming May or June. *Fruits* roundish, pale blue, ⅛ to 3/16 inch in diameter; ripening August or September.

RANGE. Coastal plain; South Carolina south to Florida.

ALTERNATE-LEAF DOGWOOD *Cornus alternifolia*

FIELD MARKS. A shrub or small tree 4 to sometimes 25 feet high, with almost horizontally spreading branches, and larger stems dark green and often streaked with white; growing in moist woods and along streams. *Branchlets* greenish and with a white pith. *Leaves* alternate, very often crowded toward the tips of the branchlets, rather long and slender stalked, oval or broadly egg-shaped, broadly pointed at base, pointed at tip, bright green and smooth above, paler and sometimes slightly downy beneath, 2 to 5 inches long. *Flowers* creamy white, in flat-topped clusters; blooming May or June. *Fruits* bluish black, roundish, ¼ to ⅜ inch in diameter, on bright-red stalks; ripening August or September.

RANGE. Newfoundland to southern Ontario, southeastern Manitoba, and eastern Minnesota; south to Florida, Alabama, and northern Arkansas.

Also called Blue or Pagoda Dogwood.

SWEET-PEPPERBUSH FAMILY (Clethraceae)

SWEET-PEPPERBUSH *Clethra alnifolia*

FIELD MARKS. A leaf-losing shrub 3 to 10 feet high, the older stems with dark gray or blackish and flaky bark; growing in wet or swampy, and usually sandy, woods and thickets. *Branchlets* slender, grayish brown, minutely downy. *Leaves* alternate, broadest above the middle, wedge-shaped at base, bluntly pointed or abruptly sharp-pointed at tip, sharply and doubly toothed on margin to somewhat below the middle, green and smooth or nearly so on both surfaces, 1½ to 3 inches long. *Flowers* white or (rarely) pinkish, 5-petalled, fragrant, in narrow end clusters from 2 to 6 inches in length; blooming June to September. *Fruits* roundish capsules about ⅛ inch in diameter, on ascending stalks; maturing September or October.

RANGE. Chiefly coastal plain; southern Maine and New Hampshire, south to Florida, west to eastern Texas.

MOUNTAIN SWEET-PEPPERBUSH *Clethra acuminata*

FIELD MARKS. A leaf-losing shrub 4 to about 15 feet high, the older stems with reddish-brown or cinnamon-colored bark which comes off in long strips; growing in rich, moist woods and along streams in the southern Appalachians. *Branchlets* slender, pale brown or ashy, more or less downy. *Leaves* alternate, rather long-stalked, oval to oblong-elliptic, pointed or roundish at base, pointed at tip, finely and sharply toothed on margin nearly to base, bright green and smooth above, paler and usually somewhat downy beneath, 2 to 7 inches long. *Flowers* white, 5-petalled, fragrant, in narrow end clusters from 4 to 8 inches in length; blooming July to September. *Fruits* roundish or slightly egg-shaped capsules about 3/16 inch in diameter, on nodding stalks; maturing September or October.

RANGE. Western Pennsylvania, western Virginia, and West Virginia south to northern Georgia.

Also known as Cinnamon Clethra.

Southern Rough-leaf Dogwood **Alternate-leaf Dogwood**

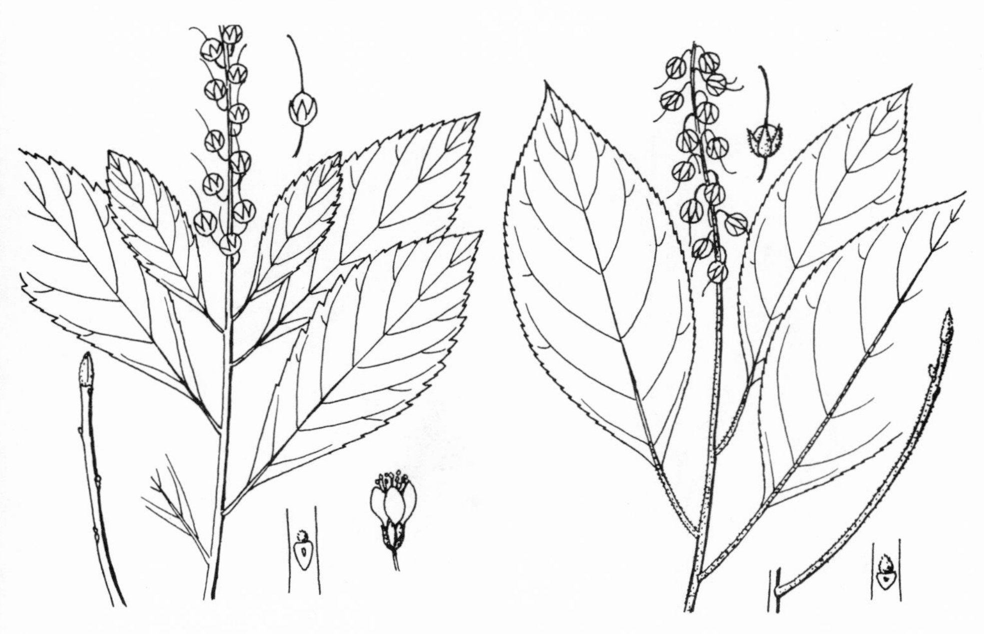

Sweet-pepperbush **Mountain Sweet-pepperbush**

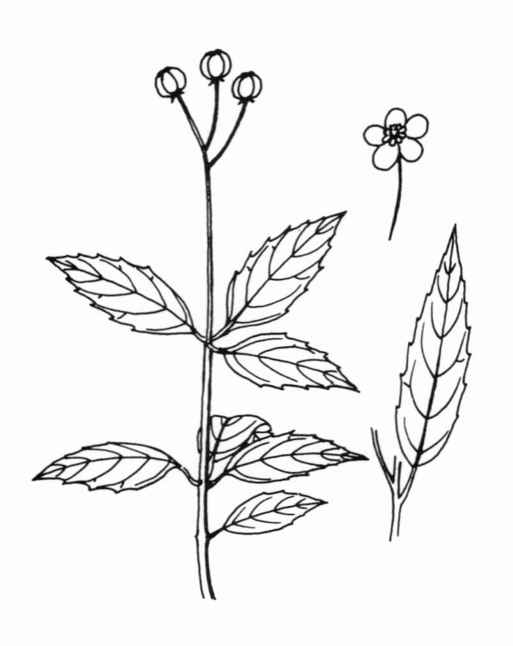

Spotted Wintergreen

Pipsissewa

Tar-flower

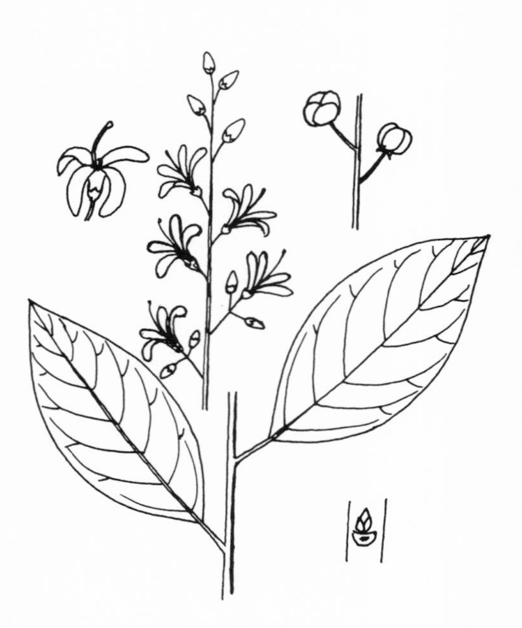

Southern-plume

WINTERGREEN FAMILY (Pyrolaceae)

SPOTTED WINTERGREEN *Chimaphila maculata*

FIELD MARKS. A low, semiherbaceous, evergreen plant with creeping underground stems and upright, leafy branches 4 to 10 inches high; growing in dry woods. *Leaves* crowded toward tips of stems or nearly in whorls of 3's, egg-shaped to lance-shaped, broadly pointed to roundish at base, pointed at tip, sharply toothed on margin, dark green mottled with white along the veins and lustrous above, usually purplish beneath, 1 to 3 inches long. *Flowers* white or pinkish, 5-petalled, about ⅝ inch across, 2 to 5 in a long-stalked end cluster; blooming June or July. *Fruits* slightly flattened, roundish, 5-parted capsules about ¼ inch in diameter.

RANGE. New Hampshire to Ontario and Michigan; south to Georgia, Alabama, and Tennessee.

Also called Spotted Pipsissewa.

PIPSISSEWA *Chimaphila umbellata*

FIELD MARKS. A low, semiherbaceous, evergreen plant with creeping underground stems and upright, leafy branches 4 to 12 inches high; growing in dry woods. *Leaves* crowded toward tips of stems and appearing to be whorled; broadest above the middle, wedged-shaped at base, blunt or broadly pointed at tip, sharply toothed on margin, bright green on both surfaces, very lustrous above, ¾ to 2½ inches long. *Flowers* white or pinkish, 5-petalled, about ½ inch across, 3 to 6 in a long-stalked end cluster; blooming July or August. *Fruits* slightly flattened, roundish, 5-parted capsules about 3/16 inch in diameter.

RANGE. Nova Scotia to Ontario; south to northeastern North Carolina, West Virginia, Ohio, northern Illinois, and Minnesota.

Also called Prince's-pine, King's-cure, and Wintergreen. Used medicinally by the Indians for a variety of ailments.

HEATH FAMILY (Ericaceae)

TAR-FLOWER *Befaria racemosa*

FIELD MARKS. An evergreen shrub 2 to about 6 feet high, with upright branches; growing in sandy pinelands. *Branchlets* long, slender, hairy. Leaves alternate, elliptic to oval, broadly pointed at base, sharply pointed at tip, untoothed on margin, thick and leathery in texture, ½ to 2 inches long. *Flowers* very showy, white or pink-tinged fragrant, usually 7-petalled, 1½ to 2 inches across; blooming June to August. *Fruits* somewhat flattened, roundish, 5-parted capsules about ¼ inch in diameter.

RANGE. Costal plain; southern Georgia and Florida.

Also called Fly-catcher, as the sticky-hairy buds and calyxes of the flowers often catch small insects.

SOUTHERN-PLUME *Elliottia racemosa*

FIELD MARKS. A leaf-losing shrub or small tree 3 to about 15 feet high; growing on oak ridges or sandhills. *Leaves* alternate, oval to oblong-elliptic, broadly pointed at base, pointed at tip, untoothed on margin, dull green and smooth above, paler and somewhat downy beneath, 2 to 5 inches long. *Flowers* showy, white, usually 4-petalled, about ½ inch long; blooming July or August. *Fruits* somewhat flattened, roundish, usually 4-parted capsules nearly ⅜ inch in diameter.

RANGE. Southern South Carolina and Georgia.

One of the rarest of American shrubs, known only from a few localities within its range.

SANDMYRTLE *Leiophyllum buxifolium*

FIELD MARKS. A much-branched evergreen shrub 6 inches to 2½ feet high, with prostrate or spreading to upright branches; growing in sandy pinelands and on rocky places in the mountains. *Leaves* mostly opposite, short-stalked, elliptic to oval, blunt at both ends, untoothed on margin, smooth, leathery, lustrous above, paler and dull beneath, smooth, ⅛ to ⅝ inch long. *Flowers* small, white, 5-petalled, in rather dense end clusters; blooming March to June. *Fruits* egg-shaped, 2- to 5-parted capsules about ⅛ inch long.

RANGE. Coastal plain, New Jersey to South Carolina; upper piedmont and mountains, eastern Kentucky, North and South Carolina.

LABRADOR-TEA *Ledum groenlandicum*

FIELD MARKS. An evergreen shrub 1 to (rarely) 3 feet high; growing in peaty soils and cold bogs. *Branchlets* rusty-woolly when young. *Leaves* alternate, short-stalked, elliptic to narrowly oblong, roundish at base, bluntly pointed at tip, margin untoothed and strongly rolled inward beneath, leathery, smooth and bright green above, densely rusty-wooly beneath, ½ to 2 inches long. *Flowers* white, 5-petalled, about ⅜ inch across, in rather dense end clusters; blooming May or June. *Fruits* narrowly oblong capsules about ¼ inch long, opening upward from base into 5 parts.

RANGE. Greenland and Labrador to Alaska; south to northern parts of New Jersey, Pennsylvania, and Ohio, Michigan, Alberta, and Washington.

RHODODENDRONS AND AZALEAS (Rhododendron)

(Key Appendix M)

These are evergreen or leaf-losing shrubs with alternate, simple leaves. They have clusters of showy flowers arising from end buds formed the previous season. The fruits are narrowly egg-shaped or oblong, 5-parted, woody capsules which contain a large number of very small seeds. Rhododendrons and azaleas are among the most beautiful of our native shrubs, and many of them are grown as ornamentals. The rhododendrons are evergreen shrubs with leathery leaves; our native azaleas are all leaf-losing shrubs.

ROSEBAY RHODODENDRON *Rhododendron maximum*

FIELD MARKS. An evergreen shrub or small tree commonly 5 to 15 feet (rarely to 40 feet) high; growing along streams and on moist, rocky, wooded slopes and sometimes forming dense thickets. *Leaves* usually broadest above the middle, tapering to a wedge-shaped base, broadly pointed at tip, untoothed on margin, very thick and leathery, lustrous above, paler and dull and sometimes downy beneath, 4 to 8 inches long. *Flowers* white to rose pink, spotted with olive green to orange, open bell-shaped with 5 rounded lobes, about 1½ inches across, in large clusters; blooming June or July. *Fruits* downy, ¾ to 1 inch long.

RANGE. Southwestern Maine to New York, southern Ontario, and Ohio; south in the mountains to northern Georgia and Alabama.

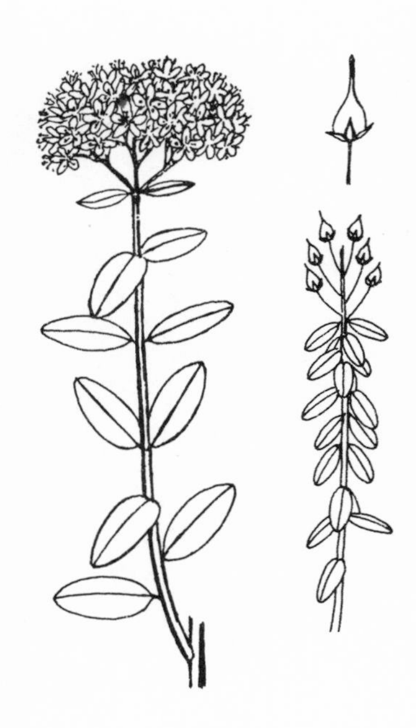

Sandmyrtle

Sandmyrtle (prostrate expression)

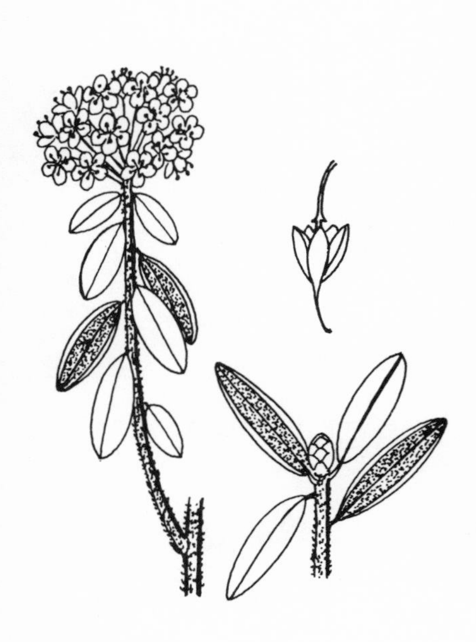

Labrador-tea

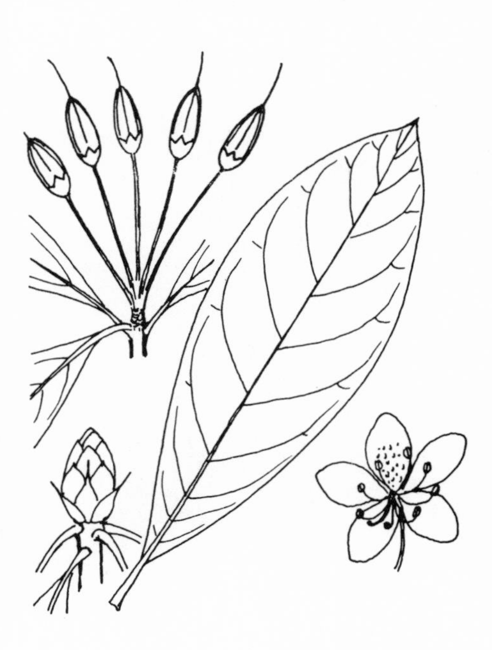

Rosebay Rhododendron

Catawba Rhododendron

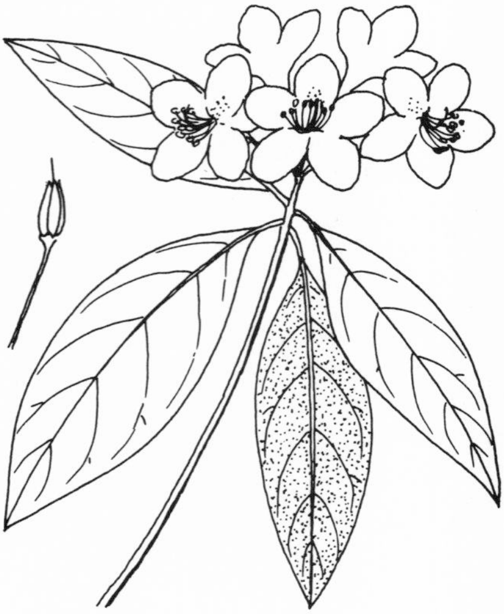

Carolina Rhododendron

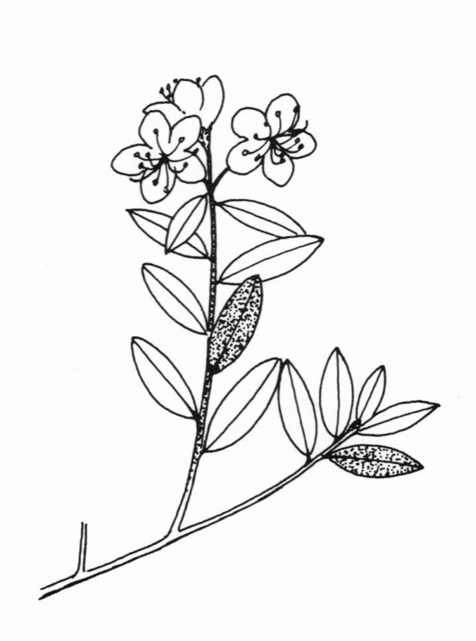

Lapland Rhododendron

Pinkshell Azalea

CATAWBA RHODODENDRON *Rhododendron catawbiense*

FIELD MARKS. A spreading and often thicket-forming evergreen shrub usually 4 to 10 (rarely 20) feet high; growing on rocky slopes, ridges, and mountain tops, usually above 3,000 feet.. *Leaves* elliptic or oblong, rounded at base, blunt or slightly pointed at tip, margin untoothed, very thick and leathery, smooth, dark green and lustrous above, paler or whitened and dull beneath, 3 to 5 inches long. *Flowers* lilac purple to rose purple spotted with olive green, open bell-shaped with 5 rounded lobes, about 2 inches across, in large and showy clusters; blooming April to June. *Fruits* narrowly oblong, rusty-hairy, ¾ to 1 inch long.

RANGE. Southwestern Virginia, southern West Virginia, and southeastern Kentucky; south to northern Georgia and Alabama; locally eastward to north-central North Carolina.

Also called Purple Rhododendron and Mountain Rosebay. Often cultivated and a parent of many cultivated hybrid rhododendrons.

CAROLINA RHODODENDRON *Rhododendron minus*

FIELD MARKS. An evergreen shrub 3 to about 8 feet high; growing on exposed or wooded slopes and along streams. *Leaves* elliptic to narrowly elliptic, pointed at both ends, margin untoothed, thick and leathery, dark green and smooth above, pale or whitened and covered with minute dotlike brown scales beneath, 2 to about 4 inches long. *Flowers* white to deep rose pink, often spotted with olive-green or orange, open bell-shaped with 5 rounded lobes, 1 to 1½ inches across, in dense or fairly open clusters; blooming April to July or later. *Fruits* oblong egg-shaped, ⅜ to ¾ inch long, rusty brown.

RANGE. South-central and western North Carolina, eastern Tennessee, and northwestern South Carolina south into Georgia and Alabama.

Also known as Piedmont and Small Rhododendron.

LAPLAND RHODODENDRON *Rhododendron lapponicum*

FIELD MARKS. A prostrate or mat-forming evergreen shrub 2 to rarely 12 inches high; growing on tundras and southward on the rocky alpine summits of high mountains. *Leaves* short-stalked, elliptic to oval, pointed or rounded at base, blunt or abruptly short-pointed at tip, margin untoothed, leathery, dark green and wrinkled in appearance above, densely covered with small brownish scales beneath, ¼ to ¾ inch long. *Flowers* purple, open bell-shaped with 5 rounded lobes, ½ to ¾ inch across, in clusters of 3 to 6; blooming June or July. *Fruits* about ¼ inch long.

RANGE. Arctic south to Newfoundland, eastern Quebec; mountains of Maine, New Hampshire, and New York; and in Dells of Wisconsin River, Wisconsin.

PINKSHELL AZALEA *Rhododendron vaseyi*

FIELD MARKS. A leaf-losing, bushy shrub 3 to 15 feet high; growing in boggy places, along streams, and in high mountain forests. *Branchlets* light reddish brown, smooth or sparingly hairy, later becoming grayish brown with flaking bark. *Leaves* elliptic, pointed at both ends, margin somewhat wavy and hairy-fringed, dark green above, paler beneath, smooth or with a few hairs along the midrib, 2 to 5 inches long. *Flowers* pale to deep rose pink spotted with yellowish orange to reddish orange, open bell-shaped with a short tube and 5 rounded lobes, 1½ to 2 inches across, 5 to 8 in a cluster; blooming in May or June. *Fruits* elliptical but somewhat lopsided, smooth or nearly so, light reddish brown, about ½ inch long.

RANGE. Jackson, Macon, and Watauga Counties, North Carolina.

Beautiful flowering shrub of very limited natural distribution but sometimes cultivated and reported as escaping from cultivation in Massachusetts.

RHODORA *Rhododendron canadense*

FIELD MARKS. A leaf-losing shrub 1 to 3 feet high; growing in cold bogs and moist rocky barrens. *Branchlets* smooth, often whitened with a bloom. *Leaves* elliptic or oblong, pointed at both ends, margin slightly rolled and hairy-fringed, dull green above, pale or whitened beneath and with some rusty hairs along the midrib, ¾ to 2 inches long. *Flowers* rose purple, 2-lipped, with a short tube, 5 to 8 in a cluster; blooming April or May. *Fruits* oblong egg-shaped, uneven at base, downy, ½ to ⅝ inch long.
RANGE. Labrador and Newfoundland to south-central Quebec; south to northern New Jersey, northeastern Pennsylvania, and central New York.

FLAME AZALEA *Rhododendron calendulaceum*

FIELD MARKS. A leaf-losing shrub 3 to (rarely) 15 feet high; growing in oak or pine woods and on mountain balds. *Branchlets* downy and with scattered spreading hairs. *Leaves* elliptic or broadest above the middle, pointed at both ends, margin hairy-fringed, bright green and often with scattered hairs above, paler and downy and with stiff hairs along midrib beneath, 1 to 3 inches long. *Flowers* yellow, orange, or scarlet, the tube with gland-tipped hairs, not fragrant, 1½ to 2 inches across, 5 to 9 in a cluster; blooming April to June, as new leaves unfold. *Fruits* narrowly egg-shaped, somewhat downy and with spreading hairs, about ¾ inch long.
RANGE. Southwestern Pennsylvania, southeastern Ohio, and West Virginia; south to northern Georgia and Alabama.
Often cultivated as an ornamental shrub.

SHOWY AZALEA *Rhododendron speciosum*

Very similar to the preceding, but flower tubes have hairs without glands, and perhaps just a variety. It grows at lower elevations in Georgia. (Not illustrated)

CUMBERLAND AZALEA *Rhododendron cumberlandense*

Much like the Flame Azalea but not blooming until the leaves are grown, and the branchlets are smooth. Oak woods, eastern Kentucky and western Virginia. (Not illustrated.

ALABAMA AZALEA *Rhododendron alabamense*

FIELD MARKS. A leaf-losing shrub 2 to about 4 feet high; growing on rich but often dry wooded slopes. *Branchlets* with scattered stiff hairs. *Leaves* elliptic or broadest above the middle, pointed at base, broadly pointed to rounded at tip, margin fringed with short hairs, dark green above, paler or slightly whitened and somewhat hairy and with stiff hairs along the midrib beneath, 1¼ to 2¼ inches long. *Flowers* white, fragrant, the tube downy and with short gland-tipped hairs on the outside; blooming in May, when the leaves are well developed. *Fruits* narrowly egg-shaped, somewhat downy and glandular-bristly, about ⅝ inch long.
RANGE. South-central South Carolina south to Florida, west to Mississippi.

PINXTER-FLOWER *Rhododendron periclymenoides*

FIELD MARKS. A leaf-losing shrub 2 to 6 (rarely 10) feet high, the stems often unbranched below but with more or less whorled branches above; growing in moist woods, clearings, and swamps. *Branchlets* smooth or with scattered spreading hairs. *Leaves* elliptic or oblong and often broadest above the middle, pointed at tip and wedge-shaped at base, margin fringed with short hairs, dull green and smooth above, paler and with stiff hairs along the midrib beneath, 2 to 4 inches long. *Flowers* pink to almost white, delicately or faintly fragrant, the tube hairy but seldom with gland-tipped hairs on the outside, about 1½ inches across; blooming March to May, before the leaves appear. *Fruits* narrowly egg-shaped, hairy, ½ to ¾ inch long.
RANGE. Massachusetts to New York and southern Illinois; south to central South Carolina and northern Georgia.
Also called Pink or Purple Azalea and Purple-honeysuckle.

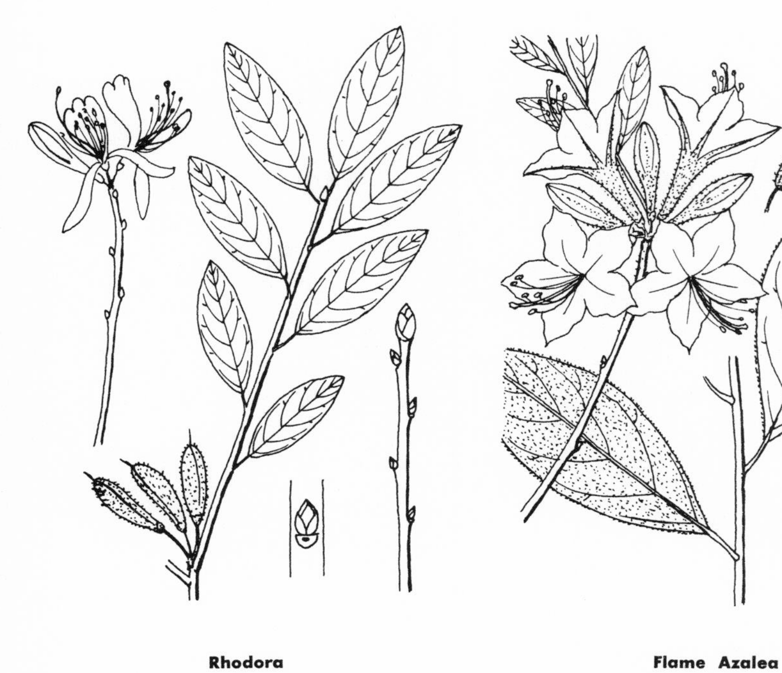

Rhodora

Flame Azalea

Alabama Azalea

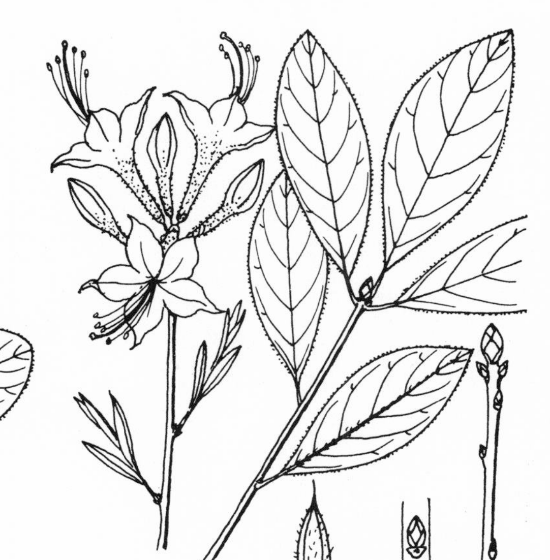

Pinxter-Flower

Mountain Azalea

Hoary Azalea

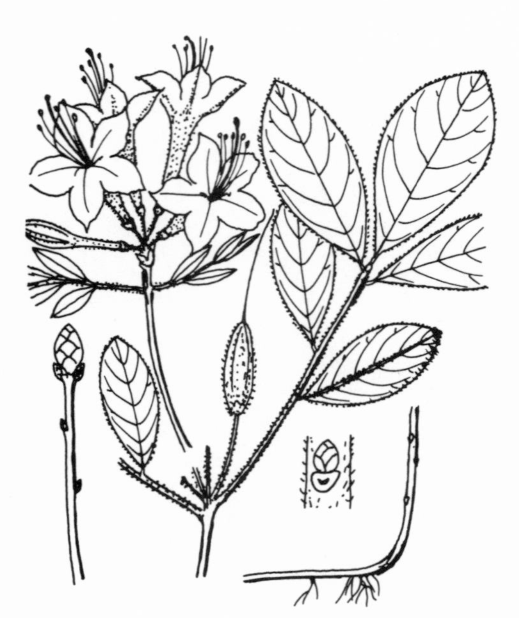

Dwarf Azalea

Swamp Azalea

MOUNTAIN AZALEA *Rhododendron prinophyllum*

FIELD MARKS. A leaf-losing shrub 3 to 10 feet high; growing in upland woods and on rocky slopes. *Branchlets* grayish-downy and with scattered stiff hairs. *Leaves* elliptic or broadest above the middle, wedge-shaped at base, broadly pointed at tip, margin hairy-fringed, upper surface often with scattered hairs, grayish-downy and with stiff hairs along midrib beneath, 1 to 3 inches long. *Flowers* usually bright pink, tube about as long as the lobes and downy and sticky-glandular, fragrant, about 1½ inches across, 5 to 9 in a cluster; blooming April to June. *Fruits* narrowly oblong, downy, glandular-hairy capsules about ¾ inch long.

RANGE. Maine to Quebec; south to western Virginia, Tennessee, and Missouri.

Also called Rose, Early, or Honeysuckle Azalea.

HOARY AZALEA *Rhododendron canescens*

FIELD MARKS. A shrub similar to the preceding; growing in moist woods, swamps, and along streams. *Branchlets* and *leaves* quite similar to those of the preceding species. *Flowers* pink or (rarely) white, the glandular-sticky and downy tube slender and about twice as long as the lobes, fragrant, about 1½ inches across, 5 to 9 in a cluster; blooming March or April, often before the leaves begin to expand. *Fruits* similar to those of the preceding species.

RANGE. Chiefly coastal plain; northern Delaware and northeastern Maryland; south to Florida, west to Texas, and north to Arkansas and Tennessee.

DWARF AZALEA *Rhododendron atlanticum*

FIELD MARKS. A leaf-losing shrub forming colonies by means of creeping underground stems, the erect branches 8 inches to 2 feet high; growing in moist to dry sandy pinelands. *Branchlets* more or less bristly-hairy. *Leaves* elliptic or broadest above the middle, wedge-shaped at base, bluntly pointed to roundish at tip, margin hairy-fringed, smooth above, pale or somewhat whitened and with stiff hairs along midrib beneath, 1¼ to 2½ inches long. *Flowers* white to pink or pale purple, tube longer than the lobes and with rows of sticky stalked glands, very fragrant, about 1½ inches across, 4 to 10 in a cluster; blooming April or May, often before the new leaves begin to appear. *Fruits* narrowly egg-shaped, bristly-hairy, about ⅝ inch long.

RANGE. Coastal plain; Delaware and southeastern Pennsylvania south to Georgia and Alabama.

SWAMP AZALEA *Rhododendron viscosum*

FIELD MARKS. A leaf-losing shrub 3 to 10 feet high; growing in swamps, bogs, along streams, and occasionally on mountain summits. *Branchlets* with scattered, stiff, brownish hairs. *Leaves* narrowly elliptic or broadest above the middle, wedge-shaped at base, blunt or short-pointed at tip, hairy-fringed on margin, dark green and smooth above, pale or slightly whitened and with stiff brownish hairs along midrib beneath, ¾ to 2½ inches long. *Flowers* white to pale pink, tube somewhat longer than the lobes and very sticky-glandular, very fragrant, about 1¼ inches across, 4 to 9 in a cluster; blooming May to July after the leaves are grown. *Fruits* narrowly egg-shaped, glandular-bristly capsules ½ to ¾ inch long.

RANGE. Southwestern Maine to northeastern Ohio; south to Georgia and Tennessee.

Also called the Clammy Azalea and White- or Clammy-honeysuckle.

SMOOTH AZALEA *Rhododendron arborescens*

FIELD MARKS. A leaf-losing shrub 5 to 20 feet high; growing in moist rocky woods, along streams, and in swamps and bogs. *Branchlets* smooth, somewhat lustrous, often with a whitish bloom. *Leaves* elliptic or broadest above the middle, wedge-shaped at base, short-pointed to blunt at tip, margin fringed with short hairs, thick and firm in texture, dark green and lustrous above, pale or whitened beneath and sometimes with a few stiff hairs along the midrib, 1 to 3 inches long. *Flowers* white or sometimes pale pink, about 1½ inches across, very sticky-hairy on the outside, the corolla tube longer than the lobes, very fragrant, 3 to 6 in a cluster; blooming late May to July. *Fruits* narrowly egg-shaped, densely glandular-hairy, ½ to ¾ inch long.

RANGE. Southern Pennsylvania to Kentucky, south to northern Georgia and Alabama.

Also called Sweet or Tree Azalea.

RED AZALEA *Rhododendron prunifolium*

FIELD MARKS. A leaf-losing shrub 4 to 10 feet high; growing in moist woods and ravines. *Branchlets* smooth. *Leaves* elliptic to narrowly elliptic or broadest above the middle, wedge-shaped at base, broadly pointed to long-pointed at tip, margin fringed with short hairs and also minutely toothed, dark green and smooth above, paler green and with some stiff hairs along the midrib beneath, 1½ to 5 inches long. *Flowers* bright red, about 1½ inches across, the corolla tube smooth or merely with occasional hairs, 3 to 5 in a cluster; blooming July and August.

RANGE. Coastal plain; southern Georgia and Alabama.

ALLEGHENY MENZIESIA *Menziesia pilosa*

FIELD MARKS. A rather straggling leaf-losing shrub 2 to 6 feet high, with bark freely shredding; growing in bogs or on thinly wooded slopes and balds in the southern Appalachians. *Branchlets* bristly-hairy and somewhat rusty-chaffy. *Leaves* alternate, often crowded toward tips of branchlets, elliptic to oval or broadest above the middle, pointed at base, abruptly pointed at tip, untoothed but hairy-fringed on margin, roughish-hairy above, pale or whitened and usually rusty-chaffy along the veins beneath, ¾ to 2 inches long. *Flowers* greenish white or greenish yellow often tinged with red, bell-shaped, about ¼ inch long, 3 to 6 in a cluster and nodding on slender stalks; blooming May to July. *Fruits* egg-shaped, bristly-hairy capsules about 3/16 inch long; maturing August to October.

RANGE. Southern Pennsylvania and West Virginia south to eastern Tennessee and northern Georgia.

Also called Minniebush.

ALPINE-AZALEA *Loiseleuria procumbens*

FIELD MARKS. A much-branched, mat-forming, evergreen shrub seldom 8 inches high; growing on tundras or mountain tops, and in cold bogs. *Leaves* opposite, crowded, elliptic to oval, blunt at tip, margin rolled and untoothed, leathery in texture, smooth, dark green above, often whitened beneath, ⅛ to ⅜ inch long. *Flowers* white or pink, bell-shaped, about ⅛ inch long, 2 to 5 in end clusters; blooming June and July. Fruits 2- to 3-parted, egg-shaped capsules about ⅛ inch long.

RANGE. Greenland and Newfoundland to Alaska; south to mountain summits in Maine and New Hampshire, Quebec, and Alberta.

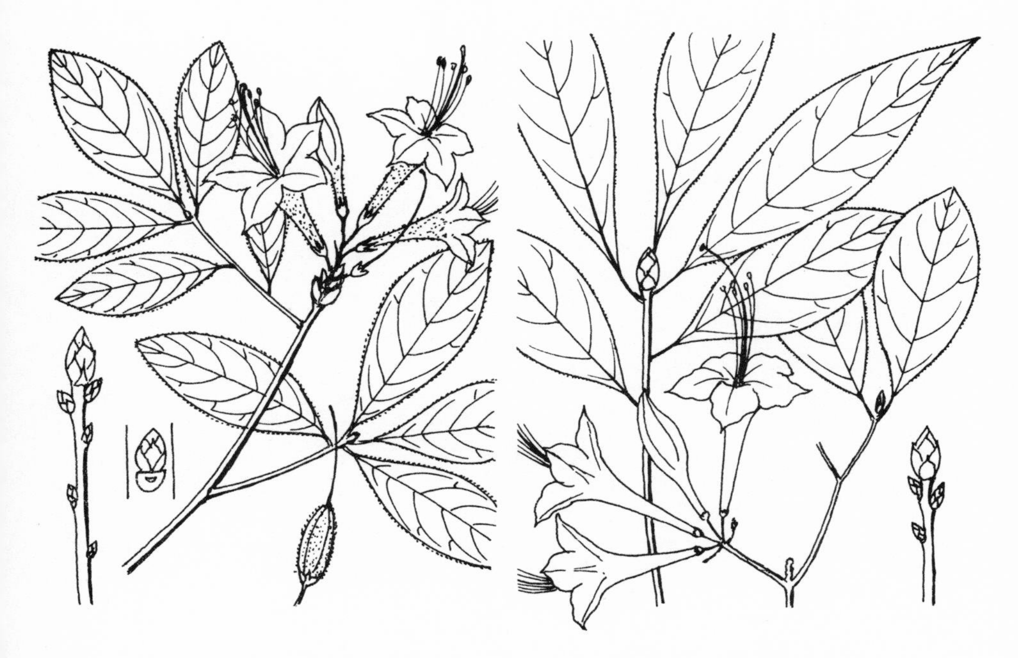

Smooth Azalea

Red Azalea

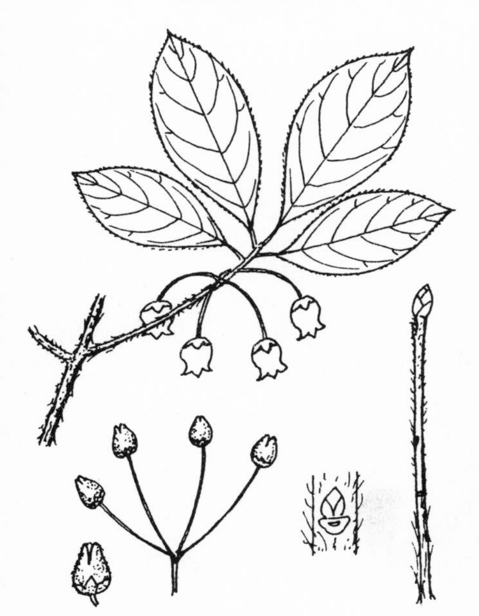

Allegheny Menziesia

Alpine-Azalea

Mountain-Laurel

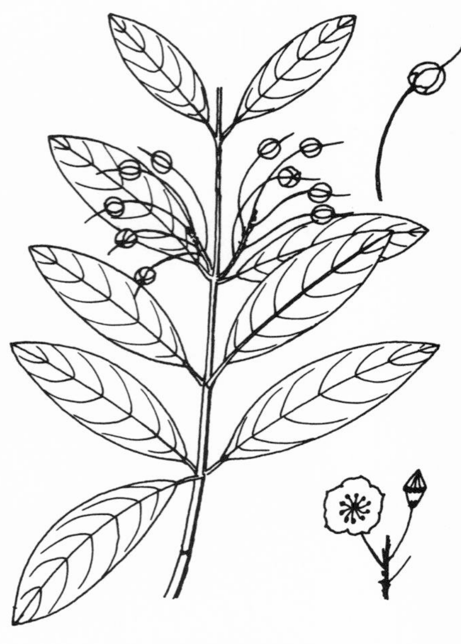

Sheep-Laurel

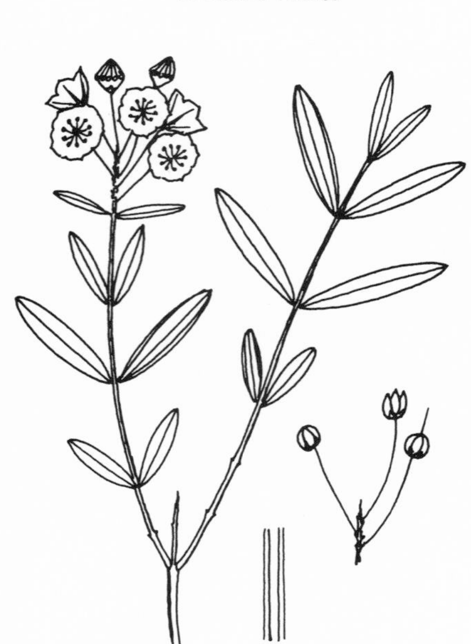

Pale-Laurel

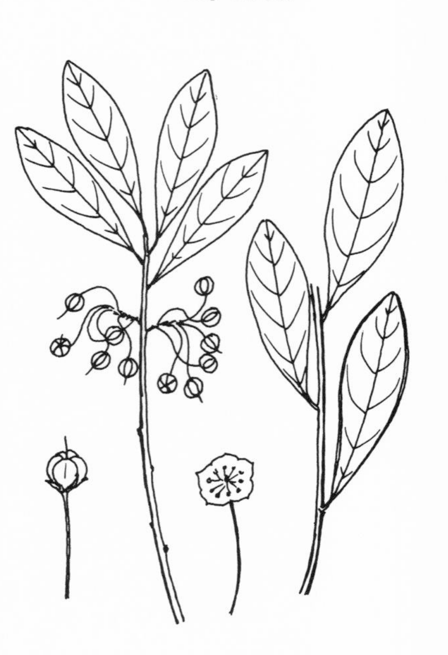

White Wicky

LAURELS (Kalmia L.)

These are shrubs with alternate, opposite, or whorled simple leaves with untoothed margins. Their showy flowers have a saucer-shaped or shallowly bowl-shaped, 5-lobed corolla. The anthers of the 10 stamens fit into small pouches in the corolla until they are released by an insect touching the springlike filaments. The fruits are 5-parted roundish or slightly flattened capsules containing a number of small seeds. Leaves of these plants are known to be poisonous to livestock, and honey made from the nectar of the flowers is said to be poisonous to humans. Although these are commonly called laurels in America, the true laurels are members of the Laurel family.

MOUNTAIN-LAUREL *Kalmia latifolia*

FIELD MARKS. An evergreen shrub or small tree 3 to (rarely) 35 feet high; growing in rocky or sandy woods. *Leaves* mostly alternate but crowded toward ends of the branchlets, elliptic, pointed at both ends, thick and leathery, smooth, lustrous above, paler yellowish green beneath, 2 to about 4 inches long. *Flowers* deep pink to whitish, about ¾ inch across, in dense end clusters; blooming April to July. *Fruits* about 3/16 inch in diameter.

RANGE. New Brunswick and Maine to Ontario, southern Indiana, and western Kentucky; south to Florida and Louisiana.

Also called Ivy (in southern Appalachians), Calicobush, and Spoonwood. Often cultivated as an ornamental shrub.

SHEEP-LAUREL *Kalmia angustifolia*

FIELD MARKS. An evergreen shrub 1 to about 3 feet high; growing in bogs and rocky or sandy woods. *Leaves* opposite or in 3's, narrowly elliptic to oblong, pointed at both ends, thin but leathery in texture, smooth above, paler green or slightly whitened and smooth beneath, 1 to 2½ inches long. *Flowers* deep pink to rose purple (rarely white), about ⅜ inch across, in clusters on branchlets of the previous year at base of the new growth; blooming April to August. *Fruits* about ⅛ inch in diameter.

RANGE. Newfoundland and Labrador to Manitoba, south to Pennsylvania and Michigan.

CAROLINA-LAUREL *Kalmia carolina*

A closely related species with permanently paler, hairy lower leaf surfaces, ranging from Virginia and Tennessee south to Georgia. (Not illustrated)

PALE-LAUREL *Kalmia polifolia*

FIELD MARKS. A straggling evergreen shrub 6 inches to 2 feet high; growing in cold northern bogs. *Branchlets* strongly 2-edged. *Leaves* opposite or in 3's, stalkless or nearly so, very narrowly elliptic, narrowed at base, blunt or short-pointed at tip, margin rolled, smooth, lustrous above, strongly whitened beneath, ½ to 1¼ inches long. *Flowers* rose purple, about ½ inch across, in end clusters; blooming May to July. *Fruits* about ⅛ inch in diameter.

RANGE. Newfoundland and Labrador to Alaska; south to New Jersey, northern Pennsylvania, Michigan, Minnesota, and Oregon.

Also called Swamp- or Bog-laurel.

WHITE WICKY *Kalmia cuneata*

FIELD MARKS. A leaf-losing shrub 8 inches to about 3 feet high; growing in wet places in the Carolina sandhills. *Branchlets* downy. *Leaves* alternate, narrowly elliptic or broadest above the middle, wedge-shaped at base, usually blunt at tip, margin slightly rolled, smooth above, downy beneath, ¾ to 2 inches long. *Flowers* white or pinkish, about ⅝ inch across, in few-flowered clusters on branchlets of the previous year at base of the new growth; blooming May or June. *Fruits* about ⅛ inch in diameter.

RANGE. North and South Carolina.

HAIRY-LAUREL *Kalmia hirsuta*

FIELD MARKS. An evergreen shrub 6 inches to about 2 feet high, with ascending branches; growing in moist, sandy pinelands. *Branchlets* hairy. *Leaves* alternate, stalkless or nearly so, elliptic to oval, broadly pointed at both ends, margin sometimes slightly rolled, dark green above, paler beneath, with spreading hairs on both surfaces, ⅛ to ½ inch long. *Flowers* pink or rose purple with hairy calyx and pointed corolla lobes, ⅜ to ⅝ inch across, usually solitary in the axils of leaves toward the ends of the branchlets; blooming May to August. *Fruits* about ⅛ inch in diameter.
RANGE. Coastal plain; southeastern Virginia south to Florida, west to Mississippi.

MOUNTAIN-HEATH *Phyllodoce caerulea*

FIELD MARKS. A prostrate evergreen shrub with ascending branches 4 to 8 inches high; growing on arctic tundras and southward on alpine summits of high mountains. *Leaves* alternate, crowded, narrow, flattened, blunt at tip, on short stalks running slightly down the branchlets, margin roughish or minutely toothed, 3/16 to ⅜ inch long. *Flowers* purplish, urn-shaped, ¼ to ⅜ inch long, solitary or 2 to 6 at ends of the branchlets; blooming June to August. *Fruits* egg-shaped, 5-parted capsules about 3/16 inch long.
RANGE. Arctic regions south to Newfoundland, Quebec, and the higher mountains of Maine and New Hampshire.

BOG ROSEMARY *Andromeda glaucophylla*

FIELD MARKS. A spreading evergreen shrub 4 inches to 2 feet high, growing in bogs and shallow pools. *Branchlets* whitened. *Leaves* alternate, very short-stalked, narrowly lance-shaped or narrowly oblong, pointed at base, short-pointed at tip, margin strongly rolled, thick and leathery in texture, dark bluish green and lustrous above, very white and minutely downy beneath, ¾ to about 2 inches long. *Flowers* white or pinkish, globe-shaped, about ¼ inch long, in end clusters; blooming May to July or later. *Fruits* turban-shaped, 5-parted capsules about 3/16 inch in diameter.
RANGE. Southwestern Greenland and Labrador to eastern Manitoba; south to northern New Jersey, West Virginia, Indiana, Wisconsin, and Minnesota.

ZENOBIA *Zenobia pulverulenta*

FIELD MARKS. A leaf-losing shrub 3 to about 10 feet high; growing in damp sandy or peaty pinelands. *Branchlets* reddish to brown, often more or less whitened with a bloom. *Leaves* alternate, elliptic to oval, pointed at base, blunt or sometimes pointed at tip, untoothed or with rather obscure wavy teeth or margin, leathery in texture, smooth on both surfaces, often whitened with a bloom but sometimes green on both surfaces, ¾ to 2½ inches long. *Flowers* white, broadly bell-shaped, about ¼ inch long, in clusters along growth of the preceding year; blooming May or June. *Fruits* roundish, 5-parted capsules nearly ¼ inch in diameter.
RANGE. Coastal plain; southeastern Virginia south to Georgia.

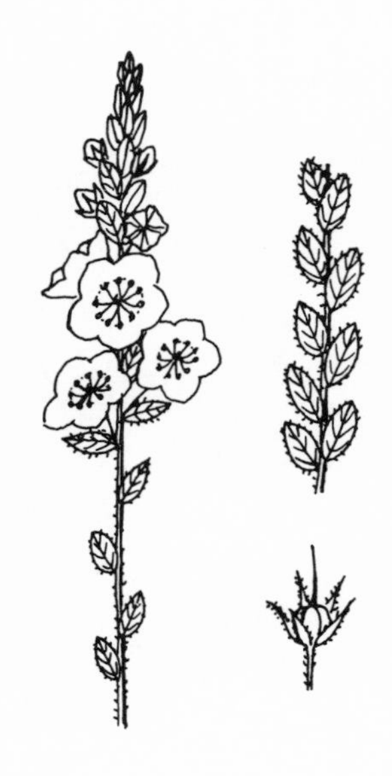

Hairy-Laurel

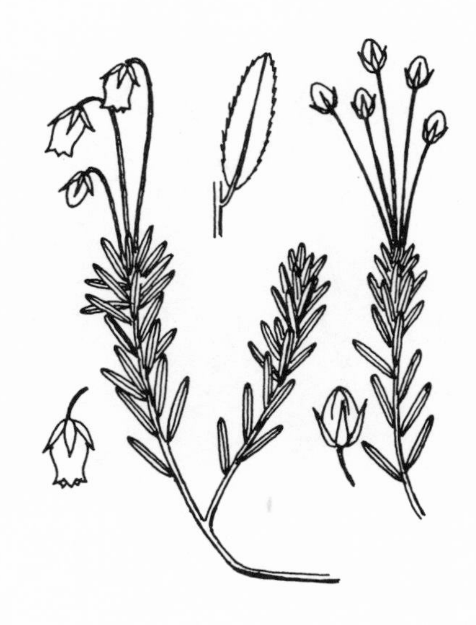

Mountain-Heath

Bog Rosemary

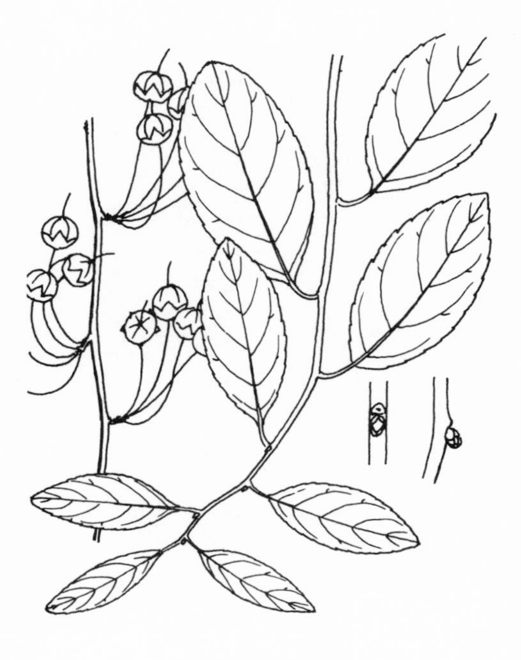

Zenobia

Drooping Leucothoë

Coastal Leucothoë

Recurved Leucothoë

Swamp Leucothoë

DROOPING LEUCOTHOË *Leucothoe fontanesiana*

FIELD MARKS. An evergreen shrub 2 to about 6 feet high, with spreading and arching, reddish branches; growing in moist mountain woodlands. *Leaves* alternate, narrowly egg-shaped to lance-shaped, rounded or broadly pointed at base, long-pointed at tip, finely toothed on margin with bristle-tipped teeth, leathery in texture, dark green and lustrous above, paler and smooth beneath, 2½ to 6 inches long; leafstalks ⅜ to ⅝ inch long. *Flowers* white, bell-shaped, fragrant, about ⅜ inch long; in many-flowered, narrow, drooping clusters arising in the axils of the leaves; blooming April and May. *Fruits* roundish, somewhat flattened, 5-lobed capsules about 3/16 inch in diameter.

RANGE. Western Virginia south to northern Georgia and northern Alabama.

Also called Dog-hobble and Switch-ivy.

COASTAL LEUCOTHOË *Leucothoe axillaris*

FIELD MARKS. An evergreen shrub similar to the preceding species but growing in wet woods and swamps in the coastal region. *Leaves* alternate, elliptic or oblong lance-shaped, pointed at base, pointed or short-pointed (rarely long-pointed) at tip, margin toothed chiefly above the middle, leathery in texture, dark green and lustrous above, paler and smooth or nearly so beneath, 1½ to 5 inches long; leafstalks ⅛ to ⅜ inch long. *Flowers* white, bell-shaped, fragrant, about ⅜ inch long; in many-flowered, narrow, drooping clusters arising in the axils of the leaves; blooming February to May. *Fruits* roundish, somewhat flattened, 5-lobed capsules about 3/16 inch in diameter.

RANGE. Coasal plain; Southeastern Virginia south to Florida, west to Mississippi.

RECURVED LEUCOTHOË *Leucothoe recurva*

FIELD MARKS. A widely branched, straggling, leaf-losing shrub 3 to 10 feet high; growing on dry, often rocky, wooded slopes in the mountains. *Leaves* alternate, egg-shaped or elliptic, pointed at both ends, finely and sharply toothed on the margin, thin in texture, bright green and smooth above, paler and sometimes slightly downy on the veins beneath, 1½ to 4 inches long. *Flowers* white, narrowly bell-shaped, fragrant; many hanging downward in 1-sided, spreading and recurved, narrow end clusters; blooming April to June. *Fruits* roundish, somewhat flattened, deeply 5-lobed capsules about 3/16 inch in diameter.

RANGE. Western Virginia and West Virginia south to northern Georgia and northern Alabama.

Also called Redtwig Leucothoe. The leaves turn bright red in fall.

SWAMP LEUCOTHOË *Leucothoe racemosa*

FIELD MARKS. A leaf-losing shrub 5 to about 12 feet high, with ascending and spreading branches; growing in swampy thickets, along streams, and about shallow ponds. *Leaves* alternate, lance-shaped or elliptic, pointed at both ends, finely and sharply toothed on margin, bright green and smooth above, paler and usually somewhat downy on the veins beneath, 1 to 3 inches long. *Flowers* white or pale pink, narrowly bell-shaped, fragrant; many standing erect in long, narrow, stiffly spreading clusters; blooming April to June. *Fruits* roundish, somewhat flattened, 5-parted capsules about 3/16 inch in diameter.

RANGE. Massachusetts, southeastern New York, and southeastern Pennsylvania; south to Florida, west to Louisiana.

FETTERBUSH *Lyonia lucida*

FIELD MARKS. An evergreen shrub 3 to about 6 feet high, with arching or drooping and sharply 3-angled branches; growing in moist pinelands, swamps, and peaty thickets. *Leaves* alternate, elliptic to oval or sometimes broadest above the middle, pointed or broadly pointed at base, abruptly pointed at tip, stiff and leathery in texture, with a prominent vein paralleling the untoothed and narrowly rolled margin, dark green and lustrous above, paler and dull and glandular-dotted beneath, 1 to 3 inches long. *Flowers* white to deep rose pink, oblong bell-shaped, about 3/8 inch long, the 5 calyx lobes narrow and spreading, 3 to 10 in clusters in the leaf axils; blooming March to May. *Fruits* roundish, 5-parted capsules about 3/16 inch in diameter.

RANGE. Coastal plain, Southeastern Virginia south to Florida, west to Louisiana.

Also called Tetterbush and Hoorah-bush (Okefenokee region).

STAGGERBUSH *Lyonia mariana*

FIELD MARKS. A leaf-losing shrub 8 inches to nearly 4 feet high, with slender upright branches; growing in peaty or sandy open woods. *Leaves* alternate, elliptic or broadest above the middle, pointed at base, sharply to bluntly pointed at tip, untoothed on margin, bright green and smooth above, paler and minutely black-dotted and sometimes downy on the veins beneath, 1 to 3 inches long. *Flowers* white or pale pink, barrel-shaped, about 3/8 inch long, the 5 calyx lobes narrow and spreading, in several whorled clusters on leafless tips of branchlets; blooming April to June. *Fruits* urn-shaped, 5-angled, point capsules about 1/4 inch long.

RANGE. Chiefly coastal plain; Southern Rhode Island, southeastern New York, and southeastern Pennsylvania; south to Florida, west to Texas, and north in the Mississippi Valley to Arkansas and eastern Tennessee.

Foliage poisonous to calves and lambs if eaten.

MALEBERRY *Lyonia ligustrina*

FIELD MARKS. A leaf-losing shrub 3 to about 12 feet high; growing in moist to dry places, often along streams or in bogs. *Leaves* alternate, commonly elliptic to narrowly egg-shaped but often broadest above the middle, more or less pointed at both ends, usually very finely toothed but sometimes inconspicuously toothed or almost untoothed on margin, smooth above but usually more or less hairy beneath, commonly 1 to 2½ (rarely 4) inches long. *Flowers* white, globe-shaped, about 1/8 inch across, in leafless or leafy-bracted clusters from buds on growth of the previous year; blooming April to July. *Fruits* roundish but somewhat flattened, 5-celled capsules about 1/8 inch in diameter.

RANGE. Central Maine to central New York and Kentucky; south to Florida, Louisiana, and Oklahoma.

Very variable as to leaf size, hairiness of branchlets, etc. Also known as Privet-andromeda, Seedy-buckberry, and Male-huckleberry.

RUSTY LYONIA *Lyonia ferruginea*

FIELD MARKS. An evergreen shrub or small tree usually 3 to 10 (rarely 30) feet high; growing in hammocks and wet pinelands. *Leaves* alternate, those near the ends of the branchlets reduced in size and rusty in appearance, oval to narrowly elliptic or broadest above the middle, pointed at base, pointed to roundish at tip, untoothed on margin, leathery in texture; the larger ones pale green above, rusty-dotted beneath, 1 to 2½ inches long. *Flowers* white, globe-shaped, about 3/16 inch across, in small clusters in the leaf axils of the previous year; blooming March to May. *Fruits* narrowly egg-shaped, 5-angled capsules about 1/4 inch long.

RANGE. Coastal plain; South Carolina south to Florida.

Fetterbush

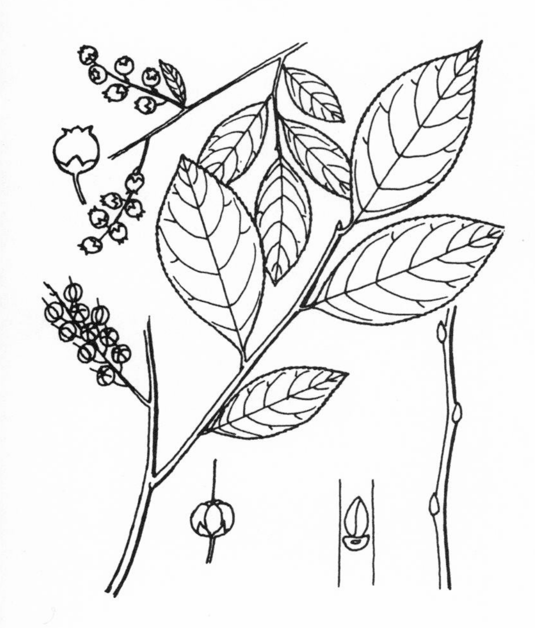

Maleberry

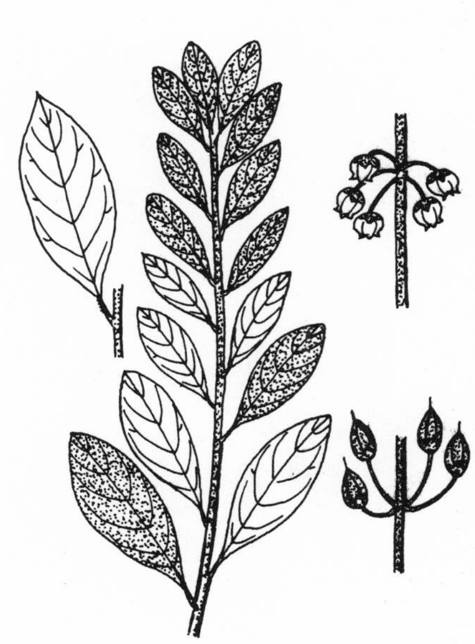

Rusty Lyonia

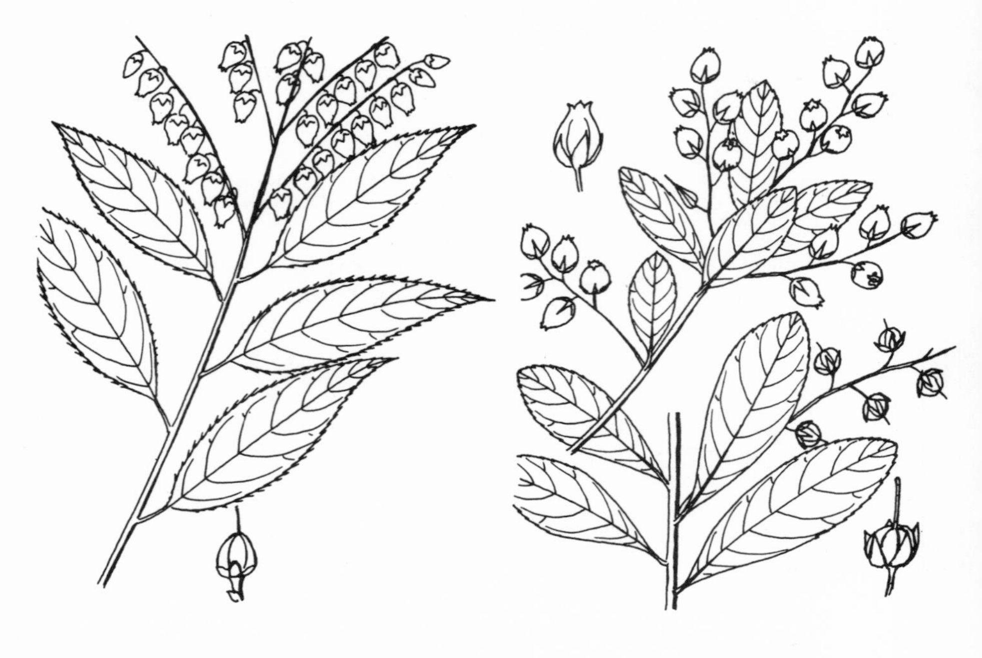

Mountain Fetterbush

Vine-wicky

Leatherleaf

Trailing-Arbutus

MOUNTAIN FETTERBUSH *Pieris floribunda*

FIELD MARKS. An evergreen shrub 3 to about 6 feet high; growing on dry to moist slopes or balds in the Appalachian Mountains. *Branchlets* slender, hairy, nearly erect. *Leaves* alternate, narrowly egg-shaped or oblong, pointed at base, sharply pointed at tip, finely toothed and bristly-hairy on margin, leathery in texture, smooth, green above, paler and minutely black-dotted beneath, 1½ to 3 inches long. *Flowers* white, vase-shaped, nodding, about 3/8 inch long, in a group of slender clusters at tips of branches; blooming March to June. *Fruits* egg-shaped, slightly 5-angled capsules about 3/16 inch long.

RANGE. West Virginia south to western North Carolina and eastern Tennessee.

Often cultivated as an ornamental shrub.

VINE-WICKY *Pieris phillyreifolia*

FIELD MARKS. An evergreen vinelike shrub creeping up the trunks of cypress trees beneath their outer bark, often to a height of 20 to 30 feet, and sending out branches every few feet; more rarely a small shrub standing alone. *Leaves* alternate, elliptic to oval or broadest above the middle, pointed at base, broadly pointed or blunt at tip, margin toothed only above the middle, leathery in texture, bright green and smooth above, paler beneath, 1 to 3 inches long. *Flowers* white, vase-shaped, about ¼ inch long, in narrow clusters from the axils of the upper leaves; blooming February to April. *Fruits* roundish, slightly flattened, 5-celled capsules about 3/16 inch in diameter.

RANGE. Coastal plain; Southeastern Georgia, northern Florida, and southern Alabama.

Also called He-huckleberry. A very unique shrub, the branches seemingly growing out of the trunks of cypress trees.

LEATHERLEAF *Chamaedaphne calyculata* var. *angustifolia*

FIELD MARKS. A much-branched, more or less evergreen shrub 1 to about 3 feet high, with spreading or horizontal branches; growing in peaty soils or in bogs. *Leaves* alternate, narrowly elliptic or broadest above the middle, pointed at base, bluntly pointed to roundish at tip, practically untoothed or with some minute teeth on the slightly rolled margin, dull green dotted with silvery scales above, brownish with minute rusty scales beneath, leathery in texture, ½ to 1½ inches long, often turning reddish in winter. *Flowers* white, bell-shaped, about ¼ inch long, in long and one-sided end clusters, each flower subtended by a leafy bract; blooming March to July. *Fruits* roundish, somewhat flattened, 5-celled capsules about ⅛ inch in diameter.

RANGE. Newfoundland to Alaska; south to the coastal plain and mountains of North Carolina, the region of the Great Lakes, northern Iowa, Alberta, and British Columbia.

TRAILING-ARBUTUS *Epigaea repens*

FIELD MARKS. A trailing evergreen shrub with slender stems which are usually bristly with rusty hairs; growing in dry rocky or sandy open woods and clearings. *Leaves* alternate, oblong-oval, roundish or heart-shaped at base, blunt to rounded at tip, untoothed but usually fringed with stiff hairs on margin, green and usually rough-hairy on both surfaces but sometimes quite smooth, ¾ to 2½ inches long. *Flowers* white to deep pink, about ½ inch long, very fragrant, in small clusters in the leaf axils or at tips of the stems; blooming March to May or June. *Fruits* roundish 5-celled capsules about ¼ inch in diameter.

RANGE. Labrador to Saskatchewan; south to Georgia, Alabama, and Iowa.

Also known as Mayflower.

TEABERRY *Gaultheria procumbens*

FIELD MARKS. An aromatic evergreen plant 2 to 6 inches high, with creeping underground stems; growing in sandy or rocky woods and clearings. *Leaves* with wintergreen odor when bruised, alternate, oval to roundish, margin with a few low and bristle-tipped teeth, leathery, smooth, lustrous above, paler green beneath, ¾ to 2 inches long. *Flowers* white, urn-shaped, about ¼ inch long, usually solitary in leaf axils; blooming June to August. *Fruits* roundish, bright red, about ⅜ inch in diameter, with spicy wintergreen odor when crushed; ripening September to November and persisting.

RANGE. Newfoundland to Manitoba; south to Georgia, Alabama, and Minnesota.

Also called Mountain-tea, Checkerberry, and Wintergreen. Oil of wintergreen used as a flavoring and in medicines is obtained from the leaves.

CREEPING-SNOWBERRY *Gaultheria hispudula*

FIELD MARKS. A creeping, aromatic, evergreen plant with very slender, brownish-hairy stems; growing in cold mossy woods and bogs. *Leaves* with a wintergreen odor when bruised, short-stalked, alternate, roundish egg-shaped, pointed or abruptly bristle-pointed at tip, broadly pointed to roundish at base, margin untoothed and rolled inward beneath, leathery, lustrous above, paler and with closely pressed brown hairs beneath, ⅛ to ⅜ inch long. *Flowers* white, bell-shaped, slightly over ⅛ inch long, solitary in leaf axils; blooming May or June. *Fruits* roundish, white, usually with some bristly brownish hairs, about ¼ inch in diameter, with wintergreen odor when crushed; ripening August or September.

RANGE. Newfoundland and Labrador to British Columbia; south to New England, West Virginia, Michigan, Minnesota, and Idaho.

MOSS-PLANT *Cassiope hypnoides*

FIELD MARKS. A prostrate or tufted, mosslike, evergreen plant 1 to 5 inches high; growing on tundras and southward on alpine mountain summits. *Leaves* narrow and needle-like, crowded in 4 rows along the branchlets, minutely downy, averaging about ⅛ inch long. *Flowers* white or pink, open bell-shaped, deeply 4- to 5-lobed, about ¼ inch long, nodding on slender stalks at tips of the branchlets; blooming June to August. *Fruits* small, roundish, 4- to 5-parted capsules.

RANGE. Arctic America south to mountains of Newfoundland, Quebec, Maine, New Hampshire, and New York.

BEARBERRY *Arctostaphylos uva-ursi*

FIELD MARKS. A trailing evergreen shrub, the long and flexible branches often rooting at the nodes; growing on tundras and rocky or sandy open areas. *Branchlets* smooth to white-woolly or sticky-downy; later becoming reddish brown to grayish brown, and with papery peeling bark. *Leaves* alternate, short-stalked, usually broadest above the middle, wedge-shaped at base, rounded at tip, untoothed on margin, leathery, smooth or nearly so on both surfaces, lustrous above, paler beneath, ½ to 1¼ inches long. *Flowers* white or pinkish, urn-shaped, almost ¼ inch long, in small end clusters; blooming May to July and rarely October or November. *Fruits* roundish, red, about ¼ inch in diameter, with mealy flesh and 5 to 10 seeds; ripening August or September and persisting.

RANGE. Labrador to Alaska; south to eastern Virginia, the region of the Great Lakes, South Dakota, New Mexico, and California.

Leaves used medicinally, for tanning leather, and for making dyes. Mixed with tobacco by Indians and called Kinnikinnik. Berries eaten by grouse and bears.

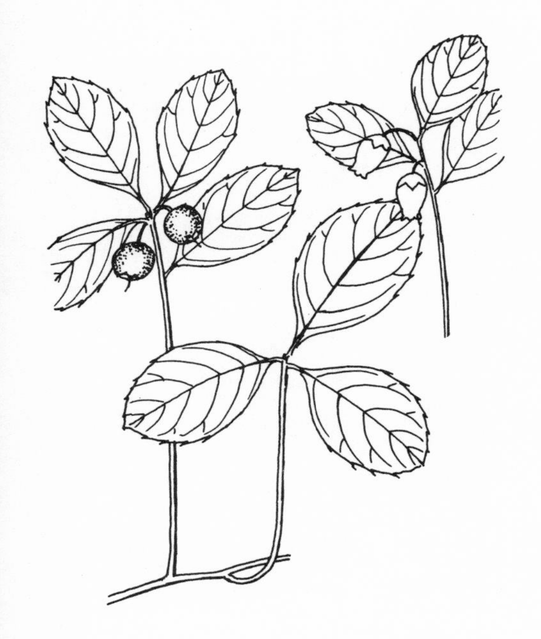

Teaberry

Creeping-snowberry

Moss-plant

Bearberry

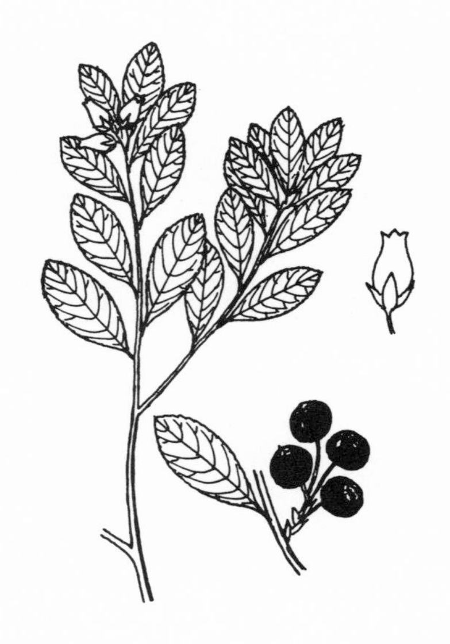

Alpine Bearberry

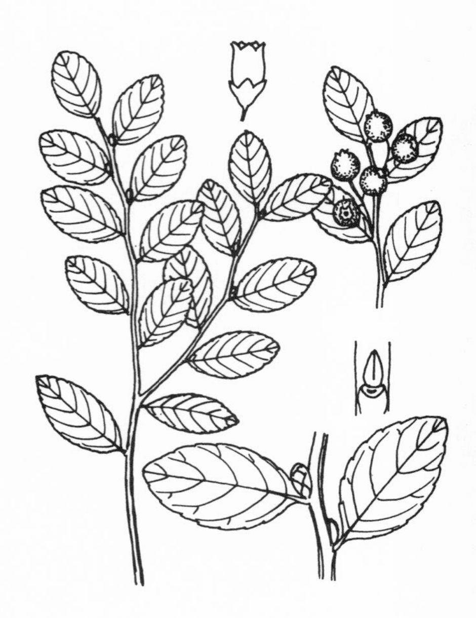

Box Huckleberry

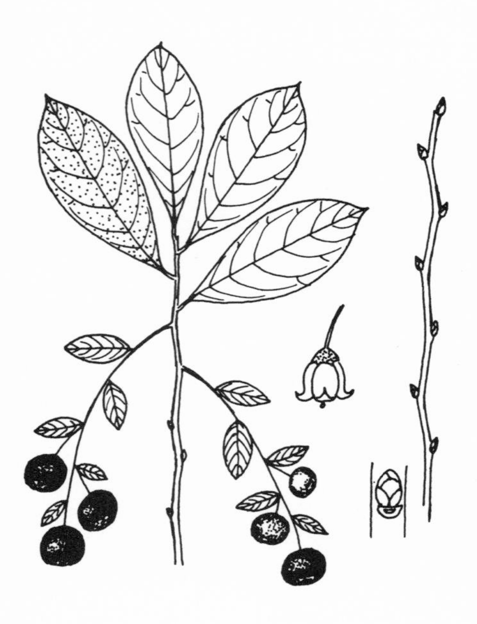

Dwarf Huckleberry

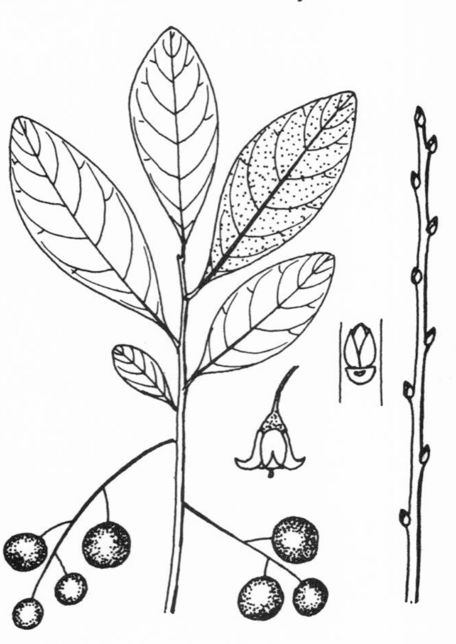

Dangleberry

ALPINE BEARBERRY *Arctostaphylos alpina*

FIELD MARKS. A trailing or mat-forming shrub with papery-barked and brittle branches; growing on tundras and southward on alpine mountains summits. *Leaves* alternate, broadest above the middle, wedge-shaped at base, pointed or blunt at tip, margin minutely toothed and often hairy-fringed toward base, wrinkled-veiny above, ⅜ to 1 inch long; turning bright red and withering in fall but persisting. *Flowers* white or pink-tinged, urn-shaped, 4- or 5-lobed, about 3/16 inch long, 2 to 4 in end clusters; blooming May or June. *Fruits* roundish, purple or purplish black, juicy, 4- or 5-seeded; ripening September or October.

RANGE. Arctic America south to Newfoundland, Maine, and New Hampshire.

HUCKLEBERRIES (Gaylussacia)

(Key Appendix N)

Huckleberries are usually leaf-losing shrubs with alternate simple leaves which are dotted with small, glistening, yellowish resin globules. The fruits are berry-like, roundish, and have 10 rather large seedlike nutlets. These leaf and fruit characteristics readily distinguish them from the blueberries.

BOX HUCKLEBERRY *Gaylussacia brachycera*

FIELD MARKS. An evergreen shrub forming colonies by creeping underground stems, branches 6 to 15 inches high; growing in sandy woods or on dry wooded slopes. *Branchlets* prominently angled, smooth. *Leaves* short-stalked, oval to egg-shaped, broadly pointed to blunt at both ends, margin with low blunt teeth and rolled inward beneath, leathery, smooth, bright green above, paler yellowish green beneath, ½ to 1 inch long. *Flowers* white or pinkish, urn-shaped, in small clusters; blooming May or June. *Fruits* blue, with a whitish bloom, about ⅜ inch in diameter; ripening June to August.

RANGE. Delaware and Pennsylvania south to South Carolina and Tennessee.

DWARF HUCKLEBERRY *Gaylussacia dumosa*

FIELD MARKS. A leaf-losing shrub 6 inches to 2 feet high, with creeping underground stems; growing in dry sandy woods. *Branchlets* slender, zigzag, more or less downy. *Leaves* broadest above the middle, wedge-shaped at base, abruptly short-pointed at tip, margin untoothed, thickish, bright green on both sides, lustrous above, somewhat glandular-hairy and resin-dotted beneath, 1 to 2½ inches long. *Flowers* whitish to greenish pink or reddish, bell-shaped, in leafy-bracted clusters; blooming March to June. *Fruits* black, sometimes glandular-hairy, ¼ to ⅜ inch in diameter; ripening June to August.

RANGE. Newfoundland and New Brunswick to eastern Pennsylvania and Tennessee, south to Florida and Louisiana.

DANGLEBERRY *Gaylussacia frondosa*

FIELD MARKS. A leaf-losing shrub 2 to 4 feet high; growing in rocky or sandy woods and bogs. *Branchlets* slender, smooth, often whitened with a bloom. *Leaves* elliptic to oval or broadest above the middle, pointed at base, bluntly pointed to roundish at tip, margin untoothed, pale green and smooth above, paler or whitened and resin-dotted beneath, 1 to 2½ inches long. *Flowers* greenish pink, bell-shaped, in rather long drooping clusters; blooming March to June. *Fruits* blue, whitened with a bloom, about ⅜ inch in diameter, sweet and juicy; ripening June to August.

RANGE. Massachusetts to southeastern New York and Ohio, south to Florida and Louisiana.

Also called Tangleberry, Blue-tangle, and Blue Huckleberry.

WOOLLY DANGLEBERRY *Gaylussacia frondosa* var. *tomentosa*

FIELD MARKS A leaf-losing shrub with stiffly erect and sparingly branched stems 2 to 4 feet high; growing in moist pinelands. *Branchlets* densely downy or woolly-hairy. *Leaves* elliptic to oval or broadest above the middle, wedge-shaped at base, broadly pointed or blunt at tip, pale green and wrinkled-veiny above, densely coated beneath with pale to brownish wool and also resin-dotted, 1 to 3 inches long. *Flowers* white or pinkish, bell-shaped, in rather long drooping clusters; blooming March to May. *Fruits* dull blue, whitened with a bloom, about ⅜ inch in diameter; ripening June to August.

RANGE. Coastal plain; southeastern South Carolina south to Florida, west to Alabama.

CREEPING DANGLEBERRY *Gaylussacia frondosa* var. *nana*

FIELD MARKS. A leaf-losing shrub with creeping underground stems and erect branches 8 inches to 2 feet high; growing in sandy oak and pine woods. *Branchlets* finely downy, often whitened with a bloom and resin-dotted. *Leaves* elliptic or broadest above the middle, pointed to roundish at base, broadly pointed or rounded at tip, margin untoothed but slightly rolled, wrinkled-veiny, smooth above, often finely downy and resin-dotted beneath, whitened on both surfaces, ¾ to 1½ inches long. *Flowers* creamy white or greenish white, often tinged with red, bell-shaped, in drooping clusters; blooming March or April. *Fruits* blue, whitened with a bloom, about ¼ inch in diameter; ripening June or July.

RANGE. Coastal plain; Georgia and Florida.

BLACK HUCKLEBERRY *Gaylussacia baccata*

FIELD MARKS. A leaf-losing shrub 1½ to 3 feet high, with erect branches and with young growth copiously dotted with sticky resin globules; growing in dry sandy or rocky woods and sometimes in bogs. *Leaves* elliptic to oval or oblong lance-shaped (rarely broadest above the middle) pointed at base, pointed to blunt at tip, margin untoothed, yellowish green above and scarcely paler beneath, resin-dotted on both surfaces but more densely so beneath, 1 to 2½ inches long. *Flowers* greenish tinged with pink or red, egg-shaped, in short clusters; blooming April to June. *Fruits usually* lustrous black, about ¼ inch in diameter, sweet; ripening July and August.

RANGE. Newfoundland to Saskatchewan, south to Georgia and Louisiana.

BUCKBERRY *Gaylussacia ursina*

FIELD MARKS. A leaf-losing shrub 2 to about 4 feet high; growing throughout woodlands of the southern Appalachians. *Branchlets* somewhat downy when young. *Leaves* elliptic or sometimes broadest above the middle, pointed to roundish at base, pointed at tip, margin untoothed but often hairy-fringed, thin in texture, bright green and smooth or nearly so except for some hairs along the midrib beneath, minutely and rather inconspicuously resin-dotted on the lower surface, 1½ to 4 inches long. *Flowers* greenish white to reddish, bell-shaped, about 3/16 inch across, in elongate clusters; blooming May or June. *Fruits* lustrous black, about ⅜ inch in diameter, insipid to rather sweet; ripening July to September.

RANGE. Southwestern North Carolina and eastern Tennessee, south into northwestern South Carolina and northern Georgia.

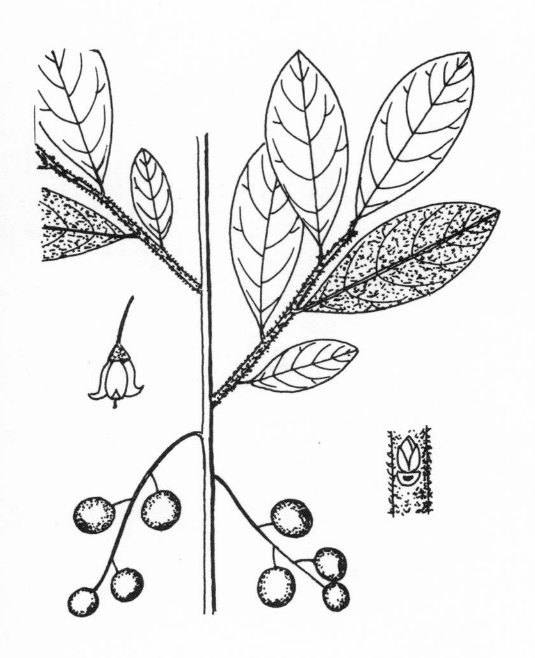

Woolly Dangleberry

Creeping Dangleberry

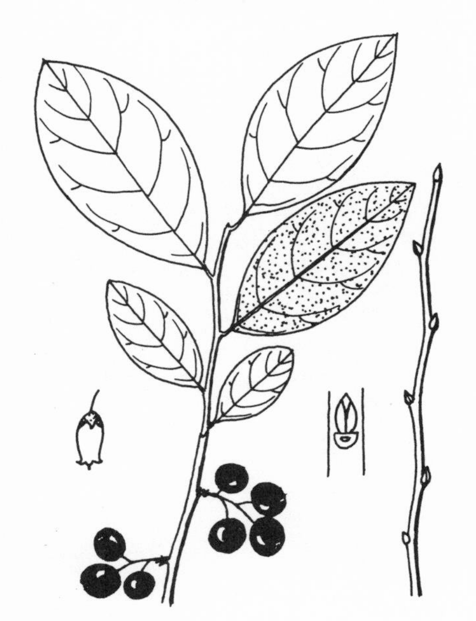

Black Huckleberry

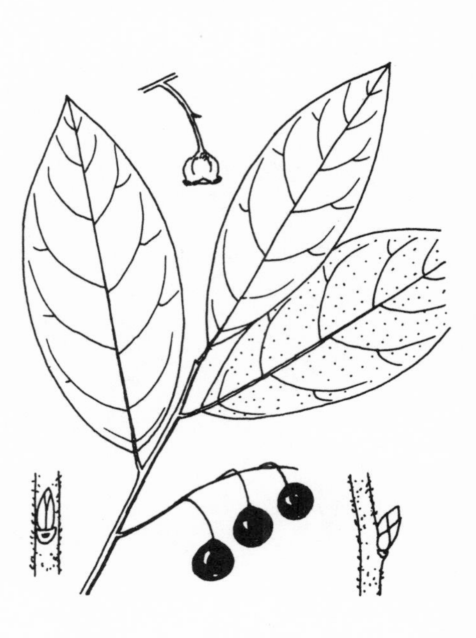

Buckberry

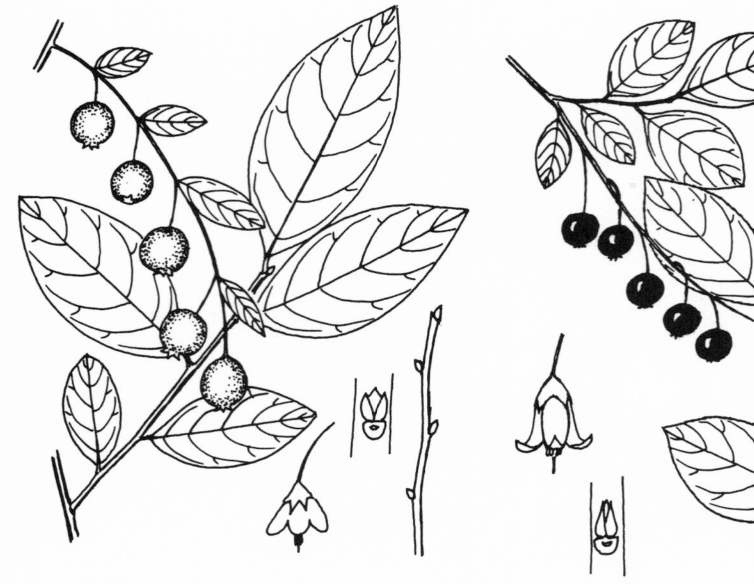

Deerberry

Sparkleberry

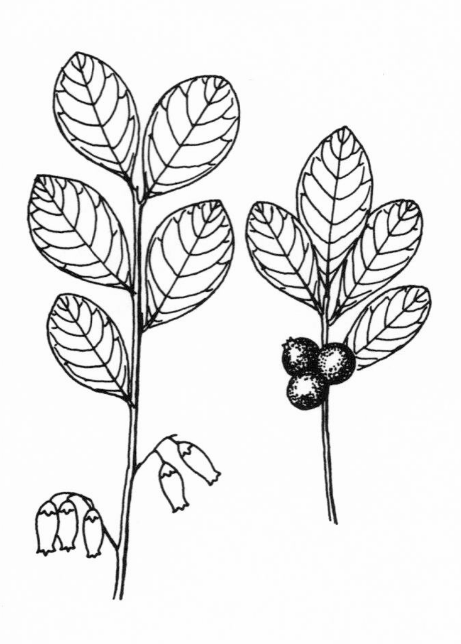

Bog Bilberry

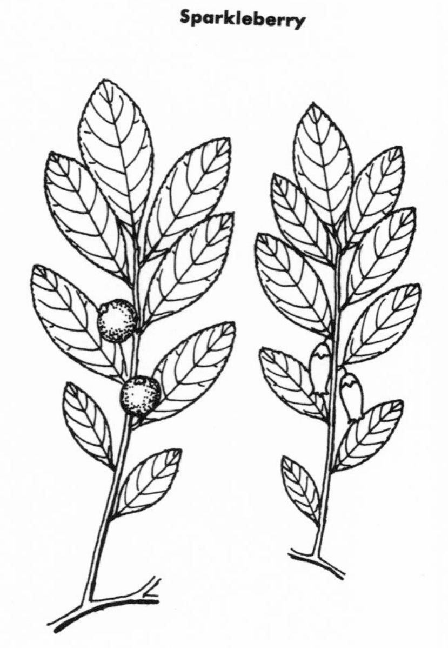

Dwarf Bilberry

BLUEBERRIES, BILBERRIES, AND CRANBERRIES (Vaccinium)

(Key Appendix O)

These are shrubs with alternate and simple leaves. The fruits are roundish berries containing a large number of small seeds. Blueberries and cranberries are well known for their edible fruits. The blueberries are a difficult group, and various botanists have described numerous species. Other botanists regard them as mere variations of the species included here.

DEERBERRY *Vaccinium stamineum*

FIELD MARKS. A leaf-losing shrub 2 to 10 feet high; growing in dry rocky or sandy woods and thickets. *Branchlets* reddish purple, often downy or with a whitish bloom. *Leaves* elliptic or egg-shaped, pointed to rounded at base, usually pointed at tip, untoothed on margin, thin, smooth above, smooth or downy and often whitened beneath, 1 to 3½ inches long. *Flowers* greenish white or purple-tinged, open bell-shaped, in leafy-bracted clusters; blooming April to June. *Fruits* greenish or pale purplish, often whitened with a bloom, 5/16 to ½ inch in diameter, rather sour but edible when cooked; ripening July to October.

RANGE. Massachusetts to southern Ontario and Kansas; south to Florida and Louisiana.

Also called Squaw-huckleberry.

SPARKLEBERRY *Vaccinium arboreum*

FIELD MARKS. A leaf-losing, or southward an evergreen, shrub or small tree 4 to 30 feet high; growing in dry sandy or rocky woods. *Leaves* elliptic to oval or broadest above the middle, more or less pointed at both ends, margin sometimes minutely toothed or slightly rolled inward on lower surface, somewhat leathery in texture, lustrous above, paler and sometimes downy beneath, ¾ to 2 inches long. *Flowers* white, bell-shaped, in leafy-bracted clusters; blooming April to June. *Fruits* black, lustrous, about ¼ inch in diameter, rather dry and insipid; ripening September or October.

RANGE. Virginia to southern Indiana and Illinois, Missouri, and Oklahoma; south to Florida and Texas.

Also called Farkleberry and Tree-huckleberry.

BOG BILBERRY *Vaccinium uliginosum*

FIELD MARKS. A spreading, much-branched, leaf-losing shrub 6 to 12 inches high; growing on tundras and southward on alpine mountain summits and lakeshores. *Leaves* almost stalkless, oval or broadest above the middle, wedge-shaped at base, rounded or blunt at tip, untoothed on margin, smooth and dull above, paler or whitened and sometimes downy beneath, 3/16 to ¾ inch long. *Flowers* pinkish, urn-shaped, 2 to 4 in a cluster; blooming June and July. *Fruits* bluish black, with whitish bloom, about ¼ inch in diameter, sweet and edible; ripening July to September.

RANGE. Arctic America south to New England, New York, Michigan, and Minnesota.

DWARF BILBERRY *Vaccinium cespitosum*

FIELD MARKS. A tufted leaf-losing shrub 2 to 12 inches high; growing in rocky woods or on gravelly shores. *Leaves* short-stalked, broadest above the middle, wedge-shaped at base, blunt or broadly pointed at tip, finely toothed on margin, thin, smooth, lustrous green on both sides, ⅜ to 1½ inches long. *Flowers* white or pink, narrowly bell-shaped, solitary in leaf axils; blooming June or July. *Fruits* light blue, with whitish bloom, about ¼ inch in diameter, sweet and edible; ripening July or August.

RANGE. Labrador to Alaska; south to New England, New York, Michigan, Wisconsin, Colorado, and California.

THINLEAF BILBERRY *Vaccinium membranaceum*

FIELD MARKS. An erect leaf-losing shrub 1 to 4½ feet high, with peeling bark; growing in cool, moist, northern woods and thickets. *Branchlets* somewhat 4-angled. *Leaves* elliptic or egg-shaped, pointed at both ends, sharply and finely toothed on margin, thin in texture, bright green and smooth or nearly so on both surfaces, ¾ to 2¾ inches long. *Flowers* greenish or purplish, solitary in leaf axils; blooming June or July. *Fruits* dark purple or black, about 5/16 inch across, rather sour; ripening July to September.

RANGE. Western Ontario, northern Michigan, and southwestern South Dakota; southern Alberta and southern British Columbia south to California.

OVAL-LEAF BILBERRY *Vaccinium ovalifolium*

FIELD MARKS. A straggling, slender, leaf-losing shrub 1 to 5 feet high; growing in cool northern woods and on peaty slopes. *Branchlets* sharply 4-angled. *Leaves* elliptical, roundish or broadly pointed at base, rounded or blunt at tip, untoothed or slightly wavy-toothed on margin, thin in texture, dull green above, pale or whitened beneath, smooth on both surfaces, 1 to 2 inches long. *Flowers* pinkish, bell-shaped, solitary in the axils of developing leaves; blooming June or July. *Fruits* blue, with a white bloom, about ⅜ inch in diameter, rather unpalatable; ripening September and October.

RANGE. Newfoundland and southeastern Labrador to western Ontario and northern Michigan; Alaska south to Idaho and Oregon.

EVERGREEN BLUEBERRY *Vaccinium myrsinites*

FIELD MARKS. A much-branched evergreen shrub 8 inches to 2 feet high; growing in open sandy woods and pinelands. *Leaves* elliptic to egg-shaped or broadest above the middle, more or less pointed at both ends, margin untoothed or with small and low bristle-tipped teeth, thickish in texture, dark green and lustrous above, paler and sometimes slightly downy or whitened beneath, ¼ to ¾ inch long. *Flowers* white or pinkish, cylindrical bell-shaped, in umbel-like end clusters; blooming late February to April. *Fruits* bluish black, sometimes with a whitish bloom; ripening April or May.

RANGE. Coastal plain; North Carolina south to Florida.

VELVET-LEAF BLUEBERRY *Vaccinium myrtilloides*

FIELD MARKS. A much-branched, leaf-losing shrub 8 inches to about 2 feet high; growing in cool, moist woods and swamps. *Branchlets* densely velvety with whitish hairs and warty-dotted. *Leaves* short-stalked, narrowly elliptic to oblong lance-shaped, pointed at both ends, untoothed on margin, thin in texture, green and somewhat downy above, densely downy beneath with pale hairs, ¾ to 1½ inches long. *Flowers* greenish white or pinkish, narrowly bell-shaped, in dense clusters; blooming May or June. *Fruits* blue, with a white bloom, ¼ to ⅜ inch in diameter, usually quite sour; ripening July and August.

RANGE. Newfoundland and Quebec to British Columbia; south to western New England, Pennsylvania, the Great Lakes region, northeastern Iowa, and Montana, and in the Appalachian Mountains to western Virginia and West Virginia.

Also called Canada Blueberry and Sourtop.

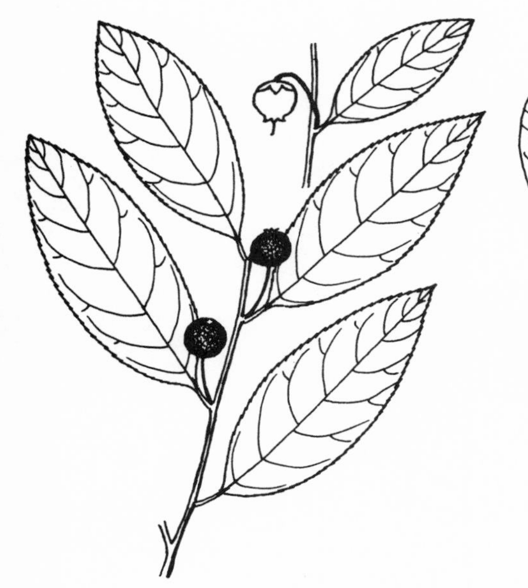

Thinleaf Bilberry

Oval-leaf Bilberry

Evergreen Blueberry

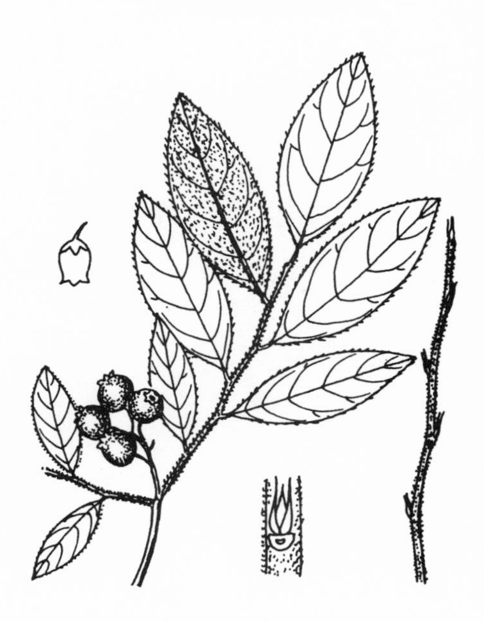

Velvet-leaf Blueberry

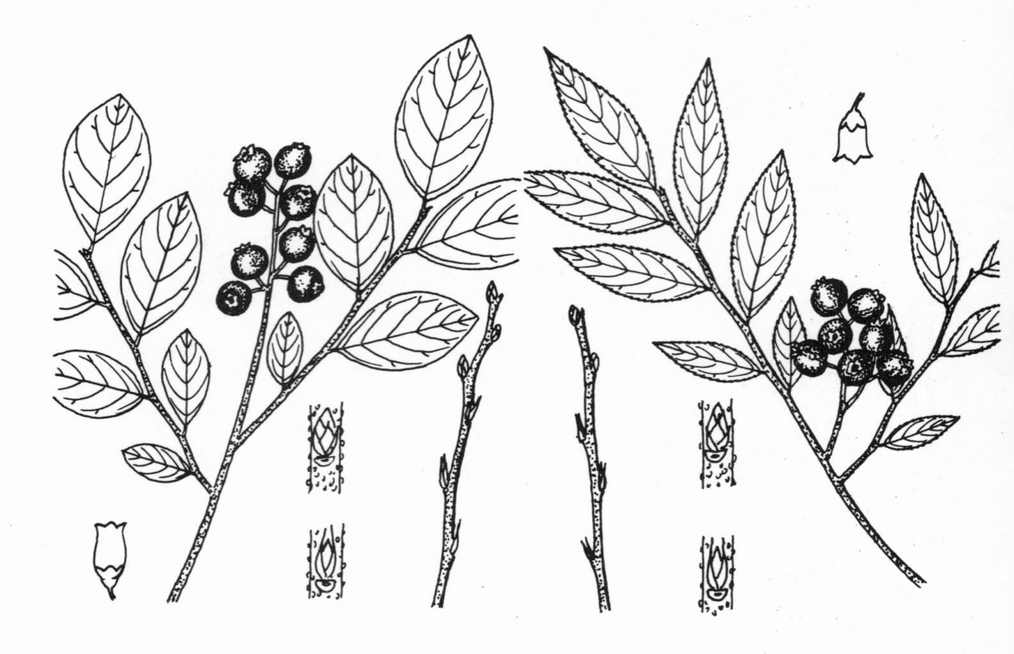

Late Low Blueberry

Early Low Blueberry

Hairy Blueberry

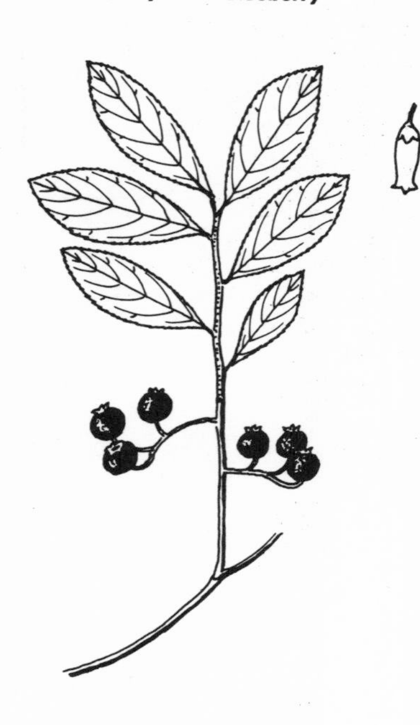

Small Black Blueberry

LATE LOW BLUEBERRY *Vaccinium pallidum*

FIELD MARKS. A spreading leaf-losing shrub 8 inches to about 2 feet high; growing in dry sandy or rocky woods and clearings. *Branchlets* yellowish green, warty-dotted, somewhat angled. *Leaves* short-stalked, oval to egg-shaped or broadest above the middle, usually broadly pointed at both ends, untoothed or sometimes finely toothed on margin, smooth on both surfaces, paler and sometimes slightly whitened beneath, ½ to 2 inches long. *Flowers* greenish white or pink-tinged, cylindrical urn-shaped, clustered; blooming March to June, when leaves are partly grown. *Fruits* dark blue, usually with a whitish bloom, ¼ to ⅜ inch in diameter, sweet and juicy; ripening June to September.

RANGE. Western Nova Scotia to southern Ontario, Michigan, and northeastern Iowa; south to Georgia, Alabama, and eastern Kansas.

EARLY LOW BLUEBERRY *Vaccinium angustifolium*

FIELD MARKS. A spreading leaf-losing shrub 8 inches to about 2 feet high; growing in open rocky woods and clearings. *Branchlets* olive green to yellowish green, somewhat grooved, warty-dotted. *Leaves* short-stalked, narrowly elliptic or lance-shaped, pointed at both ends, finely and sharply toothed on margin, bright green and smooth above, paler green and sometimes whitened or downy beneath, ½ to 1½ inches long. *Flowers* white or pink-tinged, urn-shaped, clustered; blooming April to June. *Fruits* blue, with a heavy whitish bloom, or sometimes lustrous black, about ¼ inch in diameter, sweet and juicy; ripening June to August.

RANGE. Labrador to Saskatchewan; south to western Virginia, West Virginia, Indiana, Minnesota, and northeastern Iowa.

The common commercial wild blueberry of New England.

HAIRY BLUEBERRY *Vaccinium hirsutum*

FIELD MARKS. A leaf-losing shrub 1 to 2½ feet high; growing in sandy open woods and balds in the southern Appalachians. *Branchlets* densely soft-hairy. *Leaves* elliptic or egg-shaped, usually pointed at both ends, untoothed on margin, hairy on both surfaces but paler and more densely so beneath, ½ to 2 inches long. *Flowers* cylindrical urn-shaped, greenish white or reddish-tinged, densely hairy, clustered; blooming in April or May. *Fruits* bluish black, densely hairy, ¼ inch or slightly more in diameter, sweet and juicy; ripening June or July.

RANGE. Mountains of southwestern North Carolina, western Tennessee, and northern Georgia at altitudes of 1,700 to about 5,000 feet.

SMALL BLACK BLUEBERRY *Vaccinium tenellum*

FIELD MARKS. A leaf-losing shrub with creeping underground stems and often nearly unbranched upright stems 8 inches to 2 feet high; growing in dry sandy woods and pinelands. *Leaves* broadest above the middle, wedge-shaped at base, more broadly pointed at tip, finely and rather inconspicuously toothed on margin, green and smooth above, paler and with gland-tipped hairs beneath at least when young, ½ to about 1 inch long. *Flowers* white or pink-tinged, rather narrowly cylindric, clustered; blooming March to May. *Fruits* black, about ¼ inch in diameter, rather dry but sweet; ripening June or July.

RANGE. Southeastern Virginia south to Florida and west to Mississippi.

HIGHBUSH BLUEBERRY *Vaccinium corymbosum*

FIELD MARKS. A leaf-losing shrub 3 to about 10 feet high; growing in low wet grounds, bogs, swamps, and moist rocky woods. *Branchlets* yellowish green to reddish, warty-dotted, often hairy in lines. *Leaves* short-stalked, elliptic to egg-shaped or broadly lance-shaped, pointed to rounded at base, pointed at tip, untoothed or sometimes with very fine or bristly teeth on margin, paler green and sometimes whitened or slightly downy beneath, 1 to about 3 inches long. *Flowers* white, greenish white, or pinkish, cylindrical urn-shaped, clustered; blooming late February to June, before the leaves or when the leaves are partly grown. *Fruits* blue or bluish black, whitened with a bloom, ¼ to nearly ½ inch in diameter, sweet and juicy; ripening June to August.

RANGE. Nova Scotia to southern Quebec and Wisconsin; south to Florida and Louisiana.

The parent of most cultivated varieties of blueberries. Most common southward in coastal plain but also present in the mountains.

BLACK HIGHBUSH BLUEBERRY *Vaccinium atrococcum*

FIELD MARKS. A leaf-losing shrub 3 to about 10 feet high; similar to the preceding species and growing in similar places. *Branchlets* densely downy. *Leaves* short-stalked, elliptic to egg-shaped, usually pointed at both ends, untoothed on margin, paler and densely downy beneath, 1 to 3 inches long, *Flowers* greenish white to yellowish green and often tinged with red, cylindrical urn-shaped, clustered; blooming late February to June, before or with the expanding leaves. *Fruits* lustrous black, about ⅜ inch in diameter, sweet and juicy; ripening June to August.

RANGE. New England, New York, and southern Ontario; south to Georgia and Arkansas.

Found southward chiefly in the coastal plain and piedmont.

ELLIOTT BLUEBERRY *Vaccinium elliottii*

FIELD MARKS. A leaf-losing shrub 3 to about 8 feet high; growing along streams, in sandy woods, and in swamps. *Branchlets* warty-dotted, greenish, often downy. *Leaves* almost stalkless, egg-shaped to elliptic or oval, mostly rounded at base, blunt to pointed at tip, finely toothed or almost untoothed on margin, lustrous green and smooth above, duller and often somewhat downy beneath, ⅜ to about 1 inch long. *Flowers* pink or reddish, narrowly urn-shaped or vase-shaped, clustered; blooming March or April, as the leaves are expanding. *Fruits* black or bluish black, about 5/16 inch in diameter, sweet but rather dry; ripening June or July.

RANGE. Southeastern Virginia south to Florida, west to Louisiana; north in Mississippi Valley to Arkansas.

SOUTHERN MOUNTAIN-CRANBERRY *Vaccinium erythrocarpum*

FIELD MARKS. A leaf-losing shrub 1 to about 6 feet high, with spreading branches; growing in cool moist woods or bogs in the southern Appalachians. *Branchlets* downy. *Leaves* short-stalked, egg-shaped to oblong lance-shaped, rounded to broadly pointed at base, tapering to a sharp point at tip, margin finely toothed with bristle-tipped teeth, green and often slightly downy on both surfaces, 1 to about 3 inches long. *Flowers* pale red with 4 recurved corolla lobes, solitary in the axils of leaves or leaf-like bracts; blooming May to July. *Fruits* dark purplish red, about 3/16 inch in diameter, sweetish to sour or insipid; ripening August and September.

RANGE. Western Virginia and West Virginia south to northern Georgia and eastern Tennessee.

Also called Bearberry.

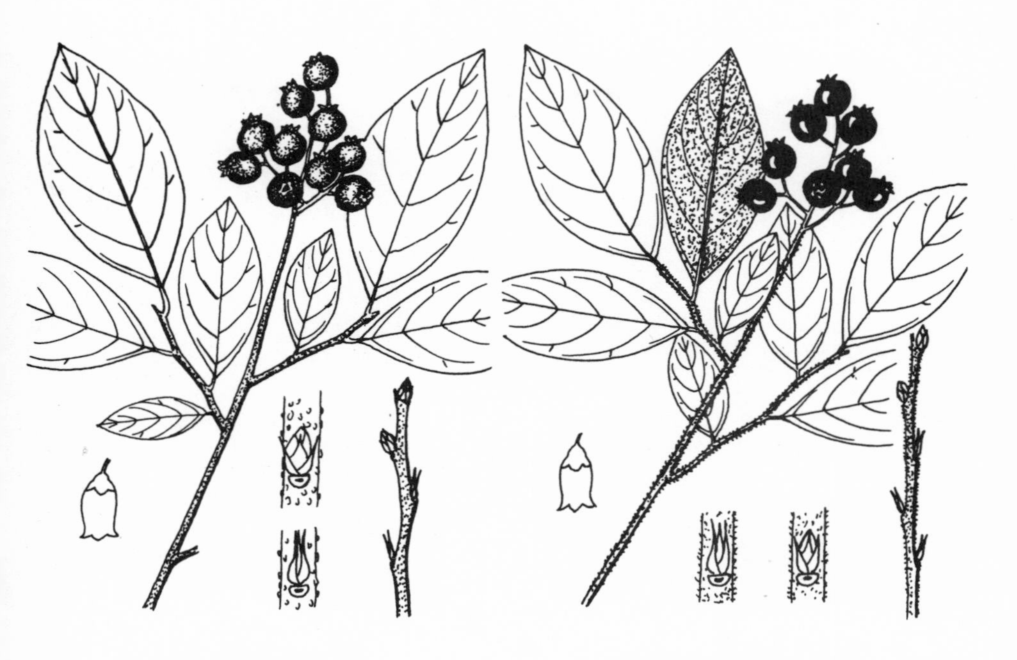

Highbush Blueberry

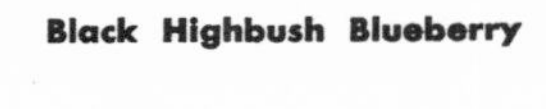

Black Highbush Blueberry

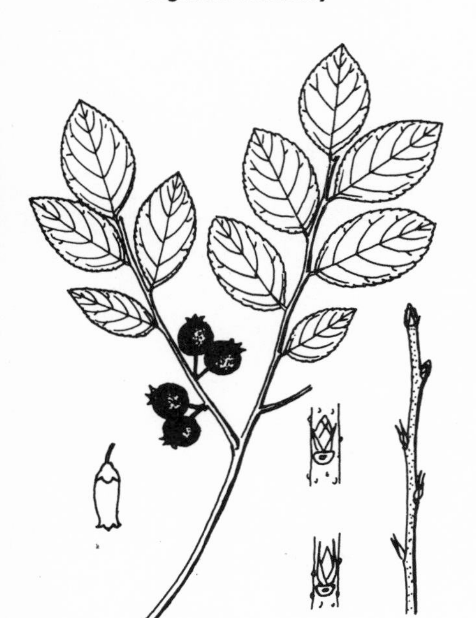

Elliott Blueberry

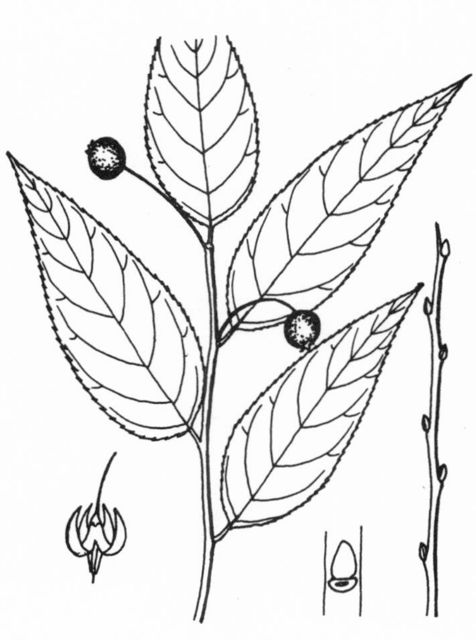

Southern Mountain-cranberry

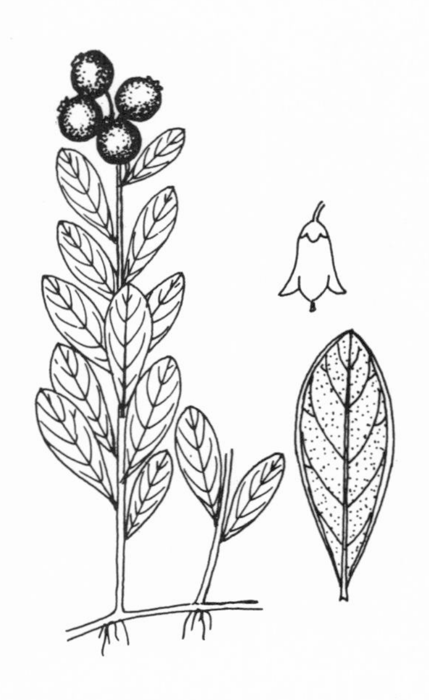

Mountain-cranberry

Creeping Blueberry

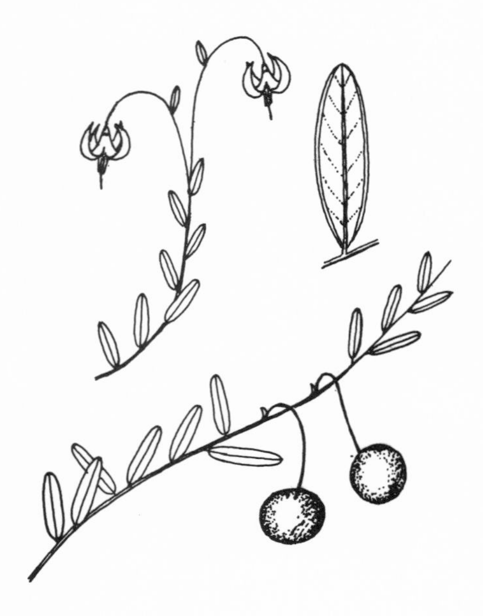

Large Cranberry

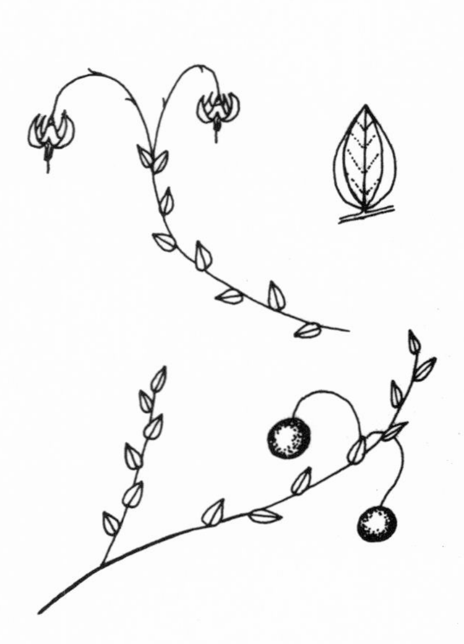

Small Cranberry

MOUNTAIN-CRANBERRY *Vaccinium vitis-idaea* var. *minus*

FIELD MARKS. An evergreen mat-forming shrub with creeping underground stems and upright branches 1 to about 6 inches high; growing in cool, moist, rocky places and bogs. *Leaves* short-stalked, oval or broadest above the middle, wedge-shaped at base, blunt or rounded at tip, margin rolled and sometimes sparingly toothed, dark green and lustrous above, pale and with minute black dots beneath, ¼ to ¾ inch long. *Flowers* white or pinkish, bell-shaped with 4 short lobes, in small end clusters; blooming June or July. *Fruits* dark red, ¼ to ⅜ inch in diameter, acid and slightly bitter but edible when cooked; ripening August and September.

RANGE. Arctic region south to Newfoundland, New England, southwestern Ontario, northern Michigan and Minnesota, Manitoba, and British Columbia.

Also called Cowberry.

CREEPING BLUEBERRY *Vaccinium crassifolium*

FIELD MARKS. A trailing evergreen shrub with slender stems up to 3 feet long from a thickened base; growing in wet peaty soils of pineland bogs. *Leaves* very short-stalked, elliptic or oval, usually broadly pointed at both ends or blunt at tip, margin thickened or slightly rolled and minutely toothed, leathery in texture, smooth, deep green above, paler green beneath, ¼ to ½ inch long. *Flowers* white or pink, globe-shaped, in small axillary clusters; blooming April or May. *Fruits* black or purplish black, lustrous, little more than ⅛ inch in diameter; ripening June or July.

RANGE. Coastal plain; southeastern Virginia south to Georgia.

In habit this plant very much resembles the cranberries.

LARGE CRANBERRY *Vaccinium macrocarpom*

FIELD MARKS. A trailing evergreen shrub with slender stems up to 3 feet long; growing in wet peaty soils and sphagnum bogs. *Leaves* very short-stalked, oblong-elliptic, roundish or blunt at both ends, margin untoothed and sometimes slightly rolled, smooth, leathery in texture, dark green and lustrous above, pale or somewhat whitened beneath, ¼ to ⅝ inch long. *Flowers* pink, with 4 recurved lobes, solitary or 2 to 4 on slender stalks in end clusters; blooming May to August. *Fruits* red, ⅜ to ¾ inch in diameter, sour; ripening August to November.

RANGE. Newfoundland to Minnesota; south to coastal plain and mountains of North Carolina, Tennessee, the region of the Great Lakes, and Arkansas.

The familiar cranberry grown commercially and sold in markets. Also called American Cranberry.

SMALL CRANBERRY *Vaccinium oxycoccos*

FIELD MARKS. A trailing evergreen shrub with almost threadlike stems up to 18 inches in length; growing in cold sphagnum bogs (sometimes along with the preceding species). *Leaves* very short-stalked, egg-shaped or triangular, broadly rounded or slightly heart-shaped at base, pointed at tip, margin untoothed and rolled, dark green and lustrous above, white beneath, ⅛ to rarely ½ inch long. *Flowers* similar to those of the preceding species but much smaller; blooming May to July. *Fruits* red, rarely ⅜ inch in diameter, very sour; ripening August to October.

RANGE. Labrador to Alaska; south to New Jersey, West Virginia, northern Ohio, Michigan, Wisconsin, Manitoba, Saskatchewan, and Oregon.

SAPODILLA FAMILY (Sapotaceae)

BUMELIAS (Sideroxylon)

Bumelias are shrubs or small trees with alternate, simple leaves which are often clustered on short lateral spurs. The branchlets are armed with short thorns and exude a milky sap when they are cut or broken. They have small, 5-parted, whitish flowers which are borne in dense clusters in the leaf axils. The fruits are berry-like in appearance but have a single large seed, surrounded by a pulp with a bittersweet taste. Also called Bullies.

BUCKTHORN BUMELIA *Sideroxylon lycioides*

FIELD MARKS. A leaf-losing shrub or small tree 5 to 20 feet or more high; growing in swampy woods or on stream banks, bluffs, and dunes. *Leaves* narrowly elliptic or broadest above the middle, wedge-shaped at base, short-pointed to rounded at tip, untoothed on margin, smooth or nearly so on both surfaces, 1 to 6 inches long. *Flowers* blooming June to August. *Fruits* black to blue-black, oblong, about ⅜ to ½ inch long; ripening September or October.

RANGE. Southeastern Virginia south to Florida, west to Texas; north in Mississippi Valley to southern Indiana, Illinois, and southeastern Missouri.

Also called Southern-buckthorn and False-buckthorn.

WOOLLY BUMELIA *Sideroxylon lanuginosum*

FIELD MARKS. A shrub or small tree 10 feet or more high, with more or less persistent leaves; growing in moist to dry sandy or rocky woods and thickets. *Leaves* narrowly elliptic, wedge-shaped at base, abruptly pointed at tip, untoothed on margin, dark green and lustrous above, sparsely to densely coated with tawny to white hairs beneath, 1 to about 4 inches long. *Flowers* blooming June or July. *Fruits* black, roundish to oval, about ½ inch long; ripening about October.

RANGE. Coastal plain, southern Georgia and northern Florida west to Texas; north in Mississippi Valley to southern Illinois, central Missouri, and southeastern Kansas.

Also called Gum-elastic, Woolly-buckthorn, and Chittamwood.

TOUGH BUMELIA *Sideroxylon tenax*

FIELD MARKS. An evergreen shrub or small tree 5 to about 20 feet high; growing in dry sandy pinelands and dunes near the coast. *Leaves* top-shaped or broadest well above the middle, wedge-shaped at base, rounded or sometimes notched at tip, untoothed on margin, lustrous green above, coated beneath with copper-colored or golden-brown silky hairs, ¾ to 2½ inches long. Flowers blooming May or June. *Fruits* black, oblong, about ½ inch long; ripening September or October.

RANGE. Coastal plain; South Carolina south to Florida.

Also called Ironwood.

NIGHTSHADE FAMILY (Solanaceae)

SALT MATRIMONY-VINE *Lycium carolinianum*

FIELD MARKS. A straggling shrub 1 to about 5 feet high, with lithe and recurving branches; growing among coastal sand dunes and about salt marshes. *Leaves* alternate, usually with smaller leaves clustered in the axils, narrow but broader toward the blunt tip, tapering to the base, fleshy in texture, 1 to 1½ inches long. *Flowers* blue or lilac purple, ½ to ¾ inch across, with 5 spreading lobes; blooming May to July. *Fruits* red, roundish egg-shapped berries, about ⅜ inch long; ripening September and October or later.

RANGE. Coastal plain; Georgia and Florida west to Texas.

Also called Christmasberry.

Buckthorn Bumelia

Woolly Bumelia

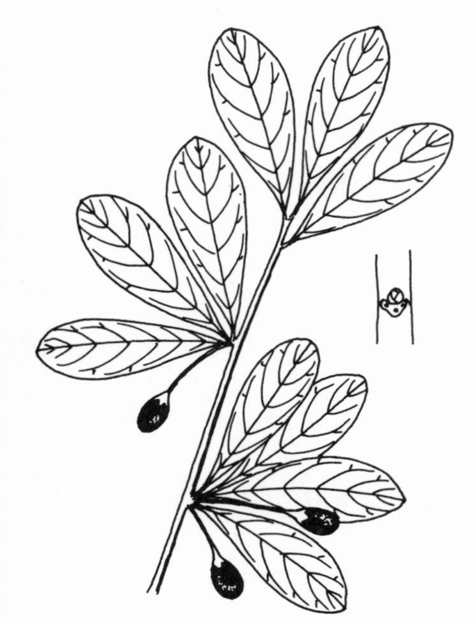

Tough Bumelia

Salt Matrimony-Vine

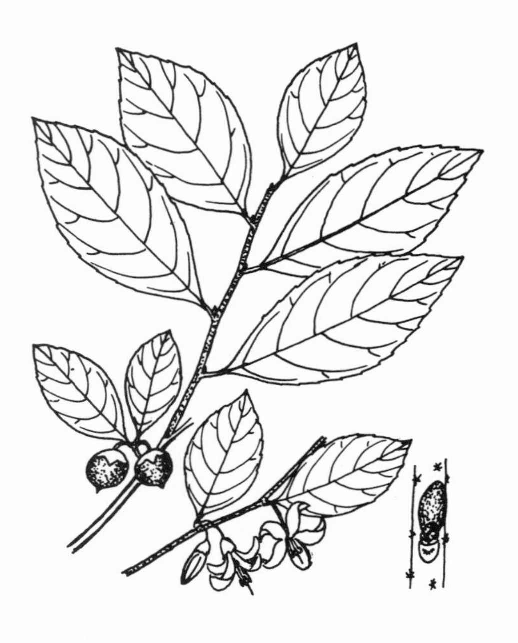

American Snowbell

Bigleaf Snowbell

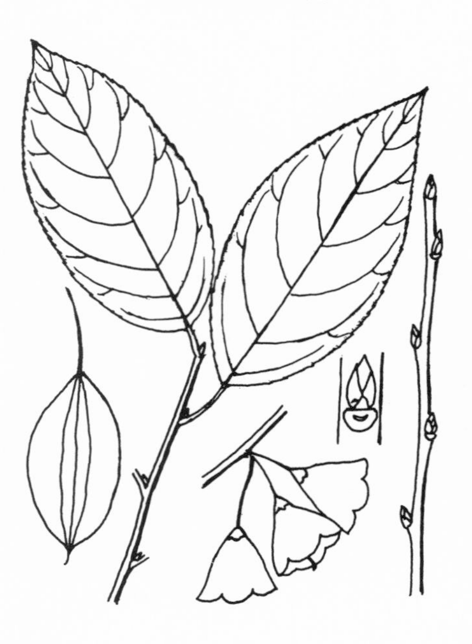

Carolina Silverbell

Little Silverbell

AMERICAN SNOWBELL *Styrax americanus*

FIELD MARKS. A leaf-losing shrub 3 to about 12 feet high; growing along streams and in swampy places. *Leaves* alternate, elliptic or sometimes broadest above the middle, pointed at both ends, margin untoothed or with rather widely spaced and low teeth, bright green and smooth above, paler and smooth or with scattered starry-branched hairs beneath (scaly and more densely hairy beneath in the variety *pulverulentus*), ¾ to 3 inches long. *Flowers* white, about ½ inch long, with 4 long and often recurved lobes, solitary or in leafy-bracted clusters of 5 to 20 in leaf axils or at ends of branchlets; blooming April to June. *Fruits* roundish, dry, 1-seeded, about ¼ inch in diameter.

RANGE. Southern Virginia to Ohio Valley region, Missouri, and Arkansas; south to Florida and Texas. Var. *pulverulentus* chiefly on Coastal Plain from Virginia and Arkansas southward.

Also called Mock-orange.

BIGLEAF SNOWBELL *Styrax grandifolius*

FIELD MARKS. A leaf-losing shrub or small tree 3 to about 12 feet high; growing in moist woods and along streams. *Leaves* alternate, broadly elliptic to oval or broadest above the middle, pointed at base, pointed or abruptly short-pointed at tip, margin untoothed or with some teeth chiefly above the middle, smooth or nearly so above, white-woolly or downy with starry-branched hairs beneath, 2 to about 7 inches long. *Flowers* similar to those of the preceding but almost 1 inch long and in fewer-flowered clusters; blooming April or May. *Fruits* roundish or oval, dry, 1-seeded, about ¼ inch in diameter.

RANGE. Southern Virginia, Tennessee, and Arkansas; south to Florida and Louisiana.

CAROLINA SILVERBELL *Halesia carolina*

FIELD MARKS. A leaf-losing shrub or small tree 5 to 30 feet or more high; growing in rich moist woods and along streams. *Leaves* alternate, elliptic to oblong or egg-shaped, broadly pointed to rounded at base, pointed or long-pointed at tip, finely toothed on margin, bright green and smooth or nearly so above, slightly paler and downy beneath, 2 to 5 inches long. *Flowers* white or pinkish, bell-shaped with 4 lobes, ⅓ to about 1 inch long, 2 to 5 in a cluster; blooming March to May. *Fruits* oblong-oval, dry, 4-winged, 1 to 1¾ inch long.

RANGE. Virginia and West Virginia to southern Ohio, southeastern and western Kentucky, and southern Illinois; south to northwestern Florida, Alabama, and western Tennessee. Also in Arkansas and southeastern Oklahoma.

Also called Snowdrop-tree and Opossumwood.

LITTLE SILVERBELL *Halesia parviflora*

FIELD MARKS. A leaf-losing shrub or small tree 4 to about 30 feet high; usually growing on dry sandy soils. *Leaves* alternate, elliptic or egg-shaped, pointed at base, pointed or long-pointed at tip, finely toothed on margin, smooth or nearly so above, downy at least on the veins beneath, 2 to 4 inches long. *Flowers* white, bell-shaped with 4 lobes, ¼ to ½ inch long; blooming March or April. *Fruits* club-shaped, rather narrowly 4-winged, dry, ⅝ to 1¼ inches long.

RANGE. Coastal plain; southern Georgia and northern Florida west to eastern Mississippi.

TWO-WING SILVERBELL *Halesia diptera*

FIELD MARKS. A leaf-losing shrub or small tree 5 to about 30 feet high; growing in low woods, swamps, or along streams. *Leaves* alternate, oval to egg-shaped or broadest above the middle, broadly pointed to rounded at both ends or abruptly pointed at tip, margin with widely spaced teeth, smooth or nearly so above, usually somewhat downy on the veins beneath, 2½ to 4½ inches long. *Flowers* white, bell-shaped and deeply 4-lobed, 3 to 6 in a cluster; blooming April and May. *Fruits* oblong or broader toward the tip, broadly 2-winged and often with narrower wings or ridges between, 1½ to 2 inches long.

RANGE. Coastal plain; Georgia and Northwestern Florida west to Texas; north in Mississippi Valley to Central Arkansas and Southeastern Oklahoma.

SWEETLEAF FAMILY (Symplocaceae)

SWEETLEAF *Symplocos tinctoria*

FIELD MARKS. A leaf-losing or semi-evergreen shrub or small tree 5 to about 30 feet high; growing in moist rocky or sandy woods and along streams. *Leaves* alternate, oblong to narrowly elliptic, pointed at both ends, untoothed or with obscure wavy teeth on margin, thickish and somewhat leathery in texture, dark yellowish green and smooth above, paler and often slightly downy beneath, 2 to 6 inches long. *Flowers* small, creamy white or pale yellow, with very conspicuous stamens, fragrant, in showy clusters along the branchlets; blooming March to May, before the leaves appear. *Fruits* elliptic, orange brown, dry, 1-seeded, about ½ inch long; maturing August or September.

Also called Horse-sugar, both horses and cattle relishing the sweetish-tasting leaves. Sometimes called Yellowwood, as it yields a yellow dye.

RANGE. Delaware, North Carolina, Tennessee, and southern Arkansas; south to northern Florida and eastern Texas.

OLIVE FAMILY (Oleaceae)

SWAMP-PRIVET *Foresteria acuminata*

FIELD MARKS. A leaf-losing shrub or small tree 4 to (rarely) 25 feet high; growing in river swamps, borders of ponds, and along streams. *Leaves* opposite, slender-stalked, oblong egg-shaped or broadly lance-shaped, tapering to a point at both ends, margin usually finely toothed above the middle, thin in texture, light green and smooth on both surfaces, 1¼ to about 4 inches long. *Flowers* very small, greenish or yellowish, in dense clusters; blooming March to May. *Fruits* dark purple, narrowly ellipsoid and pointed at both ends, 1-seeded, with a rather thin flesh, about ½ inch long; ripening May to July.

RANGE. South Carolina to southern Indiana, Illinois and Missouri, and southeastern Kansas; south to Florida and Texas.

UPLAND SWAMP-PRIVET *Foresteria ligustrina*

FIELD MARKS. A leaf-losing shrub 3 to about 10 feet high; growing on rocky slopes and bluffs and the sandy banks of streams. *Branchlets* smooth to downy. *Leaves* opposite, short-stalked, elliptic to oblong or broadest above the middle, wedge-shaped at base, bluntly pointed at tip, finely and inconspicuously toothed on margin, light green and smooth above, slightly paler and more or less downy beneath, ⅜ to 1½ inches long. *Flowers* very small, greenish or yellowish, clustered; blooming April or May. *Fruits* dark purple, oval or egg-shaped, stalkless, 1-seeded and fleshy, about ¼ inch long; ripening June or July.

RANGE. Kentucky and Tennessee south to Florida and Alabama.

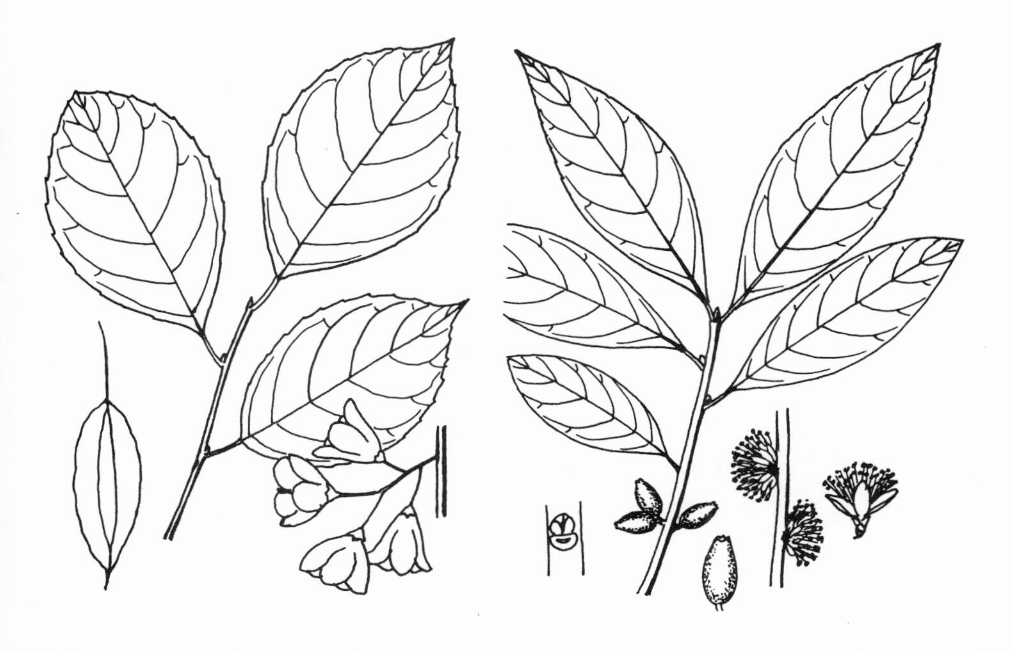

Two-wing Silverbell

Sweetleaf

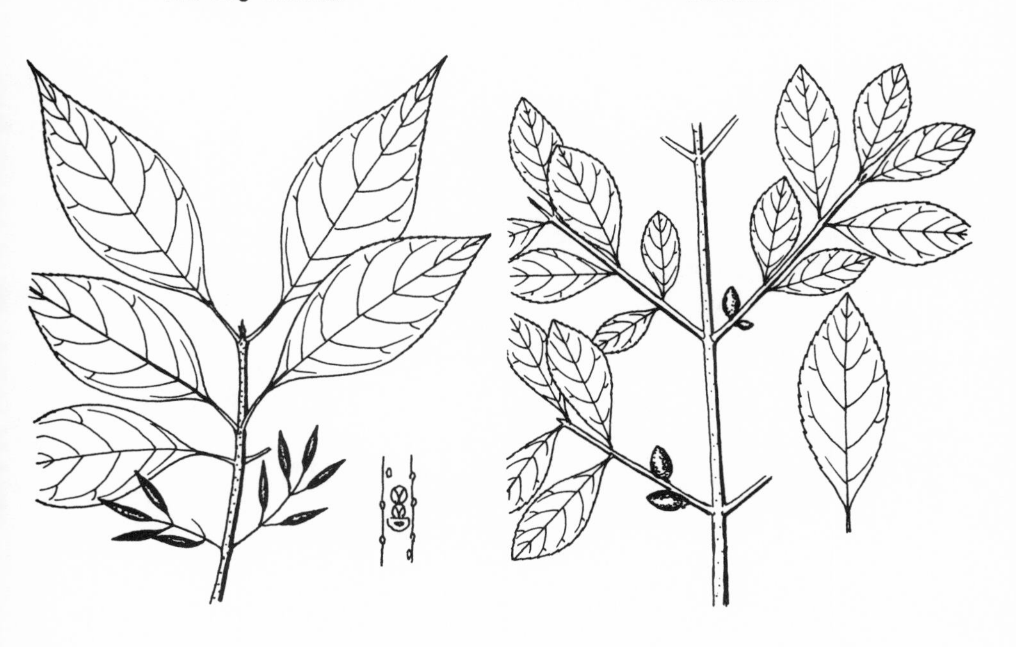

Swamp-privet

Upland Swamp-privet

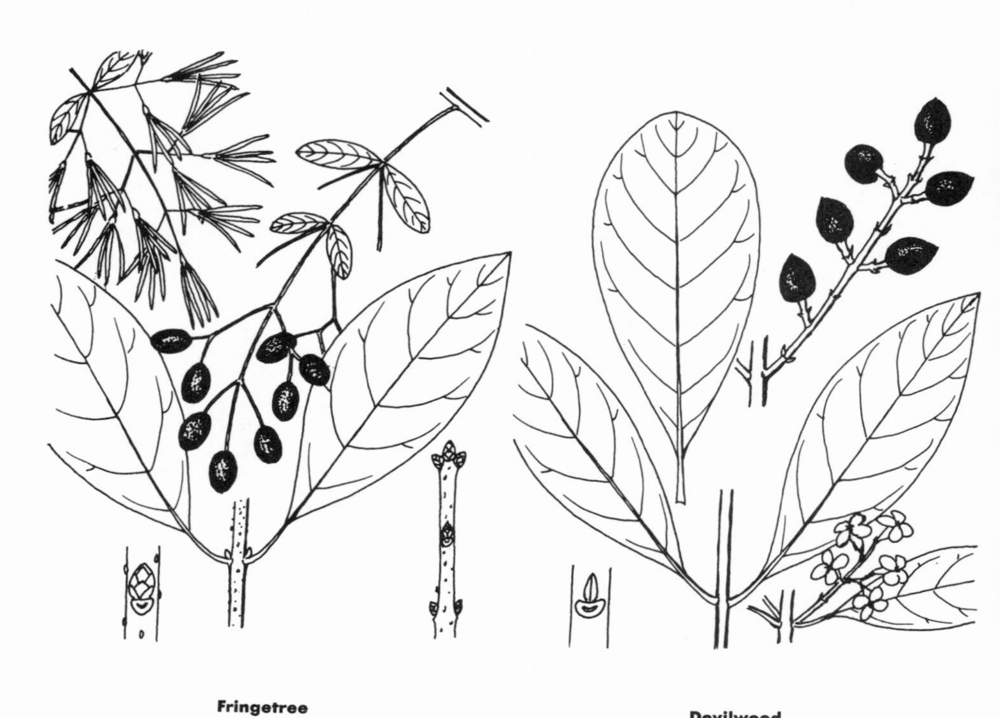

Fringetree

Devilwood

Yellow Jessamine

Beautyberry

FRINGETREE *Chionanthus virginicus*

FIELD MARKS. A leaf-losing shrub or small tree to 20 or rarely 40 feet high; growing in woods, savannahs, and along streams. *Leaves* opposite, elliptic or broadest above the middle, pointed at both ends, untoothed on margin, thickish, smooth or nearly so and paler beneath, 3 to 8 inches long. *Flowers* white, petals 4 and ribbon-like, fragrant, ¾ to 1 inch long, in rather large and drooping clusters; blooming April or May. *Fruits* oval-shaped, olive-like, 1-seeded, bluish black, usually with a whitish bloom, ⅜ to ¾ inch long; ripening July to September.

RANGE. New Jersey to southern Ohio and Missouri, and southeastern Oklahoma; south to Florida and Texas.

Also called Flowering-ash, Old-man's beard, and Grancy-graybeard.

DEVILWOOD *Osmanthus americanus*

FIELD MARKS. An evergreen shrub or small tree; growing in woods and borders of coastal swamps. *Leaves* opposite, narrowly elliptic or top-shaped, pointed at base, pointed to rounded at tip, untoothed on margin, leathery, lustrous above, paler beneath, smooth, 2½ to 4½ inches long. *Flowers* small, creamy white, in axillary clusters; blooming April or May. *Fruits* egg-shaped, olive-like, 1-seeded, dark bluish purple, ½ to ¾ inch long; ripening August to October.

RANGE. Coastal plain; Virginia south to Florida, west to Louisiana.

Also called Wild-olive.

PINKROOT FAMILY (Loganiaceae)

YELLOW JESSAMINE *Gelsemium sempervirens*

FIELD MARKS. A smooth, slender-stemmed, evergreen, twining vine; growing in borders of woods, swamps, and wayside thickets. *Leaves* opposite, lance-shaped or narrowly egg-shaped, broadly pointed to rounded at base, taper-pointed at tip, untoothed on margin, slightly leathery, lustrous above, paler beneath, smooth, 1½ to 3 inches long. *Flowers* bright yellow, trumpet-shaped, 5-lobed, very fragrant, about 1½ inches long, 1 to 3 on short stalks in leaf axils; blooming late February to May. *Fruits* oblong egg-shaped, pale-brown capsules ½ to ¾ inch long; containing flattened, winged seeds.

RANGE. Coastal plain and piedmont; Virginia south to Florida, west to Texas; north in Mississippi Valley to Arkansas.

Also called Carolina- or False-jessamine. Roots used medicinally. Children have been poisoned by sucking nectar from the flowers.

SCENTLESS YELLOW-JESSAMINE *Gelsemium rankinii*

A vine closely resembling the preceding, but the odorless flowers have pointed rather than blunt calyx lobes. The capsules are longer-beaked and seeds are wingless. Less common; southeastern North Carolina to Florida. (Not illustrated)

VERVAIN FAMILY (Verbenaceae)

BEAUTYBERRY *Callicarpa americana*

FIELD MARKS. A leaf-losing shrub 3 to about 6 feet high; usually growing in moist sandy or rocky woods. *Branchlets* ashy gray, hairy or roughish. *Leaves* opposite, elliptic to oval or egg-shaped, tapering to a point at both ends, sharply toothed on margin, roughish-hairy above, paler and woolly-hairy beneath, 3 to 6 inches long. *Flowers* small, bluish to lavender pink, funnel-shaped, clustered in leaf axils; blooming May to July. *Fruits* berry-like, violet or magenta purple, juicy, about ⅛ inch in diameter, in dense axillary clusters; ripening August to October.

RANGE. Maryland to Tennesee, Arkansas, and Oklahoma; south to Florida and Texas.

Also called French-mulberry. Handsome as an ornamental shrub.

DOGBANE FAMILY (Apocynaceae)

TRACHELOSPERMUM *Trachelospermum difforme*

FIELD MARKS. A slender, twining, high-climbing, soft-woody vine; growing in low moist woods, swamps, and along streams. *Leaves* opposite, oval to egg-shaped or lance-shaped, pointed at tip, usually pointed at base, untoothed on margin, thin in texture, 1½ to 3½ inches long. *Flowers* funnel-shaped, pale greenish yellow or cream-colored, about ½ inch long, in axillary or end clusters; blooming May to August. *Fruits* paired slender pods 5 to 9 inches long, containing a large number of small seeds with silky hairs at one end; maturing July to September.

RANGE. Delaware and southern Indiana and Illinois to Missouri and Oklahoma; south to Florida and Texas.

Also called Climbing-dogbane.

MINT FAMILY (Lamiaceae)

CONRADINA *Conradina canescens*

FIELD MARKS. An evergreen shrub 1 to 1½ feet high, with numerous stiff and ascending branches; growing in dry sandy pinelands and among coastal sand dunes. *Leaves,* commonly with small leaves in the axils, narrow and somewhat club-shaped, the margins untoothed and tightly rolled, rounded or blunt at tip, very finely downy and grayish green in color, 3/16 to about ⅜ inch long. *Flowers* pale bluish, about ⅜ inch long, in axils of the upper leaves; blooming March to May. *Fruits* small, hard, dry nutlets in groups of 4 within the hairy calyx.

RANGE. Coastal plain; northwestern Florida and Alabama.

Also called False Rosemary.

GEORGIA BASIL *Calamintha georgiana*

FIELD MARKS. An aromatic shrub 8 inches to 2 feet high; growing in rocky or sandy woods and on stream banks. *Branchlets* erect, somewhat downy. *Leaves* opposite, elliptic, more or less pointed at both ends, shallowly toothed on margin, smooth or nearly so on both surfaces, paler beneath, ⅜ to about 1 inch long, commonly with smaller leaves clustered in the axils. *Flowers* white or pink, purple-spotted, ⅜ to ½ inch long, in axillary clusters; blooming July to October. *Fruits* small, hard, dry nutlets in groups of 4; within the persistent calyx.

RANGE. Piedmont and coastal plain; North Carolina south to Florida, west to Mississippi.

Also called Calamint.

RED BASIL *Calamintha coccinea*

FIELD MARKS. An evergreen, sparingly branched shrub 1 to 3 feet high; growing in sandy or rocky open woods and on sandhills. *Branchlets* slender, erect. *Leaves* narrowly top-shaped or elliptic, wedge-shaped at base, blunt or broadly pointed at tip, untoothed on margin, smooth or nearly so on both surfaces, ¼ to ¾ inch long. *Flowers* bright red, 1 to 1½ inches long, in axillary clusters; blooming May to September, or all year southward. *Fruits* small, hard, dry nutlets in groups of 4; within the persistent calyx.

RANGE. Coastal plain; southern Georgia and Alabama, south into Florida.

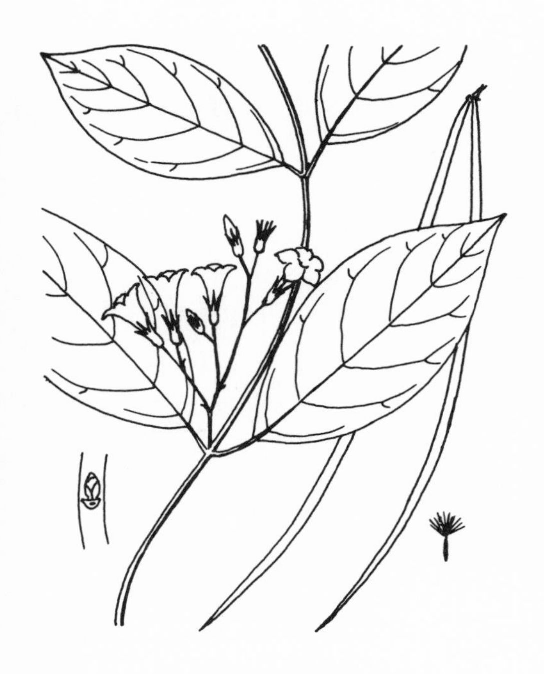

Trachelospermum

Conradina

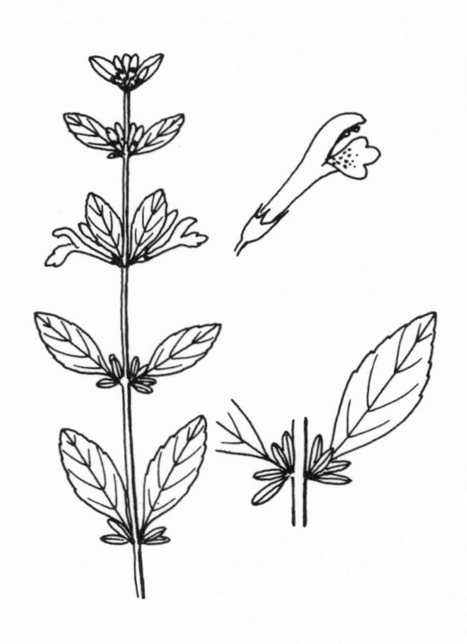

Georgia Basil

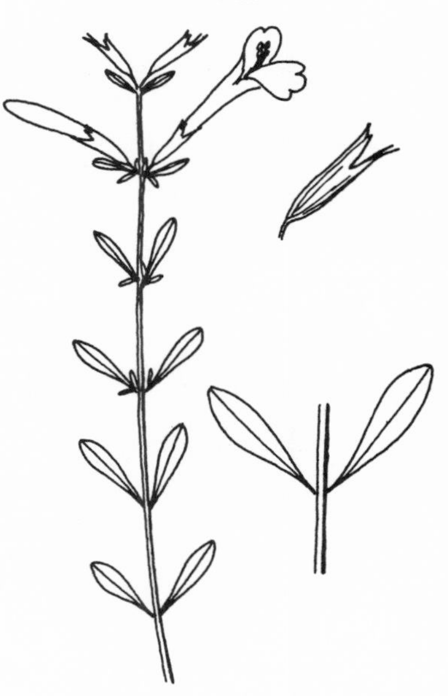

Red Basil

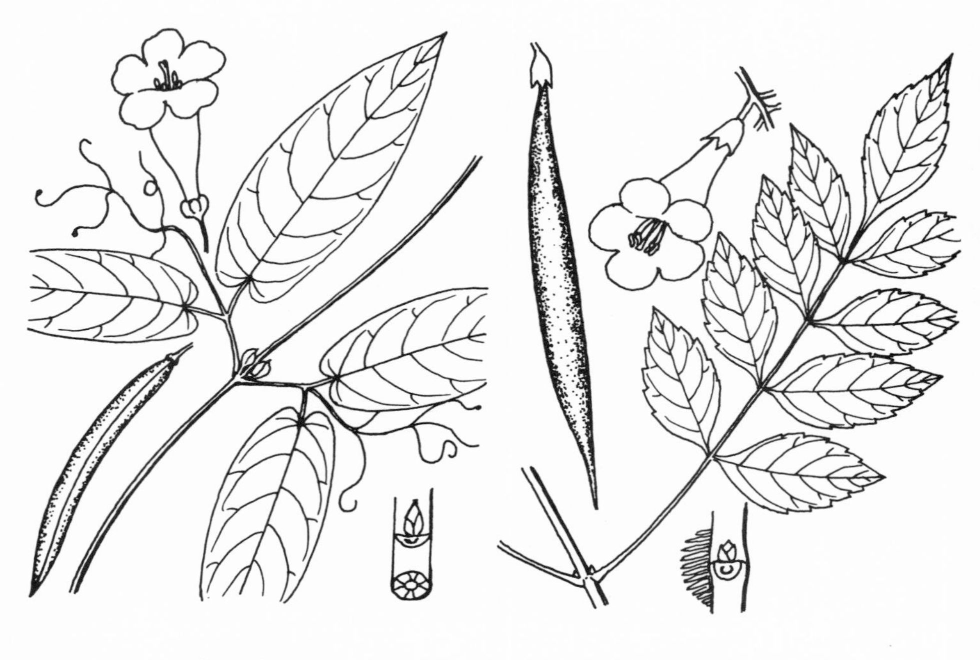

Cross-vine

Trumpet-creeper

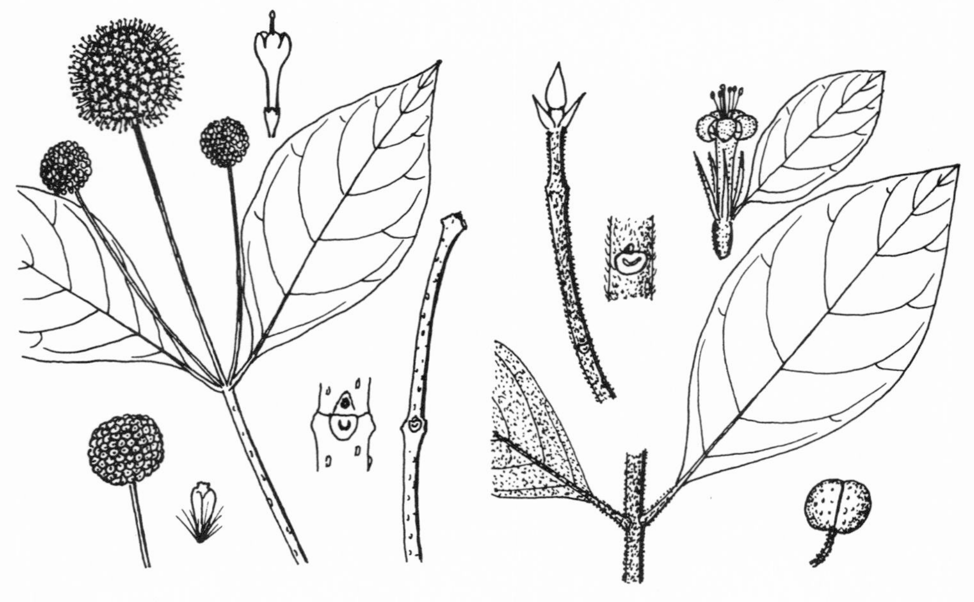

Buttonbush

Pinckneya

BIGNONIA FAMILY (Bignoniaceae)

CROSS-VINE *Bignonia capreolata*

FIELD MARKS. A high-climbing vine which is evergreen southward; growing in moist woods and swamps. *Branchlets* show cross-shaped pith in cross-section. *Leaves* opposite, divided into 2 leaflets with a branched tendril between them; leaflets stalked, oblong-egg-shaped, heart-shaped at base, pointed at tip, untoothed on margin, smooth, paler beneath, 2 to 5 inches long. *Flowers* trumpet-shaped, orange red outside and yellow within, 5-lobed, 1½ to 2½ inches long, 2 to 5 in axillary clusters; blooming April to June. *Fruits* cylindrical, somewhat flattened, 2-celled capsules 4 to 6 inches long, containing many winged and flattened seeds; maturing July or August.

RANGE. Maryland to southern Ohio and Illinois, and Missouri; south to Florida and Louisiana.

TRUMPET-CREEPER *Campsis radicans*

FIELD MARKS. A leaf-losing vine climbing by means of aerial rootlets in 2 short rows at the nodes; growing in moist woods, thickets, or along fence rows. *Leaves* opposite, compound, 8 to 15 inches long; the 9 to 11 leaflets almost stalkless, elliptic to oblong egg-shaped, pointed at both ends, coarsely and sharply toothed on margin, smooth above, slightly paler and often downy on the veins beneath, 1½ to 3 inches long. *Flowers* reddish orange, trumpet-shaped, 5-lobed, 2½ to 3½ inches long, in showy end clusters; blooming June to September. *Fruits* cylindrical, somewhat flattened, 2-ridged capsules 4 to 8 inches long, containing many winged and flattened seeds; maturing September or October.

RANGE. Connecticut and southeastern Pennsylvania to West Virginia, Kentucky, southern Illinois, and Iowa; south to Florida and Texas.

Also called Cow-itch, as the plant causes a severe skin irritation in some people. Often cultivated as an ornamental and sometimes escaping northward.

MADDER FAMILY (Rubiaceae)

BUTTONBUSH *Cephalanthus occidentalis*

FIELD MARKS. A leaf-losing shrub 3 to 10 or rarely 20 feet high; growing in swamps, shallow ponds, and along streams. *Leaves* opposite or in 3's, elliptic to egg-shaped or lance-shaped, pointed at both ends, untoothed on margin, often lustrous above, paler and sometimes downy beneath, 3 to 6 inches long. *Flowers* small, white, tubular, 5-lobed, fragrant, clustered in long-stalked ball-shaped heads 1 to 1½ inches across; blooming May to August. *Fruits* small, top-shaped, 2-seeded capsules in tight ball-shaped heads.

RANGE. Nova Scotia to southern Ontario, Minnesota, and California; south to Florida, Texas, and Mexico.

Also called Honeyballs. Often cultivated as an ornamental shrub.

PINCKNEYA *Pinckneya bracteata*

FIELD MARKS. A leaf-losing shrub or small tree 5 to rarely 25 feet high; growing in sandy swamps or along streams. *Branchlets* more or less tawny- to rusty-hairy. *Leaves* opposite, elliptic to egg-shaped, pointed to roundish at base, pointed at tip, untoothed on margin, somewhat hairy above, paler and more densely hairy beneath, 5 to 8 inches long. *Flowers* tubular, greenish yellow spotted with red, 5-lobed; with 1 or 2 calyx lobes expanded into a broad pink to whitish, petal-like blade; borne in large and showy end clusters; blooming in May. *Fruits* roundish capsules about ¾ inch in diameter, containing many small seeds with wings.

RANGE. Coastal plain; southeastern South Carolina south to Florida.

Also called Georgia-bark and Fevertree. The bark was used by early settlers for the treatment of malaria.

PARTRIDGEBERRY *Mitchella repens*

FIELD MARKS. A low evergreen plant with slightly woody, slender, creeping or trailing stems often rooting at the nodes; growing in moist to fairly dry woodlands. *Leaves* opposite, stalked, oval to broadly egg-shaped or roundish, rounded or heart-shaped at base, roundish or blunt at tip, untoothed on margin, smooth, lustrous above, paler beneath, ¼ to ¾ inch long and wide. *Flowers* funnel-shaped, 4-lobed, white (pinkish in bud), fragrant, borne in pairs at ends of branchlets; blooming April to June. *Fruits* berry-like bright red, about ¼ inch in diameter, partially joined together, edible; ripening August or September, often persisting until the following spring.

RANGE. Southwestern Newfoundland and southern Quebec to Ontario and Minnesota, south to Florida and Texas.

Also called Twinberry.

HONEYSUCKLE FAMILY (Caprifoliaceae)

BUSH-HONEYSUCKLES (Diervilla)

Bush-honeysuckles are rather low leaf-losing shrubs with opposite, simple leaves having toothed margins. Their flowers have yellow and tubular corollas with 5 short and spreading lobes at the summit. They are arranged in small clusters either at the tips of branchlets or in the axils of the uppermost leaves. The fruits are narrowly oblong or egg-shaped capsules, each ending in a beak with the remains of the 5 spreading calyx lobes at the tip. They are close relatives of the weigelias, commonly grown as ornamental shrubs.

NORTHERN BUSH-HONEYSUCKLE *Diervilla lonicera*

FIELD MARKS. A bushy-branched shrub 1 to 3 feet high; growing in dry, rocky, open woodlands and in wayside thickets. *Branchlets* nearly round with a hairy-lined ridge running down from a line connecting the bases of the leafstalks. *Leaves* obviously stalked, egg-shaped or lance-shaped, rounded or broadly pointed at base, long-pointed at tip, finely and sharply toothed and hairy-fringed on margin, smooth above, paler and sometimes downy on the veins beneath, 1½ to 5 inches long. *Flowers* pale yellow, usually 3 in a cluster; blooming June to August. *Fruits* slender, about ½ inch long.

RANGE. Newfoundland to Manitoba; south to Delaware, western North Carolina, Ohio, and Iowa.

SOUTHERN BUSH-HONEYSUCKLE *Diervilla sessilifolia*

FIELD MARKS. A bushy-branched shrub 1½ to 5 feet high; growing in moist rocky places and bogs in the southern Appalachians. *Branchlets* prominently 4-angled and with hairy lines. *Leaves* stalkless or nearly so, egg-shaped to broadly lance-shaped, rounded to somewhat heart-shaped at base, long-pointed at tip, sharply toothed and hairy-fringed on margin, smooth above, sometimes slightly downy beneath, 2 to 6 inches long. *Flowers* sulfur-yellow, in 3- to 7-flowered clusters; blooming June to August. *Fruits* oblong, ⅝ to ¾ inch long.

RANGE. Western Virginia and eastern Tennessee, south to northwestern Georgia and northern Alabama.

HAIRY BUSH-HONEYSUCKLE *Diervilla rivularis*

FIELD MARKS. A shrub 2 to about 4 feet high; growing in damp rocky woods, or on banks and cliffs in the southern Appalachians. *Branchlets* roundish, densely grayish-downy. *Leaves* short-stalked or the uppermost stalkless, egg-shaped to elliptic or lance-shaped, roundish or somewhat heart-shaped at base, long-pointed at tip, sharply and finely toothed on margin, somewhat hairy above, densely downy beneath, 1½ to about 4 inches long. *Flowers* pale yellow or greenish yellow, few to many in a cluster; blooming June to August. *Fruits* oblong, ¼ inch or slightly longer.

RANGE. Western North Carolina and eastern Tennessee, south to northwestern Georgia and northern Alabama.

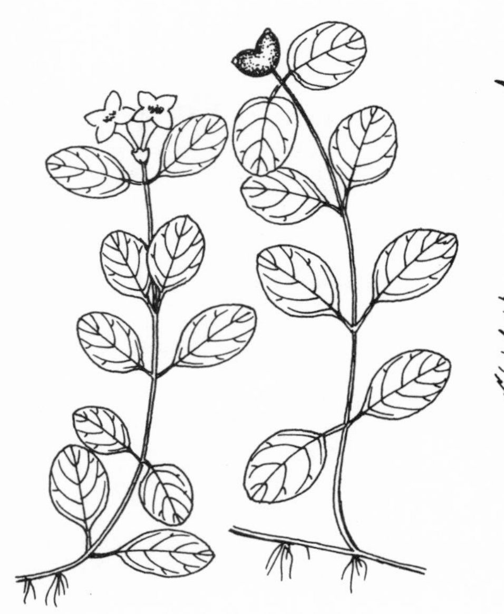

Partridgeberry

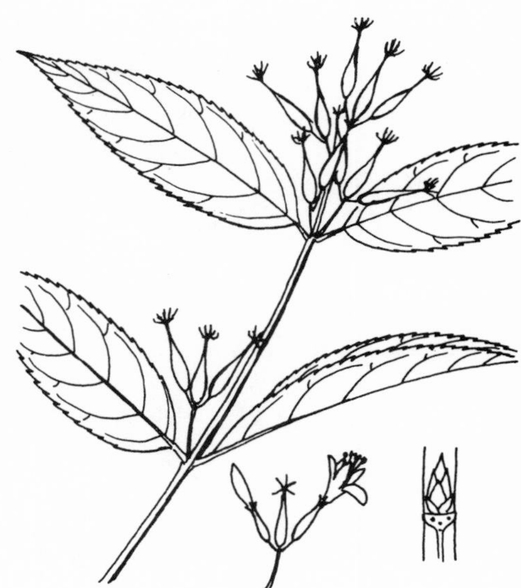

Northern Bush-Honeysuckle

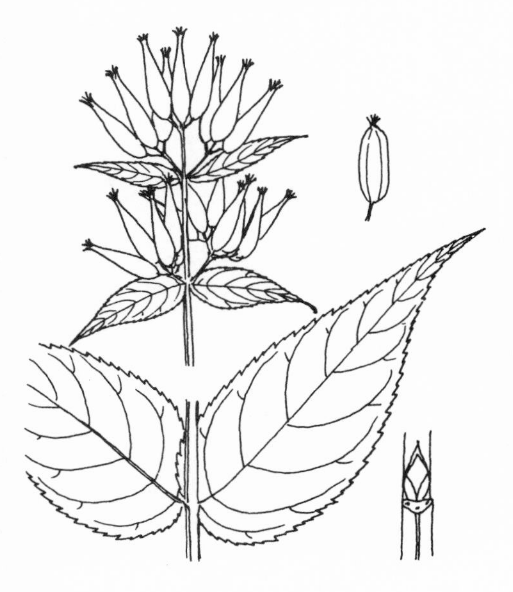

Southern Bush-Honeysuckle

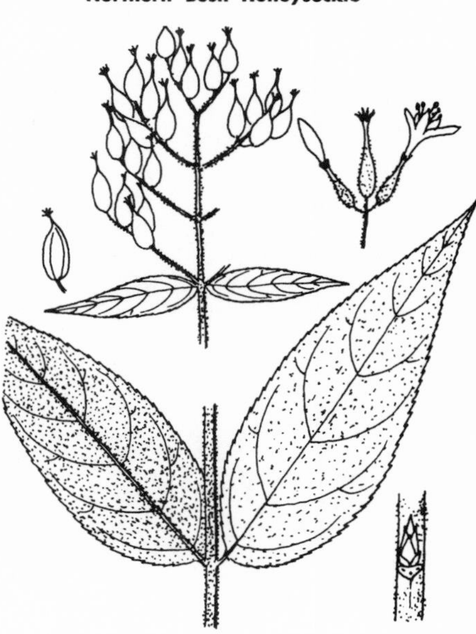

Hairy Bush-Honeysuckle

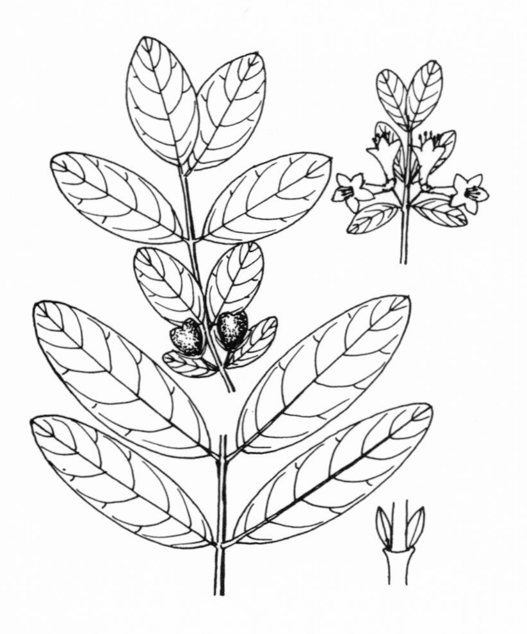

Mountain Fly Honeysuckle

American Fly Honeysuckle

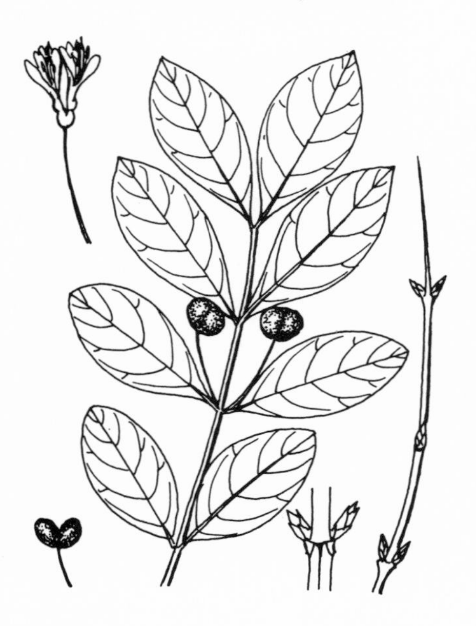

Swamp Fly Honeysuckle

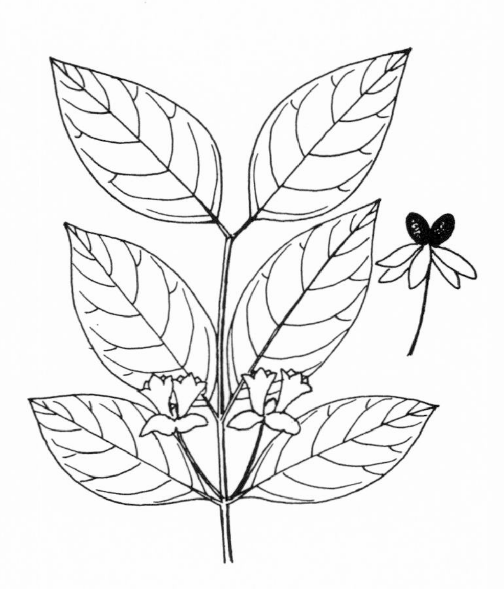

Involucred Fly Honeysuckle

HONEYSUCKLES (Lonicera)

(Key Appendix P)

Honeysuckles are erect shrubs or often vines with simple, opposite, untoothed leaves. The flowers have tubular corollas, 5-lobed at the summit or sometimes 2-lipped; borne either in axillary pairs or in end clusters. The fruits are juicy few-seeded berries which are eaten by many kinds of wild birds. The introduced Japanese Honeysuckle (*Lonicera japonica* Thunb.) often escapes cultivation and is a pernicious weed in the southeastern states.

MOUNTAIN FLY HONEYSUCKLE *Lonicera villosa*

FIELD MARKS. A leaf-losing shrub 1 to about 3 feet high, with ascending branches and shredding bark; growing in cold, moist, rocky woods or bogs. *Branchlets* may be hairy or smooth. *Leaves* short-stalked, elliptic to narrowly oblong, mostly roundish at both ends, margin untoothed but often hairy-fringed, usually hairy but sometimes quite smooth, veiny in appearance, pale or somewhat whitened beneath, ¾ to 1½ inches long. *Flowers* pale yellow, bell-shaped, 5-lobed, paired on axillary stalks less than ¼ inch long; blooming May or June. *Fruits* blue or bluish black, oval, 2-eyed berries about ¼ inch in diameter; ripening June to August.

RANGE. Newfoundland and southern Labrador to Manitoba; south to New England, northern Pennsylvania, Michigan, and Minnesota.

Also known as Blue Honeysuckle. The fruits are edible.

AMERICAN FLY HONEYSUCKLE *Lonicera canadensis*

FIELD MARKS. A leaf-losing shrub 2 to 4 feet high; growing in cool, moist woodlands. *Branchlets* smooth, flexible, spreading. *Leaves* egg-shapped, rounded or heart-shaped at base, broadly pointed or blunt at tip, margin untoothed but hairy-fringed, smooth or nearly so on both surfaces, 1 to 3½ inches long. *Flowers* greenish yellow, funnel-shaped, 5-lobed, paired on long axillary stalks; blooming April to June. *Fruits* bright red, egg-shaped, paired but distinct berries about ¼ inch long; ripening July to September.

RANGE. Nova Scotia to Saskatchewan; south to northern New Jersey, northern Georgia, Indiana, and northeastern Iowa.

SWAMP FLY HONEYSUCKLE *Lonicera oblongifolia*

FIELD MARKS. A bushy leaf-losing shrub 2 to 5 feet high; growing in cold swamps and bogs. *Branchlets* stiff, ascending, smooth. *Leaves* short-stalked, oblong or elliptic, pointed at base, pointed or blunt at tip, untoothed on margin, paler beneath and smooth or nearly so on both surfaces, ¾ to 3 inches long. *Flowers* creamy white often tinged with purple, 2-lipped, paired on long axillary stalks; blooming May to July. *Fruits* red or purplish, egg-shaped berries about ¼ inch long, often somewhat united; ripening July to September.

RANGE. Southeastern Quebec to Manitoba; south to Maine, northwestern Pennsylvania, northern Ohio, Michigan, Wisconsin, and Minnesota.

INVOLUCRED FLY HONEYSUCKLE *Lonicera involucrata*

FIELD MARKS. A leaf-losing shrub 3 to 10 feet high; growing in cool woods or along streams. *Branchlets* slightly 4-angled or 4-lined, smooth or downy. *Leaves* elliptic or oblong egg-shaped, pointed or rounded at base, pointed at tip, untoothed on margin, smooth or nearly so, 2 to 5 inches long. *Flowers* yellow, funnel-shaped, 5-lobed, sticky-hairy, paired on long stalks and with large leaflike bracts at base; blooming June or July. *Fruits* roundish or oval, blackish berries, paired on long axillary stalks and distinct; ripening August or September.

RANGE. New Brunswick to British Columbia and Alaska; south to western Ontario, Michigan, Utah, and California.

SMOOTH HONEYSUCKLE *Lonicera dioica*

FIELD MARKS. A somewhat twining to sprawling or reclining, leaf-losing shrub with stems 3 to 10 feet long. *Leaves* stalkless or short-stalked, oblong or oval, more or less pointed at base, rounded or bluntly pointed at tip, whitened and smooth or finely downy beneath, 1½ to 3½ inches long; at least the uppermost pair united at the base to form an oblong or rhombic disk. *Flowers* greenish yellow, often purplish-tinged, 2-lipped, the tube with a prominent swelling at the base, ½ to ¾ inch long, in 1 to 3 whorls in a nearly stalkless end cluster; blooming May to August. *Fruits* salmon-red, roundish, about ¼ inch in diameter; ripening July to September.

RANGE. Southwestern Maine to western Quebec and British Columbia; south to western North Carolina, Tennessee, Missouri, and Kansas.

Also called Glaucous Honeysuckle.

GRAPE HONEYSUCKLE *Lonicera prolifera*

FIELD MARKS. A somewhat twining to almost bushy leaf-losing shrub, with stems 6 to 12 feet long; growing in rocky woods or along streams. *Leaves* stalkless or short-stalked, elliptic to oval or oblong, sometimes broadest above the middle, narrowed to a pointed base, blunt or broadly pointed at tip, untoothed on margin, dark green and slightly whitened above, very much whitened and more or less downy beneath, 2 to 3½ inches long; the upper pair united at base to form a roundish disk, 2 or 3 other pairs below it often united at base. *Flowers* pale yellow, 2-lipped, the tube scarcely swollen at the base, 1 to 1¼ inches long, in 1 to 4 whorls in a nearly stalkless end cluster; blooming May to July. *Fruits* coral red, roundish, about ¼ inch in diameter; ripening July to October.

RANGE. Southern Ontario to southeastern Manitoba; south to Tennessee, Arkansas, and eastern Kansas.

HAIRY HONEYSUCKLE *Lonicera hirsuta*

FIELD MARKS. A twining and often high-climbing leaf-losing shrub; growing in woods and thickets. *Branchlets* rough-hairy. *Leaves* stalked, oval or egg-shaped, untoothed but hairy-fringed on margin, roundish at base, bluntly or broadly pointed at tip, upper surface with some appressed hairs, lower surface grayish green and densely downy, 2 to 4 inches long; uppermost 1 or 2 pairs united at the base to form a roundish to oval or rhombic disk with pointed ends. *Flowers* orange yellow, sticky-hairy, 2-lipped, ¾ to 1 inch long, in a stalked or stalkless end cluster; blooming in July. *Fruits* red, roundish, about ¼ inch in diameter; ripening in September.

RANGE. Western Quebec to Saskatchewan; south to Pennsylvania, Ohio, Michigan, Minnesota, and Nebraska.

YELLOW HONEYSUCKLE *Lonicera flava*

FIELD MARKS. A somewhat twining or reclining leaf-losing shrub; growing on rocky wooded slopes and bluffs. *Branches* smooth. *Leaves* short-stalked, oval to egg-shaped or broadly elliptic, roundish or abruptly pointed at base, broadly pointed or blunt at tip, bright green above, grayish green beneath, smooth on both surfaces, 1½ to 3½ inches long; the uppermost 1 or 2 pairs united at the base to form an oval or roundish disk. *Flowers* yellow to orange yellow, 2-lipped, fragrant, an inch or slightly longer, in a crowded and stalked end cluster; blooming April and May. *Fruits* red, roundish, about ¼ inch in diameter; ripening July and August.

RANGE. Southwestern North Carolina, northwestern South Carolina, and southeastern Tennessee; south to northern parts of Georgia and Alabama; also in the Ozark region of Arkansas, Missouri, and Oklahoma.

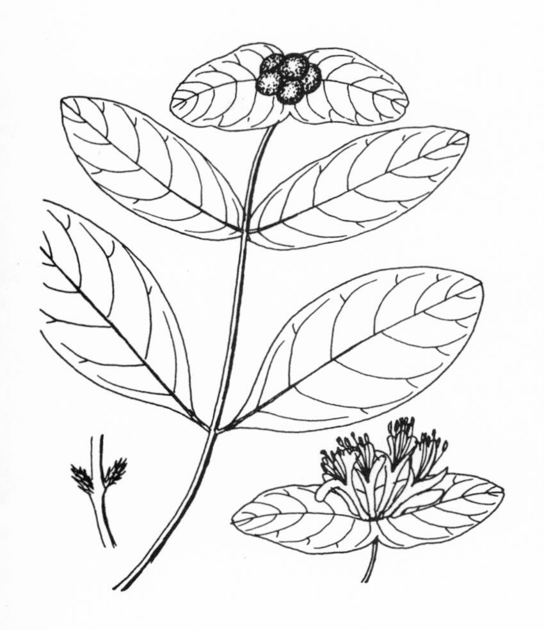

Smooth Honeysuckle

Grape Honeysuckle

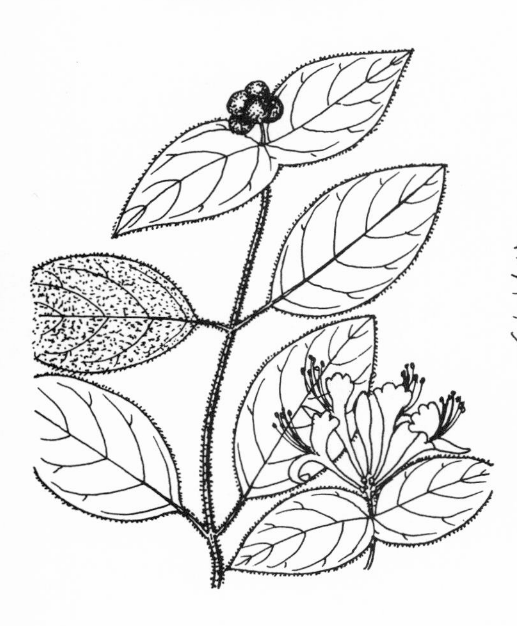

Hairy Honeysuckle

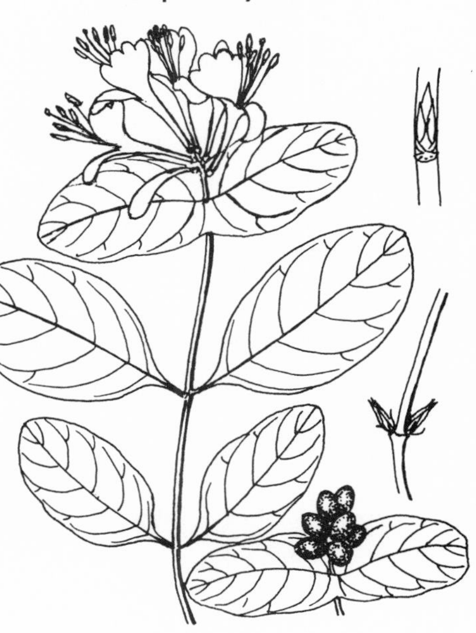

Yellow Honeysuckle

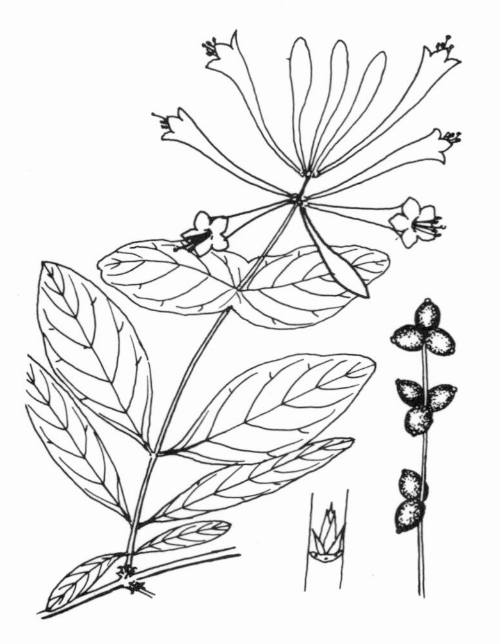

Trumpet Honeysuckle

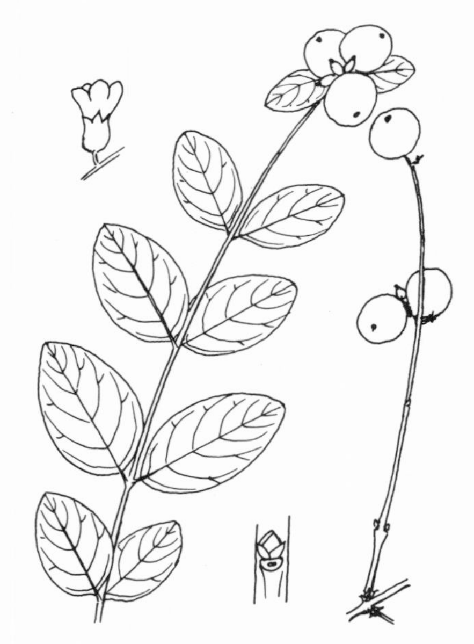

Snowberry

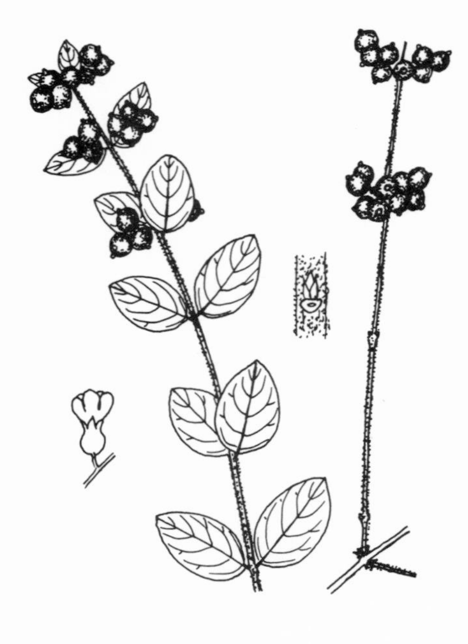

Coralberry

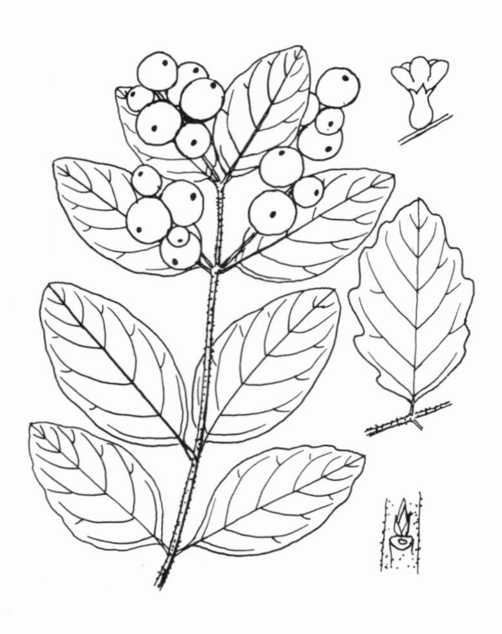

Wolfberry

TRUMPET HONEYSUCKLE *Lonicera sempervirens*

FIELD MARKS. A twining and sometimes high-climbing, leaf-losing or (southward) evergreen shrub; growing in woodlands, thickets, and along fencerows. *Leaves* oval to egg-shaped or narrowly elliptic, short-stalked, pointed to roundish at base, rounded to bluntly pointed at tip, untoothed on margin, somewhat leathery in texture, dark green and smooth above, whitened and smooth or somewhat downy beneath, 1½ to 3½ inches long; the uppermost 1 or 2 pairs united at the base to form an oval-shaped or roundish disk. *Flowers* narrowly trumpet-shaped with 5 nearly equal lobes at summit, bright red outside, yellow within, 1½ to 2 inches long, in several whorls on an end stalk; blooming March to June or later. *Fruits* bright red, egg-shaped, about ¼ inch in diameter; ripening July to October.

RANGE. Southern Maine to New York, Ohio, Iowa, and Nebraska; south to Florida and Texas.

Also called Coral Honeysuckle. Often cultivated and frequently escaping.

SNOWBERRY *Symphoricarpos albus*

FIELD MARKS. A finely branched leaf-losing shrub 1 to about 5 feet high; growing on dry, rocky, wooded slopes and banks. *Branchlets* very slender, smooth or slightly hairy. *Leaves* opposite, short-stalked, oblong-elliptic to roundish, rounded or bluntly pointed at both ends, untoothed or somewhat wavy or margin (sometimes lobed on vigorous shoots), thin in texture, green on both surfaces, smooth above and smooth or slightly downy beneath, ¾ to 2 inches long. *Flowers* pink, about ¼ inch long, corolla 5-lobed, in leaf axils or end clusters; blooming May to July. *Fruits* roundish, white berries ¼ to ½ inch in diameter; ripening August to October.

RANGE. Eastern Quebec to British Columbia; south to Massachusetts, western Virginia, Michigan, Wisconsin, Nebraska, and Colorado.

Often cultivated as an ornamental shrub.

CORALBERRY *Symphoricarpos orbiculatus*

FIELD MARKS. A finely branched leaf-losing shrub 2 to about 5 feet high; growing in low woods or dry, rocky, wooded slopes. *Branchlets* very slender, usually finely downy. *Leaves* opposite, short-stalked, egg-shaped to roundish, rounded or bluntly pointed at both ends, untoothed but often wavy on margin, dull green and smooth or nearly so above, paler and finely downy beneath, ½ to 2 inches long. *Flowers* pinkish, about 3/16 inch long, corolla 5-lobed, in short but dense clusters in the leaf axils; blooming July to September. *Fruits* roundish or slightly egg-shaped, purplish- to coral-red berries, about 3/16 inch in diameter; ripening September to November and persisting.

RANGE. Pennsylvania to Ohio, Illinois, Minnesota, South Dakota, and Colorado; south to northern Georgia, Alabama, and Mississippi.

Commonly cultivated as an ornamental and frequently escaping. Also called Indian-currant.

WOLFBERRY *Symphoricarpos occidentalis*

FIELD MARKS. A bushy-branched, leaf-losing shrub 1 to about 3 feet high, spreading freely from the roots and often forming dense colonies; growing in dry open woods and on prairies. *Branchlets* slender, often minutely downy. *Leaves* opposite, short-stalked, egg-shaped, rounded to somewhat pointed at both ends, untoothed or sometimes lobed on margin, thickish in texture, dull green and often sparingly hairy above, paler and more or less downy beneath, 1 to 4 inches long. *Flowers* pinkish, stalkless, about 5/16 inch long, corolla 5-lobed, in dense axillary or end clusters; blooming June or July. *Fruits* roundish, dull-white berries about ⅜ inch in diameter, soon becoming blackish; ripening August or September.

RANGE. Ontario to British Columbia; south to northern Illinois, Missouri, Kansas, and New Mexico.

Locally escaping from cultivation eastward.

VIBURNUMS (Viburnum)

(Key Appendix Q)

Viburnums are shrubs or small trees with opposite, simple leaves. The numerous small white or pinkish flowers are perfect and have a 5-lobed, short-tubed corolla. They are arranged in more or less flat-topped end clusters. The fruits are drupes with a large bony-covered seed surrounded by pulpy flesh. Most species are desirable ornamental plants, and the fruits provide food for wildlife.

HOBBLEBUSH *Viburnum lantanoides*

FIELD MARKS. A leaf-losing shrub 3 to 10 feet high, the forked branches often bending over and rooting at the tip; growing in cool, moist, rocky woods or along streams. *Branchlets* and large naked end buds densely coated with cinnamon-colored, starry-branched hairs. *Leaves* roundish or heart-shaped, broadly pointed at tip, finely toothed on margin, veiny, becoming smooth or nearly so above but remaining rusty-hairy on the veins beneath, 4 to about 7 inches long. *Flowers* of 2 kinds; the marginal ones about ½ inch across and sterile; blooming April to June. *Fruits* egg-shaped, bright red becoming purplish black, about ⅜ inch long; ripening July to September.

RANGE. New Brunswick to Ontario; south to northern New Jersey, Pennsylvania, Ohio, and Michigan, and in the mountains to northern Georgia.

Also called Witch-hobble.

SOUTHERN WITHEROD *Viburnum nudum*

FIELD MARKS. A leaf-losing shrub 5 to 15 feet high; growing in wet woods, swamps, and bogs. Similar to the preceding species. *Branchlets* rather lustrous. *Leaves* lustrous on the upper surface, paler and usually rusty-scurfy on midrib and veins beneath. *Flowers* all alike, the cluster on a stalk as long as or longer than its several branches; blooming April or May.

RANGE. Connecticut and southeastern Pennsylvania to Kentucky and Arkansas; south to Florida and Texas.

NORTHERN WITHEROD *Viburnum nudum* var. *cassinoides*

FIELD MARKS. A leaf-losing shrub 3 to 8 feet high; growing in cool, moist, usually rocky woods and swamps. *Branchlets* slender, dull, rather flexible. *Leaves* narrowly egg-shaped to oblong-elliptic, pointed to roundish at base, pointed to blunt at tip, sometimes obscurely toothed on margin, dull above, paler and sometimes rusty-scurfy on midrib beneath, 1½ to 4 inches long. *Flowers* all alike, the cluster on a distinct *stalk;* blooming May or June. *Fruits* roundish or slightly oval, bluish black, whitened with a bloom, about 5/16 inch long; ripening August to October.

RANGE. Newfoundland to Ontario; south to Delaware, Maryland, northern Georgia, and Alabama, and the region of the Great Lakes.

Also called Wild-raisin.

SWEET VIBURNUM *Viburnum lentago*

FIELD MARKS. A leaf-losing shrub or small tree 5 to 15 (rarely 30) feet high; growing in moist woods, thickets, and borders of swamps. *Branchlets* long, slender, flexible. *Leaves* oval or egg-shaped, rounded to broadly pointed at base, abruptly long-pointed at tip, finely and sharply toothed on margin, smooth, paler beneath, 2 to 4 inches long; leafstalks ½ to 1 inch long, prominenly winged. *Flowers* all alike, the cluster stalkless or nearly so; blooming May or June. *Fruits* oval-shaped, bluish black, whitened with a bloom, about ½ inch long; ripening September and October.

RANGE. New England and Quebec to Manitoba; south to New Jersey, West Virginia, Ohio, Missouri, and Colorado.

Also called Nannyberry and Sheepberry. The wood has a rank odor.

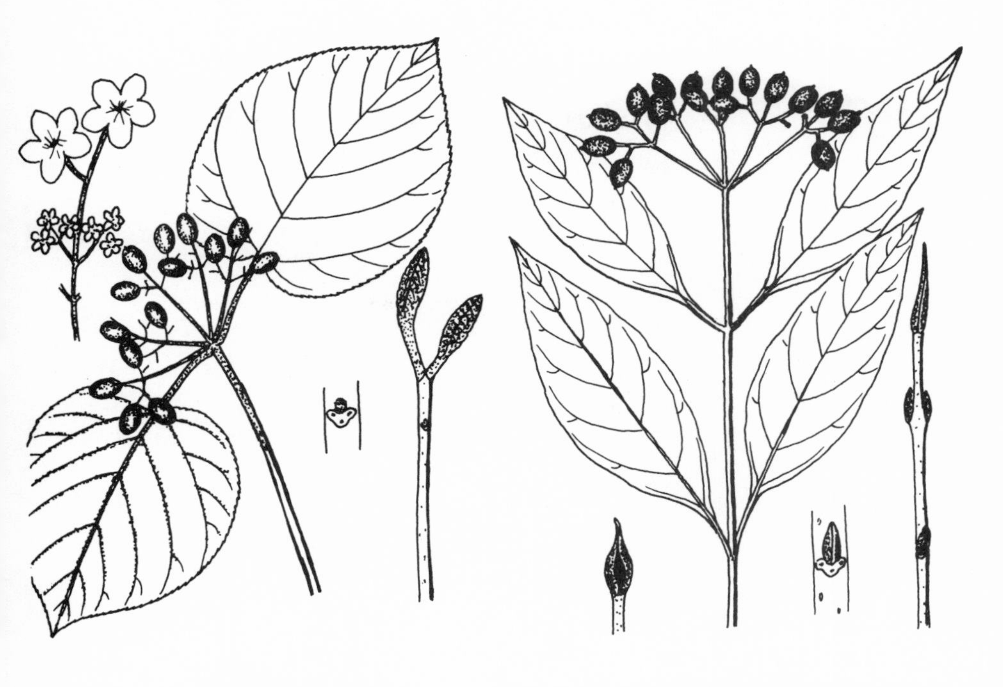

Hobblebush

Southern Witherod

Northern Witherod

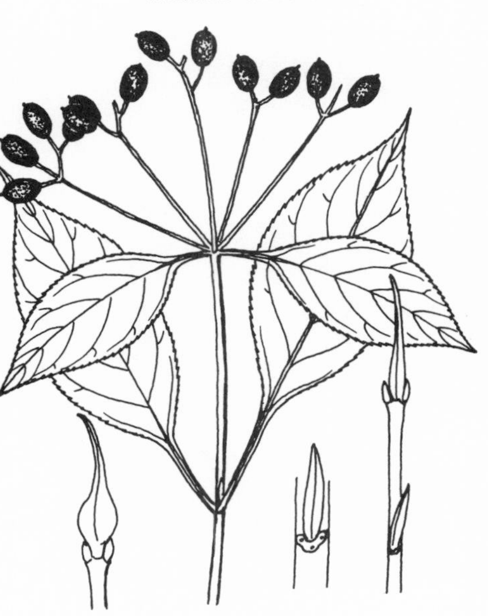

Sweet Viburnum

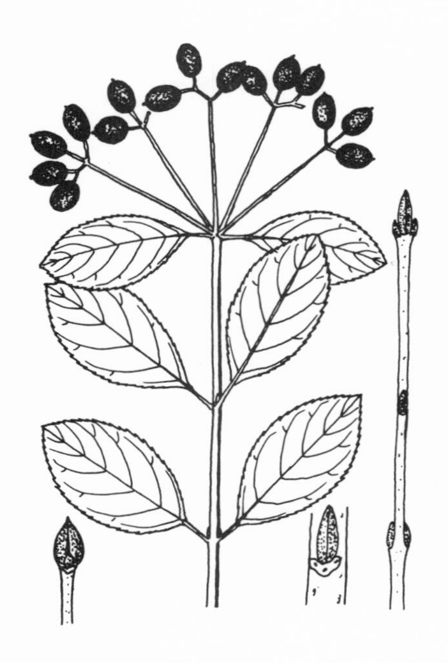

Black Haw

Rusty Black Haw

Small-leaf Viburnum

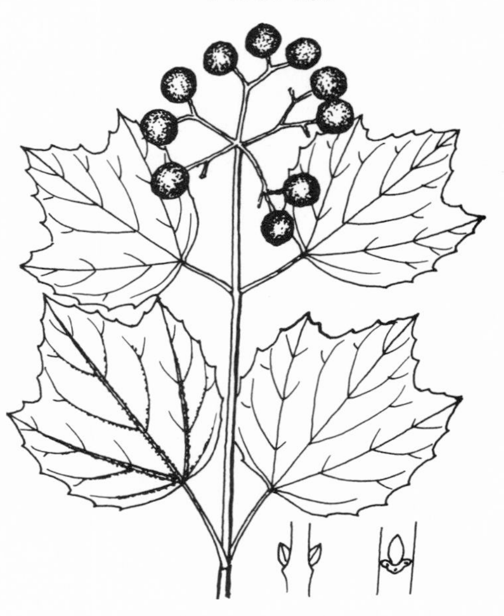

Squashberry

BLACK HAW *Viburnum prunifolium*

FIELD MARKS. A leaf-losing shrub or small tree 5 to about 15 feet high; growing chiefly in upland woods or thickets. *Branchlets* slender, rather stiff and spiky. *Leaves* elliptic to oval or egg-shaped, broadly pointed or roundish at base, blunt to broadly pointed at tip, finely and sharply toothed on margin, dull green above, paler beneath, smooth on both surfaces, 1 to 3 inches long; leafstalks grooved but not winged, ⅜ to ⅝ inch long. *Flowers* all alike, the cluster stalkless or nearly so; blooming April or May. *Fruits* oval-shaped, bluish black, sometimes with a slight whitish bloom, about ⅜ inch long, sweet and edible; ripening September or October.

RANGE. Connecticut to southern Michigan, Iowa, and eastern Kansas; south to northern Florida and Texas.

The bark of the roots is used medicinally.

RUSTY BLACK HAW *Viburnum rufidulum*

FIELD MARKS. A leaf-losing shrub or small tree 5 to about 25 feet high; growing in dry pine or deciduous woods and thickets. *Branchlets* slender, somewhat flexible, more or less rusty-scurfy. *Leaves* elliptic to oval or broadest above the middle, mostly pointed at base, blunt or abruptly short-pointed at tip, finely and sharply toothed on margin, rather thick in texture, lustrous above, paler and more or less rusty-scurfy beneath, 1½ to 3 inches long; leafstalks grooved and often slightly winged, rusty-scurfy, ¼ to ½ inch long. *Flowers* all alike, the cluster stalkless or nearly so; blooming late March to early May. *Fruits* oval-shaped, deep blue, often with a whitish bloom, about ⅜ inch long, sweet and edible; ripening September or October.

RANGE. Virginia to southern Ohio and Illinois, Missouri, and southeastern Kansas; south to central Florida and Texas.

Also called Southern Black Haw and Rusty Nannyberry.

SMALL-LEAF VIBURNUM *Viburnum obovatum*

FIELD MARKS. A shrub or small tree 5 to about 20 feet high with more or less persistent leaves; growing in swamps and low hammocks or along streams near the coast. *Branchlets* slender and rather flexible. *Leaves* stalkless or nearly so, top-shaped, broadest above the middle, wedge-shaped at base, rounded or broadly pointed at tip, untoothed or with some obscure and blunt teeth above the middle, lustrous above, paler and with some minute red hairs beneath, ¾ to 1½ inches long. *Flowers* all alike, the cluster somewhat dome-shaped and stalkless; blooming March or April. *Fruits* egg-shaped, bluish black, about ¼ inch long; ripening September or October.

RANGE. Coastal plain; South Carolina south to Florida.

Also called Walter Viburnum. Wood with a very rank odor.

SQUASHBERRY *Viburnum edule*

FIELD MARKS. A straggling or sprawling leaf-losing shrub 1 to 4 feet high; growing in cool moist woods and ravines. *Leaves* broadly oval or broadest above the middle, usually with 3 broadly pointed short lobes above the middle, rounded to somewhat heart-shaped at base, coarsely and irregularly toothed on margin, smooth above, slightly paler and more or less downy on the veins beneath, 1 to 2½ inches long; leafstalks ⅜ to ¾ inch long, often with a pair of small glands near the summit. *Flowers* all alike, the cluster less than 1½ inches broad; blooming May to August. *Fruits* roundish or slightly egg-shaped, light red or orange, about ⅜ inch in diameter; ripening August to October.

RANGE. Labrador to Alaska; south to Maine; northern New York, Michigan, and Minnesota; Colorado and Oregon.

The acid fruits are used for sauce and jelly.

HIGHBUSH-CRANBERRY *Viburnum opulus* var. *americanum*

FIELD MARKS. A leaf-losing shrub 3 to about 12 feet high; growing in cool moist woods and cold swamps. *Leaves* broadly egg-shaped, with 3 broad and rather long-pointed lobes, rounded to very broadly pointed at base, margin with coarse and rather wavy teeth, sometimes sparingly hairy above, slightly paler and often hairy along the veins beneath, 2 to 4 inches long; leaf stalks grooved, ½ to 1½ inches long, often with small stalked glands at the summit. *Flowers* of 2 kinds, the marginal ones showy, ½ inch or so across, but sterile; blooming May to July. *Fruits* roundish or slightly oval-shaped, bright red, juicy, translucent, about ⅜ inch in diameter; ripening September or October.

RANGE. Newfoundland to British Columbia; south to New England, West Virginia, the region of the Great Lakes, northeast Iowa, South Dakota, and Washington.

The acid fruits are cooked and used like cranberries.

MAPLE-LEAF VIBURNUM *Viburnum acerifolium*

FIELD MARKS. A leaf-losing shrub 2 to 6 feet high; growing in moist to dry and often rocky woods. *Branchlets* smooth or minutely downy. *Leaves* egg-shaped to roundish, usually with 3 broad and pointed lobes but sometimes almost unlobed or the lobes very short, rounded to heart-shaped at base, coarsely toothed on margin, dull green and smooth or nearly so above, usually more or less downy (rarely nearly smooth) and with minute black dots beneath, 2 to 5 inches long; leafstalks ⅜ to 1 inch long, often with a pair of narrow stipules at base. *Flowers* all alike, sometimes pink-tinged; blooming late April to July. *Fruits* roundish or slightly oval-shaped, bluish black, about ¼ inch long; ripening September or October.

RANGE. Southwestern Quebec to Minnesota south to New England, Georgia, Alabama, and Mississippi.

The pinkish to magenta leaves are very attractive in the fall.

ARROWWOOD *Viburnum dentatum*

FIELD MARKS. A bushy leaf-losing shrub 3 to about 10 (rarely 15) feet high; growing along streams or on shores of lakes, and in low wet woods and swamps. *Branchlets* more or less ridged or angled, smooth or sometimes roughish-hairy. *Leaves* with leafstalks usually more than ¼ inch long, egg-shaped to roundish, rounded or slightly heart-shaped at base, pointed to blunt at tip, prominent lateral veins ending in the large and sharp-pointed marginal teeth, smooth or nearly so on the upper surface, paler and more or less downy beneath (or smooth except for occasional tufts of down in the axils of the veins in the variety *lucidulum* Ait.), 1½ to about 4 inches long. *Flowers* all alike; blooming late March to July. *Fruits* roundish or egg-shaped, bluish black, ¼ to ⅜ inch long; ripening July to October.

RANGE. New Brunswick to southern Ontario; south to Florida and Texas.

A variable shrub often divided into several species by some botanists.

SOFTLEAF ARROWWOOD *Viburnum molle*

FIELD MARKS. A leaf-losing shrub 4 to 12 feet high, with peeling grayish bark on the older stems; growing in rocky woods and on the banks of streams. *Branchlets* smooth and roundish. *Leaves* broadly egg-shaped or roundish, heart-shaped at base, abruptly pointed at tip, prominent lateral veins ending in the large and sharp-pointed marginal teeth, smooth above, paler and softly downy beneath, 2 to 5 inches long; leaf stalks ¾ to 2 inches long,, usually with a pair of narrow stipules at the base. *Flowers* all alike; blooming May and June. *Fruits* oval-shaped, bluish black, about ⅜ inch long; ripening August to October.

RANGE. Indiana and Illinois to Missouri; south to Kentucky and Arkansas.

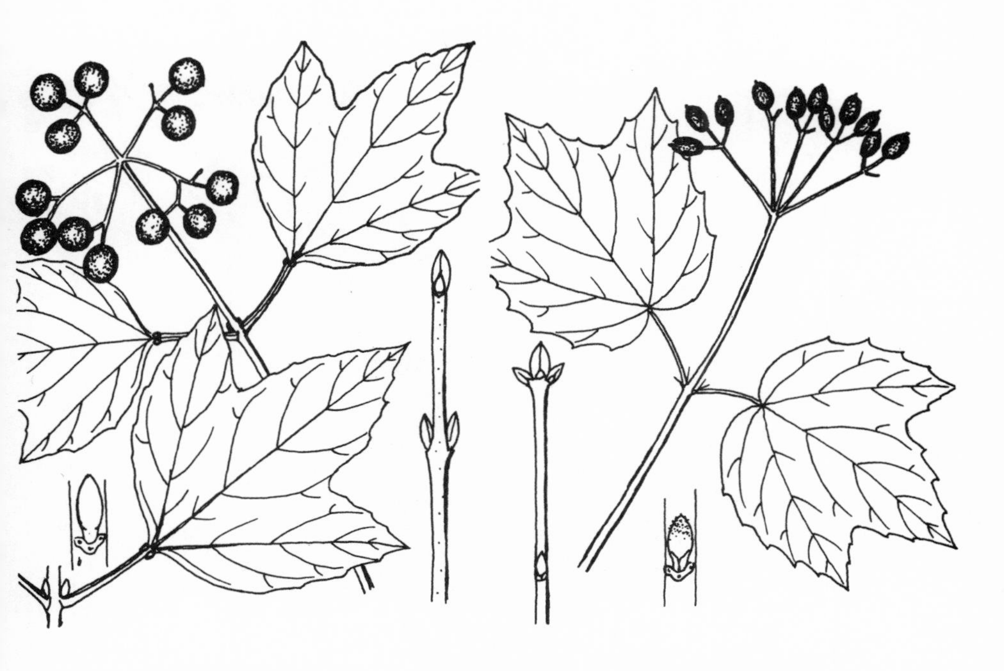

Highbush-cranberry

Maple-leaf Viburnum

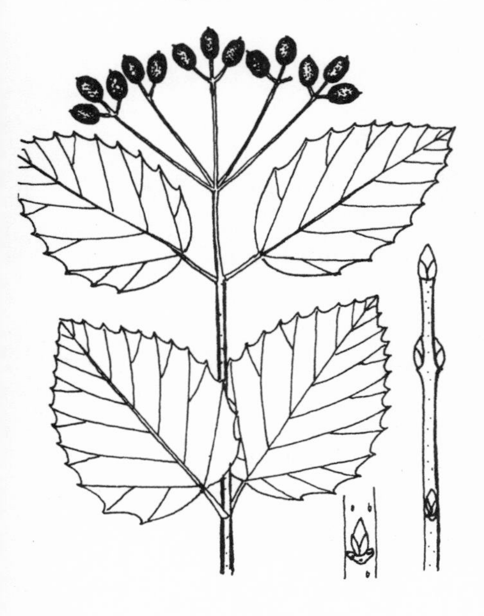

Arrowwood

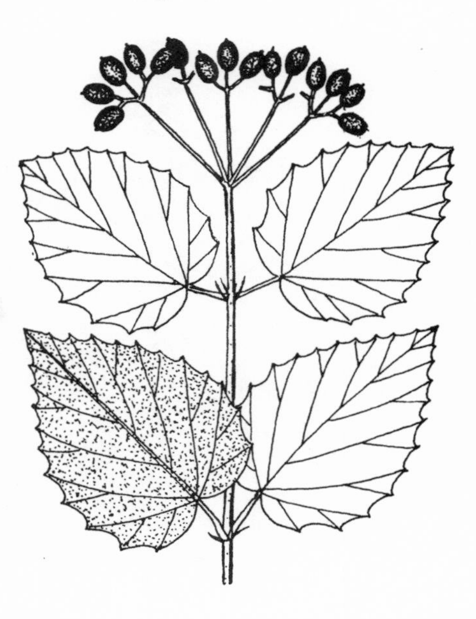

Softleaf Arrowwood

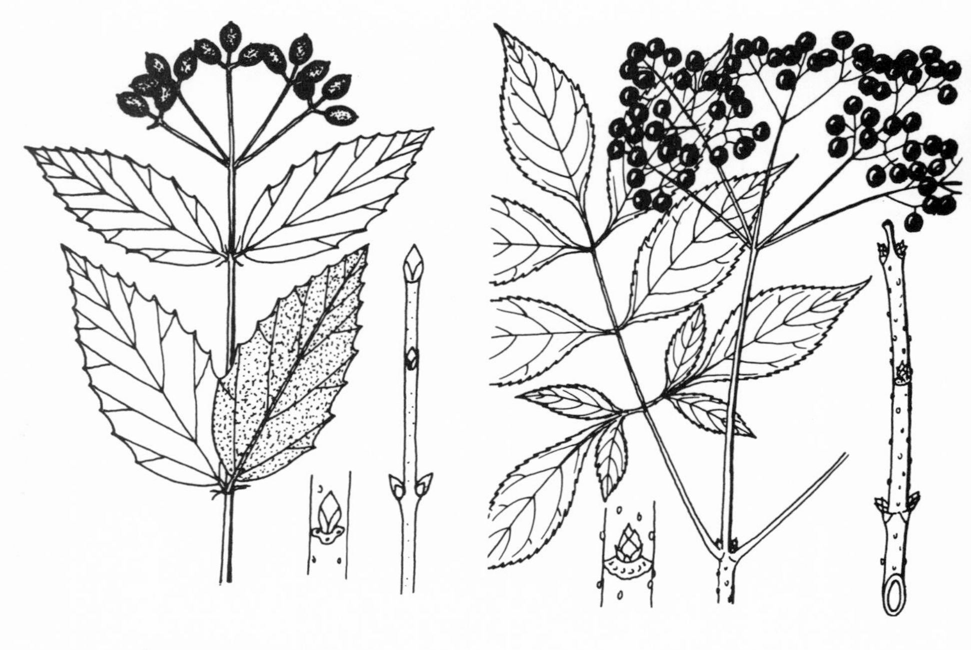

Downy Arrowwood

Common Elder

Red-berried Elder

Marsh Elder

DOWNY ARROWWOOD *Viburnum rafinesquianum*

FIELD MARKS. A leaf-losing shrub 2 to 5 feet high, with slender grayish-barked stems; growing on dry wooded slopes and banks. *Branchlets* roundish, smooth or nearly so. *Leaves* stalkless or very short-stalked and with narrow stipules, egg-shaped or oblong egg-shaped, rounded at base, pointed at tip, with prominent lateral veins ending in coarse and sharp marginal teeth, smooth or nearly so above, softly downy beneath, 1½ to 3 inches long. *Flowers* all alike; blooming April to June. *Fruits* oval-shaped, purplish black, about ¼ inch long; ripening July to September.

RANGE. Quebec to Manitoba; south to central North Carolina, northern Georgia, Kentucky, and Missouri.

COMMON ELDER *Sambucus canadensis*

FIELD MARKS. A leaf-losing shrub 4 to 12 feet high; widely distributed in moist rich soils. *Branchlets* stout, yellowish brown, with warty lenticels, large white pith, and small greenish or brown buds; odor rank when bruised. *Leaves* opposite, compound, 6 to 10 inches long; the 5 to 11 leaflets elliptic or lance-shaped, lower ones often 3-parted, mostly pointed at base and tip, margin sharply toothed, smooth above, paler and sometimes slightly downy beneath, 3 to 6 inches long. *Flowers* small, white, the short corolla tube 5-lobed, in broad flat-topped end clusters; blooming April to August. *Fruits* berry-like, round, purplish black, with 3 to 5 large seeds, about 3/16 inch in diameter; ripening July to October.

RANGE. Nova Scotia to Manitoba; south to Florida and Texas.

Fruits are eaten by birds and used for jelly, pies, and wine.

RED-BERRIED ELDER *Sambucus racemosa* var. *pubens*

FIELD MARKS. A leaf-losing shrub 3 to 10 feet high; growing in cool, moist, rocky woods and ravines. *Branchlets* stout, light brown, with warty lenticels, large brownish pith, and egg-shaped purplish-red buds. *Leaves* opposite, compound, 5 to 8 inches long; the 5 to 7 leaflets lance-shaped, pointed to unevenly rounded at base, long-pointed at tip, finely and sharply toothed on margin, smooth above, paler and usually downy beneath, 2 to 5 inches long. *Flowers* small, creamy white, the short corolla tube 5 lobed, in pyramid-shaped end clusters; blooming April to June. *Fruits* berry-like, round, bright red, about 3/16 inch in diameter; ripening June to August.

RANGE. Newfoundland to Alaska; south to New Jersey, Pennsylvania, the region of the Great Lakes, Iowa, and along the mountains to northern Georgia.

ASTER FAMILY (Asteraceae)

Members of the Aster family, formerly called composites, have small flowers grouped in heads which are surrounded by bracts. Many, like the goldenrods, asters, sunflowers, and daisies, are familiar plants. Only a few species are more or less shrubby.

MARSH ELDER *Iva frutescens*

FIELD MARKS. A woody-based plant 3 to about 10 feet high; growing in coastal salt marshes. *Leaves* mostly opposite, stalkless or nearly so, elliptic to narrowly lance-shaped, pointed at both ends, usually coarsely toothed on margin, somewhat fleshy, smooth or minutely rough-hairy, 1½ to 4 inches long. *Flowers* small, greenish, in nodding heads in axils of small leaves toward tips of branchlets; blooming August to November. *Fruits* small, dry, 1-seeded.

RANGE. Coastal plain; Newfoundland south to Florida, west to Texas.

SEACOAST MARSH ELDER *Iva imbricata*

Similar to the preceding but usually less than 2 feet high and smoother; and with alternate, very narrow, untoothed leaves. Grows on coastal sand dunes from southeastern Virginia south to Florida, west to Texas. (Not illustrated)

GROUNDSEL TREE *Baccharis halimifolia*

FIELD MARKS. A more or less resinous, soft-wooded, semi-evergreen shrub 3 to 10 feet high; growing in swampy thickets, sandy open woods or fields, and on sea beaches. *Branchlets* green, longitudinally and finely furrowed, smooth or somewhat downy. *Leaves* alternate, elliptic to roundish or broadest above the middle, the upper ones often quite narrow, wedge-shaped at base, blunt or rather pointed at tip, margin (except in uppermost leaves) coarsely toothed from about the middle to tip, pale green and smooth or nearly so on both surfaces, 1 to 3 inches long. *Flowers* small, the heads mostly stalked and in a branching open end cluster; blooming September or October. *Fruits* small, dry, 1-seeded, with a conspicuous tuft of silky whitish hairs.

RANGE. Coastal plain and piedmont; eastern Massachusetts south to Florida, west to Texas.

Also called Groundselbush and Sea-myrtle.

SESSILE-FLOWERED GROUNDSEL TREE *Baccharis glomeruliflora*

FIELD MARKS. A barely resinous, soft-wooded, more or less evergreen shrub 3 to about 10 feet high; growing in low woods and brackish marshes near the coast. *Leaves* alternate, broadest above the middle and tapering to the base, blunt or broadly pointed at tip, usually with several coarse but sharp teeth above the middle, ¾ to about 2 inches long. *Flowers* small, the heads stalkless and mostly in groups of 3 in the axils of leaflike bracts; blooming October and November. *Fruits* small, dry, 1-seeded, with a tuft of silky whitish hairs.

RANGE. Coastal plain; North Carolina south to Florida.

Also called Silverling.

FALSE WILLOW *Baccharis angustifolia*

FIELD MARKS. A more or less resinous, soft-wooded, evergreen shrub 2 to 8 feet high; growing in brackish coastal marshes. *Leaves* alternate, long and narrow with parallel sides, somewhat pointed at base and tip, untoothed or nearly so on margin, ¾ to 2 inches long, less than 3/16 inch wide. *Flowers* small, the heads mostly stalked and clustered at the ends of the branchlets; blooming late September and October, or later southward. *Fruits* small, dry, 1-seeded, with a tuft of silky whitish hairs.

RANGE. Coastal plain; North Carolina south to Florida, west to Texas.

SEA-OXEYE *Borrichia frutescens*

FIELD MARKS. A shrub 1 to 2½ feet high, with the younger parts fleshy; growing in brackish marshes along the coast. *Leaves* opposite, narrow to quite broad and widest above the middle, somewhat pointed at both ends, untoothed or with some wavy teeth on margin, thickish, grayish green with short silky hairs on both surfaces, ¾ to 2¼ inches long. *Flowers* in daisy-like heads about 1½ inches across, with bright yellow rays and a brownish disk; blooming May to September. *Fruits* small, dry, 1-seeded.

RANGE. Coastal plain; southeastern Virginia south to Florida, west to Texas.

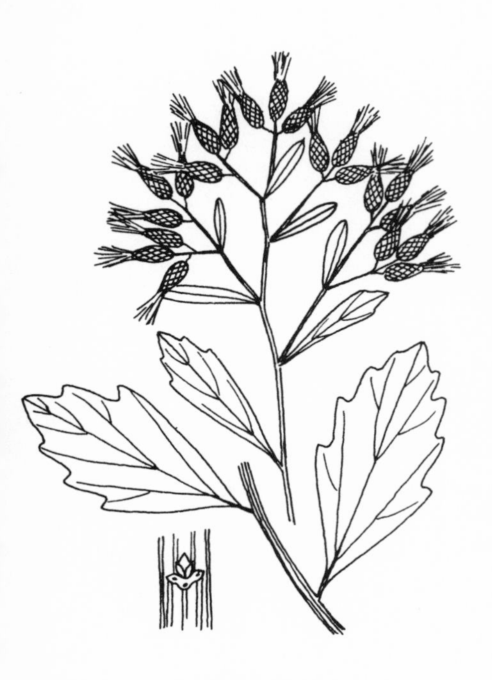

Groundsel Tree

Sessile-flowered Groundsel Tree

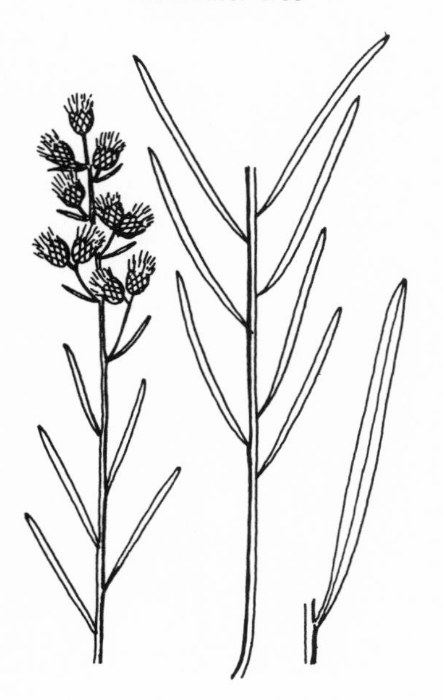

False Willow

Sea-oxeye

APPENDICES

Keys for identifying particular plants within the larger native shrub families.

A

KEY TO THE GREENBRIERS (Smilax)

1 Stems and lower leaf surfaces woolly-hairy
WOOLLY GREENBRIER *(S. pumila)* 394
1 Stems and lower leaf surfaces otherwise – 2

2 Leaves white beneath GLAUCOUS GREENBRIER *(S. glauca)* 390
2 Leaves green on both surfaces or sometimes mottled with white – 3

3 Leaves evergreen – 4
3 Leaves deciduous – 7

4 Leaves lance-shaped and pointed at both ends – 5
4 Leaves commonly fiddle-shaped or triangular – 6

5 Leaves thin, 5- to 7-veined LANCELEAF GREENBRIER *(S. smallii)* 393
5 Leaves thickish, 3-veined LAUREL-LEAF GREENBRIER *(S. laurifolia)* 393

6 Leaves usually prickly on margin, sometimes mottled with white
CHINABRIER *(S. bona-nox)* 390
6 Leaves not prickly on margin, sometimes notched at tip
WILD-BAMBOO *(S. auriculata)* 393

7 Leaves thickish; stems often angled and with stout broad-based prickles
COMMON GREENBRIER *(S. rotundifolia)* 390
7 Leaves thin; stems not angled, prickles slender when present – 8

8 Stems more or less thickly covered with needle-like blackish prickles; leaf margins roughish with minute bristly teeth; fruits black, on stalks longer than the leafstalks BRISTLY GREENBRIER *(S. hispida)* 390
8 Stems usually prickly only toward the base; leaf margins not roughish; fruits red, on stalks shorter than the leafstalks
RED-BERRIED GREENBRIER *(S. walteri)* 393

B

KEY TO THE BAYBERRIES (Myrica)

1 Floral bracts of staminate flowers exceeding stamen length; fruit surface not white-waxy; pistillate bractlets persistent SWEETGALE *(M. gale)* 394

1 Floral bracts of staminate flowers shorter than stamen length; fruit surface white-waxy; pistillate bractlets deciduous – 2

2 Leaves smooth, not aromatic when crushed, margins entire; staminate flowers with 8 or more stamens; 1 to 3 pistils in each bract ODORLESS BAYBERRY *(M. inodora)* 397

2 Leaves aromatic when crushed, scurfy, margins toothed; staminate flowers with fewer than 7 stamens – 3

3 Leaves persistent, blades oblanceolate, densely glandular on both surfaces, fruits 1/12 to 1/7 inch long SOUTHERN BAYBERRY *(M. cerifera)* 397

3 Leaves persistent or deciduous, blades obovate to elliptic, densely glandular below only, sparce or few glands above; fruits 1/8 to 1/5 inch long – 4

4 Leaves with similarly colored golden-yellow glands, margins not strongly rolled under; fruits 1/8 to 1/6 inch long, smooth, with a usually thin, gray or dark waxy cover EVERGREEN BAYBERRY *(M. heterophylla)* 397

4 Leaves with two distinctly colored glands, margins strongly rolled under; fruits 1/7 to 1/5 inch long, hairy at first, with a thick white waxy cover NORTHERN BAYBERRY *(M. pensylvanica)* 397

C

KEY TO THE WILLOWS (Salix)

1 Prostrate or creeping shrubs of arctic tundras and alpine mountain summits of the northeastern United States – 2

1 Shrubs with erect or ascending stems – 3

2 Leaves roundish, heart-shaped at base, bright green on both surfaces DWARF WILLOW *(S. herbacea)* 409

2 Leaves elliptical or broadest above the middle, wedge-shaped at base, pale beneath BEARBERRY WILLOW *(S. uva-ursi)* 409

3 Leaf margins untoothed or nearly so – 4

3 Leaf margins toothed – 10

4 Leaves smooth on both surfaces – 5

4 Leaves hairy at least beneath – 6

5 Leaves green on both surfaces, narrowly elliptic or oblong, wedge-shaped at base, margin inrolled; northern bog shrub BOG WILLOW *(S. pedicellaris)* 405

5 Leaves whitened beneath, lustrous above, elliptic or oblong, pointed or blunt at both ends TEALEAF WILLOW *(S. planifolia)* 406

6 Branchlets smooth – 7

6 Branchlets downy or woolly – 8

7 Branchlets dark green and lustrous; leaves silvery-silky beneath, the margin inrolled SILVER WILLOW *(S. argyrocarpa)* 406
7 Branchlets reddish to olive brown, usually whitened with a bloom; leaves often silky to velvety white beneath SATINY WILLOW *(S. pellita)* 406

8 Branchlets and lower leaf surfaces densely white-woolly, leaf margin inrolled; shrub of northern bogs HOARY WILLOW *(S. candida)* 406
8 Branchlets grayish-woolly; upland shrubs of wide distribution—9

9 Leaves mostly over 2 inches long, grayish-woolly to smooth and veiny beneath, margin usually wavy PRAIRIE WILLOW *(S. humilis)* 402
9 Leaves less than 2 inches long, white-woolly beneath, margin inrolled DWARF GRAY WILLOW *(S. humilis* var. *microphylla)* 402

10 Upper leaf surfaces with a wrinkled appearance, margins with some wavy teeth—11
10 Upper leaf surfaces otherwise—12

11 Leaves more or less downy on both surfaces but more densely so beneath BEBB WILLOW *(S. bebbiana)* 402
11 Leaves not hairy above, whitened and often with a few rusty hairs beneath PUSSY WILLOW *(S. discolor)* 402

12 Leaves long, narrow, and almost stalkless, the margins with shallow and widely spaced teeth SANDBAR WILLOW *(S. interior)* 398
12 Leaves rather regularly and finely toothed on margin—13

13 Leaves with short stout stalks, densely silky-hairy on both surfaces; branchlets woolly HEARTLEAF WILLOW *(S. cordata)* 401
13 Leaves otherwise—14

14 Leaves less than 3 times as long as wide—15
14 Leaves more than 3 times as long as wide—16

15 Leaves with a balsam-like odor when crushed, thin, dull above, smooth on both surfaces, margin with rather low glandular teeth; stipules small or absent BALSAM WILLOW *(S. pyrifolia)* 405
15 Leaves not fragrant, thick, firm, lustrous above, whitened and sometimes woolly beneath; margin sharply toothed; stipules large BLUELEAF WILLOW *(S. glaucophylloides)* 405

16 Branchlets downy or woolly—17
16 Branchlets smooth or nearly so—18

17 Leaves more or less grayish-hairy on both surfaces, more or less heart-shaped at base; stipules large and glandular-toothed; northern HEARTLEAF WILLOW *(S. cordata)* 401
17 Leaves smooth but dull above, whitish and with some soft hairs beneath, mostly rounded to pointed at base; stipules moderate in size; midwestern MISSOURI RIVER WILLOW *(S. eriocephala)* 401

18 Leafstalks with a pair of glands at the summit; leaves very lustrous on the upper surface—19
18 Leafstalks otherwise; leaves but slightly if at all lustrous on the upper surface—20

19 Leaves whitish and smooth beneath AUTUMN WILLOW *(S. serissima)* 398
19 Leaves green and sometimes with scattered rusty hairs beneath SHINING WILLOW *(S. lucida)* 401

20 Leaves smooth on both surfaces — 21
20 Leaves more or less silky-hairy beneath — 22

21 Leaves gradually tapered to the tip; branchlets brittle at base; southern WARD WILLOW *(S. caroliniana)* 401
21 Leaves rather abruptly pointed at the tip; branchlets not brittle-based; northern PEACHLEAF WILLOW *(S. amygdaloides)* 401

22 Leaves silvery-silky beneath, margins toothed quite to the base; branchlets brittle-based; widespread SILKY WILLOW *(S. sericea)* 398
22 Leaves whitened beneath and usually with scattered rusty or tawny hairs; margin not toothed toward the base; branchlets not brittle-based; northern SLENDER WILLOW *(S. gracilis)* 398

D

KEY TO THE CURRANTS AND GOOSEBERRIES (Ribes)

1 Branchlets and stems unarmed — 2
1 Branchlets and stems armed with spines or prickly bristles — 6

2 Main stems prostrate or creeping — 3
2 Main stems erect or ascending — 4

3 Plant with a skunklike odor when bruised SKUNK CURRANT *(R. glandulosum)* 438
3 Plant otherwise SWAMP RED CURRANT *(R. triste)* 438

4 Leaves and branchlets with small yellow resin dots WILD BLACK CURRANT *(R. americanum)* 438
4 Leaves and branchlets not resin-dotted — 5

5 Branchlets downy BUFFALO CURRANT *(R. odoratum)* 441
5 Branchlets smooth or with scattered bristles SMOOTH GOOSEBERRY *(R. hirtellum)* 437

6 Largest leaves less than 1 inch wide; southern — 7
6 Largest leaves more than 1 inch wide; northern or in mountains — 8

7 Branchlets reddish brown; fruits smooth GRANITE GOOSEBERRY *(R. curvatum)* 437
7 Branchlets grayish; fruits densely covered with gland-tipped spines FLORIDA GOOSEBERRY *(R. echinellum)* 434

8 Nodal spines less than ¼ inch long, the internodes smooth or with some short bristles — 9
8 Nodal spines more than ¼ inch long, the internodes often with prickles or bristles — 11

9 Branchlets always smooth between the nodes ROUNDLEAF GOOSEBERRY *(R. rotundifolium)* 434
9 Branchlets usually more or less bristly between the nodes — 10

10 Leaves with pointed lobes, smooth or nearly so on both surfaces SMOOTH GOOSEBERRY *(R. hirtellum)* 437
10 Leaves with bluntish lobes, somewhat hairy and glandular beneath NORTHERN GOOSEBERRY *(R. oxyacanthoides)* 437

11 Internodes thickly beset with long and prickly bristles SWAMP BLACK CURRANT *(R. lacustre)* 438

11 Internodes smooth or with rather widely scattered spines or bristles—12

12 Nodal spines less than ½ inch long; fruits prickly
PRICKLY GOOSEBERRY *(R. cynosbati)* 434

12 Nodal spines often over ½ inch long; fruits smooth or bristly—13

13 Branchlets whitish or gray, sometimes bristly; leaves smooth above but usually downy beneath; fruits smooth
MISSOURI GOOSEBERRY *(R. missouriense)* 437

13 Branchlets reddish brown, usually bristly; leaves usually hairy on both surfaces; fruits often bristly
BRISTLY GOOSEBERRY *(R. oxyacanthoides* var. *setosum)* 437

E

KEY TO THE JUNEBERRIES (Amelanchier)

1 Leaves with pointed bases and stout leafstalks less than ⅜ inch long; fruits elongate, solitary or 2 to 3 together
OBLONG-FRUITED JUNEBERRY *(A. bartramiana)* 449

1 Leaves with rounded or heart-shaped bases and slender leafstalks up to ¾ inch long; fruits roundish or nearly so, several in a cluster—2

2 Leaves coarsely toothed, the teeth about as many as the prominent straight primary veins—3

2 Leaves finely toothed, the teeth at least twice as many as the irregular and branching veins—4

3 Low shrub forming colonies by creeping underground stems; leaf margins untoothed toward the base LOW JUNEBERRY *(A. humilis)* 446

3 Tall shrub, straggling or with arching branches, not colony-forming; leaf margins toothed quite to the base
ROUNDLEAF JUNEBERRY *(A. sanguinea)* 446

4 Colony-forming shrubs with creeping underground stems—5

4 Shrubs growing in clumps, or small trees—6

5 Leaf margins untoothed toward the base; leaves smooth or slightly downy beneath; fruits with stalks of variable length
RUNNING JUNEBERRY *(A. stolonifera)* 446

5 Leaf margins toothed quite to the base; leaves downy beneath; fruits with stalks of about the same length
COASTAL JUNEBERRY *(A. obovalis)* 449

6 Leaves very blunt or rounded at the tip
OBLONG-LEAF JUNEBERRY *(A. canadensis)* 449

6 Leaves definitely pointed at the tip—7

7 Flowers and fruits in upright clusters; shrub with several stems in a clump SWAMP JUNEBERRY *(A. × intermedia)* 449

7 Flowers and fruits usually in drooping clusters; usually trees—8

8 Mature leaves more or less downy beneath, at least on the midrib and veins COMMON JUNEBERRY *(A. arborea)* 446

8 Mature leaves smooth on both surfaces
SMOOTH JUNEBERRY *(A. laevis)* 446

F

KEY TO THE ROSES (Rosa)

1 Leaflets usually 3; stems scrambling or climbing PRAIRIE ROSE *(R. setigera)* 462

1 Leaflets usually 5 or more; stems erect—2

2 Stems with broad-based, usually hooked prickles SWAMP ROSE *(R. palustris)* 461

2 Stems unarmed or with slender needle-like or bristly prickles—3

3 Stems unarmed or with a few weak or bristly prickles; fruits smooth and with persistent stipules SMOOTH ROSE *(R. blanda)* 462

3 Stems well armed—4

4 Stems with needle-like prickles at the nodes and others widely scattered; fruits glandular-bristly PASTURE ROSE *(R. carolina)* 461

4 Stems densely covered with bristly or needle-like prickles—5

5 Stems densely bristly; fruits bristly-hairy, the sepals soon shed SHINING ROSE *(R. nitida)* 461

5 Stems densely covered with needle-like prickles; fruits smooth and with persistent sepals PRICKLY WILD ROSE *(R. acicularis)* 462

G

KEY TO THE INDIGOBUSHES (Amorpha)

1 Lowest pair of leaflets close to the stem; shrubs usually less than 3 feet high—2

1 Lowest pair of leaflets ½ inch or more from the stem; shrubs usually over 3 feet high—4

2 Branchlets smooth or nearly so; leaflets conspicuously glandular-dotted beneath GEORGIA INDIGOBUSH *(A. georgiana)* 470

2 Branchlets and leaves grayish-downy or grayish-hairy—3

3 Leaflets and branchlets densely grayish-hairy; fruit pods densely hairy; midwestern shrub LEADPLANT *(A. canescens)* 469

3 Leaflets and branchlets grayish-downy; fruit pods grayish-downy and with conspicuous dark glands; southeastern coastal plain shrub PLUME-LOCUST *(A. herbacea)* 470

4 Leaflets rather broadly egg-shaped or oval and smooth on both sides MOUNTAIN INDIGOBUSH *(A. glabra)* 469

4 Leaflets rather narrow or oblong, often hairy at least beneath—5

5 Branchlets and upper surfaces of the leaflets both lustrous; fruit pods smooth or nearly so SHINING INDIGOBUSH *(A. nitens)* 470

5 Branchlets and upper surfaces of the leaflets not both lustrous; fruit pods either hairy or conspicuously glandular-dotted—6

6 Branchlets and lower surfaces of the leaflets densely brownish-hairy SCHWERIN INDIGOBUSH *(A. schwerinii)* 470

6 Branchlets and lower surfaces of the leaflets with pale to tawny hairs COMMON INDIGOBUSH *(A. fruticosa)* 469

H

KEY TO THE LOCUSTS (Robinia)

1 Branches and pods sticky from stalked or sessile glands—2
1 Branches and pods not glandular—3

2 Less than 6 feet high; branches beset with stalked glandular hairs HARTWEG LOCUST *(R. viscosa* var. *hartwegii)* 473
2 More than 10 feet high; branches beset with unstalked glandular hairs CLAMMY LOCUST *(R. viscosa)* 470

3 Plants producing abundant fruits—4
3 Plants rarely if ever producing fruits—5

4 Plants moderately to very bristly-hairy; leaves narrow, 1.5 times longer than wide LARGE-FLOWERED LOCUST *(R. hispida* var. *fertilis)* 470
4 Plants sparsely bristly-hairy or glabrous; leaves wide, 2 to 3 times longer than wide KELSEY LOCUST *(R. hispida* var. *kelseyi)* 474

5 Stems and leafstalks not prickly-hispid—6
5 Stems and leafstalks strongly prickly-hispid BRISTLY LOCUST *(R. hispida* var. *hispida)* 473

6 Plants small, less than 3 feet high, scarcely branched with short zigzig internodes; leaflets commonly less than ¾ inch long ELLIOT LOCUST *(R. hispida* var. *nana)* 473
6 Plants larger, mostly more than 3 feet high, extensively branched with straight branches; leaflets commonly more than ¾ inch long BOYNTON LOCUST *(R. hispida* var. *rosea)* 473

I

KEY TO THE HOLLIES (Ilex)

1 Leaves evergreen, thick and leathery in texture—2
1 Leaves deciduous, thin or slightly thick in texture—6

2 Leaf margins toothed quite to the base with bluntish or rounded teeth; fruits red or yellow YAUPON HOLLY *(I. vomitoria)* 490
2 Leaf margins untoothed, or toothed only above the middle—3

3 Leaf margins with some low, blunt teeth toward the tip; fruits black INKBERRY *(I. glabra)* 490
3 Leaf margins untoothed or with a few sharp, spiny, or minute teeth—4

4 Leaves narrow, less than ⅜ inch wide, very stiff and sharp at tip MYRTLE-LEAF HOLLY *(I. myrtifolia)* 493
4 Leaves broader, often with a few small spiny teeth—5

5 Leaves minutely black dotted on the lower surface; fruits black LARGE GALLBERRY *(I. coriacea)* 490
5 Leaves not black-dotted beneath; fruits red or yellow DAHOON HOLLY *(I. cassine)* 493

6 Leaf margins with very small, blunt, and inconspicuous teeth SARVIS HOLLY *(I. amelanchier)* 490

6 Leaf margins with conspicuous sharp bluntish teeth—7

7 Leaves and fruits clustered on short lateral spurlike branches—8
7 Leaves and fruits never clustered on short spurlike branches—11

8 Leaf margins with rather low blunt teeth, the leaf bases quite sharply pointed—9
8 Leaf margins with sharply pointed teeth, the leaf bases broadly pointed to roundish—10

9 Leaves less than 3 inches long, less than 1 inch wide; fruits on short stalks DECIDUOUS HOLLY *(I. decidua)* 489
9 Leaves often over 3 inches long and over 1 inch wide; fruits on stalks ½ inch long or longer LONG-STALKED HOLLY *(I. longipes)* 489

10 Largest leaves less than 2½ inches long; coastal plain shrub CAROLINA HOLLY *(I. ambigua)* 489
10 Largest leaves more than 2½ inches long; chiefly mountain shrub LARGELEAF HOLLY *(I. montana)* 489

11 Leaves thickish, dull above and downy beneath, with rather coarsely toothed margins WINTERBERRY *(I. verticillata)* 486
11 Leaves thin, lustrous above, smooth or nearly so on both surfaces, with rather finely toothed margins SMOOTH WINTERBERRY *(I. laevigata)* 486

J

KEY TO THE GRAPES (Vitis)

1 Tendrils rarely present; plant low and bushy BUSH GRAPE *(V. ruprestris)* 509
1 Tendrils always present; plants high-climbing—2

2 Tendrils simple; branchlets with pith continuing through the nodes; bark not becoming loose or shredded MUSCADINE GRAPE *(V. rotundifolia)* 509
2 Tendrils branched; branchlets with pith interrupted by a woody partition (diaphragm) at each node; bark becoming loose or shredded—3

3 A tendril or flower or fruit cluster opposite each leaf—4
3 Tendrils and flower or fruit clusters intermittent, none opposite every third leaf—5

4 Leaves with permanent, matted, tawny or rusty wool on the lower surface FOX GRAPE *(V. labrusca)* 505
4 Leaves smooth or merely with some wool on the veins of the lower surface NEW ENGLAND GRAPE *(V. × novae-angliae)* 506

5 Leaves green on the lower surface, smooth or somewhat hairy—6
5 Leaves not green on the lower surface—10

6 Lower leaf surfaces conspicuously grayish-hairy—7
6 Lower leaf surfaces smooth, or merely with axillary tufts or scattered hairs along the veins—8

7 Branchlets angled; southern POSSUM GRAPE *(V. cinerea* var. *baileyana)* 509
7 Branchlets not angled; Great Lakes region RIVERBANK GRAPE *(Vitis riparia)* 506

8 Branchlets red; leaves deeply lobed RED GRAPE *(V. palmata)* 509
8 Branchlets not red; leaves with shallow lobes if any—9

9 Leaves with broad U-shaped basal sinuses, the margins with narrowly pointed teeth; branchlets with thin nodal diaphragms
RIVERBANK GRAPE *(V. riparia)* 506

9 Leaves with narrow V-shaped basal sinuses, the margins with rather broadly pointed teeth; branchlets with thick nodal diaphragms
FROST GRAPE *(V. vulpina)* 506

10 Lower leaf surfaces smooth or nearly so and very white—11
10 Lower leaf surfaces otherwise—12

11 Branchlets with thin nodal diaphragms; midwestern
POST OAK GRAPE *(V. aestivalis* var. *lincecumii)* 509

11 Branchlets with thick nodal diaphragms; eastern
BLUELEAF GRAPE *(V. aestivalis* var. *bicolor)* 505

12 Young branchlets angled—13
12 Young branchlets not angled—14

13 Leaves whitish-woolly beneath WINTER GRAPE *(V. cinerea)* 506
13 Leaves reddish-hairy beneath PIGEON GRAPE *(V. cinerea* var. *floridana)* 506

14 Branchlets with thin nodal diaphragms; midwestern
POST OAK GRAPE *(V. aestivalis* var. *lincecumii)* 509

14 Branchlets with thick nodal diaphragms; eastern
SUMMER GRAPE *(V. aestivalis)* 505

K

KEY TO THE ST. PETER'S-WORTS AND ST. JOHN'S-WORTS (Hypericum)

1 Flowers with 4 petals; sepals 2 or 4—2
1 Flowers with 5 petals; sepals 5—5

2 Styles 3 or 4; sepals of nearly the same length but the outer pair much larger and broader ST. PETER'S WORT *(H. crux-andreae)* 510

2 Styles 2; the inner sepals very small or wanting—3

3 Flower stalks bending downward as fruits develop; bractlets near base of the flower stalk DWARF ST. PETER'S-WORT *(H. suffruticosum)* 510

3 Flower stalks remaining erect; bractlets near the base of the calyx—4

4 Plant erect ST. ANDREW'S-CROSS *(H. hypericoides)* 513
4 Plant reclining RECLINING ST. ANDREW'S-CROSS *(H. stragalum)* 513

5 Leaves very narrow (about ⅛ inch wide) or almost needle-like; flowers in elongate end clusters—6

5 Leaves broader, usually more than ⅛ inch wide; flowers relatively few or, if numerous, in flat-topped clusters—10

6 Leaves narrow but flat, the larger ones about ⅛ inch wide
BEDSTRAW ST. JOHN'S-WORT *(H. galioides)* 517

6 Leaves almost needle-like, about as wide as thick—7

7 Shrubs less than 1½ feet high—8
7 Shrubs usually over 2 feet high; growing in wet places—9

8 Largest leaves ⅜ inch or less long; shrub of wet to dry coastal plain sands SANDWEED *(H. reductum)* 517

8 Largest leaves ½ inch or more long; shrub of dry, rocky, piedmont woodlands SANDWEED *(H. lloydii)* 517

9 Stems with spongy bark peeling in thin layers; leaves with 2 longitudinal grooves beneath SANDWEED *(H. fasciculatum)* 517

9 Stems with close thin bark; leaves not distinctly 2-grooved beneath SANDWEED *(H. nitidum)* 517

10 Flowers ½ inch or less across, usually numerous in flat-topped clusters; sepals all similar, narrow and not leaflike — 11

10 Flowers larger, either not numerous or not in distinctly flat-topped clusters; sepals, or some of them, leaflike or larger — 13

11 Principal leaves over ⅜ inch wide, seldom with axillary clusters of smaller leaves; flower cluster with very small bracts if any NAKED-FLOWERED ST. JOHN'S-WORT *(H. nudiflorum)* 517

11. Principal leaves ⅜ inch or less wide, usually with axillary clusters of smaller leaves; flower cluster with narrow but prominent bracts — 12

12 Leaves pointed at base; branchlets 2-edged; capsules much longer than broad BUSHY ST. JOHN'S-WORT *(H. densiflorum)* 514

12 Leaves roundish at base; branchlets 4-angled; capsules about as broad as long ROUND-PODDED ST. JOHN'S-WORT *(H. cistifolium)* 517

13 Flowers stalkless, solitary or 2 to 3 at branchlet tips, 1 to 2 inches across GOLDEN ST. JOHN'S-WORT *(H. frondosum)* 513

13 Flowers stalked, usually more numerous, mostly less than 1 inch across — 14

14 Sepals as long as the petals, very broad and leaflike MYRTLE-LEAF ST. JOHN'S-WORT *(H. myrtifolium)* 514

14 Sepals shorter than the petals or very unequal in length — 15

15 Leaves often over ½ inch wide; flowers clustered in the axils of the upper leaves SHRUBBY ST. JOHN'S-WORT *(H. spathulatum)* 513

15 Leaves less than ½ inch wide; flowers or flower clusters only at the ends of branchlets — 16

16 Stems and branchlets roundish; flowers with oblique pointed petals; capsules almost as broad as long STRAGGLING ST. JOHN'S-WORT *(H. dolibriforme)* 514

16 Stems and branchlets angled; flowers with symmetrical petals, capsules egg-shaped but longer than broad — 17

17 Leaves less than 1 inch long; creeping plant less than 1 foot high; growing in rocky places in southern Appalachians MOUNTAIN ST. JOHN'S-WORT *(H. buckleyi)* 518

17 Leaves over 1 inch long; taller northern shrub KALM ST. JOHN'S-WORT *(H. kalmianum)* 514

L

KEY TO THE DOGWOODS (Cornus)

1 Leaves alternate, more or less crowded at tips of branchlets ALTERNATE-LEAF DOGWOOD *(C. alternifolia)* 526

1 Leaves opposite — 2

2 Leaves roughish on the upper surface — 3

2 Leaves not roughish on the upper surface — 4

3 Fruits pale blue; branchlets with white pith; southeastern
SOUTHERN ROUGHLEAF DOGWOOD *(C. asperifolia)* 526

3 Fruits white; branchlets with brownish pith; midwestern
ROUGHLEAF DOGWOOD *(C. drummondii)* 525

4 Branchlets with brownish pith—5
4 Branchlets with white pith—7

5 Leaves egg-shaped or oval, roundish at base; fruits blue; bark greenish brown to purplish
SILKY DOGWOOD *(C. amomum)* 522
5 Leaves narrowly egg-shaped to lance-shaped, pointed at base—6

6 Fruits white, in cone-shaped clusters; bark gray
GRAY-STEMMED DOGWOOD *(C. racemosa)* 525
6 Fruits blue, in flat-topped clusters; bark greenish brown to purplish PALE DOGWOOD *(C. amomum* var. *obliqua)* 522

7 Leaves with 7 to 9 pairs of lateral veins; branchlets warty-dotted; fruits lead-colored or blue ROUNDLEAF DOGWOOD *(C. rugosa)* 525
7 Leaves with 4 to 6 pairs of lateral veins; branchlets not warty-dotted—8

8 Leaves paler green beneath, smooth or nearly so; branchlets greenish or partly red, becoming gray; fruits pale blue; southeastern
STIFF DOGWOOD *(C. foemina)* 525
8 Leaves whitened or densely woolly beneath; branchlets bright red to purplish red or reddish brown; fruits white; northern
RED-OSIER DOGWOOD *(C. sericea)* 522

M

KEY TO THE AZALEAS (Rhododendron)

When in flower

1 Flower red, orange, or yellow—2
1 Flower white or pink—5

2 Flowers appearing before or with the young leaves—3
2 Flowers appearing in summer after the leaves mature—4

3 Corolla tubes with gland-tipped hairs; southern mountain shrub
FLAME AZALEA *(R. calendulaceum)* 534
3 Corolla tubes with simple, non-glandular hairs; southern lowland shrub
SHOWY AZALEA *(R. speciosum)* 534

4 Corolla tubes hairy and with some short-stalked glands; mountain shrub CUMBERLAND AZALEA *(R. cumberlandense)* 534
4 Corolla tubes smooth or nearly so; southern coastal plain shrub
RED AZALEA *(R. prunifolium)* 538

5 Corolla tube very short, less than ¼ as long as the lobes—6
5 Corolla tube as long as or longer than the lobes—7

6 Corolla strongly 2-lipped, with long and narrow lobes; stamens 10; northern shrub RHODORA *(R. canadense)* 534
6 Corolla slightly 2-lipped, the lobes broad and the upper one spotted, stamens usually 7; southern mountain shrub
PINKSHELL AZALEA *(R. vaseyi)* 533

7 Flowers appearing before or with the young leaves—8
7 Flowers appearing in summer after the leaves mature—12

8 Flower buds with stalked glands in prominent rows; colonial shrub of the coastal plain less than 2 feet high
DWARF AZALEA *(R. atlanticum)* 537
8 Flower buds otherwise; taller and non-colonial shrubs—9

9 Flowers white, the calyx glandular ALABAMA AZALEA *(R. alabamense)* 534
9 Flowers pink, the calyx not glandular—10

10 Young leaves smooth except along midrib beneath; flowers slightly if at all fragrant, the corolla rarely glandular
PINXTER-FLOWER *(R. periclymenoides)* 534
10 Young leaves densely downy beneath; flowers very fragrant—11

11 Lobes of the corolla about as long as the tube; northern and mountain shrub MOUNTAIN AZALEA *(R. prinophyllum)* 537
11 Lobes of the corolla much shorter than the tube; southern lowland shrub
HOARY AZALEA *(R. canescens)* 537

12 Stamens usually more than twice as long as the corolla tube; young branchlets and leaves smooth or nearly so
SMOOTH AZALEA *(R. arborescens)* 538
12 Stamens about twice as long as the corolla tube; young branchlets with scattered stiff hairs SWAMP AZALEA *(R. viscosum)* 537

With fruits and mature leaves

1 Leaf margins smooth or inrolled on the lower side; capsules smooth or merely downy, more or less lopsided at the base—2
1 Leaf margins hairy-fringed and sometimes minutely toothed; capsules with some scattered stiff hairs, quite symmetrical at the base—3

2 Leaves less than 2 inches long, the margins inrolled beneath; capsules downy; northern shrub RHODORA *(R. canadense)* 534
2 Leaves mostly over 2 inches long, the margin flat and somewhat wavy; capsules smooth; southern mountain shrub
PINKSHELL AZALEA *(R. vaseyi)* 533

3 Branchlets smooth; leaves smooth or with scattered hairs along midrib on lower side—4
3 Branchlets more or less hairy or downy; leaves at least with hairs along midrib on lower side—5

4 Largest leaves often over 3 inches long and long-pointed at tip; shrub of southern coastal plain RED AZALEA *(R. prunifolium)* 538
4 Largest leaves less than 3 inches long, always broadly pointed or blunt at tip, often whitened beneath
SMOOTH AZALEA *(R. arborescens)* 538

5 Colonial shrub with spreading underground branches, less than 2 feet high; coastal plain shrub DWARF AZALEA *(R. atlanticum)* 537
5 Non-colonial shrubs 3 feet or more high—6

6 Capsules and lower leaf surfaces grayish-downy and usually with some bristly hairs—7
6 Capsules and lower leaf surfaces with bristly hairs but not downy—9

7 Most long hairs of the capsules gland-tipped; northern and mountain shrub MOUNTAIN AZALEA *(R. prinophyllum)* 538
7 Most long hairs of the capsules not gland-tipped—8

8 Capsules with few or no long hairs; southern lowland shrub
HOARY AZALEA *(R. canescens)* 538

8 Capsules with numerous long hairs; southern mountain shrub
FLAME AZALEA *(R. calendulaceum)* 534

9 Young branchlets grayish-downy as well as with scattered stiff hairs; leaves with scattered hairs beneath
ALABAMA AZALEA *(R. alabamense)* 534

9 Young branchlets smooth except for scattered bristly hairs; leaves hairy only along midrib beneath—10

10 Leaves light green above; hairs of the capsules not gland-tipped
PINXTER-FLOWER *(R. periclymenoides)* 534

10 Leaves lustrous dark green above; hairs of the capsules gland-tipped
SWAMP AZALEA *(R. viscosum)* 538

N

KEY TO THE HUCKLEBERRIES (Gaylussacia)

1 Branchlets angled; leaves evergreen; toothed on margin, not resin-dotted BOX HUCKLEBERRY *(G. brachycera)* 553

1 Branchlets not angled; leaves deciduous, untoothed on margin, dotted with small yellow resin globules at least on the lower surface—2

2 Leaves resin-dotted on both surfaces but more densely so beneath; berries black, in short clusters BLACK HUCKLEBERRY *(G. baccata)* 554

2 Leaves resin-dotted only beneath; berries in rather long and more or less drooping clusters—3

3 Berries in the axils of large leaflike bracts
DWARF HUCKLEBERRY *(G. dumosa)* 553

3 Berries not in the axils of such bracts—4

4 Leaves green beneath; berries lustrous black BUCKBERRY *(G. ursina)* 554

4 Leaves pale or whitened beneath; berries blue and whitened with a bloom—5

5 Branchlets smooth; leaves smooth or nearly so beneath
DANGLEBERRY *(G. frondosa)* 553

5 Branchlets and lower leaf surfaces downy or woolly—6

6 Branchlets and lower leaf surfaces merely downy; shrub with creeping underground stems
CREEPING DANGLEBERRY *(G. frondosa* var. *nana)* 554

6 Branchlets and lower leaf surfaces conspicuously woolly; shrub otherwise WOOLLY DANGLEBERRY *(G. frondosa* var. *tomentosa)* 554

O

KEY TO THE BLUEBERRIES, BILBERRIES, AND CRANBERRIES (Vaccinium)

1 Stems trailing, creeping, or tufted and mat-forming—2

1 Stems ascending or erect—5

2 Leaves minutely black-dotted on the lower surface; stems tufted and mat-forming MOUNTAIN-CRANBERRY *(V. vitis-idaea* var. *minus)* 565

2 Leaves not black-dotted beneath, the margins more or less rolled inward on the lower side; stems very slender, creeping — 3

3 Leaves green beneath; berries blue
CREEPING BLUEBERRY *(V. crassifolium)* 565
3 Leaves pale or whitened beneath; berries bright red — 4

4 Leaves elliptical, blunt at tip, margins but slightly rolled inward beneath LARGE CRANBERRY *(V. macrocarpom)* 565
4 Leaves egg-shaped, pointed at tip, margins strongly rolled inward beneath SMALL CRANBERRY *(V. oxycoccos)* 565

5 Fruits solitary in the axils of the leaves or of leaflike bracts — 6
5 Fruits 2 to several together in a cluster; bracts, if present, very minute — 11

6 Branchlets 4-angled or 4-sided — 7
6 Branchlets round or nearly so — 8

7 Leaves pale or whitened beneath, blunt at tip, the margins untoothed; berries blue OVAL-LEAF BILBERRY *(V. ovalifolium)* 558
7 Leaves bright green on both surfaces, pointed at tip, the margins finely toothed; berries dark purple or black
THINLEAF BILBERRY *(V. membranaceum)* 558

8 Leaves toothed — 9
8 Leaves untoothed or apparently so — 10

9 Stems somewhat tufted, less than 1 foot high; fruits light blue; northern shrub DWARF BILBERRY *(V. cespitosum)* 557
9 Stems not tufted, 1 to 6 feet high; fruits purplish red; southern mountain shrub
SOUTHERN MOUNTAIN-CRANBERRY *(V. erythrocarpum)* 562

10 Leaves leathery, margins often with minute teeth or slightly rolled inward beneath, lower surface sometimes minutely downy; fruits black, about ¼ inch in diameter SPARKLEBERRY *(V. arboreum)* 557
10 Leaves thin, margins always flat and untoothed, often whitened or downy beneath; fruits over ¼ inch in diameter
DEERBERRY *(V. stamineum)* 557

11 Low shrubs less than 3 feet high — 12
11 Tall shrubs 3 or more feet high — 18

12 Largest leaves less than ¾ inch long — 13
12 Largest leaves more than ¾ inch long — 14

13 Arctic-alpine shrub; leaves dull above BOG BILBERRY *(V. uliginosum)* 557
13 Southern coastal plain shrub; leaves lustrous above
EVERGREEN BLUEBERRY *(V. myrsinites)* 558

14 Branchlets and leaves conspicuously hairy — 15
14 Branchlets minutely warty-dotted, slightly if at all hairy; leaves smooth to somewhat downy or sparingly hairy beneath — 16

15 Fruits smooth; northern shrub
VELVET-LEAF BLUEBERRY *(V. myrtilloides)* 558
15 Fruits hairy; southern mountain shrub
HAIRY BLUEBERRY *(V. hirsutum)* 561

16 Leaves, or at least some of them, with small stalked glands on the lower surface SMALL BLACK BLUEBERRY *(V. tenellum)* 561

16 Leaves without such glands beneath—17

17 Leaves pale or whitened beneath, the margins usually untoothed; widely distributed shrub LATE LOW BLUEBERRY *(V. pallidum)* 561

17 Leaves usually green beneath, the margins minutely toothed; northern shrub EARLY LOW BLUEBERRY *(V. angustifolium)* 561

18 Leaves less than 1 inch long ELLIOTT BLUEBERRY *(V. elliottii)* 562

18 Leaves more than 1 inch long—19

19 Branchlets and lower leaf surfaces densely downy; berries black BLACK HIGHBUSH BLUEBERRY *(V. atrococcum)* 562

19 Branchlets and lower leaf surfaces smooth or nearly so; berries blue HIGHBUSH BLUEBERRY *(V. corymbosum)* 562

P

KEY TO THE HONEYSUCKLES (Lonicera)

1 Scrambling shrubs or vines with flowers and fruits in end clusters, above disks formed by leaves with united bases—2

1 Erect shrubs with flowers and fruits paired on axillary stalks, the leaves all distinct—6

2 Leaves merely grayish green beneath—3

2 Leaves strongly whitened, at least on the lower surface—4

3 Branchlets rough-hairy; leaves densely downy beneath; northern HAIRY HONEYSUCKLE *(L. hirsuta)* 582

3 Branchlets and leaves both smooth; southern YELLOW HONEYSUCKLE *(L. flava)* 582

4 Leaves whitened on both surfaces; midwestern GRAPE HONEYSUCKLE *(L. prolifera)* 582

4 Leaves green on the upper surface; eastern or widespread—5

5 Branchlets whitened; flowers less than 1 inch long, 2-lipped, yellowish or purplish-tinged; fruits roundish, salmon red SMOOTH HONEYSUCKLE *(L. dioica)* 582

5 Branchlets not whitened; flowers over 1 inch long, trumpet-shaped and bright red; fruits egg-shaped and bright red TRUMPET HONEYSUCKLE *(L. sempervirens)* 585

6 Leafy bracts present at the bases of the fruits INVOLUCRED FLY HONEYSUCKLE *(L. involucrata)* 581

6 Bracts at the bases of the fruits small and narrow or none—7

7 Leaves thin, usually egg-shaped; fruits distinct, bright red; branchlets flexible and spreading AMERICAN FLY HONEYSUCKLE *(L. canadensis)* 581

7 Leaves thickish, elliptical or oblong; fruits more or less united to form a 2-eyed berry; branchlets stiff and ascending—8

8 Leaves usually rounded at the base, often hairy at least beneath; flowers and fruits with stalks less than ¼ inch long; fruits blue or bluish black MOUNTAIN FLY HONEYSUCKLE *(L. villosa)* 581

8 Leaves usually pointed at the base, smooth or nearly so; flowers and fruits with stalks over ½ inch long; fruits red or purplish SWAMP FLY HONEYSUCKLE *(L. oblongifolia)* 581

Q

KEY TO THE VIBURNUMS (Viburnum)

1 Leaves 3-lobed – 2
1 Leaves not lobed – 4

2 Lower leaf surface with minute black dots, usually softly downy; leafstalks usually with stipules at the base
MAPLE-LEAF VIBURNUM *(V. acerifolium)* 590
2 Lower leaf surfaces not black-dotted, smooth or merely downy on the main veins; leafstalks commonly with a pair of glands at the summit – 3

3 Leaves shallowly 3-lobed above the middle; low or straggling shrub
SQUASHBERRY *(V. edule)* 589
3 Leaves deeply 3-lobed; upright shrub
HIGHBUSH-CRANBERRY *(V. opulus* var. *americanum)* 590

4 Leaves usually over 4 inches wide, heart-shaped; branchlets and lower leaf surfaces with cinnamon-colored, starry-branched hairs
HOBBLEBUSH *(V. lantanoides)* 586
4 Leaves less than 4 inches wide – 5

5 Leaves with prominent straight veins running from the midrib into large marginal teeth – 6
5 Leaf margins untoothed or with fine teeth, the veins not at all prominent toward the leaf margin – 8

6 Leaves stalkless or with stalks less than 3/16 inch long
DOWNY ARROWWOOD *(V. rafinesquianum)* 593
6 Leaves with stalks usually over 1/4 inch long – 7

7 Leafstalks with stipules at the base; bases of the leaves usually heart-shaped SOFT ARROWWOOD *(V. molle)* 590
7 Leafstalks usually without stipules; bases of the leaves usually rounded
ARROWWOOD *(V. dentatum)* 590

8 Leaf margins untoothed or the teeth very inconspicuous; flower and fruit clusters either stalkless or distinctly stalked – 9
8 Leaf margins quite finely, sharply, and regularly toothed the entire length; flower and fruit clusters stalkless or nearly so – 11

9 Flower and fruit clusters stalkless; largest leaves seldom 1½ inches long, broadest toward the tip SMALL-LEAF VIBURNUM *(V. obovatum)* 589
9 Flower and fruit clusters distinctly stalked; largest leaves well over 2 inches long, usually broadest at or below the middle – 10

10 Stalk of the flower or fruit cluster shorter than its branches; leaves dull above, usually smooth beneath
NORTHERN WITHEROD *(V. nudum* var. *cassinoides)* 586
10 Stalk of the flower or fruit cluster as long as or longer than its branches; leaves lustrous above, usually rusty-scurfy on the veins beneath SOUTHERN WITHEROD *(V. nudum)* 586

11 Leaves long-pointed at the tip; leafstalks rather broadly winged
SWEET VIBURNUM *(V. lentago)* 586
11 Leaves short-pointed or blunt at the tip; leafstalks slightly if at all winged – 12

12 Leafstalks not winged, both they and the lower leaf surfaces smooth; leaves dull above BLACK HAW *(V. prunifolium)* 589

12 Leafstalks often slightly winged, both they and the lower leaf surfaces more or less rusty-scurfy; leaves lustrous above RUSTY BLACK HAW *(V. rufidulum)* 589

Glossary

ACHENE A small, dry, one-seeded fruit that does not split open at maturity.
ACUMINATE With the tip gradually tapering to an extended point.
AERIAL ROOTLETS Small, rootlike structures protruding from the stems of some vines and not penetrating the ground.
ALTERNATE Arranged singly at intervals along the stem.
ANGLED With several edges or evident ridges.
ANTHER The apical part of a stamen that produces pollen.
AQUATIC Living in water.
ARMED Bearing prickles, spines, or sharp thorns.
AROMATIC Having a pleasant odor.
ASCENDING Rising upward at an angle.
AWL-LIKE Narrow and abruptly tapering to a sharp point.
AXIL The upper angle formed between the leaf or leafstalk and the stem. Also, a similar angle formed between the principal vein of a leaf and the midrib.
AXILLARY Borne in the axil of a leaf or bract.

BEAKED Ending in a long and prominent point.
BEARD A tuft or line of hairs.
BERRY A fruit arising from a single ovary that is fleshy or pulpy throughout.
BILATERAL Divisible into two identical halves only by a single vertical axis through the center, usually of a flower.
BLADE The expanded, flat portion of a leaf or petal.
BLOOM A white, waxy, powdery substance that is easily rubbed off.
BRACT A small, leaflike structure associated with a flower cluster.
BRANCHLET A small branch of the current season; a twig.
BRISTLE A stiff hair; a weak outgrowth of a stem.
BUD An undeveloped stem, branch, flower, or flower cluster.
BULB A kind of underground bud with fleshy scales.
BUNDLE SCARS Dotlike scars within a leaf scar, representing the broken ends of ducts that led into the leafstalk.

CALYX The outer whorl of floral parts; the sepals taken collectively.
CAPSULE A dry fruit that splits into two or more parts at maturity.
CARPEL A simple pistil or one of the parts of a compound pistil.
CATKIN An inflorescence made up of small, scaly-bracted flowers arranged in a long and often drooping cluster.
CELL A compartment or division of a compound ovary.
CHAFF Small, thin, dry scales associated with the disk flowers in many composites.
CHAMBERED PITH Pith divided crosswise by woody plates or partitions.
CLAW A narrow base or stalk, usually of a petal.
CLEFT Deeply cut but not divided.
COLUMN Central structure of an orchid flower; comprised of one or two stamens united with the style and stigma.
COMPOUND LEAF A leaf in which the blade is divided into two or more leaflike parts or leaflets.
CONTINUOUS PITH Pith that is not divided into compartments by cross plates.
CORDATE Heart-shaped.
CORM An enlarged stem base that is bulblike but not solid.
COROLLA The petals of a flower taken collectively.
CORYMB A flat-topped inflorescence in which the outer or marginal flowers open first.
CRESTED Having elevated ridges or projections.
CYME A flat-topped inflorescence in which the central flowers open first.

DIAPHRAGM A zone of denser or woody tissue in the pith.
DISK FLOWER In composites, one of the tubular flowers in the center of the head.
DISTINCT Separate; not united.
DIVIDED Separated to the base.
DOWNY Covered with fine, soft hairs.
DRUPE A fleshy fruit with the seed enclosed in a hard, stony covering.
DRUPELET A tiny drupe.

EGG-SHAPED Broadest below the middle.
ELLIPTIC Widest in the middle and tapering sharply at both ends.
ELONGATE Much longer than broad.
ENTIRE Untoothed; with a smooth margin.
EPIPHYTE A plant growing upon another plant or other elevated support, but not parasitic.
EVERGREEN With leaves remaining open throughout the year.
EYE A prominent mark in the center of a flower.

FEATHER-VEINED Having veins arising along the sides of the midrib.
FERTILE STAMEN A stamen that produces pollen.
FILAMENT The long, slender, stalklike part of a stamen that supports the anther.
FLAKY Having loose scales.
FLORET A small flower, especially one in a dense cluster.
FREE Separate; not joined to other organs.
FRUIT A ripened ovary and the parts united with it; the seed-bearing structure of a plant.

GLAND A secreting surface or appendage.
GLANDULAR Bearing glands.

HEAD A dense inflorescence made up of stalkless or nearly stalkless flowers.
HOARY Grayish white with fine, close hairs.
HYBRID A cross between two closely related species.

IMPERFECT Lacking either stamens or pistils.
INFERIOR OVARY An ovary surrounded by and united with the calyx.
INFLATED Seemingly blown up; bladdery.
INFLORESCENCE A group or cluster of flowers.
INSERTED Attached to or growing out of something.
INVOLUCRE A circle (or circles) of bracts surrounding a flower or flower cluster.
IRREGULAR FLOWER A flower in which the parts are not all of the same size and shape; a bilaterally symmetric flower.

KEEL The two lower petals of a papilionaceous flower, with a central ridge like the keel of a boat.

LANCEOLATE Much longer than broad, widest near the base and gradually tapering to the tip.
LATERAL Along the sides of a branchlet.
LEAFLET One of the leaflike units of a compound leaf.
LEAF SCAR The scar left on a twig when a leaf falls.
LENTICEL A corky spot on the bark facilitating gas exchange.
LINEAR Long and narrow with the sides parallel or nearly so; like the leaves of grasses.
LIP One of the parts of an unequally divided calyx or corolla; in orchids, the petal (usually the lowest) that differs in size, shape, and often color from the other petals.
LOBE A more or less rounded extension of an organ, such as a leaf.
LOBED Having lobes or deep indentations.
LUSTROUS Shiny or glossy.

MARGIN The border or edge of a leaf.
MIDRIB The central or main vein of a leaf, often appearing like an extension of the petiole.

NAKED BUD A bud without bud scales.
NODAL Pertaining to or situated at a node.
NODE The point on a stem to which a leaf (or leaves) is attached.
NUT A hard-shelled, one-seeded fruit that does not split open at maturity.
NUTLET A small, hard, dry fruit.

OBLONG Longer than broad and with the sides nearly parallel.
OPPOSITE Occurring in pairs at the nodes.
OVARY The ovule-bearing part of a pistil.
OVULE One of the small bodies in the ovary that develops into a seed.

PALMATE Having veins, lobes, or divisions that radiate from one central point, like the spread fingers of a hand.

PANICLE An inflorescence similar to a raceme but with branching flower stalks.

PAPILIONACEOUS Butterflylike; a term used for flowers that have a standard, two wings, and a keel.

PAPPUS Hairs, bristles, or scales that form the calyx limb in composites.

PARASITE An organism that obtains its food directly from another living organism or host.

PARTED Cleft nearly, but not quite, to the base.

PARTITIONED PITH Pith divided crosswise by woody plates.

PERFECT Possessing both stamens and pistils.

PERIANTH The parts of a flower (usually sepals and petals) surrounding the stamens and pistils.

PERSISTENT Remaining attached.

PETAL A modified leaf, usually brightly colored, immediately outside the reproductive organs of a flower; one of the separate divisions of the corolla.

PETIOLE The stalk supporting a leaf.

PINNATE Having veins, lobes, or divisions of leaves arranged along the sides of a midrib.

PISTIL The female organ of a flower, which develops into fruit and seeds; composed of stigma, style, and ovary.

PITH The soft or spongy tissue in the center of a twig or stem.

POD A dry fruit that splits open at maturity.

POLLEN The minute grains produced in the anthers and containing the male reproductive cells.

POLLINIUM A mass of pollen grains that is shed as a unit from an anther.

PRICKLE A small, sharp, needlelike outgrowth of the bark.

PROSTRATE Lying flat on the ground.

RACEME A simple, long, narrow inflorescence of stalked flowers.

RADIALLY SYMMETRIC Divisible through the center on any axis into mirror images. A radially symmetric flower has identical parts, all of which radiate from the center.

RAY OR RAY FLOWER In composites, the strap-shaped, petallike flowers.

RECEPTACLE The enlarged or expanded end of a stem to which the parts of a flower are attached.

RECLINING With the lower part lying flat on the ground but the upper part curving upward.

RECURVED Curving downward or backward.

REFLEXED Abruptly bent or turned downward.

REGULAR FLOWER A flower in which all parts are alike in size and shape; a radially symmetric flower.

RESIN DOTS Minute globules of resin exuded on the surface of the plant.

REVOLUTE Curled under; rolled.

RIB A primary or prominent vein of a leaf.

ROOTSTOCK An underground stem growing more or less horizontally.

ROSETTE A cluster of leaves radiating from one point and lying more or less flat on the ground.

RUNNER A slender or trailing stem that takes root at the nodes.

SCALE A small modified leaf on the outside of a bud.

SCURFY Covered with small, branlike scales.

SEED A ripened ovule.

SEMI-HERBACEOUS Only partly woody.

SEPAL One of the outermost modified leaves surrounding the reproductive organs of a flower; collectively called the calyx.

SHREDDY Peeling off in thin, narrow, irregular strips.

SIMPLE LEAF A leaf with an undivided blade; not compound.

SIMPLE STEM A stem without branches.

SINUS The space or indentation between the lobes of a leaf blade.

SMOOTH Lacking hairs.

SPADIX A thick stem bearing crowded, small, stalkless flowers; characteristic of the Arum family.

SPATHE A large bract surrounding or partially surrounding a flower or flower cluster.

SPIKE A simple inflorescence of stalkless flowers arranged along a common stem.

SPINE A sharp-pointed, rigid, thornlike structure.

SPRAWLING Spreading irregularly.

SPUR A hollow tubular or saclike projection of a petal or sepal.

STALKED Borne on a stalk; having a narrow, necklike base.

STAMEN The male, pollen-bearing part of a flower; composed of anther and filament.

STANDARD The broad upper petal of a papilionaceous flower.

STERILE Barren or unproductive.

STERILE STAMEN A stamen lacking a pollen-producing anther.

STIGMA The part of the pistil that receives the pollen.

STIPULE A small appendage, usually paired, at the base of the leafstalk.

STIPULE SCAR A scar left on the twig by a fallen stipule.

STOLON A branch resting on the ground that may take root.

STOLONIFEROUS Having stolons or runners.

STRAGGLING Spreading irregularly but semi-upright.

STYLE The part of the pistil above the ovary and bearing the stigma.

SUPERIOR OVARY An ovary that is free from and placed above all other parts of the flower.

SUPERPOSED Placed one above the other.

TENDRIL A slender, twining outgrowth by which a plant attaches itself to some support.

TERMINAL Borne at the tip of a stem or branch.

TERNATE Arranged in threes or divided into three parts.

THORN A stiff, sharp, woody outgrowth of a stem.

THROAT The opening into the tube of a calyx or corolla having united sepals or petals.

TOOTHED Having small, toothlike projections.

TUBER A short, thick underground stem.

TUBULAR Having floral parts united to form a hollow tube, usually with lobes at the summit.

TWICE COMPOUND Divided into leaflets that are again divided into smaller leaflets.

TWIG A small shoot representing the past season's growth; a branchlet.

UMBEL An inflorescence in which the stalked flowers all arise from a common point.

WEDGE-SHAPED Narrow and tapering to a sharp point.

WHORLED Arranged in a circle about a single node.

WING A thin, flat extension. Also, one of two lateral petals of a papilionaceous flower.

WOOLLY Bearing tangled or matted loose hairs.

For Further Information

Bailey, L. H., and E. Z. Bailey. *Hortus Third,* revised and expanded by the staff of the Liberty Hyde Bailey Hortorium, 3rd ed. New York: Macmillan, 1976.

Braun, E. L. *The Woody Plants of Ohio.* Columbus: Ohio State University Press, 1961.

Core, E. L., and N. P. Ammons. *Woody Plants in Winter.* Pacific Grove, Calif.: Boxwood Press, 1973.

Duncan, W. H., and M. B. Duncan. *Trees of the Southeastern United States.* Athens: University of Georgia Press, 1988.

Eastman, J. *The Book of Forest and Thicket: Trees, Shrubs, and Wildflowers of Eastern North America.* Harrisburg, Pa.: Stackpole Books, 1992.

Fernald, M. L. *Gray's Manual of Botany,* 8th ed. New York: American Book Co., 1950.

Foote, L. E., and S. B. Jones. *Native Shrubs and Woody Vines of the Southeast.* Portland, Oreg.: Timber Press, 1989.

Gleason, H. A. *The New Britton and Brown Illustrated Flora of the Northeastern United States and Adjacent Canada* (3 volumes). New York: New York Botanical Garden, 1952.

Gleason, H. A., and A. Cronquist. *Manual of Vascular Plants of Northeastern United States and Adjacent Canada,* 2nd ed. New York: New York Botanical Garden, 1991.

Godfrey, R. K. *Trees, Shrubs, and Woody Vines of Northern Florida and Adjacent Southern Georgia and Alabama.* Athens: University of Georgia Press, 1988.

Harris, S. K. *A.M.C. Field Guide to Mountain Flowers of New England.* Boston: Appalachian Mountain Club, 1977.

Kingsbury, J. M. *Poisonous Plants of the United States and Canada.* Englewood Cliffs, N.J.: Prentice-Hall, Inc., 1964.

Magee, D. W. *Freshwater Wetlands: A Guide to Common Indicator Plants of the Northeast.* Amherst: University of Massachusetts Press, 1981.
Muenscher, W. C. *Keys to Woody Plants,* 7th rev. ed. Ithaca, N.Y.: Comstock Publishing Co., 1950.
———. *Poisonous Plants of the United States,* rev. ed. New York: Macmillan, 1951.
Newcomb, L. *Newcomb's Wildflower Guide.* Boston: Little, Brown and Co., 1977.
Niering, W. A. *The Audubon Society Field Guide to North American Wildflowers, Eastern Region.* New York: Alfred A. Knopf, 1979.
Peterson, R. T., and M. McKenny. *A Field Guide to Wildflowers of Northeastern and North-central North America.* Boston: Houghton Mifflin Co., 1968.
Petrides, G. A. *A Field Guide to Trees and Shrubs,* 2nd ed. Boston: Houghton Mifflin Co., 1982.
Radford, A. E., H. E. Ahles, and C. R. Bell. *Manual of the Vascular Flora of the Carolinas.* Chapel Hill: University of North Carolina Press, 1968.
Rehder, A. *Manual of Cultivated Trees and Shrubs.* New York: Macmillan, 1940.
Rickett, H. W. *The New Field Book of American Wild Flowers.* New York: Putnam, 1963.
———. *Wild Flowers of America.* New York: Crown Publishers, 1953.
Small, J. K. *Manual of the Southeastern Flora.* Chapel Hill: University of North Carolina Press, 1933.
Soper, J. H., and M. L. Heimburger. *Shrubs of Ontario.* Toronto: Royal Ontario Museum, 1982.
Strausbaugh, P. D., and E. L. Core. *Flora of West Virginia,* 2nd ed. Grantsville, W.Va.: Seneca Books, 1978.
Stupka, A. *Wildflowers in Color.* New York: Harper & Row, 1965.
U.S. Department of Agriculture. *Common Weeds of the United States.* New York: Dover, 1971.
Wherry, E. T. *Wild Flower Guide, Northeastern and Midland United States.* New York: Doubleday & Co., 1948.

Index